DYNAMICS OF POPULATIONS
OF PLANETARY SYSTEMS

IAU COLLOQUIUM No. 197

COVER ILLUSTRATION: ASTEROID (433) EROS

Mosaic of Eros' Northern Hemisphere

While NEAR Shoemaker orbits Eros, the asteroid appears too large for the camera's field of view. In order to get a complete view of the surface from a particular vantage point, several images are mosaicked. To do this, the digital images returned by the spacecraft are draped over a computer model of the asteroid's shape. This spectacular view –looking down on the north polar region– was constructed from six images taken February 29, 2000, from an orbital altitude of about 200 kilometers (124 miles).

Eros is the asteroid on which we have most information, thanks to the close up examination by a robotic space mission with sophisticated sensors. For this reason it is an essential benchmark against which we have to test all our theories about asteroids and other minor planetary bodies. However, Eros is just one of many thousands of similar objects. What really matters are the global properties of the asteroid population, rather than the properties of a single object which may be more or less representative.

The image of Eros has been taken by the NASA mission NEAR-Shoemaker. It has been downloaded from The National Space Science Data Center (NSSDC) (http://nssdc.gsfc.nasa.gov).

INTERNATIONAL ASTRONOMICAL UNION

UNION ASTRONOMIQUE INTERNATIONALE

DYNAMICS OF POPULATIONS OF PLANETARY SYSTEMS

PROCEEDINGS OF THE 197th COLLOQUIUM OF THE INTERNATIONAL ASTRONOMICAL UNION HELD IN BELGRADE, SERBIA AND MONTENEGRO AUGUST 31 – SEPTEMBER 4, 2004

Edited by

ZORAN KNEŽEVIĆ

Astronomical Observatory, Belgrade, Serbia and Montenegro

and

ANDREA MILANI

Department of Mathematics, University of Pisa, Pisa, Italy

CAMBRIDGE
UNIVERSITY PRESS

CAMBRIDGE UNIVERSITY PRESS
The Edinburgh Building, Cambridge CB2 2RU, UK
40 West 20th Street, New York, NY 10011 4211, USA
477 Williamstown Road, Port Melbourne, VIC 3207, Australia
Ruiz de Alarcón 13, 28014 Madrid, Spain
Dock House, The Waterfront, Cape Town 8001, South Africa

First published 2005

Printed in the United Kingdom at the University Press, Cambridge

Typeset in System LaTeX 2_ε

A catalogue record for this book is available from the British Library

Library of Congress Cataloguing in Publication data

ISBN 0 521 185203 X hardback
ISSN 1743-9213

Table of Contents

Part 1. EXTRASOLAR PLANETARY SYSTEMS

Part 2. ASTEROID FAMILIES AND STABILITY

Part 3. ASTEROID ORBIT DETERMINATION AND IMPACTS

Part 4. COMETS AND TRANS-NEPTUNIAN OBJECTS

Part 5. METEORS, METEORITES AND DUST

Part 6. SATELLITE AND STELLAR SYSTEMS

Part 7. ENDING THE COLLOQUIUM

Preface

The progress in computing results in the possibility to compute numerically the orbit of a single celestial object in almost all problems, even when there are close approaches and in other highly unstable cases. This has, however, only moved forward the frontiers of research in dynamical astronomy. Now the goal is to study the dynamical behaviour of entire populations of celestial bodies, either real (observed) or only hypothetical (simulated).

The populations discussed in the most recent research include small solar system bodies, like asteroids, comets, meteoroid streams, rings and interplanetary dust particles; artificial objects, like space debris; extrasolar planets and swarms of planetesimals, in an effort to determine the regions of phase space where such objects can be formed and where their orbits can remain stable; virtual objects, to represent the uncertainty of the orbit of a specific object as determined from the observations.

The methods developed to cope with population dynamics problems range from the analytical, semianalytical and synthetic theories of proper elements, to the improvements in performance and accuracy of numerical integration and of the approximate integrators with important qualitative properties (e.g., symplectic methods). Even the availability of efficient computational tools, giving rich but approximate information on a very large number of orbits, solves only half of the problem: how should the results be represented? The availability of advanced computer graphics, including color coded maps and animations, has made possible to present enormous amounts of information in one slide, but this might sometimes mean dowloading the difficulty of understanding the results on the reader.

Let us mention a few examples. We have now a good understanding of the dynamical structure of the asteroid main belt over intermediate time scales (few tens to a few hundreds of millions of years). To extend such understanding to the entire history of the solar system we need to account for chaotic diffusion, non gravitational perturbations (especially Yarkovsky effect) and collisional evolution. Asteroid families, formed by catastrophic disruption (and/or craterization) of parent bodies, are now a well established fact. Similarly, the sucess of Leonid predictions over recent years showed that dynamic model of meteoroid streams are reliable, although over shorter time scales. On the other hand, the very success in creating huge proper element catalogs (with $> 100,000$ objects) has created a difficult situation, in which asteroid families cannot be defined just by statistical methods, as groupings contrasting with a sparse background. Very similar problems arise in the study of meteor showers, again because of the very large data sets now available.

More difficult problems of unstable, strongly chaotic dynamics arise in the study of the transport of meteorites and Near Earth Objects (NEO). Some meteoroid streams, such as the Quadrantid, also exhibit many chaotic properties. The subject of the study is not anymore the single object, but the entire Earth-crossing population. By using simplectic integrators a large amount of qualitative and quantitative information has been gathered, but then the problem is how to understand and represent the results. Model populations of NEO are the most useful output of these studies. However, the question arises of the sensitivity of these models to assumptions and approximations used: as an example, non gravitational perturbations are relevant in this process, but a quantitative measure of this relevance is not yet well established.

The transport of Trans Neptunian Objects (TNO) and comets to the inner Solar System has also been studied by large scale numerical integrations; but, are the current models fully self consistent, taking into account all the effects, including collisions and physical 'aging' of comets? Interplenetary dust dynamics, including the formation of dust belts, is basically understood, but the relative importance of different sources is still the subject of discussion.

The dynamics of planetary systems very different from our own has been a subject of intense study in the recent years, because of the discovery of many extrasolar planetary systems with puzzling dynamical structures. While the study of a single known system is always feasible, it is not yet clear how to explore the phase space of the possible systems to constrain the observational searches, classify and understand the new discoveries. Moreover, the conservative dynamics of a planetary system as observed is the consequence of the formation of the planets from some swarm of planetesimals and of the migration of the planets under the effect of interactions with the residual planetesimals, the dust, the gas and the extended body of the central star. A unified view of the issues related to extrasolar planetary systems is a goal which has so far eluded the efforts of the astronomers: although this meeting may not be enough to achieve it, it is intended to give a significant contribution, by at least providing us with a global view of the aspects connected with dynamics.

A large populations of artificial celestial bodies in geocentric orbit has been created by the space age. To model space debris, and also the meteoroid background, in a way operationally useful to increase the security of space assets, such as telecommunication satellites and the International Space Station, is a challenge with significant economic implication.

In the effort to solve the problem to detect possibilities of future asteroid/comet impacts on our planet it was realized that the present uncertainty on the real orbit of each one of these bodies needs to be represented by means of a swarm of Virtual Asteroids (VA). Thus to compute the probability of a future impact we need to solve a population dynamics problem. Advanced methods to solve this problem involve describing the VA by means of a geometric object, either one or two dimensional, rather than using a swarm of independent points.

Although each specialist can tell an interesting tale based on the results of its own solution to the specific population dynamic problem, the main focus of the exchange of information should be on the methods. Indeed the export of methods from one problem to another can be very effective. This meeting is intended to pursue this cross-fertilization on top of presenting specific new results.

The timing of this meeting has been chosen avoding superposition with other related meetings and taking into account that a long time elapsed since the last IAU sponsored meeting not only on this subject, but even connected to the same discipline, namely Celestial Mechanics and Dynamical Astronomy.

The choice of Belgrade for the location of this Colloquium was certainly motivated by the opportunity to visit one of the historic European capital cities, and by the presence of a well established local astronomical community with long tradition and specific reputation in this field. Moreover, a meeting in this city was meant to mark the readmission of Serbia and Montenegro to the International Astronomical Union, after Yugoslavia ceased to exist, thus depriving the IAU of one of its member states.

Andrea Milani and Zoran Knežević, co-chairs SOC,
Belgrade, August 31, 2004

THE ORGANIZING COMMITTEE

Scientific

Joseph A. Burns (USA)
Rudolf Dvorak (Austria)
Sylvio Ferraz-Mello (Brasil)
Claude Froeschlé (France)
John D. Hadjidemetriou (Greece)

Zoran Knežević (co-chair, Serbia & Mo...
Anne Lemaitre (Belgium)
Andrea Milani (co-chair, Italia)
Vladimir Porubčan (Slovakia)
Giovanni B. Valsecchi (Italy)

Local

Ivan Pakvor (chair)
Milan Ćirković
Ivana Damjanov

Slavica Pavić
Rade Pavlović
Nataša Todorović

Acknowledgements

The colloquium is sponsored and supported by the IAU Divisions I (Fundamental Astronomy) and III (Planetary Systems Sciences); and by the IAU Commissions No. 7 (Celestial Mechanics and Dynamical Astronomy), No. 20 (Positions and Motions of Minor Planets, Comets and Satellites) and No. 22 (Meteors, Meteorites and Interplanetary Dust).

The Local Organizing Committee operated under the auspices of the
Astronomical Observatory, Belgrade

Funding by the
International Astronomical Union,
UNESCO-ROSTE,
Ministry of Science and Environmental Protection of Serbia,
and
Astronomical Observatory, Belgrade
is gratefully acknowledged.

Participants

Fred **Adams**, University of Michigan, Physics Department, East University, Ann Arbor, MI, USA — fca@umich.edu
Alvaro **Alvarez-Candal**, National Observatory, Rio de Janeiro, Brazil — alvarez@on.br
David **Asher**, Armagh Observatory, College Hill, United Kingdom — dja@arm.ac.uk
Daniel **Benest**, Observatoire de la Cote d'Azur, Nice, France — benest@obs-nice.fr
Stefan **Berinde**, Astronomical Observatory, Cluj-Napoca, Romania — sberinde@math.ubbcluj.ro
Carlo **Blanco**, Physics and Astronomy Department of Catania University, Italy — cblanco@ct.astro.it
Edi **Bon**, Astronomical Observatory, Belgrade, Serbia and Montenegro — ebon@aob.aob.bg.ac.yu
William **Bottke**, Southwest Research Institute (SwRI), Boulder, CO, USA — bottke@boulder.swri.edu
Miroslav **Brož**, Astronomical Institute of the Charles University, Prague, Czech Republic — mira@sirrah.troja.mff.cuni.cz
Joe A. **Burns**, Cornell University, Ithaca, NY, USA — jab16@cornell.edu
Oleg **Bykov**, Main (Pulkovo) Observatory of RAS, St-Petersbourg, Russia — oleg@OB3876.spb.edu
David **Čapek**, Astronomical Institute of the Charles University, Prague, Czech Republic — capek@sirrah.troja.mff.cuni.cz
Steven **Chesley**, Jet Propulsion Laboratory, Pasadena, CA, USA — steve.chesley@jpl.nasa.gov
Zorica **Cvetković**, Astronomical Observatory, Belgrade, Serbia and Montenegro — zcvetkovic@aob.aob.bg.ac.yu
Mihailo **Čubrović**, Department of Astronomy - Petnica Science Center, Valjevo, Serbia and Montenegro — cygnus@EUnet.yu
Milan M. **Ćirković**, Astronomical Observatory, Belgrade, Serbia and Montenegro — mcirkovic@aob.aob.bg.ac.yu
Matija **Ćuk**, Cornell University, Ithaca, NY, USA — cuk@astro.cornell.edu
Ivana **Damjanov**, Astronomical Observatory, Belgrade, Serbia and Montenegro — idamjanov@aob.aob.bg.ac.yu
Goran **Damljanović**, Astronomical Observatory, Belgrade, Serbia and Montenegro — gdamljanovic@aob.aob.bg.ac.yu
Miguel **de Val Borro**, Stockholm University, AlbaNova Center, Sweden — miguel@astro.su.se
Rudolf **Dvorak**, Institute for Astronomy, University of Vienna, Austria — dvorak@astro.univie.ac.at
Piotr **Dybczyński**, Astronomical Observatory, A.Mickiewicz University, Poznan, Poland — dybol@amu.edu.pl
Christos **Efthymiopoulos**, Research Center for Astronomy, Academy of Athens, Greece — cefthim@cc.uoa.gr
Vacheslav **Emel'yanenko**, Dept. of Comp. & Celestial Mech., South Ural Univ., Chelyabinsk, Russia — vvemel@math.susu.ac.ru
Sylvio **Ferraz - Mello**, Universidade de Sao Paulo, Brazil — sylvio@usp.br
Florian **Freistetter**, Institute for Astronomy, University of Vienna, Austria — freistetter@astro.univie.ac.at
Christiane **Froeschlé**, Observatoire de la Cote d'Azur, Nice, France — froesch@obs-nice.fr
Claude **Froeschlé**, Observatoire de la Cote d'Azur, Nice, France — claude@obs-nice.fr
Barbara **Funk**, Institute for Astronomy, University of Vienna, Austria — funk@astro.univie.ac.at
Stefan **Gajdoš**, Astronomical Institute, Faculty of Mathematics, Bratislava, Slovak Republic — Stefan.Gajdos@fmph.uniba.sk
Mikael **Granvik**, Observatory, University of Helsinki, Finland — mikael.granvik@astro.helsinki.fi
Evgeny **Griv**, Ben-Gurion University, Department of Physics, Beer-Sheva, Israel — griv@bgu.ac.il
Giovanni-Federico **Gronchi**, Department of Mathematics, University of Pisa, Italy — gronchi@dm.unipi.it
Massimiliano **Guzzo**, Dipartimento di Matematica Pura e Applicata, Universita' di Padova, Italy — guzzo@math.unipd.it
John **Hadjidemetriou**, Department of Physics, University of Thessaloniki, Greece — hadjidem@auth.gr
Jacques **Henrard**, University of Namur, Belgium — jacques.henrard@fundp.ac.be
Arika **Higuchi**, National Astronomical Observatory Japan/Kobe University, Tokyo, Japan — higuchi@th.nao.ac.jp
Sergei **Ipatov**, Catholic University of America, Greenbelt, MD, USA — siipatov@hotmail.com
Pawel **Kankiewicz**, Institute of Physics, Swietokrzyska Academy, Kielce, Poland — pawel.kankiewicz@pu.kielce.pl
Mike **Kuzmanoski**, Faculty of Mathematics, University of Belgrade, Serbia and Montenegro — mike@matf.bg.ac.yu
Jozef **Klačka**, Astronomical Institute FMPI, Comenius University, Bratislava, Slovak Republic — klacka@fmph.uniba.sk
Zoran **Knežević**, Astronomical Observatory, Belgrade, Serbia and Montenegro — zoran@aob.bg.ac.yu
Michal **Kočer**, Klet' Observatory, Ceske Budejovice, Czech Republic — kocer@klet.cz
Miroslav **Kocifaj**, Institute for Experimental Physics, University of Vienna, Austria — kocifaj@ap.univie.ac.at
Eiichiro **Kokubo**, National Astronomical Observatory of Japan, Tokyo, Japan — kokubo@th.nao.ac.jp
Thomas **Kotoulas**, Department of Physics, University of Thessaloniki, Greece — tkoto@skiathos.physics.auth.gr
Edouard **Kuznetsov**, Astronomical Observatory of Urals State University, Ekaterinburg, Russia — Eduard.Kuznetsov@usu.ru
Andjelka **Kovačević**, Faculty of Mathematics, University of Belgrade, Serbia and Montenegro — andjelka@matf.bg.ac.yu
Anne **Lemaitre**, University of Namur, Belgium — Anne.Lemaitre@fundp.ac.be
Hal **Levison**, Southwest Research Institute (SwRI), Boulder, CO, USA — hal@boulder.swri.edu
Andrzej **Maciejewski**, Institut of Astronomy,University of Zielona Gora, Poland — maciejka@astro.ia.uz.zgora.pl
Zoltan **Mako**, Babes-Bolyai University, Cluj-Napoca, Romania — zmako@math.ubbcluj.ro
Patrick **Michel**, Observatoire de la Cote d'Azur, Nice, France — michel@obs-nice.fr
Andrea **Milani**, Department of Mathematics, University of Pisa, Italy — milani@dm.unipi.it
Vasile **Mioc**, Astronomical Institute of Romanian Academy, Bucharest, Romania — vmioc@aira.astro.ro
Vjera **Miović**, Department of Physics, University of Toronto, Canada — miovic@physics.utornto.ca
Alessandro **Morbidelli**, CNRS / Observatoire de la Cote d'Azur, Nice, France — morby@obs-nice.fr
Lubos **Neslušan**, Astronomical Institute, Slovak Academy of Sciences, Tatranska Lomnica, Slovak Republic — ne@ta3.sk
blic Ernesto **Vieira Neto**, Universidade Estadual Paulista (UNESP), Guaratinguet, S.P. Brazil — ernesto@feg.unesp.br
Slobodan **Ninković**, Astronomical Observatory, Belgrade, Serbia and Montenegro — sninkovic@aob.bg.ac.yu
Benoit **Noyelles**, IMCCE - Laboratoire d'Astronomie de Lille, France — Benoit.Noyelles@imcce.fr
David **O'Brien**, Lunar and Planetary Lab, University of Arizona, Tucson, AZ, USA — obrien@lpl.arizona.edu
Dragomir **Olević**, Astronomical Observatory, Belgrade, Serbia and Montenegro — dolevic@aob.aob.bg.ac.yu
Ivan **Pakvor**, Astronomical Observatory, Belgrade, Serbia and Montenegro — jipakvor@EUnet.yu
Pavol **Pastor**, Astronomical Institute, FMPI, Comenius University, Bratislava, Slovak Republic — p.pastor@seznam.cz
Slavica **Pavić**, Astronomical Observatory, Belgrade, Serbia and Montenegro — col197@aob.bg.ac.yu
Rade **Pavlović**, Astronomical Observatory, Belgrade, Serbia and Montenegro — rpavlovic@aob.bg.ac.yu
Elke **Pilat-Lohinger**, Institute for Astronomy, University of Vienna, Austria — lohinger@astro.univie.ac.at
Dionyssia **Psychoyos**, Departement of Physics, University of Thessaloniki, Greece — dpsyc@skiathos.physics.auth.gr
Hans **Rickman**, Astronomiska observatoriet, Uppsala University, Sweden — hans@astro.uu.se
Fernando **Roig**, National Observatory, Rio de Janeiro, Brazil — froig@on.br
Alex **Rosaev**, FGUP NPC "Nedra", Yaroslavl, Russia — hegem@mail.ru
Alessandro **Rossi**, Spaceflight Dynamics Section ISTI-CNR, Pisa, Italy — Alessandro.Rossi@isti.cnr.it
Galina **Ryabova**, Research Inst. of Applied Mathematics and Mechanics of Tomsk State Univ., Russia — ryabova@niipmm.tsu.ru
Masaya **Saito**, SOKENDAI/NAOJ, Tokyo, Japan — saitohms@cc.nao.ac.jp
Natalia **Shakht**, Main (Pulkovo) Observatory of RAS, St-Petersbourg, Russia — shakht@gao.spb.ru
Magda **Stavinschi**, Astronomical Institute of Romanian Academy, Bucharest, Romania — magda@aira.astro.ro
Ferenc **Szenkovits**, Babes-Bolyai University, Cluj-Napoca, Romania — fszenko@math.ubbcluj.ro
Gonzalo **Tancredi**, Depto. Astronomia - Fac. Ciencias, Montevideo, Uruguay — gonzalo@fisica.edu.uy
Nataša **Todorović**, Astronomical Observatory, Belgrade, Serbia and Montenegro — ntodorovic@ovic@aob.bg.ac.yu
Aleksandar **Tomić**, Peoples Observatory, Kalemegdan, Belgrade, Serbia and Montenegro — aleksandartomic@hotmail.com
Giacomo **Tommei**, Department of Mathematics, University of Pisa,Italy — tommei@mail.dm.unipi.it
Kleomenis **Tsiganis**, CNRS / Observatoire de la Cote d'Azur, Nice, France — tsiganis@obs-nice.fr
Veselka **Trajkovska**, Astronomical Observatory, Belgrade, Serbia and Montenegro — vtrajkovska@aob.bg.ac.yu
Stéphane **Valk**, Faculés Universitaires Notre-Dame de la paix / FUNDP, Namur, Belgium — stephane.valk@fundp.ac.be
Giovanni **Valsecchi**, INAF-IASF, Roma, Italy — giovanni@rm.iasf.cnr.it
Harry **Varvoglis**, Department of Physics, University of Thessaloniki, Greece — varvogli@astro.auth.gr
Jenni **Virtanen**, Observatory, University of Helsinki, Finland — jenni.virtanen@helsinki.fi
David **Vokrouhlický**, Institute of Astronomy, Charles University, Prague, Czech Republic — vokrouhl@mbox.cesnet.cz
George **Voyatzis**, Department of Physics, University of Thessaloniki, Greece — voyatzis@auth.gr
Mark **Wyatt**, Royal Observatory Edinburgh, United Kingdom — wyatt@roe.ac.uk
Ji-Lin **Zhou**, Department of Astronomy, Nanjing University, China — zhoujl@nju.edu.cn

Address by the Local Organizing Committee

Dear participants, dear guests, ladies and gentlemen,

It is with pleasure that I can welcome you on behalf of the Local Organizing Committee to the 197th Colloquium of the International Astronomical Union.

17 years elapsed since the last IAU meeting held in Belgrade – the Colloquium No. 100: Fundamentals of Astrometry, held from September 8 to 11, 1987, on the occasion of the hundredth anniversary of the Astronomical Observatory, Belgrade. That meeting has been considered a great success from the scientific, as well as from the organizational point of view. We hope that we shall this time even better meet the high standards required for the organization of an international meeting, with so many participants and of such a high scientific level. Let me assure you that we in LOC will do our best to fulfill these requirements.

I would like to take this opportunity to thank all the colleagues involved with the organization of the meeting, and in particular to the members of the LOC.

We are deeply grateful to the following institutions for their financial and organizational support: the International Astronomical Union, the UNESCO-ROSTE, the Ministry of Science and Environmental Protection of Serbia, the Astronomical Observatory, Belgrade and the Society of Astronomers of Serbia.

I wish you all a successful meeting and a pleasant stay in Belgrade. Thank you very much, and welcome.

Ivan Pakvor, chair LOC

Belgrade, 31 August 2004

Address by the Scientific Organizing Committee

Within our scientific lifetime we have seen enormous changes in science and technology, perhaps the most impressive being the increase in the performance of digital computers. Over the last decades the computing power of a single chip has followed quite accurately the empirical Moore's law, increasing exponentially with a doubling time of 18 months. This corresponds to an increase by a factor of the order of one million in 30 years.

Let us consider the implication of this dramatic improvement in the available computers in our field of research, dynamical astronomy. The senior scientists, like myself, will remember that at the beginning of our carrier it was possible to explicitly compute a few orbits, e.g., of some N-body system (with small N). Now, also thanks to improvements in the algorithms, it is possible to repeat this experiment millions of times.

This has irresistibly led us to consider problems of population dynamics. We were perfectly aware that, e.g., asteroids, comets, meteoroids are not important as individual objects, but provide critical astrophysical information when considered as populations. However, only recently we have had the possibility to tackle the problems of populations dynamics which cannot be solved by comparatively simple analytical arguments. This has allowed to solve some old open problems, but results in new difficulties, of which I would like to mention the two main ones.

First, the computational complexity of the dynamics of large populations, if we use a brute force numerical approach, can defeat whatever computing power we have today and in the foreseeable future. For example, the populations of small bodies of our solar system contain billions of bodies, and if we are interested in detecting comparatively rare effects (such as an impact with our planet by an object large enough to produce damage) we simply cannot compute the orbits of all objects until the special event we are looking for is detected. Another example is the study of the possible stable configurations of extrasolar planetary systems: to model the planetary formation and migration processes requires a large population of mutually interacting planetesimals. Last but not least, we know that the most unstable orbits of the N-body type problems are strongly chaotic, thus precisely the most interesting orbits are only examples of the possible evolution, to be studied statistically, with large samples. Thus, it does not matter how much our computing power has increased, we still must use it in an intelligent way.

The second problem is that the output of a large scale integration of orbits can easily defeat our capability of understanding what the results really mean. Even the best graphical representation does show only the number densities in a parameter space, which is always 2-dimensional, although we may delude ourselves in thinking that we can show a complex multidimensional distribution. By increasing the number of data points the amount of information conveyed by a plot first increases, then decreases. Thus, the larger the number of orbits we can compute, the more difficult is the task of designing a procedure capable of extracting from the output the results which are significant and really needed that much computing power. By doing this, we may be able to find an appropriate balance between analytical and numerical methods.

We have organized this Colloquium to listen to the state of the art in the dynamics of different populations, also hoping to learn new methods, concepts and insights, which could be copied, or at least serve as inspiration, in our own struggle against these difficulties. Let us hope we come back from Belgrade with increased strength and new tools to confront the challenges of population dynamics.

Andrea Milani,
Belgrade, August 31, 2004

Part 1

EXTRASOLAR PLANETARY SYSTEMS

Dynamics of Populations of Planetary Systems
Proceedings IAU Colloquium No. 197, 2005
Z. Knežević and A. Milani eds.

© 2005 International Astronomical Union
DOI: 10.1017/S1743921304008439

Resonances and stability of extra-solar planetary systems

C. Beaugé[1], N. Callegari Jr.[2], S. Ferraz-Mello[3] and T.A. Michtchenko[3]

[1] Observatorio Astronómico, Universidad Nacional de Córdoba, Laprida 854, (X5000BGR) Córdoba, Argentina (beaugé@oac.uncor.edu)

[2] Departamento de Matemática, Universidade Federal de São Carlos, São Carlos, Brasil

[3] Instituto de Astronomia, Geofísica e Ciências Atmosféricas, Universidade de São Paulo, Cidade Univesitária, 05508-900 São Paulo, Brasil (sylvio@usp.br)

Abstract. This paper reviews recent results on the dynamics of multiple-planet extra-solar systems, including main sequence stars and the pulsar PSR B1257+12 and, comparatively, our own Solar System. Taking into account the degree of gravitational interaction of the planets, the known planetary systems may be separated into four main groups: (Ia) Planets in Mean-motion resonance (Ib) Low-eccentricity near-resonant pairs; (II) Non-resonant planets with a significant secular dynamics; and (III) Weakly interacting planet pairs. Different analytical and numerical tools can help to understand the structure of the phase space, to identify stability mechanisms and to categorize different types of motions in the cases of more significant dynamical interaction. The origin of resonant configurations is discussed in the light of the hypothesis of planetary migration.

Keywords. Exoplanets, stability, resonances

1. Introduction

Currently 12 systems with 2 or more planets around Main Sequence stars are known (see Schneider, 2004) and others are to be announced soon. Twelve of them are shown in Table 1. When the ratio of orbital periods in each planet pair is plotted (fig. 1), we see a more or less continuous distribution (at least in a logarithmic scale) between two extreme cases: In the lower end, we have planet pairs with $P_2/P_1 \approx 2$ and at the upper end, $P_2/P_1 > 300$. We may distinguish, tentatively, three different classes (one of which includes two sub-classes).

Class Ia. Planets in Resonant Orbits (MMR)

We put in Class I those planets at the lower end of the distribution shown in figure 1 with large masses and eccentricities and orbiting in relatively small orbits. These planets are liable to strong gravitational interaction and are significantly perturbed over orbital timescales. They are unable to remain stable if not tied by a mean-motions resonance (MMR). They are among the more interesting extra-solar systems for Celestial Mechanics studies.

The first two pairs are the two well-known pairs in 2:1 mean-motion resonance: HD 82943 and GJ 876 (= Gliese 876). The next two pairs appear in fig. 1 with open circles because of doubts concerning one of the planets in the pair. They are 47 Uma and the planets b,c of 55 Cnc. In these 2 pairs, the outer planet cannot yet be considered as confirmed, as there is no consensus among the observers about their existence (Naef

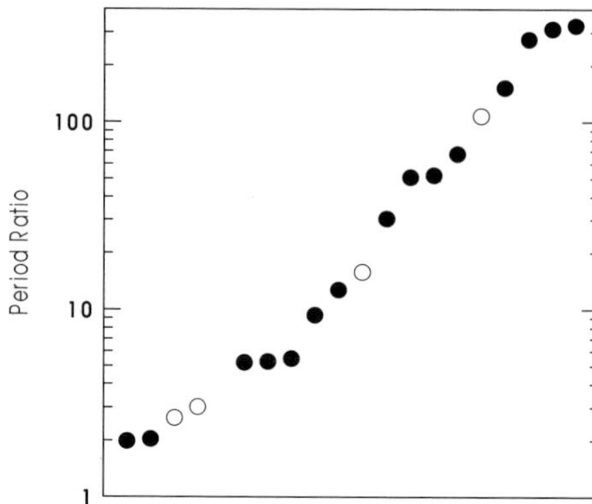

Figure 1. Ratio of the orbital periods of pairs of planets in motion around a main sequence star (see text for details). Open circles indicate pairs for which the existence of one of the components is still under discussion. The abscissas (not shown) are just sequence numbers of the pairs arranged in increasing order.

et al. 2004). If the existence of 55 Cnc c and the resonance is confirmed, this planet pair will be added to the two known resonant pairs to form a set of examples whose study may contribute to our understanding of the physics underneath the capture of exoplanets into resonance. A recent paper considers that the planetary hypothesis for the 44-day oscillations in the radial velocity is more likely because this star is a very inactive star (McArthur *et al.* 2004).

The planets of 47 UMa have large orbits and small eccentricities and, perhaps, may remain stable even in non-resonant orbits, in which case this system would rather be considered in Class Ib.

Class Ib. Low-eccentricity Near-resonant pairs

This is a special class including systems with low-eccentricity planets in successive pairs, with small period ratio, but with circumstances making the gravitational interaction between the planets less important. We put in this class the planets pairs of the pulsar PSR B1257+12. We may also include here the planets of the two sub-systems of the Solar System. In the planetary system of the pulsar PSR B1257+12 (see Table 2), as well as in the inner Solar System, the orbits are relatively close one to another, but the masses are relatively small. In the outer Solar System the masses are larger, but the distances between the planets is always large allowing this system to show long-term stability notwithstanding the fact that the planets have low period ratios.

The most conspicuous characteristic, here, is the presence of a large number of pairs in near resonance. The most conspicuous examples are the pair Jupiter-Saturn with a period ratio $\sim$2.5 (5:2 MMR) (see Michtchenko and Ferraz-Mello, 2001b) and the two outer planets of the pulsar PSR B1257 +12 with period ratio $\sim$1.5 (3:2 MMR). The closeness of the pulsar outer planets to commensurability produces perturbations large enough to be observed from Earth thus allowing the very existence of the planets to be confirmed in a few years of continuous observations (Rasio *et al.* (1992); Malhotra *et al.* (1992)).

Among the extra-solar systems, the low-eccentricity planets of 47UMa are possibly to be included in this class. The current orbits are near the resonance 8/3, but new observations are necessary to confirm the data of this system.

Table 1. Extra-solar Planetary Systems with 2 or more planets †

Star/ spectrum	planet	mass $\times \sin i$ ($m_{\rm Jup}$)	period (days)	period ratio	semi-major axis (AU)	eccentricity
HD 82943([1])	c	1.7	219.5	1.99	0.75	0.39
G0	b	1.8	436.2		1.18	0.15
GJ 876([2])	c	0.56	30.12	2.03	0.13	0.27
M4 V	b	1.89	61.02		0.21	0.10
47 UMa([2])	b	2.9	1079.2	2.64	2.1	0.05
G0 V	c (?)	1.1	2845.0		4.0	0
55 Cnc([3])	e	0.045	2.81	5.2	0.038	0.2
($=\rho^1$ Cnc)	b	0.78	14.7	2.99	0.115	0.02
G8 V	c (?)	0.22	43.9	103	0.24	0.44
	d	3.91	4517		5.26	0.3
μAra([4])	d	0.044	9.55	67.5	0.09	–
($=$HD 160691)	b	1.67	645	4.63	1.50	0.20
G3 IV-V	c	3.1	2986		4.2	0.6
υAnd([2])	b	0.64	4.617	52.2	0.058	0.01
F8 V	c	1.79	241.16	5.29	0.805	0.27
	d	3.53	1276.1		2.543	0.25
HD 12661([2])	b	2.30	263.6	5.48	0.823	0.35
G6	c	1.57	1444.5		2.557	0.20
HD 169830([5])	b	2.88	225.62	9.32	0.81	0.31
F8 V	c	4.04	2102		3.6	0.33
HD 37124([2])	b	0.86	153.3	12.7	0.543	0.1
G4 IV-V	c	1.0	1942		2.952	0.4
HD 168443([2])	b	7.64	58.1	30.5	0.295	0.53
G5	c	16.96	1770		2.873	0.20
HD 74156([2])	c	1.61	51.6	> 51	0.745	0.65
G0	b	>8.2	>265		>3.82	0.35
HD 38529([2])	b	0.78	14.309	152	0.129	0.29
G4	c	12.7	2174.3		3.68	0.36

† Update: Oct. 2004. ϵ Eri pair is too uncertain and was not included.
([1]) Ref: Ferraz-Mello *et al.* (2005).
([2]) Ref: Fischer *et al.* (2003).
([3]) Ref: McArthur *et al.* (2004).
([4]) Refs: Santos *et al.* (2004), McCarthy *et al.* (2004).
([5]) Ref: Schneider (2004).

Class II. Non-resonant Planets with a Significant Secular Dynamics

The period ratio of the planet pairs considered in this class lies above 4.5 and is large enough to make very difficult a capture into a MMR. The gravitational interaction between these planets may be strong but the conservation of the angular momentum limits the eccentricity variations, allowing them to remain stable even if not in a MMR. They present long-term variations, primarily described by secular perturbations, large

Table 2. Planetary System of the pulsar PSR B1257+12†

Star	planet	mass (m_{Earth})	period (days)	period ratio	semi-major axis (AU)	eccentricity	inclination (degrees) ‡
B1257 +12	A	0.020‡‡	25.262	2.63	0.19	0.	
	B	4.3	66.5419	1.47	0.36	0.0186	53
	C	3.9	98.2114		0.46	0.0252	47

† Ref: Konacki and Wolszczan (2003). Adopted pulsar mass: $1.4 M_\odot$
‡ over the tangent plane to the celestial sphere.
‡‡ adopting the inclination $i = 50°$

variation of the eccentricities and interesting dynamical effects such as the alignment and anti-alignment of the apsidal lines (see Michtchenko and Malhotra, 2004).

Among the extra-solar planetary systems, the most conspicuous examples in this class are three important systems: the outer planets of μ Ara, HD 12661 and the outer planets of υ And. In all cases, they do not seem to be in MMR or close to one; the outer planets of υ And are paradigms of systems showing apsidal lock due to a non-resonant secular dynamics.

Class III. Weakly interacting Planet Pairs

These are the planets given at the bottom of table 1. One example is HD 38529 where $P_2/P_1 \sim 150$. The outermost and innermost planets of 55 Cnc are such that $P_2/P_1 \sim 1700$ forming the more separated known pair (out of the scale of fig. 1). The large period ratios (> 30) allow these planets to be considered as weakly interacting planets; the mutual gravitational interaction exists, but is less important than in the previous case and the probability of capture in a MMR is negligible. We remind that in many dynamical studies of the outer Solar System, the mass of the inner planets is just added to the mass of the Sun!

Two of the candidates to this category are the planets around HD 168443 and HD 74156. However, these systems have some very massive planets and the use of hierarchical models to study these pairs should be preceded of specific analysis.

2. Chaos

Chaos is a common feature in systems with many degrees of freedom. In systems with several planets, the neighborhood of the planets is filled by MMR. In the considered low-eccentricity systems, the near resonant motion of neighboring planets gives rise to a dense set of three-planet resonances, which occur when the periods corresponding to two two-planet MMR form critical linear combinations. Jupiter and Saturn lie very close to the 5:2 MMR, Uranus is confined between the domains of the overlap of the 7:1 MMR with Jupiter and the 2:1 MMR with Neptune in one side, and the 3:1 MMR with Saturn in the other; finally, Neptune is close to the 2:1 MMR with Uranus.

The resonant structure of these systems may be known through dynamical maps constructed in the neighborhood of the actual planets (Michtchenko and Ferraz-Mello, 2001a). In these maps, the initial values of the semi-major axis and eccentricity of one planet are uniformly chosen on a rectangular grid covering the vicinity of its actual position, while the initial positions of the other planets are fixed to the actual values at the chosen epoch. The short-term oscillations (of the order of the orbital periods) are eliminated by employing a low-pass filtering procedure *on-line* with the numerical integration and the resulting semi-major axes are used to construct the maps. The spectral

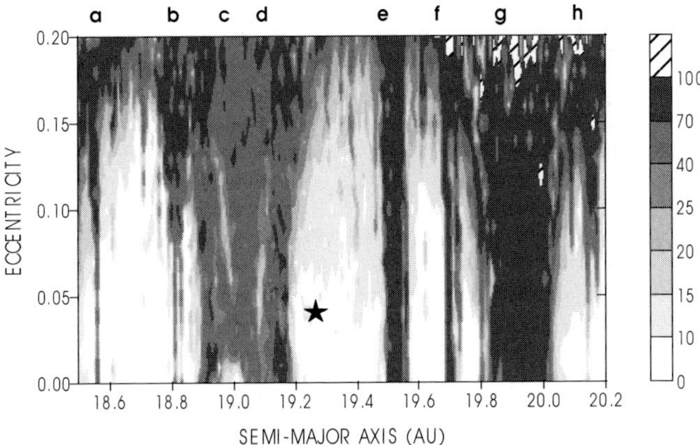

Figure 2. Dynamical map of the neighborhood of Uranus. The main apparent MMR are indicated on top of the figure by the letters a-h and identified in table 3. The star shows the current position of Uranus. (Taken from Michtchenko and Ferraz-Mello, 2001a.)

Table 3. Main MMR in Uranus' neighborhood

Position (fig. 2)	MMR†	Remarks (beat)
a	J–S–4U	beat of J–7U and S–3U
b	3J–10S+7U	beat of 2(2J–5S) and J–7U
c	U–2N	
d	J–7U	
e	2J–6S+3U	beat of 2J–5S and S–3U
f	2J–6S+3U	beat of 2J–5S and 2(S–3U)
g	S–3U	
h	2J–3S+6U	beat of 2J–5S and –2(S–3U)

† $k_1\mathrm{J} \pm k_2\mathrm{S} \pm k_3\mathrm{U} \pm k_4\mathrm{N}$ means the MMR $k_1 n_\mathrm{J} \pm k_2 n_\mathrm{S} \pm k_3 n_\mathrm{U} \pm k_4 n_\mathrm{N} \approx 0$

number N presented in the maps is defined as the number of spectral peaks above 5% of the largest peak. The counting is stopped at $N = 100$. Regular orbits are identified with small values of N, while large numerical values correspond to increasingly chaotic trajectories.

Uranus's neighborhood

We present here the dynamical map of the neighborhood of Uranus (figure 2). This neighborhood is dominated by the 3:1 resonance with Saturn (S–3U) and the 2:1 resonance with Neptune (U–2N), one on each side of the actual position of Uranus. There are also several narrow bands of chaotic motion associated with three-planet MMR (see Table 3). The small hatched domains in figure 2 are those in which collisions (i.e. disrupting close approaches) occur in the time span of the numerical integration (50 Myr).

The neighborhood of the outer pulsar planet

The dynamics of the neighborhood of the pulsar planets is not so complex as the neighborhood of the outer Solar System planets. The dynamical map constructed with minimal masses ($\sin i = 1$) is shown in fig. 3. This neighborhood is dominated by the resonance 2B–3C with narrow bands due to several higher-order resonances (indicated above the top axis of fig. 3).

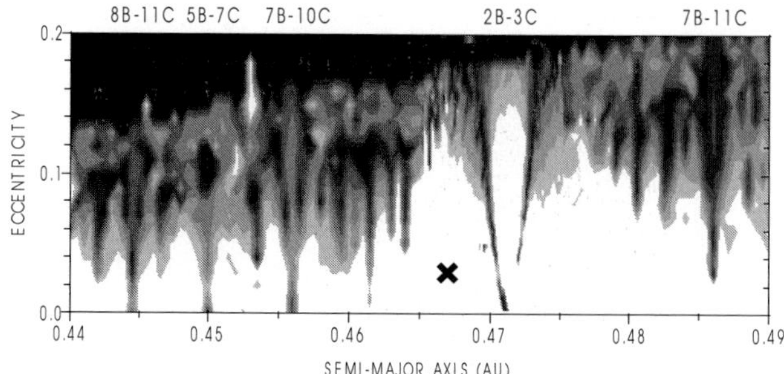

Figure 3. Dynamical map of the neighborhood of PSR B1257+12 planet C. The main apparent MMR are indicated on top of the figure. The cross shows the position of planet C. Same gray scale as fig 2. (Taken from Ferraz-Mello and Michtchenko, 2003.)

3. Dynamics of a MMR and its Neighborhood

We summarize in this section some results of the study of the low-eccentricity dynamics of the 2:1 MMR by Callegari *et al.* (2004) showing how the dynamics of a system of 2 planets evolves when it passes from outside to inside the MMR.

Fig. 4 shows three surfaces of section of the planetary motion defined by the sectioning condition $\theta_1 = 2\lambda_2 - \lambda_1 - \varpi_1 = 0$ and represented on the plane ($x = e_1 \cos \Delta\varpi$, $y = e_1 \sin \Delta\varpi$) with $\Delta\varpi = \varpi_1 - \varpi_2$, for three different energy levels†.

At the right of each surface of section are shown the corresponding dynamic power spectra of the solutions, parameterized by the value of x corresponding to $y = 0$. The points marked in these plots are the points where the frequency in the power spectrum is higher than a given level above noise level. The ordinates are the frequencies (in yr^{-1}). In dynamic power spectra, the lines have the same behavior found in frequency map analysis: frequencies remain almost constant inside resonance islands, have vertical displacements when crossing a saddle point and become erratic when a chaotic layer is reached (Laskar, 1993).

At the lower energy level (top panels in fig. 4), the system is outside the 2:1 MMR; its dynamics is dominated by secular interactions and characterized by two secular modes of motion known from the linear secular theories (see Pauwels 1983). There are two periodic solutions: one at $\Delta\varpi = 0$ and another at $\Delta\varpi = \pi$. In the surfaces of section, these solutions appear as fixed points. These two periodic orbits and the domains around them are indicated as Mode I and Mode II, respectively. The Mode I of motion is located on the right-hand side of the section while Mode II is located on the left-hand side. The curves around the fixed points are quasi-periodic solutions.

Around Mode I, the angle $\Delta\varpi$ oscillates about 0, while around Mode II, $\Delta\varpi$ oscillates about π; in both cases the eccentricity of the planets vary regularly around the eccentricity value of the periodic orbit. Between the two cases, the angle $\Delta\varpi = \varpi_1 - \varpi_2$ is in *direct* circulation. It can be shown that, in this near resonance zone, the critical angles

$$\theta_1 = 2\lambda_2 - \lambda_1 - \varpi_1$$
$$\theta_2 = 2\lambda_2 - \lambda_1 - \varpi_2$$

are in circulation. No infinite-period separatrix exist between the solutions around Mode I and Mode II and the quasi-periodic solutions shown in the surface of section form one continuous family. This fact is clearly seen in the corresponding dynamic power spectrum

† The subscripts 1 and 2 represent Uranus and Neptune, respectively.

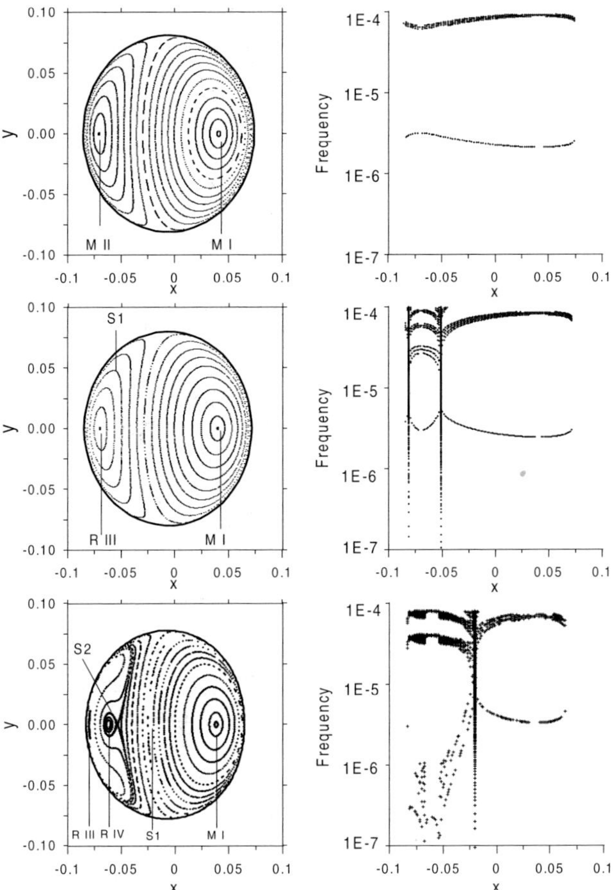

Figure 4. *Left:* Surfaces of Section of the averaged low-eccentricity planetary system in the neighborhood of the 2:1 resonance for three energy levels. Axes $x = e_1 \cos \Delta \varpi$, $y = e_1 \sin \Delta \varpi$. *Right:* Dynamic power spectrum of the solutions parameterized by the value of x corresponding to $y = 0$. Frequencies in yr^{-1}. The planets have the same masses as Uranus and Neptune. (Taken from Callegari *et al.* 2004).

where only two fundamental frequencies and a few harmonics appear. The lower line (at $\sim 3 \times 10^{-6}$ year^{-1}) corresponds to the proper frequency associated with the angle $\Delta \varpi$. We can see that the line shows a discontinuity near the fixed points, indicating that the amplitude associated to the secular frequency tend to zero at those points. The secular period is $\sim 400,000$ years in the center of Mode I and $\sim 300,000$ years in the center of Mode II. The upper line (at $\sim 10^{-4}$ year^{-1}) corresponds to the second fundamental frequency, which is associated with the circulation of the critical angles (transversal to the surface of section).

At the next energy level (middle panels in fig. 4), the resonance 2/1 is already visible. The surface of section seems similar to the previous one, but the dynamic power spectrum shows that an important difference exists. The two vertical broad lines seen in the dynamic power spectrum, at the abscissas corresponding to the curve labeled S1 in the surface of section, indicate that S1 is indeed a separatrix (even if it does not show any visible feature in the surface of section). The domain inside S1 (called RIII by Callegari *et al.*) is a resonant regime of motion in which the critical angles θ_1 and θ_2 oscillate (librate) about 0 and π, respectively. Their difference $\Delta\varpi$ librates around π. Outside S1, θ_2 and $\Delta\varpi$ alternate between libration and circulation, while in general θ_1 remains librating around zero until it reaches the domain formed by oscillations about Mode I. In this case, once again we have circulation of the critical angles. At the highest energy level (bottom panels in fig. 4), a separatrix (S2) appears inside the MMR domain, separating two distinct resonant regimes of motion. The new regime of motion (called RIV by Callegari *et al.*) fills the largest portions of the surfaces of section for higher energies (deep resonance). The dynamic power spectrum (at the right) shows that the lower frequency becomes equal to zero at the two points corresponding to the intersections of S2 with the x-axis. The direction of the phase flow inside the separatrix S2 is inverted with respect to what it is outside that separatrix. A true resonance occurs (with one proper frequency passing through zero). The dynamic power spectrum also shows that the solutions in the immediate neighborhood of S2 are chaotic. The above-discussed example (Callegari *et al.* 2004) is a <u>true</u> secular resonance inside the U-2N MMR†.

The appearance of the surface of section inside RIV is similar to that around MII, with only a different direction of the rotation. However, from the dynamical or topological point of view, these neighborhoods are very different. In the top panels, the apsidal proper frequency is always different from zero and the frequency variation from Mode II oscillations to circulations and to Mode I oscillations (from right to left in fig. 4*top*.) is smooth (continuous). In the bottom panels, the apsidal proper frequency is equal to zero at the separatrix S2; thus, the motions inside and outside S2 are separated by a true infinite-period separatrix (bifurcation) and no continuous transformation exist going across S2.

4. Mean-Motion Resonances and Apsidal Corotations

The surfaces of section in fig. 4 show that the anti-aligned periapses of the solutions in Mode II are preserved in the evolution of a system of two low-eccentricity planets towards resonance (notwithstanding the fact that the critical angles are circulating in Mode II and librating in RIII and RIV). The surfaces of section with energies intermediary between the two shown in fig. 4 (*top* and *middle*) show that the separatrix S1 emanates from the very position of the periodic orbit labeled MII in the secular dynamics and increases up to encompass the whole domain of solutions about $\Delta\varpi = \pi$ and even large amplitude oscillations around $\Delta\varpi = 0$ (see Callegari *et al.* 2004). This is a situation completely different of that occurring in the capture of a particle into a resonance with one planet (or satellite). In that case, the only effect is the capture of the critical angle‡

$$\theta_1 = (p+q)\lambda' - p\lambda - q\varpi$$

about 0 or π. The sidereal periods of particle and planet become approximately commensurable but the pericenter of the particle orbit (whose longitude is ϖ) continues to rotate. That is, $\Delta\varpi = \varpi - \varpi'$ has a monotonic time variation. However, for some well defined values of the eccentricity of the planet (satellite) orbit, it happens that not only the angle θ_1 but also the angle

$$\theta_2 = (p+q)\lambda' - p\lambda - q\varpi'$$

† Other examples of <u>true</u> secular resonance were given by Michtchenko and Ferraz-Mello (2001b) and Michtchenko and Malhotra (2004).
‡ λ and ϖ are the mean longitude and longitude of the periapsis of the trapped particle, respectively, and λ' and ϖ' are those of the trapping planet (or satellite).

is trapped in the neighborhood of 0 or π. Consequently, $\Delta\varpi$ is no longer circulating but librating (Ferraz-Mello *et al.* 1993). This is the same phenomenon known as corotation resonance in disc and ring dynamics (motion following the resonance pattern speed), extended to beyond the narrow 1:1 resonance case of the epicyclic orbits theory. The characterization of corotation resonance given by Greenberg and Brahic (1984): "resonance that depends on the eccentricity of the perturbing satellite, rather than on the eccentricity of the perturbed particle" means that we have a corotation resonance when θ_2 is in libration. But, in the trapping particle problem, θ_2 cannot be in libration if θ_1 is circulating. Therefore the simultaneous libration of $\Delta\varpi$ and θ_1 is synonymous of resonance corotation.

The resonant planar planetary three-body problem (averaged over short-period terms) is a two-degree of freedom system (see Beaugé *et al.* 2003). "Exact apsidal corotations" are solutions for which the angles

$$\theta_1 = (p+q)\lambda_2 - p\lambda_1 - q\varpi_1$$
$$\theta_2 = (p+q)\lambda_2 - p\lambda_1 - q\varpi_2.$$

and the momenta I_1, I_2 conjugated to them remain constant in time. It is important to notice that these equilibrium solutions of the averaged equations correspond to periodic orbits of the non-averaged problem. Initial conditions close to them are periodic solutions of the averaged equations (quasi-periodic solutions of the non-averaged problem): oscillations around the fixed point of the averaged system. One such solution with finite amplitude oscillations will be generically referred to as "apsidal corotation resonance ", or, for short, "apsidal corotation".

Although apsidal corotations have gained certain notoriety in exoplanetary dynamics, they are not new and can be found in our own Solar System. It has long been known (see Greenberg 1987 and references therein) that the Io-Europa pair is trapped in a 2/1 MMR and is in apsidal corotation. Both θ_1 and $\Delta\varpi$ oscillate (with very small amplitude) around fixed values. The exact apsidal corotation is defined in this case by $\theta_1 = 0$ and $\Delta\varpi = \pi$. We will refer to this case as a $(0, \pi)$-corotation, a denomination more accurate than just saying that the apsides are *anti-aligned*.

GJ 876, the first extra-solar resonant planetary system ever discovered around a main-sequence star, also exhibits an apsidal corotation, although in this case the angular variables oscillate around $\theta_1 = 0$ and $\Delta\varpi = 0$. We will denote this as an $(0, 0)$-corotation (the apsides are *aligned*).

The difference in behavior in these corotations is associated with the eccentricities of these solutions (see Lee and Peale, 2002; Beaugé *et al.* 2003): The $(0, \pi)$ solutions occur for low eccentricities of the bodies, while $(0, 0)$ corotations occur for larger eccentricities. Until recently there was little information about the exact border between both modes. The first study of families of periodic solutions of the exact equations of planets in 2/1 MMR, considering different eccentricities but constant planetary masses, is due to Hadjidemetriou (2002). Via numerical continuation of initially circular orbits for planetary masses similar to GJ 876, Hadjidemetriou (2002) found that the $(0, \pi)$ and $(0, 0)$ families are actually linked at $(e_1, e_2) = (0.097, 0)$, and it is possible to pass from one to another by a smooth variation of the total angular momentum.

Later discoveries of other extra-solar candidates in the same commensurability (as HD82943) also showed the system in apsidal corotation. The same seems to be true in other cases as the published orbits of two of the 55 Cnc planets, which lie in the 3/1 MMR. Even though some orbital fits are very imprecise, they seem to indicate that apsidal corotation resonance constitutes strong stabilizing mechanisms for high-eccentricity

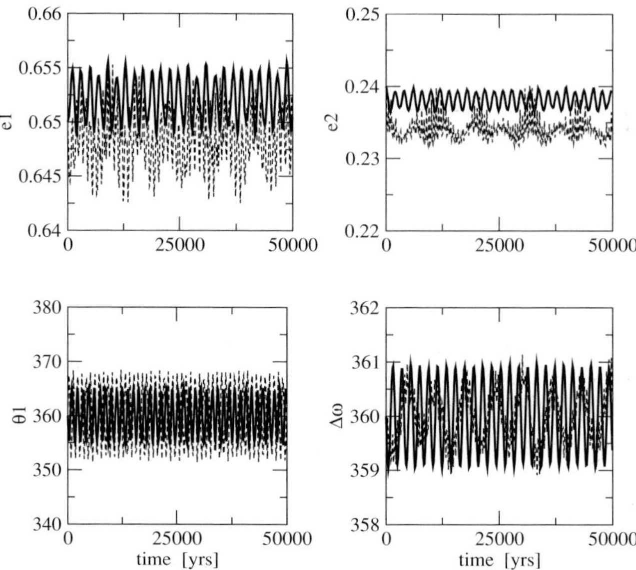

Figure 5. Two solutions in $(0,0)$ apsidal corotation. Solid lines correspond to the masses and initial conditions given in Table 4. Dashed lines are solutions obtained with same initial eccentricities and angular variables, but increasing both the masses and semi-major axes by a factor 10.

resonant planets orbiting close to the central star. These examples justify the significant effort done in recent years to study the location and stability of all apsidal corotations allowing all families of corotations to be determined as function of the planets masses and orbital elements.

Systematic searches using averaged analytical models and numerical simulations with adiabatic migration were undertaken by Beaugé *et al.* (2003) and Ferraz-Mello *et al.* (2003) aiming at finding all possible stable apsidal corotations in the 2:1 and 3:1 MMR, as functions of all the parameters of the system. The first results were that, up to second order of the masses, the position of corotations is only a function of the mass ratio m_2/m_1 of the planets, not depending on the individual masses themselves (as long as they are not large).

Similarly, the solutions are practically independent of the semi-major axes a_1, a_2, but only vary according to the value of the ratio $\alpha = a_1/a_2$. These facts are illustrated in figure 5 where solid lines correspond to a numerical simulation of a $(0,0)$-corotation (using initial conditions shown in Table 4) while dashed lines show results obtained with the same initial conditions, but increasing the planetary masses and semi-major axes by a factor 10. Although the oscillations around the exact apsidal corotation have different periods (see Beaugé *et al.* 2004), the overall behavior is practically the same. This makes clear that solutions found for given values of the six-parameter set $(m_2/m_1, \alpha, e_1, e_2, \theta_1, \Delta\varpi)$ should be valid for any resonant exoplanetary system, independent of their proximity to the star or size of the bodies.

Another important result of these investigation was the discovery of a different mode of apsidal corotation, where the equilibrium values of the angles are not equal to zero or π. They were found in both the 2/1 and 3/1 MMR and we called them "asymmetric corotations". The published orbits of the inner planets of 55Cnc seem to correspond to such an asymmetric configuration (Beaugé *et al.* 2003, Zhou *et al.* 2004).

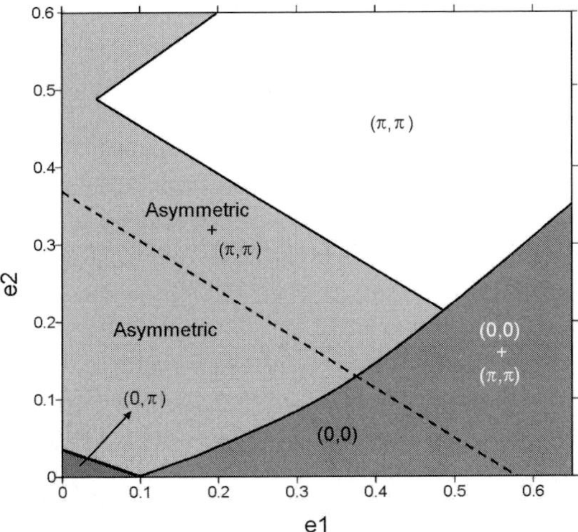

Figure 6. Domains of different types of exact apsidal corotations in the 2/1 mean-motion resonance in the e_1, e_2-plane. See text for details.

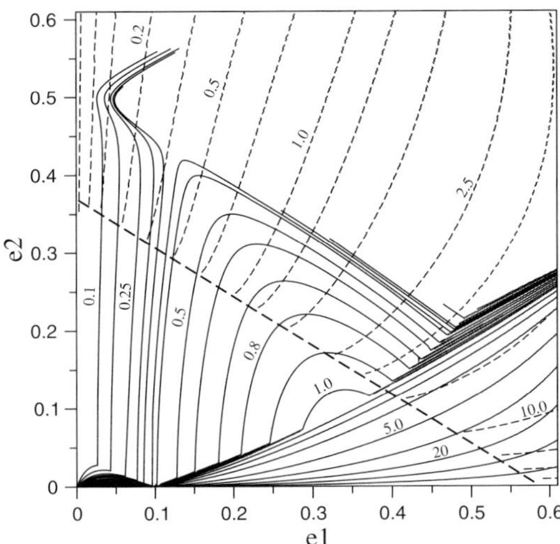

Figure 7. Level curves of constant mass ratio (i.e. $m_2/m_1 = const.$) in the eccentricitie plane for the 2/1 MMR. Broken curves show the (π, π) corotations.

However, the diversity of apsidal corotations does not stop here. Numerical studies by Hadjidemetriou and Psychoyos (2003) and by Ji *et al.* (2003) have shown a new type of solutions at very high eccentricities. Although they are symmetric, they correspond to equilibrium values $\theta_1 = \pi$, $\Delta\varpi = \pi$. We have called them (π, π)-corotations. The orbital elements initially published for HD82943 (Geneva planet search web page, July 31th, 2002) seemed to correspond to such a configuration.

Lee (2004) used numerical simulations with differential migration to map the extent of this new family, finding that stable solutions are located beyond the line defined by the collision condition of the (π, π) corotations: $a_1(1 + e_1) = a_2(1 - e_2)$. This theoretical boundary is a limit for planet masses tending to zero. In fact, averaged models are not valid if the two planets come very close one to another, the minimal distance allowed being proportional to the cube root of the planet masses (see Gladman, 1993). A similar task was also undertaken by Beaugé *et al.* (2004) with a semi-analytical method.

The main results for the 2/1 resonance are seen in Figure 6, which shows the limits of the domains of all types of apsidal corotation in the (e_1, e_2)−plane. One may also note that the (π, π)-corotations domain covers a large area at very high eccentricities; some parts of this domain overlaps with the asymmetric region, as well as the domain of $(0, 0)$-corotations, for smaller values of e_2. In the overlapping area, two distinct types of stable solutions coexist, in some cases even for the same values of the mass ratios.

Numerical simulations by Ferraz-Mello *et al.* (2003) and Lee (2004) have shown that $(0,0)$, $(0,\pi)$ and asymmetric apsidal corotations are linked by isopleths of equal mass ratio m_2/m_1 (see fig. 7). Every solution corresponds to a well-defined value of the mass ratio and, depending on the value of this ratio, a given exoplanetary system may exhibit different types of corotations. All exact apsidal corotations in these families can be reached by constant mass-ratio paths starting from initially circular orbits. This is not the case of the (π, π) apsidal corotations, which do not appear to be reachable via a smooth variation of the parameters from a path starting from the low-eccentricity domains. Even in the domain where $(0, 0)$ and (π, π) overlap it does not seem possible to have a smooth change of one into another.

Figures 8 and 9 show concrete examples of different solutions of this problem. In all cases, we can see a very stable motion around one exact apsidal corotations with a relatively small amplitude of oscillation.

Figure 8 shows two asymmetric solutions. As pointed by Lee (2004), if the mass ratio satisfies the condition $m_2/m_1 > 0.4$, the asymmetric family returns to the $(0,0)$ region for large values of e_1. Conversely, if the mass ratio is smaller than this value, the asymmetric corotations seem to converge to a thin diagonal region for large values of e_2. The two numerical simulations shown in this figure correspond to each case. In gray we can see an example of the upper high-eccentricity domain, while in black we present an asymmetric corotation in the main domain.

Figure 9 shows examples of the $(0, 0)$ and (π, π) cases taken in the rightmost area where both solutions are possible. The two orbits shown have the same planetary masses and same initial semi-major axes and eccentricities. Only the angles are different showing that both apparently opposite behaviors are possible. We point out that, in the two cases, the eccentricities do not have exactly the same averaged value (although very similar) due to the different initial phase angles chosen in each case.

The examples shown in figs. 5, 8 and 9 were obtained with numerical integrations of the exact equations with the masses and initial elements shown in Table 4.

5. Planetary Migration

Although apsidal corotation is a necessary condition for the survival of massive planets in nearby high-eccentricity orbits, they are nevertheless very particular solutions of the planetary three-body problem. The fact that they are verified by some exoplanetary systems raises interesting questions about their origin: Were the planets formed in such orbits? Or did they evolve towards them?

From early statistical studies by Roy and Ovenden (1954), it is known that it is highly improbable to find two massive bodies in an exact mean-motion resonance if they were formed independently. For instance, in the outer Solar System, the existence of several pairs of resonant satellites can only be explained by a past smooth variation of their semi-major axes due to tidal interactions; in other words, the current configuration is due to a "migration" of the primordial non-resonant orbits until a *resonance trapping* took place.

		$(0,\pi)$	Asym	U.Asym	$(0,0)$	(π,π)
m_1		1.0	1.0	1.0	1.0	1.0
m_2		0.016	0.607	0.0136	5.5	5.5
a_1		0.6295	0.6291	0.6295	0.6295	0.6295
a_2		1.0	1.0	1.0	1.0	1.0
e_1		0.03	0.20	0.13	0.65	0.65
e_2		0.02	0.28	0.57	0.24	0.24
λ_1		0.0	36.92	178.4	0.0	180.0
λ_2		0.0	216.92	358.4	180.0	180.0
ϖ_1		0.0	0.0	0.0	0.0	0.0
ϖ_2		180.0	97.17	264.14	0.0	180.0

Table 4. Initial conditions of the small amplitude corotations shown in figs. 5, 8 and 9. "Asym" = Main domain of asymmetric corotations; "U.Asym" = Upper domain of Asymmetric corotations. Planetary masses are in units of $10^{-4} M_\odot$. The mass of the central star is taken as $1 M_\odot$

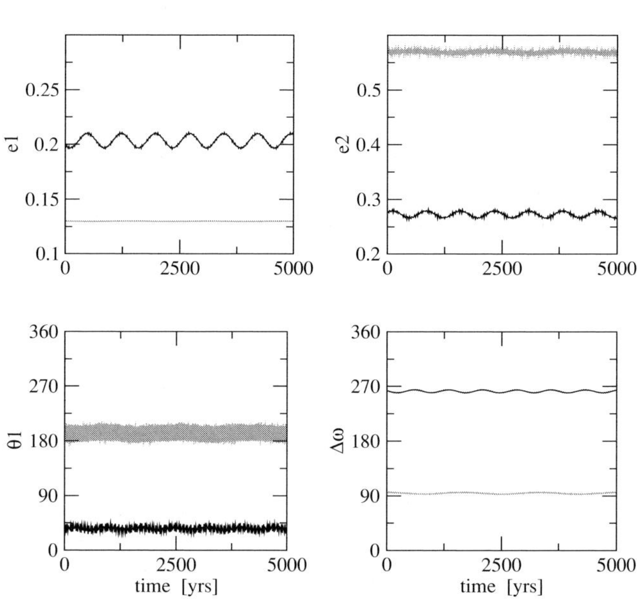

Figure 8. Numerical simulation of two initial conditions in asymmetric corotation. Black shows an example of the low-eccentricity region, while in gray we present a case of the Upper asymmetric domain. Initial conditions are given in Table 4.

After the trapping, both satellites remained locked in stable orbits with commensurable periods. If migration continued after the trapping, both bodies evolved in such a way as to maintain the resonance relationship intact. To say that a similar scenario occurred in exoplanets depends on two things: *(i)* to find a plausible driving mechanism for planetary migration, compatible with the formation process of the system, and *(ii)* to prove that this orbital evolution allows resonance capture, and yields final orbits in apsidal corotation.

In the past few years, several works have undertaken these questions. Although several migration mechanism have been initially proposed, it seems that the most probable process stems from the interaction between the planets and the gaseous primordial disk. Hydrodynamical simulations by Kley (2001, 2003), Snellgrove *et al.* (2001) and Papaloizou (2003), among others, have shown that an adequate choice of the

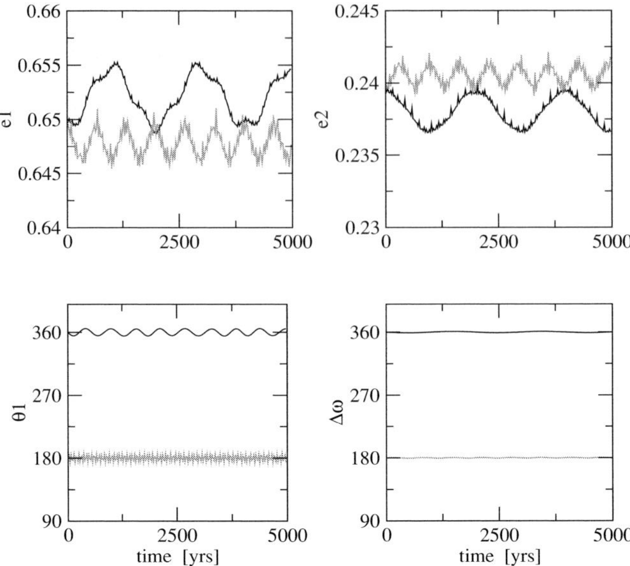

Figure 9. Numerical integration of two orbits with the same masses, semi-major axes and eccentricities. The initial angular variables were varied to place them in a stable $(0,0)$ and (π,π)-corotation. Initial conditions are given in Table 4.

parameters of the gaseous disk can favor both, a large-scale inward migration and a resonance trapping in corotation.

However, not all apsidal corotations can result from this scenario. In the case of the 2/1 resonance, we can group the different families into two distinct classes:

• Type I. Families that can be obtained through analytical continuation from initial circular orbits $e_1 = e_2 = 0$. It includes the $(0,\pi)$, $(0,0)$ and the asymmetrical families.

• Type II. The (π,π) solutions. At variance with the previous ones, this kinds of apsidal corotation does not appear to be reachable via a smooth variation of the parameters from the low-eccentricity domains.

From the point of view of migration, only Type I corotations can be attained through a smooth orbital evolution starting from quasi-circular orbits. Thus, if the planetary migration hypothesis is correct, and if all exoplanets entered the mean-motion resonance in quasi-circular orbits, then we should only expect to observe Type I solutions in real systems. The extra-solar planetary systems presently confirmed in the 2/1 resonance, GJ 876 and HD82843, show Type I apsidal corotations. In the 3/1 resonance, the published orbit of 55 Cnc also corresponds to a Type I apsidal corotation (Beaugé *et al.* 2003, Zhou *et al.* 2004).

In fact, there is only one system proposed to be in 2:1 MMR and Type II corotation: the outer planets of HD160691 (Bois *et al.* 2003). However, more recent observations rule out the possibility of a 2:1 MMR in this case (Goździewski *et al.* 2003, McCarthy, 2004). Anyway, a (π,π) apsidal corotation was unlikely, since no adiabatic evolutionary process could lead to such a situation.

6. Conclusions

In this review, we have presented a brief look at recent results on the dynamics of extra-solar planetary systems. We have separated the known planetary systems into four main groups: (Ia) Planets in Mean-motion resonance (Ib) Low-eccentricity near-resonant pairs; (II) Non-resonant planets with a significant secular dynamics; and (III) Weakly interacting planet pairs. This division reflects the different degree of interaction between the planets and, consequently, the different analytical models necessary to study their evolution and stability.

Recently, the works of Callegari *et al.* (2004) and of Michtchenko and Malhotra (2004), on the dynamics of systems with a significant secular dynamics, have shown new types of motion, many of which unknown until very recently. The same can be said for MMR, where an increasingly complex structure has been discovered. Even such simple solutions as the exact apsidal corotations (which are simply fixed points of the averaged system) show a large diversity of behavior.

A more difficult question is whether each class of extra-solar planetary systems is indicative of different past evolutions and formation scenarios. Resonant pairs are strong evidence in favor of a large-scale planetary migration before the dispersal of the stellar nebula. However, it is not very clear how systems with small period ratio but not resonant fit into this picture. Future work on disk-planet interactions, as well as better models of capture into resonance and/or scattering will help us define more clearly the relation between present day dynamics and the planetary origin.

Acknowledgements

This work has been supported by the Argentinian Research Council – CONICET –, the Brazilian National Research Council – CNPq – and the São Paulo State Science Foundation – FAPESP.

References

Beaugé, C., Ferraz-Mello, S. and Michtchenko, T.A. 2003, *Astrophys. J.*, 593, 1124

Beaugé, C., Ferraz-Mello, S. and Michtchenko, T.A. 2004, *Mon. Not. R. Astron. Soc.*, submitted (astro-ph/0404166)

Bois, E., Kiseleva-Eggleton, L., Rambaux, N. and Pilat-Lohinger, E. 2003, *Astrophys. J.*, 598, 1312

Callegari Jr., N., Michtchenko, T. and Ferraz-Mello, S. 2004, *Cel. Mech. Dyn. Astron.* 89, 201

Ferraz-Mello, S., Tsuchida, M. and Klafke, J.C. 1993, *Cel. Mech. Dyn. Astron.*, 55, 25

Ferraz-Mello, S. and Michtchenko, T.A. 2003, *Rev. Mexic. Astr. Astrof.* (serie conf.), 14, 7

Ferraz-Mello, S., Beaugé, C. and Michtchenko, T.A 2003, *Cel. Mech. Dyn. Astron.* 87, 99

Ferraz-Mello, S., Michtchenko, T.A. and Beaugé, C. 2004, *Astrophys. J.*, in press

Fischer, D.A., Marcy, G.W., Butler, R.P., Vogt, S.S., Henry, G.W., Pourbaix, D., Walp, B., Misch, A.A. and Wright, J.T. 2003, *Astrophys. J.* 586, 1394

Gladman, B. 1993, *Icarus*, 106, 247

Greenberg, R. and Brahic, A. (eds.) (1984). *Planetary Rings*, Univ. Arizona Press.

Goździewski, K., Konacki, M. and Maciejewski, A.J. 2003, *Astrophys. J.* 594, 1019

Greenberg, R. 1987, *Icarus*, 70, 334

Hadjidemetriou, J. 2002, *Cel. Mech. Dyn. Astron.*, 83, 141

Hadjidemetriou, J. and Psychoyos, D. 2003, in: G. Contopoulos and N. Voglis (eds.) *Galaxies and Chaos*, Lecture Notes in Physics. Springer-Verlag, p 412

Ji, J., Liu, L., Kinoshita, H., Zhou, J., Nakai, H. and Li, G. 2003, *Astrophys. J.*, 591, L57

Kley, W. 2001, *Mon. Not. R. Astron. Soc.*, 313, L47

Kley, W. 2003, *Cel. Mech. Dyn. Astron.*, 87, 85

Konacki, M. and Wolszczan, A. 2003, *Astrophys. J.* 591, L147

Laskar, J. 1993, *Physica D*, 67, 257

Lee, M.H. and Peale, S.J. 2002, *Astrophys. J.*, 567, 596

Lee, M.H. and Peale, S.J. 2003, *Astrophys. J.*, 592, 1201

Lee, M.H. 2004, *Astrophys. J.*, 611, 517

McArthur, B.E., Endl, M., Cochran, W.D., Benedict, G.F., Fischer, D.A., Marcy, G.W., Butler, R.P., Naef, D., Mayor, M., Queloz, D., Udry, S. and Harrison, T.E. 2004, *Astrophys. J.* 614, L81

McCarthy, C., Butler, R.P., Tinney, C.G., Jones, H.R.A., Marcy, G.W., Carter, B., Penny, A.J. and Fischer, D.A. 2004, *Astrophys. J.*, submitted (Astro-ph/0409335)

Malhotra, R., Black, D., Eck, A. and Jackson, A. 1992, *Nature* 355, 583

Michtchenko, T. A. and Ferraz-Mello, S. 2001a, *Astron. J.* 122, 474

Michtchenko, T. A. and Ferraz-Mello, S. 2001b, *Icarus* 149, 357

Michtchenko, T. and Malhotra, R. 2004, *Icarus*, 169, 237

Naef, D., Mayor, M., Beuzit, J.L., Perrier, C., Queloz, D., Sivan, J.P. and Udry, S. 2004, *Astron. Astrophys.*, 414, 351

Papaloizou, J.C.B. 2003, *Cel. Mech. Dyn. Astron.*, 87, 53

Pauwels, T. 1983, *Celest. Mech.*, 30, 229

Rasio, F.A., Nicholson, P.D., Shapiro, S.L. and Teukolsky, S.A. 1992, *Nature*, 355, 325

Roy, A.E. and Ovenden, M.W. 1954, *Mon. Not. R. Astron. Soc.*, 114, 232

Santos, N., Bouchy, F., Mayor, M., Pepe, F., Queloz, D., Udry, S., Lovis, C., Bazot, M., Benz, W., Bertaux, J.-L., LoCurto, G., Delfosse, X., Mordasini, C., Naef, D., Sivan, J.P. and Vauclair, S. 2004, *Astron. Astrophys.* (submitted)

Schneider, J. 2004, *Extra-solar planets Encyclopaedia,* `http:// www.obspm.fr/ planets`

Snellgrove, M.D., Papaloizou, J.C.B. and Nelson, R.P. 2001, *Astron. Astrophys.*, 374, 1092

Zhou, L-Y., Lehto, H.J., Sun, Y-S. and Zheng, J-Q. 2004, *Mon. Not. R. Astron. Soc.* (submitted)

Dynamics of Populations of Planetary Systems
Proceedings IAU Colloquium No. 197, 2005
Z. Knežević and A. Milani, eds.

© 2005 International Astronomical Union
DOI: 10.1017/S1743921304008440

Formation, migration, and stability of extrasolar planetary systems

Fred C. Adams

Michigan Center for Theoretical Physics, Department of Physics
University of Michigan, Ann Arbor, MI 48109 USA
email: fca@umich.edu

Abstract. This paper presents recent results concerning the planet formation, planet migration, and the long term stability of planetary systems. Most stars are found in binary systems and binary companions can disrupt both planet formation and stability. We first consider the effects of outer binary companions on the late stages of terrestrial planet formation and show how planet formation depends on the binary periastron. We then consider migration mechanisms for giant planets. In this case, planet scattering produces the full range of orbital eccentricities, but is less effective in moving planets inward (decreasing their semi-major axes). Disk torques are effective at moving planets inward, but not at increasing the eccentricities. We explore a scenario in which disk torques act in concert with planet scattering to provide the full range of orbital elements observed in extrasolar planetary systems. Finally, we consider the longer term stability of Earth-like planets in binary systems; we find that nearly 50 percent of binaries allow for Earth-like planets to remain stable over the current (4.6 Gyr) age of our solar system.

Keywords. Extrasolar planets, planetary dynamics, planet formation

---

1. Introduction

During the course of planet formation, planet migration, and the subsequent evolution of planetary systems, chaos plays an important role in the underlying dynamics. In particular, all of these systems exhibit sensitive dependence on their initial conditions, so that the result of any given process cannot be unambiguously predicted. Instead, the results must be described in terms of the distributions of possible outcomes. In this contribution, we discuss a collection of subproblems in planet formation/migration where the systems display sensitive dependence on the initial conditions. During terrestrial planet formation, the locations, masses, and numbers of planets in the resulting solar system must be described as a distribution. In this context, we consider the late stages of terrestrial planet formation in binary star systems and show that the planet formation process leads to a wide distribution of outcomes, but well-defined trends can still be determined (Figure 1). For example, we can determine the fraction of binary systems that allow the terrestrial planet formation to take place unimpeded. When planets migrate, the orbital elements of the final state display a distribution of values. Here we explore a scenario for giant planet migration in which disk torques act to move the planets inward and planet-planet scattering acts to increase the orbital eccentricities; the resulting distributions of orbital elements predicted by the theory are in reasonable agreement with those of observed extrasolar planets (Figure 2). Finally, planets on unstable orbits in binary star systems display a wide distribution of ejection times (Figure 3). Nonetheless, the distributions of ejection times exhibit well-defined dependences on the binary parameters and we can determine the fraction of binary systems that allow Earth-like planets to remain stable over the current age of the solar system.

2. Terrestrial Planet Formation in Binary Systems

The first part of this work studies the formation of terrestrial planets in binary star systems, with the goal of quantifying the effects of the binary companions on the planet formation process. We note that the extrasolar planets discovered thus far (e.g., Mayor & Queloz 1995; Marcy & Butler 1996; Butler *et al.* 1999; Marcy *et al.* 2001) have masses near that of Jupiter and are thus thought to be gaseous giant planets. Although terrestrial planets have not been detected in extrasolar systems with main-sequence primaries due to their small masses, they are expected to readily form in such systems alongside their Jovian counterparts (e.g., Lissauer 1993). Indeed, the following three lines of evidence suggest that terrestrial planets might be common: (1) Our solar system has produced terrestrial planets, and no instances of astronomical creation are known to be unique. (2) Our solar system has manufactured a large number of moons, asteroids, and other rocky bodies that presumably formed in the same manner as the planets. (3) Terrestrial planets have been discovered in orbit around the pulsar PSR 1257+12 and hence were able to form in a harsh environment (Wolszczan & Frail 1992).

In order to study the formation of terrestrial planets, we have performed a series of N-body simulations using the *Mercury* integration package (Chambers 1999), which has recently been modified to include a stellar binary companion (Chambers *et al.* 2002). These computations follow the evolution of a field of planetessimals as they evolve into a terrestrial planet system (Quintana 2004; Quintana *et al.* 2004). The simulations start with 14 planetary embryos (with mass 5.6×10^{26} g) and 140 planetessimals with a mass ten times smaller (each population of starting bodies thus accounts for half of the mass). The bodies are initially distributed on nearly circular orbits with semi-major axes in the range $0.36 \text{ AU} \leqslant a \leqslant 2.05 \text{ AU}$. This particular set of initial conditions was chosen because it tends to produce terrestrial planet systems like our own when implemented around a single star of one solar mass (Chambers 2001). The simulations run for approximately 100 Myr and evolve into terrestrial planet systems with a wide range of orbital elements and a wide range of planet masses.

These simulations are idealized in that they consider only the late stages of planetessimal accumulation, when planets are assembled from hundreds of smaller rocky bodies. As a result, these simulations do not provide a definitive determination of the full range of terrestrial planet configurations that are possible. However, the part of the problem that we understand best is the orbital dynamics and our codes properly integrate the known equations of motion. In particular, the perturbations due to the binary companion are well-modeled, so we can achieve an accurate representation of the binary effects within this idealized scenario for terrestrial planet formation.

Thus far, we have run ~ 100 simulations of terrestrial planet formation using equal mass binaries with varying orbital elements, in particular the semi-major axis a_b and orbital eccentricity ϵ_b (Quintana 2004; Quintana *et al.* 2004; Lissauer et al. 2004). One set of results is shown in Figure 1. Like most dynamical systems of this type, these systems experience chaotic behavior and exhibit sensitive dependence on the initial conditions. In the present context, for example, moving one planetessimal forward in its orbit by one meter (at the start of the simulation) can lead to a change in the number of planets produced at the end of the simulation. As a result, most quantities of interest (e.g., the number of planets produced or their orbital elements) cannot be described by a single value, but rather by a full distribution, even for effectively equivalent starting conditions. Although a study of the full parameter space for planet formation in binary systems is just beginning, the importance of presenting the results in terms of distributions is already apparent (see also Levison *et al.* 1998). As shown in Figure 1, for the set of simulations

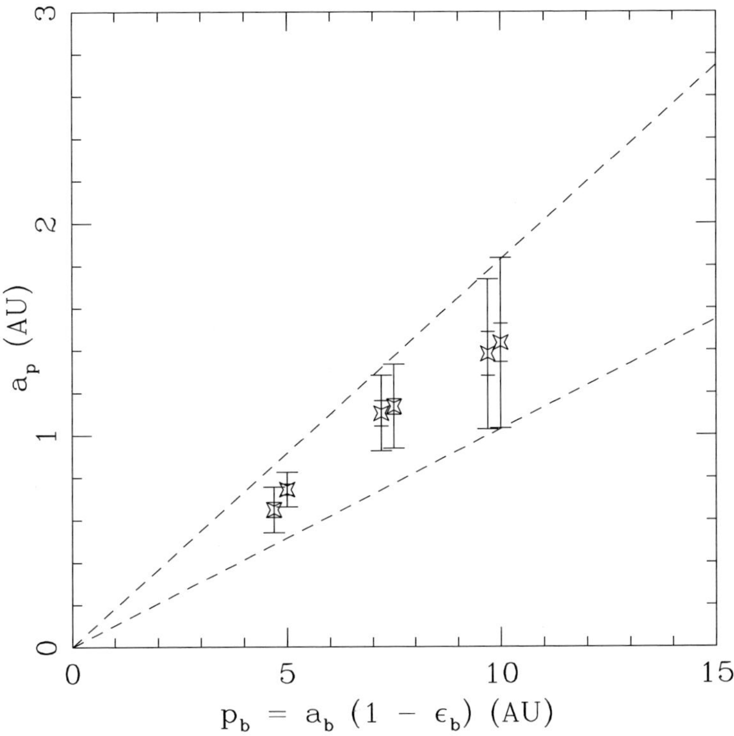

Figure 1. The effect of binary companions on the location of the outermost planet in numerical simulations of terrestrial planet formation in binary systems (Quintana *et al.* 2004). The semi-major axis a_p of the outermost terrestrial planet is plotted as a function of the periastron distance p_b for the binary. The mass ratio of the binary is taken to be unity for all cases shown; the stellar masses are 1.0 $M_\odot$ and 0.5 $M_\odot$ for the pairs of data points shown here. For this ensemble of simulations,the location of the outermost terrestrial planet takes on a distribution of values, with the width of the distribution indicated by the larger error bars. The smaller (inside) error bars indicate the effective error in determining the mean value. The dashed lines bracket the range of planet location as a function of binary periastron.

with a given value of periastron p_b, we find a wide range of values for the location a_p of the outermost terrestrial planet, where the range is depicted by the large error bars in the figure. Nonetheless, well defined trends emerge. In this case, the distribution of terrestrial system sizes (a_p) is a function of periastron: The range of planet locations (the values of a_p for the outermost planet) gets smaller and the mean value systematically decreases as the binary periastron p_b grows smaller. Further, although the initial field of rocky building blocks extends out to 2.05 AU, only a few planets are formed with semi-major axis near 2 AU and none of the simulations produce planets beyond 2 AU. The results shown here (from Quintana *et al.* 2004) thus demonstrate how binary companions can limit the growth of terrestrial planets (see also Levison & Agnor 2003) and demonstrate that many realizations of the problem must be considered in order to sample the full distribution of outcomes.

The results shown in Figure 1 indicate that when binary periastron $p_b < 10$ AU, the binary companion limits the radial extent of the terrrestrial planet region. For sufficiently wide binaries (large peristron values), however, the binary will have minimal effect on the formation of terrestrial planets. In Figure 1, the lower dashed curve, which tracks the lower end of the a_p range, can be extrapolated out to where it exceeds the extent of our terrestrial planet region ($a_p \approx 2$ AU), which occurs at $p_b \approx 19$ AU. Using this extrapolation, our results obtained to date suggest that binaries with periastron $p_b > 19$ AU have little effect on terrestrial planet formation (although more simulations and hence better statistics would be desirable). Given that more than two thirds of stars are found in binary systems (Abt 1983), what fraction of binaries are sufficiently wide, with $p_b > 19$ AU, so that terrestrial planet formation is only minimally affected?

To answer this question, we need to find the fraction of binaries with periastron greater than a set value p, as a function of p. The orbital elements of binaries have reasonably well measured distributions (Duquennoy & Mayor 1991; hereafter DM91). The binary population has a period distribution with a log-normal form; for primary stars with $M_* \sim 1 \, M_\odot$, we can convert the period distribution into the distribution of semi-major axes a_b (using the measured distribution of the mass ratio $\mu = M_C/M_*$ – again see DM91). The eccentricity distribution is also observed, and is independent of semi-major axis for the wide binaries of interest here (DM91). By integrating over the portion of binary parameter space corresponding to periastron greater than a set value p, we numerically obtain the fraction $F(p)$ of binaries with $p_b \geqslant p$ (David $et\ al.$ 2003). A fitting function to the numerically determined result can be written in the simple form

$$F(p) = F_1 \exp\left[-(a\xi + b\xi^2)\right], \tag{2.1}$$

where $F_1 = 0.711$, $a = 0.101$, $b = 0.0287$, and $\xi \equiv \ln[p/(1\mathrm{AU})]$. The function [2.1] provides a good approximation to the numerically determined result, with an absolute error less than about 0.011 and a relative error less than 4 percent. More exact fits are not warranted, as the observed distributions of binary orbital parameters are not known to this accuracy. This function can be used to estimate the fraction of binary systems with periastron greater than any specified value within the allowed range $1 < p < 10^5$. In the present context we find that about 40 percent of binaries have periastron $p_b \geqslant 19$ AU and hence have little effect on terrestrial planet formation.

3. Giant Planet Migration

The extrasolar planets detected thus far (again, see Mayor & Queloz 1995; Marcy & Butler 1996; Butler $et\ al.$ 1999; Marcy $et\ al.$ 2001) have apparently moved from their birth locations in the outer nebula and now reside in rather unusual orbits, with small semi-major axes a and large eccentricities ϵ. An important challenge for planet formation theories is to account for this migration phenomenon. During planet migration, planet-planet scattering is effective at pumping up the orbital eccentricities but is inefficient at moving planets inward (e.g., Adams & Laughlin 2003). Disk torques are effective at moving planets, but inefficient at increasing their eccentricities (for an opposing view, see Ogilvie & Lubow 2003; Goldreich & Sari 2003). A promising mechanism to account for both the semi-major axes a and eccentricities ϵ of the observed extra-solar planets combines dynamical relaxation of two planets with inward forcing driven by tidal interactions with a circumstellar disk (Adams & Laughlin 2003; Moorhead & Adams 2004). The disk exerts a torque on the outer planet, which moves inward and interacts with the inner planet. These planet-planet scattering interactions tend to pump up the eccentricities of both planets. Since these systems are highly chaotic, the outcomes depend sensitively on the initial conditions and the results must be described in terms of distributions. Nevertheless, the resulting distributions of orbital elements (a and ϵ) for the surviving planets are in reasonable agreement with the observed distributions (see Figure 2).

We have conducted a comprehensive numerical study of this migration scenario using two planets and an exterior disk (with $\sim$8000 simulations thus far). In these experiments, the two planets are placed on widely spaced orbits around a solar mass star. The inner planet starts with a period of 1900 days; the initial period of the outer planet is larger by an irrational factor so the planets do not start out in resonance. The planet masses are drawn independently from two different planetary mass distributions (denoted here as the IMF). The first IMF is a random distribution with the planet mass sampled from the range $0 \leqslant m_P \leqslant 5m_J$; the second IMF is a log-random distribution sampled from the range $-1 \leqslant \log_{10}[m_P/m_J] \leqslant 1$. The outer planet is tidally influenced by the circumstellar disk and is gradually driven inwards. The simulations include three additional effects: (1) The angular momentum exchange between the disk and planet damps the orbital eccentricity (e.g., Agnor & Ward 2002) of the outer planet. Since current estimates of this damping time scale give divergent results, we adopt a parametric approach and study the effect of different eccentricity damping time scales drawn from the range 0.1 Myr $\leqslant \tau_{\rm ed} \leqslant 1$ Myr. (2) The force equations include relativistic corrections, which drive the periastron of both planetary orbits to precess and tend to move the planets away from resonance. (3) The simulations include energy loss due to tidal interactions between the planets and the star (e.g., Papaloizou & Terquem 2001).

The simulations are integrated until only one planet remains or the integration time reaches a fiducial scale of 1 Myr. The evolutionary trend is as follows: The planets start out of resonance, but the outer planet is forced inward by the dissipative term until the planets enter into a mean motion resonance (usually the 3:1 resonance because of the starting conditions). The two planets then migrate inwards together, near resonance, but planetary interactions tend to increase the eccentricity of both orbits. The growing eccentricities drive the planets to exhibit ever-larger departures from the resonant condition. Sometimes the outer planet passes through the 3:1 resonance and then becomes held up with a period ratio of 2:1. The eccentricities increase in chaotic fashion until the system (usually) becomes unstable, i.e., until one of the planets is ejected or accreted by a star (or the planets collide). After a planet is lost, however, the orbit of the surviving planet continues to evolve as long as the disk is present. To account for this additional evolution, we assume that the disk has a randomly chosen lifetime (with $\tau_{\rm disk} \leqslant 1$ Myr) and correct the orbital elements of the surviving planet for energy dissipation and eccentricity damping over this time scale.

The orbital elements of the surviving planets display a distribution of values and this distribution is in reasonable agreement with observations. Figure 2 shows the resulting orbital elements in the $a - \epsilon$ plane for two ensembles of theoretical simulations and for the observed extrasolar planets (with the data taken from the California and Carnegie Planet Search website www.exoplanets.org). The open triangles depict the results for planet masses chosen according to the linear IMF, the open squares show the results for the log-random IMF, and the stars represent the orbital elements of the observed planets. To leading order, all three distributions fill the entire portion of the $a - \epsilon$ plane shown here. Upon closer inspection, however, one sees that the theoretical simulations tend to overproduce eccentricity relative to the observed distribution. Over much longer time scales (the $\sim$3 Gyr lifetimes of the primary stars, rather than the $\sim$1 Myr migration times integrated here), tidal interactions between the planets and their parental stars can lead to circularization of the orbits. The solid lines in Figure 2 delimit the portion of the $a - \epsilon$ plane for which this effect is important (where the two lines represent the

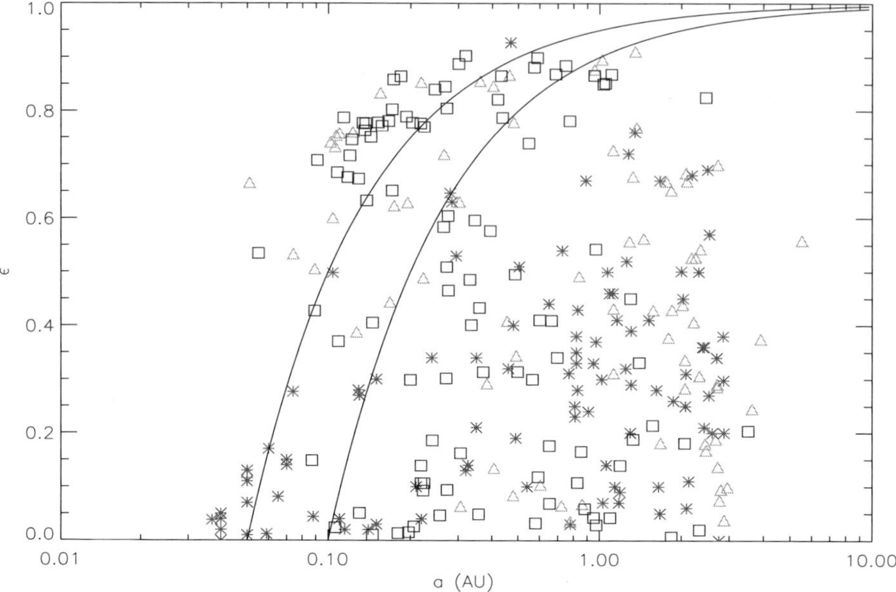

Figure 2. The $a - \epsilon$ plane for the observed population of extra-solar planets and the surviving planets in solar systems starting with two planets surrounded by a circumstellar disk (from Moorhead & Adams 2004). The star symbols represent the observed planetary orbits. The open triangles show the surviving planets for theoretical simulations using a linear IMF. The open squares show the surviving planets for simulations using a log-random IMF. As shown here, this planet migration process results in a distribution of orbital elements. The theoretical model (with multiple planets and a disk) produces distributions of orbital parameters in reasonable agreement with the observed distribution.

variation due to the possible range in the planet's Q-value: see, e.g., Wu & Goldreich 2002). With this correction for tidal circularization, the theoretical distributions are in decent agreement with the observed distribution of orbital elements.

The observed distribution of planetary mass is close to a log-random distribution. Although the planets of lowest mass tend to be those that are ejected, we find that the final mass distribution for simulations started with a log-random IMF are in good agreement with the observed mass distribution. For this log-random IMF, the numerical experiments end with about 33% ejections, 33% accretion events, and 3% collisions. The remaining simulations reach the stopping time of 1 Myr, corresponding to depletion of the circumstellar disk, without losing a planet. As a result, about one third of the systems are predicted to have multiple planets, a fraction that is somewhat higher than that of the observational sample (although additional planets might be found in the future). As expected, the outer planets are more likely to be ejected, whereas the accreted planets are almost exclusively the inner planets. As mentioned above, the ejected planets tend to be the lighter ones: The mean mass of the ejected planets is about $\langle m_P \rangle \approx 1.2\ m_J$, nearly four times smaller than the mean mass of the ejectees (with $\langle m_P \rangle \approx 4.4 m_J$). The average time for the first planet to be ejected is about 0.5 Myr, comparable to the viscous damping time of 0.3 Myr. Finally, we note that the mean ejection speed is about 5 km/s, a result that has consequences for future observations of freely floating planets (Moorhead & Adams 2004).

4. Stability of Earth-like Planets in Binary Systems

Most stars have companions: The majority of solar type stars reside in binary systems (Abt 1983) and thus have stellar companions. In addition, recent discoveries of extrasolar planets show that Sun-like stars often have planetary mass companions (e.g., Butler *et al.* 1999; Marcy et. al 2001). The distributions of orbital parameters for binary star systems are relatively well known (DM91) and the orbital parameters for extrasolar planets are under intense study. If terrestrial planets form in these solar systems, the companions can affect their prospects for long term orbital stability. This section addresses the issue of planetary stability with the overall goal of estimating the fraction of solar systems that allow an Earth-like planet to remain stable over the current age of our Sun (we note that a great deal of previous work has been done, e.g., Pilat-Lohinger & Dvorak 2002).

To study the stability of Earth-like planets in binary systems, we have performed large numbers ($\sim$100,000 to date) of three-body simulations (David *et al.* 2003; Fatuzzo *et al.* 2004). Specifically, we consider the possible ejection of an Earth-like planet that starts in a circular orbit with radius 1 AU around its parent star, with a stellar or planetary companion that acts as a source of gravitational perturbations. Through long term dynamical interactions with the companion, the orbital elements of the Earth-like planet evolve, generally in chaotic fashion (Wisdom & Holman 1991), until the planet is ejected from the system. For the sake of definiteness, we use a solar mass primary and a planet with one Earth mass. The mass M_C, eccentricity ϵ, and semi-major axis a of the companion body are then specified for each run. We consider numerical experiments with the same binary orbital parameters (M_C, a, ϵ) and a random distribution for the remaining (angular) orbital elements. In our initial set of simulations (David *et al.* 2003), the inclination angle was set to $i = 0$ so the systems are co-planar; ongoing work is exploring the effects of $i \neq 0$ (Fatuzzo *et al.* 2004; see also Pilat-Lohinger *et al.* 2003). For each set of initial conditions, we integrate the system forward in time until Earth is ejected, or it collides with either star.

For simulations with given binary properties (M_C, a, ϵ), the ejection time $\tau_{\rm ej}$ varies with the choices of the remaining orbital elements. The systems are highly chaotic and this variation is not smooth. Figure 3 shows that the ejection time displays a log-normal distribution for an ensemble of different realizations of the same underlying problem, i.e., the same (M_C, a, ϵ) and a random sampling of the remaining orbital parameters. These systems thus display a distribution of ejection times with two important properties: The width of the distribution is substantial and the distribution is log-normal (so that $\log \tau_{\rm ej}$ is the relevant variable for doing statistics). The distribution of ejection times shown in Figure 3 corresponds to one point in the parameter space (M_C, a, ϵ). Since each point in the space corresponds to a distribution of ejection times, one must perform multiple realizations of the three-body problem for each point in question. Suppose, for example, that each point (M_C, a, ϵ) has the same width $\sigma \approx 0.5$ for its distribution of $\log \tau_{\rm ej}$. To determine the ejection time itself to 1 percent accuracy, one needs about 2625 different realizations of the problem.

As mentioned earlier, nearly two-thirds of solar-type stars live in multiple systems (Abt 1983). To estimate the fraction of binaries that allow for stable Earth-like orbits, we must combine our exploration of planetary stability with the observed distributions of binary parameters (DM91). Our initial survey of parameter space (David *et al.* 2003) indicates that the ejection time of an Earth-like planet is a steeply increasing function of the binary periastron p (see also Chambers *et al.* 1996, Holman & Wiegart 1999). To a good working approximation, the ejection time varies with periastron according to the exponential law $\tau_{\rm ej} = \tau_0 \exp[(p - 1)/\ell_0]$, where the length scale ℓ_0 and time scale

τ_0 depend on the companion mass M_C and where are lengths are given in AU (David *et al.* 2003). Although the numerical simulations are only carried out to $10 - 100$ Myr, the steepness of this time scale function allows us to extrapolate the ejection times to the age of the solar system. To obtain a conservative extrapolation, we use the lower end of the range of ejection times for each value of periastron p and use the results for the largest companion mass M_C. This procedure indicates that an Earth-like planet can remain stable in a binary system if the periastron $p > 7$ AU, so we need to estimate the fraction of binaries with $p > 7$ AU. By integrating over the relevant fraction of binary parameter space (M_C, a, ϵ) using the results from DM91 and equation [2.1], we estimate that 50 percent of binary systems allow Earth-like planets to remain stable for the current age of the solar system (taken to be 4.6 Gyr).

5. Conclusions

This paper presents a collection of recent results concerning planet formation (Quintana *et al.* 2004), planet migration (Moorhead & Adams 2004; Adams & Laughlin 2003), and the longer term stability of planetary systems (David *et al.* 2003). Our findings can be summarized as follows:

During terrestrial planet formation in binary systems, we have shown that effectively equivalent starting conditions lead to a distribution of final solar system properties. In spite of the wide distribution of outcomes, however, we can extract general trends. For example, we find that the extent of terrestrial planet systems, as measured by the semi-major axis of the outermost planet, depends primarily on binary periastron p_b (Figure 1). Combining our simulation results with the observed distribution of binary parameters, we find that about 40 percent of binary systems are wide enough (specifically, with periastron $p_b \geqslant 19$ AU) so that terrestrial planet formation is relatively unaffected.

Next we have explored a promising mechanism for giant planet migration which includes a circumstellar disk and multiple planets. The disk exerts torques on the outermost planet, drives it inward, and thereby decreases its semi-major axis a; the inner planet often becomes locked in resonance with the outer planet and is also driven inward. During the migration epoch, scattering interactions between the planets are effective at increasing the eccentricities of both orbits. In these systems, the resulting distributions of orbital elements (e.g., in the $a - \epsilon$ plane) for the surviving planets are in reasonable agreement with the observed distributions (Figure 2).

Finally, we have considered the longer term stability of planetary systems. Binary companions can disrupt planetary orbits, and the ejection time for Earth-like planets in habitable orbits displays a wide distribution with nearly a log-normal form (Figure 3). Nonetheless, the mean ejection time varies (almost) exponentially with the binary periastron and sufficiently wide systems allow planets to remain stable over long spans of time. Again combining our numerical results with the observed distributions of binary parameters, we find that nearly 50 percent of binary systems allow Earth-like planets to survive over the current (4.6 Gyr) age of the solar system.

The results from this set of numerical investigations emphasize a general aspect of solar system dynamics. For a given set of initial conditions, none of the subproblems studied here actually has an answer. Instead, every set of effectively equivalent starting conditions leads to a full distribution of possible outcomes. The process of terrestrial planet formation leads to a distribution of planet number, planet masses, and planetary orbits. Similarly, the result of planet migration is a distribution of orbital elements for the surviving planets. Even the seemingly simple question – how long can an Earth-like planet survive in a particular binary system – has a distribution of answers (due to the

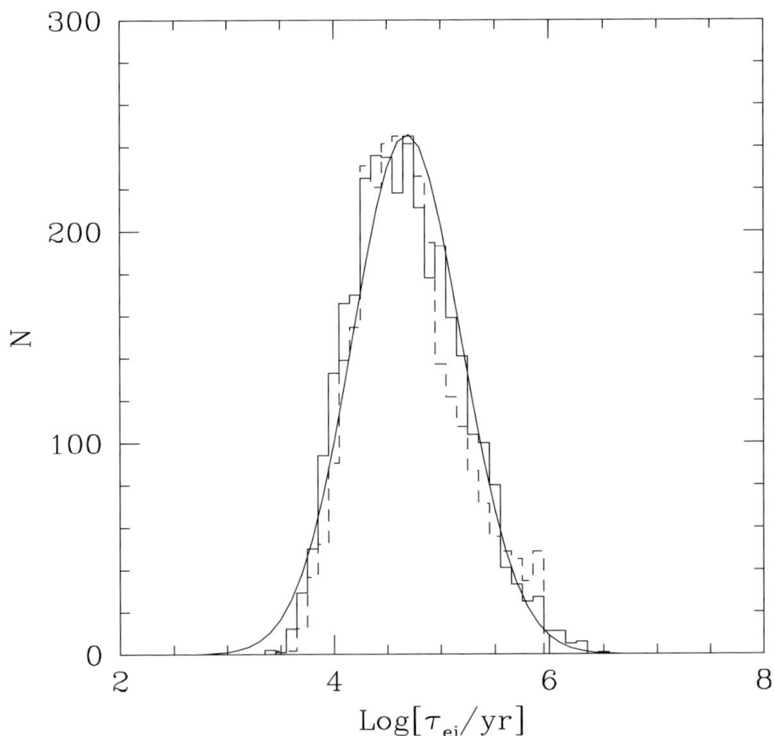

Figure 3. The distribution of ejection times for an Earth-like planet in a binary, using different realizations of the same system (from David *et al.* 2003). In this set of the experiments, the binary companion has mass $M_C = 0.1\ M_\odot$, eccentricity $\epsilon = 0.5$, semi-major axis $a = 5$ AU, and inclination angle $i = 0$. The solid histogram shows the distribution of ejection times resulting from the B-S code (for a random sampling of the remaining orbital elements). The dashed histogram shows the corresponding distribution of ejection times resulting from the symplectic code (again, for a random sampling of orbital elements). The smooth curve shows a log-normal distribution with the same peak value and width as the computed distributions. Notice that the distributions predicted by the two numerical codes are similar and that both have a log-normal form (with the same width and peak location). Because the width of this distribution is substantial, any determination of ejection times must consider a large number of realizations of the underlying problem in order to adequately sample the distribution.

dependence of the ejection time on the starting angular phases of the bodies). As a result, the dynamical systems involved in planet formation, migration, and longer term evolution display sensitive dependence on their initial conditions, one of the signposts of chaotic behavior. In the face of such chaos, the outcome of the planet formation/migration process must generally be described in terms of distributions of possible results. In spite of this complication, however, astronomical results can be obtained: In the context of planet migration, we have shown that our theoretical mechanism provides roughly the same distribution of orbital elements as the observational sample. In addition, we have shown here that 40 percent of binary systems are wide enough to not disrupt the formation of terrestrial planets and that about 50 percent of binary systems are wide enough to allow Earth-like planets to remain stable for the age of the solar system.

Acknowledgements

This contribution describes work done with a number of colleagues and I would like to especially thank the following collaborators: Eva David, Marco Fatuzzo, Greg Laughlin, Jack Lissauer, Althea Moorhead, and Elisa Quintana. In addition I would like to thank the conference organizers and local hosts, especially Zoran Knežević and Milan Ćirković, for their hospitality. This work was supported at the University of Michigan by the Michigan Center for Theoretical Physics and by NASA through the Terrestrial Planet Finder Mission (NNG04G190G) and the Astrophysics Theory Program (NNG04GK56G0).

References

Abt, H. 1983, *ARA&A* 21, 343

Adams, F. C., & Laughlin, G. 2003, *Icarus* 163, 290

Agnor, C. B., & Ward, W. R. 2002, *Astrophys. J.* 567, 579

Butler, R. P., *et al.* 1999, *Astrophys. J.* 526, 916

Chambers, J. E. 1999, *Mon. Not. R. Astron. Soc.* 304, 793

Chambers, J. E. 2001, *Icarus* 152, 205

Chambers, J. E., Quintana, E. V., Duncan, M. J., & Lissauer, J. J. 2002, *Astron. J.* 123, 2884

Chambers, J. E., Wetherill, G. W., & Boss, A. P. 1996, *Icarus* 119, 261

David, E.-M., Quintana, E. V., Fatuzzo, M., & Adams, F. C. 2003, *Publ. Astron. Soc. Pacific* 115, 825

Duquennoy, A., & Mayor, M. 1991, *Astron. Astrophys.* 248, 485 (DM91)

Fatuzzo, M., Adams, F. C., Doering, C. R., & Gaubin, R. 2004, in preparation

Goldreich, P., & Sari, R. 2003, *Astrophys. J.* 585, 1024

Holman, M. J., & Wiegart, P. A. 1999, *Astron. J.* 117, 621

Levison, H. F. & Agnor, C. 2003, *Astron. J.* 125, 2692

Levison, H. F., Lissauer, J. J., & Duncan, M. J. 1998, *Astron. J.* 116, 1998

Lissauer, J. J. 1993, *ARA&A* 31, 129

Lissauer, J. J., Quintana, E. V., Chambers, J. E., Duncan, M. J., & Adams, F. C. 2004, in: *Gravitational Collapse: From Massive Stars to Planets*, in press

Marcy, G. W., & Butler, R. P. 1996, *Astrophys. J.* 464, L147

Marcy, G. W., *et al.* 2001, *Astrophys. J.* 556, 296

Mayor, M., & Queloz, D. 1995, *Nature* 378, 355

Moorhead, A. V., & Adams, F. C. 2004, *Icarus*, submitted

Ogilvie, G. I., & Lubow, S. H. 2003, *Astrophys. J.* 587, 398

Papaloizou, J.C.B., & Terquem, C. 2001, *Mon. Not. R. Astron. Soc.* 325, 221

Pilat-Lohinger, E., & Dvorak, R. 2002, *Cel. Mech. Dyn. Astron.* 82, 143

Pilat-Lohinger, E., Funk, B., & Dvorak, R. 2003, *Astron. Astrophys.*, 400, 1085

Quintana, E. V. 2004, PhD Thesis, Physics Department, University of Michigan

Quintana, E. V., Adams, F. C., & Lissauer, J. J. 2004, in preparation

Wisdom, J., & Holman, M. 1991, *Astron. J.* 102, 1528

Wolszczan, A., & Frail, D. A. 1992, *Nature* 355, 145

Wu, Y., & Goldreich, P. 2002, *Astrophys. J.* 564, 1024

Dynamics of Populations of Planetary Systems
Proceedings IAU Colloquium No. 197, 2005
Z. Knežević and A. Milani, eds.

© 2005 International Astronomical Union
DOI: 10.1017/S1743921304008452

Dynamical evolution of extrasolar planetary systems

Ji-Lin Zhou and Yi-Sui Sun

Department of Astronomy, Nanjing University, Nanjing 210093, China
email: zhoujl@nju.edu.cn,sunys@nju.edu.cn.

Abstract. To date, more than 130 extrasolar planets around main sequence stars are revealed mainly by the Doppler radial velocity measurements. Due to the observational biases, most of the detected planets are moving in orbits close to the host stars, with some in highly eccentric orbits. Dynamical processes during the late stage of planet formation are important to account for the present orbital properties. These processes include: planet migrations and resonance trappings caused by gravitational interactions between protostellar disk and planets, dynamical scattering due to interactions between planets, etc. In this paper, we review the major effects of these dynamical processes on the orbital characteristics of the planet systems.

Keywords. Celestial mechanics, stars: planetary systems

---

1. Introduction

The study of dynamical evolution of planets under mutual gravitational interactions is a classic subject for celestial mechanics, which can be backdated to the works of Laplace and Lagrange in the late eighteen's century. However, as the solar system was the only paradigm of planetary system before 1990's, the classical perturbation theory for planet dynamics was based on near-circular orbits with large separations. The discovery of the extrasolar planets around a neutron star PSR 1257+12 by Wolszczan & Frail (1992) and a solar-type star 51 Peg by Mayor & Queloz (1995) opened a new era of planetary science. To date, more than 130 extrasolar planets around solar-type stars have been discovered (†), among them 11 multiple planetary systems are confirmed. As most of the discovered extrasolar planets are moving in orbits close to the host stars, with some in high eccentricity orbits, both the classical planet formation theories and the classical perturbation and stability theories are facing great challenges.

There have been many excellent reviews on the dynamical evolution of extrasolar planet systems during the past years, e.g., Lin *et al.* (2000), Ward & Hahn (2000), *et al.*. In this paper, we collect some of the current knowledge on the this subject. First we make some statistics on the orbital distributions of the discovered extrasolar planets, then we review some important dynamical processes as well as some recent progresses that are helpful to understand these characteristics.

2. Statistics of Extrasolar Planets

We use the elements of 114 planets, where 111 planets are taken from the web (exoplants.org), and the rest 3 are from the recently discovered "very hot Jupiters": OGLE-TR-56b (Torres *et al.* 2004), OGLE-TR-113b (Konacki *et al.* 2004), OGLE-TR-132b (Bouchy *et al.* 2004). The orbital distributions of these planets show the following statistical properties (see also Udry, Mayor & Queloz 2003, Marcy *et al.* 2003):

† See http://www.obspm.fr/encycl/encycl.html, http://exoplanets.org/

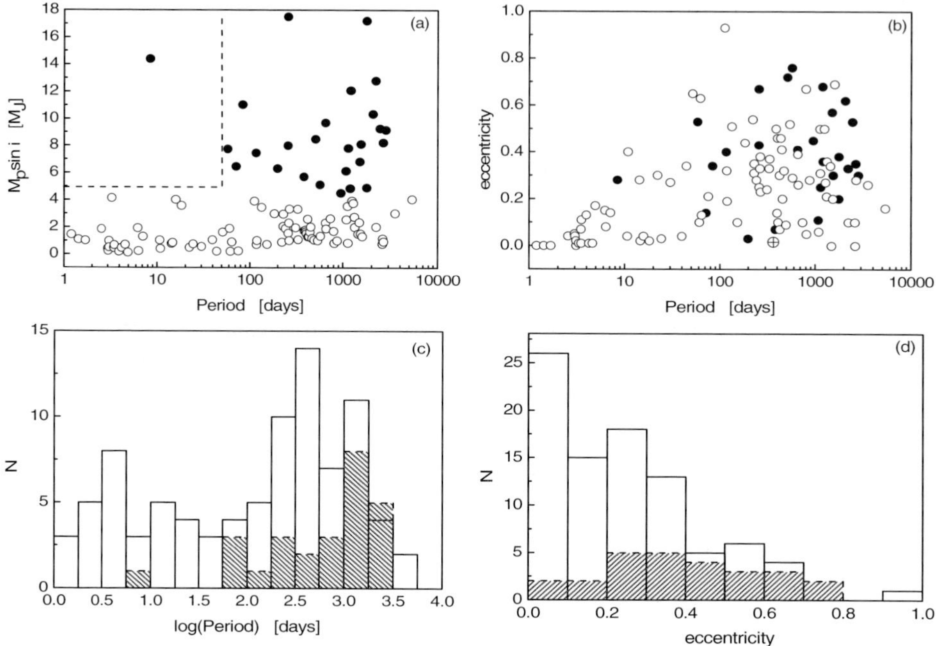

Figure 1. Statistical graphes for the dynamical parameters of the observed 114 planets (see text). In (a) and (b): open circles are for planets with $M_p \sin i < 5M_J$ (total No.=88), and filled circles for those with $M_p \sin i > 5M_J$ (total No.=26). The dotted horizontal (vertical) line in (a) corresponds to $M_p \sin i = 5M_J$ (period= 50days, resp.). In (c) and (d): the distributions with empty (shaded) bars are for planets with $M_p \sin i < 5M_J$ ($M_p \sin i > 5M_J$,resp.).

Period-mass graph. Due to the limitation of the observations, most of the planets are located in short period orbits with masses comparable to Jupiter(Fig.1a). If we classify the planets into two groups: large planets ($M_p \sin i > 5M_J$) and small planets ($M_p \sin i \leqslant 5M_J$), then almost all large planets are located in orbits with $P > 50$days (~ 0.27AU) (Fig.2b). The only one exception at present time is the planet of HD162020, with $M_p \sin i = 14.4M_J$, and a period of 8.4 days. This phenomenon may be the consequence of either the inefficient accretion of planets at small radius during the late stage of formation, or the dependence of migration speed on planet masses after their formation. We will discuss this later.

Period-eccentricity graph. Both larger planets and small planets can reach high eccentricities, indicating that there are some robust mechanisms to pump planet eccentricities (Fig.1b). The tidal circularization time scale is ~ 5.4Gyr for a Jupiter mass planet around a solar mass star in a 3-day period orbit (Lin *et al.* 2000), thus the eccentricities of planets in orbits with periods < 3days were most possibly damped. The detection of the three "very hot Jupiters" reduces the "cutoff" period to 1day (~ 0.02 AU).

Period distribution. There is a slight shortage of small planets at intermediate period ($\sim 10 - 200$ days) orbits (Fig.1c). Large planets locate somewhat evenly in orbits with period $100 \sim 3000$ days, except a small peak at period ~ 1000days.

Eccentricity distribution. The eccentricity distribution of the 26 large planets has an average at 0.39, with a variance 0.2 (Fig.1d). For the 88 small planets, the average value is 0.23, with a variance 0.19. The large difference of the average eccentricities between large and small planets shows that they might have passed some different dynamical histories.

3. Dynamics of Single Planet Systems

We begin with a planet system which is in the late stage of planet formation, so at least one protoplanet (or planet core) was formed in a gaseous disk, together with a planetesimals disk in the equatorial plane of the host star. It is generally believed that short-period planets were formed several AU away from their host stars and subsequently migrated to their present locations (Lin, Bodenheimer & Richardson 1996). On the other hand, orbital migration is a natural consequence of the interactions between gaseous disk and planets.

When the protoplanets formed through core accretion scenario, dynamical frictions with the residual populations of smaller bodies reduce the orbital eccentricities and inclinations of the protoplanets, which results nearly circular, coplanar orbits for the protoplanets (Kokobu & Ida, 1998). Thus the eccentricities of the observed planets could, most probably, be excited after the protoplanets were formed.

3.1. *planet-Gaseous disk interactions*

A protoplanet perturbs its nascent gaseous disk through gravitation. According to the linear theory, the protoplanet exerts torques on a dynamically cold disk ($c \ll r\Omega$, where c, r, Ω are the sound speed of gas, the orbital radius and angular velocity of planet motion, respectively) mainly at the Lindblad and corotation resonances (Goldreich and Tremaine, 1979, 1980). In a Keplerian disk(and hereafter), these resonances occur at locations with $\Omega(r) : \Omega_p = l : (m + \epsilon)$, where Ω_p is the angular velocity of the protoplanet, l, m are integers, $|l-m|$ is the order of the resonance, $\epsilon = -1, 0, 1$ correspond to the inner Lindblad resonances (ILRs), corotation resonances(CRs), and outer Lindblad resonances (OLRs), respectively. The torque of a protoplanet exerts on the disk at a Lindblad resonance (LR) is given as (Goldreich and Tremaine, 1980),

$$T_{l,m}^{L} \approx \frac{\pi^2}{3\epsilon}\left[\sigma\Omega^{-2}\left(\frac{rd\psi_{l,m}}{dr} - \frac{2m}{\epsilon}\psi_{l,m}\right)^2\right]_{r=r_L} \tag{3.1}$$

and that at a corotation resonance (CR) is given as,

$$T_{l,m}^{C} = -\frac{4}{3}m\pi^2\left[\left(\frac{r}{\Omega}\right)\psi_{l,m}^2\frac{d}{dr}\left(\frac{\sigma}{\Omega}\right)\right]_{r=r_C} \tag{3.2}$$

where σ is the surface density of the disk, $\psi_{l,m} \sim O(\mu e^{|l-m|})$ is the coefficient of the protoplanet potential under Fourier expansion, and $\mu = M_p/M_*$ where M_* is the mass of the star. Thus the torques at LRs have the same sign of ϵ, while those at CRs depend on the gradient of σ/Ω.

3.1.1. *Orbital migrations*

There are two main types of migrations for a protoplanet under tidal interaction with the gaseous disk:

Type I migration. Suppose the protoplanet is too small to make large modifications to the surface density of the disk, and the disk has surface density like the minimum-mass solar nebula, where $\sigma \sim r^{-3/2} \sim \Omega$. in such disks the torques at CRs (3.2) vanish in an almost unperturbed disk (Ward 1988, Artymowicz 1993). The differential torques generated by LRs between the inner and outer part (relative to the protoplanet orbit) of the disk drive the protoplanet to migrate inward, which is called type I migration (Ward 1997a). The time scale for the orbital decay, $\tau_I = r/\dot{r}$, is given by (Ward 1997b)

$$\tau_I \sim \frac{1}{2\pi c_1}\left(\frac{M_*}{M_p}\right)\left(\frac{M_*}{\sigma r^2}\right)\left(\frac{c}{r\Omega_p}\right)^3 T_p \tag{3.3}$$

Where c_1 is a constant of order unit, $T_p = 2\pi/\Omega_p$ is the orbital period of the planet. Setting $M_* = M_\odot$, $c/\Omega_p \sim h = 0.07r$, where h the scale height of the disk, $\sigma r^2 = 10^{-3} M_\odot$, thus for a planet with mass $10 M_\oplus$ at $r = 5\text{AU}$, one has $\tau_I \approx 2 \times 10^4$ years.

Gap formation. As the torques at ILRs (OLRs) are negative (positive,respectively), the inner disk loses angular momentum while the outer disk gains it. Hence, a protoplanet tends to form a gap near it, while disk viscosity tends to close it. The critical mass of the protoplanet that can open a gap is (Lin and Papaloizou 1979, Lin and Papaloizou 1993),

$$\mu_c = \frac{40\nu}{r_p^2 \Omega_p} = 40\alpha \left(\frac{c}{r_p \Omega_p} \right)^2 \tag{3.4}$$

where we have used the α as prescribed by the kinematic viscosity (Shakura-Sunyaev 1973), $\nu \sim \alpha c^2/\Omega$. With the above parameters, and set $\alpha \sim 10^{-4}$, we have $m_{gap} = \mu_c M_\odot \approx 7 M_\oplus$. Hydrodynamical simulations show that the above criterion is accurate within a factor of 2 (Bryden *et al.* 1999). It was believed that gap formation will reduce greatly the mass accretion of the protoplanet (Lin and Papaloizou 1993). Hydrodynamical simulation found the mass accretion rate through the gap onto the planet is reduced markedly when the disk viscosity is small($\alpha \leqslant 5 \times 10^{-4}$), but when the disk has large viscosity ($\alpha \sim 10^{-3}$), the protoplanet is still able to accrete lots of mass even after the formation of a nearby gap (Kley 1999).

Type II migration. When a gap is maintained near a protoplanet, the tidal torques at resonances balance the viscosity torque. Thus the protoplanet is locked to the disk, and will drift inward with a time scale determined by the viscosity of the disk:

$$\tau_\nu = \frac{r^2}{\nu} \sim \frac{1}{2\pi\alpha} \left(\frac{r\Omega_p}{c} \right)^2 T_p \tag{3.5}$$

Adopting $\alpha = 10^{-4}$ and $c/(r\Omega) = 0.07$, we find a planet at 5AU will drift inward with a time scale of $\tau_\nu \approx 3 \times 10^6$ years.

3.1.2. *Problems on planet migration*

Time scale of migration. Both type I and type II migrations predict fast inward migrations for a protoplanet after it is formed, especially the type I migration has a time scale of $\sim 10^4$ years, which is much shorter than the age of the gaseous disk ($\sim 10^7$ years). To reduce the risk of migrating planets swallowed by the host star, a possible solution is trying to reduce the minimum mass for gap formation, thus reduce the time of a planet in type I drift. Considering the nonlinear evolution of the density waves in a thick disk could slightly reduce m_{gap} to $2 - 15 M_\oplus$ (Rafikov 2002). Koller, Li & Lin (2003) propose that secondary shearing instabilities will occur near the streamline separatrix at $3.5 - 5$ Hill's radius away form the planet, which can lead to the formation of large vortices and thus results in fluctuating (positive and negative) torques instead of the negative ones. Laughlin, Steinacker and Adams (2004), Nelson and Papaloizou (2004) argue that, in the locations where the magnetorotational instability is active, type I migration can be readily overwhelmed by turbulent perturbations, and gravitational torques arising from magnetohydrodynamics turbulence will contribute a random walk component to the migratory evolution of the planets, thus increasing the drift time scale.

The inward migration of a planet could be stopped near the star as a result of tidal interaction with the host star, or through the evacuation of the inner disk by some processes such as interactions with stellar magnetosphere (Lin, Bodenheimer & Richardson 1996), magnetorotational instability (Kuchner & Lecar 2002), photoevaporation by irradiation from the central star (Matsuyama, Johnstone and Murray 2003), etc. The negligible eccentricity of all extra-solar planets with periods less than 6 days can be accounted for by tidal dissipation induced by their host stars. While the coexistence of planets with periods 7-21 days on both circular and eccentricity can be accounted for by the variations in spin-down rates of the young stars (Dobbs-Dixon, Lin, Mardling, 2004)

Lack of large planets on small period orbits. The observed lack of large planets on small period orbits, if it is a true phenomenon, needs an explanation (Fig.1a). Tidal interactions between host stars and the protoplanets cannot account for this, since the orbital decay timescale would be $\sim 10^{12}$ years for a Jupiter mass planet in an orbit with period of 10 days (Terquem *et al.* 1998, Lin *et al.* 2000). Type II migration alone can not account for this either, since the migration speed is independent of the protoplanet mass. Ivanov, Papaloizou & Polnarev (1999) found that, when the mass of the planet is larger than the characteristic mass with which it tidally interacts, the inertia of the planet becomes important in slowing down the orbital evolution. This model predicts that a protoplanet with mass substantially larger than M_{d0}, the mass of disk inside the planet orbit, should not increase its mass significantly before migrating to the center of the disc. Thus the planet mass should be limited at about $M_{d0} \sim \sigma r^2$, which implies that planets with comparatively large period orbits could have larger masses. Numerical results of Nelson *et al.* (2000) show that the maximum mass of a protoplanet that can accretion while migrating to the neighborhood of the central star would be $\sim 5 M_J$.

Direction of planet migrations. While most of the theories and numerical results predict inward migrations, an outward scenario is required by the existence of planets with large semi-major axis (like Uranus and Neptune). This because, according to the core formation scenario, the time scale for the planets to form at larger semi-major axis (several tens of AU) would be prohibitively long. Outward migrations can be caused also by interactions with gas disk in some circumstances. Veras & Armitage (2004) proposed that, a strong mass loss from the disk can cause a planet to migrate outward. The strong mass loss from the disk can be caused by photoevaporation from the host star if it is more massive than the Sun, or result from removal by other perturbations from nearby stars within a cluster. Other mechanisms such as interactions between planet and planetesimal disc (see Sec.3.2) and protoplanets scattering can also cause outward migrations.

3.1.3. *Eccentricity evolution*

The linear theory predicts that LRs can excite the eccentricity of a protoplanet in the disk, while CRs damp the eccentricity. Thus the final evolution of the eccentricity depends on the competitions between these two effects. However, as the magnitudes of the CRs depend on the gradient of the local surface density of the disk, the final evolution of planet eccentricity may be quite elusive (Goldreich & Tremaine 1980, Lin and Papaloizou 1993). Briefly speaking, it depends on the circumstances of the disk and planet:

Small planet case. If a protoplanet is too small to open a gap near it($\mu \sim 10^{-6}$), the CRs are not important. The role of the LRs on the protoplanet depends on their locations. According to Ward (1988), LRs outside (both interior and exterior) it's orbit excite planet eccentricity, while those co-orbiting with the protoplanet damp the eccentricity more effectively. Thus the eccentricity of the protoplanet ultimately suffers decay.

Moderate planet case. If the protoplanet is large enough to open a gap near it ($\mu = 10^{-5} \sim 10^{-4}$), the gradient of σ/Ω is large near the gap, so the CRs dominate by a slight margin, and the eccentricity is damped (Goldreich & Tremaine 1980).

Massive planet case. When the protoplanet is massive ($\mu \sim 10^{-3}$), the gap near it is wide enough so that all the important resonances (with large m) fall into the gap, thus both LRs and CRs can be severely weakened, in this case the increase of eccentricity is possible. Lin and Papaloizou (1993) show both analytically and numerically that, for a massive planet with its gap truncated near the 2 : 1 resonance, it's eccentricity may increase with a rate,

$$\tau \sim \left(\frac{1}{\mu}\frac{M_*}{\pi\sigma r_p^2}\right)T_p \tag{3.6}$$

So for a Jupiter mass planet at 5AU, suppose $\pi\sigma r_p^2 \sim 10^{-3}M_*$, its eccentricity could increase in 10^7 years, the same time scale as the age of the gaseous disk. Goldreich & Sari (2003) shows the evolution of the disk and a massive planet could lead to a increase of planet eccentricity. Moreover, the depletion of the disk can also cause changes of planet eccentricities (Nagasawa, Lin & Ida 2003).

Numerical simulations give similar qualitative results, i.e., the eccentricity of planets with large masses will increase under planet-disk interactions, but those with small masses may have their eccentricity damped. With two-dimensional hydrodynamical simulations, Papaloizou, Nelson and Masset (2001) shows that, there exist a transition mass ($\sim 20M_J$) such that, for a planet with lower mass the eccentricity would be damped, otherwise the eccentricity is excited to values of 0.1 $\sim$ 0.25. They suggested that the transition mass might be reduced into the range of the observed extrasolar planets at a very low disc viscosity.

3.2. *Interactions between planets and planetesimal disk*

Migration due to interactions between planets and planetesimal disk is well studied in the evolution of outer solar system (e.g., Malhotra 1993, Hahn & Malhotra, 1999). The direction of the planet migrations is mainly determined by the averaged angular momentum $H = \sqrt{a(1-e^2)}$ of the planetesimals. For planetesimals with H greater than that of the planet's H_p, they have a larger average tangential velocity relative to the orbital motion of the planet, thus the result of close encounters between the planetesimals and the planet is to increase the angular momentum of the planet, which leads to an outward migration of the planet (Ida *et al.* 2000, Gomes, Morbidelli, Levison 2004). Since the encounters between planetesimals and planets are more or less erratic, the migration caused by interactions between planets and planetsimal disk is not smooth(Ida et al. 2000, Zhou *et al.* 2002).

Murray *et al.* (1998) proposed that resonant interactions between the planet and the planetesimals can remove angular momentum from the planetesimals, increasing their eccentricities. Subsequently, the planetesimals either collide with or are ejected by the planet, reducing the semi-major axis of the planets. If the surface density for the planetesimals is above some critical values, an instability would occur so that the planet can migrate inward with by a large amount. However, the critical surface density for the planetesimals should be $\sim 200 gcm^{-3}r^{-3/2}$, which is an order of magnitude larger than that of the minimum solar nebula.

4. Dynamics of Multiple Planet Systems

The evolution of a multiple planet system under their mutual attractions is a paradigm of the classical N body problem, which is not integrable for $N \geqslant 3$. The study of the stability of planet systems can give constrains on the parameters of the planets, especially the mass of planets, which are determined up to a $\sin i$ factor from the Doppler survey. For the convenience of stability study, multiple planet systems can be classified in three types (e.g., Barnes & Quinn 2004): resonant, interacting and hierarchical systems. Resonant systems contain planets in lower order mean motion resonances (MMR). Interacting system contain planets that are not in lower order mean motion resonances but are separated with period ratio less than 10 : 1. Hierarchical systems contain planets with

Table 1. Parameters of the planets in Resonances

star	M_* ($M_\odot$)	$M_p \sin i$ (M_{Jup})	a (AU)	Period (days)	eccentricity	comment
55 Cnc b	1.03	0.84	0.115	14.65	0.02	3:1 MMR+AC
55 Cnc c	1.03	0.21	0.24	44.28	0.34	3:1 MMR+AC
GJ876 c	0.32	0.56	0.13	30.1	0.27	2:1 MMR+AC
GJ876 b	0.32	1.89	0.21	61.0	0.10	2:1 MMR+AC
HD82943 c	1.05	1.85	0.75	219.4	0.54	2:1 MMR+AC
HD82943 b	1.05	1.84	1.18	435.1	0.41	2:1 MMR+AC
47 UMa b	1.03	2.54	2.09	1089	0.06	ASR
47 UMa c	1.03	0.76	3.73	2594	0.1	ASR
Ups And c	1.30	1.89	0.829	241.5	0.28	ASR
Ups And d	1.30	3.75	2.53	1284	0.27	ASR
HD12661 b	1.07	2.3	0.82	263.6	0.33	ASR
HD12661 c	1.07	1.5	2.46	1530	0.2	ASR

Note : MMR-mean motion resonance; AC-apsidal corotation; ASR-apsidal secular resonance. Data of HD82943 from Mayor *et al.* 2004, others from http://exoplanets.org/

a period ratio grater than $10:1$, and can be approximated as dynamically decoupled systems, provided that their eccentricities are not very large.

According to the above definition, our solar system is an interacting system. For the 11 confirmed multiple extra-solar planetary systems to date, 3 pairs of planets are confirmed in mean motion resonances (Fischer *et al.* 2003): HD82943, GJ876, 55 Cancri b and c, see Table 1. The HD160691 system, with uncertain parameters, could be most probably in a 2:1 resonance (Bois *et al.* 2003). Interacting systems include 3 systems: Ups And (c-d), HD12661, 47 Uma. 5 systems are hierarchical: HD38529, HD74156, HD168443, HD37124, HD169830. The last two systems are marginal hierarchical, with period ratios 9.8 and 9.3, respectively. As the eccentricities of planet orbits in these two systems are large, their stability need inspection. The outmost planet of 55 Cancri and the innermost planet of Ups And are hierarchical with other planets in the same system.

4.1. *Resonant Systems*

Mean motion resonance is an important mechanism that leads to stable configurations for nearby planets. Among the three confirmed pairs of planets in resonances, 5 planets are in orbits with large eccentricities, which is in contradiction with our knowledge about stability. Thus a natural question is, can stable periodic solutions exist with high eccentricities, corresponding to equilibriums of mean motion resonances? Hadjidemetriou (2002) gives a positive answer. Let us take the $2:1$ resonance as an example. Define the corresponding two resonance angles as

$$
\begin{aligned}
\theta_1 &= 2\lambda_2 - \lambda_1 - \varpi_1, \\
\theta_2 &= 2\lambda_2 - \lambda_1 - \varpi_2, \\
\sigma &= \theta_2 - \theta_1 = \varpi_1 - \varpi_2 = \Delta\varpi \ .
\end{aligned}
\tag{4.1}
$$

Resonant solutions with $\theta_1 = 0, \sigma = 0$ are symmetrical periodic solutions with apsidal corotation (AC) in a coordinate system comoving with one planet. Using the continuation method of the periodic orbits, Hadjidemetriou (2002), Hadjidemetriou & Psychoyos (2003) studied the families of periodic orbits near the 2:1 and 3:1 resonances in a 2-planet systems with applications to the GJ876 and HD82943 systems. Stable resonance orbits are found at large eccentricities with $e_1 > e_2$, in accordance with the two systems.

The existence of asymmetric libration solutions ($\theta_1 \neq 0, \pi, \sigma \neq 0, \pi$) has been shown by Beaugé, Ferraz-Mello, and Michtchenko (2003), with a new perturbation method developed by Beaugé and Michtchenko (2003). The method allows us to study analytically the evolution of extrasolar planets on high eccentricity orbits. They find that the existence of stable periodic orbits depends on the mass ratio of the two planets m_1/m_2, thus it gives a constraint for the mass of the observed planets. They also find that the evolutionary track of GJ876 and HD82943's eccentricity coincide with the lines of stable 2:1 resonance solutions for the observed mass ratio. This gives a strong evidence of a migration history for these systems. This method is applied to the 55 Cancri system: the two inner planets are in 3:1 resonance and also in apsidal corotation (Zhou *et al.* 2004). From hydrodynamical simulations, tidal migrations of planets pairs in 55 Cancri can lead to a trapping of both 3:1 resonance and apsidal corotation (Kley 2002).

One of the remaining problems concerning the migration evolution of the resonant systems is the damping of their eccentricities. Suppose the planets are initially in circular orbits, when they migrate under the tidal interaction of either gaseous disk or planetesimal disk, their eccentricities will increase inevitably (e.g., Lee and Peal 2002). So a damping mechanism for the eccentricity are required to fit the observed values. Lee and Peal (2002) find that a damping rate with

$$\frac{\dot{e}_i}{e_i} = -K \left| \frac{\dot{a}_i}{a} \right|, \quad (i = 1, 2) \tag{4.2}$$

and $K \approx 100$ is required to account for the observed eccentricities of the two planets in GJ876 systems. Though this type of damping rate is coincident in form with that from disk-planet interactions, the required value of K is much larger than that obtained from the density wave theory.

4.2. *Interacting systems*

For the interacting systems, apsidal secular resonance (ASR) could be one of the dynamical effects that enforce the stability of planetary systems. An ASR occurs when the secular motion of two planets' longitudes of the perihelion (ϖ_1 and ϖ_2) have almost the same frequency, thus $d(\Delta\varpi)/dt \approx 0$. Planets in aligned ASR can reduce the possibility of close encounters, thus ASR can stabilize the interacting planets. An impulsive perturbation (e.g. by ejection of other planets) which increases the eccentricity of one planet can possibly lead two planets into ASR (Malhotra 2002). All the 3 pairs of planets in the present observed interacting systems are in ASR (see Table 1): Ups And c-d, 47UMa, HD12661. According to the linear secular perturbation theory, whether two planets in ASR depends only on m_1/m_2, a_1/a_2, e_1/e_2 and $\Delta\varpi$ (Zhou and Sun 2003).

The HD12661 system is the first extra-solar planetary system found to have two planets in anti-aligned ASR ($\Delta\varpi = \pi$). Lee and Peale (2003) use octopole-level secular perturbation theory to study the secular evolution of a planetary system. They find that, when the angular momentum ratio $L_1/L_2 \approx (m_1/m_2)(a_1/a_2)^{1/2}$ is equal to some critical values, the phase space of the secular system is occupied by apsidal librations. They find HD12661 system is close to this critical configurations, thus the secular system of HD12661 is stable. However, as a general three-body system, the present observed HD12661 system is on the border between chaos and regular motion (Kiseleva-Eggleton *et al.* 2002, Zhou & Sun 2003).

4.3. *Stable region*

Most works are devoted to the stability study of the observed multiple planet systems from the fitted initial conditions. The general N-body model is used during the numerical integrations, with the auxiliary computation of Lyapunov characteristic numbers (LCN), Fast Lyapunov Indicator (FLI) developed by Froeschlé, Lega & Gonczi (1997) and the MEGNO technique by Cincotta & Simó (2000). Barnes & Quinn find that the stable regions of the two resonant systems, HD82943 (with old data from 'Extrasolar Planets Encyclopaedia') and GJ876, are very narrow (Barnes & Quinn 2004). This indicates that the HD82943 and GJ876 systems are protected only by the mean motion resonances, and most likely the present large eccentricities are the consequence of planet migrations. For the unconfirmed HD160691 system, the planets can be stable only in a 2:1 MNR (Bois *et al.* 2003), which shows HD160691 system should be a member of the resonant systems as well.

For the interacting systems, two systems (Ups And , 47 Uma) have border stable regions (Barnes & Quinn 2004), thus they could be stable even without the apsidal secular resonances, and the observed eccentricity could be the results of planet scattering. The HD12661 system is on the border of chaotic and regular motions as mentioned before (Zhou & Sun 2003,Barnes & Quinn 2004).

The stability of the marginal hierarchical system HD169830 is studied by Goździewski & Konacki (2004), which shows the present system is in a wide stable zone, thought the planets have large eccentricities. For another marginal hierarchical system HD37124, Kiseleva-Eggleton *et al.* (2002) found the present configuration is near the border of a wide stable region. Dvorak *et al.* (2003) find there is difficult to host planets in the habitable zone of the hierarchical system HD74156. A systematical study of dynamics of binary stars planetary system have been done by their group (e.g., Pilat-Lohinger, Funk, Dvorak, 2003).

4.4. *Onset of instability*

Since generally a Hamiltonian system with more than two degrees of freedom is not integrable, instability will arise for a planet system with two planets or more. Fortunately, for a planet system with sufficient small planet mass and sufficient large separations, the speed of orbital diffusion in the action space (corresponds to angular momentum and energy, etc.) could be very small. Knowing the time scale over which instability may occur would be helpful to understand the time scale of the excitation of the orbital eccentricities by scattering among planets.

For a planet system with two planets initially on circular orbits, the system will be Hill stable (stable against close approaches for all time) if the relative orbital separation of the two planets ($\delta = (a_2 - a_1)/a_1$) fulfils, $\delta > 2.4(\mu_1 + \mu_2)^{1/3}$, where $\mu_i = m_i/m_*$(i=1,2) (Gladman, 1993). However, when the planet number is larger than 2, things become quite different. Suppose N planets with equal mass m initially on circular orbits, and the separations are equal when scaled with their mutual Hill's radius, $R_H = (2\mu/3)^{1/3} (a_i + a_{i+1})/2,(i = 1, ...n - 1)$. Chambers, Wetherill and Boss (1996) found the crossing time T_c, defined as the minimum time that orbital crossing is expected, could be approximated by $\log(T_c) = b(\Delta a_0/R_H) + c$, where $b(> 0), c$ are constants, $\Delta a_0 = a_{i+1} - a_i$, and T_c is in unit of years. Thus whenever the planet number is larger than 2, there is no Hill-stable region in the phase space so that $T_c \to \infty$. An important result is that b, c seem to have little dependence on N when $N \geqslant 5$. Yoshinaga, Kokubo & Makino (1999) extended the study to the case when planets are initially in elliptic and inclined orbits. An extension to a much wider initial separations shows that, the orbital crossing time of planets in initially circular orbits can be expressed as (Zhou, Lin, Sun, in preparation):

$$\log(T_c) \approx -2 - \frac{2}{7}\log\mu + (19 + 1.2\log\mu)\log\left(\frac{\Delta a_0}{2.3R_H}\right), \quad (\Delta a_0 > 2.3R_H) \qquad (4.3)$$

where T_c is in the unit of years. The crossing time for planet initially in circular and non circular orbits are shown in Fig.2 (the initial eccentricity e_0 is scaled by $h = (a_{i+1} - a_i)/(a_{i+1} + a_i)$). According to (4.3) and Fig.2, there exist systems that are dynamically unstable but the time scale over which such instability occurs is much longer than the physical age of the system, thus it is "effectively stable".

On the other hand, for planets with small separations, the orbital crossing time can be very short. According to (4.3), the instability time scale will $\sim 10^{6.7}$ years for $N(N > 3)$ protoplanets with mass $\mu = 10^{-7}$ and $\Delta a_0/R_H = 10$. Thus the excitation of planet eccentricities can occur within a short time scale. But as the protoplanets have larger masses, the excitation of the eccentricities require a longer time.

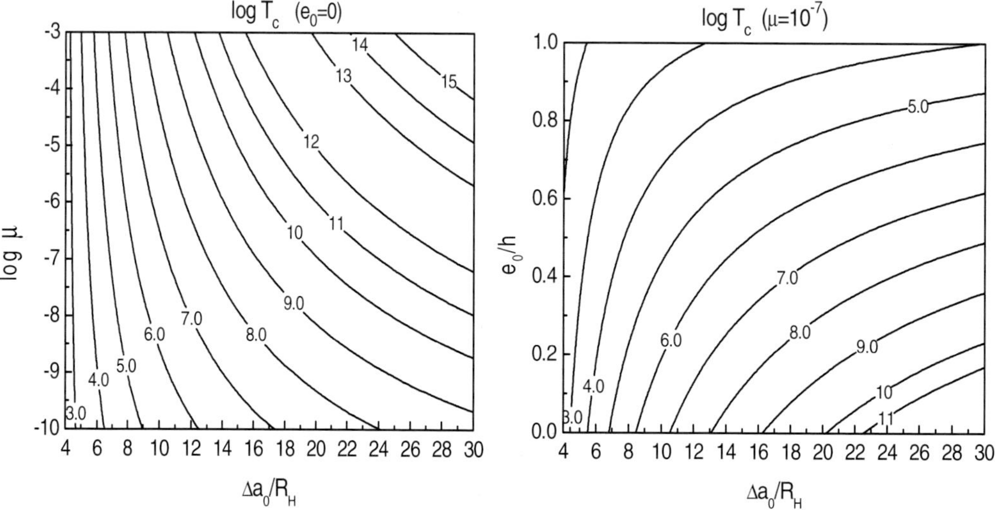

Figure 2. The orbital crossing time of multiple planet systems with equal planet masses and scaled separations (see text). (a) planets are initially in circular orbits. (b) Planets with $\mu = 10^{-7}$.

5. Summary

The dynamical evolution of the extra-solar planetary systems is a complicate process involving planet-disk interactions and interplanetary scattering. According to the present knowledge, we have the following conclusions:

(1) For the observed extra-solar planets with short period orbits, migration is believed to occur after the planets are formed. The type I migration rate given by the linear density wave theory would be too fast. The lack of large planets with short period orbits may be due to either the inefficient mass accretion for planets at small period orbits, or due to the inefficient (or too efficient) migration of large planets. If the latter scenario is right, we need a mechanism to generate a mass-dependent migration.

(2) For single planet systems with massive planet mass (of the order of Jupiter mass), the interactions with gaseous disk can excite the eccentricities of the planet orbits, However, the required minimum mass of planet is too large as found by some numerical simulations. Other dynamical processes, e.g., scattering with smaller, unseen planets, can also generate the eccentricity.

(3) For multiple planet systems, the excitation of eccentricity can either by interplanetary scattering, which may lead to an erratic change of the eccentricity, or by

adiabatic migration due to interactions with remnant gaseous disk, which leads to a gradual increase of eccentricity. The eccentricities of planets in interacting systems and hierarchical systems can be excited by the former mechanism. On the other hand, the eccentricities of planets in resonant systems are most probably excited by the latter mechanism during the migration, being trapped into resonance. However, in the resonance trapping case, we still need a mechanism to keep the eccentricities as the observed values after they are excited.

Our current knowledge is limited due to the insufficient observations, and it's still too early to say that our solar system is very special among the known planet systems. With the development of the observation technics we can get more knowledge of the possible configurations of planetary systems, and celestial mechanics will become more and more important in the study of dynamics of the extrasolar planetary systems.

Acknowledgements

This work is supported by the Natural Science Foundation of China under Grant No.19903001 and the Special Funds for Major State Basic Research Project(G200077303).

References

Artymowicz, P., 1993, *Astrophys. J.*, 419, 155

Barnes, R. & Quinn, T., 2004, *Astrophys. J.*, 611, 494

Beaugé, C., Ferraz-Mello, S., & Michtchenko, T.A., 2003, *Astrophys. J.*, 593, 1124

Beaugé, C. & Michtchenko, T.A., 2003, *Mon. Not. R. Astron. Soc.*, 341, 760

Bois, E., Kiseleva-Eggleton, L., Rambaux, N., & Pilat-Lohinger, E., 2003, *Astrophys. J.*, 598, 1312

Bouchy, F., Pont, F., Santos, N.C., Melo, C., Mayor, M., Queloz, D., & Udry, S., 2004 *Astron. Astrophys.* 421, L13

Bryden,G., Chen, X., Lin, D.N.C., Nelson, R.P., & Papalozou, J.C.B., 1999 *Astrophys. J.*, 514, 344

Chambers, J.E., Wetherill, G.W., & Boss, A.P., 1996, *Icarus*, 119, 261

Cincotta, P.M. & Simó, C., 2000, *A&AS*, 147, 205

Dobbs-Dixon, I, Lin, D.N.C. & Mardling, R. A., 2004, *Astrophys. J.*, 610, 464

Dvorak, R., Pilat-Lohinger, E., Funk, B., & Freistetter, F., 2003 *Astron. Astrophys.*, 410, L13

Froeschle, C., Lega, E., & Gonczi, R., 1997, *Celest. Mech. Dyn. Astron.* 67, 41

Fischer, D.A., Marcy, G.W., Butler, R.P., Vogt, S.S., Henry, G.W., Pourbaix, D., Walp, B., Misch, A.A., & Wright, J.T., 2003, *Astrophys. J.*, 586, 1394

Gladman, B., 1988, *Icarus*, 106, 247

Goldreich, P. & Sari, R.,2003, *Astrophys. J.*, 585, 1024

Goldreich, P. & Tremaine, S., 1979, *Astrophys. J.*, 233, 857

Goldreich, P. & Tremaine, S., 1980, *Astrophys. J.*, 241, 425

Gomes, R.S., Morbidelli, A., & Levison, H.F., 2004, *Icarus*, 170, 492

Goździewski, K. & Konacki, M., 2004, *Astrophys. J.*, 610, 1093

Hadjidemetriou, J.D. & Psychoyos, D., 2003, in: G. Contopoulos, N. Vogglis (eds.), *Galaxies and Chaos*, Lecture Notes in Physics, Vol. 626, p. 412

Hadjidemetriou, J.D., 2002, *Cel. Mech. Dyn. Astron.*, 83, 141

Hahn, J.M. & Malhotra, R., 1999, *Astron. J.*, 117, 3041

Ida, S., Bryden, G., Lin, D.N.C., & Tanaka, H., 2004, *Astrophys. J.*, 534, 428

Ivanov, P.B., Papaloizou, J.C.B., & Polnarev, A.G., 1999, *Mon. Not. R. Astron. Soc.*, 307, 79

Kley, W., 1999, *Mon. Not. R. Astron. Soc.* 303, 696

Kley, W., 2003, *Cel. Mech. Dyn. Astron.* 87, 85

Koller, Josef; Li, Hui; Lin, & Douglas, N.C., 2003, *Astrophys. J.*, 596, L01

Kokubo, E.K. & Ida, S., 1998, *Icarus*, 131, 171

Konacki, M., Torres, G., Sasselov, D.D., Pietrzynski, G., Udalski, A., Jha, S., & Ruiz, M.T., Gieren, W., & Minniti, D., 2004, *Astrophys. J.*, 609, L37

Kuchner, M.J. & Lecar, M., 2002, *Astrophys. J.*, 574, L87

Kiseleva-Eggleton, L., Bois, R., Rambaux, N., & Dvorak, R., 2002, *Astrophys. J.*, 578, L145

Laughlin, G., Steinacker, A., & Adams, R., 2004, *Astrophys. J.*, 608, 489

Lee, M.H. & Peale, S.J. 2002, *Astrophys. J.*, 567, 596

Lee, M.H. & Peale, S.J. 2003, *Astrophys. J.*, 592, 1201

Lin, D.N.C, Bodenheimer, P., & Richardson, D.C., 1996, *Nature*, 380, 606

Lin, D.N.C. & Papaloizou, J., 1979, *Mon. Not. R. Astron. Soc.*, 186, 799

Lin, D.N.C. & Papaloizou, J.C.B., 1993, in: E.H. Levy, & J.I. Lunine (eds.), *Protostars and Planets III*, Univ. Arizona Press, p. 749

Lin, D.N.C., Papaloizou, J.C.B., Terquem, C., & Bryden, G., 2000 in: V. Mannings, A.P. Boss, & S.S. Russell (eds.), *Protostars and Planets IV*, Univ. Arizona Press, p. 1111

Malhotra, R., 1993, *Nature*, 365, 819

Malhotra, R., 2002, *Astrophys. J.*, 575, L33

Marcy, G.W., Bulter, R.P., Fischer, D.A. *et al.* 2003, in: D. Deming, & S. Seager (eds.), *Scientific Frontiers in Research on Extrasolar Planets*, ASP Conferences Series, Vol. 294, p. 1

Mayor, M. & Queloz, D., 1995, *Nature* 378, 355

Mayor, M., Udry, S., Naef, D., Pepe, F., Queloz, D., Santos, N.C., & Burnet, M., 2004, *Astron. Astrophys.*, 415, 391

Matsuyama, I., Johnstone, D., & Murray, N., 2003, *Astrophys. J.*, L143

Murray, N., Hansen, B., Holman, M., & Tremaine, S., 1998, *Science*, 279, 69

Nagasawa, M., Lin, D.N.C., & Ida, S., 2003, *Astrophys. J.*, 586, 1374

Nelson, R.P. & Papaloizou, J.C.B., 2004, *Mon. Not. R. Astron. Soc.*, 350, 849

Nelson, R.P., Papaloizou, J.C.B., Masset, F., & Kley, W., 2000, *Mon. Not. R. Astron. Soc.*, 318, 18

Papaloizou, J.C.B., Nelson, R.P., & Masset, F., 2001, *Astron. Astrophys.*, 366, 263

Pilat-Lohinger, E., Funk, B., & Dvorak, R. 2003, *Astron. Astrophys.*, 400, 1085

Rafikov, R.R., 2002, *Astrophys. J.*, 572, 566

Shakura, N.I. & Sunyaev, R.A., 1973, *Astron. Astrophys.*, 24, 337

Terquem, C., Papaloizou, J.C.B., Nelson, R.P., & Lin , D.N.C., 1998, *Astrophys. J.*, 502, 788

Torres, G., Konacki, M., Saaaelov, D.D., & Jha, S., 2004, *Astrophys. J.*, 609, 1071

Udry, S., Mayor, M., & Queloz, D. 2003, in: D. Deming, & S. Seager (eds.), *Scientific Frontiers in Research on Extrasolar Planets*, ASP Conferences Series, Vol. 294, p. 17

Veras, D. & Armitage, P.J., 2004, *Mon. Not. R. Astron. Soc.*, 347, 613

Ward, W.R., 1988, *Icarus*, 73,330

Ward, W.R., 1997a, *Icarus*, 126,261

Ward, W.R., 1997b, *Astrophys. J.*, 482, L211

Ward, W.R. & Hahn, J.M., 2000, in: V. Mannings, A.P. Boss, & S.S. Russell (eds.), *Protostars and Planets IV*, Univ. Arizona Press, p. 1111

Wolszcan, A. & Frail, D.A., 1992, *Nature* 355, 145

Yoshinaga, K., Kokubo, E., & Makino, J., 1999, *Icarus*, 139, 328

Zhou, J.L. & Sun, Y.S., 2003, *Astrophys. J.*, 598, 1290

Zhou, L.Y., Sun, Y.S., Zhou, J.L., Zheng, J.Q., & Valtonen, M., 2002, *Mon. Not. R. Astron. Soc.*, 336, 520

Zhou, L.Y., Lehto, H.J., Sun, Y.S., & Zheng, J.Q., 2004, *Mon. Not. R. Astron. Soc.*, 350, 1495

Dynamics of Populations of Planetary Systems
Proceedings IAU Colloquium No. 197, 2005
Z. Knežević and A. Milani, eds.

© 2005 International Astronomical Union
DOI: 10.1017/S1743921304008464

Dynamics of planetesimals:
the role of two-body relaxation

Eiichiro Kokubo

National Astronomical Observatory of Japan, Osawa, Mitaka, Tokyo, Japan
email: kokubo@th.nao.ac.jp

Abstract. In the standard scenario of planet formation, solid planets are formed through accretion of small bodies called planetesimals. The dynamics of planetesimals is important since it controls their growth mode and timescale. Here, I briefly explain the basic dynamics of planetesimals due to the two-body gravitational relaxation process. The important roles of two-body relaxation in a planetesimal system are viscous stirring and dynamical friction. Due to viscous stirring, the random velocities (eccentricities and inclinations) of planetesimals increase, while dynamical friction realizes the energy equipartition of the random energy. I also explain the orbital repulsion of protoplanets which is the coupling effect of two-body scattering and dynamical friction.

Keywords. Solar system: formation, celestial mechanics

---

1. Introduction

In the standard scenario for planet formation, terrestrial planets and the cores of Jovian planets are formed through the accretion of small bodies called planetesimals. This process is called as planetary accretion. Planetary accretion is an important stage of planet formation since it determines the timescale of planet formation and the spatial structure of the planetary system.

The dynamics of planetesimals controls the planetary accretion. For a planetesimal, the solar gravity is a dominant force. Thus, the orbit is nearly Keplerian. The deviation velocity v of the planetesimal from the Keplerian velocity v_K of the local non-inclined circular orbit is called random velocity and is given by

$$v \simeq (e^2 + i^2)^{1/2} v_K, \tag{1.1}$$

where e and i are the eccentricity and inclination of the planetesimal. Note that the random velocities in the radial and vertical directions are proportional to e and i, respectively. The random velocity is an important factor that controls planet formation. For example, the growth timescale of planetesimals depends on the random velocity as $T_{\mathrm{grow}} \propto v^2$ when gravitational focusing is effective in collisions.

The orbit evolves through mutual gravitational interaction among planetesimals, i.e., two-body gravitational relaxation. In terms of stellar dynamics, a planetesimal system is a collisional system where two-body encounters play key roles in the dynamical evolution of the system. The two basic effects of two-body relaxation in a planetesimal system are viscous stirring and dynamical friction. Viscous stirring increases the random velocity of planetesimals, while dynamical friction realizes the energy equipartition of the random energy. In the present paper, I briefly demonstrate the basics of viscous stirring and dynamical friction in a planetesimal system. I also describe the orbital repulsion process in which two large bodies expand their orbital separation in a swarm of planetesimals.

For details of the velocity evolution of planetesimals, see e.g., Ida (1990), Ohtsuki (1999), and Stewart & Ida (2000).

2. Two-Body Relaxation

The equation of motion for a planetesimal is given as

$$\frac{d^2\boldsymbol{x}_i}{dt^2} = -GM_\odot \frac{\boldsymbol{x}_i}{|\boldsymbol{x}_i|^3} + \sum_{j=1, j\neq i}^{N} Gm_j \frac{\boldsymbol{x}_j - \boldsymbol{x}_i}{|\boldsymbol{x}_j - \boldsymbol{x}_i|^3} + \boldsymbol{f}_{\rm gas}, \tag{2.1}$$

where the terms on the r.h.s. express the solar gravity, the mutual gravitational interaction, and the gas drag, from left to right. For the standard minimum-mass disk model, the solar gravity is dominant except for the rare cases of close encounters of planetesimals and thus the orbit of planetesimals is almost Keplerian. The mutual interaction of planetesimals is the main perturbing force in a planetesimal system.

The orbit of planetesimals gradually changes due to the mutual gravitational interaction. This process is equivalent to the two-body relaxation process in star clusters. In terms of stellar dynamics, a planetesimal system is a collisional system like globular clusters, in the sense that the system evolves through two-body encounters.

The mutual interaction increases the random velocity of planetesimals on average. On the other hand, the gas drag damps the random velocity. In planetary accretion, there are equilibrium values of the eccentricity and inclination, at which the effects of the mutual interaction of planetesimals and the gas drag balance. In the present paper, we focus on the effect of the mutual interaction. The timescale of two-body relaxation for an equal-mass (m) many-body system is given by

$$T_{\rm 2B} \equiv \frac{\sigma^2}{d\sigma^2/dt} \simeq \frac{1}{n\pi r_{\rm G}^2\sigma \ln \Lambda} = \frac{\sigma^3}{n\pi G^2 m^2 \ln \Lambda}, \tag{2.2}$$

where n is the number density of constituent particles, σ is the velocity dispersion, $r_{\rm G}$ is the gravitational radius given by $r_{\rm G} = Gm/\sigma^2$, and $\ln \Lambda$ is the Coulomb logarithm (e.g., Binney & Tremaine 1987).

The two major roles of two-body relaxation in a planetesimal system are viscous stirring and dynamical friction. In the following, we demonstrate both roles by showing examples of N-body simulations. The initial conditions of planetesimals we use in the following sections are summarized as follows: 1000 equal-mass ($m = 10^{24}$g) planetesimals are distributed in a ring of the radius $a = 1$AU with width $\Delta a = 0.07$AU. The surface density of the ring is almost consistent with the minimum-mass disk model. The initial distributions of eccentricities and inclinations are set by the Rayleigh distributions with dispersions $\sigma_e = \sigma_i = 2r_{\rm H}/a$, where $r_{\rm H}$ is the Hill radius of planetesimals given by

$$r_{\rm H} = \left(\frac{2m}{3M_\odot}\right)^{1/3} a. \tag{2.3}$$

Note that σ_e and σ_i are proportional to the velocity dispersions in thr radial and vertical directions, respectively. For the numerical integration, the fourth-order Hermite integrator (Kokubo *et al.* 1998) is used with a special-purpose computer GRAPE-6 (Makino *et al.* 2003). Gas drag and collisions are not included for simplicity.

3. Viscous Stirring

Viscous stirring is the process in which velocity dispersions of planetesimals increase due to two-body encounters. Here we consider an equal-mass planetesimal system for simplicity. In the dispersion-dominated regime ($\sigma_e, \sigma_i \gtrsim r_{\rm H}/a$) where the relative velocity

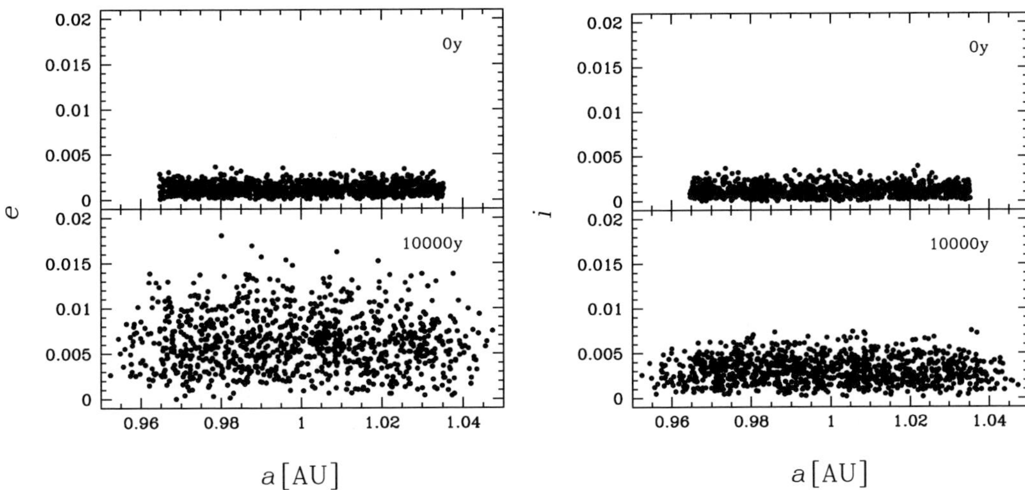

Figure 1. Snapshots of the planetesimal system on the a-e (left) and a-i (right) planes at $t = 0$ year (top) and 10000 year (bottom).

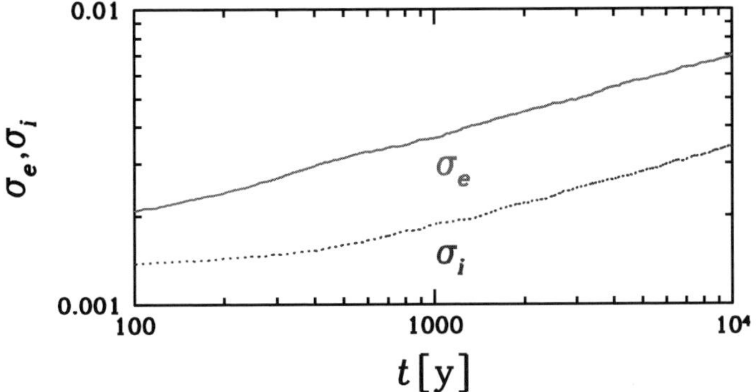

Figure 2. Time evolution of σ_e (solid) and σ_i (dashed).

of planetesimals is mainly determined by the random velocity, the rate of viscous stirring for σ_e is given by

$$\frac{d\sigma_e^2}{dt} \sim \frac{\sigma_e^2}{T_{2\mathrm{B}}}.$$ (3.1)

We can obtain the rate of viscous stirring for σ_i by replacing σ_e with σ_i in (3.1).

Figures 1 show the snapshots of the system at $t = 0$ year and 10000 year on the a-e and a-i planes. The eccentricities and inclinations of most planetesimals significantly increase in 10000 years. On average, the increase of e is larger than that of i. The distributions of e and i relax into the Rayleigh distributions. We also see the diffusion of planetesimals in a, which is the result of random walk in a due to two-body scattering. These increases of e and i with the diffusion in a are the basics of viscous stirring.

The time evolutions of σ_e and σ_i are shown in figure 2. It is clearly shown that σ_e and σ_i increase with $t^{1/4}$, and $\sigma_e/\sigma_i \simeq 2$ that corresponds to the anisotropic velocity dispersions in radial and vertical directions. These properties of σ_e and σ_i are the characteristics of two-body relaxation in a disk-shaped system (Ida *et al.* 1993).

As the number density of planetesimals is inversely proportional to the scale height of the system that is proportional to the velocity dispersion, in other words, $n \propto \sigma^{-1}$, we

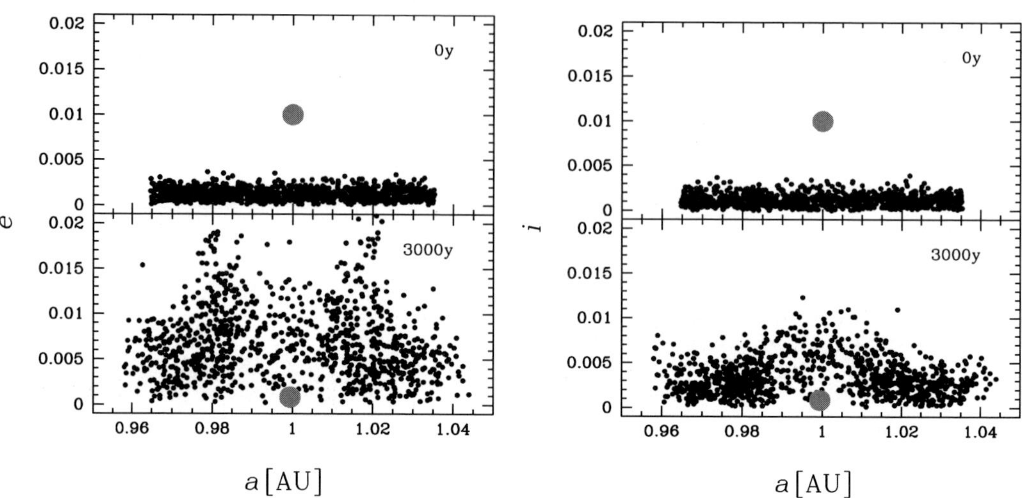

Figure 3. Snapshots of the planetesimal system on the a-e(left) and a-i (right) planes at $t = 0$ year (top) and 3000 year (bottom). The large circle indicates the protoplanet.

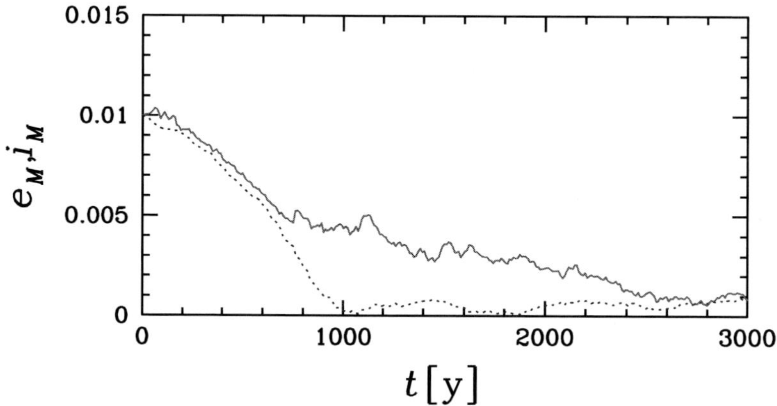

Figure 4. Time evolution of e_M (solid) and i_M (dashed) of the protoplanet.

have $T_{2B} \propto \sigma^4$. Thus, we have $\sigma \propto t^{1/4}$. The origin of the anisotropy of the velocity dispersion is the shear velocity between planetesimals with different a due to the differential rotation in the Kepler potential. This type of anisotropy is also known for the galactic stellar disk (e.g., Kokubo & Ida 1992).

4. Dynamical Friction

Dynamical friction is the process of the equipartition of the random energy, $(1/2)mv^2$, of planetesimals. In other words, the random velocity becomes $v \propto m^{1/2}$. As an illustration of dynamical friction, we consider a simple case with a protoplanet (large planetesimal) with $M = 100\, m$ embedded in a swarm of planetesimals. We focus on the orbital evolution of the protoplanet. The initial orbital elements of the protoplanet are $a_M = 1\mathrm{AU}$ and $e_M = i_M = 0.01$.

For $M \gg m$ and $e_M > e$, the rate of dynamical friction for e_M is approximated as

$$\frac{de_M^2}{dt} \sim -\frac{m}{M}\frac{e_M^2}{T_{2B}}.$$

(4.1)

The rate for i_M is obtained by replacing e_M with i_M in (4.1). Figures 3 show the snapshots

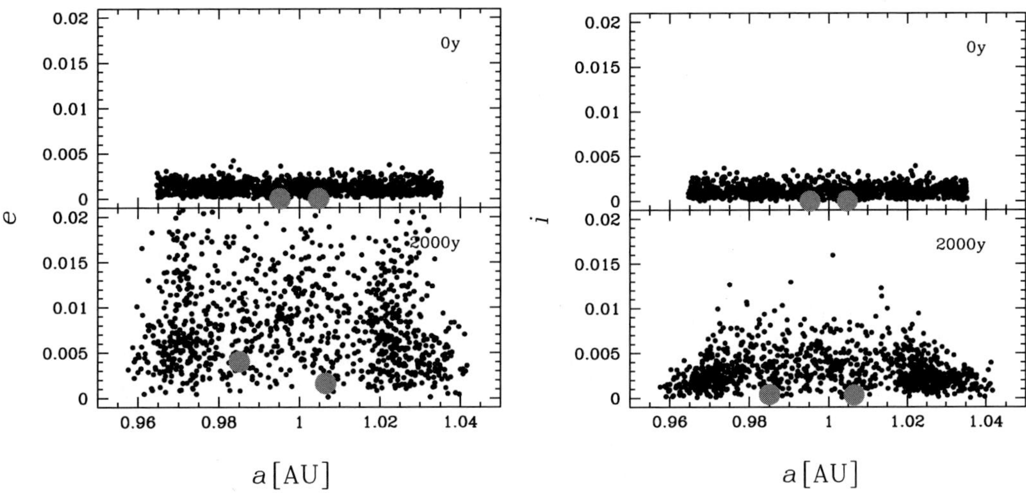

Figure 5. Snapshots of the planetesimal system on the a-e(left) and a-i (right) planes at $t = 0$year (top) and 2000year (bottom).The large circles indicate the protoplanet.

of the system at $t = 0$ year and 3000 year on the a-e and a-i planes. We see that the eccentricity and inclination of the protoplanet decrease to almost 0 in 3000 years. However, its semimajor axis is kept almost constant. On the other hand, the eccentricities and inclinations of the neighbor planetesimals are raised by reaction. The V-like structure around the protoplanet on the a-e plane corresponds to the constant Jacobi energy curve. This "heating of neighbor planetesimals by a protoplanet" leads to the decrease of the growth rate of the protoplanet (Ida & Makino 1993).

The time evolution of e_M and i_M of the protoplanet is shown in figure 4. In 3000 years, the eccentricity and inclination of the protoplanet are reduced to ~ 0.001 due to dynamical friction from small planetesimals, in other words, the orbit of the protoplanet becomes a non-inclined nearly circular orbit.

One of the important features of dynamical friction is that the rate of dynamical friction does not depend on the mass of individual particles but on the density of the system. For protoplanet-planetesimal scattering, T_{2B} for the protoplanet is $T_{2B} \propto nM^2$ from (2.2). Therefore, (4.1) leads to $|de_M^2/dt| \propto nm = \rho$, where ρ is the density of the system.

5. Orbital Repulsion

As a simple application of two-body relaxation, let us think the orbital evolution of two interacting protoplanets embedded in a swarm of planetesimals. Kokubo & Ida (1995) found that if the orbital separation, b, of the two protoplanets is smaller than a few times their mutual Hill radius, they expand their orbital separation to $b \gtrsim 5r_{\mathrm{H}}$. This phenomenon is called as orbital repulsion. In figures 5, the protoplanets with $M = 100m$ initially on non-inclined circular orbits with the orbital separation of $b = 3r_{\mathrm{H}}$ expands the orbital separation to $b \simeq 8r_{\mathrm{H}}$ keeping nearly non-inclined circular orbits in 2000 years.

By orbital repulsion, protoplanets grow keeping their orbital separation $b \gtrsim 5r_{\mathrm{H}}$. The typical orbital separation is $b \sim 10r_{\mathrm{H}}$ on the course of protoplanet growth (Kokubo & Ida 1998). This is one of the important factors that realize oligarchic growth of protoplanets.

Orbital repulsion is a coupling effect of two-body scattering and dynamical friction. Figure 6 schematically explains the two stage mechanism of orbital separation:

(a) Scattering between two protoplanets on nearly circular non-inclined orbits increases their eccentricities and orbital separation.

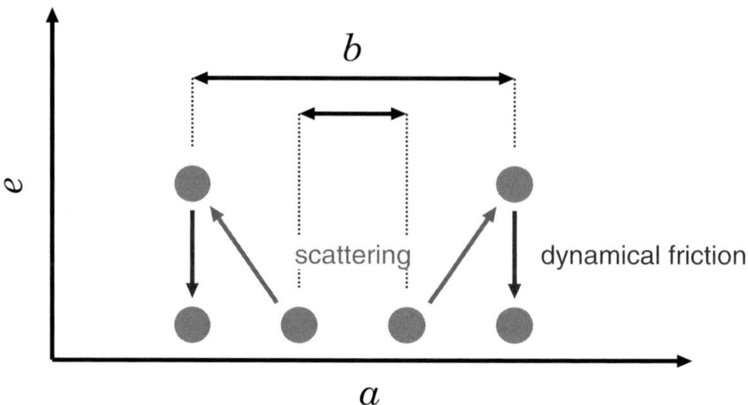

Figure 6. Schematic illustration of orbital repulsion on the a-e plane.

(b) Dynamical friction from planetesimals decreases the eccentricities of the proto-planets keeping their semimajor axes.

We can analytically show the behavior of the first stage, based on the conservation of energy and angular momentum in two-body scattering under the solar gravity.

6. Summary

We have demonstrated how two-body relaxation plays key roles in the velocity evolution of planetesimals by showing the examples of N-body simulations. The important processes are viscous stirring and dynamical friction. Viscous stirring increases σ_e and σ_i in proportion to $t^{1/4}$, keeping the ratio $\sigma_e/\sigma_i \simeq 2$, which is the characteristics of two-body relaxation in a disk-shape system. Dynamical friction realizes the equipartition of the random energy, in other words, $e, i \propto m^{-1/2}$. This equipartition is a sufficient condition of runaway growth of planetesimals (e.g., Kokubo & Ida 1996). Orbital repulsion of two protoplanets embedded in a swarm of planetesimals is also explained. The orbital separation of the two protoplanets is kept $b \gtrsim 5r_{\rm H}$ due to orbital repulsion. Orbital repulsion is one of the key processes for the oligarchic growth of protoplanets. All these elementary processes control the basic mode, timescale, and spatial structure of planetary accretion.

Acknowledgements

I would like to thank Shigeru Ida and Junichiro Makino for the useful discussion on the N-body simulation of planetesimals.

References

Binney, J. & Tremaine, S. 1987, *Galactic Dynamics* (Princeton: Princeton University Press)
Ida, S. 1990, *Icarus* 88, 129
Ida, S. & Makino, J. 1993, *Icarus* 106, 210
Ida, S., Kokubo, E., & Ida, S. 1993, *Mon. Not. R. Astron. Soc.* 263, 875
Kokubo, E. & Ida, S. 1992, *Publ. Astron. Soc. Japan* 44, 601
Kokubo, E. & Ida, S. 1995, *Icarus* 114, 247
Kokubo, E. & Ida, S. 1996, *Icarus* 123, 180
Kokubo, E. & Ida, S. 1998, *Icarus* 131, 171
Kokubo, E., Yoshinaga, K., & Makino, J. 1998, *Mon. Not. R. Astron. Soc.* 297, 1067
Makino, J., Fukushige, T., Koga, M., & Namura, K. 2003, *Publ. Astron. Soc. Japan* 55, 1163
Ohtsuki, K. 1999, *Icarus* 137, 152
Stewart, G. R. & Ida, S. 2000, *Icarus* 143, 28

Dynamics of Populations of Planetary Systems
Proceedings IAU Colloquium No. 197, 2005
Z. Knežević and A. Milani, eds.

© 2005 International Astronomical Union
DOI: 10.1017/S1743921304008476

Fitting orbits

**Andrzej J. Maciejewski[1], Krzysztof Goździewski[2]
and Szymon Kozłowski[1]**

[1] Institute of Astronomy, University of Zielona Góra, Podgórna 50, PL-65–246 Zielona Góra,
Poland
email: maciejka@astro.ia.uz.zgora.pl

[2] Toruń Centre for Astronomy, N. Copernicus University, Gagarina 11, PL-87–100 Toruń,
Poland

Nowadays, more than one hundred extra-solar planets are known, and about a dozen of multi-planetary systems have been discovered. Most of them have been detected by the radial velocity (RV) method. The recovery of orbital parameters from RV data leads to several problems. Usually RV data cover irregularly a short time interval which is frequently shorter than the orbital period of the most distant planet. Moreover, observations contain a noise due to the instabilities of the star. The distribution of this noise is unknown. A precise determination of the dynamical state of a multi-planetary system is important for understanding its stability and evolution. In most cases observers determine the orbital parameters for multi-planetary systems just fitting a sum of Keplerian orbits. The parameters obtained in such a way are in most cases the only accessible data about an extra-solar system because the observes very rarely publish their observations. However, the parameters from a multi-Keplerian fit as it has already been observed by many authors, cannot be interpreted as the osculating elements for actual planetary orbits. Moreover, these parameters can be considered as Keplerian elements of: relative, barycentric or Jacobi orbits. One can find arguments that the interpretation of parameters from a multi-Keplerian fit as elements of Keplerian orbits in the Jacobi coordinates is the most proper one, see [Lee and Peale, 2002; Goździewski *et al.* 2003].

Our first aim was to determine how badly a multi-Keplerian fit determines osculating orbits. To this end, we performed several numerical simulations. For a chosen planetary system with two planets we generated a synthetic RV observations using the Newtonian three body problem. Then we fitted to these observations the Keplerian model and compared the obtained Keplerian elements with the true osculating elements of orbits. Then we changed the semi-major axis and the eccentricity of one planet and repeated all calculations. In this way we obtained maps of differences between the true and the fitted Keplerian elements (relative, barycentric and Jacobi) for a given system with two planets.

The conclusions from these experiments are following. Even for a quite big separation of planets (2 AU), multi-Keplerian fits are bad. The errors appear mainly in the positions of planets in their orbits and can achieve 60 deg and more. The errors in eccentricities and semi-major axes achieve a few percent, but they can be bigger for bigger masses of planets, or when the observations cover only a part of the period of the external planet. Moreover, the errors are maximal for systems close to a mean motions resonance. All the above conclusions do not depend on how we interpret the parameters of a Keplerian fit: relative, barycentric, as well as Jacobi elements are equally bad if we look at the overall results.

References

Lee, M. H. and Peale, S. J. 2002, *ApJ*, 567, 596

Goździewski, K., Konacki, M. and Maciejewski, A. J. 2003, *ApJ*, 594, 1019

Dynamics of Populations of Planetary Systems
Proceedings IAU Colloquium No. 197, 2005
Z. Kneževic and A. Milani, eds.

© 2005 International Astronomical Union
DOI: 10.1017/S1743921304008488

The secular planetary three body problem revisited

Jacques Henrard and Anne-Sophie Libert

Département de mathématique FUNDP, 8, Rempart de la Vierge, B-5000 Namur, Belgique
e-mail: jacques.henrard@fundp.ac.be

Abstract. We analyze the secular interactions of two coplanar planets based on a high order (order 12) expansion of the perturbative potential in powers of the eccentricities. The model depends on only two parameters (the ratio of semi-major axis and the mass ratio of the planets) and can be reduced to a one degree of freedom system, allowing for an exhaustive parametric analysis. Following Pauwels (1984) we map the phase space on a sphere, avoiding in this way the artificial singularities introduced by other mappings. We show that the twelve order expansion is able to describe correctly most of the exosolar planetary systems discovered so far, even if the eccentricities of these planets are considerably larger than the eccentricities of our own solar system. The expansion is even able to reproduce, at moderate eccentricities, the secular resonances discovered numerically by Michtchenko and Malhotra (2004) at moderate to large eccentricities.

Keywords. Secular planetary problem, high order expansions, exosolar planetary systems

---

1. Introduction

The number of discovered exoplanets is growing fast. A few of them, but their number is also growing, form genuine planetary systems, being planets of the same star. Due possibly to observational bias, these systems are quite different from our own planetary system: large planetary masses, very large eccentricies are common. Some of the pairs of planets are locked in a mean motion resonance and their orbits are significantly perturbed; but even in the non-resonant case, long term secular effects may produce interesting dynamical phenomena and large amplitude variation of eccentricities.

Due to the large values of the eccentricities, the applicability of the classical Laplace-Lagrange linear perturbation theory is very limited. But it is worth investigating if a *non-linear* analytical theory, based on a high order expansion of the secular perturbation may not be applicable, at least for some of the discovered systems.

This is the aim of this contribution. We expand the secular interactions of a couple of coplanar planets up to order 12 in the eccentricities, and check that this expansion modelizes correctly several of the observed systems.The analytical modelization, valid not only for low, but for moderately high eccentricity, facilitates the parametric analysis of the secular interaction of a couple of planets. Essentially, two parameters are involved, the ratio $\alpha = a_1/a_2$ of the semi-major axis and the mass ratio $\mu = m_1/(m_1 + m_2)$. We restrict ourself to planar systems because no or very little information on inclination is available, but, of course, a similar analytical modelization of inclined systems is possible although its analysis would be more involved.

To describe the phase space of the problem we use the representation introduced by Pauwels (1983) which is free of the artificial singularities which mar the traditional representations. We believe that, with this representation, the phase-space diagrams are simpler and much more readable.

As a further indication of the interest of our analytical approach, we reproduce, at moderate eccentricities, the numerical results of Michtchenko and Malhotra (2004) concerning the existence of a secular resonance at moderate to large eccentricities.

2. The Planar Secular Three Body Problem

Let us consider a central star of mass m_0 and two coplanar planets of mass m_1 and m_2. Using the usual Jacobi coordinates the Hamiltonian of the dynamics of this system is (up to the second powers in the mass ratios m_1/m_0 and m_2/m_0)

$$\mathcal{H} = -\frac{Gm_0 m_1}{2a_1} - \frac{Gm_0 m_2}{2a_2} - Gm_1 m_2 \left[\frac{1}{|\vec{r}_1 - \vec{r}_2|} - \frac{(\vec{r}_1|\vec{r}_2)}{r_2^3} \right] \tag{2.1}$$

where a_i, $\vec{r}_i$ and r_i are, respectively, the osculating semi-major axis, the position vector and the norm of the position vector of the mass m_i (see for instance Brouwer and Clemence 1961 or Laskar, 1990). We assume that the mass m_1 is the one closest to the central star. To the second order in the mass ratios, the classical modified Delaunay's elements (see references above) are:

$$\begin{aligned}
\lambda_i &= \quad \text{mean longitude of } m_i & L_i &= \quad m_i \sqrt{Gm_0 a_i} \\
p_i &= \quad - \text{ the longitude of the pericenter of } m_i & P_i &= \quad L_i \left[1 - \sqrt{1 - e_i^2} \right].
\end{aligned} \tag{2.2}$$

The perturbation (the last term in equation (2.1), can be expanded in powers of the eccentricities (see for instance Murray and Dermott 1999) to yield:

$$\begin{aligned}
\mathcal{H} = &-\frac{Gm_0 m_1}{2a_1} \quad -\frac{Gm_0 m_2}{2a_2} \\
&-\frac{Gm_1 m_2}{a_2} \sum_{k,i_1,i_2,j_1,j_2} A^k_{i_1,i_2,j_1,j_2} \, e_1^{|j_1|+2i_1} \, e_2^{|j_2|+2i_2} \cos \Phi \;,
\end{aligned} \tag{2.3}$$

where $\Phi = [(k+j_1)\lambda_1 - (k+j_2)\lambda_2 + j_1 p_1 - j_2 p_2]$. The coefficients $A^k_{i_1,i_2,j_1,j_2}$ depend only on the ratio a_1/a_2 of the semi-major axis. As we plan to use extensively the canonical variables defined in (2.2), we prefer to use the expressions $E_i = \sqrt{2P_i/L_i}$ rather than e_i. Notice that for small to moderate eccentricities $E_i \approx e_i$. Hence the Hamiltonian reads:

$$\begin{aligned}
\mathcal{H} = &-\frac{Gm_0 m_1}{2a_1} \quad -\frac{Gm_0 m_2}{2a_2} \\
&-\frac{Gm_1 m_2}{a_2} \sum_{k,i_1,i_2,j_1,j_2} B^k_{i_1,i_2,j_1,j_2} \, E_1^{|j_1|+2i_1} \, E_2^{|j_2|+2i_2} \cos \Phi \;.
\end{aligned} \tag{2.4}$$

As we are interested in the long term dynamics and as we assume that we are not close to a mean motion commensurability, we average the Hamiltonian function over the "fast variables" λ_i. To the first order in the mass ratios, the averaging amount to a "averaging by scissors" (dropping the fast periodic terms from the function). We are left with:

$$\mathcal{K} = -\frac{Gm_0 m_1}{2a_1} - \frac{Gm_0 m_2}{2a_2} - \frac{Gm_1 m_2}{a_2} \sum_{k,i_1,i_2} C^k_{i_1,i_2} \, E_1^{k+2i_1} \, E_2^{k+2i_2} \cos k(p_1 - p_2) \;, \tag{2.5}$$

where a_i, E_i and p_i designate now averaged values. The "fast variables" λ_i are now ignorable which means that the L_i and thus the a_i are constant. The first two terms of (2.5), being constant, can be dropped from the Hamiltonian. Also the Hamiltonian depends only on the difference $(p_1 - p_2)$; hence the sum $P_1 + P_2$ is a first integral of the problem which can be reduced to one degree of freedom.

Before analyzing further this simplified model, it is useful to check whether it may be applied to exosolar systems. The eccentricity of many of the observed exoplanets is rather high and one may wonder if the above expansion in powers of the eccentricities can represent their orbits with enough accuracy. We have computed the expansion (up to order 12) for several of the exosystems of planets and report the numerical convergence in Table I.

Table I: Convergence of the expansion (2.5) for some exosystems.

	(c-d) v And.	Gliese 876	HD168443	HD169830	HD37124
α	0.332	0.62	0.10	0.22	0.22
μ	0.335	0.23	0.30	0.42	0.38
e_1	0.28	0.10	0.53	0.31	0.10
e_2	0.27	0.27	0.23	0.33	0.69
$p_1 - p_2$	$-10°$	$3°$	$110°$	$-104°$	$-200°$
ord. 2	$5.9\ 10^{-3}$	$1.4\ 10^{-2}$	$7.8\ 10^{-4}$	$7.3\ 10^{-3}$	$7.5\ 10^{-3}$
ord. 4	$4.7\ 10^{-4}$	$-2.5\ 10^{-4}$	$-1.2\ 10^{-5}$	$2.4\ 10^{-3}$	$3.1\ 10^{-3}$
ord. 6	$2.8\ 10^{-5}$	$-3.6\ 10^{-6}$	$-6.1\ 10^{-7}$	$7.1\ 10^{-4}$	$1.1\ 10^{-3}$
ord. 8	$1.4\ 10^{-6}$	$5.9\ 10^{-8}$	$-1.2\ 10^{-8}$	$2.0\ 10^{-4}$	$3.8\ 10^{-4}$
ord. 10	$7.2\ 10^{-8}$	$1.8\ 10^{-9}$	$-1.4\ 10^{-10}$	$5.9\ 10^{-5}$	$1.4\ 10^{-4}$
ord. 12	$3.8\ 10^{-9}$	$-1.4\ 10^{-11}$	$-4.4\ 10^{-13}$	$1.8\ 10^{-5}$	$5.0\ 10^{-5}$

Listed in Table I are the physical parameters of the systems: $\alpha = a_1/a_2$ the ratio of semi-major axis, $\mu = m_1/(m_1 + m_2)$, the mass ratio, the eccentricities of the two components of the systems and the contributions to the value of the Hamiltonian, from order 2 to order 12 in E_1 and E_2, of the terms in the expansion (2.5). The numerical convergence is excellent for the first three systems, marginal for the last two.

3. Reduction to One Degree of Freedom and Display of the Phase Space

In order to bring forward the fact that the problem may be reduced to a one degree of freedom problem, it is enough to introduce the new canonical variables:

$$u = p_2 \quad ; \quad U = (P_1 + P_2)/[(m_1 + m_2)\sqrt{Gm_0 a_2}]$$
$$v = p_1 - p_2 \quad ; \quad V = P_1/[(m_1 + m_2)\sqrt{Gm_0 a_2}]. \tag{3.1}$$

We have introduced the factor $(m_1 + m_2)\sqrt{Gm_0 a_2}$ in order to make non dimensional the new actions U and V. Notice that the constant U is a weighted mean of the square of the two eccentricities when they are small:

$$U = \mu\sqrt{\alpha}\left[1 - \sqrt{1 - e_1^2}\right] + (1 - \mu)\left[1 - \sqrt{1 - e_2^2}\right]. \tag{3.2}$$

Up to a constant (the first two terms of (2.5)) and to a factor (the factor $-Gm_1m_2/a_2]$), which can be absorbed by redefining the time scale, the Hamiltonian reads

$$\mathcal{K}^* = \sum_{n\geqslant 1} U^n \sum_{k=0}^{n} \sum_{i_1=0}^{n-k} C_{i_1,i_2}^k \left[\frac{2V}{U\mu\sqrt{\alpha}}\right]^{(k+2i_1)/2} \left[\frac{2(U - V)}{U(1 - \mu)}\right]^{(k+2i_2)/2} \cos kv, \tag{3.3}$$

where $i_2 = n - k - i_1$. The quantity U being a constant, the trajectories of the one

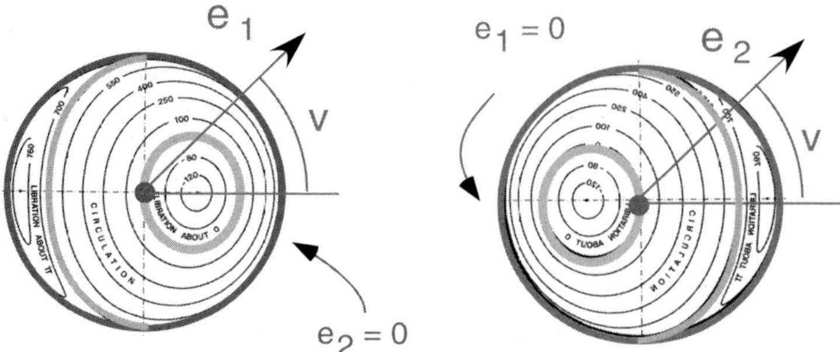

Figure 1. The usual representation of the level curves of the Hamiltonian (3.3) on a plane parametrized by polar coordinates with one of the eccentricities as distance and the difference in longitude of the pericenter as angular variables. The drawback of this representation is that the outer circle represents a single point; a fact that introduce artificial singularities.

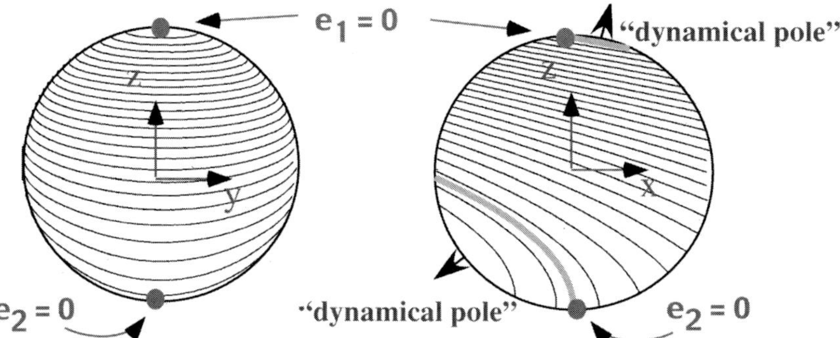

Figure 2. The projection on the $[x, z]$ and the $[y, z]$ plane of the Pauwel's sphere. The thick curves are the ones passing through the geographical poles (where one of the eccentricities vanishes). In the traditional representation they are singular, but nothing set them apart in this representation of the phase space.

degree of freedom system, (v, V), are the level curves of the Hamiltonian function on a two dimensional manifold. Usually the level curves are drawn on a plane the polar coordinates of which are v and e_1 or v and e_2 (see Figure 1). The fact that the outer circles correspond to a single point ($e_1 = 0$ to the left, $e_2 = 0$ to the right) introduces uncalled for artificial singularities in the problem and makes difficult the interpretation of the figures.

We much prefer to use the representation of the manifold $U = $ constant introduced by Pauwels (1983). Let us define an angle ϕ by the equation:

$$\frac{P_1}{P_1 + P_2} = \frac{V}{U} = \frac{1}{2}[1 - \sin \phi] . \qquad (3.4)$$

As V/U belongs to the interval $[0, 1]$ ($V/U = 0$ when $e_1 = 0$ and $V/U = 1$ when $e_2 = 0$), the angle ϕ belongs to the interval $[-\pi/2, \pi/2]$. At the extreme values the angular variable $v = p_1 - p_2$ is undefined. Hence the phase space has the topology of a sphere with ϕ as latitude and v as longitude. The particular values $e_1 = 0$ and $e_2 = 0$ represent the poles of this sphere. Of course it is not as easy to draw level curves on a sphere than on a plane, but we believe that the projections of the sphere on some plane (we have chosen, in Figure 2, the $[x, z]$ plane and the $[y, z]$) is informative enough. The really interesting points are the equilibria which we call *dynamical poles*.

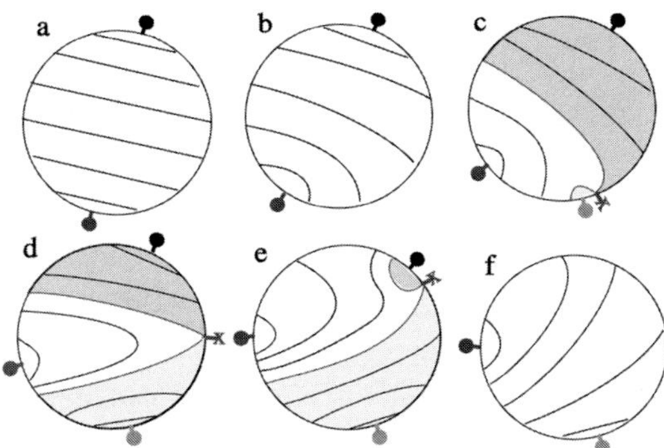

Figure 3. The projection on the $[x, z]$ plane of the Pauwel's spheres for $\alpha = 0.166$ and $m_1/m_2 = 1$. From (a) to (f) the constant U, and thus the weighted mean eccentricity, increases. The secular resonance appears from (c) to (e) – from $U = 0.12$ and $U = 0.151$. See the text for further comments.

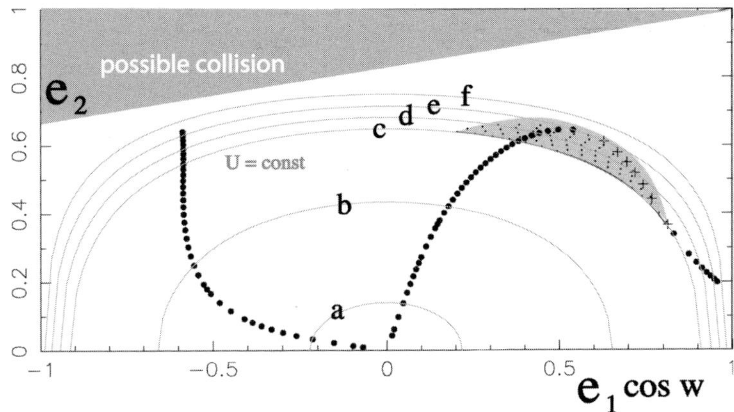

Figure 4. The singular points on the Pauwel's sphere for $\alpha = 0.166$ and $m_1/m_2 = 1$. The right half of this plane (the part with $\cos w = 1$) should be compared with the lower left panel of Figure 8 of Michtchenko and Malhotra paper (2004). See the text for further comments.

4. The Non-Linear Resonance of the Planar Secular Planetary Three-Body Problem

Michtchenko and Malhotra (2004) have shown recently that there exists, at moderate-to-high eccentricity, a secular resonance between the motion of the pericenter of the two planets. Their results are based on a semi-numerical approach which employs a numerical averaging of the short period interactions of the planets and a subsequent numerical integration of the averaged differential equations. It is a challenge for our analytical approach to reproduce at least part of their results. Of course we are unable to reproduce them when they appear at very high eccentricities; but, as shown in Figures 3 and 4, the analytical modelization is able to detect and describe correctly the secular resonance when it appears at moderate eccentricity. In that case, the analytical approach leads to a much easier analysis. The Hamiltonian function and its derivatives are rather simple functions of the spherical coordinates (ϕ, v). Trajectories are level curves of $\mathcal{K}^*$; equilibria are roots of a simple polynomial; maximum and minimum values of the eccentricities along a trajectory are easily computed, etc.

Figure 3 shows, for $\alpha = a_1/a_2 = 0.166$ and $\mu = m_1/(m_1 + m_2) = 0.5$, the evolution of the phase space when the parameter U increases. For $U = 0.005$ (Figure 3a) the dynamical problem is almost linear and the level curves are almost circles centered on a line $(y, z) = (\sin \beta, \cos \beta)$. Actually, for the linear problem obtained by keeping only the quadratic terms in the expansion of the Hamiltonian, a rotation of the sphere by the angle β (which is a simple function of the parameters α and μ) maps the level curves on the lines of constant latitude. The poles of this transformed sphere are the equilibria. This is the reason why we have called them *dynamical poles*, in contrast to the *geometrical poles* which correspond to the phases $e_1 = 0$ and $e_2 = 0$.

For $U = 0.05$, the southern dynamical pole goes toward the geometrical equator much faster than the northern dynamical pole and the level curves are distorted. For a value of U slightly smaller than 0.12 a cusp appears on one of the level curves and for $U = 0.12$ (see Figure 3c) a small loop is created containing a third equilibrium. An unstable equilibrium generates two homoclinic orbits which divide the sphere in three domains associated each with one of the equilibria. For larger values of U (see Figure 3d for $U = 0.135$) the small loop grows and the domain associated with the northern pole shrinks. At $U = 0.151$ (see Figure 3e) this domain is reduced to a small loop surrounded by one of the homoclinic orbits. For a slightly larger value of U this dynamical pole and the unstable equilibrium merge and disappear. For larger values (see Figure 3f for $U = 0.17$), the topology of the phase space is the same as before the apparition of the equilibrium. The sphere is undivided by homoclinic orbits. What was the northern dynamical pole has disappeared,; what was the southern dynamical pole has moved in the northern hemisphere and the third equilibrium, promoted to the role of a dynamical pole, has taken a position close to the southern geometric pole (i.e. for $e_2 \approx 0$).

Figure 4 summarize the location of the dynamical poles on the circles $y = 0$ of the Pauwel's spheres. For the value of U of each of the sphere of Figure 3, the curve $U = $ constant is drawn in the plane e_1, e_2. Notice that in order to distinguish between the left and right sides of the maps in Figure 3, we attribute a negative value to e_1 when x is negative or equivalently when $v = \pi$. Dots indicate the position of dynamical poles (stable equilibria). Between $U = 0.12$ and $U = 0.151$ they are three of them. A cross indicates the position of the unstable equilibria responsible for the homoclinic trajectories which divide the spheres in three domains. Small dots indicates the extend of the secular resonance (dark areas in Figure 3). The right half of Figure 4 reproduce almost exactly the lower left panel of Figure 8 of Michtchenko and Malhotra (2004) paper.

References

Brouwer, D. and Clemence, G.M 1961, *Methods of Celestial Mechanics*, Academic Press

Laskar, J. 1990, in: D. Benest and C. Froeschlé (eds.), *Les méthodes modernes de la mécanique céleste*, Editions Frontière, p. 63

Michtchenko, T.A. and Malhotra, R. 2004, *Icarus*, 168, 237

Murray, C.D. and Dermott, S.F. 1999, *Solar System Dynamics*, Cambridge Univ. Press

Pauwels, T. 1983, *Celest. Mech.*, 30, 229

Dynamics of Populations of Planetary Systems
Proceedings IAU Colloquium No. 197, 2005
Z. Knežević and A. Milani, eds.

© 2005 International Astronomical Union
DOI: 10.1017/S174392130400849X

Dynamics of extrasolar systems at the 5/2 resonance: application to 47 UMa

Dionyssia Psychoyos and John D. Hadjidemetriou

Department of Physics, University of Thessaloniki, 54124 Thessaloniki, Greece
email:hadjidem@auth.gr

Abstract. A complete study is made of the 5/2 resonant motion of two planets revolving around a star, in the model of the general planar three body problem. Families of 5/2 resonant symmetric periodic orbits are computed numerically, for the masses of the extrasolar system 47 UMa. The phase of the two planets (alignment or antialignment of perihelia and position of each planet at perihelion or aphelion) plays an important role, and the change of the phase, other things being the same, may destabilize the system. Stable motion exists even in the case where the two planetary orbits intersect. A small value of the eccentricities, for the same phase, stabilizes the system. The above results are applied to the study of 47 UMa, which according to some observations is close to the 5/2 resonance.

Keywords. periodic orbits - resonances - extrasolar systems - 47 UMa

1. Introduction

The stability and the long term evolution of a planetary system is determined from the topology of its phase space. It is clear that the topology of the phase space of any dynamical system depends on the position and the stability properties of the periodic orbits, or equivalently, on the fixed points of the Poincaré map. In particular, the mean motion resonances in a planetary system correspond to periodic motion, in a rotating frame. This is the reason why the resonances play an important role in the study of the long term evolution of a planetary system, although the corresponding periodic orbits are a set of measure zero.

There are several planetary systems at different resonances, and we shall study in the present paper the 5/2 resonance. This resonance appears in our own planetary system, between Jupiter and Saturn. Also, according to Fisher *et al.* 2002, the extrasolar planetary system 47 UMa has a ratio of the planetary periods equal to $T_2/T_1 = 2.38$, which can be considered as close to the 5/2 resonance (but also close to the 7/3 resonance). A more recent analysis of the observational data for 47 UMa (Fisher *et al.* 2003) revise these values and give a new value for the planetary periods, $T_2/T_1 = 2.64$, which is close to the 8/3 resonance (but not far from the 5/2 resonance). Altough the planetary masses are also revised, the ratio m_1/m_2 is in all cases larger than unity.

There are several papers that study the dynamics of a planetary system at the 5/2 resonance and of the system 47 UMa in particular, by numerical simulations and analytic work (Ji *et al.* 2003, Barnes and Quinn, 2004, Gozdziewski, 2002, Laughlin *et al.* 2002, T.A.Michtchenko and S. Ferraz-Mello 2001, Beaugé *et al.* 2003). In the present study we make a global analysis of the 5/2 resonant motion of a planetary system. In this global analysis, we used for the planetary masses, the (minimum) masses given for the 47 UMa system by Fisher *et al.* (2002). Note that for these masses $m_1/m_2 \approx 0.30$, which is very close to the ratio of the masses of the Jupiter-Saturn system and consequently the dynamics is also applicable to our own planetary system.

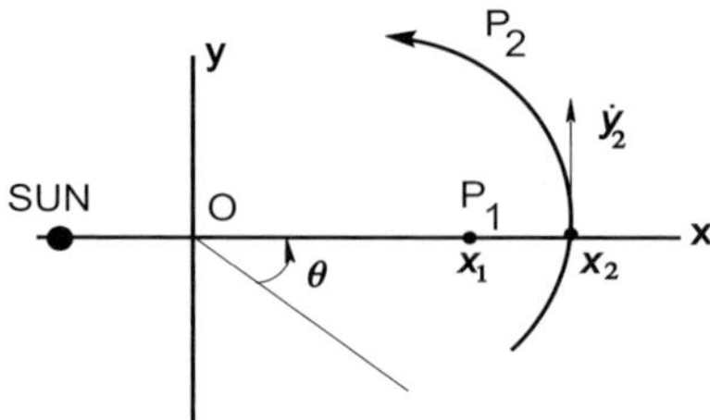

Figure 1. The rotating frame xOy

We find all the basic families of resonant periodic orbits at this resonance and in this way we have a complete knowledge of the regions of the phase space where stable motion exists. The orbits are periodic in a *non-uniformely* rotating frame, which means that the *relatative* configuration is repeated in space, and are symmetric with respect to the rotating x-axis, defined in the next section. The stability analysis of these families gives the regions of the phase space where a planetary system at the 5/2 resonance can exist, and also the regions of the phase space where a planetary system could be trapped, if it had followed in the past a migration process. In addition, the motion close to a stable periodic orbit is the motion with small variation of the orbital elements, a condition which may play an important role in the appearance of life.

The general study of the 5/2 resonance that we made is applied to the observed extra-solar planetary system 47 UMa, using the data given by Fisher *et al.* 2002. The stability of this system, for different initial phases, is compared with the exact resonant periodic motion at the 5/2 resonance.

In all the following the central star will be called the *sun*, the inner planet will be called P_1 and the outer planet P_2.

2. The dynamical Model

The model we used in the study of periodic motion of the planetary system is the general three body problem, for planar motion.

The center of mass of the planetary system is considered as fixed in an inertial frame, and the study is made in a non-uniformely rotating frame of reference xOy, whose x-axis is the line *sun* - P_1, the origin O is the center of mass of these two bodies and the y-axis is perpendicular to the x-axis (Figure 1). In this rotating frame P_1 moves on the x-axis and P_2 in the xOy plane. The coordinates are the position x_1 of P_1, the position x_2, y_2 of P_2 and the angle θ between the x-axis and a fixed direction in the inertial frame. The coordinates x_1, x_2, y_2, define the position of the system in the rotating frame and the angle θ defines the orientation of the rotating frame, so these four coordinates determine the position of the system in the inertial frame. This is a system of four degrees of freedom, but it turns out that the angle θ is ignorable, and consequently the angular momentum integral $L = \partial \mathcal{L}/\partial \dot{\theta} =$ constant, where $\mathcal{L}$ is the Lagrangian of the system. So the study is reduced to a system of three degrees of freedom, in the rotating frame only, and the angular momentum L is a fixed parameter (Hadjidemetriou 1975).

Symmetric periodic orbits exist in the above rotating frame, such that the planet P_2 starts perpendicularly from the x-axis ($y_2 = 0$, $\dot{x}_2 = 0$) and at that time $\dot{x}_1 = 0$, and after some time $t = T/2$ the planet P_2 crosses again the x-axis perpendicularly and at that time it is $\dot{x}_1 = 0$. This means that the non zero initial conditions of a symmetric periodic orbit, in the rotating frame, are x_{10}, x_{20}, $\dot{y}_{20}$. So, a family of symmetric periodic orbits is represented as a smooth curve in the three dimensional space x_{10} x_{20} $\dot{y}_{20}$. The

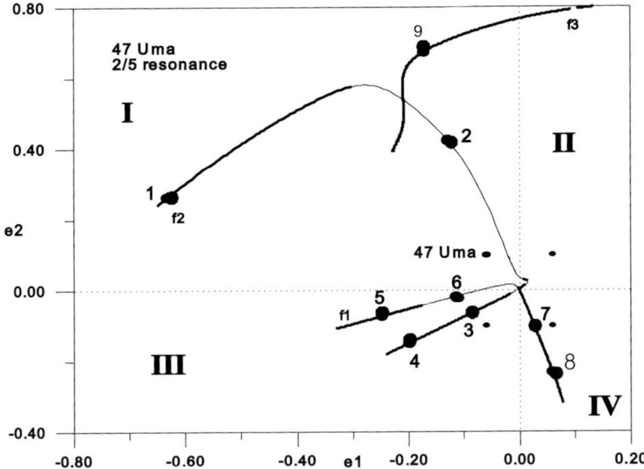

Figure 2. The three families of 5/2 resonant periodic orbits. The unstable sections are indicated by a thicker line. The sectors I-IV, defined by the sign of e_1 and e_2, correspond to the different phases given in Figure 3. The position of the 47 UMa is also shown, for all four possible phases.

symmetry implies that the perihelia of the two planets are either in the same direction or in opposite directions when the two planets and the sun are aligned. Viewed from the inertial frame, the line of apsides of the two planets precesses slowly, in such a way that $\Delta\omega$ is equal to 0 or 2π.

In order to avoid duplication of the results we fix the units of mass, length and time. This is achieved by taking the total mass of the system as the unit of mass, the gravitational constant equal to unity and also by keeping a fixed value of the angular momentum L for all the orbits of a family of periodic orbits. So, the normalizing conditions are $m_0 + m_1 + m_2 = 1$, $G = 1$, $L = constant$. In practice, we made the integration of the planetary system in the inertial frame (where the center of mass is fixed) and the reduction to three degrees of freedom, in the rotating frame, was made by a coordinate transformation. The method of integration was based on Taylor series expansion, and the accuracy was 10^{-14}.

3. The 5/2 resonance

The extrasolar planetary system 47 UMa is close to a resonant system, but not exactly resonant. The elements of this system that we used in the present study are (Fisher *et al.* 2002): $m_0 = 1.03\ M_{SUN}$, $m_1 \sin i = 2.54$ MJ, $m_2 \sin i = 0.76$ MJ, $a_1 = 2.09$ AU, $a_2 = 3.73$, $T_1 = 1089 \pm 3$ d, $T_2 = 2594 \pm 90$ d, $e_1 = 0.061 \pm 0.014$, $e_2 = 0.1 \pm 0.1$, $\omega_1 = 172^0$, $\omega_2 = 127^0$. Note that $\omega_1 - \omega_2 = 45^0$. For later updates see the web page maintained by Jean Schneider (*http://www.obspm.fr/encycl/catalog.html*). The planetary masses are multiplied by $\sin i$, where i is the inclination of the planetary orbit and are therefore the minimum masses. The observed values give a ratio of the planetary periods equal to $T_2/T_1 = 2.38$, which is close to the 5/2 resonance (but also close to the 7/3 resonance). In this paper we restrict the study to the computation of families of periodic orbits for the 5/2 resonance and for the minimum masses of the system 47 UMa. The study of the 7/3 and 8/3 resonances will be included in a forthcoming paper. In all the computations we used normalized values of the (minimum) masses, which are $m_0 = 0.996942$, $m_1 = 0.002354$, $m_2 = 0.000704$.

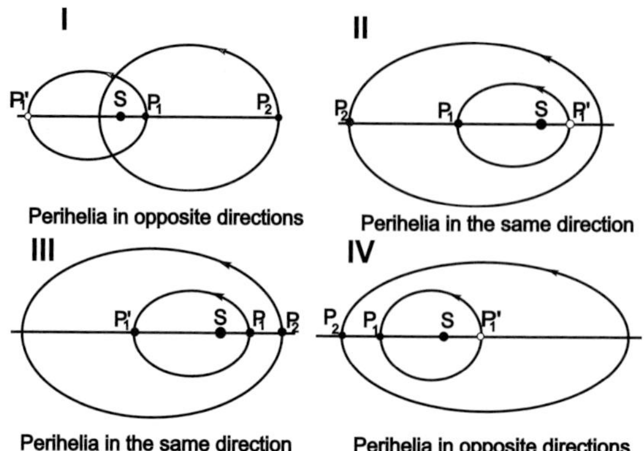

Figure 3. The four different configurations at $t = 0$ and $t = T/2$ (denoted by prime). The planet P_2 is at the same position after half a period, but the planet P_1 shifts from perihelion to aphelion (or vice versa). The panels I-IV correspond to the sectors of Figure 2 (for positive or negative eccentricities), as indicated.

The 5/2 resonance is of a different nature than the 2/1 resonance, studied by Psychoyos and Hadjidemetriou, 2004. In the present case there are two distinct resonant families, that bifurcate from the family of circular orbits from two points close to each other. One more family also exists, which does not bifurcate from the circular family.

3.1. *Periodic orbits*

In Figure 2 we present three families of 5/2 resonant periodic orbits, for the (normalized) values of the masses of the planetary system 47 UMa. To make the presentation clearer from the physical point of view, we present these families in the eccentricity space $e_1\, e_2$, instead of the initial condition space $x_{10}\, x_{20}\, \dot{y}_{20}$. To avoid artificial discontinuities in the presentation of the families, we use the notation $e_i > 0$ for position of the planet at aphelion and $e_i < 0$ for position at perihelion ($i = 1, 2$). Note that for a symmetric periodic orbit the planets are either at perihelion or at aphelion at $t = 0$ and at $t = T/2$, where T is the period.

Along the families 1 and 2 (denoted by $f1$ and $f2$, respectively, in Figure 2), the eccentricities of the planets increase, starting from zero values, because these two families bifurcate from the circular family, where $e_1 = e_2 = 0$. For the family 3 (denoted by $f3$ in Figure 2), only the eccentricity e_1 of the first planet crosses the zero point, while e_2 stays at high values. We also computed the linear stability. The unstable sections of these families are indicated by a thicker line. In Figure 3 we give the four possible positions of a 5/2 resonant planetary system, corresponding to the four different configurations I-IV of Figure 2. It is simple geometry to see that there are eight different initial positions of the two planets at $t = 0$: The perihelia of the two planetary orbits can be aligned or antialigned and in each case the planets can be either at perihelion or at aphelion. These eight configurations are however equivalent in pairs, due to the fact that we are at a symmetric resonance, and consequently there are only four different configurations. Consider for example the case where the two planets are both at perihelia at $t = 0$ (cases III and IV in Figure 3). After half a period, at $t = T/2$, P_2 will be at perihelion but P_1 will be at aphelion.

3.2. *Stability*

In order to find the region of stability around a periodic orbit, we perturbed it and computed the Poincaré map on the surface of section $y_2 = 0$. We considered as perturbation

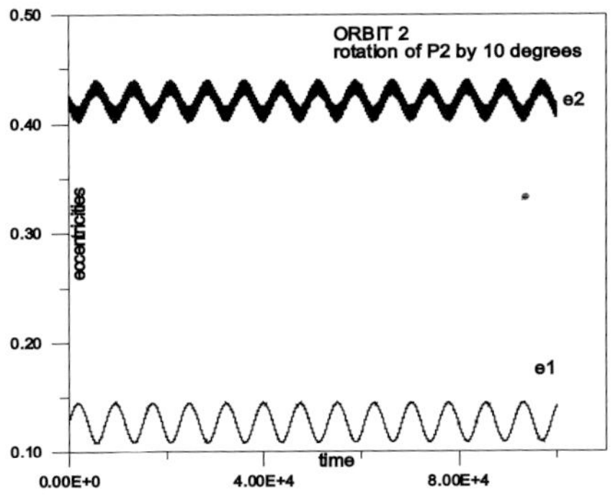
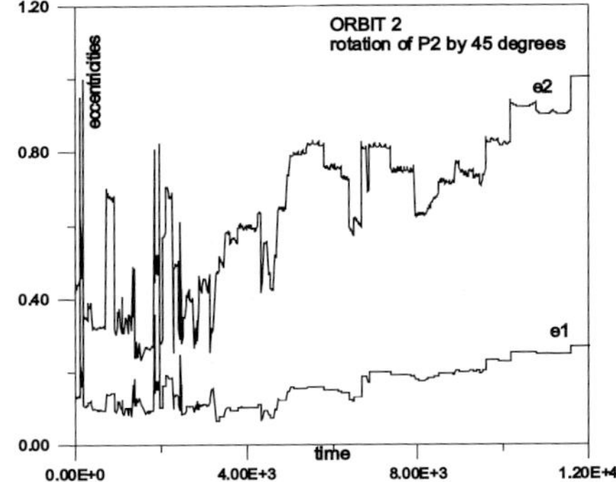

Figure 4. The evolution of the eccentricities of an orbit close to the stable periodic orbit 2 of the family 2, when the orbit of P_2 is rotated by 10^0 (panel a) and by 45^0 (panel b). The motion is bounded for a small perturbation, but for a larger perturbation the system is destabilized.

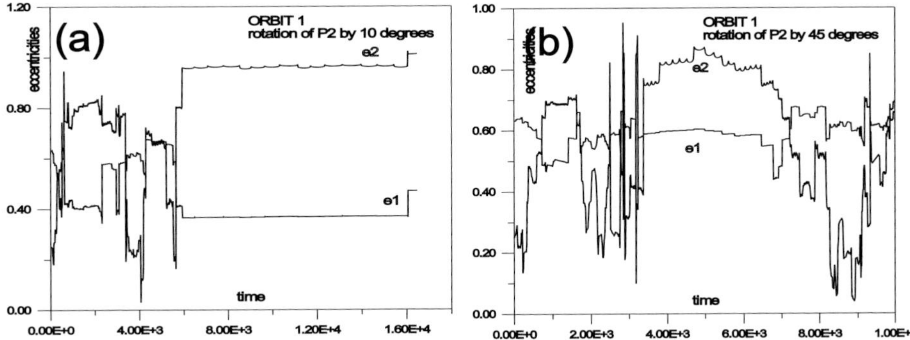

Figure 5. The evolution of the eccentricities of an orbit close to the unstable periodic orbit 1 of the family 2, when the orbit of P_2 is rotated by 10^0 (panel a) and by 45^0 (panel b). The motion is chaotic, implying that large values of the eccentricities destabilize the system.

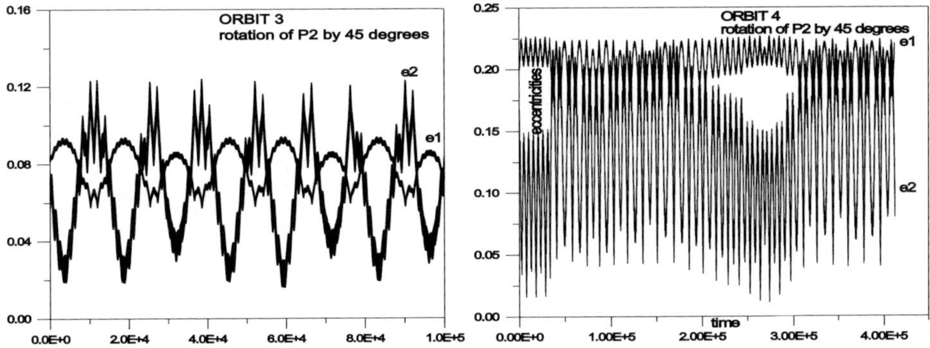

Figure 6. The evolution of the eccentricities of an orbit close to the unstable periodic orbits 3 (panel a) and 4 (panel b) of the family 2, when the orbit of P_2 is rotated by 45^0. The motion is bounded, but the amplitude of the variation is much larger for the orbit 4, which has larger eccentricities than the orbit 3.

the rotation of the orbit of P_2, at $t = 0$, by a certain angle, thus destroying the symmetry of the system. Some typical results are given bellow.

We considered first the stable periodic orbit 2 on the family 2 (Figure 4a), which corresponds to the phase where the perihelia are antialigned and P_1 is at perihelion and P_2 at aphelion (phase I). This orbit is linearly stable and a small perturbation, which in

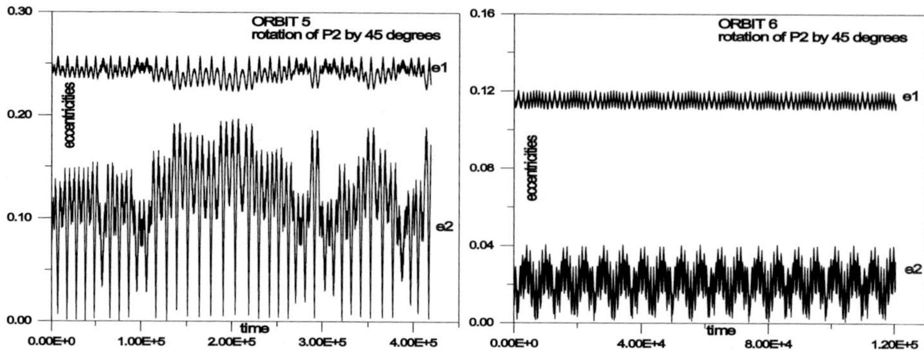

Figure 7. The evolution of the eccentricities of an orbit close to the unstable periodic orbit 5 (panel a) and the stable periodic orbit 6 (panel b) of the family 1, when the orbit of P_2 is rotated by 45^0. The motion is bounded, but the amplitude of the variation is much larger for the unstable orbit 5 which has larger eccentricities than the orbit 6.

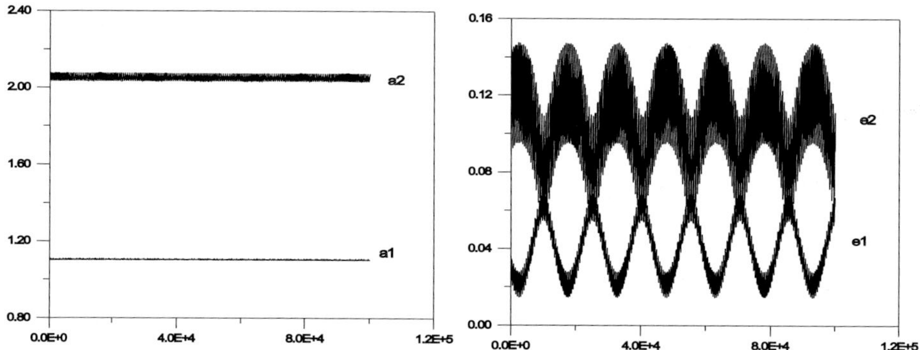

Figure 8. The evolution of the semimajor axes (panel a) and of the eccentricities (panel b), of an orbit close to the unstable periodic orbit 7 of the family 1, when the orbit of P_2 is rotated by 10^0. The motion is bounded.

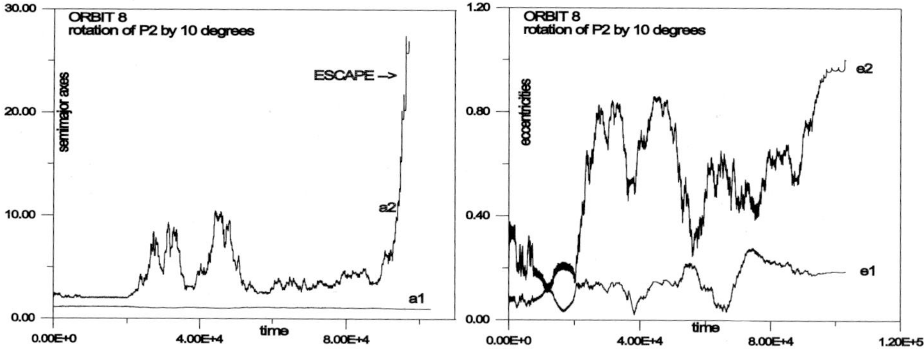

Figure 9. The evolution of the semimajor axes (panel a) and of the eccentricities (panel b), of an orbit close to the unstable periodic orbit 8 of the family 1, when the orbit of P_2 is rotated by 10^0. The motion is chaotic.

this case is the rotation of P_2 by 10^0, gives bounded motion. We remark that the two planetary orbits intersect, and still the motion is stable. A larger perturbation however (rotation by 45^0), destabilizes the system (Figure 4b). Next, we consider the unstable orbit 1 of the same family 2, which has the same phase but larger eccentricities than the orbit 2. A rotation of the orbit of P_2 by 10^0 and also by 45^0 results to instability

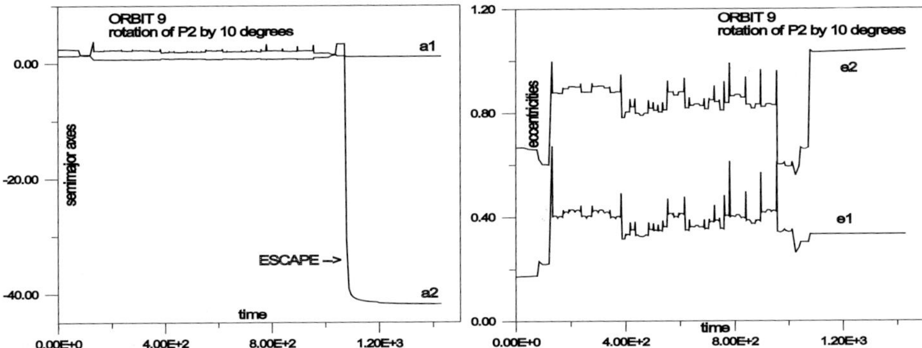

Figure 10. The evolution of the semimajor axes (panel a) and of the eccentricities (panel b), of an orbit close to the unstable periodic orbit 9 of the family 3, when the orbit of P_2 is rotated by 10^0. The motion is chaotic.

(Figure 5). We note that the orbit 1 has the same phase as the orbit 2, which implies that small values of the eccentricities are essential, in the 5/2 resonance, for the stability of the system.

The third orbit that we considered is orbit 3, on family 2, which corresponds to the phase where the perihelia are aligned and both planets are at perihelia (phase III). This orbit is linearly unstable, but a rotation of P_2 by 45^0 gives bounded motion (Figure 6a). The fourth orbit on same family that we consider is the unstable orbit 4, which has larger eccentricities than the orbit 3. A rotation of the orbit of P_2 by 10^0 in this latter case results to bounded motion, but now the variation of the eccentricities is much larger, compared to the orbit 3 on this family, which has the same phase. Again we see that a small value of the eccentricities is necessary for the stability of the system.

We repeated the same procedure as for the orbits 1-4 of the family 2, to the orbits 5-8 on the family 1. The orbits 5 and 6 have the same phase (phase III). The motion in their neighborhood is bounded, when the system is perturbed by rotating the orbit of P_2 by 45^0, but the variation of the eccentricities is much larger in the unstable orbit 5 (Figure 7a) than in the stable orbit 6 (Figure 7b).

The orbits 7 and 8 are linearly unstable, and correspond to the same phase (phase IV), but close to the orbit 7, which has small planetary eccentricities, the motion is bounded, as shown if Figure 8. On the contrary, the same perturbation to the orbit 8 results to destabilization of the system, (Figure 9). Again we note that a small value of the eccentricities is necessary for a stable planetary system at the 5/2 resonance.

All orbits of the family 3 are unstable and a small perturbation results to the destabilization of the system. A typical example is shown in Figure 10 for the orbit 9.

We remark that in all bounded orbits of the system, the amplitude of the variation of the semimajor axes is small. This is in agreement with Barnes and Quinn (2004), even in the cases where the variation of the eccentricirties is quite large.

3.3. *The system 47 UMa*

In Figure 2 we show the position of the system 47 UMa for all the four different phases, which are symmetric with respect to the origin and also symmetric with respect to the e_1 and e_2 axes. We note that these positions are close to periodic motion which has a stable region in its vicinity. In order to study the stability of these systems, we computed the evolution of these four different configurations, by the Poincaré map on a surface of section, by rotating the orbit of P_2 by 45^0 (because in the observed system it is $\omega_1 - \omega_2 = 45^0$). In all cases the numerical computations showed that we have a well defined bounded motion. Note that for these small eccentricities we have bounded motion both for alignment and for antialignment of the apsides. This is in agreement with Ji *et al.* (2003).

4. Discussion

From all the above results we come to the conclusion that small planetary eccentricities are necessary for stable motion at the 5/2 resonance. We remark that stable motion exists even in the case where the planetary orbits intersect (orbit 2), and even in the case where the system is linearly unstable (orbits 3, 7), provided that the phase is III (alignment of perihelia and position of both planets at perihelion) or IV (antialignment of perihelia and position of both planets at perihelia). We also note that the system 47 UMa lies close to the regions in the e_1 e_2 space of Figure 2, where we have bounded motion. This system would be unstable if the eccentricities were higher (see also Laughlin *et al.* 2002, Barnes and Quinn, 2004), contrary to the 2/1 resonance, where a large value of the eccentricities stabilizes the system (Psychoyos and Hadjidemetriou, 2004), because, in this latter case, the minimum distance between the planets is increased when the eccentricities increase (for the same phase). We also remark that the symmetry of the orbit is a stabilizing factor, and the system is destabilized if the symmetry is destroyed, other parameters being the same, since a rotation of the orbit of P_2 results to instability in some cases.

Since the ratio of the masses of the Jupiter-Saturm system is very close to the ratio we used for 47 UMa, the same remarks for the stability apply to our own solar system.

Acknowledgement

This work was supported by the research programme *Pythagoras*, Nr.21878 of the Greek Ministry of Education and the E.U.

References

Barnes, R. and Quinn Th. 2004, *The (in)stability of planetary systems*, preprint.
Beaugé, C., Ferraz-Mello, S. and Michtchenko, T.A. 2003, *Astrophys.J.*, 593, 1124
Michtchenko T.A. and Ferraz-Mello S. 2001, *Icarus*, 149, 357
Fischer, D., Marcy, G., Buttler, P., Laughlin, G. and Vogt, S. 2002, *Astrophys. J.* 564, 1028
Fischer, D., Marcy, G., Buttler, P., Vogt, S., Henry, G.W., Pourbaix, D., Walp, B., Misch, A.A. and Wright, J.T. 2003, *Astrophys. J.* 586, 1394
Gozdziewski, K. 2002, *Astron. Astrophys.* 393, 997
Hadjidemetriou J.D. 1975, *Cel. Mech.* 12, 155
Hadjidemetriou J.D. and Psychoyos D. 2004, *Cel. Mech. Dyn. Astron.*, in press.
Ji J., Liu L., Kinoshita H., Zhou J., Nakai H. and Li G. 2003, *Astrophys. J.* 591, L57
Laughlin, G., Chambers, J. and Fisher, D. 2002, *Astrophys. J.* 579, 455

Dynamics of Populations of Planetary Systems
Proceedings IAU Colloquium No. 197, 2005
Z. Knežević and A. Milani, eds.

© 2005 International Astronomical Union
DOI: 10.1017/S1743921304008506

Our solar system as model for exosolar planetary systems

Rudolf Dvorak[1], Áron Süli[2] and Florian Freistetter[1]

[1]Institute for Astronomy, University of Vienna, Türkenschanzstrasse 17, A-1180 Vienna,
Austria
email: dvorak@astro.univie.ac.at

[2]Department of Astronomy, Eötös University H-1117 Budapest, Pázmány Péter sétány 1/A,
Hungary
email: a.suli@astro.elte.hu

Abstract. We investigate the dynamical behaviour of a simplified model of our planetary system (Mercury and the planets Uranus and Neptune were excluded) when we change the mass of the Earth via a mass factor $\kappa_E \in [1, 300]$. This is done to study the motions in this "model planetary system" as an example for extrasolar systems. It is evident that the new systems under consideration can only serve as a model for a limited number of exosystems because they have massive planets sometimes with large orbital eccentricities. We did these numerical experiments using an already well tested numerical integration method (LIE-integration) in the framework of the Newtonian equations of motions. We can show that these planetary systems are very stable up to several hundred earth masses, but for some specific values of κ_E they show a typical chaotic behaviour already in the semi-major axis. It is know from the inner Solar System that the planets move in a small region of weak chaos, but this behaviour (close to $\kappa_E = 5$) was quite unexpected. We then use a 1st order secular theory to explain the appearance of chaos. The results may serve for a better understanding of the dynamics of some extrasolar planetary systems.

Keywords. Solar system: dynamical evolution, extrasolar planetary system

1. Introduction

Although we do not have an analytical proof up to now, all numerical and semi-analytical results indicate, that our planetary system is dynamically stable for the lifetime of the Solar System (e.g. Ito & Tanikawa 2002). From their and other results we can see that the large planets exhibit remarkable stability of their orbital parameters on gigayear time scales. On the contrary for the inner Solar System with the 4 terrestrial like planets we have evidence that they are in a zone of weak chaos (Laskar 1990, 1994, 1996). The cause of this chaotic behaviour is now partly understood as a consequence of secular resonances between the motions of the perihelia and the nodes.

With the detection of extrasolar planetary systems (we now know more than 120 extrasolar planets†) the investigations of the dynamics of our own planetary systems became more important. Although the orbital parameters are quite different in what concerns the eccentricity (some planets move on very eccentric orbits) our own planetary system may serve as a model for the dynamics of stable planetary systems.

In a previous paper (Dvorak & Süli 2002) the dynamical evolution of a simplified Solar System consisting of the Sun, Venus, Earth, Mars, Jupiter and Saturn was studied; the

† http://www.obspm.fr/encycl/catalog.html

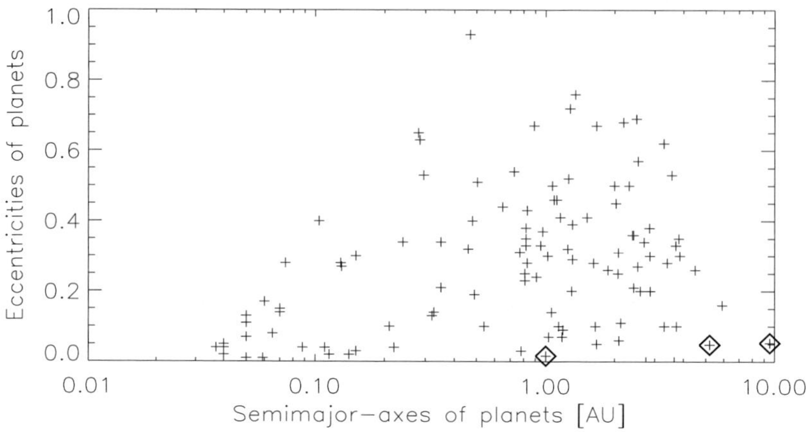

Figure 1. Eccentricity vs. the semi-major axis for extrasolar planets. The x-axis is logarithmic. The position of the Earth, Jupiter and Saturn are also indicated as diamonds with a plus sign in the middle.

masses of the inner planets were uniformly magnified via a mass factor κ. The surprising result was that in this model all the orbits were still stable for time scales of 10^7 years up to $\kappa < 245$. In this new investigation we used the same dynamical model, but only the mass of the Earth was multiplied by $\kappa_E \in [1, 300]$.

In Section 2 we discuss some characteristics of extrasolar planetary systems, Section 3 presents the dynamical model, in Section 4 some results of the dynamics of our planetary system together with the interesting motion of the node of Mars are shown. In Section 5 we show the results of the numerical experiments with a larger mass of the earth together with more detailed results for $\kappa_E = 5$ and finally we summarise the investigation shortly in Section 6.

2. Some characteristics of extrasolar planetary systems

The stability studies got a new impulse because of the recent findings of other planetary systems around single and even double stars. About 120 exoplanetary systems (exosystems) are now known. The so far discovered exoplanets have a mass range $(m \cdot \sin(i_p))$ from 0.11 m_J to 17.5 m_J where m_J is Jupiter's mass and i_p is the inclination of the orbital plane with respect to the plane of the sky. More than 90 % of these planets are orbiting 'their sun' well inside Jupiter's orbit ($a \leqslant 3.3$ AU). Their semi-major axes accumulate around 1 AU which is a matter of the analysis of observations of the radial velocity curve over periods in the order of years. We can see from Fig. 1 that most of them have significant eccentricities (some 70 % of their orbital eccentricities are larger than $e = 0.2$). Nevertheless our planetary system may serve as model case for the ones with small eccentricities which are also the ones where we may expect stable terrestrial planets moving in habitable zones (e.g. Ashgari *et al.* 2004).

3. The dynamical model and the methods of investigation

The dynamical model which we studied consisted of the Sun, Venus, Earth, Mars, Jupiter and Saturn. We have chosen this model to minimise the required CPU time for the integration of the – purely Newtonian – equations of motion. Because of the small mass of Mercury, it only slightly perturbs the motion of the other planets, but it would require a time step four times smaller than without the innermost planet. Although Uranus and Neptune are massive planets, they orbit the Sun more than twice as far as Saturn, and do not influence the motion of the inner planets significantly. This model was tested already for other numerical computation of the orbits of the near earth asteroids (e.g. Dvorak and Freistetter 2000) and turned out to be quite accurate.

Our method used to solve the equations of motion was the Lie-integration, a method which uses the property of recurrence formulae for the Lie-terms (up to the order 14); one of its advantages is that also high eccentric orbits are integrated accurately due to the automatic step size. A detailed description can be found in Hanslmeier and Dvorak (1984) and Lichtenegger (1984). Many numerical results were derived with this integrator (e.g. Dvorak and Tsiganis 2000, Tsiganis *et al.* 2000) where the method was also compared to other numerical integrators. The total CPU time used on a high end PC (AMD 2600+ and comparable) was more than 200 days for our study. The interval of data output was set to 100 years which led to a total amount of data of approximately 6 GBs. Complementary we used a secular perturbation theory up to the first order which was implemented in MAPLE (Süli 2003). The comparison of analytical method and numerical results showed quite a good agreement†.

4. Description of the our actual planetary systems

It is well known that the character of the variation of planetary orbital elements does not change significantly even over the course of long-term simulations over hundreds of million years. The variations of the semi-major axes are very small, for the inner planets it is less than a 0.001 AU, for the outer ones it is less than 0.01 AU. We know from the work of Laskar that the inner Solar System behaves in a chaotic manner and that there are critical angles which change from circulation to libration and vice-versa on time scales of tens of million years.

We do not have results for such long times in our recent study but we found a very interesting behaviour of Ω, the longitude of the ascending node of Mars. If we compare the time evolution of the orbital element i with Ω, it turns out, that whenever $i_{Mars} < i_{cr}$, then Ω_{Mars} begins to librate around $105°$. The center of the temporarily libration of Ω_{Mars} coincides with those of Jupiter ($90° < \Omega_{Jupiter} < 120°$) and of Saturn ($75° < \Omega_{Saturn} < 140°$). The same behaviour can be observed in the case of Venus and Earth (which we do not show here) although there it is not so obvious in the respective plots.

The above phenomena is very well visible in Figs. 2 and 3, where Ω and i of Mars are plotted versus the time. It is already reproduced by the first order secular theory. We have used the well known Laplace-Lagrange theory for this system (Süli 2003): a comparison of Figs. 2 and 3 with Fig. 4 shows that the periods of the appearance of libration between numerical and analytical results agree qualitatively but not quantitatively ($P_{num} \sim 2P_{theory}$). The reason for the above phenomena is the following: if the reference frame is chosen such that the x, y plane is orthogonal to the total angular momentum vector (i.e. it coincides with the invariant plane) than all the nodes circulate. If the reference frame is chosen sufficiently inclined with respect to the invariant plane, some of the nodes appear to librate. This is the case of Jupiter's and Saturn's node in our ecliptic reference frame, and the reason for the temporarily libration of Mars' node. This phenomena is very well visible in Fig. 4, where Ω, and $i \cdot 10$ of Mars are plotted versus

† For the initial planetary orbital elements we have chosen for JD 2449200.5 with respect to the mean ecliptic and equinox J2000 from the JPL ephemerides DE200. (Standish 1990).

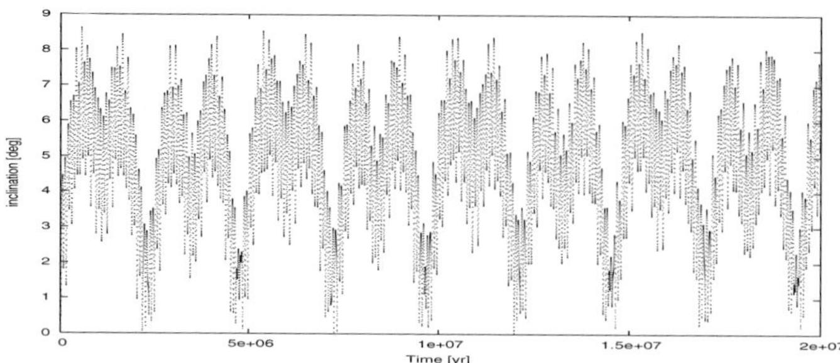

Figure 2. Inclination for Mars for our planetary system versus the time for 20 million years.

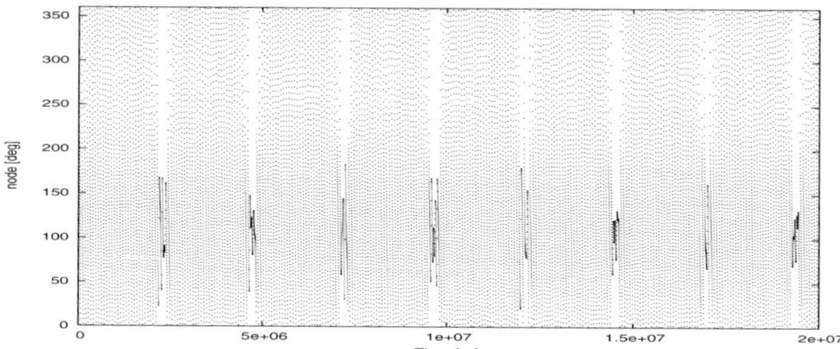

Figure 3. Longitude of the node of Mars for our planetary system for 20 million years; note that the libration modes coincide exactly with the minimum values of the inclinations.

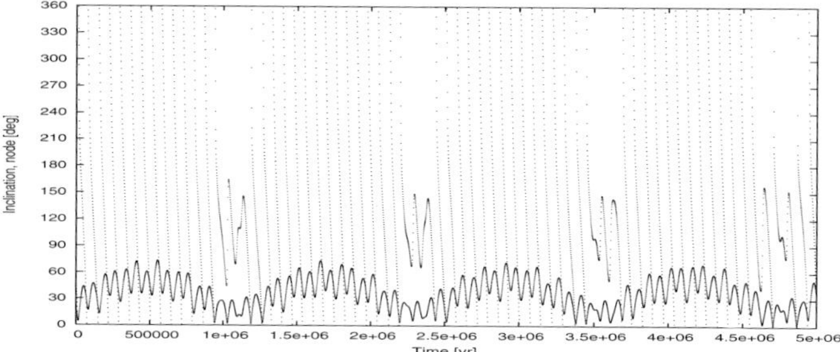

Figure 4. Longitude of the node and inclination (ten times larger) of Mars for 5 Million years by a first order secular theory. Note that the periodic changes from circulation to libration have only half the period of the numerically derived results.

the time. The i_{cr} is related to the inclination of the x, y plane relative to the invariant plane.[†]

We emphasize that the same features are also present when integrating the equations of motion with all 8 massive planets; it characterises the motion of Mars for at least 100 million years into the future and into the past (Dvorak and Gamsjaeger 2003). In the respective plot for Mars (Fig. 5) we see the evolution of the inclination for the first 10 million years together with the evolution of the node. The period is quantitatively the same as in the simplified model. For the time between 90 and 100 million years we found that the change from libration to circulation and vice versa is NOT periodic any more as a consequence of the chaotic nature of the inner planets.

[†] In the first order theory high order terms are ignored and the change from libration to circulation is sensitive with respect to the periods in the inclination of Mars; this causes the difference in the periods between numerical results and the theory.

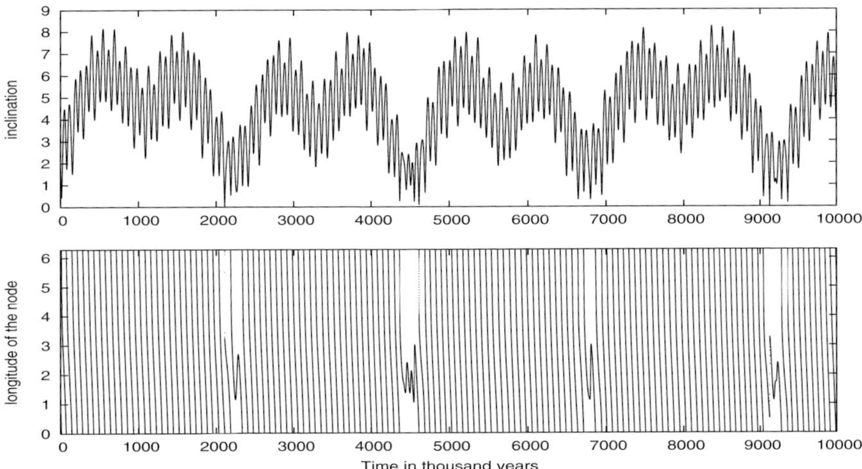

Figure 5. Inclination (upper graph) and longitude of the node (in radians) (lower graph) of Mars for 10 million years after an accurate numerical integration of the complete planetary system. Note that the periodic changes from circulation to libration have the same period as the one in the simplified system.

5. Numerical experiments with a very massive Earth

Our aim was to study the stability of the system with respect to the mass of the Earth which may be taken as an adequate model for some of the known exoplanetary system. This kind of a model planetary system is of prior interest for systems where Trojans planets may move on low eccentric orbits in habitable zones; we may even take it then as a reference planetary system! We therefore magnified the mass of the Earth by a mass factor between $1 \leqslant \kappa_E \leqslant 300$. Between $1 \leqslant \kappa_E \leqslant 15$ the step in κ_E was one, then up to $\kappa_E = 100$ the step size was 5, then up to $\kappa_E = 300$ the step size was 10. For $\kappa_E \sim 90$ the mass of Earth is comparable to the gas giant Saturn, for $\kappa_E \sim 300$ to Jupiter; then Venus and Mars can be considered as being protoplanets with relatively small masses. These resulting systems may be considered as models for 47 Ursae Majoris with its three massive planets, when we multiply the semi-major axes of this extrasolar system by 2.5. In Fig. 6 we show how the masses of the planets compare when we introduce the mass factor κ_E for the Earth. In Fig. 7 also the mass distribution of the known exoplanetary systems is shown as a function of the mass parameter $(M_{Jup} \sin i)$.

We expected that Venus but especially Mars will suffer more and more from the perturbations due to a more massive Earth. To quantify these effects we checked the orbital element *eccentricity* during the whole integration time of $2 \cdot 10^7$ years for every model and every planet. We determined the maximum value of the eccentricity (=ME) which is an important parameter for the dynamical evolution and stability of the system. It is a reliable indicator of chaos and the maximum eccentricity method (MEM) was already used successfully in many numerical investigations concerning the dynamics of planetary systems (e.g. Dvorak *et al.* 2003). The respective plot of ME for the whole range is shown in Fig. 7 (left panel), where no effect on Jupiter and Saturn is visible. We note that the ME of Earth for $\kappa_E = 1$ is 60% greater than for $\kappa_E = 2$, and it decreases further as the mass factor increases. The ME of Venus shows a similar behaviour as a consequence of the strong coupling between Venus and Earth; we observe two larger values of the ME of Venus, which are still rather small and do not influence the stability of the system.

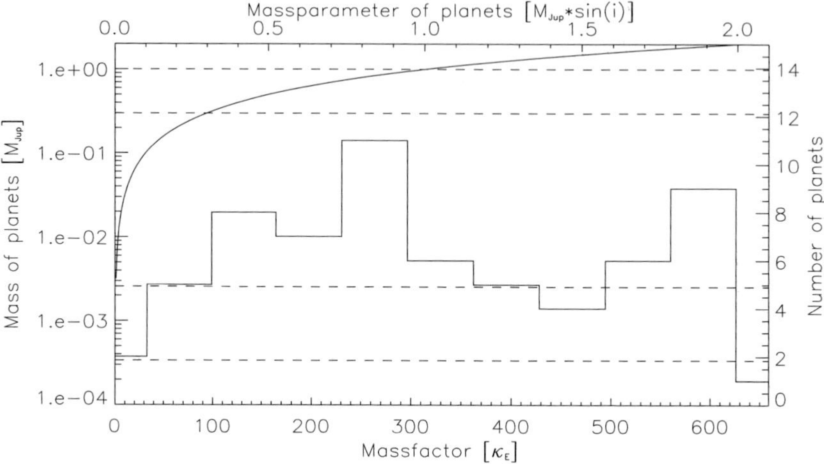

Figure 6. The mass distribution of the known exoplanetary systems are plotted together with the mass of the Earth as a function of the mass factor (κ_E) (solid line). The masses of the other planets were unchanged, and are represented by straight dashed lines (units are Jupiter masses). On the lower x-axis the mass factor is scaled; the upper x-axis shows the scaling of the mass parameter of the exoplanetary system. The left y-axis is the mass of the planets in a logarithmic scale; the right y-axis is the number of planets. The bin size is 0.2 $m_{Jupiter}$.

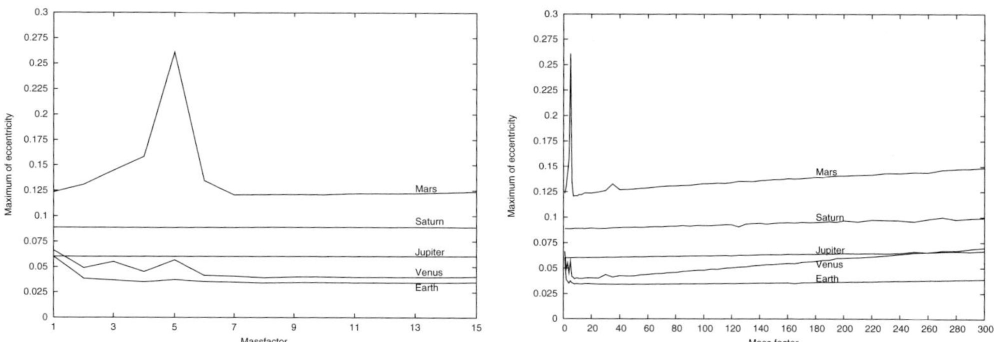

Figure 7. Results of the MEM: the maximum eccentricity as a function of κ_E. On the left graph $1 \leqslant \kappa_E \leqslant 15$; on the right graph $1 \leqslant \kappa_E \leqslant 300$

Then, for larger values of $\kappa_E > 15$ one can see a gradual but small increase of Venus' ME (see Fig. 7, right panel).

Mars is playing a special role (Fig. 7, left panel): the ME of Mars steadily increases with κ_E, and at $\kappa_E = 5$, it suddenly reaches a very high value ($e_{Mars} = 0.26$, perihelion distance $q = 1.13$ AU). After this peak the ME drops down to its starting value, and stays around 0.123. Beyond $\kappa_E = 15$ Mars' ME linearly increases with the mass factor.

Checking the elements of Mars for $\kappa_E = 5$ we can see that the semi-major axes shows an interesting non periodically modulation, which is also visible from the plot of the eccentricity (Fig. 8). The eccentricity increases rapidly to $e = 0.26$ and shows a periodic large variation $0.07 < e < 0.26$, but after several million years stays rather large $0.19 < e < 0.26$. To explain this special behaviour we used again the first order secular theory, although we know that for larger eccentricities it will fail to provide precise results. Nevertheless we found the reason for this irregular behaviour: Table 1 shows the analysis of the fundamental frequencies involved. We observe the closeness of the two frequencies

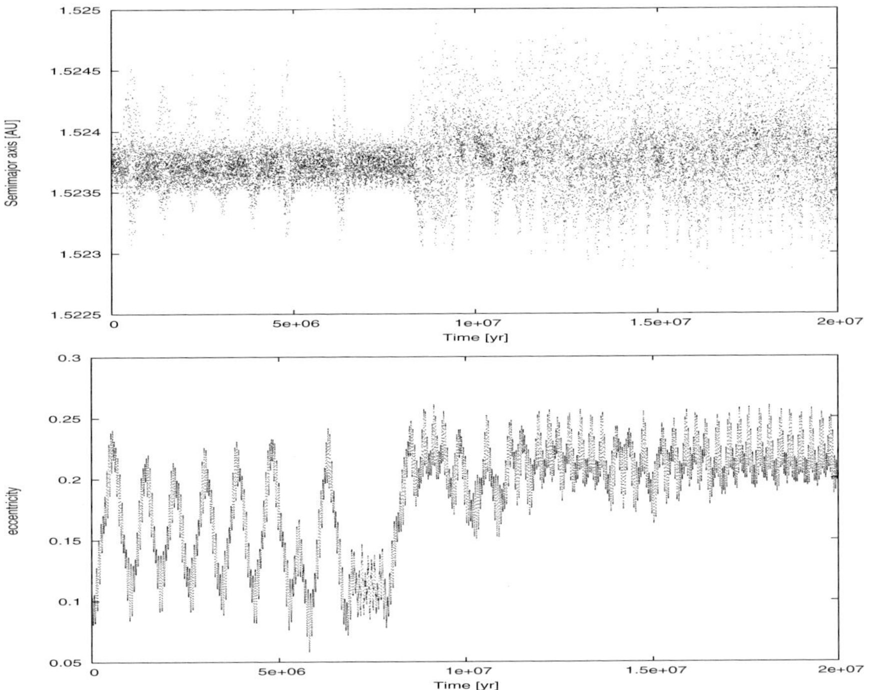

Figure 8. Semi-major axis (top graph) and eccentricity (bottom graph) of the system with $\kappa = 5$ for $2 \cdot 10^7$ years

f_2 and f_3 which shows that they are almost in a 1:1 resonance! A thorough investigation of the planetary systems around the value of $\kappa_E = 5$ using numerical and analytical tools is in progress. We hope to give then an even more detailed explanation of this interesting and – up to now – not known dynamical behaviour of our slightly modified planetary system.

Table 1. The secular frequencies of the model for $\kappa_E = 5$ calculated using the Lagrange-Laplace secular theory. The units are in arcsec/yr, year and in degrees.

j	g_j ["/yr]	P_j [yr]	β_j [deg]	f_j ["/yr]	P_j [yr]	γ_j [deg]
1	3.540609	366038.71	290.97	-47.160423	27480.67	76.93
2	8.679525	149316.93	109.16	-25.838731	50157.26	306.49
3	22.284358	58157.39	23.72	-25.682030	50463.30	129.59
4	25.686232	50455.05	116.03	-7.155340	181123.47	297.68
5	45.645800	28392.54	304.59	0	–	–

6. Conclusions

We studied the dynamics of a simplified dynamical model of the Solar System to have a kind of general model of exoplanetary systems with planets on not too eccentric orbits. We found an interesting periodic change from circulation to libration of the Mars' longitude of the node (and vice versa) what we have explained as a consequence of the inclination of the ecliptic reference frame with respect to the invariant plane.

We then studied in the same model the orbits of the planets when we increase the mass of the Earth by a mass factor between $1 \leqslant \kappa_E \leqslant 300$. These model systems were still stable even when the Earth had Jupiter's mass; older, not published results, showed that from approximately $m_{Earth} \sim 3 M_{Jupiter} \sim \kappa_E = 1000$ on close approaches of Mars and Venus to the Earth made the system dynamically unstable.

For $\kappa_E = 5$ signs of chaotic motion appeared in the semi-major axis of Mars, whose orbital eccentricity reached values up to $e = 0.256$ during the integration time of $2 \cdot 10^7$ years. Using the results of the Lagrange-Laplace secular theory to the first order for $\kappa_E = 5$, we found a secular resonance acting between the motions of the nodes of the Earth and Mars. According to these results, the stability of the Solar System depends strongly on the masses of the planets, and small changes in these parameters result in a different dynamical evolution of the planetary system. Therefore we think that this planetary system may be used as dynamical reference model for a better understanding of the orbits in extrasolar systems.

Acknowledgements

Most of the numerical integrations were accomplished on the NIIDP (National Information Infrastructure Development Program) supercomputer in Hungary. F. Freistetter acknowledges the support by the Austrian FWF project P16024. Á. Süli is grateful to the Hungarian NRF (grant no. OTKA T043739).

References

Asghari, N., Broeg, C., Carone, L., Casas-Miranda, R., Palacio, J. C. Castro Csillik, I., Dvorak, R., Freistetter F. and 33 more authors 2004, *Astron. Astrophys.*, in press.

Dvorak, R. and Gamsjäger, C. 2003, in: F. Freistetter, R. Dvorak, B. Érdi (eds.), *Proceedings of the 3rd Austrian Hungarian workshop on trojans and related topics*, p. 49

Dvorak, R., Pilat-Lohinger, E., Funk, B. and Freistetter, F. 2003, *Astron. Astrophys.*, 398, L1

Dvorak, R. and Süli, Á. 2002, *Cel. Mech. Dyn. Astron.*, 83, 77

Dvorak, R. and Tsiganis, K. 2000, *Cel. Mech. Dyn. Astron.*, 78, 125

Dvorak, R. and Freistetter, F. 2000, *Planet. Space Sci.*, 49, 803

Hanslmeier and A., Dvorak, R. 1984, *Astron. Astrophys.*, 132, 203

Ito, T. and Tanikawa, K. 2002, *Mon. Not. R. Astron. Soc.*, 336, 483

Laskar, J. 1990, *Icarus*, 88, 266

Laskar, J. 1994, *Astron. Astrophys.*, 287, L9

Laskar, J. 1996, *Cel. Mech. Dyn. Astron.*, 64, 115

Lichtenegger, H. 1984, *Celest. Mech.*, 34, 357

Süli, Á. 2003, in: F. Freistetter, R. Dvorak, B. Èrdi (eds.), *Proceedings of the 3rd Austrian Hungarian workshop on trojans and related topics*, p. 85

Tsiganis, K., Dvorak, R. and E. Pilat-Lohinger. 2000, *Astron. Astrophys.*, 354, 1091

Standish, E.M. 1990, *Astron. Astrophys.*, 233, 252

Dynamics of Populations of Planetary Systems
Proceedings IAU Colloquium No. 197, 2005
Z. Knežević and A. Milani, eds.

© 2005 International Astronomical Union
DOI: 10.1017/S1743921304008518

Planetary motion in double stars: the influence of the secondary

Elke Pilat-Lohinger

Institute for Astronomy, University of Vienna, Türkenschanzstrasse 17, A-1180 Vienna,
Austria
email: lohinger@astro.univie.ac.at

Abstract. Among more than 120 discovered exo-planets, less than 20 were found in double
star systems. Out of this sample we studied the planetary motion in those systems that can be
regarded as close binaries, i.e. HD41004 AB, γ Cephei and Gliese 86. In this study we concentrate
on the first two systems, where the secondaries are M4 V dwarfs at about 20 AU from the host-
star. A comparison of previous studies – where the dynamical behavior was studied in the (semi-
major axis, inclination) plane (see Dvorak *et al.* 2003a) for γ Cephei and Pilat-Lohinger & Funk
(2004) for HD41004 A) – shows significant differences in the stability maps. Our numerical
investigation examines the region between 0.5 and 1.2 AU, which is influenced mainly by mean
motion resonances when the initial position of the detected planet $a_{gp} < 1.5$ AU. If we move
the planet farther away from the host-star (to distances > 1.5 AU) we observe an arc-shaped
chaotic structure in the dynamical map.

Keywords. Binaries: close, celestial mechanics, methods: numerical

---

1. Introduction

Dynamical investigations of detected exo-planets in binaries are mainly restricted to
the study of S-type motion (where the planet moves around one stellar component),
whereas the P-type motion (where the planet revolves around both stars on a distant
orbit) can not be applied yet, as no planet has been found in a very close binary. Moreover,
there are only three binaries with distances of about 20 AU, all others have distances
between 100 and 6000 AU (cf. Eggenberger *et al.* 2004). Previous studies of planetary
motion in close binaries – i.e. γ Cephei (cf. Dvorak *et al.* 2003a) and HD41004 AB (cf.
Pilat-Lohinger & Funk 2004) led to significantly different results although the systems
seem to be similar, since in both cases the distance of the stellar components is about
20 AU and the perturbing secondary is a M4 main sequence star of 0.4 solar-masses. The
numerical stability study of fictitious planets in the examined (semi-major axis (a_{fp}),
inclination (i)) parameter space showed for γ Cephei an influence of the secondary at
low inclinations (see Section 3 and Dvorak *et al.* 2003a), which was not found for the
system HD41004 AB, where this region is divided by mean motion resonances into several
stable stripes (see Section 4 or Pilat-Lohinger & Funk 2004).

In our numerical study we used for both exo-solar planetary systems the restricted
four body problem (R4BP), that describes the motion of a massless planet in the gravitational field of three massive bodies, i.e. the binary and the detected giant planet.
Furthermore, to see a potential effect of the secondary on the region between 0.5 and 1.2
AU, we studied the same massless bodies in the restricted three body problem (R3BP) –
where we ignored the secondary. The orbital behavior was determined by means of the
Fast Lyapunov Indicators (FLIs) (cf. Froeschlé *et al.* 1997). Moreover, in some cases we

Table 1. Orbital parameters of the binary γ Cephei (Cochran *et al.* 2002)

	primary	secondary	planet
mass [solar masses]:	1.6	0.4	.00168
semi-major axis [AU]:		21.36	2.15
eccentricity:		0.44	0.209
period [years		70	2.47]

computed additionally the maximum eccentricity of the orbits. Based on the results of previous studies (cf. Dvorak *et al.* 2003a and Pilat-Lohinger & Funk 2004) we examined how the variation of the initial semi-major axis of the giant planet (a_{gp}) will influence the region between 0.5 and 1.2 AU – this is shown in Section 5, where one can see the stability maps for different a_{gp} in both systems.

2. Numerical methods

To define the dynamical state of the planetary orbits we used the Fast Lyapunov Indicator (cf. Froeschlé *et al.* 1997), which distinguishes qualitatively between regular and chaotic motion. This chaos indicator is defined as the norm of the largest tangent vector: $\psi(t) = \sup_i \|v_i(t)\|$ with $i = 1, \ldots n$ (n denotes the dimension of the phase space), which grows exponentially for orbits in a chaotic region. The FLI program uses the Bulirsch-Stoer integration method; the integration time was 10^4 periods of the binary.

Additionally, we carried out some straightforward orbital computations using the LIE integration method (see e.g. Lichtenegger 1984, or Hanslmeier & Dvorak 1984) and determined the maximum eccentricity (MEC) of the fictitious planets over the whole integration time span, which was between 10^5 and 10^6 years. Such a supplementary study has already been conducted in previous investigations, where the so-called habitable zone (HZ)† of several exo-solar planetary systems was analyzed (see e.g. Dvorak *et al.* 2003a and 2003b, Pilat-Lohinger & Funk 2004, Erdi *et al.* 2004).

3. Planetary motion in the binary γ Cephei

The double star system γ Cephei can be found at a distance of about 11.8 pc from the Sun, where the detected Jupiter-like planet moves on an eccentric orbit around its host-star (a K1 IV star). The computations have been carried out using the orbital parameters of Cochran *et al.*, who discovered this planet in 2002 (see Table 1).

On the basis of a general stability study of S-type motion (cf. Pilat-Lohinger & Dvorak 2002), which defines the border of the stable zone for γ Cephei at about 3.6 AU, we were able to establish the stable motion for the discovered planet.

To determine regions where other fictitious planets might exist we used the R4BP as dynamical model, and found a stable zone between the host-star and the detected planet. The results of these computations are summarized in Fig. (1.a), where we varied the initial semi-major axis of the fictitious planet (a_{fp}) from 0.5 to 1.8 AU (x-axis) and increased its initial inclination from 0° to 40° on the y-axis. The dark region in Fig. (1.a) indicates chaotic motion and long-term stability is referred to the white region. Moreover, one can see a significant "chaotic path", which appears due to the fact that (i) we expect stronger perturbations of the secondary for low inclined orbits, and (ii) at about 1 AU

† The habitable zone of a Sun-like star is, roughly speaking, located at a distance from the star where liquid water can exist on the surface of a terrestrial-like planet.

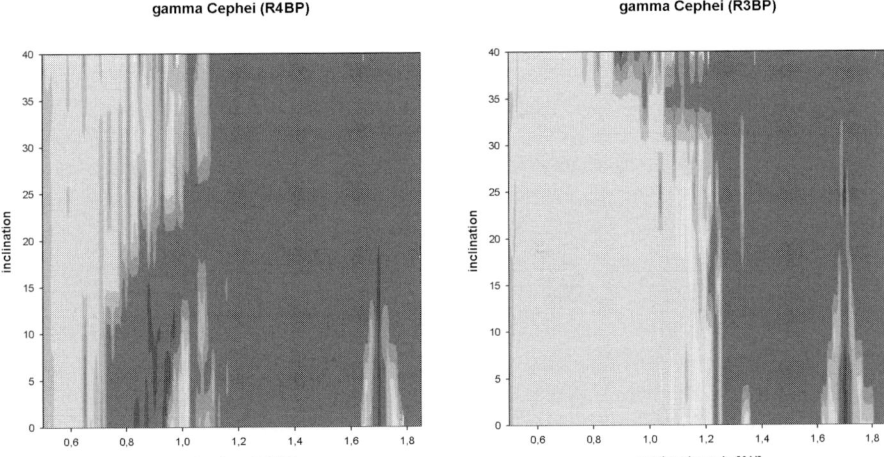

Figure 1. Stability maps for a fictitious planet in the vicinity of γ Cephei: left panel (a) shows the result in the R4BP (i.e. γ Cephei + secondary + detected planet + fictitious planet) and right panel (b) shows the result in the R3BP (i.e. γ Cephei + detected planet + fictitious planet). The dark region shows the chaotic zone and the white area the stable one; for details see the text.

a small stable island remains, which is close to the 3:1 mean motion resonance. This stable island appears for all chosen mean anomalies $(0^o, 45^o, 90^o, 135^o, 180^o, 225^o, 270^o$ and $315^o)$ of the fictitious planets, so that we suspect a stabilization of this region due to the resonance. Fig. (1.b) shows the same dynamical map for the computations in the R3BP (where we neglected the secondary). In this model one can see a much larger stable region with a well defined border. Only for high inclinations $(> 30^o)$ a reduction of the stable zone occurs due to the Kozai resonance. A comparison of the two dynamical maps led to the assumption that the arc-shaped chaotic structure could result from secular perturbations, but this has still to be proved. The two stable islands between 1.6 and 1.8 AU seem to be influenced also by the secondary, since they are smaller in size in the R4BP (Fig. 1.a).

4. Planetary motion in HD41004 AB

The study of planetary motion in the close binary HD41004 AB was carried out to examine the dynamical state of terrestrial-like planets in the habitable zone (HZ) of HD41004 A†, which is according to Kasting *et al.* (1993) between 0.48 and nearly 1 AU. A planetary companion for HD41004 A was announced by M. Mayor in July 2003 during the XIXth IAP Colloquium on "Extra-solar planets today and tomorrow" in Paris, but the system parameters (see Table 2) show that the study of this system is actually more complicated, since the secondary is accompanied by a brown dwarf of 19 m_{Jup}. However, similar dynamical studies for this exo-solar planetary system like the ones for γ Cephei in the R4BP and R3BP allowed a simplification of the investigation of the HZ of HD41004 A, since the two dynamical maps (Figs. (2.a) and (2.b)) do not indicate a significant difference in the (a,i)-plane, which probably means that this region is not under a strong influence of the secondary. In both figures one can see the same structure of the HZ, which is divided into stable stripes by mean motion resonances (1:4 near 0.52 AU, 7:2 near 0.57 AU, 3:1 around 0.63 AU and 8:3 near 0.68 AU), like the asteroid

† A detailed study of the HZ of HD41004 A will be presented in Pilat-Lohinger & Funk 2004)

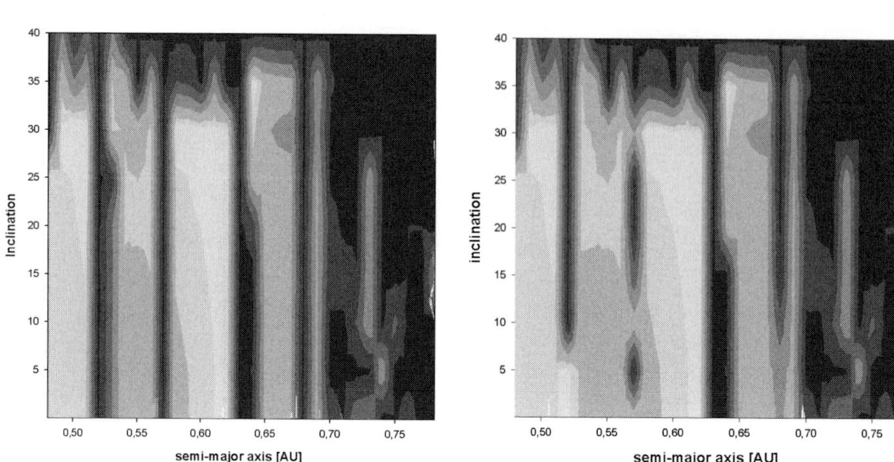

Figure 2. Stability maps for a fictitious planet in the HZ of HD41004 A: left panel (a) shows the result in the R4BP (i.e. binary HD41004 AB + detected planet + fictitious planet) and right panel (b) shows the result in the R3BP (i.e. HD41004 A + detected planet + fictitious planet). The different gray shades indicate the zones of different eccentricities: $e < 0.2$ (white), $0.2 < e < 0.3$ (light gray), ..., $e > 0.8$ (black, i.e. unstable region). Note that only the part of the HZ, where stable motion occurs, is plotted.

Table 2. Orbital parameters of the binary HD41004 AB (Zucker *et al.* 2004)

	primary	secondary	brown dwarf	planet
mass [solar masses]:	0.7	0.4	0.01814	0.002196
semi-major axis [AU]:		~ 22	~ 22	1.31
eccentricity:		?	0	0.39 ± 0.17
period [years]		~ 70	~ 70	~ 1.8

main-belt in our solar system. The left panel shows the stable motion (white region) in the R4BP, where the binary's eccentricity was 0.2, and the right panel summarizes the motion in the R3BP, where the secondary was neglected. In both cases we determined the dynamical behavior by means of the FLIs and studied additionally the maximum eccentricity (MEC), to cancel out high eccentric motion ($e > 0.5$), in order to be sure that most of the orbits will be in th HZ. Even if the simplified system of HD41004 AB (i.e. without brown dwarf) seems to be quite similar as γ Cephei, we found the contrary from our results (compare Figs. (1.a and b) and (2.a and b)). The orbital parameters of Tables 1 and 2 show that the most significant difference of the two exo-planetary systems is the position of the detected giant planet. Therefore, we studied both systems in the model of the R4BP, where we varied the initial a_{gp}, in order to find an explanation for this difference in the results.

5. Stability study for different a_{gp}

We computed the motion of fictitious bodies in the region between 0.5 and 1 AU for the following positions of the giant planet: 1.3 AU, 1.5 AU, 1.7 AU, 1.9 AU and 2.1 AU. The most important results of the FLI computations are summarized in Figs. (3.a and b) and (4.a and b), where the white region denotes stable motion and the dark

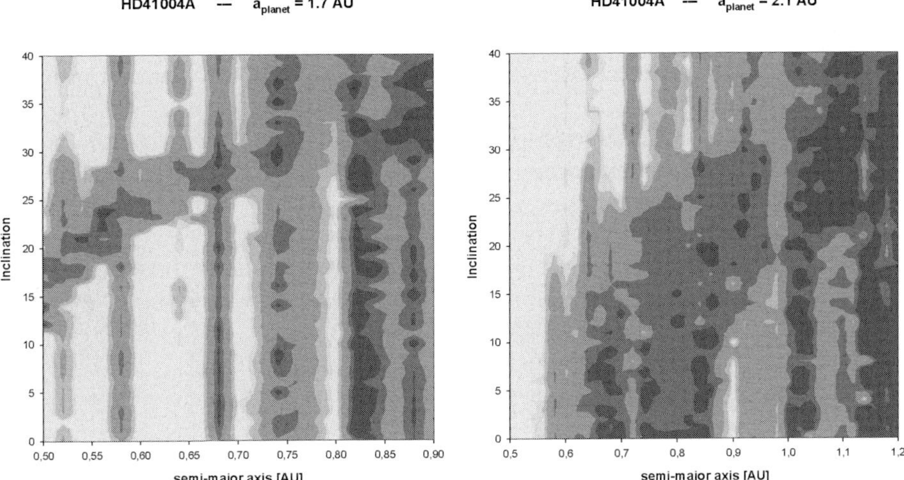

Figure 3. Stability maps for the HZ of HD41004 A with the detected giant planet moving at different distances from its host-star: (a) $a_{gp} = 1.7$ AU and (b) $a_{gp} = 2.1$ AU. The dynamical behavior was determined by means of the FLIs, where white zones denote stable motion and dark region corresponds to chaotic motion (for more details see Section 5).

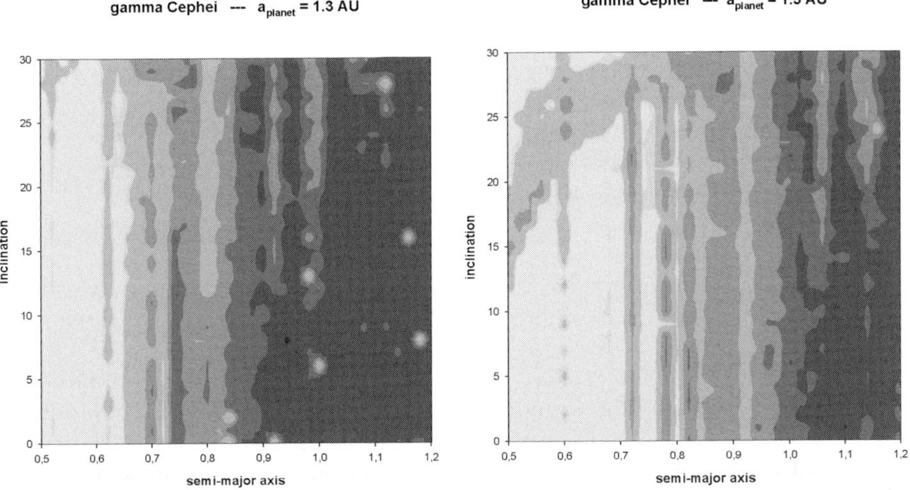

Figure 4. Stability maps for the HZ of γ Cephei, with the detected giant planet moving at different distances from its host-star: (a) $a_{gp} = 1.3$ AU and (b) $a_{gp} = 1.5$ AU. The dynamical behavior was determined by means of the FLIs, where white zones denote stable motion and dark region corresponds to chaotic motion (for more details see Section 5).

region indicates chaotic motion. One can see, that for $a_{gp} = 1.3$ AU (see Figs. (2.a) for HD41004 A and (4.a) for γ Cephei) the region in the (semi-major axis, inclination)-map is influenced mainly by mean motion resonances (dark straight lines), while the chaotic arc-like structure appears only in the dynamical maps, where $a_{gp} \geqslant 1.5$ AU. Since we expect stronger perturbations from the secondary on the giant planet, if it is moved toward the secondary, we assume that this curved chaotic shape results from secular perturbations (which has still to be proved). Following the sequence of Figs. (4.b), (3.a) and (3.b) one can see how the chaotic arc changes if a_{gp} is closer and closer to the value corresponding to the secondary.

6. Conclusions

Summarizing the numerical study of planetary motion in the region between 0.5 and 1 AU from a host-star, where we have to expect planetary and stellar perturbations, we can conclude the following:

(a) when the giant planet is close enough to the host-star (i.e. < 1.5 AU for the two binary systems) the region seems to be only perturbed by the giant planet of the system and the stable zone is divided into several stable striped due to to mean motion resonances;

(b) when the position of the giant planet is $\geqslant 1.5$ AU from the host-star we assume perturbations of the secondary on the planet, which causes an arc-shaped chaotic structure probably due to secular perturbations. Since all massive bodies move in the same plane, we expect a precession of the perihelion of the giant planet.

Even if we use the same position of the giant planet in the two binary systems, we cannot find exactly the same dynamical map:

(i) for $a_{gp} = 1.3$ AU one can see that the splitting of the stable zone into several stripes due to mean motion resonances with respect to the giant planet is much more clearly visible for HD41004 AB;

(ii) the arc-like chaotic structure appears earlier with the binary γ Cephei (when $a_{gp} = 1.5$ AU), than for HD41004 A (when $a_{gp} = 1.7$ AU);

(iii) for $a_{gp} = 2.1$ AU the arched chaotic structure (also called chaotic path in Dvorak *et al.* 2003a) is more distinct in the binary γ Cephei.

Therefore, we conclude that we have to study also other parameters like:

- the eccentricity of the binary;
- the mass-ratio of the binary;
- the mass of the giant planet.

which might change the dynamical maps as well.

A more detailed study is in progress and will be presented elsewhere.

Acknowledgements

The author wishes to acknowledge the support by the Austrian FWF (Hertha Firnberg Project T122) and wants to thank Dr. Z. Knezevic for his critical reading of the manuscript.

References

Cochran, W.D., Hatzes, A.P., Endl, M., Paulson, D.B., Walker, G.A.H, Campbell, B. and Yang, S. 2002, *Astron. Astrophys. Suppl. Ser.* 34, 916

Dvorak, R., Pilat-Lohinger, E., Funk, B. and Freistetter, F. 2003a *Astron. Astrophys.* 398, L1

Dvorak, R., Pilat-Lohinger, E., Funk, B. and Freistetter, F. 2003b, *Astron. Astrophys.* 410, L13

Eggenberger, A., Udry, S. and Mayor, M. 2004, *Astron. Astrophys.* 417, 353

Erdi, B., Dvorak, R., Sandor, Z., Pilat-Lohinger, E. and Funk, B. 2004, *Mon. Not. R. Astron. Soc.* 351, 1043

Froeschlé, C., Lega, E. and Gonczi, R. 1997, *Cel. Mech. Dyn. Astron.* 67, 41

Hanslmeier, A. and Dvorak, R. 1984, *Astron. Astrophys.* 132, 203

Lichtenegger, H. 1984, *Celest. Mech.* 34, 357

Pilat-Lohinger, E. and Dvorak, R. 2002, *Cel. Mech. Dyn. Astron.* 82, 143

Pilat-Lohinger, E. and Funk, B. 2004, "A stability study of the habitable zone of HD41004 A", in preparation

Zucker, S., Mazeh, T., Santos, N., Udry, S. and Mayor, M. 2004, *Astron. Astrophys.* 426, 695

Dynamics of Populations of Planetary Systems
Proceedings IAU Colloquium No. 197, 2005
Z. Kneževic and A. Milani, eds.

© 2005 International Astronomical Union
DOI: 10.1017/S174392130400852X

Planetary orbits in double stars: influence of the binary's orbital eccentricity

Daniel Benest[1] and Robert Gonczi[1,2]

[1] C.N.R.S. U.M.R. 6202 Cassiopée, O.C.A. Observatoire de Nice,
B.P. 4229, F-06304 NICE Cedex 4, FRANCE

[2] U.N.S.A. U.M.R. 6525 L.U.A.N.,
Parc Valrose, F-06108 NICE Cedex 2, FRANCE

Abstract. Regularity and chaos of "quasi-stable" (i.e. appearing stable during a finite interval of time) planetary orbits around one component of binary stars is investigated for different values of the binary's mass ratio and orbital eccentricity e. The behavior of fictitious planetary orbits around 16 Cyg B-like stars is presented. Among the quasi-stable orbits we found that there exists a ("stability zone") for every values of e, but that the existence of nearly-circular planetary orbits is restricted to values of e less than 0.8. However, not all the quasi-stable orbits are regular: emergence of chaos when e increases is shown in two sets of quasi-stable orbits (each set has fixed initial conditions for the planet, but varying values for e from 0 to 0.99). In the first set, which lies in the "heart" of the stability zone, chaos appears only when e approaches 1. The second one is near the border of the stability zone, and chaos appears as soon as e reaches 0.7. The influence of the binary's orbital eccentricity on the limit between regularity and chaos is therefore stronger in the latter case (wider planetary orbits) than for the former one (closer planetary orbits).

Keywords. Celestial mechanics, methods: n-body simulations, stars: binaries: individual (16 Cyg B), planetary systems

---

1. Introduction

Cosmogonical theories as well as recent observations allow us to expect the actual existence of numerous exo-planets, including in binaries (for example around 16 Cyg B and in the 16 Cyg system; see Gorshanov *et al.* 2005). Then arises the dynamical problem of stability for planetary orbits in double star systems. Modern computations (Dvorak *et al.* 1989, 2004; Hale 1994; Holman & Wiegert 1999; Lohinger *et al.* 1993; Udry *et al.* 2004) have shown that many such stable orbits do exist, among which we consider orbits around one component of the binary (called S-type orbits; see Pilat-Lohinger & Dvorak 2002; Pilat-Lohinger 2005).

The model is the planar restricted three-body problem, where the parameters reduce to e, the orbital eccentricity of the binary, and μ, the reduced mass of the primary of the planet, and where the initial conditions of the massless body may be reduced to two: $X_o > 0$ and V_o. The equations of motion are written in a rotating-pulsating frame, and are integrated using a classical 4th order Runge-Kutta scheme with improved coefficients and with variable time step. We call an orbit stable – or, more rigorously, "quasi-stable" (but hereafter called nevertheless "stable" for simplification) – if there has been neither collision with a star nor escape during a given time, as stated in previous papers (see, e.g., Benest 1988a, which contains a large list of references).

Within this framework, the phase space of initial conditions for fictitious S-type planetary orbits in given binaries is systematically explored. The subset of stable initial conditions – the "stability zone" – is established, and easily visualised on the plane $(X_o > 0, V_o)$.

2. Stable planetary orbits in binaries

In previous papers (Benest 2003, and references therein), one of us showed that stable planetary orbits, around either the lightest (B) or the heaviest (A) component of Sirius and of 4 nearby binary systems for which $\mu = 0.45$ and 0.55, exist up to distances from their primary of the order of more than half the binary's periastron separation. The results obtained in the circular case ($e = 0$) were then confirmed in the more realistic elliptic case. Among these stable orbits, nearly-circular ones exist for the binaries having a not too high orbital eccentricity.

Besides, we use Lyapunov indicators as indicators of chaos, computed with a Bulirsch & Stoer integrator with variable time step, and we study the stochasticity of some such stable planetary orbits. Then, we have found an example of a chaotic orbit which, nevertheless, lasted during 10^{10} years in a quasi-stable state: a longer integration would be needed to determine its actual nature, bounded chaos (i.e. the motion is chaotic but stays inside a finite volume for an infinite time; see, e.g., Laskar 1990) or sticky orbit (i.e. the motion looks like a bounded chaos during a long, sometimes very long, time – which may be called "confined chaos" – but finally escapes out to infinity; see Dvorak *et al.* 1998).

3. Orbiting 16 Cygni B-like stars

More recently, we have extended this study to planetary orbits around 16 Cygni B-like stars, i.e. around the lightest component of binaries having same value of μ that the actual 16 Cygni B ($\mu = 0.485$) but varying e: the orbital eccentricity of the double star 16 Cyg is not well known, its value being generally evaluated between 0.54 and 0.96, and we have extended this interval to [0,1] for our study.

The first results (Benest & Gonczi 2004) have confirmed the existence of stable (during 100 revolutions of the binary) nearly-circular S-type planetary orbits (very abundant for low values of e, see Figure 1; fairly abundant for $e = 0.7$, see Figure 2), excluding the case of very high eccentricity of the binary (Figure 3, $e = 0.95$). Moreover, investigating the chaotic behaviour of two sets of stable planetary orbits (one of close orbits in the very "heart" of the stability zone, and the other of wide ones just inside its border), we have shown that the stability of the first set is not destroyed when the binary's eccentricity increases even to very high values: for $e = 0.95$, the planetary orbit of this first set is chaotic but stays confined up to 1 billion years. On the contrary, the stability of the second set is destroyed as soon as the eccentricity e reaches the value 0.8, and the planetary orbit of this second set for $e = 0.7$ stays in confined chaos up to 1 billion years as well.

In the next section, we put the light on these two chaotic orbits which lie very near the limit between regularity and chaos, on the chaotic side. Both stay confined during the computation time span cited above, but determining the kind of confinement (bound or stickiness) would need more computation.

4. Bounded or sticky?

Figures 4 and 5 show, for the two orbits mentioned above, the evolution with time of the greatest Lyapunov indicator (top) together with the two main planetary orbital elements – eccentricity e and semi-major axis a in a.u. – (bottom). The time span of the computation

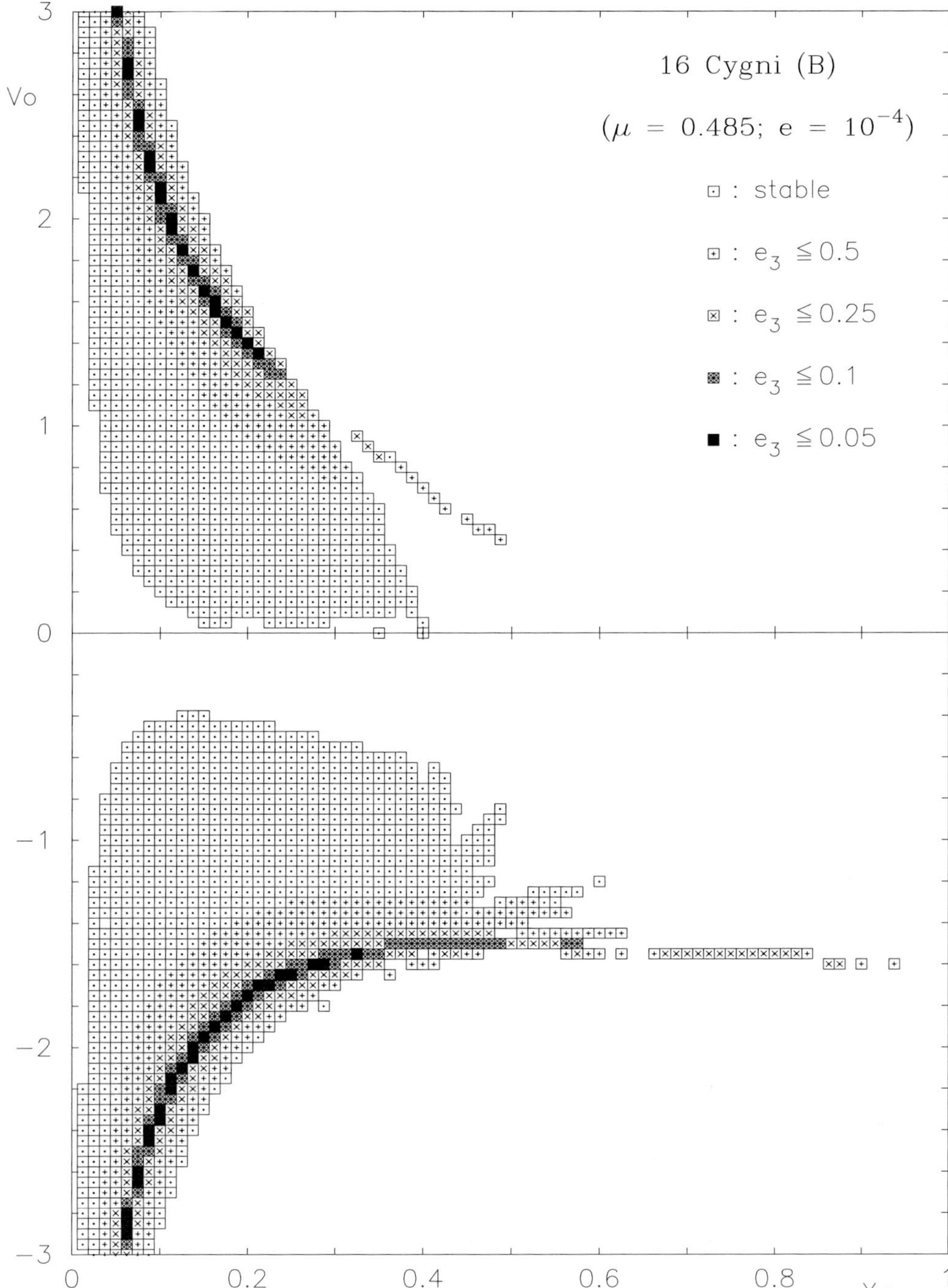

Figure 1. The subset of initial conditions in the (X_0,V_0) plane (dimensionless coordinates in the rotating-pulsating frame) for stable planetary orbits (stability zone) around a 16 Cyg B-like star when $e = 10^{-4}$: the pointed (resp. crossed, ×-ed, asterisked and filled) squares indicate planetary orbits whose eccentricity e_3 stays under the respective values given in the graph.

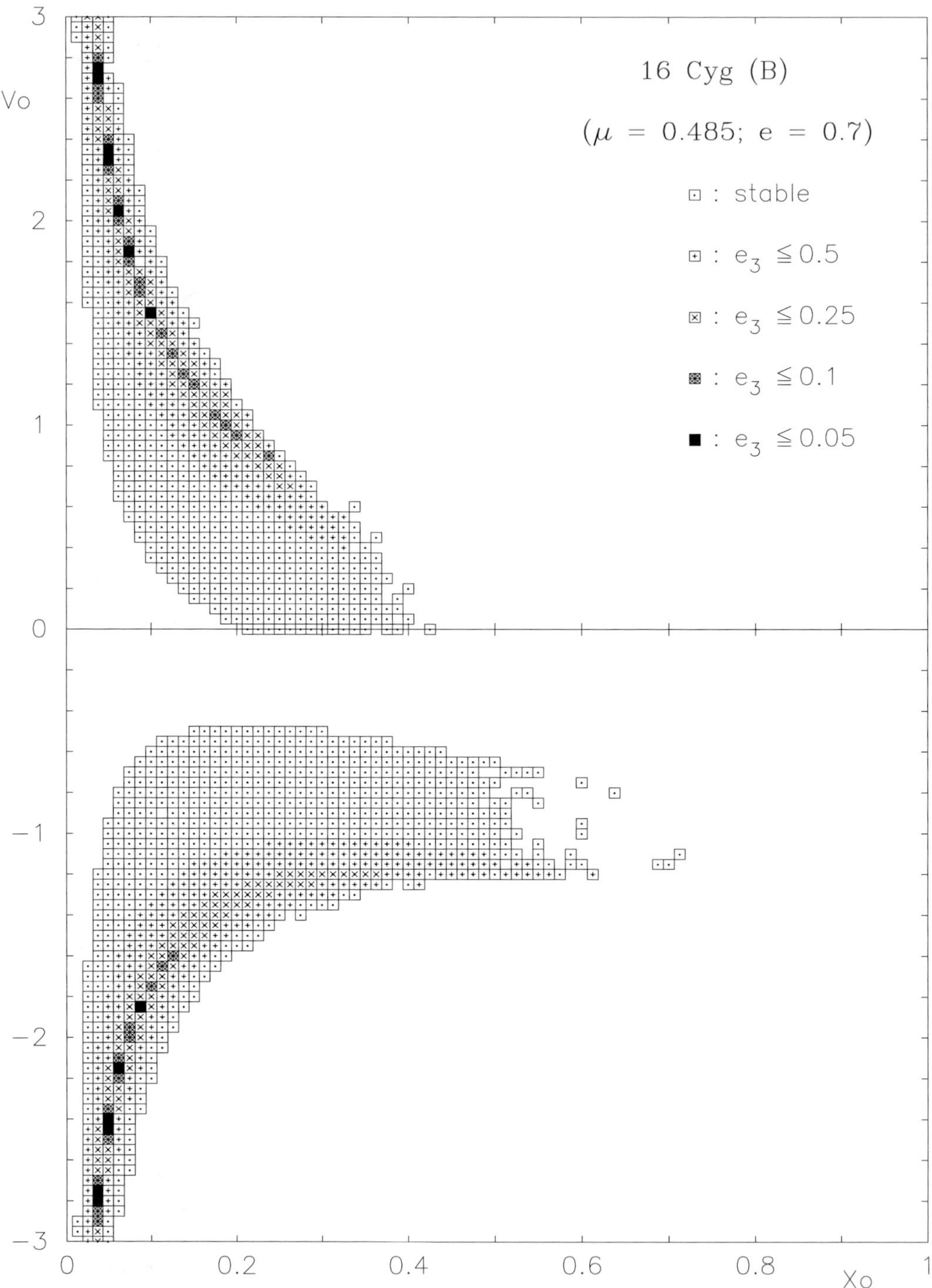

Figure 2. The stability zone for planetary orbits around a 16 Cyg B-like star when $e = 0.7$; same notations as in Fig. 1.

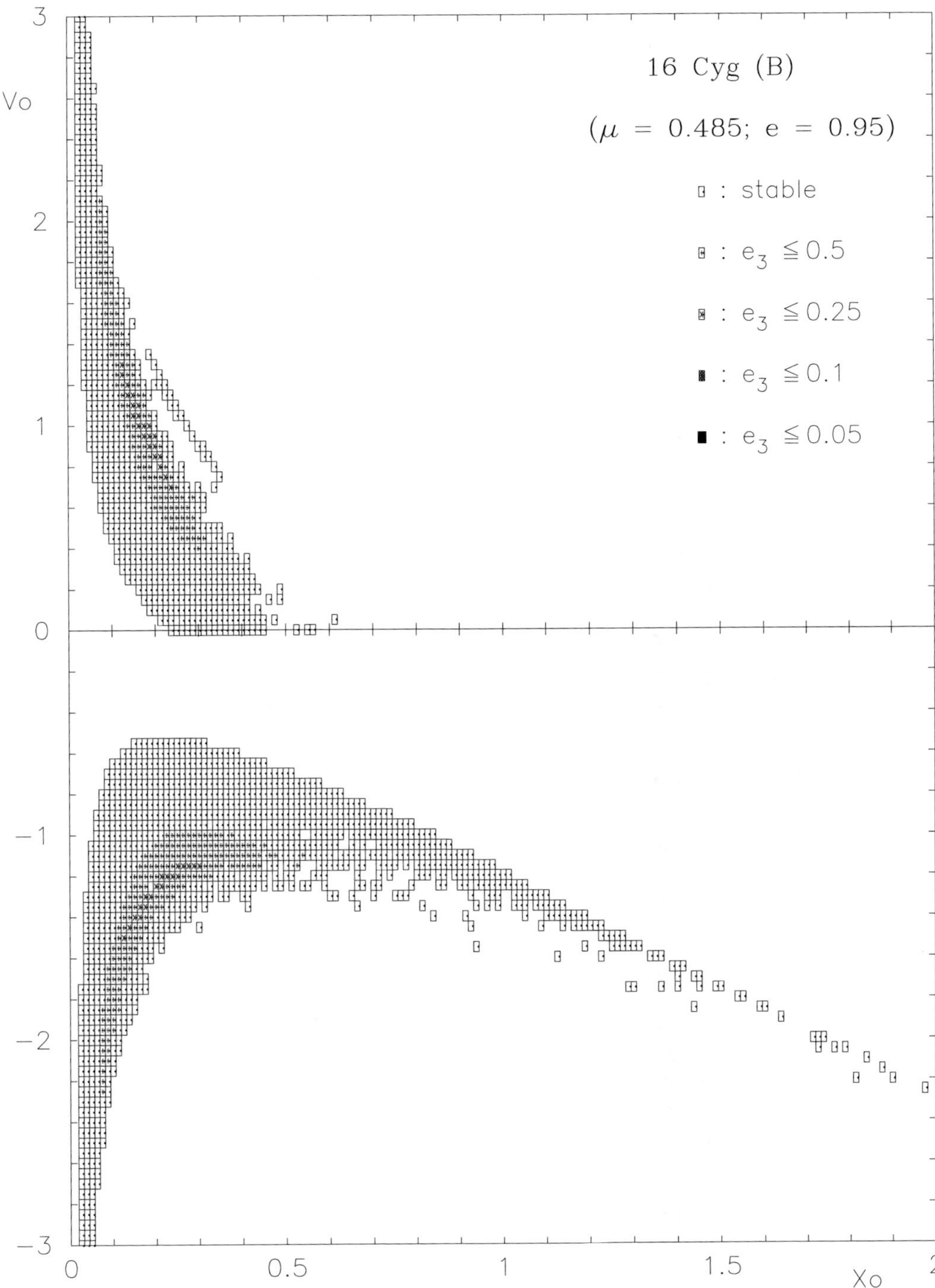

Figure 3. The stability zone for planetary orbits around a 16 Cyg B-like star when $e = 0.95$; same notations as in Fig. 1.

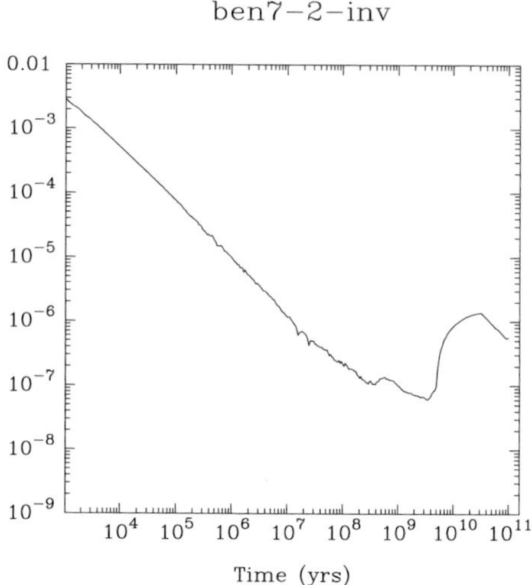

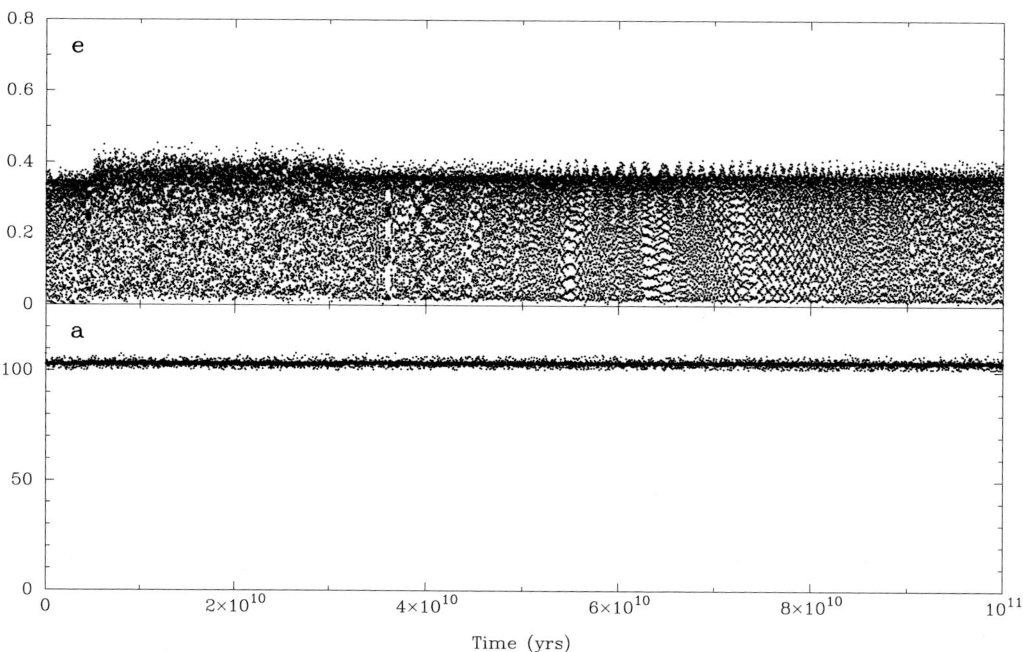

Figure 4. Evolution vs. time of the greatest Lyapunov indicator (top) and of two orbital elements – eccentricity **e** and semi-major axis **a** – (bottom) for a fictitious planet widely orbiting a 16 Cyg B-like star when $e = 0.7$ (initial conditions given in the text).

is 100 billion years for orbit 1 (figure 4, $e = 0.7$, wide planetary orbit, $X_o = 0.325$ and $V_o = $ -1.45 – see fig. 2), and only 25 billion years for orbit 2 (figure 5, $e = 0.95$, close planetary orbit, $X_o = 0.075$ and $V_o = $ -2.0 – see fig. 3), the latter value being shorter due to the integration variable time step which is much more little for a close orbit than for a much wider one.

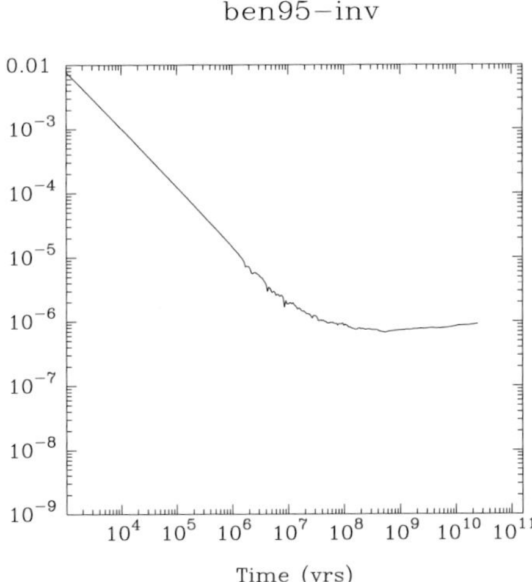

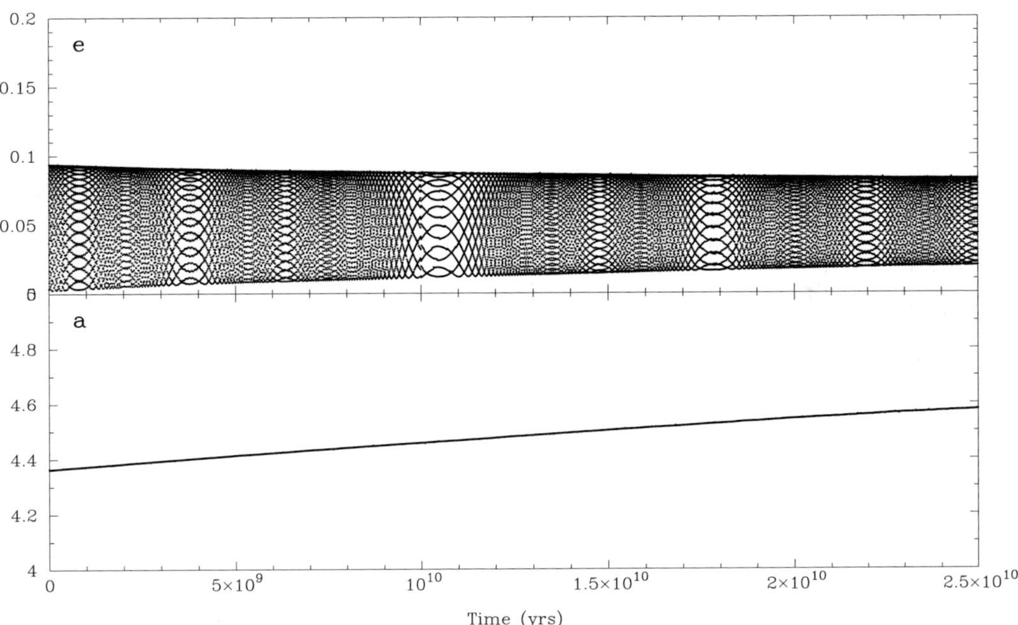

Figure 5. Greatest Lyapunov indicator, e and a for a fictitious planet closely orbiting a 16 Cyg B-like star when $e = 0.95$ (initial conditions given in the text).

For the two orbits, the greatest Lyapunov indicator, after having decreased regularly during almost 1 (Fig. 5) to 10 (Fig. 4) millions years, begins to deviate and even increase – and further decrease again for orbit 1 –, which is clearly an indicator of chaos; nevertheless, the planetary orbital elements keep a regular behaviour: e oscillates with a relative large amplitude but without any dangerous jump; besides, the amplitude of a is low – very low for orbit 2, which shows moreover a slow drift. Therefore, the chaos should be fairly weak and seems to stay confined during the integration time span. However, the question: do we have bounded chaos or stickiness? is still open dynamically speaking. But practically, after the death of the star, and its passage through the red giant stage, no interesting planet could probably survive (although the detection of exoplanets around pulsars could incline us to be cautious about such an assertion).

5. Conclusion

Our main result is that the influence of the binary's orbital eccentricity on the limit between regularity and chaos is therefore stronger in the latter case (wider planetary orbits) than for the first one (closer planetary orbits). Of course, the existence of stable orbits does not mean that there are actual planets on these trajectories, for this is the problem of formation of planets in a young binary's environment (see, e.g., Santos *et al.* 2004).

Let us conclude by two remarks. Firstly, our model is a planar one, but it has been shown since the seventies and by modern calculations (Pilat-Lohinger *et al.* 2003) that instability arises only for high values of the inclination. Finally, the existence of the gravitational field of other celestial bodies, such as stars or giant interstellar clouds, as well as galactic tides, may disturb this stability (regularity or confined chaos). But the present study is made within the frame of the three-body dynamical model and any more-body problem is, as Kipling said, another story.

Acknowledgements

This research is supported by "Action spécifique Planètes Extra-solaires" of the "Programme National de Planétologie" (C.N.R.S.: french "Centre National de la Recherche Scientifique"). Part of this work has been performed using the computing facilities provided by the program "Simulations Interactives et Visualisation en Astronomie et Mécanique (SIVAM)". This research has made use of the SIMBAD database, operated at CDS, Strasbourg, France.

References

Benest, D. 1988, *Celest. Mech.* 43, 47

Benest, D. 2003, *Astron. Astrophys.* 400, 1103

Benest, D. & Gonczi, R. 2004, *Earth, Moon, Planets* in press

Dvorak, R., Contopoulos, G., Efthymiopoulos, C. & Voglis, N. 1998, *Planet. Space Sci.* 46, 1567

Dvorak, R., Froeschlé, Ch. & Froeschlé, C. 1989, *Astron. Astrophys.* 226, 335

Dvorak, R., Pilat-Lohinger, E., Bois, E., Funk, B., Freistetter, F. & Kiseleva-Eggleton, L. 2004, *Rev. Mex. Astron. Astrofis., Serie de Conf.* 21, 222

Gorshanov, D.L., Shakt, N.A., Kisselev, A.A. & Polyakov, E.V. 2005, *in these proceedings*

Hale, A. 1994, *Astron. J.* 107, 306

Holman, M.J. & Wiegert, P.A. 1999, *Astron. J.* 117, 621

Laskar, J. 1990, *Icarus* 88, 266

Lohinger, E., Froeschlé, C. & Dvorak, R. 1993, *Cel.Mech.Dyn.Astr.* 56, 315

Pilat-Lohinger, E. 2005, *in these proceedings*

Pilat-Lohinger, E. & Dvorak, R. 2002, *Cel. Mech. Dyn. Astron.* 82, 143

Pilat-Lohinger, E., Funk, B. & Dvorak, R. 2003, *Astron. Astrophys.* 400, 1085

Santos, N., Israelian, G. & Mayor, M. 2004, *Astron. Astrophys.* 415, 1153

Udry, S., Eggenberger, A., Mayor, M., Mazeh, T. & Zucker, S. 2004, *Rev. Mex. Astron. Astrofis., Serie de Conf.* 21, 207

Dynamics of Populations of Planetary Systems
Proceedings IAU Colloquium No. 197, 2005
Z. Knežević and A. Milani, eds.

© 2005 International Astronomical Union
DOI: 10.1017/S1743921304008531

Astrometric observations of 51 Peg and Gliese 623 at Pulkovo observatory with 65 cm refractor

N.A. Shakht

Central (Pulkovo) Astronomical Observatory,
Russian Academy of Sciences, St.Petersburg, Russia
email: shakht@gao.spb.ru

Abstract. The photographic observations of 51 Peg with the known planetary companion which has been discovered on the basis of radial velocities (Mayor and Quelos 1995), are being performed at Pulkovo since 1995 by means of 65 cm refractor. So far 46 plates with 170 individual positions have been obtained. The mean error of one exposure is $0.''031$ and the error of one plate is $0.''020$. The external error of one plate or the error of unit weight is $0.''033$, while the error of one yearly position equals to $0.''010$. The aim of our observations is to investigate whether this star has some satellites of low mass (stellar or substellar) with periods of rotation from some years to a decade, or more. At present we have the possibility to study its motion over a time span of 8 years and to learn of probable perturbations due to the presence of possible satellites with the periods 0.5–6.0 years and with masses more than $0.2M_\odot$. The absence of satellites with such periods and masses has been shown by means of our observations. The results are compared with the Pulkovo series of the star Gliese 623, whose satellite with the period of 3.76 yr and mass of $0.09M_\odot$, has been confirmed by observations.

Keywords. Photographic observations, individual: 51 Peg, Gliese 623, planetary satellites

---

1. Introduction

51 Peg [$m_{\rm vis} = 5.^m5$, $G5V$, $\alpha_{(2000.0)} = 22^h57.^m5$, $\delta_{(2000.0)} = +20°46'$, $\pi = 0''.074$] with the known planetary satellite (Mayor and Quelos 1995) is inluded the Pulkovo program of single and double stars with suspected unseen components. The plate measurements have been made with Pulkovo automatic machine "Fantasy". The relative positions with respect to six stars with small proper motions, obtained in the period 1995–2000, have been published in Shakht *et al.* (2002). Our aim was to check the residuals O–C remained after the excluding of proper motion and parallactic displacement of this star.

2. Limitations of our astrometric observations

It is rather straightforward to determine the limits for the periods and masses of the components whose gravitational influence we could reveal from our observations with Pulkovo 65 cm refractor and from the comparison of the precision of observations with the gravitational influence of real and suspected satellites of the star.

A number of selected stars from our program are presented in Table 1. Here D denotes the distance of the star from the Sun in parsecs, M is its mass in $M_\odot$, σ is the error of the mean yearly position given in milliarcsecs. R1–R5, expressed in milliarcsecs, are the angular displacements of the principal star with respect to the centre of mass due to a satellite with corresponding mass and period of rotation (the third and fourth lines in the columns 7–11). ΔT is the duration of regular observations expressed in years (61 Cyg was observed with 65 cm refractor for 42 years and with normal astrograph for more than 100 years).

The angular displacements which can be detected with 65 cm refractor are marked in bold. The displacements which have been discovered in our series are marked by asterisks. In this case, we suppose that they are caused by suspected dark satellites of low masses. The displacements caused by real satellites, confirmed for Gliese 623 and discovered for δ Gem on the basis of Pulkovo observations (Shakht 1995, 1997, 2000) are marked by double asterisks. R1 corresponds to the influence of a satellite with the Jupiter-like mass and period.

Using Scargle's method of periodograms (Black and Scargle 1982) and taking into account the specific distribution of 51 Peg observations in time and the precision of our observations, we conclude that our series allow to discover the possible perturbations from the satellites with periods from 0.5 to 6.0 years and with amplitudes more than $0.''020$ and hence with masses more than $0.2 M_\odot$.

Table 1. Detectable displacements for Pulkovo program stars

1	2	3	4	5	6	7	8	9	10	11
No	Name	D	M	ΔT	σ	R1 0.001 12	R2 0.01 15	R3 0.01 3.5	R4 0.1 3.5	R5 0.2 6.0
		[pc]	$[M_\odot]$	[yr]	[mas]					
1	Lal21185	2.5	0.33	32	16	4	**50**	**18**	**160**	**400**
2	61 CygA	3.4	0.70	42	7	2	**20*	**8*	**65**	**200**
3	ADS11632	3.5	0.34	36	12	4	**34*	**13*	**110**	**280**
4	Gliese 623	7.2	0.34	20	8	15	**16**	6	****55**	**140**
5	51 Peg	13.5	1.0	8	10	0.4	4	2	16	44
6	δ GemA	16.4	1.8	32	18	0.2	2	1	9	****25**

3. Results

At the first stage, we have obtained the values of the parallax and proper motion of 51 Peg and then the residuals O–C have been analysed. In Table 2, the relative proper motion and parallax of 51 Peg with their errors are given. The errors of unit weight σ_x and σ_y obtained by least-squares solution are given too.

Table 2. Relative proper motion and parallax of 51 Peg

μ_x	$0.''2159 \pm 0.''0030$	μ_y	$0.''0695 \pm 0.''0032$
π_x	$0.''0645 \pm 0.''0083$	π_y	$0.''0721 \pm 0.''0134$
σ_x	$\pm 0''.032$	σ_y	$\pm 0.''034$

One can see that the precision is insufficient for the direct determination of parallaxes so that we excluded the parallactic displacements from the relative positions of star using the catalog values of parallaxes. Moreover, one can notice that the errors for π_x and for π_y are of comparable size, because this star is close to the pole of the ecliptique. With larger observational material we can use these values for the estimation of the influence of possible satellites with low masses, if we would find some correlations between perturbations and parallactic displacements.

We can compare O–C obtained for 51 Peg with the O–C obtained for Gliese 623 [$10.^m5$, $M3$, $\alpha_{(2000.0)} = 16^h24^m.0$, $\delta_{(2000.0)} = +48°21'$, $\pi = 0.''138$]. The latter star has been observed at Pulkovo since 1979 (Shakht 1995, 1997) with the same precision as 51 Peg. Residuals O–C (see Figs. 1 and 2). have been obtained in the same way, that is after the exclusion of proper motion and parallactic displacements.

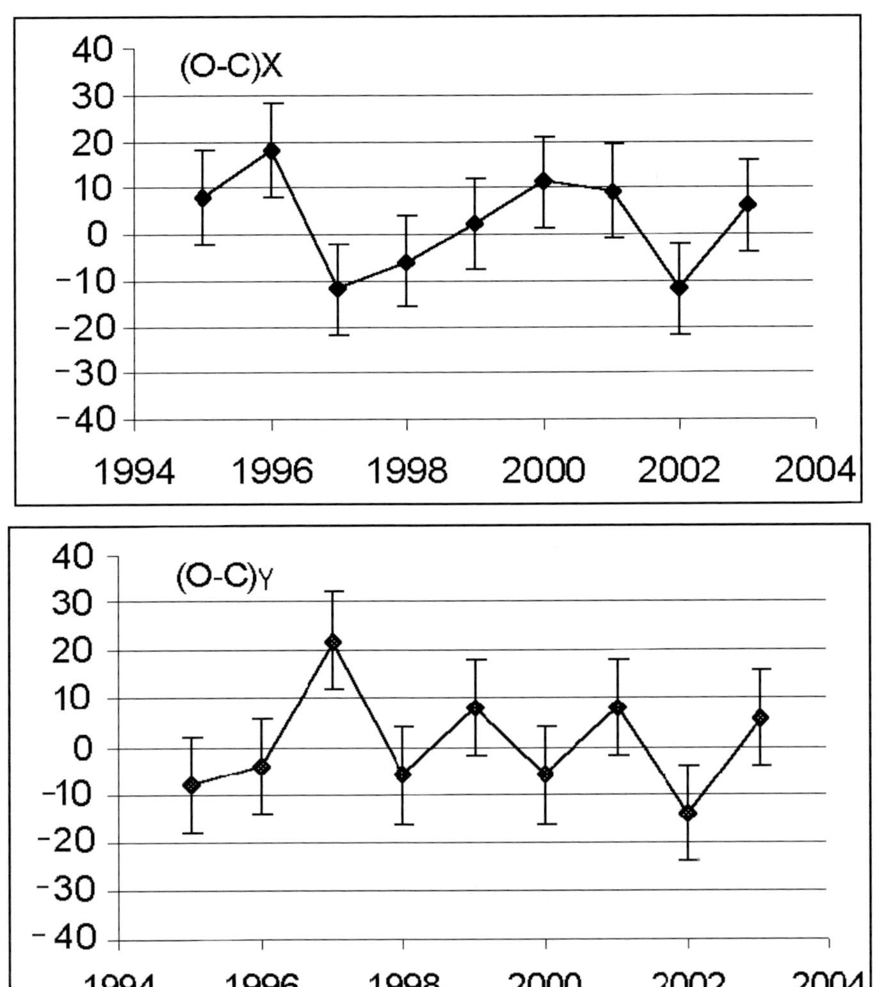

Figure 1. Residuals in the motion of 51 Peg (in mas).

The comparison has shown that the error of unit weight for Gliese 623 calculated on the basis of the residuals O–C, equals $0.''042$ for both coordinates because it contains the orbital motion of the photocenter, whereas the corresponding error for 51 Peg equals to $0.''033$.

Fig. 3 shows the power spectrum $S(\omega)$, obtained on the basis of residuals in the motion of 51 Peg (I) and for Gliese 623 (II). The latter is presented for comparison. The frequency is equal to 0.26, and corresponds to the period of 3.76 years for a satellite of the mass $0.09 M_\odot$, which belongs to Gliese 623 and which has been discovered by astrometric method (Lippincott and Borgman 1978) and confirmed by means of other astrometric

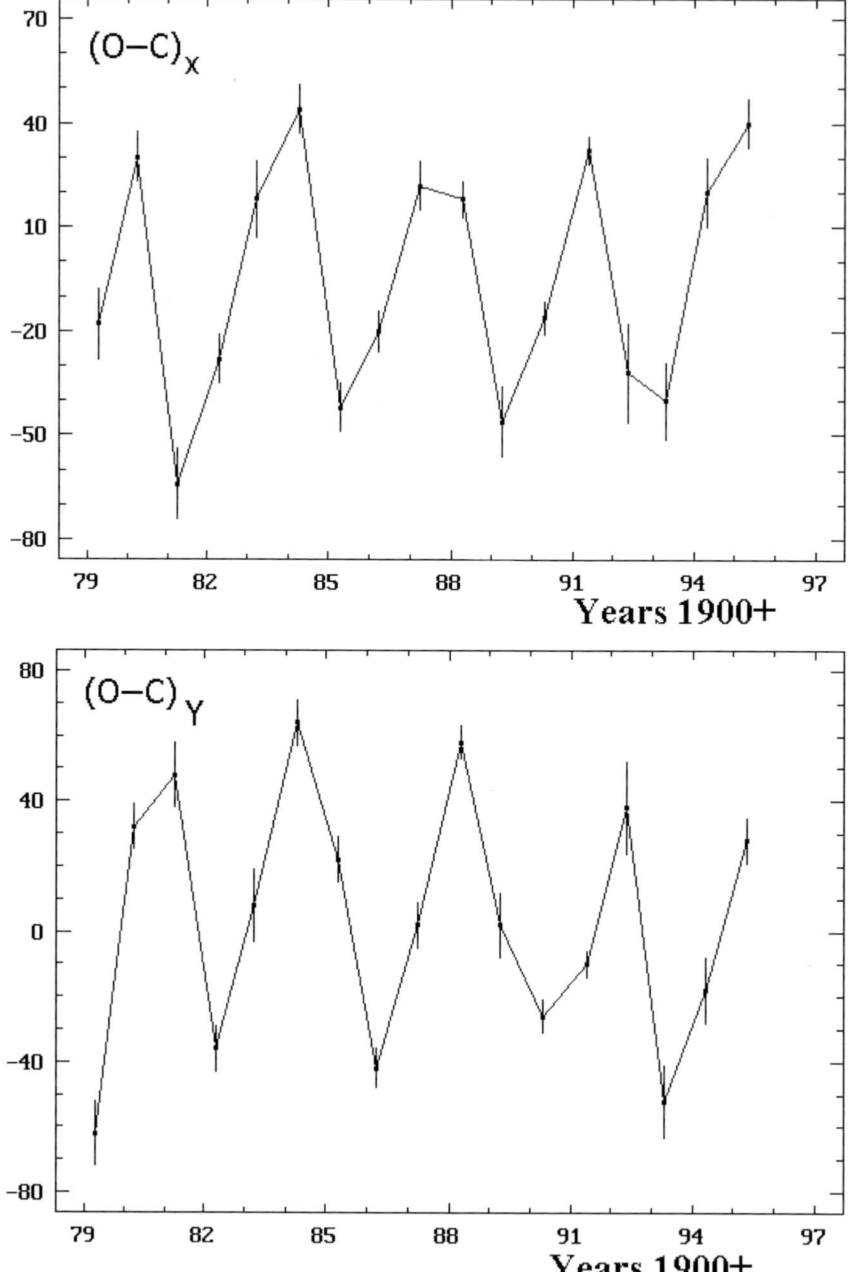

Figure 2. Residuals in the motion of Gliese 623 (in mas).

observations, including Pulkovo data. One can also see that $S(\omega)$ for 51 Peg does not show any noticeable frequencies and periods.

We used the method of the periodogram by Scargle to study the behavior of residuals and for the calculations of possible perturbations from the satellites with periods and masses shown in Table 3. Table 3 gives calculated displacements of the photocenter of 51 Peg caused by theoretically possible satellites.

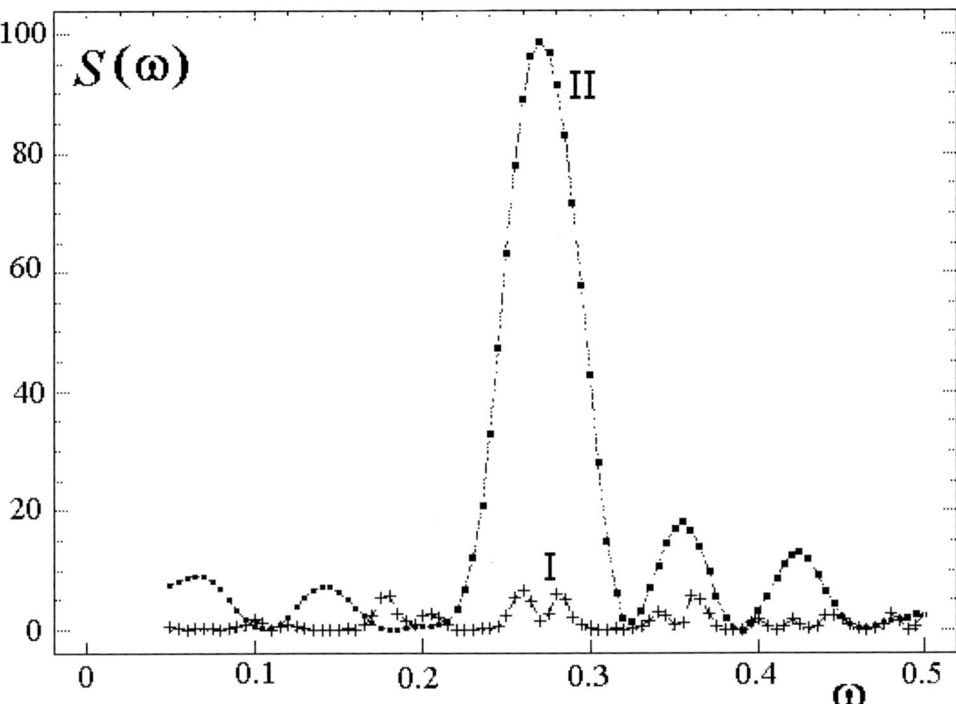

Figure 3. The periodogram obtained on the basis O–C in the motion of 51 Peg (crosses) and in the motion of Gliese 623 (dots). The ordinate is the power $S(\omega)$, where $\omega = 1/P$ and P is period in years. The great peak corresponds to the period 3.76 yr of the satellite of Gliese 623.

Table 3. Calculated displacements in the motion of 51 Peg due to the possible satellites

P [yr]	$1^y.5$	$2^y.5$	$3^y.0$	$3^y.5$	$6^y.0$
$0.5 M_\odot$	$0''.037$	$0''.050$	$0''.060$	$0''.065$	$0''.095$
0.4	30	40	50	55	78
0.3	24	35	40	45	62
0.2	17	24	27	30	44
0.1	9	13	14	16	23
0.08	7	10	12	13	18

Table 4. Parameters of the possible satellites of 51 Peg

P [yr]	1.5	2.5	3.0	3.5	3.5	4.0	4.0	6.0	6.0
$M_\odot$	0.2	0.2	0.2	0.1	0.2	0.2	0.1	0.2	0.08
A in a.e.	1.2	1.6	1.8	2.2	2.0	2.2	2.4	3.0	3.1

In the Table 4 the parameters of the possible satellites of 51 Peg accessible to Pulkovo 65 cm refractor after 12 years of observations are given. Among these there are satellites with masses less than $0.1 M_\odot$.

4. Conclusions

We present our first regular series of astrometric observations of 51 Peg performed by means of a long-focus astrograph, on the basis of which we were able to analyse a limited range of proposed periods and masses of possible satellites of 51 Peg. So far no perturbations due to the presence of possible satellites with the periods 0.5–6.0 years and with masses more than $0.2 M_\odot$ were found. Nevertheless, we have photographic plates which permit to continue the observations of this star. The precision of the observations and measurements by automatic Pulkovo complex "Fantasy" will provide the possibility to estimate the influence of satellites with masses less than $0.1 M_\odot$, near to the limit mass of the brown dwarfs $0.08 M_\odot$, and over a rather large range of rotation periods.

One of the important problems of the Pulkovo 65 cm refractor remains the necessity to have large field CCD to be used for observations in the frame of our program.

References

Black D.C. and Scargle J.D 1982, *Astron. J.* 263, (2), 854
Lippincott S.L. and Borgman E.R. 1978, *Publ. Astron. Soc. Pacific*, 90, 226
Mayor M. and Quelos D. 1995, *Nature*, 378, 355
Shakht N.A. 1995, *Proc. Symp. 166 IAU*, 359
Shakht N.A. 1997, *Astr. Astroph. Trans.*, 13, 327
Shakht N.A. 2000, *Izvestiya GAO* 214, 77 (in Russian).
Shakht N.A., Grosheva E.A., Kisselev A.A., Polyakov E.V. and Rafalsky V.B. 2002, *Izvestiya GAO*, 216, 363 (in Russian)

Dynamics of Populations of Planetary Systems
Proceedings IAU Colloquium No. 197, 2005
Z. Knežević and A. Milani, eds.

© 2005 International Astronomical Union
DOI: 10.1017/S1743921304008543

Observations of 61 Cyg at Pulkovo

Denis L. Gorshanov, N. A. Shakht, A. A. Kisselev and E. V. Poliakow

Central(Pulkovo) Astronomical Observatory,
Russian Academy of Sciences, St.Petersburg, Russia
email: dengorsh@mail.ru

Abstract. Results of the photographic observations of 61 Cyg are given. The orbital elements and mass ratio of the components of the pair are calculated. The preliminary orbit of a hypothetical satellite with a period of 6.5 yr is given.

Keywords. Photographic observations, 61 Cyg, dark satellite.

1. Introduction

The photographic observations of double star 61 Cyg [$21^h00^m.6$, $+38°45'$, $\pi = 0''.296$; $5.^m4$, $6.^m1$; K5V, K7V; $\rho = 30''.5$, $\theta = 150°(2000)$], have been collected at Pulkovo since 1895 by means of the Carte du Ciel (CdC) astrograph, and since 1958 by means of 65 cm refractor ($D = 65$ cm, $F = 10.4$ m, $M_0 = 19''.81/mm$). As it is known, the assumption on the existence of a satellite near a star of 61 Cyg was first made by Wilsing back in a 1893. Strand (1943) published a short message on the elements of orbit, providing also data on a dark satellite of 61 Cyg, with mass of $0.016 M_\odot$ and with the period of the rotation 4.9 years. A supposition about the presence of two dark companions of this star with periods of 6 and 12 years has been made on the basis of the deviations in its orbital motion by A. N. Deutsch in 1960's. The most prominent period of 6.0 years enabled to construct a model of an orbit of the photocentre described under the influence of this hypothetical satellite with elements: $T_0 = 1957.0, e = 0.2, a = 0''.006 \pm 0''.002, i = 34°, \omega = 108°, \Omega = 301°$. The lower mass limit of the probable satellite appeared to be $0.004 M_\odot$ (Deutsch & Orlova 1977).

Recently all the observations from the period 1958–1997 have been remeasured by means of the automatic machine "Fantasy" (see Gorshanov *et al.* 2002), and some preliminary results have been discussed. The final outcome of processing of the forty-year series of observations of 61 Cyg with 65 cm refractor is presented in this report.

We would like to point out that the history of research of 61 Cyg shows how with increase of the accuracy of observations and with new theoretical work the previous conclusions had to be reconsidered, since they have not always been supported by new observations. There are papers devoted to the study of this star in which the presence of satellites was not confirmed at all (see, for instance, V_r observations in Campbell *et al.* 1988 and astrometric ones in Jostis 1983, as well as some theoretical conclusions by Marcy & Chen 1992).

The period of deviations in orbital motion of 61 Cyg from Pulkovo observations is found to be 6.5 ± 0.5 years, that is close to the main period found by Deutsch. We tried to estimate it accurately and to draw some conlusions on the possible origin of this periodic wave, taking into account the corresponding observations of other authors and some alternative explanations.

We considered in particular:

(*a*) the sum of masses and the mass-ratio of components,

(*b*) the change of geometrical scale over some decades,

(*c*) the behaviour of residuals O–C obtained after the reduction to the relative orbit for 61 Cyg (ADS 14636) with respect to the control stars ADS 14710 and ADS 7251, and also the behaviour of the residuals obtained after the removal of relative proper motion, the parallax and the orbital motion with respect to the center of masses for the components A and B of 61 Cyg, separately,

(*d*) possible periodic deviations in the motion of reference stars.

2. Observations

There are at present two series of observations of 61 Cyg in Pulkovo plate archive: I – the plates obtained with 65 cm refractor, and II – the ones from Pulkovo Carte du Ciel astrograph. In this paper we consider only the series I. This series of observations with 65 cm refractor have been collected in the period 1958–1997. It contains 326 plates and about 6000 individual positions with mean error for one position from $0.''028$ to $0.''042$. The mean errors of annual positions are $0.''007$ for distance and $0.°01$ for position angle.

For the purpose of control, a wide pair (located far enough from the program star) – ADS 14710 [21^h $10^m.5$, $+22° 27'$, $6.^m9$, $7.^m8$; A1V, A0; $\rho = 18''.1$; $\pi = 0''.002$] has been observed after 61 Cyg each night since 1976. The detection of identical periodic variations with both pairs of stars thus indicates some instrumental problems or astro-climate variations, excluding the misinterpretation in terms of the presence of invisible satellites. So far we collected and measured 170 plates (2600 individual positions) of ADS 14710.

For the additional control we used a series of observations of another double star ADS 7251 [$9^h 14^m.4$; $+52°41'$; $7^m.8, 7^m.9$; K2, K2; $\pi = 0.''166$; $\rho = 17.''3$]. The series consists of about 220 plates (3300 individual positions) with equal number of the nights and it covers nearly the same period of time (1962–1998) as the series of 61 Cyg.

3. Results

The first stage of our procedure was the determination of the preliminary relative orbit of the visual components and analysis of the residuals O–C in the distances and position angles, or in coordinates X and Y. The data processing followed in general the technique of astrometric reduction described in the Pulkovo Catalogue of Double Stars (Kisselev *et al.* 1988). On the basis of these measurements, the orbit of the visible components was determined (Kisselev *et al.* 1997), yielding: $a = 24.''4., P = 659$ years, $e = 0.48, i = 146°, T_0 = 1697.0, \omega = 146°, \Omega = 176°$. To use the method of the Apparent Visible Parameters (see, Kisselev *et al.* 1997), we had to take into account the relative radial velocity ($\Delta V_r = 1.1$ km/s) of the components. Also for the best fitting of the orbit the old observations far enough in time from present ones are used. The total mass of the visible components was found to be $1.3 M_\odot$, and this value is in the best agreement with the mass – luminosity relation ($M - L$).

In the next stage we have determined the mass ratio $B = M_B/(M_A + M_B)$ using the measurements of two components B and A of the visual pair with respect to reference star. The time interval 1958–1997 was sufficient for the determination of the mass ratio B with an error less than 10%. The following equations for 240 positions have been solved in X and Y coordinates:

$$X_i = C_X + \mu_X(t_i - t_0) - B\Delta X_i$$
$$Y_i = C_Y + \mu_Y(t_i - t_0) - B\Delta Y_i$$

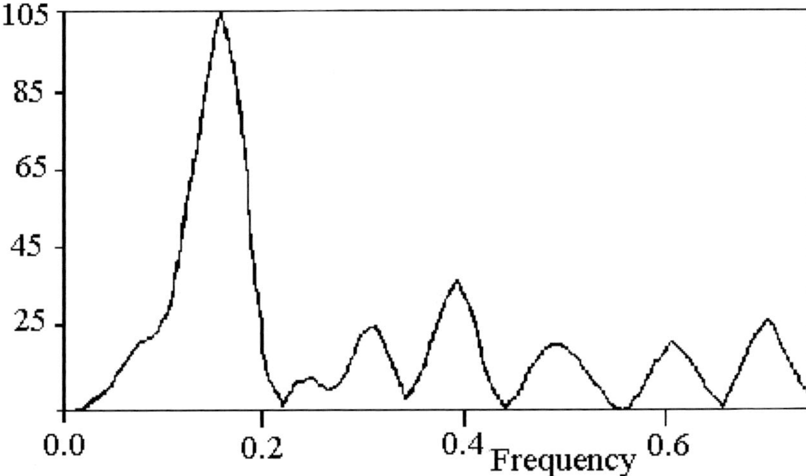

Figure 1. The periodogram of O–C in X projection of B–A distance. The main peak corresponds to the period of 6.5 yr. The power P on y-axis is in relative units.

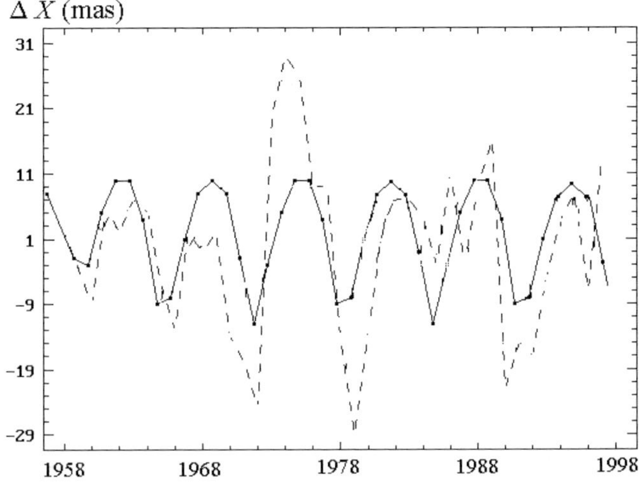

Figure 2. The comparison of residuals in orbital motion of 61 Cyg (dashed line) with the ephemerides calculated on the basis of our new data for photocentric orbit (solid line)

where X_i, Y_i are the positions of the component A on the current plate, with respect to standard plate at the middle epoch t_0. The parallactic displacement and the secular acceleration have been derived from X_i, Y_i;

C_X, C_Y are the positions of the center of the mass at zero epoch;

ΔX_i, ΔY_i are relative positions B–A of the components.

The value of B from the solutions of these equations is equal to 0.38 ± 0.03, and this mass ratio and the sum do not contradict to $M - L$ regularity and to spectral classes of the components.

As a result, the residuals O–C obtained after removal of the preliminary orbital motion from the relative positions B–A of 61 Cyg have shown a small periodic deviations with 6.5 ± 0.5 yr period, which were repeated in two parts of the observation interval, in

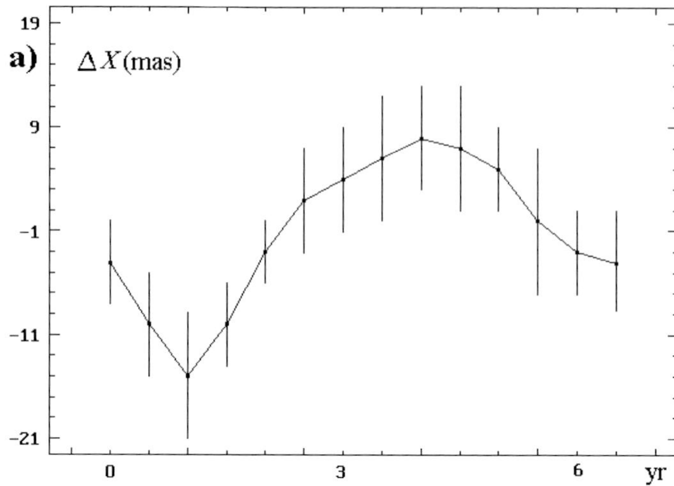

Figure 3. Residuals ΔX averaged over the period of 6.5 years in mas.

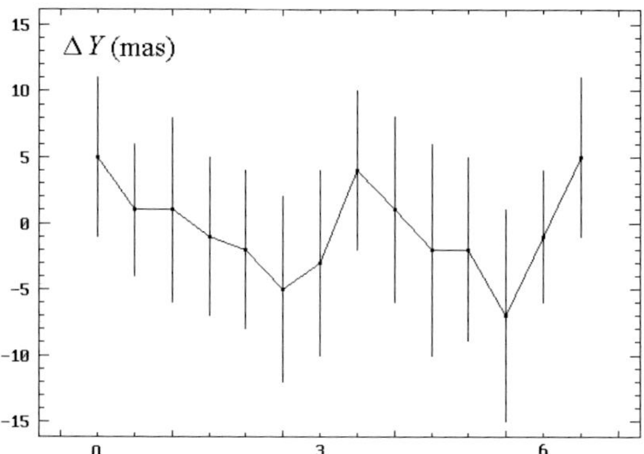

Figure 4. Residuals ΔY averaged over the period of 6.5 years in mas.

1958–1976, and in 1976–1997. However, these periodic deviations were found only in the projection on the RA axis (see Figs. 1–4), but not in the individual motion of each component with respect to the reference stars.

The residuals obtained after the removal of preliminary orbital motion in the relative positions B–A of ADS 7251 and ADS 14710 have also not shown any noticeable periodic deviations in the preliminary orbital motion of the visual components on the all interval of observations.

4. Conclusions

The long term series of observations of 61 Cyg obtained with 65 cm refractor allow to determine the relative orbit and the mass ratio for visual components. The sum and the mass ratio did not show the presence of a noticeable hidden mass of any stellar or substellar component type.

Nevertheless our long series show the periodic deviations in $(B - A)_X$ differences, with periods ranging from 6.2 to 6.7 years, with an average amplitude from 0.″015 to 0.″010, and with mean error of 0.″006 for both, the first part (1958–1974), and the second part (1975–1997) of the series, which were repeated over all the interval of the observations.

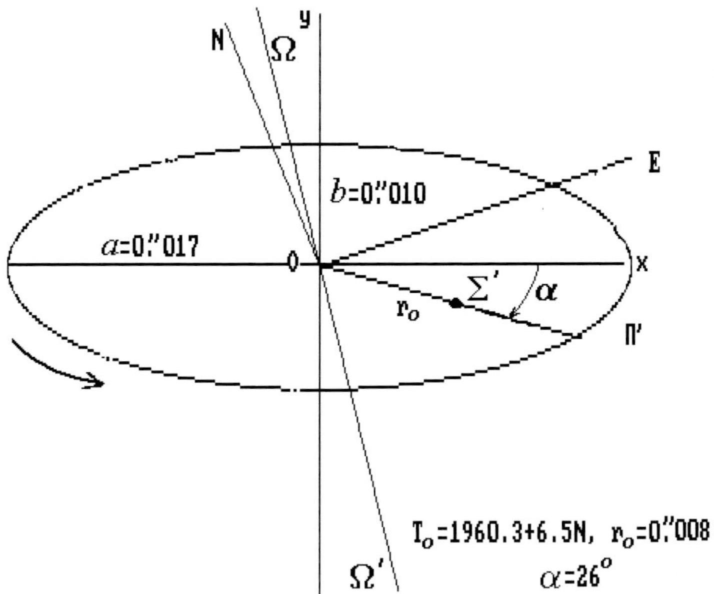

Figure 5. The visible ellipse constructed as a model of the preliminary photocentric orbit for 61 Cyg with orientation to semi-axis and with the period of 6.5 yr. The elements of visual orbit a, b, r_o, T_o are obtained by graphical method of Kisselev (1997) (Σ' is the projection of the center of masses; Π' is the projection of the periastron). The elements of true orbit are $A = 0.''018, E = 0.55, i = 126°, \omega = 50°, \Omega = 9°$. The lower limit of mass of the hypothetical satellite is $0.014 M_\odot$.

The construction of the visible ellipse by using these results (see Fig. 5) was quite uncertain. But on the basis of this ellipse and using the method of Kisselev *et al.* 1997 we have calculated a preliminary orbit of the hypothetical satellite with the lower limit of the mass of about $0.014 M_\odot$. Probably, we can accept that there are no satellites with masses greater than $0.015 M_\odot$ and with periods in the range 3.5–20 years which is accessible for our estimations. But it is possible that some complicated phenomenon created by few satellites with possibly unstable orbits may perturb the observed motion.

We have investigated also the motion of control stars and we studied the geometrical scale, the probable temperature factors and the reduction (in particular, whether any small perturbations exist in the motion of reference stars) etc., looking for alternative explanations of the observed deviations. But our control stars did not exhibit any similar periodic behaviour and the change of the scale allows to consider only the period of 12.0 years as a consequence of some instrumental or astroclimatic causes.

We have taken into account all results of the observers of this star and the other authors, and all arguments against the planetary satellites of this star. But we are aware that the problem of possible existence of one or more unseen satellites of 61 Cyg with the long periods of revolutions and maybe with unstable orbits is not yet solved neither from the astrophysical, nor from the celestial mechanics points of view.

Acknowledgements

Authors are grateful to the IAU Coll. 197 Organizing Committee for the opportunity to present the paper.

References

Campbell B., Wolker G.A.H. and Yang S. 1988, *Astrophys. J.*, 331, 902

Deutsch A.N. & Orlova O.N. 1977, *Astron. Zhurnal*, 54(2), 327 (in Russian)

Gorshanov D.L., Shakht N.A., Polyakov E.V., Kisselev A.A. and Kanaev I.I. 2002, *Izvestiya GAO*, 216, 100 (in Russian)

Jostis P.J. 1983, *Low. Obs. Bull.*, 167, 16

Kisselev A.A. 1997, *Workshop Visual Double Stars, Kluwer Acad. Publ.*, (Santiago de Compostella, Spain), 357

Kisselev A.A., Kiyaeva O.V. and Romanenko L.G. 1997, *Workshop Visual Double Stars, Kluwer Acad. Publ.*, (Santiago de Compostella, Spain), p. 377

Kisselev A.A. *et al.* 1988, *Nauka Publ.*, Leningrad, p. 39 (in Russian)

Marcy G.W. & Chen G.H. 1992, *Astrophys. J.*, 390, 550

Strand K.A. 1943, *Publ. Astron. Soc. Pacific*, 55, 322

Dynamics of Populations of Planetary Systems
Proceedings IAU Colloquium No. 197, 2005
Z. Knežević and A. Milani, eds.

Formation of the solar system by instability

Evgeny Griv and Michael Gedalin

Department of Physics, Ben-Gurion University of the Negev, Beer-Sheva 84105, Israel
email: griv@bgu.ac.il

Abstract. The early gas and dust protosolar nebula of the solar composition is considered analytically. A simultaneous formation of the sun and all the planets around it ($\approx 5 \times 10^9$ yr ago) through a local gravitational Jeans-type instability of small-amplitude gravity perturbations in the nebula disk is suggested. It is shown that a collective process, forming the basis of the disk instability hypothesis, solves with surprising simplicity the two main problems of the dynamical characteristics of the system, which are associated with its observed spacing and orbital momentum distribution, namely, Bode's law on planet spacing and the concentration of angular momentum in the planets and mass in the sun. Besides, the analysis is found to imply the existence of new planets or other Kuiper-type belts of comets at mean distances from the sun of 87 AU, 151 AU, 261 AU, 452 AU, 781 AU (Mercury, Venus, ..., Asteroid belt, ..., Neptune, Kuiper belt, new planets or other Kuiper-type belts).

Keywords. Planetary systems, planetary formation, solar system, instabilities and waves

1. Introduction

The "standard" theory of the multistage accretionary formation of planets, or the so-called core accretion mechanism (Safronov 1972; Pollack *et al.* 1996) remained the most popular until recently, when it was criticized by Boss (2002, 2003) and others. The main problem is the timescale, which is longer than estimates of the lifetime of many planet-forming disks (Taylor 1992, §1.9.1; Feigelson & Montmerle 1999).

We suggest that the sun and all the planets around it were created simultaneously by disk instability. That is, at an early stage, the protosolar nebula refers to a fragment that separated from a molecular cloud. Planetary formation is thought to start with inelastically colliding gaseous and dust particles settling to the central plane of this rotating nebula to form a thin layer around the plane. During the early evolution of the disk it is believed that the dust particles coagulate into kilometer-sized rocky comets–"planetesimals." On attaining a certain critical thickness (and, correspondingly, very low temperature) small in comparison with the radius of the system R, as a result of a local gravitational collapse the nebula disintegrated into the central body and a number of separate protoplanets.† Boss (2004) already demonstrated that convective cooling is able to cool the disk midplanes at the desired rate to produce bound clumps in marginally unstable disks. Following Boss *et al.* (2002), the hypothesis of disk instability envisions coagulation and settling of dust grains within the protoplanets to form ice and rock cores. A protoplanet accreted an additional amount of gas subsequently from the solar nebula after accumulating a solid core, followed by the loss of the light elements of the terrestrial planets through the thermal emission of the early sun. After its formation and

---

† Destabilizing self-gravity in much more "dangerous" in thin disks than in thick disks. If a rotating gaseous disk has a large vertical thickness owing to a high internal temperature, then it is stabilized against gravitational instabilities. Instabilities arise as the thickness of the disk is reduced (Shu 1970; Safronov 1980).

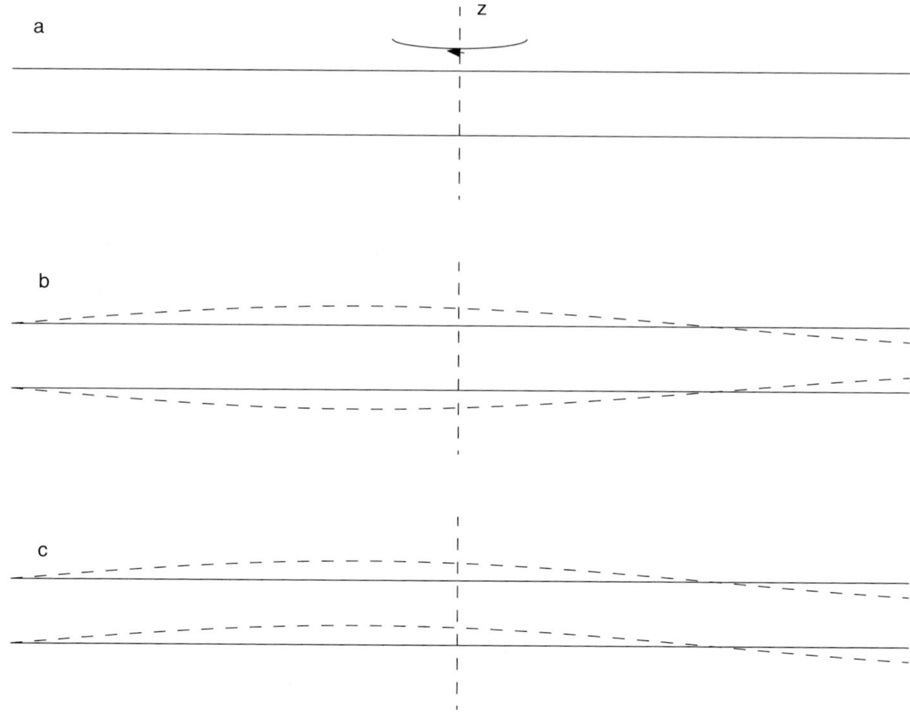

Figure 1. Sketch of perturbations of a three-dimensional protoplanetary disk. In (a) a section of the disk is shown edge-on. In (b) an even (Jeans-type) perturbation is shown (the dashed line). In (c) an odd (bending firehose-type; Kulsrud *et al.* 1971; Bertin & Casertano 1982; Griv & Chiueh 1998) perturbation is illustrated (the dashed line).

the loss of the light elements, the terrestrial planets had suffered a "late planetesimal bombardment" (Wetherill 1989).

The basic idea of planet formation through gravitational instability has been pioneered by Kuiper (1951), Urey (1958), and Cameron (1978) by considering the formation of giant planets in the solar nebula. Gravitational instability appears to be capable of forming giant planets with modest cores of ice and rock faster than the core accretion mechanism can (Boss 2002). Apparently, Cassen et al. (1981) were the first to simulate the planet formation in gravitationally unstable gaseous protostellar disks by N-body experiments. Recently, Mayer *et al.* (2002), Johnson & Gammie (2003), and Rice *et al.* (2003) used hydrodynamic simulations and N-body orbit integrations to study the long-term evolution of a fragmenting disk with realistic cooling. The advantages of the disk instability model are that (1) the instability process itself is quite fast, and could form planets in $\lesssim 10^4$ yr (Boss 2003), (2) in unstable, nonaxisymmetric disks differential rotation can simultaneously transfer angular momentum outward and mass inward through gravitational torques (Larson 1989; Taylor 1992, §2.9.2), and (3) such a model obviates the requirements for turbulent viscosity, frequently appealed to as a physical mechanism for outward transfer of angular momentum. This paper has precedents in earlier studies of gravity disturbances in galactic disks and Saturn's ring disk (Lin & Shu 1966; Shu 1970; Griv *et al.* 2001, 2002, 2003; Griv & Gedalin 2003) and uses a largely previously published analysis (Griv *et al.* 1999; Griv & Gedalin 2004) of the linear growth of perturbations in a rotating disk to draw conclusions about the formation of the solar system.

2. Oscillation spectrum

The dynamics of the gaseous component in the presence of the collective self-gravitational field is considered. A Langrangian description of the motion of a gas element under the influence of a perturbed gravity field is used, looking for time-dependent waves which propagate in a differentially rotating, spatially inhomogeneous, and two-dimensional disk. This approximation of an infinitesimally thin disk is a valid approximation if one considers perturbations with a radial wavelength that is greater h, the typical disk thickness (Shu 1970; Safronov 1980).

The time dependent surface density $\sigma(\vec{r}, t)$ is splited up into a basic and a developing (perturbation) part, $\sigma = \sigma_0(r) + \sigma_1(\vec{r}, t)$ and $|\sigma_1/\sigma_0| \ll 1$, where r, φ, z are the cylindrical coordinates and the axis of the disk rotation is taken oriented along the z-axis. The gravitational potential of the disk $\aleph(\vec{r}, t)$ and the gaseous pressure $P(\vec{r}, t)$ are also of this form. These quantities σ, $\aleph$, P are then substituted into the equations of motion of a gas element, the continuity equation, the Poisson equation, and the second order terms of the order of σ_1^2, $\aleph_1^2$, P_1^2 may be neglected with respect to the first order terms. In our study we restrict the analysis to a treatment of "sausage-like" Jeans perturbations (Bertin & Casertano 1982) which are symmetric with respect to the $z = 0$ equatorial plane of the disk (which do not cause it to bend). See Fig. 1 for an explanation. These perturbations are associated with such phenomena as, for example, the appearance of the spiral structure of disk galaxies (Lin & Shu 1966; Lin *et al.* 1969; Shu 1970; Lin & Lau 1979; Griv *et al.* 1999). The resultant equations of motion are cyclic in the variables t and φ, and hence by applying the widely used local WKB method (Alexandrov *et al.* 1984) one may seek solutions in the form of normal modes by expanding

$$\sigma_1(\vec{r}, t) = \sum_{\vec{k}} \tilde{\sigma}_{\vec{k}} \exp\left(ik_r r + im\varphi - i\omega_{\vec{k}} t\right) + \text{c.c.}, \tag{2.1}$$

where $\tilde{\sigma}_{\vec{k}} = \text{const}$ is the real amplitude, $k_r(r)$ is the real radial wavenumber, m is the nonnegative (integer) azimuthal mode number, $\omega_{\vec{k}} = \Re\omega_{\vec{k}} + i\Im\omega_{\vec{k}}$ is the complex frequency of excited waves, suffixes **k** denote the $\vec{k}$th Fourier component, and "c.c." means the complex conjugate. In the linear theory, one can select one of the Fourier harmonics: $\tilde{\sigma} \exp\left(ik_r r + im\varphi - i\omega t\right) + \text{c.c.}$. The solution in such a form represents a spiral plane wave with m arms or a ring ($m = 0$). The imaginary part of ω corresponds to a growth ($\Im\omega > 0$) or decay ($\Im\omega < 0$) of the components in time, $\sigma_1 \propto \exp(\Im\omega t)$, and the real part to a rotation with angular velocity $\Omega_{\rm p} = \Re\omega/m$. Thus, when $\Im\omega > 0$, the medium transfers its energy to the growing wave and oscillation buildup occurs.

It is important to note that in the WKB method, the radial wavenumber is presumed to be of the form

$$k_r(r) = \mathcal{A}\Psi(r), \tag{2.2}$$

where $\mathcal{A}$ is a large parameter and $\Psi(r)$ is a smooth, slowly varying function of the radial distance r, i.e., $d\ln k_r/d\ln r = O(1)$, and in the WKB approximation $|k_r|r \gg 1$.

Paralleling the analysis leading to Eq. (34) in Griv *et al.* (1999), we obtain that

$$\sigma_1 = \frac{\sigma_0 \beth}{\omega_*^2 - \kappa^2}\left(k_r^2 + \frac{4\Omega^2 - \kappa^2 + \omega_*^2}{\omega_*^2}\frac{m^2}{r^2} + \frac{2\Omega}{\omega_*}\frac{m}{rL}\right) + \text{c.c.}, \tag{2.3}$$

where $\sigma_1(t \to -\infty) = 0$, so by considering only growing perturbations we neglected the effects of the initial conditions, $\omega_* = \omega - m\Omega$ is the Doppler-shifted (in a rotating frame) wavefrequency, $\Omega(r)$ is the angular velocity of rotation at the distance r from the center, $\kappa \approx \Omega$ is the epicyclic frequency, $\beth = \aleph_1 + c^2\sigma_1/\sigma_0$, and $c = (\partial P/\partial\sigma)^{1/2}$ is the sound

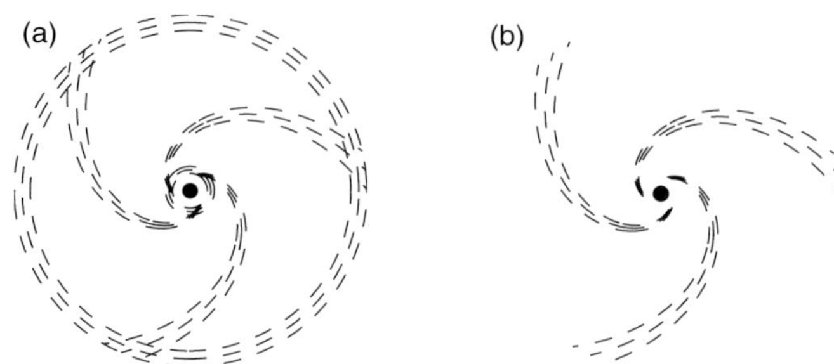

Figure 2. A schematic model of a Jeans-unstable disk with $m = 3$ spiral arms; (a) the Safronov–Toomre unstable disk ($c < c_{\rm T}$) and (b) the Safronov–Toomre stable disk ($c \gtrsim c_{\rm T}$ but $c < (2\Omega/\kappa)c_{\rm T}$ and in the protosolar nebula $2\Omega/\kappa \approx 2$).

speed. In Eq. (2.3) only the most important low-frequency ($|\omega_*^2| < \kappa^2$) perturbations developing in the $z = 0$ plane between the inner and outer Lindblad resonances are considered (Lin & Shu 1966; Shu 1970; Griv *et al.* 1999, 2002).

Equating the perturbed density σ_1 (Eq. (2.3)) to the perturbed density given by the asymptotic ($k_r^2 \gg m^2/r^2$) solution of the Poisson equation $\sigma_1 = -|k|\aleph_1/2\pi G + {\rm c.c.}$ (e.g., Griv *et al.* 1999, 2003), the solution of the generalized Lin–Shu-type dispersion relation is easily obtained

$$\omega_{*1,2} \approx \pm p|\omega_{\rm J}| - 2\pi G\sigma_0 \frac{\Omega}{\omega_{\rm J}^2} \frac{m}{r|k|L}, \tag{2.4}$$

where $p = 1$ for gravity-stable perturbations with $\omega_*^2 \approx \omega_{\rm J}^2 > 0$, $p = i$ for gravity-unstable perturbations with $\omega_*^2 \approx \omega_{\rm J}^2 < 0$, $|L| = |\partial \ln \sigma_0/\partial r|^{-1}$ is the radial scale of spatial inhomogeneity, $|kL| \gg 1$, the term involving L^{-1} is the small correction,

$$\omega_{\rm J}^2 = \kappa^2 - 2\pi G\sigma_0(k_*^2/|k|) + k_*^2 c^2 \tag{2.5}$$

is the squared Jeans frequency, $k = \sqrt{k_r^2 + m^2/r^2}$ is the total wavenumber, $k_*^2 = k^2 \left\{ 1 + [(2\Omega/\kappa)^2 - 1]\sin^2\psi \right\}$ is the squared effective wavenumber, and $\psi = \arctan(m/rk_r)$ is the perturbation pitch angle. Equation (2.4) determines the spectrum of oscillations. This equation differs from the standard Lin–Shu expression (Lin & Shu 1966; Lin *et al.* 1969; Shu 1970) by the appearance of the total k and effective k_* wavenumbers, which originate from the consideration of the nonaxisymmetrical modes $\propto \psi$, and by the factor $\propto L^{-1}$, which originates from the consideration of the effects of spatial inhomogeneity (see Morozov 1980, 1981 and Griv *et al.* 2002, 2003 for a discussion).

From Eq. (2.5), the disk is Jeans-unstable ($\omega_{\rm J}^2 < 0$) to both axisymmetric (radial) and nonaxisymmetric (spiral) perturbations if $c < c_{\rm T}$, where $c_{\rm T} = \pi G\sigma_0/\kappa$ is the Safronov–Toomre (Safronov 1960; Toomre 1964) critical sound speed to suppress the instability of axisymmetric $m = 0$ perturbations. Nonaxisymmetric (m, or $\psi \neq 0$) instabilities in a differentially rotating disk is more difficult to stabilize; stability is achieved only for sufficiently large sound speed $c \approx (2\Omega/\kappa)c_{\rm T} \approx 2c_{\rm T}$ (Morozov 1980, 1981; Griv et al. 1999, 2002). Thus, if the disk is thin, $c \ll r\Omega$, and dynamically cold, $c < c_{\rm T}$ (or Toomre's stability parameter $Q \equiv c/c_{\rm T} < 1$, respectively), then such a model will be gravitationally unstable, and it should almost instanteneously (see below for a time estimate) taken on the form of a cartwheel, that is, a structure of spirals and rings (Fig. 2a). One concludes that Toomre's Q-parameter that is < 1 suggests that the disk is likely subject to radial and azimuthal gravitational instabilities and might therefore be clumpy. The instability is driven by a strong nonresonant interaction of the gravity fluctuations (e.g., those produced by a spontaneous perturbation or a satellite system) with the bulk of the particle population: in Eq. (2.3), $\omega_* - l\kappa \neq 0$, where $l = 0, \pm 1$.

The growth rate of the instability is relatively high, $\Im\omega_* \approx \sqrt{2\pi G\sigma_0(k_*^2/|k|)} \sim \Omega$, that is, the instability develops rapidly on a dynamical time scale (on a time of 3-4 disk rotations, or $\lesssim 10^4$ yr in the solar nebula). An important feature of the instability under consideration is the fact that it is almost aperiodic ($|\Re\omega_*/\Im\omega_*| \ll 1$). From Eq. (2.5), the growth rate of the instability has a maximum at the wavelength $\lambda \approx 4c^2/G\sigma_0$. At the boundary of instability ($c \approx c_{\mathrm{T}}$, or $Q \approx 1$, respectively), $\lambda \approx \lambda_{\mathrm{crit}} = 4\pi^2 G\sigma_0/\kappa^2 \sim (2-4)\pi h$; thus $\lambda_{\mathrm{crit}} \gg h$. It means that of all harmonics of initial gravity perturbation, one perturbation with $\lambda_{\mathrm{crit}} \approx 3\pi h$, with the associated number of spiral arms m_{crit}, and with the pitch angle ψ_{crit} will be formed asymptotically. For the parameters of the solar nebula ($R \sim 1\,000$ AU, $\kappa = 2\pi/T_{\mathrm{orb}} \sim 10^{-10}$ s^{-1}, and the total mass of the disk $M_{\mathrm{d}} \sim M_\odot$), one obtains the typical mass of a protoplanet $M_{\mathrm{c}} \sim \lambda_{\mathrm{crit}}^2\sigma_0 \sim 10^{-3}\,M_\odot \sim 300\,M_\oplus$. The latter is coincident in order of magnitude with the masses of giant planets. (The detection of a number of extrasolar planets with minimum masses ranging from 0.5 to $4\,M_{\mathrm{J}}$, where $M_{\mathrm{J}} = 318\,M_\oplus$ = the mass of Jupiter, has removed much of the concern that giant planets might be rare in our Galaxy; Boss 2002.)

The larger part of the initial mass of protoplanets of the Earth's group was probably blown away due to intensive thermal emission of the early sun. Such a point of view is not unnatural since the planets of the Earth's type consist mainly of elements with a high melting temperature and are almost lacking light elements. By adding to the present masses of the terrestrial planets the amount of light gases which is necessary to restore the chemical composition of giant planets, one obtains masses larger by a factor of several hundreds, coincident with the masses of giant planets.

3. Spacing of the planets

There exists the empirical Titius–Bode (TB) rule which gives the mean orbital distances of the planets and which can be written in the Blagg–Richardson formulation:

$$r_{\mathrm{n}} = r_0 A^{\mathrm{n}}, \tag{3.1}$$

where r_{n} is the distance of the nth planet from the Sun (in AU), n = 1 for Mercury, 2 for Venus, ..., and 9 for Neptune, $A \approx 1.73$ is the mean ratio between two consecutive planetary distances, and $r_0 \approx 0.21$. (The pure geometric sequence in the Blagg–Richardson form is better than the original TB form $r_{\mathrm{n}} = 0.4 + 0.3 \times 2^{\mathrm{n}}$, where n $= -\infty, 0, 1, \ldots, 5$ and r_{n} are the distances (in AU), respectively, to Mercury, Venus, ..., Neptune. In particular, the skip in index from minus infinity to zero in the older form is quite unnatural. The mean Asteroid belt distance is considered as a regular planetary distance, $r_5 \approx 2.8$ AU. Unlike Polyachenko (Polyachenko & Fridman 1972), we do not treat Pluto as a planet but as a member of a "resonant" population of the Kuiper belt objects. The recently discovered Kuiper belt constitutes one of the few fossil records of the formation of planets and planetesimals in the early solar system.) Also, one cannot overlook the fact that many of the regularities which are found in the planetary system are also to be seen in the regular satellite systems of Jupiter, Saturn, and Uranus, e.g., the spacing of the regular satellites is a variation of the TB rule (Fig. 3). This suggests that the same cosmogonic process must have been responsible for the origin of both types of systems. Lynch (2003) has argued that it is not possible to conclude unequivocally that laws of TB type are, or are not, significant. Therefore, the possibility of a physical explanation for the observed distributions remains open. Using the Lynch method, Neslušan (2004) has continued the discussion on the statistical significance of agreement between planetary

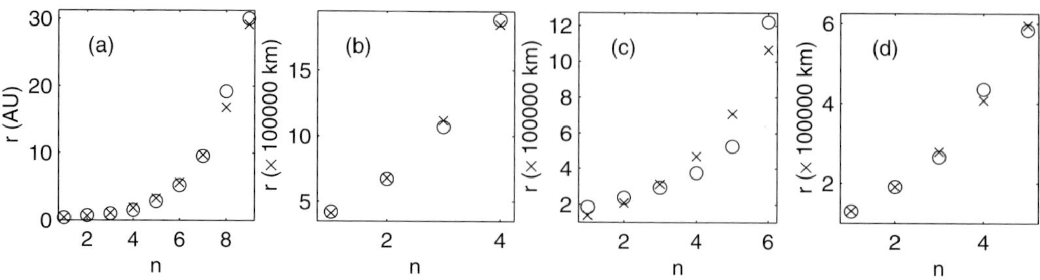

Figure 3. Relation between distances of planets (satellites) from the Sun (giant planets) r and their numbers n. Data observed are represented by circles: (a) the solar system, $r_0 = 0.21$ and $A = 1.73$, (b) the satellite system of Jupiter, $r_0 = 249.679$ and $A = 1.649$, (c) the satellite system of Saturn, $r_0 = 92.416$ and $A = 1.503$, (d) the satellite system of Uranus, $r_0 = 89.737$ and $A = 1.46$. The crosses represent the TB rule, Eq. (3.1). The organization of both types of systems does not contradict to the TB rule.

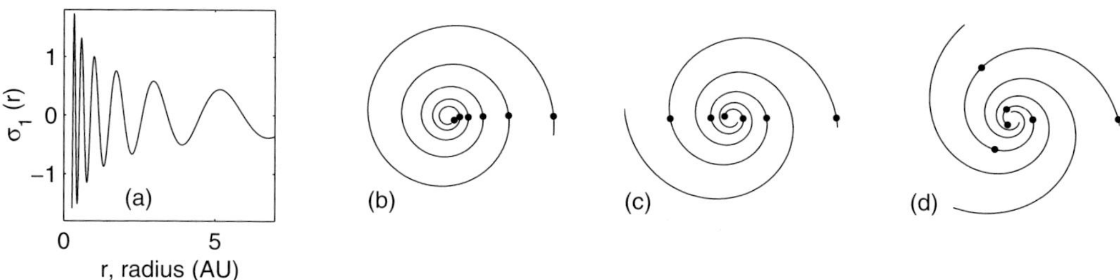

Figure 4. (a) Dependence of the perturbed surface density of the protoplanetary disk $\sigma_1(r)$ (arbitrary units) on the radius r, Eq. (3.3). The maxima of the perturbed density coincide with locations of the planets. (b) Spiral density waves with $m = 1$ arm in the (r, φ)–plane, (c) density waves with $m = 2$ arms, and (d) density waves with $m = 3$ arms. The filled circles represent the maxima of the perturbed density (protoplanets) of Jeans waves, which are unstable to both axisymmetric and nonaxisymmetric perturbations.

distributions and a power law. Interestingly, the mean orbital distance to the recently discovered Kuiper belt objects, $r \approx 46$ AU (Luu & Jewitt 2002), is in fair agreement with that given by the TB rule for the solar system's 10th planet, $r_{10} \approx 50$ AU. Liboff (2004) has predicted the existence of a planet at a mean radius from the sun of ≈ 51 AU.

Equation (3.1) can be rewritten:

$$(2\pi/\ln 1.73)\ln(r_\mathrm{n}/0.21) = 2\pi\mathrm{n}. \tag{3.2}$$

Next, the surface density of the disk may be represented in the form of the sum of the basic equilibrium surface density $\sigma_0(r)$ and the perturbed surface density

$$\sigma_1(r) = \tilde{\sigma}(r)e^{\Im\omega t}\cos\left[11.46\ln(r/0.21) + m\varphi\right], \tag{3.3}$$

where $\tilde{\sigma}(r)$ is the amplitude varying slowly with radius, and $[11.46\ln(r/0.21) + m\varphi]$ represents the phase varying rapidly with r, that is,

$$|k_r|r \equiv 11.46\left|(\mathrm{d}/\mathrm{d}r)\ln(r_\mathrm{n}/0.21)\right|r \gg 1.$$

Equation (3.2) and the condition $\tilde{\sigma}(r) > 0$ on the initial phase imply that the maximum values of the perturbed density in Eq. (3.3) coincide with the positions of all the planets (Fig. 4a).

Interestingly, and this is the central part of our theory, the TB rule (Eq. (3.2)) *satisfies* the conditions of the WKB wave with the effective TB radial wavenumber (Eq. (3.3))

$$k_{\text{eff}} = 11.46\, r^{-1}\,, \tag{3.4}$$

$d \ln k_{\text{eff}}/d \ln r = O(1)$, and $k_{\text{eff}} r \gg 1$ (cf. Eq. (2.2)). Note that Polyachenko (Polyachenko & Fridman 1972) has considered this analogy in his investigation of the possibility of the explanation of the law of planetary distances by the gravitational instability, but evidently without success. In particular, Polyachenko studied only axisymmetric perturbations, which do not carry angular momentum (see §4 below). Because $k_{\text{eff}} r \gg 1$, the short-wavelength ($\lambda_{\text{crit}}/R \ll 1$) WKB approximation used in the theory does not fail.

Thus, if the space dependence of the perturbed surface density of the protoplanetary disk has the form of Eq. (3.3) with $\Im \omega > 0$, the maxima of both radially and azimuthally unstable gravity perturbations are located in places of the solar system's planets (Figs 4b, c, d). Let us define conditions under which the density maxima are localized on planetary orbits. If the disk is inhomogeneous with respect to equilibrium parameters, the radial wavelength of a perturbation with a maximum growth rate λ_{crit} will be a function of the radius r. From the above, the radial wavelength $\lambda_{\text{crit}} \approx 4\pi^2 G \sigma_0/\kappa^2$, corresponding to the minimum on the dispersion curve (2.4) (see also Griv *et al.* 2002, their Fig. 1). On the other hand, the radial wavelength is $\lambda_{\text{eff}} = 2\pi/k_{\text{eff}}$. Comparing λ_{crit} with λ_{eff}, we see that in the case where the disk density is dependent on radius according to the law

$$\sigma_0(r) \approx 0.014\, G^{-1}\, \kappa^2\, r\,, \tag{3.5}$$

the maxima of time-increasing, both radially and azimuthally Jeans-unstable density perturbations are arranged in it according the TB rule. Interestingly, both optical and near-infrared observations of pre-main-sequence stars of intermediate mass have revealed the structure of spirals and rings, and thus presumably the Jeans instability of radial and spiral perturbations, in the circumstellar disks (Fukagawa *et al.* 2004).

One concludes, therefore, that if the surface density of a protoplanetary disk falls according to the law (3.5), the increasing maxima of density perturbations of a Safronov–Toomre unstable disk ($c < c_{\text{T}}$, or $Q < 1$, respectively) are located between the Lindblad resonances in places of the planets. We believe to have obtained a theoretical interpretation of the TB rule: the distance between planets is the wavelength of the most Jeans-unstable perturbations λ_{crit} at the given point of the protoplanetary disk.

4. The transfer of angular momentum

We next turn to the question of how to account for the concentration of angular momentum in the planets and of mass in the sun. The collective torque exerted by the gravity perturbations on the disk is $\Gamma = -\int \int d^2 r (\vec{r} \times \vec{\nabla}\aleph_1)\sigma_1$ or

$$\Gamma = -\int_{r_1}^{r_2} r dr \int_0^{2\pi} \sigma_1(r,\varphi) \frac{\partial \aleph_1(r,\varphi')}{\partial \varphi'} d\varphi'\,. \tag{4.1}$$

The points r_1 and r_2 in which $\omega_* \pm \kappa = 0$ are called the points of inner and outer Lindblad resonances. They play an important role in the theory: the solution of spiral type (2.1) rapidly oscillating in the radial direction lies between r_1 and r_2. Outside the resonances, $r < r_1$ and $r > r_2$, the solution decreases exponentially. A special analysis of the solution near spatially limited corotation ($\omega_* = 0$) and Lindblad ($\omega_* \pm \kappa = 0$) resonances is required. Resonances of a higher order, $\omega_* \pm l\kappa = 0$ and $l = 2, 3, \cdots$, are dynamically of less importance (Shu 1970). The present analysis is restricted to consideration of only the main part of a disk between the inner and outer Lindblad resonances. Lynden-Bell & Kalnajs (1972), Goldreich & Tremaine (1980), Meyer-Vernet & Sicardy (1987), and Griv *et al.* (2000) have investigated the wave–particle resonances.

Using Eq. (2.3), in terms of the Fourier components defined in Eq. (2.1), $\Gamma = \sum_{m=1}^{\infty} \Gamma_m$, from Eq. (4.1) one finds ($|\omega_*^2| < \kappa^2$)

$$\Gamma_m \approx -\frac{8\pi m^2}{\Omega \Im \omega_*} \aleph_1 \aleph_1^* \int_{r_1}^{r_2} dr \frac{\partial \sigma_0}{\partial r} \approx -8\pi \frac{m^2 \sigma_0}{\Omega \Im \omega_*} |\tilde{\aleph}|^2 e^{2\Im \omega_* t} \quad \text{if} \quad \Im \omega_* > 0, \qquad (4.2)$$

or $\Gamma_m = 0$ if $\Im \omega_* \leqslant 0$. In Eq. (4.2), $\aleph_1^*$ is the complex conjugate potential, and the values of $\aleph_1$, $\aleph_1^*$, σ_0, Ω are evaluated at $r = r_1$. Four physical conclusions can be deduced from Eq. (4.2). *(a)* The distribution of the angular momentum of a disk will change under the action of only the nonaxisymmetric forces $\propto m$. The latter is obvious: axially symmetric motions of a system, studied by Polyachenko, produce no gravitational couplings between the inner parts and the outer parts. *(b)* The distribution of the angular momentum will change with time only under the action of growing, that is, Jeans-unstable perturbations ($\Im \omega_* > 0$). In the opposite limiting case, $\Im \omega_* \to 0$, absorption and emission of angular momentum are confined only to resonate particles (Lynden-Bell & Kalnajs 1972). *(c)* Unstable perturbations can transfer angular momentum only in an inhomogeneous disk ($\partial \sigma_0 / \partial r \neq 0$). And *(d)* $\Gamma_m < 0$: the spiral perturbations remove angular momentum from the disk. This takes place in the main part of the disk between the Lindblad resonances where spiral density waves are self-excited by means of a nonresonant wave–"fluid" interaction. This in turn cannot be done for all masses because the total orbital momentum must remain constant. As a result, the bulk of angular momentum is transferred outward whereas the bulk of mass is correspondingly transported inward; a relatively small group of outer particles with radii $r > r_2$ moves outward, taking almost all of the angular momentum. We speculate that a large portion of the initial mass of the nebula was transported toward the sun about 5×10^9 yr ago by the gravitational torque.†

Finally, let us evaluate the gravitational torque for a realistic model of the protoplanetary disk. In accordance with the theory developed above, the fastest growing spiral mode with $m \gtrsim 1$, $k_* = k_{\mathrm{crit}}$, and $\Im \omega_* \sim \Omega$ is considered. Taking into account that $8\pi m^2 \aleph_1 \aleph_1^* \sim \aleph_0^2$ (an astrophysicist might well consider a perturbation with $\aleph_1/\aleph_0$ of $1/10$ or even $1/3$ to be quite small) and $\aleph_0 \sim r^2 \Omega^2$, where $\aleph_0$ is the basic potential, from Eq. (4.2) one obtains $|\Gamma| \sim \sigma_0 r^4 \Omega^2$. The angular momentum of the disk $\mathcal{L} \sim \sigma_0 r^4 \Omega$. Then the characteristic time of the angular momentum redistribution is $t \sim \mathcal{L}/|\Gamma| \sim \Omega^{-1}$. Thus, already in the first three to four disk revolutions, in, say, about 10^4 yr, the gas–dust protoplanetary disk sees almost all of its angular momentum transferred outward and mass transported inward. With this efficient transport of angular momentum the sun would be able to grow in mass. We conclude that the Jeans instability studied here can give rise to torques that can help to clear the nebula on timescales of $10^5 - 10^6$ yr, in accord with astronomical requirements (Taylor 1992, §2.9.2). Besides, the analysis is found to imply the existence of new *planets* or other Kuiper-type *belts* of comets at mean distances from the sun of $r_{11} \approx 87$ AU, $r_{12} \approx 151$ AU, $r_{13} \approx 261$ AU, $r_{14} \approx 452$ AU, $r_{15} \approx 781$ AU (Mercury, Venus, ..., Asteroid belt, ..., Neptune, Kuiper belt, new planets or other Kuiper-type belts).

---

† Lynden-Bell & Kalnajs (1972, p. 6) have proved that the gravitational torques can only communicate angular momentum outward if the spirals trail. However, in contrast to the present study, Lynden-Bell & Kalnajs considered a model perturbation propagating in a gravitationally stable disk, $\Im \omega = 0$.

5. Discussion

With the paradigm of the standard accretion theory in crisis, the theory of "disk instability," proposed back in the 1950s, is regaining popularity. It postulates the disintegration of a gas-and-dust protosolar nebula under the influence of local gravitational instabilities into massive fragments, which then collapse into planets (and the sun). In this work, we considered the rather neglected in the theory and numerical simulations Safronov–Toomre unstable disks of gas and dust with Toomre's Q-values less than unity. Clearly, more detailed studies, in particular, simulations which include a realistic treatment of the system under study, are needed to definitely distinguish between different mechanisms of the solar system formation. In turn, observations may provide an indication on whether the Q-value for protostellar disks is indeed comparable to or less than unity. (Interestingly, observations have already indicated that the outer regions of accretion disks in both active galactic nuclei and young stellar objects are close to gravitational instability, e.g., Johnson & Gammie 2003).

Even though the analysis presented here shows that there is a dominant nonaxisymmetric Fourier mode of maximum instability with the wavelength λ_{crit}, the number of spiral arms m_{crit}, and the pitch angle ψ_{crit} in the protosolar nebula, at the present we cannot explain these quantities in the *local* WKB version of our theory. In this work we investigate the collective instabilities of the self-gravitating protoplanetary disk on the following fundamental assumption, i.e., the local WKB analysis. This assumption may has essential defects for the wave phenomena and instabilities of self-consistent systems though the physical mechanisms of the instabilities are well clarified (Alexandrov *et al.* 1984). Therefore we have to investigate the effects of nonlocality in the next step (Alexandrov *et al.* 1984, p. 249). In the local WKB approximation it is assumed that the wave vector and the wavefrequency vary continuously. We will show in the next paper of the series by utilizing the more accurate nonlocal WKB approximation that in fact the characteristic oscillation frequencies of an inhomogeneous disk must be "quantized," i.e., must pass through a discrete series of values. According to the WKB method the spectrum of frequencies ω is determined by the quasiclassical rules of Bohr–Sommerfeld quantization:

$$\int_{r_1}^{r_2} k_r(r')\mathrm{d}r' = \left(n + \frac{1}{2}\right)\pi, \tag{5.1}$$

where r_1, r_2 are the "reflection" points (say, the inner and outer Lindblad resonances) and $n = 0, 1, 2, \cdots$. Equation (5.1) implies that ω is independent of r; that is, the spiral-wave front is not distorted by differential rotation. The pattern is composed of one or more modes, and discrete modes are stationary in a rotating frame and hence do not wind up. It seems likely that such an approach will alow us to determine the critical wavelength λ_{crit}, the critical number of spiral arms m_{crit}, and the critical pitch angle ψ_{crit}. The weakly inhomogeneous approximation ($|kL| \gg 1$) used throughout this paper has the meaning that the discrete spectrum will differ little from a continuous spectrum, and in the zero approximation may be regarded as continuous. Further, the nonlocal theory provides a formal basis for the main idea that formed the basis for the local description, namely, that of short-wavelength approximation $|k_r|r \gg 1$.

Acknowledgements

We thank David Eichler, Yury Lyubarsky, Valery Polyachenko, Frank Shu, Raphael Steinitz, and Chi Yuan for valuable discussions. This work was supported in part by the Israel Science Foundation and the Israeli Ministry of Immigrant Absorption.

References

Alexandrov, A.F., Bogdankevich, L.S., & Rukhadze, A.A. 1984, *Principles of Plasma Electrodynamics* (Berlin: Springer-Verlag)

Bertin, G., & Casertano, S. 1982, *Astron. Astrophys.* 106, 274

Boss, A.P. 2002, *Astrophys. J.* 576, 462

Boss, A.P. 2003, *Astrophys. J.* 599, 577

Boss, A.P. 2004, *Astrophys. J.* 610, 456

Boss, A.P., Wetherill, G.W., & Haghighipour, N. 2002, *Icarus* 156, 291

Cameron, A.G.W. 1978, *Moon Planets* 18, 5

Cassen, P.M., Smith, B.F., Miller, R.H., & Reynolds, R.T. 1981, *Icarus* 48, 377

Feigelson, E.D., & Montmerle, T. 1999, *ARA&A* 37, 363

Fukagawa, M., Hayashi, M., Tamura, M., *et al.* 2004, *Astrophys. J.* 605, L53

Goldreich, P., & Tremaine, S. 1980, *Astrophys. J.* 241, 435

Griv, E., & Chiueh, T. 1998, *Astrophys. J.* 503, 186

Griv, E., & Gedalin, M. 2003, *Planet. Space Sci.* 51, 899

Griv, E., & Gedalin, M. 2004, *Astron. J.* 128, 1965

Griv, E., Gedalin, M., & Eichler, D. 2001, *Astrophys. J.* (Letters) 555, L29

Griv, E., Gedalin, M., Eichler, D., & Yuan, C. 2000, *Phys. Rev. Lett.* 84, 4280

Griv, E., Gedalin, M., & Yuan, C. 2002, *Astron. Astrophys.* 383, 338

Griv, E., Gedalin, M., & Yuan, C. 2003, *Mon. Not. R. Astron. Soc.* 342, 1102

Griv, E., Yuan, C., & Gedalin, M. 1999, *Mon. Not. R. Astron. Soc.* 307, 1

Johnson, B.M., & Gammie, C.F. 2003, *Astrophys. J.* 597, 131

Kuiper, G.P. 1951, *Proc. Natl. Acad. Sci.* 37, 1

Kulsrud, R.M., Mark, J.W.-K., & Caruso, A. 1971, *Astrophys. Space Sci.* 14, 52

Larson, R.B. 1989, in: H.A. Weaver & L. Danly (eds.), *The Formation and Evolution of Planetary Systems*, Cambridge: Cambridge Univ. Press, p. 31

Liboff, R.L. 2004, *Astron. J.* 126, 3132

Lin, C.C., & Lau, Y.Y. 1979, *SIAM. Stud. Appl. Math.* 60, 97

Lin, C.C., & Shu, F.H. 1966, *Proc. Natl. Acad. Sci.* 55, 229

Lin, C.C., Yuan, C., & Shu, F.H. 1969, *Astrophys. J.* 155, 721

Luu, J.X., & Jewitt, D.C. 2002, *ARA&A* 40, 63

Lynch, P. 2003, *Mon. Not. R. Astron. Soc.* 341, 1174

Lynden-Bell, D., & Kalnajs, A.J. 1972, *Mon. Not. R. Astron. Soc.* 157, 1

Mayer, L., Quinn, T., Wadsley, J., & Stadel, J. 2002, *Science* 298, 1756

Meyer-Vernet, N., & Sicardy, B. 1987, *Icarus* 69, 157

Morozov, A.G. 1980, *Soviet Astron.* 24, 391

Morozov, A.G. 1981, *Soviet Astron.* 25, 421

Neslušan, L. 2004, *Mon. Not. R. Astron. Soc.* 351, 133

Pollack, J.B., Hubickyj, O., Bodenheimer, P., *et al.* 1996, *Icarus* 124, 62

Polyachenko, V.L., & Fridman, A.M. 1972, *Soviet Astron.* 16, 123

Rice, W.K.M., Armitage, P.J., Bonnell, I.A., *et al.* 2003, *Mon. Not. R. Astron. Soc.* 346, L36

Safronov, V.S. 1960, *Ann. d'Astrophys.* 23, 979

Safronov, V.S. 1972, *Evolution of the Protoplanetary Cloud and Formation of the Earth and Planets* (Jerusalem: Israel Program for Scientific Translations)

Safronov, V.S. 1980, in: D. Lal (ed.), *Early Solar System Processes* (Amsterdam: North-Holland), p. 73

Shu, F.H. 1970, *Astrophys. J.* 60, 99

Taylor, S.R. 1992, *Solar System Evolution.* Cambridge Univ. Press

Toomre, A. 1964, *Astrophys. J.* 139, 1217

Urey, H.C. 1958, *Proc. Chem. Soc. London,* 67 (March)

Wetherill, G.W. 1989, in: H.A. Weaver & L. Danly (eds.), *The Formation and Evolution of Planetary Systems*, Cambridge Univ. Press, p. 1

Dynamics of Populations of Planetary Systems
Proceedings IAU Colloquium No. 197, 2005
Z. Knežević and A. Milani, eds.

© 2005 International Astronomical Union
DOI: 10.1017/S1743921304008567

Behaviour of a two-planetary system on a cosmogonic time-scale

Konstantin V. Kholshevnikov[1] and Eduard D. Kuznetsov[2]

[1]Sobolev Astronomical Institute, St.Petersburg State University, Universitetsky pr., 28,
St.Petersburg, Stary Peterhof, 198504 Russia
email: kvk@astro.spbu.ru

[2]Astronomical Observatory, Urals State University, Lenin pr., 51, Ekaterinburg, 620083 Russia
email: Eduard.Kuznetsov@usu.ru

Abstract. The orbital evolution of planetary systems similar to our Solar one represents one of the most important problems of Celestial Mechanics. In the present work we use Jacobian coordinates, introduce two systems of osculating elements, construct the Hamiltonian expansions in Poisson series for all the elements for the planetary three-body problem (including the problem Sun–Jupiter–Saturn). Further we construct the averaged Hamiltonian by the Hori–Deprit method with accuracy up to second order with respect to the small parameter, the generating function, the change of variables formulae, and the right-hand sides of the averaged equations. The averaged equations for the Sun–Jupiter–Saturn system are integrated numerically over a time span of 10 Gyr. The Liapunov Time turns out to be 14 Myr (Jupiter) and 10 Myr (Saturn).

Keywords. Celestial mechanics, methods: analytical, methods: numerical, planets and satellites, Jupiter, Saturn

---

1. The Hamiltonian of the planetary three-body problem

The orbital evolution of a planetary system similar to our Solar one represents one of the most important problems of Celestial Mechanics. The secular behaviour of the planetary three-body problem has been extensively investigated both from the mathematical and numerical point of view by many researchers. The stability of the spatial planetary three-body problem using KAM theory has been investigated by Robutel (1993a), Robutel (1993b), Laskar & Robutel (1995), Robutel (1995). These results have been obtained for extremely small planetary eccentricities and masses. The motion of the Jupiter–Saturn planetary system near the 5 : 2 mean-motion resonance has been modeled analytically by Michtchenko & Ferraz-Mello (2001) in the framework of the planar general three-body problem.

In this paper, we continue researches of the spatial planetary three-body problem Sun–Jupiter–Saturn begun in Kholshevnikov, Greb, & Kuznetsov (2001), Kholshevnikov, Greb, & Kuznetsov (2002) and Kuznetsov & Kholshevnikov (2004). We use Jacobian coordinates as the best suited. Let us assume $m_0, \mu m_0 m_1, \mu m_0 m_2$ as masses of the Sun, Jupiter and Saturn respectively. The small parameter μ is equal to 10^{-3}. In this case the dimensionless masses m_1 and m_2 are of order unity ($m_1 \approx 1, m_2 \approx 1/3$).

Let us represent the Hamiltonian as a sum of the unperturbed part h_0 and the perturbed one μh_1:

$$h = h_0 + \mu h_1. \tag{1.1}$$

The first term depends on the semi-major axes only

$$h_0 = -\frac{Gm_0m_1}{2a_1} - \frac{Gm_0m_2}{2a_2} \,,$$

G being the gravitational constant. Here and below the subscripts 1 and 2 for coordinates and elements correspond to Jupiter and Saturn respectively. The second term of (1.1) may be thought of as a constant factor having the dimension of velocity squared and a dimensionless part h_2

$$h_1 = \frac{Gm_0}{a_0} h_2 \,, \qquad h_2 = h_3 + h_4 \,, \tag{1.2}$$

where a_0 is an arbitrary parameter with dimension of the length,

$$h_3 = \frac{m_2 a_0}{\mu}\left(\frac{1}{r_2} - \frac{1}{\rho}\right) = \frac{m_2 a_0 \left[2\frac{m_1}{1+\mu m_1}\mathbf{r}_1\mathbf{r}_2 + \mu\left(\frac{m_1}{1+\mu m_1}\right)^2 r_1^2\right]}{r_2\rho(r_2+\rho)} \,,$$

$$h_4 = -\frac{m_1 m_2 a_0}{\Delta} \,, \qquad \rho = \left|\mathbf{r}_2 + \frac{\mu m_1}{1+\mu m_1}\mathbf{r}_1\right| \,, \qquad \Delta = \left|\mathbf{r}_2 - \frac{1}{1+\mu m_1}\mathbf{r}_1\right| \,.$$

2. Systems of osculating elements

Let us introduce two systems of osculating elements. The first system is close to the Keplerian one

$$x^{(1)}_{3s-2} = \widetilde{a}_s, \quad x^{(1)}_{3s-1} = e_s, \quad x^{(1)}_{3s} = \widetilde{I}_s \,, \quad y^{(1)}_{3s-2} = \alpha_s, \quad y^{(1)}_{3s-1} = \beta_s, \quad y^{(1)}_{3s} = \gamma_s. \tag{2.1}$$

Here $\widetilde{a} = (a - a^0)/a^0$, $\widetilde{I} = \sin(I/2)$, $\alpha = l + g + \Omega$, $\beta = g + \Omega$, $\gamma = \Omega$ are expressed in terms of Keplerian elements a, a^0, e, I, l, g, Ω: semi-major axis and its mean value, eccentricity, inclination, mean anomaly, argument of pericenter, longitude of ascending node. The index s changes from 1 to the number of planets $N = 2$.

The second system realizes simplifications due to the homogeneity of the perturbation function with respect to the semi-major axes. In this system the denominators arising in a process of averaging transforms are extremely simple. On the other hand, it introduces a complication, by mixing the orbital elements of different planets

$$x^{(2)}_{3s-2} = z_s, \quad x^{(2)}_{3s-1} = e_s, \quad x^{(2)}_{3s} = \widetilde{I}_s \,, \quad y^{(2)}_{3s-2} = \alpha_s, \quad y^{(2)}_{3s-1} = \beta_s, \quad y^{(2)}_{3s} = \gamma_s \,, \tag{2.2}$$

where $z_1 = \omega_1^0/\omega_1 - 1$, $z_2 = (\omega_1^0\omega_2)/(\omega_2^0\omega_1) - 1$. Here $\omega_s = \kappa_s a_s^{-3/2}$ are mean motions of the planets, ω_s^0 are constants close to mean values ω_s, $\kappa_s^2 = Gm_0m_s/M_s$ are gravitational parameters of the planets, reduced masses are $M_s = m_s(1 + \mu m_1 + \cdots + \mu m_{s-1})/(1 + \mu m_1 + \cdots + \mu m_s)$, $s = 1, 2$.

3. Expansion of disturbing Hamiltonian into Poisson series

The disturbing Hamiltonian h_2 is presented as Poisson series

$$h_2 = \sum A_{kn}x^k \cos ny. \tag{3.1}$$

Here $x = \{x_1, \ldots, x_6\}$ are action-like elements, $y = \{y_1, \ldots, y_6\}$ are angular ones, A_{kn} are numerical coefficients, $k = \{k_1, \ldots, k_6\}$ and $n = \{n_1, \ldots, n_6\}$ are multi-indices. The summation is taken over non-negative k_s and n_1 and integer $n_2, \ldots, n_6$.

It is well known (see for example Charlier 1927, Subbotin 1968) that

$$n_1 + n_2 + n_3 + n_4 + n_5 + n_6 = 0\,, \qquad n_3 + n_6 = \text{even}\,,$$

$$k_s = |n_s| + \text{non-negative even} \quad (s = 2, 3, 5, 6) \tag{3.2}$$

The simplest restrictions on k and n are

$$k_1 + k_2 + \cdots + k_6 \leqslant d\,, \qquad n_1 \leqslant c\,, \qquad |n_4| \leqslant c. \tag{3.3}$$

The other n_s are less than d according to D'Alembertian properties of the Hamiltonian h_2 (Kholshevnikov 1997, 2001).

Designate b the order of approximation with respect to μ. To evaluate $d(b)$ it is sufficient to take into account that Jupiter's and Saturn's eccentricities $e_1 = 0.05$, $e_2 = 0.05$ and the sines of the half-angles of inclinations $I_1 = 0.01$, $I_2 = 0.02$ are small values of the same order $\mu_1 \sim \mu^{1/2}$. The choice of $c(b)$ is controlled by the rate of convergence at $\mu_1 = 0$. According to Kholshevnikov, Greb, & Kuznetsov (2002):

$$d(2) = 6\,, \quad d(3) = 11\,, \quad d(4) = 16\,, \qquad c(2) = 13\,, \quad c(3) = 25\,, \quad c(4) = 37. \tag{3.4}$$

The parameters a_0, m_1, m_2, a_1^0, a_2^0 (necessary to calculate the coefficients A_{kn} and the multiplier to convert the Hamiltonian h_1) are given in Kholshevnikov, Greb, & Kuznetsov (2002). In this paper it is shown also that the expansions up to μ^4 has only one small divisor $2\omega_1^0 - 5\omega_2^0$. The constant $F = |2\omega_1^0 - 5\omega_2^0|/\omega_1^0$ describing commensurability degree is equal to 0.023331 for the chosen values of the parameters: this indicates that a weak resonance $F \sim \sqrt{\mu}$ is present.

The Poisson series processor PSP (Brumberg 1995, Ivanova 1996) is used to construct the expansion of disturbing Hamiltonian h_2 into Poisson series (3.1). The rational version of the PSP is used to decrease round-off errors during calculations of the coefficients A_{kn}. The expansion of the disturbing Hamiltonian is processed up to μ^2. The summation is taken over $k_1 + \cdots + k_6 \leqslant 6$, $|n_s| \leqslant 15$ $(s = 1, \ldots, 6)$. For each of the osculating elements system two variants of the expansion are constructed. The first variant deals with numerical values of parameters (masses, mean values of semi-major axes, $\ldots$) corresponding to the Sun–Jupiter–Saturn system. The second one deals with the litteral expressions depending upon the parameters of the system.

4. Averaged planetary three-body problem

The Hori–Deprit method (Lie transformation method) is used to construct the averaged Hamiltonian H. This method is based on Poisson brackets that allow us to use non-canonical elements, by writing down the Poisson brackets in the corresponding system of phase variables (Kholshevnikov & Greb 2001).

For a stationary change of variables without mixing impulses $p = (p_1, \ldots, p_6)$ and coordinates $q = (q_1, \ldots, q_6)$ the Poisson bracket is written down through partial brackets (Kholshevnikov & Greb 2001) as

$$\{f, g\} = V_{jk}(f, g)_{jk}\,, \qquad V_{jk} = \frac{\partial x_j}{\partial p_i}\frac{\partial y_k}{\partial q_i}\,, \qquad (f, g)_{jk} = \frac{\partial f}{\partial x_j}\frac{\partial g}{\partial y_k} - \frac{\partial f}{\partial y_k}\frac{\partial g}{\partial x_j}.$$

Summation is made by repeating indices from 1 to 6.

The matrices $\mathcal{V}$ with elements V_{jk} for both the first and the second systems of elements (2.1), (2.2) was obtained in Kholshevnikov & Greb 2001, Greb (2002). For the first system (2.1) the Poisson matrix is block diagonal. For the second one (2.2) it is triangular.

The Hamiltonian h (1.1) is averaged over the fast variables α_1 and α_2. The averaged Hamiltonian H is represented by a power series in the small parameter μ. The Hamiltonian H and the generating function are echeloned Poisson series (Rom 1971).

For calculations we use the rational version of the echeloned Poisson series processor EPSP (Ivanova 2001) to reduce the round-off errors. Transformations are made for both systems of elements with numerical parameters corresponding to the Sun–Jupiter–Saturn system.

The Hamiltonian H is represented by an echeloned Poisson series upto μ^2. Two approximations of Hori–Deprit method are made for the first system of elements (2.1). For the second system (2.2) only the first approximation is realized. The generating function, the change of variables formulae between averaged and osculating elements, and the right-hand sides of the averaged equations of motion are thus obtained.

5. Behaviour of the Sun–Jupiter–Saturn system

The averaged equations are integrated numerically over a time span of 10 Gyr. The equations for slow variables are integrated by 15 order Everhart and 11 order Runge-Kutta methods. The equations for fast variables are integrated by spline interpolation method.

The accuracy of the integration is controlled by computation of the integrals of energy and area. The absolut value of the relative error for the energy integral calculation is less than $5.2 \cdot 10^{-13}$ for both integrators over a time span of 10 Gyr. The mean value of the relative error is a constant over all integration time.

To calculate the area integrals the properties of the form conservation in Jacobian coordinates (Charlier 1927) and under averaging transform (Kholshevnikov 1991) are used. The area integrals are determined with respect to the Laplace plane. In this case the vector $\sigma = (\sigma_x, \sigma_y, \sigma_z)$ is directed along the z-axis. Hence $\sigma_x = \sigma_y = 0$. The absolute value of the relative error for the σ_z area integral calculation is $3.5 \cdot 10^{-10}$ for both integrators over a time span of 10 Gyr. The amplitude and mean value of the relative error are constant over all integration time. We find out that the area integrals σ_x, σ_y are preserved with low accuracy: $|\sigma_x/\sigma_{z0}| < 8.2 \cdot 10^{-7}$, $|\sigma_y/\sigma_{z0}| < 5.2 \cdot 10^{-7}$. The non-conservation of the σ_x and σ_y integrals has the following reason. As it is proved in Kholshevnikov (1991), they are preserved in the system determined by the averaged Hamiltonian H. But *they are not preserved in a system determined by a finite sequence of Poisson expansion* of the averaged Hamiltonian H taking into account the restrictions (3.3), (3.4) on k and n.

Table 1 presents the low and upper limits, mean values and amplitudes of oscillations for the averaged eccentricities and inclinations obtained from the numerical integration of the averaged equations for the first and second approximations over 10 Gyr. The results obtained by the two integrators show a good agreement. The relative differences between the first and second approximations are given in the two last columns of the table 1. They are more than the small parameter $\mu = 1 \cdot 10^{-3}$, but they are generally less than the quotient $\mu/F = 4.4 \cdot 10^{-2}$. The only exception is Jupiter's eccentricity amplitude e_a for Runge–Kutta method.

The evolution of the ascending node longitudes γ of Jupiter and Saturn orbits depends upon the choice of the reference plane. In the first approximation the evolution of the ascending nodes longitudes with respect to the ecliptic plane turns out to be in libration with amplitudes $12.9°$ and $32.8°$ for Jupiter and Saturn respectively. The libration amplitudes are in agreement with those obtained by Smart (1953). The evolution of the ascending nodes longitudes with respect to the Laplace plane is secular. On the Laplace

Table 1. Range of the eccentricity and inclination changes for Jupiter and Saturn orbits on time interval 10 Gyr.

| | First approximation | | Second approximation | | Relative difference | |
	15 order Everhart method	11 order Runge–Kutta method	15 order Everhart method	11 order Runge–Kutta method	15 order Everhart method	11 order Runge–Kutta method
	Jupiter					
e_{min}	0.0184	0.0187	0.0171	0.0171		
e_{max}	0.0510	0.0508	0.0511	0.0511		
e_{mean}	0.0347	0.0348	0.0341	0.0341	$1.8 \cdot 10^{-2}$	$2.1 \cdot 10^{-2}$
e_a	0.0163	0.0161	0.0170	0.0170	$4.1 \cdot 10^{-2}$	$5.3 \cdot 10^{-2}$
i_{min}	1.2657	1.2698	1.2705	1.2679		
i_{max}	2.0011	2.0002	2.0005	2.0006		
i_{mean}	1.6334	1.6350	1.6355	1.6342	$1.3 \cdot 10^{-3}$	$4.9 \cdot 10^{-4}$
i_a	0.3677	0.3652	0.3650	0.3664	$7.4 \cdot 10^{-3}$	$3.3 \cdot 10^{-3}$
	Saturn					
e_{min}	0.0212	0.0212	0.0194	0.0194		
e_{max}	0.0772	0.0771	0.0780	0.0780		
e_{mean}	0.0492	0.0492	0.0487	0.0487	$1.0 \cdot 10^{-2}$	$1.0 \cdot 10^{-2}$
e_a	0.0280	0.0280	0.0293	0.0293	$4.4 \cdot 10^{-2}$	$4.4 \cdot 10^{-2}$
i_{min}	0.7240	0.7346	0.7344	0.7340		
i_{max}	2.5366	2.5319	2.5317	2.5349		
i_{mean}	1.6303	1.6333	1.6331	1.6344	$1.7 \cdot 10^{-3}$	$6.7 \cdot 10^{-4}$
i_a	0.9063	0.8987	0.8987	0.9005	$8.5 \cdot 10^{-3}$	$2.0 \cdot 10^{-3}$

plane the difference between the ascending nodes longitudes of Jupiter and Saturn orbits is equal to $180°$ exactly. This property is used as a test of the integration accuracy.

In the second approximation the evolution of the ascending nodes longitudes with respect to the ecliptic plane turn out to be a large amplitude oscillations with a slow secular motion. The mean values over one period of the ascending node longitudes decrease by $6°$ per Gyr.

The evolution of the pericentre longitudes β of Jupiter and Saturn orbits turns out to be secular for both approximations.

Table 2 gives estimations of the mean squared norm

$$||f||_2 = \left(\frac{1}{N} \sum_{j=1}^{N} (f_j)^2 \right)^{1/2}$$

and the uniform norm

$$||f||_\infty = \max_{j=1,\dots,N} |f_j|$$

of the variable change functions describing, for each element, the short periodic perturbations, i.e. the differences between mean and osculating elements. Here N is the number of the orbital elements values on the time interval.

The short-period perturbations of Jupiter and Saturn semi-major axes a do not exceed 0.0022 a.u. and 0.0121 a.u. respectively. The tables 1 and 2 comparison shows that the maximum values of the short-period perturbation norms for the eccentricities e and inclinations i are much less than the amplitudes of the corresponding long-period perturbations. The variable change function norms for the longitudes α, β and γ are much less than the amplitudes of the long-period perturbations also.

Table 2. Mean squared and uniform norms of the variable change functions.

Planet	Orbital elements					
	a, a.u.	e	i, degrees	α, degrees	β, degrees	γ, degrees
Mean squared norm						
Jupiter	0.0006	0.0005	0.0005	0.0113	0.9738	0.0182
Saturn	0.0038	0.0012	0.0013	0.0402	1.7033	0.0522
Uniform norm						
Jupiter	0.0022	0.0011	0.0013	0.0304	3.1538	0.0616
Saturn	0.0121	0.0027	0.0032	0.1127	7.6316	0.2603

Estimates of the Liapunov Exponents for the Sun–Jupiter–Saturn system have been obtained. The corresponding Liapunov Time turns out to be 14 Myr (for Jupiter) and 10 Myr (for Saturn).

6. Conclusion

A numerical integration of the averaged equations for the Jupiter–Saturn system shows that the motion has a quasiperiodic character over the time-scale of 10 Gyr. The eccentricities and inclinations of Jupiter and Saturn orbits remain small and their values are separated from zero. The lines of nodes and apsides have a secular motion. The short-periodic perturbations remain small during the considered time interval.

Acknowledgements

This work was partly supported by the RFBR, Grant 02-02-17516, and the Leading Scientific School, Grant NSh-1078.2003.02.

References

Brumberg, V. A. 1995, *Analytical Techniques of Celestial Mechanics*. (Springer, Heidelberg)

Charlier, C. L. 1927, *Die Mechanik des Himmels*. (Walter de Gruyter & Co., Berlin und Leipzig)

Greb, A. V. 2002, PhD thesis, St. Petersburg State University, St. Petersburg

Ivanova, T. V. 1996, in: S.Ferraz-Mello, B.Morando and J.-E.Arlot (eds.), *Dynamics, Ephemerides and Astrometry of the Solar System*, IAU Symp. 172, (Kluwer Academic Publishers), p. 283

Ivanova, T. 2001, *Cel. Mech. Dyn. Astron.*, 80, 167

Kholshevnikov, K. V. 1991, *Astron. Zh.*, 68, 660

Kholshevnikov, K. V. 1997, *Astron. Rep.* 41, 135

Kholshevnikov, K. V. 2001 *Astron. Rep.*, 45, 577

Kholshevnikov, K. V., Greb, A. V. and Kuznetsov, E. D. 2001, *Solar System Res.*, 35, 243

Kholshevnikov, K. V. and Greb, A. V. 2001, *Solar System Res.*, 35, 415

Kholshevnikov, K. V., Greb, A. V. and Kuznetsov, E. D. 2002, *Solar System Res.*, 36, 68

Kuznetsov, E. D. and Kholshevnikov, K. V. 2004, *Solar System Res.*, 38, 147

Laskar, J. and Robutel, P. 1995, *Cel. Mech. Dyn. Astron.*, 62, 193

Michtchenko, T. A. and Ferraz-Mello, S. 2001, *Icarus*, 149, 357

Robutel, P. 1993a, *Cel. Mech. Dyn. Astron.*, 56, 197

Robutel, P. 1993b, *Cel. Mech. Dyn. Astron.*, 57, 97

Robutel, P. 1995, *Cel. Mech. Dyn. Astron.*, 62, 219

Rom A. 1971, *Celest. Mech.*, 3, 331

Smart, W. M. 1953, *Celestial mechanics*. (Longmans, Green and Co., London, New York, Toronto)

Subbotin, M.F. 1968, *Introduction to theoretical astronomy*. (Nauka, Moscow)

Dynamics of Populations of Planetary Systems
Proceedings IAU Colloquium No. 197, 2005
Z. Knežević and A. Milani, eds.

© 2005 International Astronomical Union
DOI: 10.1017/S1743921304008579

Boundaries of the habitable zone: unifying dynamics, astrophysics, and astrobiology

Milan M. Ćirković

Astronomical Observatory Belgrade,
Volgina 7, 11160 Belgrade-74, Serbia and Montenegro
email: mcirkovic@aob.aob.bg.ac.yu

Department of Physics, University of Novi Sad,
Trg Dositeja Obradovića 4, 21000 Novi Sad,
Serbia and Montenegro

Abstract. We are witnessing tremendous progress in the nascent multidisciplinary field of astrobiology, encompassing the origin and evolution of life in the cosmic context. One of the key concepts recently introduced in this field is the Galactic Habitable Zone (GHZ): an interval of galactocentric distances convenient for formation of stars possessing habitable planets. The boundaries of the GHZ are still poorly understood, however. Here we present a comparative analysis of various proposals for the mechanisms determining the GHZ boundaries, as well as different numerical values obtained. When joined with the models of Galactic stellar distribution, this gives us a better handle on the number of potential life-bearing sites.

Keywords. Astrobiology, Galaxy: structure, Galaxy: kinematics and dynamics, extraterrestrial intelligence

1. Introduction

We are lucky enough to live in an epoch of great progress in the nascent discipline of astrobiology, which deals with the three canonical questions: How does life begin and develop? Does life exist elsewhere in the universe? What is the future of life on Earth and in space? A host of important discoveries have been made during the last decade or so, the most important certainly being a large number of extrasolar planets, but also the existence of many extremophile organisms possibly comprising "deep hot biosphere" of the late Thomas Gold; the discovery of subsurface water on Mars and the huge ocean on Europa, and possibly also Ganymede and Callisto; the unequivocal discovery of amino-acids and other complex organic compounds in meteorites; modelling organic chemistry in Titan's atmosphere; the quantitative treatment of the Galactic habitable zone; the development of a new generation of panspermia theories, spurred by experimental verification that even terrestrial microorganisms easily survive conditions of an asteroidal or a cometary impact; progress in methodology of SETI studies, etc. (for recent beautiful reviews see Des Marais and Walter 1999; Darling 2001; Grinspoon 2003).

One of the most important and fruitful concepts recently introduced in astrobiology is Galactic habitable zone (henceforth GHZ). Proposed by Gonzalez, Brownlee and Ward (2001), it represents an annular ring comprising the potential sites for genesis of complex metazoans in the Milky Way. It is easy to see intuitively why is that so: locations at either large or very small galactocentric distances will be inhospitable to life as we know it. At very large galactocentric distance, for example, the metallicity of too low, and terrestrial planets will be very difficult to form (not to mention suppression of the complicated chemical evolution necessary for supporting biochemistry). Close to the Galactic Center, the number of close stellar encounters, as well as threats of supernovae or even nuclear outbursts are likely to decrease habitability.

But the exact boundaries of GHZ remain somewhat mysterious: different authors suggested not only different particular values, but also different criteria for the zone boundary. It is the purpose of this contribution to try to sort out the confusion and evaluate different estimates. We shall also briefly discuss the issue of relevance of the GHZ concept for the SETI projects.

2. Outer boundary

Outer boundary is, rather incontrovertibly, set by the metallicity gradient of the Milky Way disk (Hou, Prantzos, and Boissier 2000; Tadross 2003). The original study of Gonzalez *et al.* (2001) was not very quantitative on this issue, but it is quite easy to see how a reliable estimate can be achieved using the appropriate scaling relationship.

According to most observations, $\nabla Z = 0.09$ dex kpc^{-1} (e.g. Tadross 2003). In order to determine the outer boundary, we need an assumption on the mass scaling of terrestrial planets, as well as the minimal mass of the terrestrial planet viable from the astrobiological point of view. The former is suggested by Gonzalez *et al.* (2001) to be the scaling (we assume that [Fe/H] represents overall metallicity sufficiently well). The latter is a topic of considerable debate in astrobiological circles, but we can take as a prototype often expressed opinion that an object of the size of Mars ($M_{\mathrm{min}} \approx 0.1\,M_{\oplus}$) is the minimum, for several reasons (development of plate tectonics, retention of O_2 and O_3, etc.). In principle, we shall have

$$R_{\mathrm{out}} = R_{\odot} - \frac{1}{\nabla Z} \log\left(1 + \frac{2}{3}\log\frac{M_{\mathrm{min}}}{M_{\oplus}}\right), \qquad (2.1)$$

where $R_{\odot} = 8.5$ kpc is the Solar galactocentric distance. In this approximation, we obtain $R_{\mathrm{out}} = 13.5$ kpc, which is significantly less than the Holmberg radius of the Milky Way.

Of course, since the overall metallicity grows in time, the net effect on the outer boundary is to move slowly outward, which is fairly uncontroversial (compare §3 below): $\frac{dR_{\mathrm{out}}}{dt} > 0$. Thus, regions of the Galaxy devoid of habitable planets will slowly become habitable over time, and this process can, in principle, continue until the asymptotic maximal metallicity is reached in the physical eschatological context (Adams and Laughlin 1997). However, due to the exponential stellar distribution, the net gain in number of habitable sites due to the expansion of the GHZ outer boundary will become less and less important as time passes.

3. Inner boundary

The inner boundary of GHZ is subject to controversy, since at least three possible mechanisms for its determination have been proposed: 1. dynamical instability of planetary orbits under perturbations from stellar encounters, 2. high radiation environment due to frequent SNe and gamma-ray bursts, and 3. cosmogonical problems linked with high metallicity and UV-induced evaporation of protoplanetary material. The first two are obviously contingent mechanisms (i.e. those whose operation affect habitability in a random manner), while the third one can be considered deterministic. The first two has been proposed in a qualitative manner by Gonzalez *et al.* (2001), and subsequently quantified by Lineweaver *et al.* (2004). The third has been proposed by Pena-Cabrera and Durand-Manterola (2004); they have considered only suppression of terrestrial planets due to very high Z. Photoevaporation of protoplanetary disks have been considered in a different context by Adams *et al.* (2004); its application to GHZ remains to be done.

Clearly, the true GHZ inner boundary will be the value driven by whichever mechanism is dominant at particular epoch: $R_{\mathrm{inn}} = \max[R_{\mathrm{rad}}, R_{\mathrm{dyn}}, R_{\mathrm{cosm}}]$. For example, the dynamical radius can be estimated in a following crude manner. The rate of stellar encounters is $\Gamma = \langle nv\sigma \rangle \approx \langle n \rangle \langle v \rangle \langle \sigma \rangle$, where n, v, and σ are stellar density, velocity and encounter cross-section respectively. We take the distribution of stars to be exponential, with the disk lengthscale of $R_d = 3$ kpc (Drimmer and Spergel 2001). Let us define the critical value of Γ as $\Gamma_{\mathrm{crit}} = 1$ (age of the Earth)$^{-1}$.

According to most of recent calculations and simulations, an encounter disruptive enough to destroy terrestrial planets either directly or indirectly (through perturbing orbits of outer planets sufficiently to create high eccentricities driving the habitable planets into their sun) occurs with an average cross-section of $\langle \sigma \rangle = 100$ AU^2 (Laughlin and Adams 2000). At the Solar circle, the average relative stellar velocity is $\langle v_0 \rangle \simeq 40$ km s^{-1}; however, in the interior of the Galactic bulge, where the inner boundary is located, we should use larger value, say $\langle v \rangle \simeq 122$ km s^{-1} suggested by Freeman *et al.* (1988). In this approximation, we obtain

$$R_{\mathrm{inn}} = R_{\odot} - R_d \ln \frac{\Gamma_{\mathrm{crit}}}{\langle n_0 \rangle \langle v \rangle \langle \sigma \rangle}. \qquad (3.1)$$

Inserting numerical values, we obtain negative values for R_{inn}, which indicates that the dynamical inner limit is, in fact, irrelevant, if not non-existent. Even using the most radical assumption that $\langle \sigma \rangle = \pi(30 \text{ AU})^2$ we obtain that for purely dynamical constraints $R_{\mathrm{inn}} = 1.3$ kpc. Calculations of Lineweaver, Fenner, and Gibson (2004)—which greatly improve on the previous study of Lineweaver (2001)—indicate that an inner bound due to probability of supernovae exploding too close and too often to life-bearing planetary systems is much more conservative (see Fig. 2). Their study indicate that the probability of a planet being undisturbed for sufficient amount of time to evolve complex life is rather small within an inner boundary of about 3 kpc (today). If that is the case, the conclusions of this section are strongly reinforced, since the Galactic disk is still farther from being uniform in the astrobiological sense. (Although, of course, the exact value depends on still poorly understood effects of close supernovae on planetary environments.) The same general conclusion follows from the estimate of Pena-Cabrera and Durand-Manterola (2004), who find $R_{\mathrm{inn}} = 4$ kpc, on the basis of the assumption that metal-rich planets in the inner Galaxy would grow too fast to be inhabitable; this assumption, however, is highly questionable since another kind of habitable planets could plausibly form under high-Z conditions (see, e.g. Léger *et al.* 2004). Additional effect not taken into account by Pena-Cabrera and Durand-Manterola (2004) is that in the currently preferred theories of planet formation, protoplanet growth rate scales as the stellar mass. This can lead to an astrobiologically interesting effect, namely that in metal-rich systems, the preferred planet-forming systems (for terrestrial planets) become those with lower mass. Since it has been argued that planets around K and M dwarfs can indeed be habitable (e.g. Heath *et al.* 1999), this indicates the breakdown of this criterion for the inner GHZ boundary.

As far as the evolutionary changes in the inner boundary with time go, the confusion is compounded by different trends exhibited by candidates for the dominant mechanism. Since the overall metallicity increases with time, the maximal mass for a terrestrial plan*et* *also* increases, thus yielding $\frac{dR_{\mathrm{cosm}}}{dt} > 0$. However, the decrease in SNe frequency observable in external galaxies (and following predicted star formation history), will lead to the $\frac{dR_{\mathrm{rad}}}{dt} < 0$ behavior, while on the timescales smaller than physical eschatological ones, we expect to encounter $\frac{dR_{\mathrm{dyn}}}{dt} \approx 0$. Thus, the direction of the temporal evolution

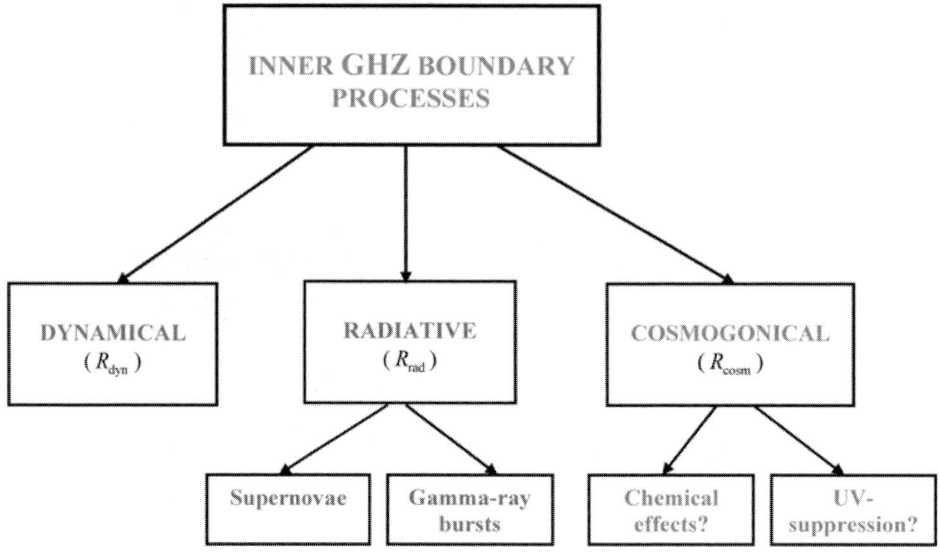

Figure 1. A scheme of various physical mechanisms determining the inner boundary of GHZ. Contingent mechanisms are listed in blue and deterministic in red color. Note that both cosmogonical mechanisms are still quite controversial, at least when applied to GHZ stratification.

of the inner GHZ boundary is not clear at present. In future research, it will be important to explicate the assumption of "biological invariance": that intrinsic probability of emergence/complexification of life at life-friendly sites does not vary with time.

4. An interlude: GHZ and SETI studies

There are at least two conclusions following from the studies of GHZ so far which are relevant to SETI studies. First and foremost, the 4-D structure of GHZ is such that it comprises a large number of Sun-like stars and habitable planets much older than the Solar System. According to Lineweaver (2001), Earth-like planets around other stars in the Galactic habitable zone are, on average, 1.8 ± 0.9 Gyr older than our planet, and mostly located at somewhat smaller galactocentric distance than the Sun. Applying the Copernican assumption naively, we would expect that correspondingly complex lifeforms on those others to be on the average 1.8 Gyr older.† This makes Fermi's paradox (for a review, see Webb 2002) more serious than hitherto thought. Obviously, this spells failure for any meaningful SETI-type communication; instead, the attention should be directed toward the youngest members of the GHZ-family, notably the outer rim of it.

The second issue of some practical importance in SETI targeting is the size of the volume to be search in radio eavesdropping and similar projects (e.g. Tarter 2001). Suppose that somehow (using a crystal ball or Drake equation) we ascertain the number of

† Strictly speaking, we need to take into account the width of the age distribution as well; it is conceivable, for instance, that most habitats in the Galaxy last shorter than the required biological timescale, Earth being exceptionally long-lived compared to the average (non-Copernican!). This leads us to the so-called Carter argument, very interesting topic in astrobiology and SETI studies, which is beyond the scope of the present paper.

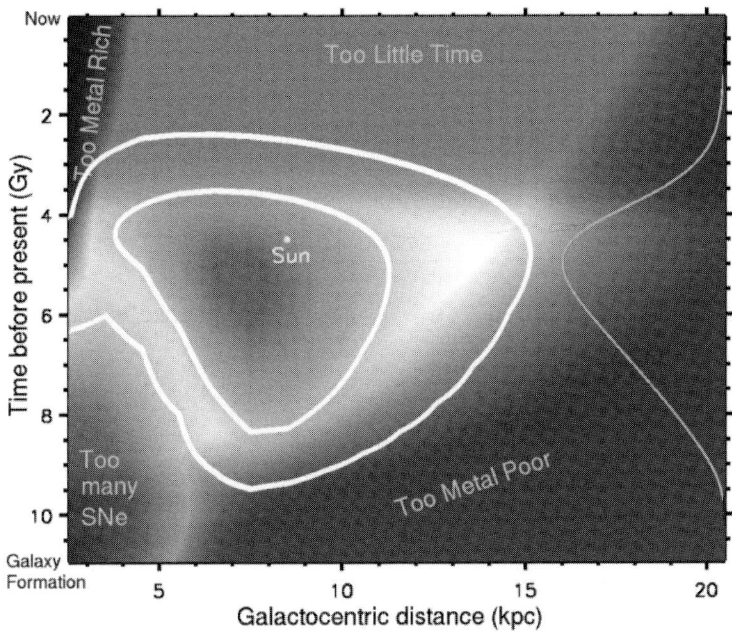

Figure 2. GHZ according to Lineweaver *et al.* (2004) in the disk of the Milky Way based on the star formation rate, metallicity, sufficient time for evolution, and freedom from life-extinguishing supernova explosions. The white contours encompass 68% (inner) and 95% (outer) of the origins of stars with the highest potential to be harboring complex life today (reproduced with kind permission of Prof. Charles Lineweaver).

civilizations in the "communication window" in the Galaxy as N_c (Duric and Field 2003). Efficiency of any particular SETI project depends on the density of prospective targets (i.e. inhabited sites). In the first approximation, then, taking into account GHZ acts to increase the real density of possible SETI targets by a factor $V_{MW}/V_{GHZ} \approx 5$. However, since (although it is not entirely clear from the derivation for Duric and Field 2003 study) N_c is perhaps derived from an assumption about the total stellar census of the Milky Way, the reduction of the area of interest to GHZ will have smaller effect. It will not entirely cancel the change in volume, since the distribution of stars is inhomogeneous, notably decreasing exponentially with the galactocentric radius. Roughly speaking, the number of stars within radius R scales as $N \propto R^2 \exp(-R/R_d)$, with maximum at $R \sim 6$ kpc, within GHZ. Thus, the average stellar density inside GHZ is—for the same normalization constants and the same value of the disk radius—larger than the average Galactic stellar density by the factor of $\langle n \rangle_{GHZ}/\langle n \rangle_{MW} \approx 1.4$. In future, we may expect significant improvement in precision of this type of estimate.

5. Prospects

Clearly, much work remains to be done in order to pin-point the extent of GHZ. In particular, research on the influence of close supernovae and local GRBs on Earth-like biospheres or prebiotic environments seems a promising way to go. But the purely dynamical constraints, especially in the long-term view, including the secular evolution of the Galactic gravitational potential, are no less important, particularly at the heuristical level. A unified modelling of dynamical, astrochemical and astrophysical constraints, both deterministic and contingent ones, is likely to be within reach of state-of-the-art astrobiological research within a decade. Of course, this should not preclude further research toward limitations of the concept of the habitable zone itself. Here, the stellar case again serves as a useful prototype: astrobiologists are investigating mechanisms acting to define the boundaries of the circumstellar habitable zones (like the carbon-silicate cycle, etc.), while keeping an eye open on the possibility of life outside of it (for instance, life in subglacial oceans on Europa, etc.).

Acknowledgements

The author acknowledges the useful comments of a referee, helping to significantly improve the final version of the manuscript. It is a pleasure to thank Nick Bostrom, Karl Schroeder, Robert J. Bradbury, and Srdjan Samurović for many helpful discussions. This manuscript has benefited from the technical help of Vesna Milošević-Zdjelar, Olga Latinović, Nick Bostrom, Nikola Milutinović, and Milan Bogosavljević. Ministry of Science and Environment of the Republic of Serbia supports the author through Projects #1196 "Astrophysical Spectroscopy of Extragalactic Objects" and #1468 "Structure and Kinematics of the Milky Way". Charles Lineweaver is acknowledged for kindly allowing reproducing of Figure 2.

References

Adams, F. C. and Laughlin, G. 1997, *Rev. Mod. Phys.* 69, 337

Adams, F. C., Hollenbach, D., Laughlin, G., and Gorti, U. 2004, *ApJ* 611, 360

Ćirković, M. M. 2004, *J. Br. Interplan. Soc.* 57, 53

Darling, D. 2001, *Life Everywhere* (Basic Books, New York)

Des Marais, D. J. and Walter, M. R. 1999, *Annu. Rev. Ecol. Syst.* 30, 397

Drimmer, R. and Spergel, D. N. 2001 *ApJ* 556, 181

Duric, N. and Field, L. 2003, *Serb. Astron. J.* 167, 1

Freeman, K. C., de Vaucouleurs, G., Wainscoat, R. J., and de Vaucouleurs, A. 1988, *ApJ* 325, 563

Gonzalez, G., Brownlee, D., and Ward, P. 2001, *Icarus* 152, 185

Grinspoon, D. 2003, *Lonely Planets: The Natural Philosophy of Alien Life* (HarperCollins, New York)

Heath, M. J., Doyle, L. R., Joshi, M. M., and Haberle, R. M. 1999, *Orig. Life Evol. Biosph.* 29, 405

Hou, J. L., Prantzos, N., and Boissier, S. 2000, *Astron. Astrophys.* 362, 921

Irion, R. 2004, *Science* 303, 27

Laughlin, G. and Adams, F. C. 2000, *Icarus* 145, 614

Léger, A., Selsis, F., Sotin, C., Guillot, T., Despois, D., Mawet, D., Ollivier, M., Labéque, A., Valette, C., Brachet, F., Chazelas, B., and Lammer, H. 2004, *Icarus* 169, 499

Lineweaver, C. H. 2001, *Icarus* 151, 307

Lineweaver, C. H., Fenner, Y., and Gibson, B. K. 2004, *Science* 303, 59

Pena-Cabrera, G. V. Y., and Durand-Manterola, H. J. 2004, *Adv. Space Res.* 33, 114

Scalo, J. and Wheeler, J. C. 2002, *ApJ* 566, 723

Tadross, A. L. 2003, *New Ast.* 8, 737

Tarter, J. 2001, *Annu. Rev. Astron. Astrophys.* 39, 511

Ward, P. D. and Brownlee, D. 2000, *Rare Earth: Why Complex Life Is Uncommon in the Universe* (Springer, New York)

Webb, S. 2002, *Where is Everybody? Fifty Solutions to the Fermi's Paradox* (Copernicus, New York)

Part 2

ASTEROID FAMILIES AND STABILITY

Dynamics of Populations of Planetary Systems
Proceedings IAU Colloquium No. 197, 2005
Z. Knežević and A. Milani, eds.

© 2005 International Astronomical Union
DOI: 10.1017/S1743921304008580

Asteroid proper elements: recent computational progress

Fernando Roig[1] and Cristian Beaugé[2]

[1]Observatório Nacional, Rio de Janeiro, Brazil
email: froig@on.br

[2]Observatorio Astronómico Córdoba, Córdoba, Argentina
email: beauge@oac.uncor.edu

Abstract. In this work, we review the analytical and semi-analytical tools introduced to deal with resonant proper elements and their applications to the Trojan asteroids, the numerical computation of synthetic proper elements for resonant and non resonant asteroids, and the introduction of proper elements for planet crossing asteroids. We discuss the applications and accuracy of these methods and present some comparisons between them.

Keywords. Minor planets, asteroids, proper elements, perturbation theory

1. Introduction

Proper elements play a major role in the characterization of the long term stability of asteroid orbits, as well as in the identification and definition of asteroidal families. In recent years, the development of new analytical and numerical tools have allowed to extend the computation of proper elements to huge sets of asteroid orbits including main belt, resonant and planet crossing asteroids. This has had a deep impact on our knowledge of the dynamical structure of the asteroid belt, and on the dynamical and collisional processes taking place there.

Among these new tools, the semi-analytical model introduced by Beaugé and Roig (2001) to deal with resonant proper elements represents a major advance in the field. Their method allowed to determine the proper elements of the Trojan asteroids and to confirm the existence of families among these bodies. A purely analytical method to deal with resonant proper elements has been recently introduced by Miloni, Ferraz-Mello and Beaugé (in preparation), who also presented a preliminary application of their method to the Hilda asteroids. On the other hand, Knežević and Milani (2000) elaborated a synthetic theory of the long term asteroidal motion that allowed the numerical computation of highly accurate proper elements for non resonant and resonant asteroids, without the typical limitations of the analytical or semi-analytical models previously used (e.g. Milani and Knežević 1990; Lemaitre and Morbidelli 1994). Another major improvement concerns the computation of proper elements for planet crossing orbits. This method has been introduced by Gronchi and Milani (2001), and has been successfully applied to predict planet collisions of Near-Earth Asteroids (NEAs).

Besides the theoretical development of all these techniques, the access to more powerful computational resources at lower costs has allowed the computation of proper elements (either analytical, semi-analytical or numerically) for very huge sets of orbits and for very different populations, from the main asteroid belt to the trans-Neptunian region. It is also possible to keep large databases periodically updated at the same rhythm of discovery of new asteroids (Knežević and Milani 2003). This has had a major impact in our knowledge of the dynamical structure of the asteroid belt, and especially in the detection of asteroid families.

In this contribution we review these issues. The paper is organized as follows: in Sect. 2 we provide the basic theoretical background about the computation of proper elements. Section 3 is devoted to describe the method of Beaugé and Roig (hereafter B-R). Section 4 summarizes the method of Miloni, Ferraz-Mello and Beaugé The synthetic theory of Knežević and Milani (hereafter K-M) is presented in Sect. 5. In Sect. 6 we present an application of the B-R and K-M methods to the Trojan asteroids. Finally, Sect. 7 describes the method of Gronchi and Milani.

2. Theoretical background

In a strict sense, proper elements should be integrals of motion of the dynamical system representing the motion of an asteroid under the perturbation of the planets. Since this system is not integrable, integrals of motion do not exist at all, but in most cases it is possible to compute quantities that are close to these integrals in the sense that they vary very little over very long time scales. These quasi-integrals of motion are referred to as "the proper elements".

The idea behind the computation of proper elements is to perform a canonical transformation (or a set of canonical transformations) such as to reduce the original Hamiltonian of the system to an integrable approximation. Schematically, suppose that the Hamiltonian can be separated as follows

$$F(\theta, J) = F_0(J) + \varepsilon F_1(\theta, J)$$

where F_0 is an integrable part, F_1 is a perturbation of order $\varepsilon \ll 1$, and θ, J are the angle-action variables of F_0. We search for a canonical transformation

$$(\theta, J) \to (\theta^*, J^*)$$

such that the new Hamiltonian becomes

$$F^*(\theta^*, J^*) = F_0^*(J^*) + \varepsilon^n F_1^*(\theta^*, J^*)$$

with $n > 1$. If the reminder of $\mathcal{O}(\varepsilon^n)$ can be neglected, then the new actions J^* are the proper elements we are looking for.

In practice, this procedure is accomplished by the computation of a time averaging that eliminates the angular dependence of the Hamiltonian. The final result of this averaging method strongly depends on the choice of the averaging "kernel" (the Hori's kernel), that is F_0, which determines how the angles actually vary with time. This choice must be done in such a way that F_0 accounts for the basic dynamical features of the system, or in other words, for the basic topology of the phase space. Thus, the key problem when dealing with the computation of proper elements is how to split the Hamiltonian for a suitable averaging.

In the asteroidal problem, it is possible to separate the angular dependence of the Hamiltonian according to the different time scales of the perturbations, leading to a set of "fast" and "slow" angles. The first ones are related to the mean longitudes of the asteroid (λ) and the planets (λ_i), while the second ones are related to the longitudes of perihelia (ϖ, ϖ_i) and nodes (Ω, Ω_i). Thus, for example, the classical definition of asteroids proper elements, based on Yuasa's theory (Milani and Knežević 1990), involves two averaging: the first to eliminate the fast angles and the second to eliminate the slow ones. Schematically, we first write

$$F = F_0(L, L_i) + \varepsilon F_1(\lambda, \varpi, \Omega, \lambda_i, \varpi_i, \Omega_i, L, W, Z, L_i, W_i, Z_i)$$

where F_0 basically represents the two body problem, and $\varepsilon \sim m_i$, which are the masses of the perturbing bodies. Here, L, W, Z are the canonical momenta conjugated to λ, ϖ, Ω, respectively. Then, we introduce a canonical transformation

$$(\varpi, \Omega, \varpi_i, \Omega_i, W, Z, W_i, Z_i) \rightarrow (\bar{\varpi}, \bar{\Omega}, \bar{\varpi}_i, \bar{\Omega}_i, \bar{W}, \bar{Z}, \bar{W}_i, \bar{Z}_i)$$

from "osculating elements" to "mean elements" through a *first* averaging that eliminates λ, λ_i, assuming that these angles vary with time following the solution of F_0. The "averaged" Hamiltonian takes the form

$$\bar{F} = \bar{F}_0 + \varepsilon \bar{F}_1(\bar{\varpi}, \bar{\Omega}, \bar{\varpi}_i, \bar{\Omega}_i, \bar{W}, \bar{Z}, \bar{W}_i, \bar{Z}_i)$$

where $\bar{F}_0$ is a constant that can be disregarded. The averaged perturbation can the be re-written as

$$\bar{F}_1 = \bar{F}_{10}(\bar{\varpi}, \bar{\Omega}, \bar{\varpi}_i, \bar{\Omega}_i, \bar{W}, \bar{Z}, \bar{W}_i, \bar{Z}_i) + \epsilon \bar{F}_{11}(\bar{\varpi}, \bar{\Omega}, \bar{\varpi}_i, \bar{\Omega}_i, \bar{W}, \bar{Z}, \bar{W}_i, \bar{Z}_i)$$

where, again, $\bar{F}_{10}$ is an integrable part basically represented by an harmonic oscillator with a forced term, and ϵ is a small parameter somehow related to the high powers of the eccentricities and inclinations of the bodies. The *second* averaging is performed assuming that the mean angles vary linearly with time with frequencies given by the fundamental frequencies of $\bar{F}_{10}$. After the averaging, we arrive to an integrable Hamiltonian and the proper elements are given by the actions of $\bar{F}_{10}$ plus a correction of order ϵ arising from $\bar{F}_{11}$. The proper frequencies are also given by the fundamental frequencies of $\bar{F}_{10}$ plus a correction of order ϵ. It is then usual to proceed in an iterative way by repeating the average using these corrected frequencies until their values converge.

Other problems in asteroidal dynamics are treated in a similar way. The only differences arise from the form in which the integrable part of the Hamiltonian is separated at the different stages of the procedure, and also the form in which the average is done. Three cases are of particular interest:

• When the asteroid orbit is in a mean motion resonance, some linear combination of the mean longitudes λ, λ_i has a frequency close to zero. This linear combination constitutes the resonant angle, which has to be isolated, so the first average is performed only over the non resonant angles. This procedure leads to an averaged Hamiltonian, where $\bar{F}_0$ has the basic features of a pendulum (actually, an Andoyer Hamiltonian). This must be taken into account when performing the second average, which introduce additional complexity to the problem.

• When the asteroid orbit is in a secular resonance, the corresponding resonant angle has to be isolated from $\bar{F}_1$, and the second average is performed only over the non resonant angles. This procedure leads to an averaged Hamiltonian, where $\bar{F}_{10}$ has the basic features of a pendulum (Morbidelli 1993).

• When the asteroid is in a largely eccentric or largely inclined non resonant orbit, the separation of the averaged Hamiltonian in $\bar{F}_{10}$ and $\bar{F}_{11}$ is no longer valid because ϵ is not small. Other re-arrangements are possible in this case (e.g. Lemaitre and Morbidelli 1994), but the corresponding results are always restricted to limited ranges of the eccentricity and inclination.

• When the asteroid is in a planet crossing orbit, the first average cannot be performed because there is a singularity along the integration path.

In any of these cases, the classical theory to compute proper elements will fail. Therefore, specific techniques have to be developed to treat them, as we will show in the following.

3. Proper elements for resonant orbits

When dealing with resonant orbits in the framework of the restricted three body problem, the Hamiltonian obtained after the elimination of the fast angles has the form

$$\bar{F} = \bar{F}_0(\bar{\sigma}, \bar{L}, \bar{W}, Z, \bar{W}', \bar{Z}') + \varepsilon \bar{F}_1(\bar{\sigma}, \bar{\varpi}, \bar{\Omega}, \bar{\varpi}', \bar{\Omega}', \bar{L}, \bar{W}, \bar{Z}, \bar{W}', \bar{Z}') \tag{3.1}$$

where $\bar{F}_0(\bar{\sigma}, \bar{L}, \bar{W}, \bar{Z}, \bar{W}', \bar{Z}')$ is a pendulum-like Hamiltonian, ε is proportional to the eccentricity and inclination of the perturbing body, $\bar{\sigma}$ is the resonant angle that librates around a certain value $\bar{\sigma}_c$, $\bar{L}$ is the canonical momentum conjugated to $\bar{\sigma}$, and primed variables refer to the perturber. In order to apply a second averaging to eliminate all the angles, we have to take into account that $\bar{\varpi}, \bar{\Omega}, \bar{\varpi}', \bar{\Omega}'$ are linear functions of time but $\bar{\sigma}$ is not. Therefore, the time averaging cannot be directly replaced by an average over $\bar{\sigma}$. A possible solution is to introduce a canonical transformation to find the action-angle variables of $\bar{F}_0$. This is usually accomplished by solving the equations of motion for $\bar{F}_0$ numerically, substituting this solution in $\bar{F}_1$, and computing the time average with a numerical quadrature. However, this has the drawback of being very CPU-time consuming, and does not explicitly yield the proper element associated to the pair $\bar{\sigma}, \bar{L}$.

Another possibility has been introduced by Beaugé and Roig (2001), based on ideas by Jupp (1969) for the Ideal Resonance Problem. This consists into find a canonical transformation from $(\bar{\sigma}, \bar{\varpi}, \bar{\Omega}, \bar{L}, \bar{W}, \bar{Z})$ to new variables $(\theta, \varpi^*, \Omega^* J, W^*, Z^*)$ where all the angles are non resonant. The idea can be summarized as follows: Let us think about the libration region of a resonance as a set of invariant curves around the libration point $\bar{\sigma}_c$. If we only concentrate on this region and disregard the structure of qthe phase space outside the separatrix, we can think of these orbits as distorted circulations around a center which is displaced from the origin of the coordinate system. Now, if we find a canonical transformation $(\bar{L}, \bar{\sigma}) \to (J, \theta)$ that is simply a translation of the origin to the libration center, we will obtain a new angle θ having a frequency different from zero, and the integral of J along any orbit will be the action of that trajectory. In other words, we will have an angle $\bar{\sigma}$ that librates transformed into another angle θ that circulates with frequency $\nu_\theta = \nu_{\bar{\sigma}}$. These new variables will have properties of being "non-resonant" (even though they are a simple translation), and we can use any classical averaging method, such as Hori's method, to determine the corresponding action-angle variables.

A simple way to determine (J, θ) is based on the following series of transformations:

$$(\bar{L}, \bar{\sigma}) \to (K, H) = \sqrt{2\bar{L}}(\cos \bar{\sigma}, \sin \bar{\sigma})$$
$$(K, H) \to (X, Y) = (K - K_c, H - H_c)$$
$$(X, Y) = \sqrt{2J}(\cos \theta, \sin \theta) \to (J, \theta) \tag{3.2}$$

where $(K_c, H_c) = \sqrt{2\bar{L}_c}(\cos \bar{\sigma}_c, \sin \bar{\sigma}_c)$ marks the center of libration. This center is nothing but the equilibrium point of $\bar{F}_0$ and can be easily obtained numerically. Beaugé and Roig (2001) introduced a slightly different procedure, in the sense that it can no longer be thought of as a simple translation. The transformation in their case is represented by the relationship:

$$X = \Gamma^{-1/2}\big(\widehat{K} - (K_c^2 - \widehat{H}^2)^{1/2}\big); \qquad\qquad Y = \Gamma^{1/2}\widehat{H} \tag{3.3}$$

where $(\widehat{K}, \widehat{H}) = \sqrt{2\bar{L}}\,(\cos(\bar{\sigma} - \bar{\sigma}_c), \sin(\bar{\sigma} - \bar{\sigma}_c))$ and $\Gamma = \Gamma(K_c)$ is a scaling factor which

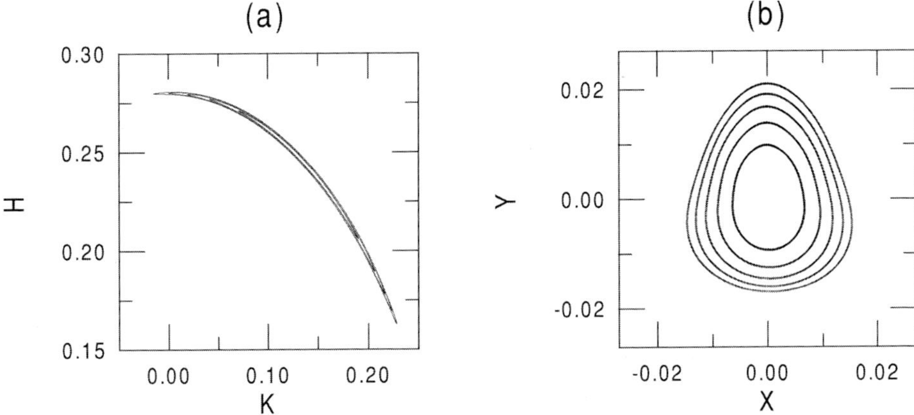

Figure 1. (a) A set of invariant curves representing Trojan-type librations. (b) The corresponding transformation to local variables given by Eq. (3.3).

modifies the shape of the trajectories. This transformation is canonical and is valid as long as $|\bar{\sigma}_{\max} - \bar{\sigma}_c| < \pi/2$ (with both $\bar{\sigma}_{\max}$ and $\bar{\sigma}_c$ defined between $\pm\pi$). Since the transformation is an explicit function of the center of libration, Beaugé and Roig called Eq. (3.3) the *local variables*. An example of this transformation is shown in Fig. 1.

By means of this very simple and purely geometrical "banana-to-pear" transformation, it is possible to bypass the difficulties generated by the libration of σ, and to define variables suitable for the application of Hori's averaging method. The averaged Hamiltonian can then be written in the form:

$$\bar{F} = \bar{F}_0(\theta, J, W^*, Z^*, W^{*\prime}, Z^{*\prime}) + \varepsilon \bar{F}_1(\theta, \varpi^*, \Omega^*, \varpi^{*\prime}, \Omega^{*\prime}, J, W^*, Z^*, W^{*\prime}, Z^{*\prime}) \quad (3.4)$$

and it would be ready to proceed with the second average.

3.1. *Averaging Methods with Adiabatic Invariance*

In order to treat the second average, Beaugé and Roig (2001) introduced a general procedure to analyze multi-dimensional Hamiltonian systems having a "hierarchical" separation in time of the different degrees of freedom. The procedure can be summarized as follows: Suppose a generic two degrees of freedom system defined by a Hamiltonian

$$F \equiv F(J, \theta) = F_0(J_1, J_2) + F_1(J_1, J_2, \theta_1, \theta_2)$$

where (J, θ) are action-angle variables of F_0. Assuming that neither θ_1 nor θ_2 are resonant angles and that there are no significant commensurabilities between them, this system can be solved using Hori's averaging method. The transformation $(J, \theta) \rightarrow (J^*, \theta^*)$ up to first order, is given by the equations:

$$J_k = J_k^* + \frac{\partial B_1}{\partial \theta_k^*}; \qquad \theta_k = \theta_k^* - \frac{\partial B_1}{\partial J_k^*} \qquad (k = 1, 2) \qquad (3.5)$$

where B_1 is the first-order generating function. The idea is to think about Eqs. (3.5) as a system of 4 algebraic equations corresponding to two different sets of variables (the degrees of freedom). Instead of taking all equations simultaneously, the system is broken in two parts and a hypothesis of adiabatic invariance is adopted, assuming that the unperturbed frequencies of each degree of freedom satisfy the condition $\nu_1 \gg \nu_2$. In this way, the two equations corresponding to the first degree of freedom ($k = 1$) can be solved separately, assuming fixed values for the second degree of freedom and writing the solution in terms of these values.

It is easy to show that this procedure is equivalent to solve the "one degree of freedom" Hamiltonian

$$F \equiv \widetilde{F}(J,\theta) = \widetilde{F}_0(J_1; J_2,\theta_2) + \widetilde{F}_1(J_1,\theta_1; J_2,\theta_2) \tag{3.6}$$

where (J_2,θ_2) are fixed parameters. Note that $\widetilde{F}$ is nothing but the original Hamiltonian F split in a different way. The solution of $\widetilde{F}$ by Hori's method will provide certain values $(\widetilde{J}_1^*,\widetilde{\theta}_1^*)$ of the action and angle, and the Theory of Adiabatic Invariants guarantees that the difference between this solution and the solution (J_1^*,θ_1^*) of the real system in which (J_2^*,θ_2^*) are slowly varying with time is such that

$$\widetilde{J}_1^* - J_1^* \propto \epsilon\, \mathcal{K}(\widetilde{J}_1^*,\widetilde{\theta}_1^*; J_2^*,\theta_2^*); \qquad\qquad \widetilde{\theta}_1^* - \theta_1^* \propto \epsilon\, \mathcal{L}(\widetilde{J}_1^*,\widetilde{\theta}_1^*; J_2^*,\theta_2^*)$$

where $\epsilon = \nu_2/\nu_1$, and $\mathcal{K}$ and $\mathcal{L}$ are functions of order unity (Henrard and Roels 1974). Since $\epsilon \ll 1$, both sets of solutions are approximately the same. In other words, using the adiabatic approximation, the new action-angle variables are determined up to order ϵ, and this parameter defines the precision of the method.

The action $\widetilde{J}_1^*$ is an invariant of the "frozen" Hamiltonian $\widetilde{F}$, but not of the full Hamiltonian F. Since (J_2,θ_2) vary slowly with time, so does $\widetilde{J}_1^*$, and according to the Adiabatic Theory, this variation is such that

$$\frac{d\widetilde{J}_1^*}{dt} \sim \epsilon^2.$$

For very small values of ϵ, this second order variation can be neglected and the resulting "constant" value of $\widetilde{J}_1^*$ is called an *adiabatic invariant* of Hamiltonian F. It is worth noting that these second order corrections to the adiabatic invariant are periodic with the same period of (J_2,θ_2). Thus, they could be eliminated by a suitable averaging of $\widetilde{J}_1^*$ over a period of (J_2,θ_2). In most cases, averaging the corrections provides a better approach to the adiabatic invariant than neglecting them.

Once the action-angle variables for the first degree of freedom have been determined (up to order ϵ), it is possible to solve the equations for the second degree of freedom. The idea is to introduce the solution $J_1^* = J_1^*(J_2^*,\theta_2^*)$ and $\theta_1^* = \theta_1^*(J_2^*,\theta_2^*)$ into the generating function B_1 and to solve the sub-system of Eqs. (3.5) corresponding to $k = 2$. This is equivalent to solve a one-degree of freedom non-autonomous Hamiltonian, since θ_1^* is a linear function of time. Actually, this is equivalent to take the original Hamiltonian F, introduce the solution for the first degree of freedom $J_1 = J_1(t, J_2,\theta_2)$, $\theta_1 = \theta_1(t, J_2,\theta_2)$, and average the resulting expression with respect to θ_1. The procedure leads to a new "one degree of freedom" Hamiltonian $\widehat{F}(\langle J_2\rangle_{\theta_1^*}, \langle\theta_2\rangle_{\theta_1^*})$, where $\langle .\rangle_{\theta_1^*}$ represents the average over θ_1^*, whose solution by Hori's method provides the corresponding action-angle variables (J_2^*,θ_2^*). In other words, we can average the original Hamiltonian F over a reference orbit of the first degree of freedom (which is obtained by adiabatic approximation assuming that the second degree of freedom is fixed), and then, we can use this averaged Hamiltonian to solve the second degree of freedom.

The whole procedure can be easily extended to the general case with N degrees of freedom in which the unperturbed frequencies ν_i of each angular variable θ_i are finite and large, and satisfy the condition $\nu_1 \gg \nu_2 \gg \ldots \gg \nu_N$. Thus, introducing the small parameters $\epsilon_{i,j} = \nu_j/\nu_i$ $(j > i)$ the system can be solved in a hierarchical form, solving one degree of freedom at a time.

4. Resonant averaging theory

The main limitation of the B-R method is that it can be applied only under the hypothesis of adiabatic invariance. Unfortunately, this situation does not hold in other mean motion resonances, like the 3/2 with Jupiter or the 2/3 with Neptune, both associated to large populations of minor bodies. For these cases, a generalized resonant averaging theory has been formally introduced by Ferraz-Mello (1997, 2002), and has been recently applied by Miloni, Ferraz-Mello and Beaugé (in preparation).

Their method can be summarized as follows: Consider the Hamiltonian of the restricted planar three body problem

$$F = F_0(L, L') + m'F_1(\lambda, \varpi, \lambda', \varpi', L, W, L', W')$$

where F_0 is the Keplerian part, F_1 the disturbing function, and primed variables refer to the perturber. The disturbing function is further expanded using the expansion of Beaugé (1996), which does not have the convergence limitations of the classical expansions, nor the phase space domain limitations of the asymmetric expansions. The Hamiltonian is then given by a harmonic series with constant coefficients and can be manipulated in a fully analytical way.

Introducing the resonant angles

$$\sigma = \frac{p+q}{q}\lambda' - \frac{p}{q}\lambda - \varpi; \qquad \sigma' = \frac{p+q}{q}\lambda' - \frac{p}{q}\lambda - \varpi'$$

(p, q integers) and averaging (up to first order) over the short period angle $\lambda - \lambda'$, the Hamiltonian takes the form

$$\bar{F} = F_0(\bar{L}) + m'\bar{F}_1$$

In order to split this Hamiltonian for further averaging, the authors expand F_0 around the reference value $\bar{L}_0$ corresponding to the exact mean motion resonance, and assume that $\bar{L} - \bar{L}_0 \sim \mathcal{O}(m'^{1/2})$. This assumption is crucial since it allows to re-arrange terms of the same order in m' so as to write:

$$\bar{F} = m'\widetilde{F}_0(\sigma, S, S') + m'^{3/2}\widetilde{F}_1(\sigma, \sigma', S, S')$$

where $m'\widetilde{F}_0$ is a pendulum Hamiltonian, basically constituted by a quadratic term in $S \sim \bar{L} - \bar{L}_0$ plus a term $m'\cos\sigma$. The next step is to find the actions of the pendulum, which is accomplished by expanding the solution by means of elliptic integrals. This allows to explicitly compute the actions analytically in terms of σ, S. The resulting Hamiltonian takes the form:

$$F^* = m'F_0^*(J, J') + m'^{3/2}F_1^*(\theta, \theta', J, J')$$

and is suitable for the application of Hori's method to totally solve it. Note that in this case, the perturbation equations of Hori's method will be grouped in orders of $m'^{3/2}, m'^{5/2}$, and so on. In practice, the averaging is carried out up to the "first" order only. After finding the proper actions of F^*, it is possible to analytically go back with the transformation to compute the proper amplitude of libration S^* and the remaining proper elements.

The method has been successfully applied by the authors for a preliminary computation of proper elements of the Hilda asteroids in the planar case, and at present it is being extended to include the inclinations.

5. Synthetic proper elements

The idea underneath the computation of synthetic proper elements is to fit the time series of the asteroid orbital elements to some predefined function. This function usually has the form of a harmonic series with a given number of harmonics. The fit consists of determining the frequencies, amplitudes and phases of the different harmonics by linear regression, which is nothing but to decompose the time series through a Fourier transform. After this decomposition, the signal can be easily filtered to remove all the periodic terms, leaving just the constant terms of the series which constitute the proper elements.

At variance with the usual averaging methods, which deals with the *solution of the averaged equations* of motion

$$\left\langle \frac{dq_i}{dt} \right\rangle = \left\langle \frac{\partial F}{\partial p_i} \right\rangle ; \qquad\qquad \left\langle \frac{dp_i}{dt} \right\rangle = - \left\langle \frac{\partial F}{\partial q_i} \right\rangle ,$$

the filtering deals with the *average of the solution*, that is $(\langle q_i \rangle , \langle p_i \rangle)$. The equivalence between both approaches is given by the condition

$$\frac{d \langle q_i \rangle}{dt} = \left\langle \frac{dq_i}{dt} \right\rangle ; \qquad\qquad \frac{d \langle p_i \rangle}{dt} = \left\langle \frac{dp_i}{dt} \right\rangle .$$

However, this equivalence is strictly valid only if the averages are made over the "perturbed" solution. While the filtering fulfills this condition, the usual averaging methods don't because they are always made over an "unperturbed" or intermediate solution (e.g. the solution of the Hori's kernel). Therefore, the filtering always provides a more accurate approach to the proper elements than the usual averaging theories (both analytical and numerical).

Knežević and Milani (2000) used a synthetic theory to compute asteroid proper elements. They numerically integrated the orbit of the asteroid over intervals of time ranging from 2 to 10 Myr and performed a Fourier analysis of the output. The original output is represented by the time series of the equinoctal elements

$$(k, h) = e(\cos \varpi, \sin \varpi) \qquad\qquad (q, p) = \sin \frac{I}{2}(\cos \Omega, \sin \Omega)$$

These are filtered on-line in order to remove the short period variations related to the mean anomalies, which also allows to decimate the output and to reduce the data storage size. The filtered output is then processed in three steps:

(*a*) The forced secular perturbations with known frequencies $(g_5, g_6, ...)$ and $(s_5, s_6, ...)$ are removed from the filtered series by identifying the corresponding harmonics in the Fourier transform.

(*b*) The time series of the free angles (ϖ_f, Ω_f) are fitted by straight lines and the proper frequencies are determined from the slope of the fits.

(*c*) The components with period 2π are extracted from the data series $k(\varpi_f), h(\varpi_f)$ and $q(\Omega_f), p(\Omega_f)$. These constitute the proper modes and their amplitudes define the proper elements e_p and $\sin I_p/2$.

(*d*) The proper semi-major axis a_p is computed as the average of the filtered semi-major axis.

Simultaneously to the above procedure, the maximum Lyapunov Characteristic Exponent (LCE) is also computed by numerically solving the variational equations of the orbit. This is used as an indicator of chaos and provides an indication of the reliability of the computed proper elements.

The K-M procedure requires to perform a numerical integration of the orbits over long time scales. This has an advantage in the sense that it automatically provides a stability test for the proper elements: Indeed, with the same output it is possible to compute proper elements over a running box with a shorter time width and to see how these values vary with time. However, the main disadvantage is that updates of the data to incorporate new asteroids are hard to perform, since they are very time consuming. In spite of this, the method provides very precise values of the proper elements.

5.1. *Synthetic resonant proper elements*

The extension of the synthetic theory to the case of resonant orbits is straightforward. In fact, synthetic theories for the computation of proper elements were first developed for resonant orbits, more specifically, for the Trojan case. The origins of the method go back to Bien and Schubart (1984) and Schubart and Bien (1987), and it was fully developed by Milani (1993).

The idea is to consider the time series of $a \exp \iota\sigma$, $e \exp \iota\varpi$ and $\sin I \exp \iota\Omega$. These time series are first filtered on-line to remove the short period variations and then it is applied the same kind of harmonic decomposition as explained above. However, the (filtered) time series of $a \exp \iota\sigma$ cannot be processed as such because σ is librating rather than circulating. Therefore, a transformation $(a, \sigma) \rightarrow (D, \theta)$ to local variables, like Eq. (3.2), is introduced:

$$D(\cos \theta, \sin \theta) = \left(\sigma - \sigma_c, \frac{a - a_c}{\gamma} \right) \tag{5.1}$$

where (σ_c, a_c) is the center of libration and γ is a scaling factor that relates D, i.e. the semi-amplitude of libration in σ, with the semi-amplitude of libration in a. The angle θ is no longer librating but circulating with a fundamental frequency equal to the frequency of libration. The synthetic theory is then applied to the time series of $D \exp \iota\theta$, and the proper semi-amplitude of libration D_p is obtained together with the corresponding proper frequency.

6. An application to the Trojan asteroids

The B-R method and the K-M method described above have been applied to compute proper elements for the Trojan asteroids. Many dynamical properties of these asteroids complicate the elaboration of an analytical model for their long-term motion, so the application of the K-M synthetic theory seems to be a better option. The K-M procedure involves the transformation Eq. (5.1) with $\sigma_c = \pm\pi/3$ and $a_c = a_{\text{Jupiter}}$. This is a major limitation of their method because it is well known that, at large eccentricities and inclinations, the true center of libration may be significantly displaced from its "standard" location at $\pm\pi/3$. Thus, the method will produce fake estimates of D_p and, especially, of the proper frequency of libration, whenever D is smaller that the difference between the true center of libration and $\pi/3$. Fortunately, these cases are very rare among the real Trojan asteroids, and the K-M method can be safely applied in most cases.

On the other hand, there is a major dynamical feature that may be exploited by the B-R method: The different degrees of freedom of the system are well separated with respect to their periods. In fact, while the period of libration of the resonant angle σ is typically about 150 yr, the period of oscillation of ϖ is of the order of 3,500 yr, and the period of Ω is even longer: $10^5 - 10^6$ yr. In this way, it is possible to introduce the adiabatic approach to the problem defining the small parameters $\epsilon_{12} = \nu_\varpi/\nu_\sigma, \epsilon_{23} = \nu_\Omega/\nu_\varpi$ and $\epsilon_{13} = \nu_\Omega/\nu_\sigma$.

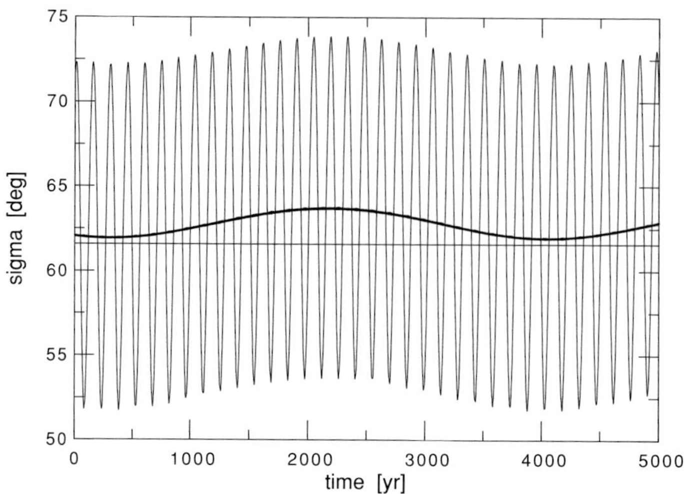

Figure 2. Time evolution of σ from the full Hamiltonian F (thin curve), from the circular planar Hamiltonian F_0 (thin horizontal line), and from the frozen Hamiltonian $\widetilde{F_0}$ (thick curve). Note that the solution of $\widetilde{F_0}$ follows the long term evolution of the center of libration while the solution of F_0 doesn't.

The B-R method starts with the averaged restricted three body problem for the $1/1$ mean motion resonance, similar to Eq. (3.1):

$$F = F_0(\sigma, L, W, W') + \varepsilon F_1(\sigma, \varpi, \Omega, \varpi', \Omega', L, W, Z, W', Z')$$

where F_0 corresponds to the circular planar problem. Then, the transformation to local variables (Eq. 3.3) is introduced, which leads to a Hamiltonian like Eq. (3.4)

$$F = F_0(\theta, J, W, W') + \varepsilon F_1(\theta, \varpi, \Omega, \varpi', \Omega', J, W, Z, W', Z').$$

This Hamiltonian is further expanded using an *asymmetric* expansion around the center of libration, that is, a Taylor-Fourier expansion around $J = 0$, $W = 0$, and $Z = 0$†. The secular variation of Jupiter's orbit is introduced through the synthetic planetary theory LONGSTOP 1B (Nobili *et al.* 1989). The direct gravitational effects of Saturn, Uranus and Neptune on the asteroid are also included, assuming that these planets move on fixed circular orbits with zero inclination.

In order to average over the libration period, the Hamiltonian is re-arranged like in Eq. (3.6):

$$F = \widetilde{F_0}(J; \varpi, \Omega, \varpi', \Omega', W, Z, W', Z') + \mu \widetilde{F_1}(\theta, J; \varpi, \Omega, \varpi', \Omega', W, Z, W', Z') \qquad (6.1)$$

where $\varpi, \Omega, \varpi', \Omega', W, Z, W', Z'$ are taken as fixed parameters. Here, $\widetilde{F_0}$ is no longer the Hamiltonian of the circular planar problem and μ is a small parameter somehow related to the amplitude of libration. The key point is that $\widetilde{F_0}$ contains much more information than F_0, since it has embedded the slow variation of the other degrees of freedom. The advantage of this can be appreciated in Fig. 2. The "first order" solution of Hamiltonian Eq. (6.1) provides the proper action and angle (J^*, θ^*).

The averaging over the libration period leads to a new Hamiltonian that corresponds to a two degrees of freedom non autonomous system. This Hamiltonian is split so that

---

† Note that an expansion around $J = 0$ is indeed asymmetric

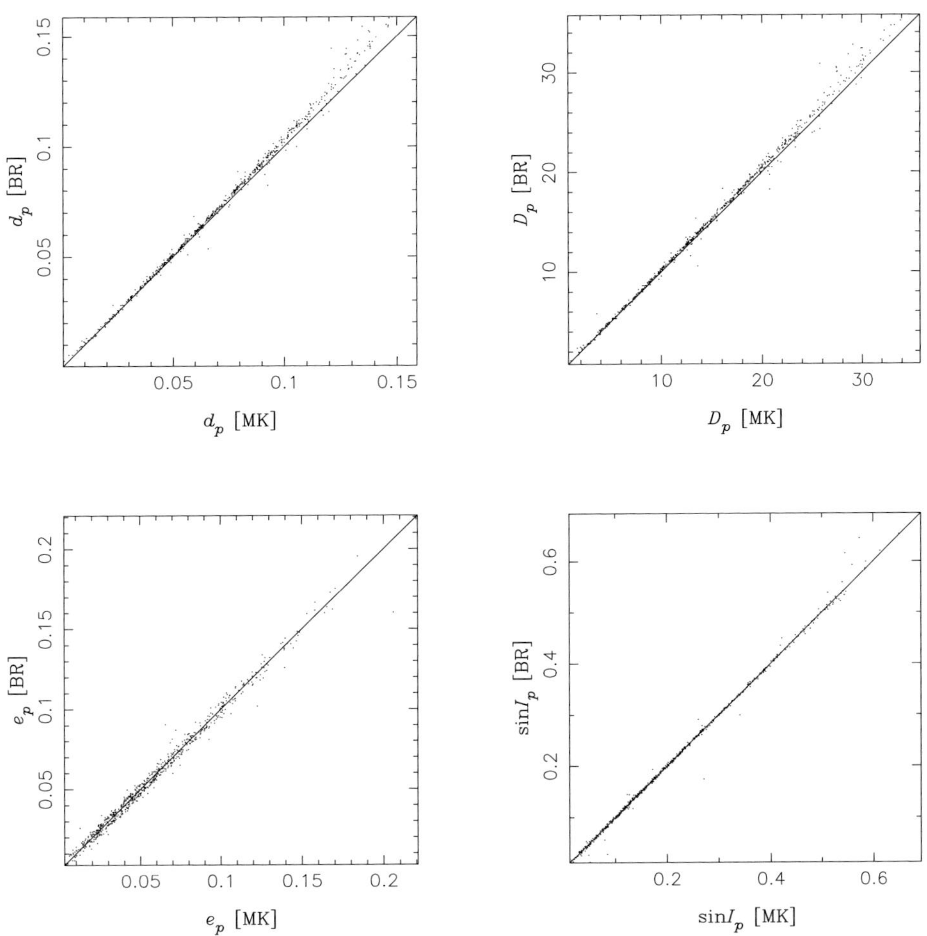

Figure 3. Comparison between the Trojan proper elements obtained with the B-R method (in the ordinates) and the K-M method (in the abscissas). The straight line in each plot represents the identity function. d_p is the proper semi-amplitude of libration in a.

all the terms depending on the angles $(\varpi, \Omega, \varpi', \Omega')$ are grouped in the "perturbation", and the "first order" solution by Hori's method provides the proper actions and angles $(W^*, Z^*, \varpi^*, \Omega^*)$. In the end, the set of proper actions (J^*, W^*, Z^*) depend solely on the initial conditions and on the orbital elements of the perturber, and are transformed back to the set (D_p, e_p, I_p).

A comparison between the results of the B-R method and those of the K-M synthetic theory is shown in Fig. 3. In the case of e_p, I_p, there is practically no differences between both sets. The agreement is also very good for small values of D_p, but there is a systematic bias of the B-R proper elements with respect to the K-M set for large amplitudes of libration. This bias is probably related to the early truncation of the asymmetric expansion of the disturbing function. However, it may also be related to the chaotic character of the large amplitude orbits that precludes the possibility of defining accurate proper elements. This is clearly shown in Fig. 4. In spite of this, we have verified that the systematic character of the bias does not significantly affect the identification of asteroid families.

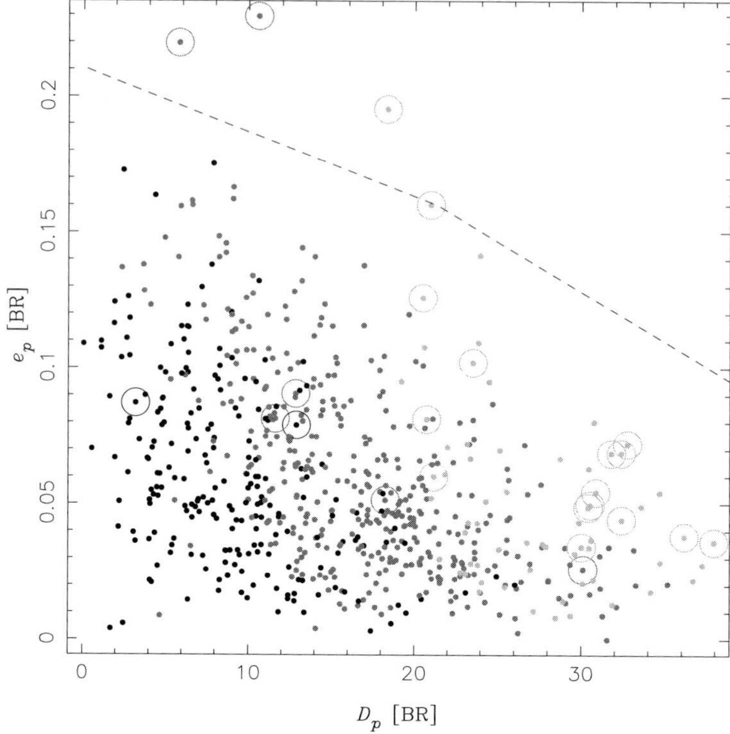

Figure 4. Trojan proper elements computed with the B-R method. The points are coded in a gray-scale according to their LCE (computed from the K-M method): the darker, the more stable. A circle around the dots indicates those cases for which the differences between the B-R and K-M proper elements are larger than 1%. Most of these cases are related to chaotic orbits.

7. Proper elements for planet crossing orbits

As we already said, the application of a perturbative method based on an averaging principle to solve the equations of motion of an asteroid under the perturbation of the planets will fail whenever the perturbation has a singularity along the domain of the solution. To avoid this problem, Gronchi and Milani (1999, 2001) introduced a generalized averaging principle, that can be summarized as follows.

Consider the restricted circular N-body problem, where the planets are assumed to move in circular and coplanar orbits. The Hamiltonian can be written as

$$F = F_0(L, L_i) + \varepsilon F_1(\lambda, \varpi, \lambda_i - \Omega, L, W, Z)$$

where F_1 is the disturbing function expanded in terms of the (modified) Delaunay variables and depends on the combination $\lambda_i - \Omega$. The secular motion of the system, under the absence of any mean motion resonances is given by the equations

$$\left\langle \frac{d\varpi}{dt} \right\rangle = \frac{\partial \langle F \rangle}{\partial W} \qquad\qquad \left\langle \frac{dW}{dt} \right\rangle = -\frac{\partial \langle F \rangle}{\partial \varpi}$$

$$\left\langle \frac{d\Omega}{dt} \right\rangle = \frac{\partial \langle F \rangle}{\partial Z} \qquad\qquad \left\langle \frac{dZ}{dt} \right\rangle = -\frac{\partial \langle F \rangle}{\partial \Omega} = 0 \qquad (7.1)$$

where $\langle X \rangle = \frac{1}{4\pi^2} \int_0^{2\pi} \int_0^{2\pi} X \, d\lambda \, d\lambda_i$. These expressions have embedded the property that

$$\left\langle \frac{\partial F}{\partial x} \right\rangle = \frac{\partial \langle F \rangle}{\partial x} \tag{7.2}$$

for any x provided that F were continuous and differentiable in all the domain of interest. The averaged Hamiltonian is a one degree of freedom Hamiltonian usually known as the Kozai Hamiltonian.

When the orbits of the asteroid and one planet intersect each other, F_1 has a first order pole, arising from the direct perturbation, and the $\partial F_1 / \partial x$ have a second order pole. In such case $\langle F_1 \rangle$ is a complete improper integral, but $\langle \partial F_1 / \partial x \rangle$ is a divergent integral and the equations for the secular motion make no sense. The strategy of Gronchi and Milani was to realize that this is true only at the point of intersection of the orbits, where the pole exists. At all the other points along the orbit the relation (7.2) is still valid, and so are Eqs. (7.1).

The idea is then to numerically integrate Eqs. (7.1) until arriving very close to the intersection point, then to suitably manipulate the jump across the singularity and to continue the integration. In order to manipulate the singularity, the authors introduce a method of singularity extraction proposed by Kantorovich which consists into decompose the direct part of the disturbing function in two terms whose averages can be computed in terms of convergent integrals. In short, they write

$$\frac{\partial}{\partial x} \int_0^{2\pi} \int_0^{2\pi} \frac{1}{\Delta_i} \, d\lambda \, d\lambda_i = \frac{\partial}{\partial x} \int_0^{2\pi} \int_0^{2\pi} \frac{1}{\delta_i} \, d\lambda \, d\lambda_i + \frac{\partial}{\partial x} \int_0^{2\pi} \int_0^{2\pi} \left(\frac{1}{\Delta_i} - \frac{1}{\delta_i} \right) d\lambda \, d\lambda_i$$

where Δ_i is the actual mutual distance between the asteroid and the i-th planet, and δ_i is the mutual distance computed by assuming that, close to the mutual node, the bodies are moving along straight lines tangent to the orbits at that point (Wetherill 1967). With this method, the principal part of the singularity of $\frac{\partial}{\partial x} \frac{1}{\Delta_i}$ is removed, and the reminder $\frac{\partial}{\partial x} \left(\frac{1}{\Delta_i} - \frac{1}{\delta_i} \right)$ has now a first order pole, so that

$$\int_0^{2\pi} \int_0^{2\pi} \frac{\partial}{\partial x} \left(\frac{1}{\Delta_i} - \frac{1}{\delta_i} \right) d\lambda \, d\lambda_i = \frac{\partial}{\partial x} \int_0^{2\pi} \int_0^{2\pi} \left(\frac{1}{\Delta_i} - \frac{1}{\delta_i} \right) d\lambda \, d\lambda_i$$

The principal term $\frac{\partial}{\partial x} \frac{1}{\delta_i}$ still has a second order pole, but the computation of

$$\frac{\partial}{\partial x} \int_0^{2\pi} \int_0^{2\pi} \frac{1}{\delta_i} \, d\lambda \, d\lambda_i$$

can be performed analytically, giving rise to analytical expressions for the solution of the equations of motion close to the mutual nodes (Gronchi 2002). These expressions allow to jump over the singularity during the numerical integration of the equations.

The proper elements are then obtained from the solution of the Eqs. (7.1) by computing the average semi-major axis, the maximum and minimum eccentricity and the maximum and minimum inclination over one period of $\omega = \varpi - \Omega$ (either if ω is circulating or librating). The solution also provides the proper frequencies and the encounter circumstances with each planet (radiant, planetocentric velocity and date). This latter information allows to predict the occurrence of node crossings and is particularly interesting in the case of Earth crossing asteroids, since it provides a way to link meteor streams with NEAs and to evaluate potential Earth impactors.

Let's say to close this review that these proper elements do not have the same meaning as the usual proper elements since they are not quasi-integrals of motion in a strict sense.

Their stability is guaranteed only over a short time scale, either of the order of the period of a complete oscillation of ω, or until the next very close approach to a planet.

References

Beaugé, C. 1996, *Cel. Mech. Dyn. Astron.* 64, 313
Beaugé, C. & Roig, F. 2001, *Icarus* 153, 391
Bien, R. & Schubart, J. 1984, *Celest. Mech.* 34, 425
Ferraz-Mello, S. 1997, *Cel. Mech. Dyn. Astron.* 66, 39
Ferraz-Mello, S. 2002, *Cel. Mech. Dyn. Astron.* 83, 275
Gronchi, G. F. 2002, *Cel. Mech. Dyn. Astron.* 83, 97
Gronchi, G. F. & Milani, A. 1999, *Cel. Mech. Dyn. Astron.* 71, 109
Gronchi, G. F. & Milani, A. 2001, *Icarus* 152, 58
Henrard, J. & Roels, J. 1974, *Celest. Mech.* 10, 497
Jupp, A. H. 1969, *Astron. J.* 74, 35
Knežević, Z. & Milani, A. 2000, *Cel. Mech. Dyn. Astron.* 78, 17
Knežević, Z. & Milani, A. 2003, *Astron. Astrophys.* 403, 1165
Lemaitre, A. & Morbidelli, A. 1994, *Cel. Mech. Dyn. Astron.* 60, 29
Milani, A. 1993, *Cel. Mech. Dyn. Astron.* 57, 59
Milani, A. & Knežević, Z. 1990, *Cel. Mech. Dyn. Astron.* 49, 347
Morbidelli, A. 1993, *Icarus* 105, 48
Nobili, A. M., Milani, A. & Carpino, M. 1989, *Astron. Astrophys.* 210, 313
Schubart, J. & Bien, R. 1987, *Astron. Astrophys.* 175, 299
Wetherill, G. W. 1967, *J. Geophys. Res.* 72, 2429

Proceedings Dynamics of Populations of Planetary Systems
Proceedings IAU Colloquium No. 197, 2005
Z. Knežević and A. Milani, eds.

© 2005 International Astronomical Union
DOI: 10.1017/S1743921304008592

Asteroid family classification from very large catalogues

Anne Lemaitre

Department of Mathematics, University of Namur
Rempart de la Vierge, 8 - B-5000 Namur Belgium
email: anne.lemaitre@fundp.ac.be

Abstract. The paper presents a review of the recent contributions and open questions concerning the families of asteroids. Due to the availability of very large catalogues (synthetic and analytical proper elements of the asteroids and large observational surveys of their spectra) and to the introduction of non gravitational forces in their determination, the concept of *static* family has disappeared, to be replaced by this of *dynamical* families. The proper elements are not constant anymore but are ageing on very long timescales. The size distributions of the populations of asteroids, in and out the families, their ages, the ejection velocities of the fragments after an impact, have been reconsidered by several teams of research, with this new approach. Parallel numerical simulations of collisions and fragmentations of bodies have showed that most of the asteroids are likely rubble piles or agglomerates than monolithic blocks. The methods of classification have been refined and combine, in their newest versions, the dynamics and the observations, working now on 5 dimensional space instead of 3. A series of sub families of the large well-known families have been recently identified, using catalogues with more than 100 000 asteroids (the cluster Karin for example).

Keywords. Asteroid family, proper elements, surveys, catalogues, micro family.

1. Introduction

The so-called families of asteroids were detected by Hirayama (1918), on a set of a few hundreds of planets. The problem of the missing planet between Mars and Jupiter seemed then to have found a partial answer ; several quite big bodies had first the opportunity of growing up in this region, but repetitive collisions between them or systematic bombardments led to their fragmentation, and reduced them, chock after chock, to bodies with a much smaller size and volume. The asteroids are obviously a population of bodies whose evolution is deeply and extensively influenced by collisional processes, and families are direct proofs that these collisions really occur in the asteroid belt. However their origin is not a single large body destroyed by collisions.

Up to the last decade, and before the new large catalogues, the research in that particular field was very well organized, in three separate steps, concerning different people and knowledge. Let us summarize these three sequential steps and their main past characteristics :

- The calculation of proper elements for the asteroids.

Based on a double elimination of periodic terms, of short (orbital scale) and long (secular scale) periods, the purpose is to calculate three invariants of motion, supposed to be very close to each other for the members of a same family. This calculation is very technical; after a pioneer work of Kozai (1962) let us mention two classical approaches : a completely analytical one, developed by Yuasa, Knežević and Milani (see Yuasa 1973, and Milani and Knežević 1994), and a semi-numerical one, introduced by Williams (1989) and adapted by Lemaitre and Morbidelli (1994). Specific regions, crossed by mean motion or secular resonances require a special treatment, by adding the resonant angle to the slow variables. The dynamics is therefore more complex; the definition of the corresponding invariants of motion is purely local, and not compatible with the non resonant one. The work of Schubart for the Hilda's (jovian resonance 3/2), is summarized on the web site http://www.rzuser.uni-heidelberg.de/s24/hilda.htm. Let us also mention the paper of Morbidelli (1993) on the calculation of secular proper elements.

- The application of a clustering method.

On the three dimensional set of proper elements and taking into account a local background density of population, clusters and groups of asteroids are identified and considered as candidates for a common origin, unless the mineralogical compositions show incompatibilities. Two approaches were developed and compared: a Hierarchical Clustering Method (HCM) developed by Zappala *et al.* (1990) and a Wavelet Analysis Method (WAM) due to Bendjoya *et al.* (1991).

The principle of HCM is the use of a specific and adequate metric in the proper elements $(a, e, \sin i)$ space to search the nearest neighbour of each body :

$$d = na(C_a(\delta a/a)^2 + C_e(\delta e)^2 + C_i(\delta \sin i)^2)^{\frac{1}{2}}$$

where $C_a = \frac{5}{4}$, $C_e = 2$ and $C_i = 2$ is a possible and usual choice for these coefficients ; d is given here in m/sec, represents the "distance", while δa, δe and $\delta \sin i$ are the differences in semimajor axis, eccentricity and sine of inclination between the two bodies.

A diagram called *dendrogram* connecting all the objects with increasing distances is drawn, the results are given in the form of stalactites. Systematic comparisons with stalactites obtained from quasi-random simulated populations are performed and a cut off of about 100 m/sec is used to identify the main families, on a sample of a few thousands objects.

WAM idea is to superimpose a network to the set of data (in 2 and 3 dimension proper elements space); at each node of the network, a wavelet coefficient is calculated, as an indicator of local increase in density in the vicinity of the node. Comparisons with a random distribution of proper elements (with the same local density) are performed, and the significant coefficients are used to model the cluster responsible for these high values. Bendjoya *et al.* (1993) showed that both methods (HCM and WAM) were able to identify families (even with potentially odd structures like filaments in proper elements space) : they used simulations in which synthetic families were put inside background populations of increasing spatial density.

- The spectral analysis of the candidates to a common origin.

From the beginning, the specialists in proper elements and families have claimed that it was necessary to test the mineralogical composition of the different bodies, to be sure of their common origin. A catalogue of spectra was then elaborated by the observers, mapping the position, brightness, and the 5-color CCD photometry of the concerned objects (see Cellino *et al.* 2002 for references).

A classification of the asteroids was proposed by Tholen (1984) on a sample of 405 objects; 14 classes were distinguished, called: C (for dark carbonaceous objects), S for Silicate, A, B, D, F, G, T for some spectrally distinct curves, E, M, P, three classes reduced to X in the absence of the albedos, and 3 classes with a unique object : Q (Apollo), R (Dembowska) and V (Vesta).

The definition of a family, given in the glossary of Asteroids III in 2002 corresponds to this static and sequential approach of the phenomenon : *a family is a statistically significant cluster of asteroids in proper elements space that may share a common origin perhaps by the collisional disruption of a larger body.*

The situation has evolved quite rapidly, thanks to the availability of very large catalogues of asteroids, and spectacular new surveys, in parallel with several bright theoretical results. The proper elements are not really constant, once non gravitational forces are taken into account, the clustering methods have to mix dynamics and observations so to recover obvious new families (since the recent available results of large surveys). Moreover, as a consequence of the use of CCD cameras, a new taxonomy is introduced for the classification, more adapted to the data.

In the following sections those important contributions of the last years in each of the mentioned fields, are listed, still considered as independent. In the last section I shall show how all these features are related to each other and should be considered together in the future.

2. The new synthetic proper elements

A new systematic way of calculating the proper elements has been developed by Knežević and Milani (2003); it consists of a purely numerical procedure, based on time integrations over periods of 2 Myr (10 Myr for some cases). The elimination of the short-periodic effects is performed through digital filters. A Fourier analysis allows to identify proper modes. A whole portrait of the belt can be drawn, with systematic and individual checking of the accuracy (standard deviations and maximum excursions). The maximum Lyapunov characteristic exponent is given so to measure the chaotic diffusion. The relative velocities at breakup are estimated with an error of about 5 m/s (through the Gauss equations - see for example Zappala *et al.* 1996) and the accuracy has been globally improved by more than a factor 3.

Full information is available on the site http://hamilton.dm.unipi.it/cgi-bin/astdys/astibo, with complete catalogues regrouping more than 100 000 planets, with analytical and synthetic proper elements.

Proper elements are also provided (using the synthetic technique) for very specific groups like the Trans-Neptunian Objects (TNO) or the Trojans.

3. The Trojan and Hilda specific proper elements and corresponding families

Besides the synthetic catalogues, an interesting semi-analytical work has been developed for the Trojans by Beaugé and Roig (2001). They provided a set of proper elements for the two Lagrange points, L4 and L5, and identified several obvious families (or groups) about L4, while the situation about L5 is less structured, with only one or two potential groups. Let us also mention the agreement with the purely numerical approach of Burger *et al.* (1998).

For the Hilda group, a new semi-numerical and promising approach has been proposed by Miloni *et al.* (2004).

4. The surveys

Many more data concerning lightcurves are now available, due to systematic surveys of regions of the sky. Besides the Space Watch Survey and several Subaru japanese surveys, let us describe three of the most often quoted in the search for families :

• SMASS : Small Belt Asteroid Spectroscopy Survey divided in two phases : SMASS I (Xu *et al.* 1995) and SMASS II (Bus and Binzel 2002a); they catalogue the spectra of

1447 asteroids of small and medium size, mainly from the inner main belt, or from the NEA population for SMASS I, and from the middle-region of the belt, for SMASS II.

• S3OS2 : Small Solar System Objects Spectroscopy Survey : Lazzaro *et al.* (2001) with more than 734 objects

• SDSS MOC : Sloan Digital Sky Survey Moving Object Catalog (Ivezic *et al.* 2000 or Stoughton *et al.* 2002) which has already covered one eighth of the entire sky, which means 125 283 objects (amongst them, 35 401 known asteroids were identified).

5. The official families

The specialists of the clustering methods (see Zappala *et al.* 1995) carried out a new comprehensive search for families based on the largest catalogues ever used for this purpose, and using both methods of identification developed by Torino and Nice teams. A review of these methods and the results of these calculations is also summarized in Bendjoya and Zappala (2002).

The simulations were based on a much larger set of proper elements (12 487 minor planets) than in their first comparison . 20 official families were identified with the criterion of sharing at least 75 % of objects, and less than 10 families could be added to this list if we accept a criterion of 50 % of common members for the two theories. Except for a few exceptions, those families are not very different from the previous ones; the number of planets in each family is just much larger, gathering a quasi constant percentage of the total number of objects (about 35% of the whole asteroid population belong to families).

6. The taxonomy by Bus

Based on the SMASS survey, and due to the intensive use of CCD cameras, a new taxonomy was recently proposed by Bus (1999) or Bus and Binzel (2002b) based on 1447 asteroids, and respecting the following properties :

• It is still connected to Tholen's classification, easy to use and applicable by others

• It is based only on spectral features and on multivariate analysis

• It shows agreement with natural groups, for the sizes and boundaries of the taxonomic classes

• It provides 3 spectral principal components : the spectral slope, and two others, mixing the data coming from the five colors observations with different weights (see Bus and Binzel 2002a for complementary information).

The result gives 26 classes, in which we find Tholen's classes; new subdivisions of the classes C, S, X appear (C, Cg, Cgh, Ch for example). This taxonomy is already adopted by other groups, as shown by the results of Carvano *et al.* (2003) or Mothé-Diniz *et al.* (2003).

7. The background and family density of population

A very significant discussion for the understanding of the families concerns the densities of population inside and outside the families. Since official families have been identified, the size distribution of families have been analyzed and similar curves have been drawn for the background population of asteroids (objects not belonging to a family).

These analyses (for example, Zappala and Cellino 1996) concluded that families likely dominated the asteroid inventory., due to their steep size distribution. A simple extrapolation law was used to extend this result to very small objects, not yet observed, and it tended to prove that most of the objects should belong to families. Different authors completed this theory by adding new elements to the discussion : for example, the geometry of the family offers an explanation of this steep distribution of sizes (see Tanga *et al.* 1999).

A possible explanation was recently given in a paper of Morbidelli *et al.* (2003); the idea is to debiase the distribution of families relative to that of the background, using the data coming from the new surveys as calibration parameters. Their conclusion is that, after treatment of the samples by this procedure, for the objects with a magnitude $H > 13$ (which roughly means a radius less than 10 km), most families have a magnitude distribution even *shallower* than that of the background. The population of 1 km size objects could then be dominated by background bodies and not by families members.

8. The collisions and the fragmentation phase

The birth of a family is usually due to a collisional event, with another asteroid or a comet. This kind of mechanisms were simulated in laboratories, where projectiles of different size and compositions were crashed on meter size rocks. The fragments were described, followed, classified, the ejection velocities were measured and several scenarios were build, following all these features. At that point occurs a problematic step : are we sure that everything really observed in laboratories corresponds effectively to the reality in the space, after application of a simple scale factor ? An affirmative answer means that the fragmentation models do not depend on the size of the bodies but only on their relative scales. The values obtained by this technique for the fragment dispersion velocities were systematically smaller than the ones calculated by the theory (for a review, see Holsapple *et al.* 2002).

Fortunately, the collisional dynamics was considerably improved and understood those last years, in particular through the work of Michel and collaborators (Michel *et al.* 2001, 2002 and 2003). These authors performed a very technical and complete numerical simulation of collisions, introducing a fragmentation phase in the process, thanks to an hydrocode (software 3DSPH) inspired by the fluid dynamics concepts. They simulated the dispersion and reaccumulation of the fragments (1 km size minimum) after a collision. The process also allows diffusion, and for an asteroid size body, the catastrophic event (from the impact to the end of the partial reaccumulation) takes a few hours or days.

Tests have been made on the classical families Eunomia, Koronis, Flora and for the new cluster Karin.

We can summarize the conclusions as follows : the reaccumulation process differs from one case to the other one, due to the internal structures and masses of the bodies. For Eunomia (284 km size body) the largest fragment after reaccumulation has reassembled 70 % of the initial mass before impact. For Flora (154 km size body) the percentage is of 56 %, and for Koronis (119 km size body) it is only of 4 %. It seems quite obvious that the largest family members must be made of reaccumulated fragments and be closer to rubble piles than to monolithic rocks. Again, as in the laboratories experiments, the ejection velocities are much smaller than the 100 m/sec predicted by independent estimates of Dell'Oro *et al.* (2004).

9. The ejection of the fragments and the ageing of the proper elements

As already mentioned in the previous sections, the ejection velocities of the fragments after an impact seem to be sur-estimated by the theories, in comparison with the data coming from the laboratories experiments and the hydrocode numerical simulations.

The theoretical calculation of these velocities refers to papers of Zappala *et al.* (1996) and Zappala *et al.* (1997) and aims to rebuild the velocities field at the impact with the present orbital motions of the members of the family. The method was based on the likely existence of an axial symmetry.

The first step is the determination of the proper elements of the different members, the second one is the calculation, in the proper elements space, of the parameters δa, δe and δi (respectively the differences in semi major axis, eccentricity and inclination between the member and the estimated center of mass of the group). The third step is to replace the quantities δa, δe and δi in the lefthand side of the Gauss equations, so to extract, from the righthand side, the three components of the velocity, tangential, V_T, radial V_R and perpendicular V_W. The principle is simple, except that the equations also depend on f, the true anomaly, for V_T and V_R and on f and ω (the argument of pericenter), for V_W. The proper elements sets is three dimensional; the values of f and ω at the impact time, are not known.

They are then determined through an optimization process, so to maximize the axial symmetry of the reconstructed field. The results (the reconstructed fields) were convincing, except that the ejection velocities took very high values, often greater than 100 m/sec.

Different authors completed this theory, for example (Cellino *et al.* 1999) and (Pisani *et al.* 1999).

However, these classical studies consider that proper elements are constants for ever; it is now commonly accepted that it is not the case. Several features could play a role in the process. Dell'Oro *et al.* (2004) introduced a systematic noise on the proper elements, so to simulate this ageing of the constants of motion; they re-applied the same technique for the fragment velocities, corrected the estimators and obtained different results, in better agreement with the data.

10. The non gravitational forces and evolving families

In the paper of Dell'Oro *et al.* (2004) a list of effects potentially responsible for the ageing of the families is given :

- The intrinsic noise in proper elements
- The presence of resonances
- The incompleteness of the families (there are biases in the observations)
- The close encounters with large asteroids like Ceres
- The Low-velocities collisions between asteroids
- The presence of non gravitational forces, like Yarkovsky
- The family structure itself

The non-gravitational forces, combined with resonances, have already proved their efficiency; a series of papers from Vokrouhlický *et al.* (2001), Nesvorný *et al.* (2002a) and Bottke *et al.* (2001) showed how the addition of the thermal Yarkovsky effect could affect the stability of the proper elements.

Let us recall the principle of the Yarkovsky force : the surface of an asteroid is heated by the Sun during its day and cools off during its night; the asteroid emits more heat from its afternoon sight. This unbalanced thermal radiation produces a tiny acceleration and acts on any asteroid with a diameter smaller than 20 km. The numerical simulations make use of synthetic families (less dispersed than the real ones by a factor 3), and do not consider planets with diameters D smaller than 2 km. They are based on isotropic ejection fields of velocities (not exceeding 100 m/sec at infinity).

The following families were selected by the authors mentioned above : Flora, Maria, Eunomia and Koronis for the S-Type, Themis and Dora for the C-type; the thermal parameters are chosen with respect to the taxonomy because they are linked to the conductivity of the matter. Integrations were performed over periods of time between 300 and 500 Myr, with the integrator SWIFT of Levison and Duncan (2000), with the perturbation of 4 outer planets (except for Flora, where a more complete set of planets was used). Numerical proper elements were calculated for each member of the synthetic family. The results show an excellent concordance between the simulations and the expected positions of real bodies. The role of the resonances, even the three-body ones, appears clearly, cutting some families in a sharp way and exciting the eccentricities (see also Tsiganis *et al.* 2003). Some of the most spectacular figures can be found on the following web sites :

- http://www.boulder.swri.edu/davidn/yarko/yarko.html
- http://www.boulder.swri.edu/bottke/Animation/koronis.gif.

The Yarkovsky effect introduces a kind of random walk, pushing the objects in or close to the resonances; it allows the members of the families to diffuse in the proper elements space, giving a non uniform spreading shape to the cluster. The concept of *dynamical* families, in opposition with *static*, is really introduced here. The results nicely reproduce the proper semi-major axis distributions, but a fit of the eccentricities and of the inclinations of the members of the different families is more difficult to achieve.

With the same philosophy, the close encounters with the largest asteroids (like Ceres) were tested for some families, like Gefion and Adeona (more concerned, because of their relative position) by Carruba *et al.* (2003). Their simulations showed a slightly modification of the proper elements due to Ceres, but nothing comparable with Yarkovsky effect, for which they reproduced similar qualitative results as the previous studies, even with different parameters and adjustment of constants.

The Yarkovky effect was measured for the first time thanks to several radar observations of the NEA asteroid 6489 Golevka. This 0,5 km size asteroid was observed in 1991, 1995, 1999 and 2003, and a shift of 15 km on the semi major axis was clearly detected on this period of 12 years by Chesley *et al.* (2003).

This gives a first real measurement of the Yarkovsky effect on the semi-major axis : more than 10^{-4} AU per Myr, for a 1 km size body, which is, as mentioned by Dell'Oro *et al.* (2004), smaller than the values used by Bottke *et al.* (2001) or Morbidelli and Vokrouhlický (2003), but larger than those by Spitale and Greenberg (2001) or Carruba *et al.* (2003).

11. A new cluster : Karin

A new cluster was discovered by Nesvorný *et al.* (2002b) using the numerical integration methods mentioned previously, including Yarkovsky forces; they started from a subtle filament-like structure appearing in the Koronis family, identified by looking at the results obtained by the extended database of asteroids proper elements. It is called Karin and could be composed of 39 potential members, with confirmation for 13 of them.

The family contains two large bodies, respectively of 19 and 14 km size : Karin (832) and 1990FV (4507,) and a few others in the range of 2 to 7 km size. Backwards integrations of the 13 members of the family converged to one single orbit, this of the pre-breakup parent object. A collision between an initial body of 25 km with another one of 3 km, at about 5 km/sec, could be at the origin of the breakup, 5.8 Myr ($\pm$ 0.2 Myr) ago. Karin is then a young family and represents a very interesting event, especially for space weathering, because it is quite recent event, precisely dated.

12. The new families

A very convincing approach to the classification has been introduced very recently by Nesvorný *et al.* (2004). This is a real interaction between *dynamical* proper elements, *mixed* clustering methods, *taxonomic* components coming from *new* surveys and *large* catalogues, so to get a sample of 106 264 minor planets in which 40 families are clearly identified, regrouping 38 625 asteroids (about 36,3 % of the whole population).

The authors use the data coming from the SSDC MOC survey (more than 125 000 objects), classified in the 5 colors bands (u, g, r , i, z),. They run the PCA algorithm (Principal Component Analysis) , to calculate (among all the combinations of the 5 possibly correlated variables), the 2 principal components, maximizing the de-correlation. They obtain the following linear combinations :

$$PC_1 = 0.396(u - g) + 0.553(g - r) + 0.567(g - i) + 0.465(g - z)$$
$$PC_2 = -0.819(u - g) + 0.017(g - r) + 0.090(g - i) + 0.567(g - z)$$

(PC standing for Principal Component).

They generalize the notion of distance used in the HCM method; the distance between two objects A and B is now calculated in a 5 dimensional space and is given by the expression :

$$d(A, B) = n\,a\big(C_a(\delta a/a)^2 + C_e(\delta e)^2 + C_i(\delta \sin i)^2 + C_{PC}(\delta PC_1)^2 + C_{PC}(\delta PC_2)^2\big)^{\frac{1}{2}}$$

always expressed in m/sec where $C_a = \frac{5}{4}$, $C_e = 2$, $C_i = 2$ and $C_{PC} = 106$.

The cut off values are chosen between 0 and 150 m/sec. The numerical coefficients are given here as examples; other choices are also relevant. Depending on the chosen values, large families are re-discovered or smaller sub-structures appear; for example, with a cut off of 23 m/sec, Koronis is still a unique block, but with 12 m/sec, Karin is identified. Of course, to distinguish the sub-structures of the main families (the *micro families*) a large number of bodies is necessary and such studies have to be performed with a minimum of 100 000 bodies to give significant results.

13. The age of the families and the space weathering

Concerning the attempts to estimate the age the families, let us recall the precursor approach concerning the family of Veritas, performed by Milani and Farinella (1994), using a chaotic chronology. For the first time, the formation of a family appeared as a very recent event : between 50 and 100 Myr for the first approach, confirmed by Knežević and Pavlović (2002). Following Nesvorný *et al.* (2004) we can list a short review of past and present methods for dating a family : the SFD (Size - Frequency Distribution) developed by Marzari *et al.* (1995), the Global Main Belt SFD, by Durda and Dermott (1997), the family spreading via thermal forces and numerous tiny resonances, the collisional ejection of fragments corrected by Dell'Oro *et al.* (2004) and the backward numerical integration of orbits (for young families like Karin) used by Nesvorný *et al.* (2002b).

Many of the catastrophic events at the origin of the breakup of a family can now be dated and, surprisingly, some of them are very young (less than 1 Myr). These families should be used as a fundamental constraint in order to derive the original ejection velocities, since in one Myr, Yarkovsky has not had the time to modify, in a relevant way, the family structure.

The age of the families has a very important consequence : it allows to measure the traces of space weathering, by comparative study of the surfaces of members of old and young families. By space weathering we refer to all the processes that alter optical properties of surfaces of airless bodies.

The first studies have shown (Chapman 2004) evidence of space weathering, with first estimations of the timescales; it is now recognized that an age dependent component alters asteroid colors.

14. The spin vectors

Let us mention a very interesting result obtained by the observers and experts in photometry.

In the Koronis family, systematic higher lightcurves amplitudes were observed, in comparison with the other regions. Mechanisms were proposed to explain this situation : either it was a mechanism linked to the process of formation of the family, or it was due to the alignment of the spin vectors of the different bodies. Slivan (2002) and Slivan *et al.* (2003) showed a bimodality in the spin orientation of Koronis family members, and proved that at least 10 members of Koronis had aligned spin vectors (using only the photometric data of the family). Morbidelli and Vokrouhlický (2003) presented a first tentative explanation of this peculiar situation, based on Yarkovsky effect and the presence of a resonance with Saturn.

15. Conclusions

The important point to mention is that dynamical and observational features can now be treated together and not separately; the new clusters or the new members of the known families will be probably discovered with those combined approaches.

Acknowledgements

I would like to acknowledge Alessandro Morbidelli and David Nesvorný for useful discussions, precious documents and preprints, and very kind help. I also thank A. Cellino for his comments and complementary information.

References

Beaugé, C. and Roig, F. 2001, *Icarus* 153, 391

Bendjoya, Ph., Slézak, E. and Froeschlé, C. 1991, *Astron. Astrophys.* 272, 651

Bendjoya, Ph., Cellino, A., Froeschlé, C. and Zappala, V. 1993, *Astron. Astrophys.* 272, 651

Bendjoya, Ph. and Zappala, V. 2002, in: Bottke, W.F., Cellino, A., Paolicchi, P., Binzel, R.P. (eds.), *Asteroids III*, Univ. Arizona Press, Tucson, p. 613

Bottke, W.F., Vokrouhlický, D., Broz, M., Nesvorný, D. and Morbidelli, A. 2001, *Science* 294, 1693

Burger, Ch., Pilat-Lohinger, E., Dvorak, R. Christaki, A. 1998, in: J. Henrard and S. Ferraz-Mello (eds.), *Proceedings of the IAU Colloquium 172, Impact of Modern Dynamics on Astronomy*, Namur.

Bus, S.J. 1999, *Compositional structure in the asteroid belt : results of a spectroscopic survey.*, Ph.D Thesis, MIT, **DAI-B 61-01**, p. 311.

Bus, S.J. and Binzel, R.P. 2002a, *Icarus* 158, 106

Bus, S.J. and Binzel, R.P. 2002b, *Icarus* 158, 146

Carruba, V., Burns, J.A., Bottke, W., and Nesvorný, D. 2003, *Icarus* 162, 308

Carvano, J.M., Mothé-Diniz, T. and Lazzaro, D. 2003, *Icarus* 161, 356

Cellino, A., Michel,P., Tanga,P., Zappal,V., Paolicchi,P. and Dell'Oro, A. 1999, *Icarus* 141, 79

Cellino, A., Bus, S.J., Doressoundiram, A. and Lazzaro, D. 2002, in: Bottke, W.F., Cellino, A., Paolicchi, P., Binzel, R.P. (eds.), *Asteroids III*, Univ. Arizona Press, Tucson, p. 633

Cellino, A., Dell'Oro, A and Zappala, V. 2004, *Planet. Space Sci.*, In press.

Chapman, C.R. 2004, *Annual Rev. Earth Planet. Sci.* 32, 539

Chesley, S.R., Ostro, S.J., Vokrouhlický, D., Capek, D., Giorgini, J.D., Nolan, M. C., Margot, J., Hine, A. A., Benner, L. A. and Chamberlin, A.B. 2003, *Science* 302, 1739

Dell'Oro, A., Bigongiari, G., Paolicchi, P. and Cellino, A. 2004, *Icarus* 169, 341

Durda, D.D. and Dermott, S.F. 1997 *Icarus* 130, 15

Hirayama, K. 1918 *Astron. J.* 31, 185

Holsapple, K., Giblin, I., Housen, K., Nakamura, A. and Ryan, E. 2002, in: Bottke, W.F., Cellino, A., Paolicchi, P., Binzel, R.P. (eds.) *Asteroids III*, Univ. Arizona Press, Tucson, p. 443

Ivezic, Z. and 32 colleagues, 2001, *Astron. J.* 122, 2749

Knežević, Z. and Pavlović R. 2002, *Earth, Moon, Planets* 88, 155

Knežević, Z. and Milani, A. 2003, *Astron. Astrophys.* 403, 1165

Kozai, Y. 1962, *Astron. J.* 67, 591

Lazzaro, D., Carvano, J.M., Mothé-Diniz, T., Angeli, C.A., Duffard, R., Florczak, M. 2001, in: Claria, J.J., Lambas, D.G. and Levato, H. (eds.), *Proceedings of the 10th regional meeting of Latin America, Rev. Mexicana Astron. Astrofys., Serie Conf.* 14, 104.

Lemaitre, A. and Morbidelli, A. 1994, *Cel. Mech. Dyn. Astron.* 60, 29

Levison, H.F. and Duncan, M.J. 2000, *Astron. J.* 120, 2117

Marzari, F., Davis, D. and Vanzani, V. 1995, *Icarus* 113, 168

Michel, P., Benz, W., Tanga, P. and Richardson, D.C. 2001, *Science* 294, 1696

Michel, P., Benz, W., Tanga, P. and Richardson, D.C. 2002, *Icarus* 160, 10

Michel, P., Benz, W. and Richardson, D.C. 2003, *Nature* 421, 608

Milani, A. and Farinella, P. 1994, *Nature* 370, 40

Milani, A. and Knežević, Z. 1994, *Icarus* 107, 219

Miloni, O., Ferraz-Mello, S. and Beaugé, C. 2004, *Cel. Mech. Dyn. Astron.*, submitted.

Morbidelli, A. 1993, *Icarus* 105, 48

Morbidelli, A., Nesvorný, D., Bottke, W.F., Michel P., Vokrouhlický, D. and Tanga, P. 2003, *Icarus* 162, 328

Morbidelli, A. and Vokrouhlický, D. 2003, *Icarus* 163, 120

Mothé-Diniz, T., Carvano J.M., Lazzaro, D. 2003, *Icarus* 162, 10

Nesvorný, D., Morbidelli, Vokrouhlický, D. A., Bottke, W.F., and , Broz, M. 2002a, *Icarus* 157, 155

Nesvorný, D., Bottke, W.F., Dones, L. and Levison, H.F. 2002b, *Nature* 417, 720

Nesvorný, D. , Bottke, W.F., Levison, H. F. and Dones, L. 2003, *Astrophys. J.*, 591, 486

Nesvorný, D., Jedicke, R., Whiteley, R.J. and Ivezic, Z. 2004, *Icarus*, in press.

Pisani, E., Dell'Oro, A.,and Paolicchi, P. 1999, *Icarus* 142, 78

Slivan, S. M. 2002, *Nature* 419, 49

Slivan, S. M., Binzel, R. P., Crespo da Silva, L. D., Kaasalainen, M., Lyndaker, M.M. and Krco, M. 2003, *Icarus* 162, 285

Spitale J. and Greenberg R, 2001, *Icarus* 149, 222

Stoughton, C. and 191 colleagues 2002, *Astron. J.* 123, 485

Tanga, P., Cellino, A., Michel, P., Zappala, V., Paolicchi, P. and Dell'Oro, A. 1999, *Icarus* 141, 65

Tholen, D.J. 1984, *Asteroid taxonomy from cluster analysis of photometry*, Ph. D. Thesis, Arizona University, Tucson.

Tsiganis, K., Varvoglis, H. and Morbidelli, A. 2003, *Icarus* 166, 131

Vokrouhlický, D., Broz, M., Farinella, P. and Knežević, Z. 2001, *Icarus* 150, 78

Williams, J. G. (1989), in: R.P. Binzel, T. Gehrels and M.S. Matthews (eds.), *Asteroids II*, Univ. Arizona Press, Tucson, p. 1034

Xu, S., Binzel, R.P., Burbine, T.H. and Bus, S.J. 1995, *Icarus* 115, 1

Yuasa, M. 1973, *Publ. Astron. Soc. Japan* 25, 399

Zappala, V., Cellino, A., Farinella, P. and Knežević, Z. 1990, *Astron. J.* 100, 2030

Zappala, V., Bendjoya, Ph., Cellino, A., Farinella, P. and Froeschlé, C. 1995, *Icarus* 116, 291

Zappala, V. and Cellino, A. 1996, *Astron. Soc. Pacific Conf. ser.*, 107, 29

Zappala, V., Cellino, A., Dell'Oro, A., Migliorini, F. and Paolicchi, P. 1996, *Icarus* 124, 156

Zappala, V., Cellino, A., di Martino, M., Migliorini, F. and Paolicchi, P. 1997, *Icarus* 129, 1

Dynamics of Populations of Planetary Systems
Proceedings IAU Colloquium No. 197, 2005
Z. Knežević and A. Milani, eds.

© 2005 International Astronomical Union
DOI: 10.1017/S1743921304008609

Non-gravitational perturbations and evolution of the asteroid main belt

David Vokrouhlický[1], M. Brož[1], W.F. Bottke[2], D. Nesvorný[2] and A. Morbidelli[3]

[1]Institute of Astronomy, Charles University, Prague, V Holešovičkách 2, 18000 Prague 8, Czech Republic
email: vokrouhl@mbox.cesnet.cz, mira@sirrah.troja.mff.cuni.cz

[2]Southwest Research Institute, 1050, Walnut St., Suite 400, Boulder, CO-80302, USA
email: bottke@boulder.swri.edu, davidn@boulder.swri.edu

[3]Observatoire de Nice, Dept. Cassiopee, BP 4229, 06304 Nice Cedex 4, France
email: morby@obs-nice.fr

Abstract. Gravity is the most important force to affect the motion of bodies in the Solar system. At small sizes, however, additional forces must be taken into account to explain fine details of their translational and rotational motion, as well as parameters of their populations. This is because the strength of the non-gravitational perturbations typically increases as $\simeq 1/D$ toward small sizes D. The principal perturbation acting on macroscopic main-belt bodies (sizes up to several kilometers for timescales of about billion years) is due to the anisotropic thermal re-radiation of the absorbed sunlight. In orbital dynamics this is known as the Yarkovsky effect, while in rotational dynamics the same physical phenomenon is called the YORP (Yarkovsky-O'Keefe-Radzievskii-Paddack) effect. We review the main observational implications of the Yarkovsky/YORP effects as understood and evidenced today.

Keywords. minor planets, asteroids; interplanetary medium; radiation mechanisms: thermal

---

1. Introduction

The motion of celestial bodies has been a domain of discovery, application and testing of the gravity force for a long time. Less than two centuries ago, the motion of comets and dust particles in their tails required, for the first time, to account for the recoil effect of sublimated gas/dust and the solar radiation pressure. Asteroids and their fragments down to microscopic sizes, although inactive, are susceptible to pressure due to the absorbed, scattered and thermally re-processed solar radiation. At very small sizes, effects of solar wind drag and Lorentz force due to the interplanetary magnetic field become important. Here we review evidence that non-gravitational perturbations affect the orbital and rotational dynamics of asteroids in the main belt. We restrict to this topic, leaving the issue of cometary dynamics and, for the most part, interplanetary dust dynamics to reviews by S.R. Chesley and M.C. Wyatt (this volume).

Common to all non-gravitational effects relevant for our discussion is their $\propto 1/D$ fading for bodies with large diameter D. This is because their strength is directly proportional to the amount of photons or solar wind particles intercepted by their surface that scales as $\propto D^2$. Since their mass increases as $\propto D^3$, the non-gravitational acceleration has dependence inversely proportional to size. With a finite observation accuracy, and a finite possible orbital evolution period (limited by the Solar system age), this implies that a given non-gravitational perturbation is relevant for bodies up to some largest size only. Similarly, the strength of non-gravitational effects fades at small sizes. This occurs for various reasons: e.g. at sub-micron sizes solar visible radiation ceases to interact with dust particles and does not impart momentum to them; similarly, thermal conduction through a centimetre-sized meteoroid becomes efficient enough to produce a uniform temperature and the Yarkovsky force diminishes. This lower limit means that efficiency of non-gravitational perturbations is usually restricted to some finite interval of sizes. Poynting-Robertson (PR) drag appears to be important for particles between $\simeq 1\,\mu$m and $\simeq 1$ cm, while the Yarkovsky effect dominates in the size range between $\simeq 10$ cm till $\simeq 10$ km.

Unlike gravitational perturbations, which can be often approximated using a minimum-parameter point-like model, non-gravitational effects depend on a number of parameters, e.g. albedo and other physical constants, rotation state, etc. This makes models involving these effects less than optimum constrained. Still, insufficiency of the gravitational-only models to explain a number of observational facts and the right order-of-magnitude arguments make them credible. The models involving non-gravitational effects thus usually fit statistical parameters of a large sample of objects (such as the near-Earth asteroids or the asteroid families). However, the Yarkovsky effect has also been directly detected in the motion of a single object (Vokrouhlický *et al.* 2000, 2004b,c; Chesley *et al.* 2003) – in that case it is possible to estimate the unknown physical parameters of the body.

Given our tight page limit, we restrict ourselves to review recent results on the Yarkovsky and YORP effects, with only a brief comment on how radiation forces affect interplanetary dust particles. More detailed reviews on these topics can be found in: (i) Dermott *et al.* (2001) who discuss the radiative effects, including the PR drag, for asteroid dust particles and (ii) Bottke *et al.* (2003) who discuss the thermal effects. A textbook which reviews both subjects is Bertotti *et al.* (2003).

2. Solar radiation pressure: dust band orbital evolution

Space-borne infrared tools of modern astronomy have significantly changed our understanding of the zodiacal cloud, primarily showing it is complex and highly structured. Among these coherent and localized features, the most outstanding are the bands discovered by the IRAS spacecraft (Low *et al.* 1984), which are associated with dust produced in the main asteroid belt (Dermott *et al.* 1984). Having been able to match the mid-latitude of the high-inclination dust band at $\simeq 9.35°$, Nesvorný *et al.* (2002) convincingly argued these particles were produced by recent asteroid disruption events like Karin and Veritas (Sec. 5). Yet, the age of these events is significantly more than the dynamical transport time of $10-200\,\mu$m particles from the main belt. This means the currently observed dust population cannot originate from the parent break-up directly. Since the estimated collisional lifetime of interplanetary dust particles with sizes smaller than $\simeq 150\,\mu$m is longer than the depletion time by the PR effect (e.g. Grün *et al.* 1985; Dermott *et al.* 2001), the dust band population should be significantly affected by the PR drag. A testable prediction from this model is a shallow size distribution because small particles are more efficiently removed from the system by PR drag (e.g. Dermott *et al.* 2001).

Using numerical integrations, Grogan *et al.* (2001) built a model of such dynamically evolving dust bands and compared the predicted shape and amplitude of their structures in the several IRAS wavebands†. The best fitted value of the power-law index of the incremental mass distribution was $q \simeq -1.4$, significantly less than what is expected

---

† Note that each of the wavebands preferentially samples radiation emitted by particles of different sizes, allowing thus to characterize their size distribution.

for the collisionally evolved population of particles ($q \simeq -11/6 \simeq -1.8$; e.g. Dohnanyi 1969; O'Brien & Greenberg 2003). The difference between the power-law index of the observed population and that of the collisionally evolved, but otherwise dynamically static, population is about -1 for the size distribution‡, and this matches very well the model of strong dynamical depletion by PR drag. Less evident, but often cited, observational argument for PR-drag-dominated depletion of the main asteroid belt grain population in the size range $\simeq (1 - 100)\,\mu$m is the measurement of the LDEF spacecraft (Love & Brownlee 1993). The observed shallow mass power-law index ($q \simeq -1.2$) is interpreted similarly to the Grogan *et al.* (2001) analysis of the dust band particles. Here, however, a certain fraction of LDEF detected particles might be of cometary origin, which makes interpretation of the results less clear.

3. Yarkovsky and YORP effects: principle and early applications

Solar radiation heats cosmic bodies in an asymmetric way: the noon side receives sunlight and gets slightly hotter, while the midnight side is cooling. The temperature distribution on the surface is affected by the ability of the material to conduct heat (thermal inertia), by rotation rate and by its ability to re-emit radiation. It almost always keeps some of irradiation asymmetry (with an exception of very small bodies, $\leqslant 10$ cm for plausible materials, which are kept isothermal by efficient internal heat conduction). As the body re-emits thermally, it experiences a recoil force proportional to the linear momentum carried away by thermal photons; assuming the emitted radiation is in a local equilibrium with the surface material, the infinitesimal force applied on the body is proportional to the fourth power of the surface temperature. The resulting force applied on the whole body is non-zero because of the persistent temperature asymmetry. In this way the body is self-accelerated by its own thermal radiation; this perturbation is called the Yarkovsky force (e.g. Bottke *et al.* 2003). Since asteroids and their fragments typically have an irregular shape, the very same thermal effect results in a net torque; Rubincam (2000) coined an acronym YORP for the first letters of scientists (Yarkovsky-O'Keefe-Radzievskii-Paddack) whose work was directed toward recognizing the importance of this phenomenon in the long-term evolution in asteroid and meteoroid rotation.

The finite value of the surface thermal inertia produces a secular drift of the body's semimajor axis. Direct solar radiation pressure on a body of an arbitrary shape cannot produce such an effect, and even if albedo variations are taken into account, the net semimajor axis change is small (e.g. Vokrouhlický & Milani 2000). Difficulties in estimating the Yarkovsky-induced semimajor axis drift rate for a given body stems from its dependence on material (thermal) properties and rotation states. Asteroid observations in the infrared and lightcurve analyses can constrain these parameters, but not precisely.

The YORP effect produces a steady change in an asteroid obliquity and rotation rate and can thus significantly modify the distribution of these parameters for small asteroids and meteoroids† (e.g. Rubincam 2000; Vokrouhlický & Čapek 2002). Unlike in the Yarkovsky effect case, surface thermal inertia is not necessary for YORP to operate secularly; moreover, solar radiation pressure produces a similar perturbation (e.g. Rubincam 2000). It has been however found that thermal inertia does affect the net result (e.g.

---

‡ Note the power law index b of the incremental size distribution is related to the power law index q of the incremental mass distribution via $b = 3q + 2$.

† Because the Yarkovsky effect depends also on rotation parameters, there is a complicated interplay between both thermal effects.

Čapek & Vokrouhlický 2004). In addition to the Yarkovsky effect parameters, the YORP torque depends sensitively on the body's shape.

Meteoroid transport from the main belt to Earth was the original motivation to bring the Yarkovsky effect into planetary science (e.g. Burns *et al.* 1979). That was also the topic of primary interest when Afonso *et al.* (1995), Rubincam (1995, 1998) and Farinella *et al.* (1998) revived the Yarkovsky effect from years of hibernation. These works put the idea into a modern context where part of the transport scenario is due to gravitational resonant effects. Thus Bottke *et al.* (2000) numerically studied the interaction of the Yarkovsky migrating meteoroid orbits with a grid of weak resonances in the inner part of the main asteroid belt, while Vokrouhlický & Farinella (2000) developed a semi-analytical model to fit the observed distribution of cosmic ray exposure ages of chondrites by applying Yarkovsky-driven delivery to the resonances.

In what follows, we review recent Yarkovsky and YORP effect results.

4. Yarkovsky effect: sustaining mechanism for unstable populations

A population of asteroids residing on planet-crossing orbits, and on Earth-crossing orbits in particular, must be continously resupplied from some vast reservoir of objects. This is because their orbits are dynamically unstable, with their typical fates being they fall into the Sun, are ejected from the Solar system, or they hit one of the terrestrial planets. The nature of this process, and thus the origin of near-Earth asteroids (NEAs), has been debated for years. Progress in NEAs discovery programs along with new computational facilities over the last decade allowed the construction of a model in which majority of NEAs are former escapees from the main asteroid belt (e.g. Bottke *et al.* 2000, 2002). Two major constrains on the delivery mechanism(s) follow from this work: (i) there are about 220 multi-kilometre asteroids to be resupplied into the planet-crossing orbits every My, a majority of which comes from the main asteroid belt, and (ii) the cumulative size distribution of the de-biased multi-kilometre population of NEAs can be approximated as $N(< H) \propto 10^{\gamma H}$ with $\gamma \simeq 0.35$ (in the absolute magnitude range $15.5 < H < 18$). This latter finding should be compared with the estimated size distribution of the main belt population in the same size range that leads to $\gamma \simeq 0.26$ (e.g. Ivezić *et al.* 2001).

A standard model for NEA origin, with regard the asteroid source, assumed collisional ejecta being thrown into the major orbital resonances acting as the immediate routes to deliver material onto planet-crossing orbits. However, Zappalà *et al.* (2002) reviewed population of plausible parent bodies of multi-kilometre fragments near the major resonances and found the mechanism short to explain the constraint (i) above. Moreover, the value 0.35 of the γ-exponent in this model appears arbitrary and unrelated to the similar exponent of the population in the main asteroid belt. On the other hand, Morbidelli & Vokrouhlický (2003) developed a model where the bulk of the NEAs is delivered from the main asteroid belt to the adjacent resonance zones by the Yarkovsky effect. They noticed the high rate of 220 multi-kilometre asteroids per My is affordable within this model (with about right proportion of the three resonant routes; Bottke *et al.* 2000, 2002). Additionally, the difference of the γ exponents in the NEA and in the main belt populations appears naturally explained by the fundamental $\propto 1/D$ size dependence of the Yarkovsky effect strength. Guillens *et al.* (2002) analysed the observed numbered asteroids on low-inclination orbits in the unstable zone surrounding the 3/1 resonance. Indeed they found about 40% of these asteroids, with sizes up to $\simeq 15$ km, have short dynamical lifetime and will fall in the resonance within 100 My; these objects are the "next generation" of NEAs. Though their work does not prove these asteroids were placed onto

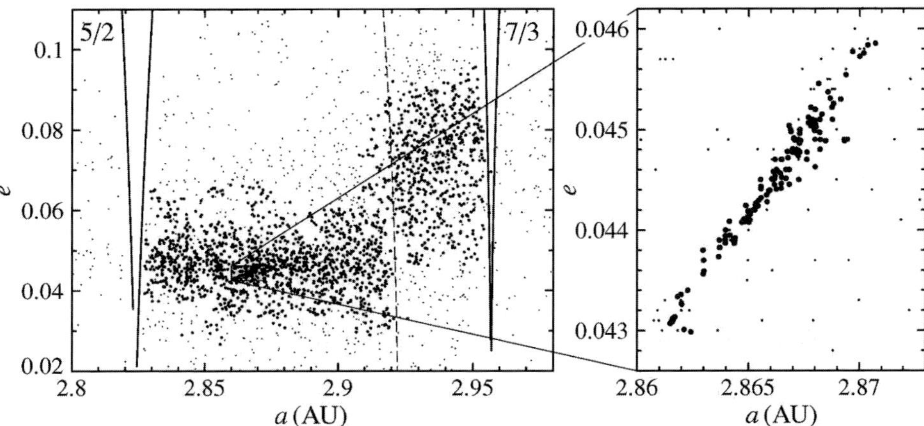

Figure 1. Koronis family and Karin sub-cluster in proper semimajor and eccentricity space (dots are background asteroids and full symbols are the family members). In the left figure, the major mean motion resonances with Jupiter – 5/2 and 7/3 – are shown to bracket the family. Dashed line is a location of the high-order secular resonance $g + 2g_5 - 3g_6$, that makes mean eccentricity in the Prometheus clan (semimajor axes larger than $\simeq 2.92$ AU) higher than for the remaining part of the family. In the right figure we show a zoom on the Karin sub-cluster.

their unstable orbits by the Yarkovsky effect, they believe this possibility is the most plausible scenario.

A similar approach has been used to explain the observed dynamically unstable resonant populations via Yarkovsky delivery of small asteroids from the main belt. In particular, Tsiganis *et al.* (2003) showed a population of objects in the 7/3 mean motion resonance with Jupiter had to be former Koronis and Eos members delivered into their current orbits by the Yarkovsky effect (see below for more details). Brož *et al.* (2004), using a similar model, succeeded in matching the parameters of asteroids on unstable orbits within the 2/1 mean motion resonance with Jupiter.

5. Yarkovsky effect: dynamical spreading of asteroid families

Farinella & Vokrouhlický (1999) were the first to point out that a large semimajor axis dispersion of small members in the Astrid family might be caused by the spreading of an initially more compact family by the Yarkovsky effect. Though not the best example, this work paved the idea of "Yarkovsky contribution" to the asteroid families evolution. The analysis of the non-gravitational effects in the families is important since that they represent a constrained sample of bodies which were, for the most part, created at the same instant. Though we often do not know when, the common time origin is an important circumstance.

A majority of recent studies have assumed the Yarkovsky effect is the driving mechanism producing the semimajor axis spreading in the families. We note however the work of Dell'Oro *et al.* (2004) who, by analysing current configuration of families, independently checked their width and determined they must have expanded by a factor of $\simeq 2$ in semimajor axis over time. Here we briefly review the results for three families.

Koronis.– While studying a bizarre shape of this family (Fig. 1), Bottke *et al.* (2001) brought the first "definitive" proof of Yarkovsky spreading within the asteroid families. First, these authors pointed out that the family identified with a standard clustering methods sharply terminates at the location of 5/2 and 7/3 mean motion resonances with Jupiter. Since these resonances efficiently remove asteroids from the main belt (e.g. Gladman *et al.* 1997), the shape of the family suggests a permanent flow in the semimajor axis from the centre of the family toward the resonances. This is exactly what the Yarkovsky model would predict. However, a major puzzle was the displacement in the mean eccentricity value of Koronis members with $a \geqslant 2.92$ AU (the so called Prometheus clan); no such effect is seen in mean inclination. No reasonable initial velocity field can create such a distorted distribution of fragments. Bottke *et al.* (2001) proved that the fragments, initially created near the center of the family, migrated outward from the Sun and reached a high-order secular resonance $g + 2g_5 - 3g_6$ that forced their orbital eccentricity e to increase. The amount of this quasi-instantaneous step in e is just right needed to "lift" the asteroids to the Prometheus clan. Since the nodal frequency s does not take part in the resonance frequency, the inclination remains unperturbed.

The Yarkovsky-spreading model implies some fraction of family members should be located at the edge of the major resonances. This was indeed known even earlier, in particular Milani & Farinella (1995) noticed the asteroid Vysheslavia, a Koronis family member, is located tightly near the 5/2 resonance so that its dynamical lifetime is $\simeq 10$ My[†]. This is about two orders of magnitude less than the estimated age of the family. As was concluded by Milani & Farinella (1995), Vysheslavia is too large ($\simeq 15$ km size) to be a secondary collisional fragment. A more robust scenario is that the body was transported via Yarkovsky effect from stable regions to the 5/2 resonance; Vokrouhlický *et al.* (2001) determined a $\sim$ Gy timescale is needed.

Similarly, Tsiganis *et al.* (2003) recently analysed population of asteroids inside the 7/3 mean motion resonance that brackets the Koronis family (Fig. 1). They noted the population of 23 objects is subdivided in two groups distinct by their orbital inclinations: (i) the low-inclination orbits match the typical values of Koronis members, while (ii) the high-inclination orbits match the typical values of Eos members. With a median dynamical lifetime of several tens of My, this resonant population must be resupplied from the two adjacent families. Importantly, Tsiganis *et al.* (2003) prove the pace at which the Yarkovsky effect could deliver asteroids from the available Koronis population near the 7/3 resonance is about right to explain the observed low-inclination population (and similarly for the Eos family and the high-inclination resonant population).

Karin.– Nesvorný *et al.* (2002) noticed an exceptionally compact cluster of asteroids inside the Koronis family (Fig. 1). With a direct integration of their orbits, they proved that the Karin members originated from a single disruptive event ~ 5.8 My ago (a similarly spectacular result has been also achieved for the Veritas family; Nesvorný *et al.* 2003). For the first time, the age of a past disruptive event in the main asteroid belt had been estimated with an uncertainty smaller than few percents. It opened a broad variety of studies. Among them, Nesvorný & Bottke (2004) dwell on the fact that the proper frequencies s and g, with which the proper values of node and perihelion longitudes circulate, depend on the semimajor axis. Since the originally determined convergence of nodes and perihelia ($\simeq 40°$; Fig. 2a) was nearly an order of magnitude larger than expected just after the disruption event[‡], secular drift da/dt of the semimajor axes may help

---

† Knežević *et al.* (1997) found two more asteroids with similarly short dynamical lifetime; later Brož & Vokrouhlický (2001) continued searching in this region and found additional four asteroids.

‡ From the semimajor axis, eccentricity and inclination dispersion of Karin members one can easily derive the maximum relative speed of ejection $V_{\mathrm{ej}} \simeq 15$ m/s (Nesvorný *et al.* 2002); as an example, Gauss equations then imply the initial node dispersion was $\delta\Omega \simeq 2(V_{\mathrm{ej}}/V_{\mathrm{orb}})/(\sqrt{6}\sin I) \simeq 1°$ ($V_{\mathrm{orb}} \simeq 18$ km/s is the orbital velocity and $I \simeq 2°$ is the inclination).

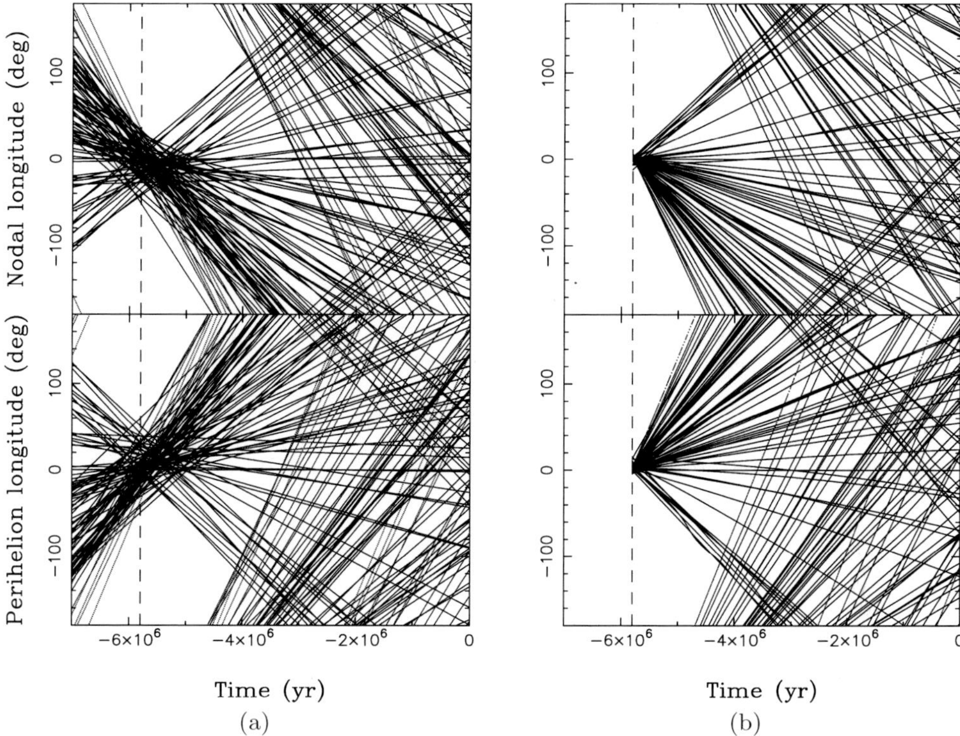

Figure 2. Convergence of the proper longitude of node and the proper longitude of perihelion for 90 members of Karin cluster at -5.75 My: (a) without the Yarkovsky effect, (b) with the Yarkovsky effect (Nesvorný & Bottke 2004). We note a much better convergence of the orbits occurs in the case (b) with a scatter of $\simeq 5°$ at the initial time (compared to $\simeq 40°$ in the case (a)); a better result is prevented by the gravitational influence of large main belt asteroids whose mass is not known exactly and whose relative motion with respect to the Karin members cannot be modeled deterministically on the My timescale.

align them more compactly via additional terms such as $\Delta\Omega_{\text{Yark}} \simeq \frac{1}{2}(\partial s/\partial a)(da/dt)\tau^2$ (Fig. 2b). Here the values of da/dt appear as a free parameter for each of the asteroids making the exercise apparently over-parametrized. However, Nesvorný & Bottke (2004) noted both proper node and proper perihelion must be aligned for all asteroids and this removes any arbitrariness. Moreover, derived values of da/dt are of the right order of magnitude expected for the estimated sizes, which further strengthens the case. Extreme values of da/dt imply the corresponding asteroids must have spin axes nearly perpendicular to their orbits and this is a well-testable prediction. Another important conclusion from this work is that several asteroids whose node and perihelion could not be found convergent with Karin members were proven to be interlopers (among them, the large asteroid (4507) 1990 FV).

Eos.– Like the Koronis case, the Eos family presented a hard problem for the static family model with its puzzling shape. Morbidelli *et al.* (1995) pointed out this family is intercepted by the 9/4 mean motion resonance with Jupiter. Numerical integration of fake asteroids inside this resonance indicated their median lifetime of $\simeq 100$ My and showed they should be distributed to high eccentricity and inclination values. This finding motivated Zappalà *et al.* (2000) to search for such asteroids. Spectroscopic similarity with the Eos members should identify them as escaping former family members. Indeed, Zappalà *et al.* succeeded in discovering 7 positive cases. In fact, this was an initial hint

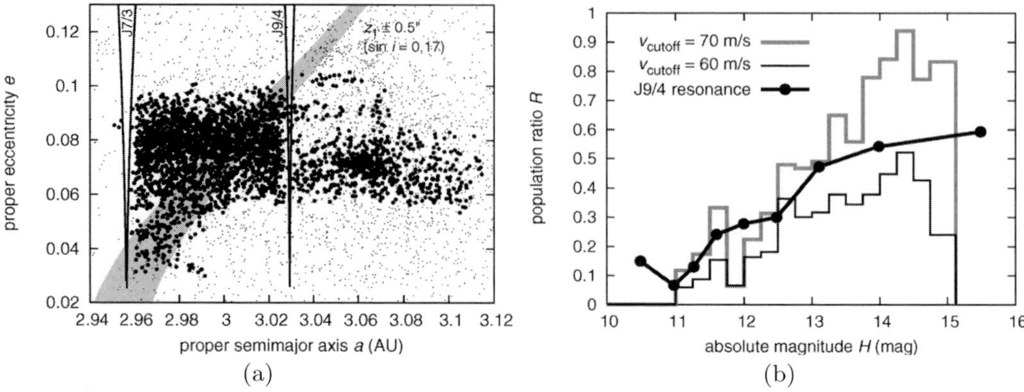

Figure 3. The Eos family asteroids (full circles) in the projection of proper semimajor axis and eccentricity (left part (a); background asteroids are dots). The shaded zone depicts location of the high-order secular resonance z_1 for the low inclination edge of the family. On the right – part (b) – a relative fraction R of Eos members with higher value of the semimajor axis than the location 9/4 mean motion resonance with respect to the population of asteroids with lower value of the semimajor axis shown as a function of absolute magnitude (the two histograms for two cutoff velocities of the family identification procedure, 60 and 70 m/s). The thick solid curve indicates the numerically determined fraction of Yarkovsky migrating orbits that crossed the resonance.

for the ongoing dynamical spreading in this family, since it appears unlikely that the Eos family would be $\simeq 100$ My old.

Further systematic work has been done by Vokrouhlický *et al.* (2004d; to be submitted) who found four main lines of evidence for Yarkovsky spreading in this family†. First, similar to the Koronis case, the Eos family sharply terminates at the 7/3 mean motion resonance (Fig. 3a). Second, objects with larger semimajor axis value than that of the 9/4 mean motion resonance are under-populated as compared to those with smaller semimajor axis value. This is true even when corrected for the miscentered position of the 9/4 resonance. Vokrouhlický *et al.* (2004d), however, prove that the observed under-population corresponds to the probability to cross the 9/4 resonance with outward migrating orbits due to the Yarkovsky effect (Fig. 3b). Third, small members in the family are more dispersed in semimajor axis than large members in a way consistent with Yarkovsky evolution. Although one would also assume that the small members acquired initially larger relative velocities, detailed analysis supports a model where the initial semimajor axis dispersion was doubled by the subsequent Yarkovsky migration of small family members. This is because small asteroids tend to populate regions with extreme values of the semimajor axis and under-populate the central zone of the family. Assuming the YORP effect preferentially tilts the obliquity of these objects toward extreme values of $0°$ and $180°$ (Čapek & Vokrouhlický 2004), the Yarkovsky drift becomes maximum and produces the observed central depletion. Performing a fit to the observed distribution of the Eos members and using a combined Yarkovsky/YORP model, Vokrouhlický *et al.* (2004d) determine the age of this family of (1.3 ± 0.2) Gy. The initial semimajor axis dispersion was found to be about half of the current value which agrees with results in Dell'Oro *et al.* (2004). Finally, the shape of the Eos family is found to adhere to the high-order secular resonance $z_1 = s + g - s_6 - g_6$ at small inclinations (Fig. 3a). This appears anomalous in a static model, while it is expected within the Yarkovsky dispersal

† Preliminary analysis has been reported by us at the ACM 2002 meeting in Berlin, see abstract 12-01, p. 115 of the ACM Abstract Book.

model because orbits migrating towards smaller values of the semimajor axis are found to be efficiently trapped in this resonance (see also Vokrouhlický & Brož 2002).

6. YORP effect: transport into resonant states

Compared to the Yarkovsky effect, we currently have less evidence about the YORP effect, in part because its revival came years later (Rubincam 2000). Efforts toward its direct detection via precise observations of the rotation rate of small NEAs have been so far unsuccessful[†]. With that, the most convincing evidence about a long-term YORP perturbation comes from an analysis of the spin axis distribution of Koronis family members.

Concluding a long-term observation survey, Slivan (2002) and Slivan *et al.* (2003) reported astonishing results about the rotation state of several small members of the Koronis family. First, and the most strange, four asteroids with a prograde sense of rotation all have spin axes nearly parallel in space, with obliquity near $55°$, and have all rotation periods clustered in a tight range of values between 7.5 hr and 9.5 hr. Second, five asteroids with a retrograde sense of rotation have obliquities pushed near the maximum value of $180°$ and their rotation periods are anomalously short or long. Since these results are not a random fluke, they call out for a robust theoretical model.

The YORP renaissance came right on time to meet this interesting problem. Vokrouhlický *et al.* (2003) proved the prograde group of asteroids resides in a secular spin-orbit resonance. This means their spin axes follow a particular mode (here related to the s_6 frequency) of the mean orbital plane precession in space; in classical astronomy this is the Cassini state 2 (e.g. Colombo 1966). Because this mode is forced by planetary perturbations, it acts the same on all asteroids and this naturally explains the alignment of their spin axes in ecliptic longitude (and thus their true collinearity in space). With that, however, the problem is not solved; we still need to explain why the prograde asteroids are captured by this fine resonance. Here the YORP effect became an instrumental part of the model. Vokrouhlický *et al.* (2003) proved that in $2 - 3$ Gy of evolution, a generic initial prograde rotation state (with rotation periods $\simeq 4 - 9$ hr) evolves toward the s_6 secular resonance. The YORP torque first tilts the obliquity to a small value, where the capture probability becomes unity; then YORP continues to push evolution toward the stable point of the resonance. The generalized Cassini state, with YORP effect included, has necessarily a resonant obliquity near $55°$[‡]; this explains the observed obliquities and also provides evidence that the YORP effect is the underlying dynamical mechanism. The resonant state was found weakly unstable, such that the future evolution will drive the asteroidal spin axes out of its proximity. This has likely already happened to smaller unobserved asteroids, where the strength of the YORP effect is larger. Spin axes of the retrograde group of asteroids evolve toward extreme values in both obliquity and rotation period by YORP since there is no obvious spin-orbit resonance in this zone. Hence, the rotation states of the retrograde asteroids also point to past YORP evolution. Note

† With a near certainty the YORP and the Yarkovsky effects will be detected using Hayabusa observations as a slight deceleration of (25143) Itokawa's rotation and about a 20 km orbit displacement in the late 2005 (e.g. Vokrouhlický *et al.* 2004a; Ostro *et al.* 2004). The analysis of an early photometry of (433) Eros by J. Ďurech, p. 95 of the Abstracts book, suggests YORP might have been affecting rotation of this asteroid. This empirical result for the Eros rotation period increase compares surprisingly well with the prediction by Čapek & Vokrouhlický (2004), but large observational errors prevent a clear-cut conclusion.

‡ In spite of the solid numerical evidence about this result, it has never been theoretically explained why the YORP torque on rotation speed vanishes at this obliquity value. It remains an interesting problem for the future work.

that this model was further supported by new observations in 2004 (S. Slivan, personal communication).

7. YORP effect: rotation period and obliquity distribution

As pointed out above, a fundamental advantage in analyzing asteroid family members is their common age. The YORP effect perturbs the rotation of any small asteroid in the main belt, but their "evolution clocks" are not synchronized to bring them, at a certain time, to a similar evolution state. An interesting exception are NEAs, because most of them recently ($\simeq 1 - 50$ My ago; Bottke *et al.* 2002) left the main asteroid belt via three major routes: (i) the 3/1 mean motion resonance with Jupiter, (ii) the ν_6 secular resonance, and (iii) the intermediate Mars-crossing zone. Assuming the Yarkovsky effect is the fundamental force that drives asteroids into these escape routes, as suggested by Morbidelli & Vokrouhlický (2003), one tentatively expects roughly 2/3 of NEAs should preferentially rotate retrograde and 1/3 prograde. This is because with retrograde rotation the Yarkovsky effect causes the asteroids to migrate inwards to the Sun, while prograde rotation makes them to migrate outward from the Sun. Only one of the three options for the transport resonant routes, the 3/1 resonance for inner belt asteroids, is populated by the outward migrating objects. This argument assumes the same number of asteroids leaks through the three escape routes (i) → (iii); when corrected for their relative weight (derived by Bottke *et al.* 2000, 2002), one would predict 71% of NEAs have prograde rotation and 29% of NEAs have retrograde rotation.

This rather rough prediction has been confirmed by the work of La Spina *et al.* (2004) who find that the ratio of retrograde vs. prograde rotating NEAs is indeed 2:1 with a high statistical confidence. Though more observations are certainly needed to make the current sample of 21 asteroids more sound, this result complements previous lines of evidence for the ongoing Yarkovsky perturbation of orbits of the main belt asteroids.

8. Conclusions

The last decade provided ample evidence about ongoing dynamical evolution in the main asteroid belt, with several examples mentioned above. Still, a number of topics remain for further work:

- make use of the Yarkovsky and YORP effects to build a chronology of asteroid families (e.g. Carruba *et al.* 2003);
- make use of the Yarkovsky and YORP effect to investigate the role of dynamical depletion for analysis of the size distribution function of both the main belt and NEA populations (see W.F. Bottke *et al.*, this volume);
- do the Yarkovsky and YORP effects significantly limit lifetime of (compact) asteroid binaries?
- further detections of the Yarkovsky effect as a systematic tool for asteroid mass determination (e.g. Vokrouhlický *et al.* 2004b,c);
- searching for more opportunities for a direct detection of the YORP effect (e.g. Vokrouhlický *et al.* 2004a);
- does YORP produce a significant fraction of slowly tumbling asteroids?
- does YORP produce a significant fraction of compact binaries among the near-Earth asteroids?

The overall importance of these problems for planetary science promise the analysis of the non-gravitational effects will continue to be an active research field over the next decade.

Acknowledgements

The work of DV and MB was supported by the Grant Agency of the Czech Republic.

References

Afonso, G.B., Gomes, R.S. & Florczak, M.A. 1995, *Planet. Sp. Sci.* 43, 787

Bertotti, B., Farinella, P. & Vokrouhlický, D. 2003, *Physics of the Solar System* (Kluwer)

Bottke, W.F., Rubincam, D.P. & Burns, J.A. 2000, *Icarus* 145, 301

Bottke, W.F., Jedicke, R., Morbidelli, A., Petit, J.-M. & Gladman, B.J. 2000, *Science* 288, 2190

Bottke, W.F., Vokrouhlický, D., Brož, M., Nesvorný, D. & Morbidelli, A. 2001, *Science* 294, 1693

Bottke, W.F., Morbidelli, A., Jedicke, R., Petit, J.-M., Levison, H.F., Michel, P. & Metcalfe, T.S. 2002, *Icarus* 156, 399

Bottke, W.F., Vokrouhlický, D., Rubincam, D.P. & Brož, M. 2003, in: Bottke, W.F., Cellino, A., Paolicchi, P. & Binzel, R.P. (eds.), *Asteroids III* (The University of Arizona Press, Tucson), p. 395

Brož, M. & Vokrouhlický, D. 2001, in: Pretka-Ziomek, H., Wnuk, E., Seidelmann, P.K. & Richardson, D. (eds.), *Dynamics of Natural and Artificial Celestial Bodies* (Kluwer Academic), p. 307

Brož, M., Vokrouhlický, D., Roig, F., Nesvorný, D., Bottke, W.F. & Morbidelli, A. 2004, *Mon. Not. R. Astron. Soc.*, submitted

Burns, J.A., Lamy, P.L. & Soter, S. 1979, *Icarus* 40, 1

Čapek, D. & Vokrouhlický, D. 2004, *Icarus*, in press

Carruba, V., Burns, J.A., Bottke, W.F. & Nesvorný, D. 2003, *Icarus* 162, 308

Chesley, S.R., Ostro, S.J., Vokrouhlický, D., Čapek, D., Giorgini, J.D., Nolan, M.C., Margot, J-L., Hine, A.A., Benner, L.A.M. & Chamberlin, A.B. 2003, *Science* 302, 1739

Colombo, G. 1966, *Astron. J.* 71, 891

Dell'Oro, A., Bigongiari, G., Paolicchi, P. & Cellino, A. 2004, *Icarus* 169, 341

Dermott, S.F., Nicholson, P.D., Burns, J.A. & Houck, J.R. 1984, *Nature* 312, 505

Dermott, S.F., Grogan, K., Durda, D.D., Jayaraman, S., Kehoe, T.J.J., Kortenkamp, S.J. & Wyatt, M.C. 2001, in: Grün, E., Gustafson, B.A.S., Dermott, S.F. & Fechtig, H. (eds.), *Interplanetary Dust* (Springer), p. 569

Dohnanyi, J.W. 1969, *J. Geophys. Res.* 74, 2531

Farinella, P. & Vokrouhlický, D. 1999, *Science* 283, 1507

Farinella, P., Vokrouhlický, D. & Hartmann, W.K. 1998, *Icarus* 132, 378

Giorgini, J.D., Ostro, S.J., Benner, L.A.M., Chodas, P.W., Chesley, S.R., Hudson, R.S., Nolan, M.C., Klemola, A.R., Standish, E.M., Jurgens, R.F., Rose, R., Chamberlin, A.B., Yeomans, D.K. & Margot, J-L. 2002, *Science* 296, 132

Gladman, B.J., Migliorini, F., Morbidelli, A., Zappalà, V., Michel, P., Cellino, A., Froeschlé, C., Levison, H.F., Bailey, M. & Duncan, M. 1997, *Science* 277, 197

Guillens, S.A., Vieira Martins, R. & Gomes, R.S. 2002, *Astron. J.* 124, 2322

Grogan, K., Dermott, S.F. & Durda, D.D. *Icarus* 152, 251

Grün, E., Zook, H.A., Fechtig, H. & Giese, R.H. 1985, *Icarus* 62, 244

Ivezić, Z. and 31 coauthors 2001, *Astron. J.* 122, 2749

Knežević, Z., Milani, A. & Farinella, P. 1997, *Planet. Sp. Sci.* 45, 1581

La Spina, A., Paolicchi, P., Kryszczynska, A. & Pravec, P. 2004, *Nature* 428, 400

Love, S.G. & Brownlee, D.E. 1993, *Science* 262, 550

Low, F.J., Young, E., Beintema, D.A., Gautier, T.N., Beichman, C.A., Aumann, H.H., Gillett, F.C., Neugebauer, G., Boggess, N. & Emerson, J. P. 1984, *Astrophys. J. Lett.* 278, 19L

Morbidelli, A. & Vokrouhlický, D. 2002, *Icarus* 163, 120

Morbidelli, A., Zappalà, V., Moons, M., Cellino, A. & Gonczi, R. 1995, *Icarus* 118, 132

Nesvorný, D. & Bottke, W.F. 2004, *Icarus* 170, 324

Nesvorný, D., Bottke, W.F., Dones, L. & Levison, H.F. 2002, *Nature* 417, 720

Nesvorný, D., Bottke, W.F., Levison, H.F. & Dones, L. 2003, *Astrophys. J.* 591, 486

O'Brien, D.P. & Greenberg, R. 2003, *Icarus* 164, 334

Ostro, S.J., Benner, L.A.M., Nolan, M.C., Magri, C., Giorgini, J.D., Scheeres, D., Broschart, S., Kaasalainen, M., Vokrouhlický, D., Chesley, S.R., Margot, J.-L., Jurgens, R., Rose, A., Yeomans, D., Suzuki, S. & DeJong, E. 2004, *Meteor. Planet. Science* 39, 407

Rubincam, D.P. 1995, *J. Geophys. Res.* 100, 1585

Rubincam, D.P. 1998, *J. Geophys. Res.* 103, 1725

Rubincam, D.P. 2000, *Icarus* 148, 2

Slivan, S.M. 2002, *Nature* 419, 49

Slivan, S.M., Binzel, R.P., Crespo da Silva, L.D., Kaasalainen, M., Lyndaker, M.M. & Krčo, M. 2003, *Icarus* 162, 285

Tsiganis, K., Varvoglis, H. & Morbidelli, A. 2003, *Icarus* 166, 131

Vokrouhlický, D. & Farinella, P. 2000, *Nature* 407, 606

Vokrouhlický, D. & Milani, A. 2000, *Astron. Astrophys.* 362, 746

Vokrouhlický, D. & Čapek, D. 2002, *Icarus* 159, 449

Vokrouhlický, D. & Brož, M. 2002, in: Celletti, A., Ferraz-Mello, S. & Henrard, J. (eds.), *Modern Celestial Mechanics: from Theory to Applications* (Kluwer Academic), p. 467

Vokrouhlický, D., Milani, A. & Chesley, S.R. 2000, *Icarus* 148, 118

Vokrouhlický, D., Nesvorný, D. & Bottke, W.F. 2003, *Nature* 425, 147

Vokrouhlický, D., Brož, M., Farinella, P. & Knežević, Z. 2001, *Icarus* 150, 78

Vokrouhlický, D., Čapek, D., Kaasalainen, M. & Ostro, S.J. 2004a, *Astron. Astrophys.* 414, L21

Vokrouhlický, D., Čapek, D., Chesley, S.R. & Ostro, S.J. 2004b, *Icarus*, in press

Vokrouhlický, D., Čapek, D., Chesley, S.R. & Ostro, S.J. 2004c, *Icarus*, submitted

Zappalà, V., Cellino, A. & Dell'Oro, A. 2002, *Icarus* 157, 280

Zappalà, V., Bendjoya, P., Cellino, A., Di Martino, M., Doressoundiram, A., Manara, A. & Migliorini, F. 2000, *Icarus* 145, 4

Dynamics of Populations of Planetary Systems
Proceedings IAU Colloquium No. 197, 2005
Z. Knežević and A. Milani, eds.

© 2005 International Astronomical Union
DOI: 10.1017/S1743921304008610

Diffusion in the asteroid belt

Harry Varvoglis

Section of Astrophysics, Astronomy and Mechanics, Department of Physics, University of
Thessaloniki, GR-541 24 Thessaloniki, Greece
email: varvogli@physics.auth.gr

Abstract. In the beginning we review briefly the evolution of the ideas on the motion of the
bodies in our solar system, from Newton's clockwork Universe to the presently accepted ubiquity
of chaotic transport in the asteroid belt. Then we discuss the result of chaotic motion, which
is transport in phase space, and we introduce the concept of diffusion of an asteroid in action
space. We proceed by reviewing recent work on numerical as well as analytical study of asteroids
following chaotic trajectories and we summarize the main results. We present several applications
of the theoretical modelling of asteroid motion as diffusion in action space, to problems of specific
interest.

Keywords. Asteroid belt, chaos, diffusion

---

1. Introduction

Celestial Mechanics, the first branch of Astronomy where mathematical modelling
managed to interpret and predict celestial phenomena, has a record of impressive achievements. In less than three hundred years, starting from Newton's law of gravitation and
his three laws of dynamics, Celestial Mechanics managed finally to accurately describe,
in practice, the trajectories of all planets and minor bodies of our solar system. The
theory, behind all this, was based on the implicit assumption that the functions describing all these trajectories had all the "nice" mathematical properties and, in particular,
that they were analytic in all variables and parameters, including the initial conditions.
Astronomers were aware, of course, since the work of Poincaré, that the simplest model
used in Celestial Mechanics to describe perturbed motion, the planar circular restricted
three-body problem, does not posses any other analytic integrals of motion, besides the
Jacobi integral. The general consensus was, however, that the homoclinic and heteroclinc
tangles, described so vividly by Poincaré, were of little importance to the actual problem
of solar system dynamics, at least for time scales of practical interest.

Things started to change in the last half of the twentieth century, after the seminal work
by Hénon and Heiles (1965) revealed that chaos in the homoclinic tangle, as described
by Poincaré, may dominate the phase space of very simple dynamical systems. Soon,
through the work of Giffen (1973) in the 2:1 and Scholl & Froeschlé (1974) in the 3:1
resonance, it was understood that the simplest model used in Celestial Mechanics, the
planar restricted three-body problem, possesses non-ignorable chaotic regions in phase
space. Subsequently Wisdom (1980) showed that chaos in the outer belt is due to the
overlap of first-order resonances and derived the law

$$\Delta a = 1.5 a_J \mu^{2/7} \tag{1.1}$$

which, in the model of the circular three-body problem, relates the width, Δa, of the
resonance-overlap region around the orbit of the secondary body, to its mass, μ, and semi-
major axis, a_J. Later it was shown that a similar effect is produced by the overlapping
of the components of high-order resonant multiplets (Murray & Holman (1997)) in the

model of the elliptic three-body problem. In this way it became widely accepted that chaos is an important phenomenon in the solar system. Therefore Newton's idea of a clockwork universe, which, once started to move, would continue moving in the same way "ad perpetuum", suddenly was proved to be wrong. The bodies of the solar system, especially the minor ones, may follow trajectories that change secularly in time, so that "collisions" (either true or just close encounters) and ejections play an important role as sinks (both) and sources (the former) of bodies, even in the present era. As a result, the question now is not anymore whether chaos affects the dynamics of planetary systems, but on which time-scales it does so and on whether it is the rule or the exception. Recent calculations put the percentage of main belt asteroids on chaotic orbits at 30% (proper elements and LCE computations, Milani & Knežević (2003)).

However it was not so easy to understand how to treat mathematically the dynamical evolution of bodies on chaotic trajectories, since this was a novel situation in Celestial Mechanics. Laplace's theory and its continuation, proper elements theory, are not valid in chaotic regions. As numerical experiments clearly show, the values of proper elements in chaotic regions change in a secular, non-quasi-periodic, way. This is how the concepts and methods of Statistical Physics, in particular transport theory, were introduced in Celestial Mechanics. In the rest of this article we will review these methods and show how transport theory may be used in order to obtain useful information in real problems of Celestial Mechanics.

2. Transport vs. diffusion

The "classical" way to treat a system, in which certain quantities change in an erratic way with time, is the statistical approach. Statistical Physics has developed a wide range of tools that may be used in the description of such a system. Basic concept in this respect is the concept of diffusion. Here, however, one should be very careful, since the term *diffusion* has been sometimes used in a liberal way in Celestial Mechanics. In Statistical Physics the term *diffusion* has a restrictive meaning. Speaking in simple terms, it describes the evolution of an ensemble of n "particles", undergoing a *classical random walk* in some state variable(s), i.e. a process described by a random variable with finite second moments. Even more restrictive is the set of available tools for the mathematical description of such a system. They consist, essentially, in a set of partial differential equations, named *transport equations*. Principal among these equations is the *Fokker-Planck* equation (e.g. see Lichtenberg & Lieberman (1992))

$$\frac{\partial f}{\partial t} = -\frac{\partial (Bf)}{\partial x} + \frac{\partial^2 (Df)}{\partial x^2} \tag{2.1}$$

which gives the evolution of the probability density function, $f(x, t)$, of an ensemble of "particles" that undergo a random walk in x-space.† The first term in Eq. (2.1) describes drift, i.e. a systematic motion, through the *frictional coefficient B*, and the second diffusion, i.e. a random motion, through the *diffusion coefficient D*, in the one-dimensional $x-$space. The equation can be easily generalized for more than one degrees of freedom.

In the case of conservative dynamical systems, the drift term can be incorporated into the diffusion term, due to the conservation of phase space volume, so that in the simplest possible case of a 1-D conservative system the Fokker-Planck equation takes the simple

---

† I.e. $f(x, t)dt$ gives the probability that, at time t, the co-ordinate x of a "particle" lies within the interval x and $x + dx$.

form,

$$\frac{\partial P}{\partial t} = \frac{\partial}{\partial I}\left(\frac{D}{2}\frac{\partial P}{\partial I}\right) \tag{2.2}$$

in which enters only one parameter, the *diffusion coefficient* $D(I)$.† Note that in Eq. (2.2) we have used action-angle variables, $I - \theta$, and this is not accidental, since this selection has two important consequences. First in the case of a perturbed integrable system, where the action is an integral of the integrable part, any "random" change in the action may be considered, under some conditions which will be discussed later, as a diffusion. The second is that the "randomness" of the process can be easily put in mathematical terms, by invoking the *random phase approximation*, in which we assume that the angle variable is evolving in a random way. The above mathematical formulation seems to fit very well to the problems encountered in Celestial Mechanics, where the overwhelming majority of the models used are conservative perturbations of the Keplerian motion and they are very often cast in Hamiltonian form and action-angle variables (Delauné, Poincaré, or any modified form). However, there is a small, but crucial, point, that makes the application of the Fokker-Planck equation to problems of Celestial Mechanics less than straightforward. This point is that, due to the degeneracy of the Hamiltonian describing planetary motion in general, the osculating elements of a regular orbit in the three-body problem are not constant, even in the integrable (Laplace) approximation, but undergo secular oscillations. Therefore one has to separate the "regular" change of the elements, due to secular motion, from the "diffusive" one, due to the chaotic nature of the motion. This was pointed out by Murray & Holman (1997) and can, in principle, be accounted for by using in Eq. (2.2) *free* elements, instead of osculating. However in practice this imposes a lot of supplementary computational work, which makes cumbersome the numerical solution of the problem.

In the derivation of the Fokker-Planck equation several assumptions are implicit, so that, before using it to describe transport in a real problem, one has to ensure that all these assumptions really hold. Otherwise there is a danger that one could end up by solving correctly a wrong model! In particular, one should check the following conditions.

• First we should confirm that the set of trajectories, whose evolution we are trying to calculate, lie in a chaotic region of phase space. The chaotic nature of the region should by confirmed by calculating the Lyapunov number(s) (LCNs).

• The second condition is that the chaotic region, in action space, should be *simply connected.*

• The third condition is that the motion could be considered as a Markovian process, which means that there should be a fast decay of correlations.

• The fourth condition is that the "time-step" of the random walk model should be larger than the autocorrelation time, $\Delta t \gg t_c$, while the "typical" change in action should be small, $\Delta I \ll I$.

• The fifth condition is that the variance of the random-walk probability density function should be finite, i.e. $\sigma \neq \infty$.

Now it is instructive to give some examples of cases of interest in Celestial Mechanics, which cannot be described mathematically by a Fokker-Planck equation, at least in a straightforward way.

• Real dynamical systems posses a *divided phase space*, i.e. the chaotic regions are closely intermingled with ordered ones. If the measure of the ordered regions is an

---

† The diffusion coefficient might depend on time as well, i.e. $D = D(I,t)$, but the discussion of this case lies outside the scope of the present article.

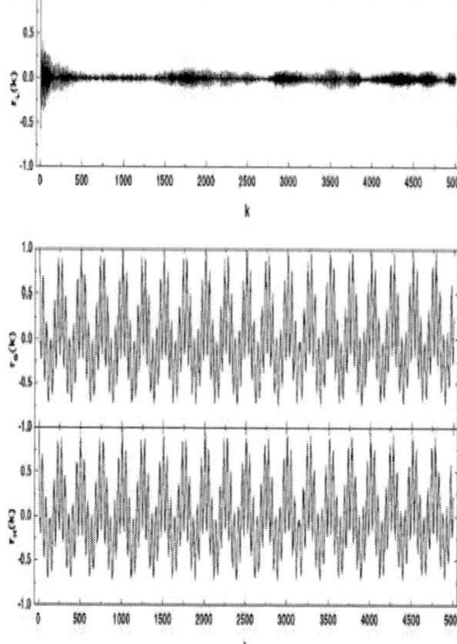

Figure 1. The autocorrelation functions of the three Delaunay variables of a stable chaotic asteroid. The one corresponding to the semi-major axis drops almost exponentially, an indication that the variations of the semi-major axis are chaotic. The autocorrelation functions of the other two actions, however, oscillate, roughly, between, $+1$ and -1, an indication that the variation of G and H are quasi-periodic, so that they cannot be modelled as diffusion in the corresponding action space. Taken from Tsiganis, Varvoglis & Hadjidemetriou (2002a).

important fraction of the total available phase space, the simple statistical description (either theoretical or numerical) fails.

• When the second condition is not satisfied, a few "holes" in phase space may be accounted for by introducing absorbing or reflecting barriers. But a complicated self-similar topology, like the one existing at the border between ordered and chaotic regions, might lead to Lévy-like statistics (see fifth item below).

• The third condition excludes several transport processes that appear in chaotic systems. E.g. in systems with more than one degrees of freedom some of the degrees might be strongly chaotic, so that the maximal Lyapunov exponent has a large value, while others might be so mildly chaotic, as to appear ordered on physically important time-scales. In this way the particles diffuse, in the space of the strongly chaotic degrees, on a much shorter time-scale than in the space on the mildly chaotic degrees. A typical example is the case of *stable chaos* in the solar system (Milani & Nobili (1992)). In a stable chaotic trajectory, the autocorrelation function of one action (corresponding to the semi-major axis) decays on a time scale of the order of the Lyapunov time (i.e. the inverse of the maximal Lyapunov number). However, the autocorrelation functions of the other two actions (corresponding to the eccentricity and the inclination) vary quasi-periodically for thousands of Lyapunov times (Fig. 1). Therefore correlations in the eccentricity do not decay fast and the process cannot be considered as diffusion in eccentricity and/or inclination space (although it might be considered as diffusion in semi-major axis space).

• The variation of the "normal", non-resonant action variables in the vicinity of a low-order resonance cannot be considered as diffusion, since it fails to fulfill the third and the fourth conditions, hence the need to use "free" elements.

• Finally the fifth condition is a subtle one. The cornerstone of classical Statistical Mechanics is the Central Limit Theorem, which holds if the variance of a random variable is finite. If this is true for the random variable related to the random walk of the diffusing particles, the process is *classical diffusion* and one arrives naturally from the Fokker-Planck equation, Eq. (2.2), to Fick's law

$$\sigma^2(t) = \langle [I(t) - \langle I(t) \rangle]^2 \rangle = Dt \tag{2.3}$$

where the average is taken over all particles and over the total time interval. If this is not true, then

$$\sigma^2(t) \propto t^b \tag{2.4}$$

with $b \neq 1$. This is exactly the regime of the *non-classical* Lévy diffusion (Metzler & Klafter (2000)), named after the French mathematician who first described it. In particular the dependance of the variance might be either faster than linear in time ($b > 1$, super-diffusion) or slower ($b < 1$, sub-diffusion). An interesting example of super-diffusive behavior in Celestial Mechanics has been found recently by Carruba et al. (2003), who showed that close encounters with Ceres disperse the asteroids semi-major axes with a power-law exponent $b > 1$, i.e. faster than Fick's law.

It has been postulated that, at least some cases of slow chaos in the asteroid belt, are manifestations of sub-diffusion. This situation appears, in particular, in the regions of a dynamical system close to the border dividing order from chaos. Presently there is no widely accepted mathematical description of this type of transport. Several authors† have proposed transport equations with fractional derivatives, which might account not only for a Markovian process with infinite second moments, but for processes with memory as well. However these equations are very hard, in general, to solve, either analytically or numerically. The present practice is to model these processes by a Fokker-Planck equation, whose diffusion coefficient is selected ad hoc, so that its solutions mimic a Lévy probability density function.

3. Statistical approach and Numerical Experiments

The first authors to attempt a statistical description of chaotic motions in the asteroid belt was the group of Lecar (e.g. Lecar, Franklin & Murison (1992), Lecar, Franklin & Soper (1992), Murison, Lecar & Franklin (1994)). They performed numerical experiments, by integrating the equations of motion for a large number of asteroids and they calculated the Lyapunov time of each trajectory as well as a "characteristic" time, which was meant to describe the time needed for the eccentricity of an asteroid to reach a planet crossing value, after which it would soon be ejected from the main asteroid belt due to a close encounter. They arrived at a very surprising conclusion: the plot of the "characteristic" times vs. the Lyapunov times showed a linear trend in a log-log plot. They interpreted this result as a "universal law" in the asteroid belt. As it turned out, the trend is real but the correlation coefficient is low, so that it cannot be considered as a law. Lecar and his co-workers found a *strong* correlation, only because they assigned to Jupiter a mass ten times larger than its real one. But in this way they changed the dynamics of the system, as it was pointed out by Murray & Holman (1997). In particular, from eq. (1.1) it is evident that in their model all mean motion resonances overlapped, creating a simply connected chaotic region. However this unexpected result triggered a number of papers by other authors, who tried to understand and interpret it. To my opinion this fact was in the heart of the development of the statistical description of chaotic asteroid motion, in general, and of the diffusive approach, in particular.

† e.g. see Metzler & Klafter (2000) and references therein

Morbidelli & Froeschlé (1996) arrived at the conclusion that in the resonance overlap regime there is, indeed, a polynomial dependence between the escape and the Lapunov times, but there is no universal value of the exponent. Shevchenko (1998), working in the modulated pendulum approximation, assumed that the diffusion is taking place near the border of a separatrix and arrived at a the conclusion that such a relation exists and it is, moreover, quadratic. Varvoglis & Anastasiadis (1996) introduced to this problem a purely statistical approach. They assumed that the asteroid motion is a Markovian process and they wrote down a modified Fokker-Planck equation. Then they attempted to solve the equation in the limiting case of strong perturbation, when the surviving islands of stability are of negligible measure. Assuming, moreover, a diffusion coefficient depending only on the value of the Lyapunov number and not on the action, they solved the equation analytically and showed that a relation similar to the one found by the Lecar group is naturally recovered. These authors suggested as well that the motion of Helga, the first example of stable chaos found by Milani & Nobili (1992), could not be modelled as diffusion in eccentricity space.

4. Diffusion of asteroids - Mathematical formulation

In the framework of the three-body problem, asteroids can follow chaotic trajectories when in the vicinity of a mean motion resonance. Due to the ellipticity of Jupiter's orbit, each resonance is in fact a multiplet of close-by harmonics. Their overlapping is the main source of chaos.

Two different approaches have been presented so far, with the aim to (i) understand the cause of chaotic diffusion and (ii) provide analytic estimates for the relevant quantities, i.e. the diffusion coefficient and the escape time of chaotic orbits. The Hamiltonian describing the motion of an asteroid in a mean motion resonance of order q has the form (in the 2-D elliptic three-body problem)

$$\mathcal{H} = \frac{1}{2}\alpha\Psi^2 + \beta\Phi + \varepsilon\sum_{p=0}^{q} c_p \cos(\psi - p\,\phi) \tag{4.1}$$

where the (ψ, Ψ) degree of freedom describes the libration inside the resonance and the (ϕ, Φ) degree of freedom describes the motion of the free pericenter. The last term of the Hamiltonian is the resonant multiplet and $c_p \sim \Phi^{p/2} \sim e^p$, where e is the free eccentricity of the asteroid. The harmonics are separated by the amount $\delta\Psi = \beta/\alpha$, while their width is a function of the eccentricity, given by $\Delta\Psi = \sqrt{2\varepsilon c_p/\alpha}$. What controls the dynamics in the resonance is the ratio of the distance between the harmonics, with respect to their widths, $K = \Delta\Psi/\delta\Psi$.

The results of Murray & Holman (1997) support that $K = \Delta\Psi/\delta\Psi \sim 1$ in most resonances in the outer asteroid belt and, hence, the situation is close to the one shown in Fig. 2(b). In this case we can consider that the variable Ψ performs repeated uncorrelated random jumps of order $\Delta\Psi$ within a time equal to the Lyapunov time. The diffusion in Ψ is bounded within a region of size $\sum\Delta\Psi(\Phi)$, which depends on the eccentricity. At the same time it results to a diffusion in $\Phi \sim e^2$, with a diffusion coefficient of the form

$$D \sim \varepsilon\Phi^p \sim \varepsilon c_p^2 \tag{4.2}$$

Since, in the resonances where $K = \Delta\Psi/\delta\Psi \sim 1$ it turns out that $c_p \sim \varepsilon$, the value of the diffusion coefficient turns out to be of order ε^3. Murray & Holman (1997) used this result to solve the Fokker-Planck equation and derive analytic estimates for the

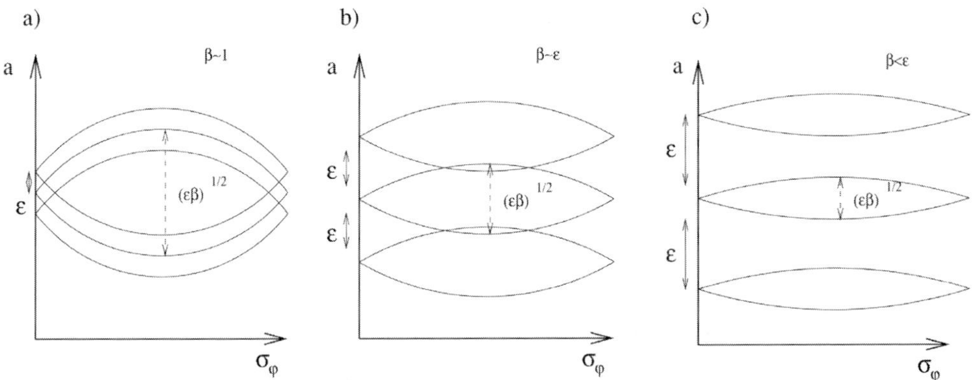

Figure 2. The three different possible arrangements of secondary resonances. Either they completely overlap, in which case the dynamics resemble to that of a modulated pendulum (a), or they overlap partially, in which case the dynamics are those described by Murray & Holman (1997) (b) or they do not overlap at all, in which case there is no diffusion (c). Taken from Morbidelli & Guzzo (1996).

escape time of asteroids. Their estimates agree, in general, with numerical integrations conducted in the frame of more advanced models. Although the applicability of this theory in most resonances of the asteroid belt can be questioned, it is an elegant theory providing meaningful results.

The results of Wisdom (1985) for the 3:1 resonance and those of Neishtadt (1987) focus on the situation $K \gg 1$ described in Fig. 2 (a). We note that this approximation holds in low- and medium- order resonances and can take into account important secular effects that appear inside mean motion resonances, as e.g. is the 3:1 case. In fact this approximation, although more laborious than the one presented above, may be more suitable for most of the resonances in the asteroid belt.

When $K \gg 1$, the resonant dynamics resemble those of a slowly modulated pendulum. The variable Ψ is *not* an appropriate action, as it cannot be considered as having repeated uncorrelated jumps of order $\Delta\Psi$ within a Lyapunov time. The relevant action, which is random walking in the case, is the area of libration of the pendulum, which suffers small changes when the separatrix of the pendulum sweeps by. Neishtadt (1987) has calculated the change in this adiabatic invariant over a secular cycle (i.e. $t \sim 1/\varepsilon$), finding $\Delta\mathcal{J} \sim \varepsilon$. This gives a diffusion coefficient for $\mathcal{J}$ which is of order $D \sim \frac{\langle\Delta\mathcal{J}\rangle^2}{\langle\Delta t\rangle} \sim \varepsilon^3$. Thus, we find that the diffusive time-scale for resonances of this type is as long as the one estimated through the Murray & Holman theory, provided (as we discuss in the next paragraph) that the resonance does not support a periodic orbit in the simplest non-trivial approximation of the elliptic planar three-body problem. This result agrees with what numerical integrations show, i.e. test-particles may take more than 1 Gyr to leave a high-order resonance (Tsiganis, Varvoglis & Hadjidemetriou (2002b), Tsiganis & Morbidelli (2003)).

However, if the secular dynamics are such that low-eccentricity regions are smoothly connected to high-eccentricity regions, the situation is different. This is the case for e.g. the 3:1 resonance, which possesses, in the the planar elliptic restricted problem, a resonant periodic orbit of order 1. The separatrix of the pendulum intersects the homoclinic orbit and slow diffusion of $\mathcal{J}$ allows the two eccentricity regions to communicate. In this way the eccentricity suffers large intermittent jumps, while $\mathcal{J}$ is slowly diffusing and changes only by small amounts. Using his calculations for $\Delta\mathcal{J}$, Neishtadt (1987) estimated the

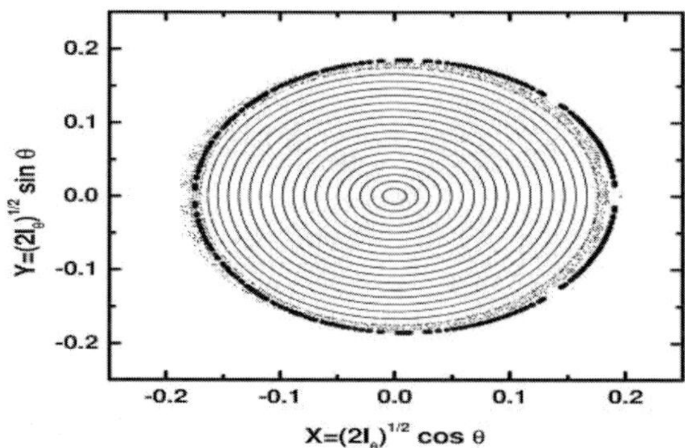

Figure 3. The secular dynamics, in the case of complete overlap, for the resonance 8/3, where there is no periodic orbit. The thick curve, superimposed on the graph, denotes the points of intersection of the (X, Y) plane with the separatrix of the modulated pendulum. All orbits intersecting this curve become chaotic. Notice the small width of the chaotic strip. Taken from Tsiganis & Morbidelli (2003).

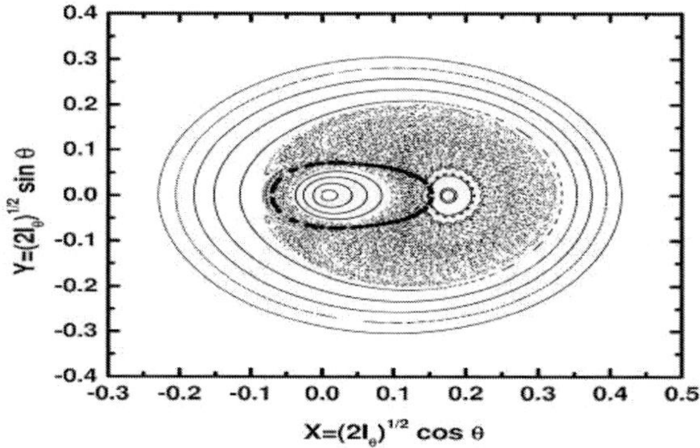

Figure 4. Same as Fig. 3, except that the resonance portrayed here is the 3/1, where there exists a periodic orbit. Notice the extended width of the chaotic region. Taken from Tsiganis & Morbidelli (2003).

time scale between two eccentricity jumps to be of order 10^5 yrs, as was found earlier by the numerical integrations of Wisdom (1985). This process is the cause of fast, non-diffusive transport in the asteroid belt.

We note that the regime of applicability of the two different approximations for diffusive transport has not been clarified yet. Although a unique theory would be more preferable, it seems that a case-by-case study should be made for mean motion resonances in the asteroid belt.

5. Applications to specific astronomical problems

Besides the work on the development of the basic ideas and tools, the statistical approach to problems that involve diffusive motions of asteroids has found already several applications.

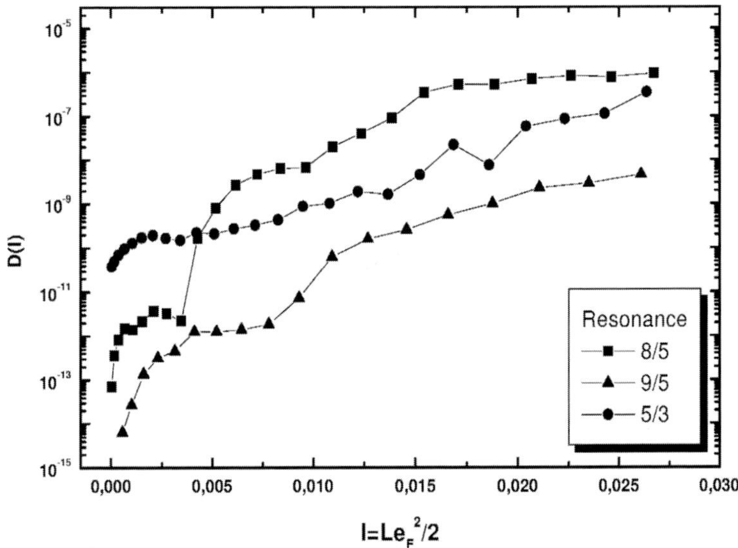

Figure 5. The diffusion coefficient calculated numerically for three different resonances, as a function of the eccentricity of the diffusing body. Taken from Tsiganis, Varvoglis & Anastasiadis (2003).

5.1. *Numerical calculation of $D(I)$*

Tsiganis, Varvoglis & Anastasiadis (2003) calculated numerically (Fig. 5) the functions $D(I)$ for the 5/3, 8/5 and 9/5 resonances in the framework of the elliptic restricted three-body problem. They found that the diffusion coefficient indicates differences not only between low and high eccentricity values, but also between resonances that support a periodic orbit in this model and those that do not.

- (a) The 8/5 and 9/5 resonances, which *do not* support a periodic orbit, have distinctively different functional behavior at low and high eccentricity regions. In the low eccentricity region they have a "stair-like" form, which is predicted by the theory of Murray & Holman (1997) and indicates that the dynamics are those of partial resonance overlap. The stair-like form is due to the fact that, for different values of e_f, the strongest term of the resonant multiplet, which determines the exponent p in Eq. (4.2), is different. The small-eccentricity part of these curves can, indeed, be fitted by curves of the form $D(I) \sim I^b$. However for $e_f \gtrsim 0.20$ this is not possible, because at intermediate eccentricities the dynamics are those of a modulated pendulum.

- (b) The functional form of the diffusion coefficient for the case of the 5/3 resonance cannot be fitted by a power law at all, because in this resonance there exists a periodic orbit in the framework of the elliptical planar restricted three-body problem, the dynamics are similar to those of the 3/1 resonance, and the theory of Neishtadt (1987) should be more appropriate.

Finally, at high eccentricities (say, $e_f \gtrsim 0.25$) all curves converge to the same limit, $D = D_{QL}$, the *quasi-linear* value, which is approached as all mean motion resonances begin to overlap.

5.2. *Chronology of the Veritas family*

It has been found (Milani & Farinella (1994)) that several members of the Veritas family, including the largest body in the family, asteroid (490) Veritas itself, are located inside

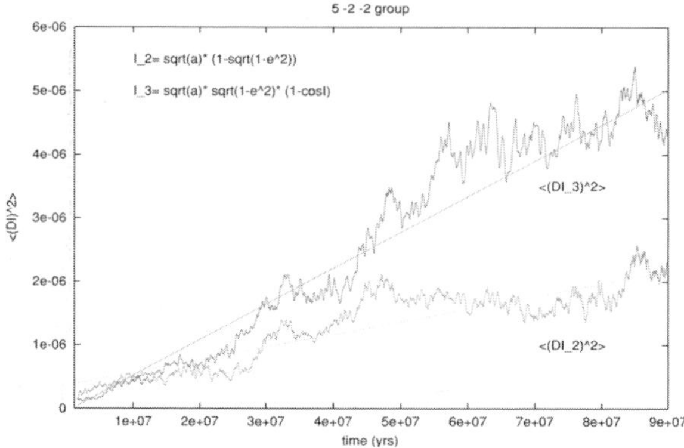

Figure 6. Morbidelli & Nesvorný (1998) have shown that some members of the Veritas family are in the $(5 - 2 - 2)$ resonance. The evolution of the variance of the I_2 and I_3 actions, as a function of time, for the chaotic members of this group is linear in time. The action I_2 is related to the eccentricity and the action I_3 is related to the inclination. Note the different value of the diffusion coefficient in the two actions. Taken from Tsiganis, Knežević & Varvoglis (2004).

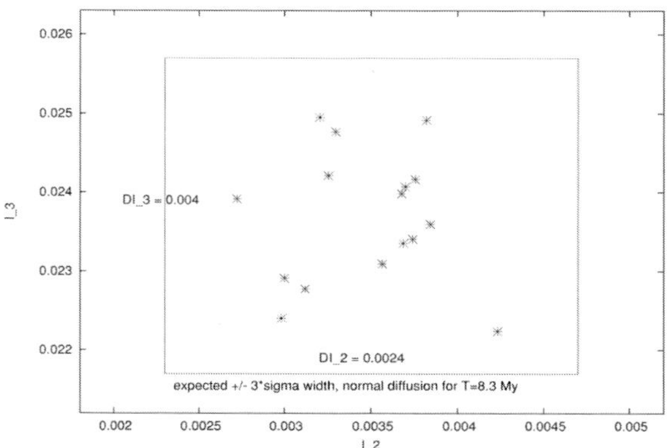

Figure 7. The chaotic members of the Veritas family in the (5 -2 -2) resonance plotted on the I_2 - I_3 space. The frame of the figure corresponds to $3\sigma(t)$, for $t = 8.3$ Myrs, in both variables. From the fact that all asteroids in this group lie within the frame, we can infer that the age of the family is ≈ 8.3 Myrs. Taken from Tsiganis, Knežević & Varvoglis (2004).

the chaotic zone and exhibit a typical chaotic behavior. Lyapunov times for these bodies are of the order of 10,000 yrs and they all appear to be in the resonance overlap regime. Chaotic diffusion brings these bodies outside the family boundaries on a 100 Myrs time-scale, which can be interpreted as an upper bound to the age of the family (Milani & Farinella (1994)). Knežević *et al.* (2002) confirmed that the motion of the chaotic family members is indeed diffusive, by showing that the evolution of the variance of the eccentricity and inclination obeys Fick's law and, from the slope of the curve $\sigma^2(t) = Dt$, estimated the diffusion coefficient (Fig. 6). Then, by assuming that the initial probability density function was a δ-function, they compared the analytical solution of the Fokker-Planck equation to the observed distribution of the actual chaotic members of the family. From this they estimated (Tsiganis, Knežević & Varvoglis (2004)) that the age of the family is ≈ 8.3 Myrs (Fig. 7), a value that agrees nicely with the result found recently by Nesvorný *et al.* (2003), by back integrating the members of the family on ordered orbits.

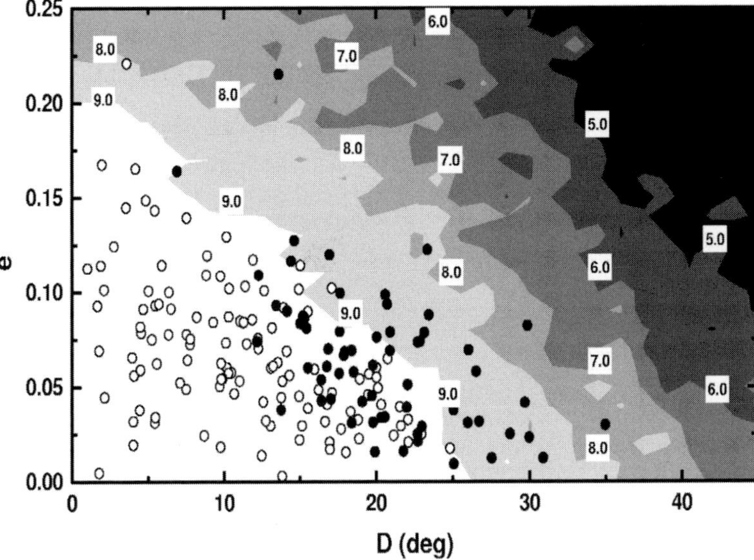

Figure 8. The location of ordered and chaotic real Trojans in the eccentricity (e)-libration width (D) space. The real bodies (white dots ordered, black dots chaotic) are superimposed on the map of the logarithm of escape time (in yrs), derived by integrating fictitious bodies. Taken from Tsiganis, Varvoglis & Dvorak (2004).

5.3. *Chaotic diffusion of Jupiter Trojans*

We now know that many Jupiter Trojans follow chaotic orbits (Milani (1993)). Many types of secondary resonances exist inside the 1:1 libration region (see Robutel et al. 2004), resulting in a slow chaotic motion, which resembles a diffusive evolution of the width of libration. Levison et al. (1997) had already shown, by means of numerical experiments, that the Trojans are slowly dispersing with time. However, the question whether this slow depletion is the result of chaotic diffusion or of a non-conservative phenomenon (e.g. collisions or Yarkovsky) remained open for some years.

Tsiganis, Varvoglis & Dvorak (2004) extended the numerical results of Levison et al. (1997). They derived the distribution of regular and chaotic bodies in the Trojan swarms, as a function of the proper elements. They also defined the "effective stability region", in which a Trojan has an escape time greater than the age of the solar system (see Fig. 8). For objects escaping within 1 Gyr, they showed that a power-law trend, relating the Lyapunov and escape time scales, exists (Fig. 9). As the boundary of stability is approached, this power-law trend is severely distorted, a fact that indicates a non-classical diffusive behavior. Finally, using the observations they estimated the diameters of the Trojans and derived the distributions of diameter for (i) the population of ordered and (ii) the population of chaotic Trojans. As shown in Fig. 10, the two distributions were found to be nearly identical, which shows that the transport of Trojans from the outskirts of the effective stability region towards the chaotic domain is a size-independent process, i.e. chaotic diffusion instead of radiation forces or collisions.

6. Discussion and conclusions

Proper elements theory is considered as the ultimate tool for studying the long time evolution of the Solar System, in general, and the asteroid belt, in particular. It is

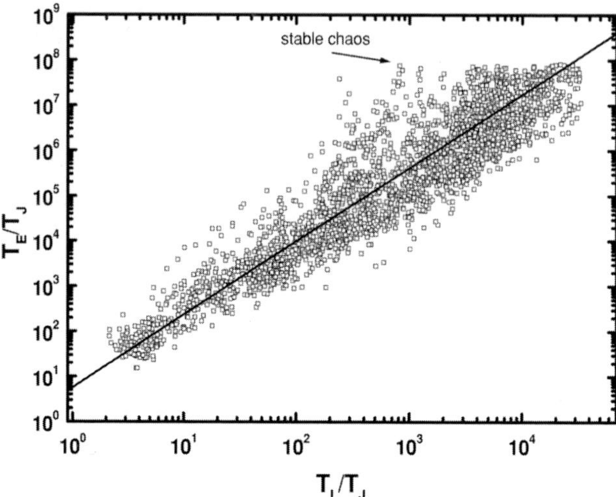

Figure 9. The relation between the Lyapunov time of Trojans and their escape time (through a collision with Jupiter). Several cases of stable chaos are noted. Taken from Tsiganis, Varvoglis & Dvorak (2004).

interesting to note that proper elements turn out to be of use not only in the case of asteroids on ordered motion, when the corresponding theory is valid, but even in some cases of mildly chaotic trajectories. However the theory has its limitations and becomes meaningless for asteroids with Lyapunov times less than 10^4 years. For these cases it might be more appropriate to change the way of approach and, instead of trying to make accurate predictions for the motion of individual bodies, it could be more instructive to seek the statistical evolution of a whole ensemble. According to the theory of Laplace-Lagrange, the semi-major axis does not show important secular changes, so that we focus our attention to the evolution of eccentricity and inclination. If the process in the corresponding actions can be considered as *classical diffusion*, the appropriate mathematical tool that can help us in this task is the Fokker-Planck equation, which in the case of conservative systems contains only one parameter, the *diffusion coefficient*. Therefore all information that can be drawn from the point of view of Statistical Physics depends only on the knowledge of the diffusion coefficient and the initial distribution of the ensemble of asteroids.

However even the functional form of the diffusion coefficient is not the same throughout the asteroid belt. Presently there exist two different theories for this form, both for the simplest possible model of the planar elliptic three-body problem, each one applying to a different topological structure of the chaotic region near an orbital resonance. If the various harmonics of the resonant multiplet overlap partially, a situation that holds usually in high-order resonances, the theory by Murray & Holman (1997) assumes that the action corresponding to the eccentricity is diffusing, with a diffusion coefficient of the form $D \propto \varepsilon c_p^2$, where p is the order of the harmonic. If the harmonics of the resonant multiplet are located almost on top of each other, a situation that appears usually in low- and medium-order resonances, the theory of Neishtadt (1987) assumes that the action diffusing is the one corresponding to the adiabatic invariant $\mathcal{J}$, with a diffusion coefficient that is of order ε^3, where $\varepsilon \sim 10^{-3}$ is the mass of Jupiter (in solar masses). In the cases where low-order resonant periodic orbits in the ERTBP or the 3DRTBP do

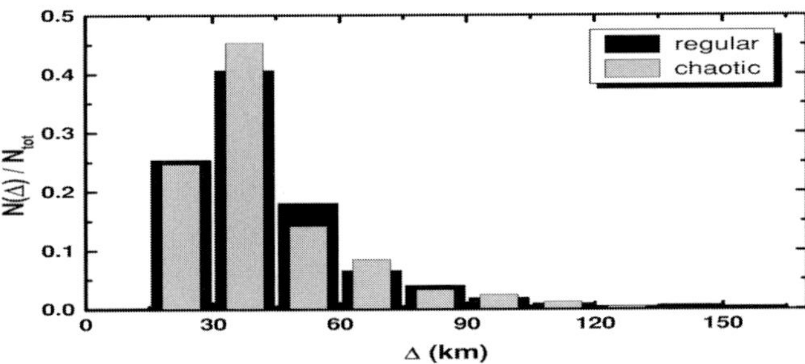

Figure 10. The distribution of regular and chaotic Trojans, with respect to their diameters. Taken from Tsiganis, Varvoglis & Dvorak (2004).

not dominate the dynamics, the diffusion described by the two theories gives the correct diffusive time-scale, although the physical processes are totally different. Of course in the real world the topology should be, as a rule, an intermediate case between the two limiting ones, something that complicates the theoretical treatment of any real problem.

Things become even more complicated by the fact that the above theories do not address all the possible *simple* cases. If there exists a resonant periodic orbit in the framework of the model used to describe the dynamical evolution of the ensemble, i.e. the ERTBP, something that in general is true for resonances of low order, transport in eccentricity space is not anymore purely diffusive and it cannot be described solely by any diffusion coefficient. Besides this case, there exists another phenomenon that complicates the statistical study of asteroid evolution, *stable chaos*. Stable chaotic trajectories are characterized by autocorrelation times that vary by orders of magnitude between each degree of freedom, the autocorrelation time of the semi-major axis being usually the shortest. In this way the evolution of the eccentricity cannot be considered as a random walk and the diffusive approach fails completely. Therefore the diffusive approach has to be applied with caution on a case by case base.

Despite all the above complications, the diffusive approach and its tools turned out to be of use in attacking several real problems. In the present article we presented three such applications. In the first, the numerical calculation of the diffusion coefficient is used in order to differentiate between the two mathematically developed models of diffusion in the specific cases of orbital resonances. In the second, the age of the Veritas family is estimated by assuming a constant diffusion coefficient, which was calculated numerically. In the third case it is concluded that the evolution of Jupiter Trojans is diffusive, on two indications: first that their Lyapunov times have a positive correlation to their escape times, a fact that characterizes diffusion in the border between order and chaos, and second that the distributions of both ordered and chaotic Trojans with respect to their diameters are effectively identical, a fact that precludes the importance of nongravitational forces. It is hoped that in the near future the theoretical understanding of diffusion in the asteroid belt will enable the tackling of more sophisticated problems.

Acknowledgements

Part of this article is based on work that has been done in collaboration with Profs. R. Dvorak and J. Hadjidemetriou and Drs. A. Anastasiadis, A. Morbidelli and

K. Tsiganis. I would like to thank Drs. A. Morbidelli and K. Tsiganis, who read a first draft of the article and made many useful comments.

References

Carruba, V., Burns, J.A., Bottke, W. & Nesvorný, D. 2003, *Icarus* 162, 308

Giffen, R., 1973, *Astron. Astrophys.* 23, 387

Holman, M. & Murray, N. 1996, *Astron. J.* 112, 1278

Knežević, Z., Tsiganis, K. & Varvoglis, H. 2002, in: B. Warmbein (ed.), *Asteroids, Comets, Meteors - ACM2002* (ESA SP-500), p. 335

Lecar, M., Franklin, F. & Murison, M. 1992, *Astron. J.* 104, 1230

Lecar, M., Franklin, F. & Soper, P. 1992, *Icarus* 96, 234

Levison, H., Shoemaker, E.M. & Shoemaker, C.S. 1997, *Nature* 385, 42

Lichtenberg, A.J. & Lieberman, M.A. 1992, *Regular and Chaotic Dynamics*, (New York: Springer)

Metzler, R. & Klafter, J. 2000 *Phys. Rep.* 339, 1

Milani, A. 1993 *Cel. Mech. Dyn. Astron.* 57, 59

Milani, A. & Knežević, Z. 2003, *http://hamilton.dm.unipi.it/cgi-bin/astdys/astibo*

Milani, A. & Farinella, P. 1994 *Nature* 370, 40

Milani, A. & Nobili, A. 1992, *Nature* 357, 569

Morbidelli, A. 2002, *Reg. Chaotic Dyn.* 111, 1718

Morbidelli, A. & Froeschlé, C. 1996, *Cel. Mech. Dyn. Astron.* 63, 227

Morbidelli, A. & Guzzo, M. 1996, *Cel. Mech. Dyn. Astron.* 65, 107

Morbidelli, A. & Nesvorný, D. 1998, *Astron. J.* 116, 3029

Murison, M.A., Lecar, M. & Franklin, F.A. 1994, *Astron. J.* 108, 2323

Murray, N. & Holman, M. 1997, *Astron. J.* 114, 1246

Neishtadt, A. 1987 *J. Appl. Math. Mech.* 51, 586

Nesvorný, D., Bottke, W.F., Levison, H.F. & Dones, L. 2003, *Astrophys. J.* 591, 486

Robutel, P., Gabern, F. & Jorba, A. 2004, *Cel. Mech. Dyn. Astron.*, in press

Scholl, H. & Froeschlé, C. 1974, *Astron. Astrophys.* 33, 455

Shevchenko, I.I. 1998, *Phys. Lett.* A 241, 53

Tsiganis, K., Knežević, Z. & Varvoglis, H. 2004, in preparation

Tsiganis, J., & Morbidelli, A. 2003, *Ann. MCFA* 3, 999

Tsiganis, K., Varvoglis, H. & Anastasiadis 2003, in: A. Celletti, S. Ferraz-Mello & J. Henrard (eds.), *Modern Celestial Mechanics: from Theory to Applications*, (Dordrecht: Kluwer), p. 451

Tsiganis, K., Varvoglis, H. & Dvorak, R. 2004, *Cel. Mech. Dyn. Astron.* in press

Tsiganis, K., Varvoglis, H. & Hadjidemetriou, J. 2000, *Icarus* 146, 240

Tsiganis, K., Varvoglis, H. & Hadjidemetriou, J. 2002, *Icarus* 155, 454

Tsiganis, K., Varvoglis, H. & Hadjidemetriou, J. 2002, *Icarus* 159, 284

Tsiganis, K., Varvoglis, H. & Morbidelli, A. 2003, *Icarus* 166, 131

Varvoglis, H. & Anastasiadis, A. 1996, *Astron. J.* 111, 1718

Wisdom, J. 1980, *Astron. J.* 85, 1122

Wisdom, J. 1985, *Icarus* 63, 272

Dynamics of Populations of Planetary Systems
Proceedings IAU Colloquium No. 197, 2005
Z. Knežević and A. Milani, eds.

© 2005 International Astronomical Union
DOI: 10.1017/S1743921304008622

Accurate model for the Yarkovsky effect

David Čapek and David Vokrouhlický

Institute of Astronomy, Charles University, V Holešovičkách 2, CZ-18000 Prague 8,
Czech Republic
email: capek@sirrah.troja.mff.cuni.cz, vokrouhl@mbox.cesnet.cz

Abstract. Yarkovsky effect (YE), a tiny nongravitational force due to radiative recoil of the anisotropic thermal emission, is known to secularly affect the orbital semimajor axis. Therefore, angular phases such as longitude in orbit or proper longitude of node undergo a quadratic perturbation. This is fast enough to allow direct detection of the YE. The first positive case was obtained for (6489) Golevka in 2003 and prospects are very good for many more detections in the near future. To make productive scientific use of the YE detections, we need to accurately compute its strength for a given body. Simple models, available so far, will likely not be adequate in many of the forthcoming YE detection possibilities. We thus developed a complex numerical approach capable of treating most of them. Here we illustrate its power by discussing the cases of: (i) Toutatis, with a tumbling (non-principal-axis) rotation state, and (ii) 2000 DP107, a binary system.

Keywords. Minor planets, asteroids: individual (Toutatis, 2000 DP107).

---

1. Introduction

The Yarkovsky effect (YE), and its consequences for planetary science, has attracted a considerable attention during the past decade (e.g. Bottke *et al.* 2003; Vokrouhlický *et al.*, this volume). It became a vital part of models for meteorite and asteroid delivery to the planet-crossing region (e.g. Farinella *et al.* 1998; Farinella & Vokrouhlický 1999; Vokrouhlický & Farinella 2000; Morbidelli & Vokrouhlický 2003), dynamical aging of the asteroid families (e.g. Bottke *et al.* 2001; Vokrouhlický *et al.* 2002; Nesvorný & Bottke 2004) or populating metastable asteroidal orbits (e.g. Vokrouhlický *et al.* 2001; Tsiganis *et al.* 2003; Brož *et al.*, this volume). Though important, these applications assume large samples of bodies and do not allow direct detection of the YE (with the unusual exception of the Karin family; Nesvorný & Bottke 2004).

Since the YE continues to perturb accurately known orbits of the planet-crossing asteroids, Vokrouhlický *et al.* (2000) suggested a direct detection can follow from their precise tracking (see also Vokrouhlický & Milani (2000) who discuss effects of other radiative forces on the motion of planet-crossing asteroids). This is because the YE makes a steady perturbation of the orbital semimajor axis, producing a quadratic advance along the orbit; in a number of cases the resulting displacement exceeds ephemerides uncertainty and allows YE detection. With that goal, Chesley *et al.* (2003) conducted a successful experiment by radar ranging to the near-Earth asteroid (6489) Golevka. Their analysis also proved the YE detection contains a significant scientific information, most importantly it has the capability to constrain an asteroid's mass. Vokrouhlický *et al.* (2004a,b) recently reviewed future possibilities for YE detection and noted about a dozen cases might be obtained in the next decade, with more possibly later on. Several of these candidate objects present unforseen difficulties in terms of the YE computation.

This situation motivated us to develop dedicated software for accurate YE computation: the purpose of this paper is to discuss its properties. Our goal is to tackle most of the "real-world" cases, including bodies of unusual shape, orbit and/or rotation state. Here we discuss two spectacular objects: (i) 4179 Toutatis, a body with the most accurately known tumbling state, and (ii) 2000 DP107, a binary system. If the YE is detected in the Toutatis' motion in October 2004 (see Vokrouhlický *et al.* 2004a), Toutatis might become a landmark case in several respects: (i) this will be the first multi-kilometre asteroid for which YE would be detected, and (ii) with further observation possibilities till 2012 this might be the first case for which the YE will be repeatedly measured. Similarly, if YE signal is too weak for 1998 RO1, the system 2000 DP107 might be the first binary for which YE will be detected (see also Vokrouhlický *et al.* 2004b).

2. Numerical model

Analytical expression of the Yarkovsky force components have been obtained so far for a spherical body residing on a low-eccentricity orbit (e.g. Vokrouhlický 1998, 1999; Vokrouhlický & Farinella 1999); moreover, these results assume linearization of the boundary condition (2.2). Though largely simplified, this formulation was successfully used by Vokrouhlický *et. al.* (2000) for low-accuracy, but reliable, predictions and is available at `http://newton.dm.unipi.it/` as a Fortran source within the `OrbFit` software package.

Apart from a non-linear nature of the heat diffusion problem (HDP), computation of the Yarkovsky force for near-Earth asteroids (NEAs) frequently brings some, or a combination, of the following complexities: (i) large orbital eccentricity, (ii) highly irregular shape (such as a part of the surface may cast shadow on another part), (iii) temperature-dependent thermal constants and/or (iv) unusual rotation state (including free motion of the rotation axis in the body, i.e. the "tumbling state"). Moreover, a fair fraction of NEAs are not solitary but compose binary systems (e.g. Merline *et al.* 2003). All these factors could invalidate the very simplified analytical approach and need to be considered for a high-accuracy YE computation.

Formulation of the heat diffusion problem.– In general, a fully 3D formulation of the HDP is needed to characterize the temperature inside and on the surface of a body. However, since we assume external energy sources only (such as impinging sunlight), in the most relevant situations the body consists of an isothermal core with temperature variations occurring in a thin surface slab. In that case, one can adopt a simplified, 1D approach with temperature $T(t, z)$ dependent on the depth z below the surface and time t (for an early formulation see Wesselink 1948). This is justified when the penetration depth of the most important thermal wave (diurnal or seasonal) is significantly smaller than the size of the body. Bodies larger than $\simeq 20$ m generally meet this condition, unless a very high thermal inertia†. The HDP is thus solved for each of the (infinitesimal) surface elements separately, as if there were no thermal communication between them through latitudinal thermal gradients.

The heat diffusion equation now reads

$$\rho C \, \frac{\partial T}{\partial t} = \frac{\partial}{\partial z} \left(K \, \frac{\partial T}{\partial z} \right) \, , \qquad (2.1)$$

where ρ is the density, C is the specific heat capacity and K is the thermal conductivity, all of which might be temperature dependent. If this effect is taken into account, we

† An exceptional group of very small NEAs, such as 1998 KY26 or 2003 YN107, may require a full-fledged 3D analysis as in Spitale & Greenberg (2000).

use empirical fits to laboratory and/or space measurements (e.g. Wechsler *et al.* 1972; Yomogida & Matsui 1983).

The system (2.1) must be supplemented with boundary conditions to make the solution unique. In the space coordinate this means (i) energy input on the surface, and (ii) constancy of the temperature at large depth; put in mathematics we have

$$\varepsilon\sigma T^4(t,0) = K \frac{\partial T}{\partial z}(t,0) + E(t) , \tag{2.2}$$

$$\frac{\partial T}{\partial z}(t,\infty) = 0 , \tag{2.3}$$

where we explicitly made clear depth z of the boundary. Here ε is the surface infrared emissivity, σ the Stephan-Boltzmann constant and $E = (1-A)\Phi(\mathbf{n}\cdot\mathbf{n}_0)$ is the radiative energy flux through the surface element; A is the albedo value in optical, Φ the incident solar radiation flux, $\mathbf{n}$ is the external unitary normal vector to the surface facet and $\mathbf{n}_0$ is the local direction to the Sun. We note E is nil, when $\mathbf{n}\cdot\mathbf{n}_0 < 0$ and also when another part of the body casts a shadow onto the chosen surface element (see below).

In the time coordinate we impose periodicity after interval P, thus $T(t,z) = T(t+P,z)$ for all grid nodes. After the period P the body must be brought into the same conditions, namely experience the same exterior radiation field. In practice this means to be at the same phase of revolution about the Sun and to have the same orientation in space. Though most asteroids of interest are in the principal-axis rotation mode, their rotation and revolution periods are not necessarily commensurate. However, the rotation period $P_{\rm rot}$ is usually much shorter than the revolution period $P_{\rm rev}$ and it is without loss of accuracy in evaluation of the YE to slightly modify $P_{\rm rot}$ in order to become commensurate with $P_{\rm rev}$. Then $P = P_{\rm rev}$. A more tricky situation occurs for a special class of tumbling asteroids (e.g. Pravec *et al.*, 2004), for which their orientation in space might not repeat at any time. Luckily, near-repetitions are usually found and they could be made commensurate with $P_{\rm rev}$; see Sec. 3 for an example.

We note that scaled, rather than physical, variables are best suitable in our problem. Depth z is expressed in terms of the penetration depth $h_T = \sqrt{K P_{\rm rot}/2\pi\rho C}$ of the diurnal thermal wave, thus introducing $z' = z/h_T$. Time t is replaced with the mean anomaly ℓ of the orbital motion. The "isothermal-core" condition (2.3) is applied at typically $10-15$ penetration depths of the seasonal thermal wave ($= \sqrt{P_{\rm rev}/P_{\rm rot}}\, h_T$), and the solution is (multiply) 2π periodic in the ℓ variable. Standard discretization methods are used to represent the heat diffusion equation (2.1) and Spencer *et al.* (1989) scheme is used for the non-linear surface boundary condition (2.2). An isothermal initial seed in the whole mesh quickly converges to the desired solution, though faster convergence is achieved when analytical approximation are used (such as in Wesselink 1948). We stop iterations of the numerical solution when a fractional change in temperature of all surface elements between two successive iterations is smaller than 10^{-4}.

Rotation state.– The surface energy input function $E(t)$ in (2.2) is computed from the known position of the Sun with respect to the surface element and it is a function of the asteroid orbit and its orientation in space. The latter is expressed using a rotation matrix $\mathbf{R}$ that refers body-fixed frame to an inertial frame. In general $\mathbf{R}$ may be parametrized by three Euler angles; for principal-axis rotators those depend on pole position, rotation period and epoch of local meridian,† while for tumbling asteroids the Euler

† In fact the result only weakly depends on the phase of local meridian at a given time, so that this information may be waived and replaced with an arbitrary zero value.

equations are numerically integrated with given initial data (e.g. Landau and Lifschitz 1976; Kryszczyńska *et al.* 1999).

Shape/shadowing.– We use polyhedron representation of the asteroid shape with typically several thousands triangular surface elements. These models are mostly due to radar sensing analysis, to lesser extend due to direct satellite reconnaissance and/or lightcurve inversion (data are generally available at the PDS node `http://www.psi.edu/pds/archive/rshape.html`). Solution of the HDP is preceded with a preliminary analysis, where we store in computer memory all combinations of mutual shadowing effects of different parts of the asteroid. This information is used for evaluation of the energy source function $E(t)$ in (2.2).

Yarkovsky force.– Once the surface temperature is determined by the numerical analysis described above, we compute components of the Yarkovsky force. For an oriented surface facet $d\mathbf{S} = \mathbf{n}\,dS$ their infinitesimal values read (see e.g. Milani *et al.* 1987)

$$d\mathbf{f}(\ell) = -\frac{2}{3}\frac{\varepsilon\sigma T^4(\ell,0)}{c}\,\mathbf{n}\,dS\,, \tag{2.4}$$

where the isotropic (Lambert) thermal emission is used. Total force† components are expressed as a sum of partial results for all surface elements and they are exported in an output file (with an appropriate header describing model parameters). In complex situations, like those discussed below, we export the resulting force components once every fraction of the diurnal cycle (typically $20 - 200$ times per asteroid rotation). In the case of solitary asteroids with principal-axis rotation, we further locally average over a diurnal cycle, making roughly $100 - 500$ normal points of the Yarkovsky force components per asteroid's revolution about the Sun. This procedure makes then the orbit determination faster.

Data and their availability.– Examples of our results are available through the `http://sirrah.troja.mff.cuni.cz/~davok/` web site where we also maintain a page coordinating efforts for the future YE detections.

3. Two examples

In what follows we briefly discuss results for two cases that require a high-accuracy YE computation. More details can be found in Vokrouhlický *et al.* (2004a,b).

Toutatis: a tumbling asteroid.– Toutatis was the first asteroid for which the non-principal-axis (tumbling) rotation state was discovered and accurately determined (Hudson & Ostro 1995). With the orbit residing near the 1/4 exterior mean motion resonance with the Earth, Toutatis undergoes frequent close Earth encounters during a couple of decades and this might permit YE to be detected (Vokrouhlický *et al.* 2004a). Accurate radar astrometry was acquired in 1992 and 1996 (Ostro *et al.* 1999), and a single Doppler measurement from 2000 is less useful but still makes a valuable constraint on the orbit. A spectacularly close encounter which occurs late September 2004 may give the first opportunity to detect YE (Vokrouhlický *et al.* 2004a), with further refinements during 2008 and 2012 encounters (all within the reach of the current radar systems; see `http://echo.jpl.nasa.gov/`).

As noted in Sec. 1, a productive use the YE measurement requires ability of a high

---

† We also standardly compute total thermal torque $d\mathbf{t} = \mathbf{r} \times d\mathbf{f}$ ($\mathbf{r}$ is the position vector of the surface element) affecting body's rotation, the so called YORP effect; e.g. Bottke *et al.* (2003). As an example, this has been used for prediction of the YORP observability in the case of asteroid (25143) Itokawa (Vokrouhlický *et al.* 2004c).

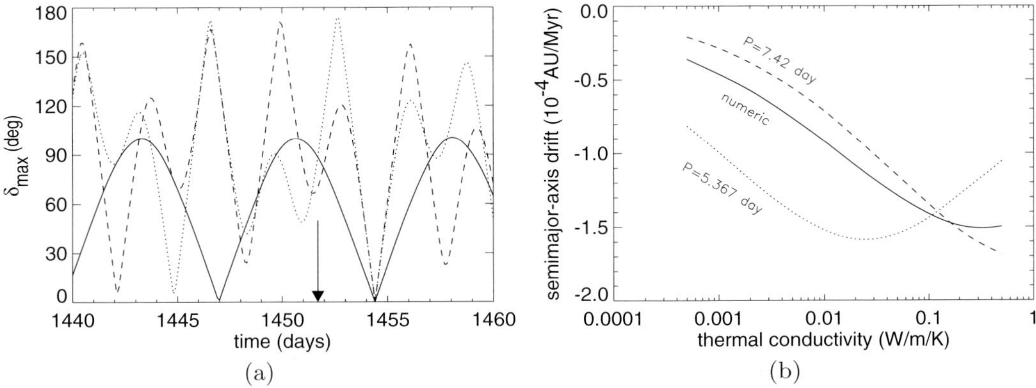

Figure 1. *Part a*: The angle between body principal axes at the initial epoch and time t (abscissa; in days): (i) solid for the longest axis, (ii) dotted for the middle axis, and (iii) dashed for the shortest axis. There is a sharp minimum in all angles at time $\simeq 1454.4$ d, meaning a near coincidence with the initial-epoch orientation (better than $0.5°$; initial epoch from Ostro *et al.* 1999). The arrow indicates orbital period. *Part b*: Estimated mean drift rate of the semimajor axis of Toutatis orbit due to the YE as a function of the surface thermal conductivity. Solid curve from the high-accuracy model, dashed and dotted curves from a simplified analytic approach assuming a spherical body with two characteristic periods: (i) 5.367 d (dotted), and (ii) 7.42 d (dashed); this model assumes spin axis along the total angular momentum of Toutatis.

accuracy Yarkovsky force computation. This appears non-trivial for elongated and tumbling Toutatis. The particular trouble for this body is its non-axial rotation: spin vector wobbles around the longest body axis in 5.367 d (in the body-fixed frame) and the longest body axis precesses around the nearly conserved total angular momentum in 7.42 d (in the inertial frame). Both motions are slow, which means the diurnal thermal lag is small and this strengthens requirements on accurate prediction of the YE magnitude. As for the boundary condition issue we note Toutatis undergoes a near repetition of its space orientation in $\simeq 1454.4$ d, remarkably close to the orbital period $P_{\rm rev} \simeq 1451.7$ d (Fig. 1a). Except an unlikely case of random coincidence, we do not have explanation for this interesting commensurability that may warrant future theoretical work. It appears important for our work, since we can take $P = P_{\rm rev}$ for the periodicity of the temperature solution.

We use a high-quality polyhedral model with 12 796 triangular facets adopted from `http://www.psi.edu/pds/archive/rshape.html`. Surface parameters are as follows: the mean density $\rho = 2\,{\rm g/cm}^3$, the mean specific thermal capacity $C = 800\,{\rm J/kg/K}$, the surface albedo $A = 0.1$ and the mean thermal conductivity varied in the interval $K = 0.0005 - 0.5$ W/m/K (though we consider $\simeq 0.01$ W/m/K the most likely value, compatible with the estimated thermal inertia reported by Howell *et al.* (1994)). When converting the Yarkovsky force to acceleration components, we adopt a bulk density $\rho_b = 2.6\,{\rm g/cm}^3$, slightly higher than ρ (presumably affected by surface microporosity).

Figure 1b shows the resulting mean rate of change of Toutatis' semimajor axis due to the YE as a function of the poorly constrained surface conductivity K. Because of the slow rotation the YE strength drops for small values of K. For interest, we also show prediction of the linearized analytical theory that would assume an equivalent spherical body uniformly rotating about the direction of Toutatis angular momentum with two characteristic periods. Interestingly, the 7.42 d period does a fairly good job, especially for high conductivity values. Adopting $K = 0.01$ W/m/K we predict the YE displacement should exceed during the early October 2004 a 3σ formal orbit-determination error due to uncertainty in observations, thus being possibly detectable at a statistically significant level (more details in Vokrouhlický *et al.* (2004a)). It is, however, yet to be verified that

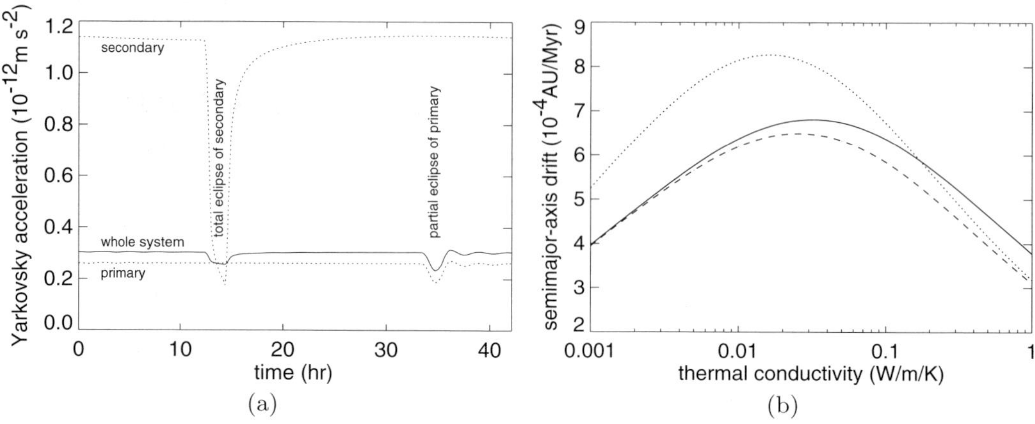

Figure 2. *Part a:* The effect of mutual eclipses between components of 2000 DP107 system: amplitude of the Yarkovsky acceleration during one revolution about their common center of mass. Eclipses produce dips in the signal; smooth variation during the shadow/eclipse entry and exit is due to a finite value of the surface thermal inertia (note the different effects for the fast rotating primary and the slowly rotating secondary asteroid). Solid curve is the effective Yarkovsky acceleration as it appears in the translational motion of the center of mass about the Sun. *Part b:* Drift rate of the orbital semimajor axis of the 2000 DP107 system due to the YE as a function of surface thermal conductivity K: (i) solid curve is for the whole system, (ii) dashed curve is for the primary component only, as if it were a solitary asteroid (no eclipses), and (iii) the dotted curve is for the primary component only and with the analytic formulation of the YE. A fair agreement indicates the effect of the secondary is minor, except for large value of K.

the orbit uncertainty due to gravitational perturbation by asteroids does not prevent the YE detection. This concern is mainly because of a low inclination of Toutatis' orbit $(0.44°)$, thus leading to frequent encounters with many asteroids in the main belt. Assuming the encounter probability scales inversely proportinally with inclination (e.g. Öpik 1951, 1976), we might expect about 5 times larger orbit-uncertainty due to gravitational effects of asteroids than in the case of (6489) Golevka. This latter has been estimated to be about $15\,\mu$s in delay measurement (Chesley *et al.* 2003; note the predicted delay displacement due to the YE is in between $15 - 30\,\mu$s in October 2004).

2000 DP107: a binary system.– 2000 DP107 belongs to the $10 - 15\%$ population of binary asteroids among NEAs (e.g. Margot *et al.* 2002; Merline *et al.* 2003). It consists of two components, a primary with estimated size of $\simeq 800\,$m and a secondary of $\simeq 300\,$m. The primary component exhibits a fast rotation with $P_1 \simeq 2.775\,$h, while the secondary component is likely orbit-synchronous with a period of $P_2 \simeq 1.755\,$d (in our model we slightly tweaked these values to become commensurable with the orbital period of the system about the Sun, namely P_1 be 1/5034 and P_2 be 1/332 part of that value). The mutual orbit of the two asteroids is quasi-circular with radius of $\simeq 1310\,$m. Current data do not allow shape resolution, so that we use spherical models for both components, represented in our model by 1004-facet polyhedra, with spin axes perpendicular to their mutual orbital plane (all data from Margot *et al.* (2002)).

In compact binaries, such as 2000 DP107, mutual eclipses produced by the two asteroids play important role and must be taken into account (Fig. 2a). We accordingly adapted our software to compute simultaneously Yarkovsky force for both asteroids in the system. The C-type classification for the primary component and tracking of mutual motion of the two components suggest lower density of $1.7\,$g/cm^3 (Margot *et al.* 2002; we assume this value for both surface and bulk density). The specific thermal capacity is taken to be

$\simeq 800\,\mathrm{J/kg/K}$, the surface optical albedo $A = 0.1$, while we again let the surface thermal conductivity to change in a broad range of values $0.001 - 1\ \mathrm{W/m/K}$.

Figure 2b shows the mean drift rate of the semimajor axis of the center of mass orbital motion about the Sun due to the Yarkovsky effect (one easily shows that the effective YE for the center of mass heliocentric motion is given by a mass-weighted mean of the YE on the two asteroids). We note the contribution of the secondary component is small, but not entirely negligible. Not shown here, however, is the role of the YE for motion of the two components about their common center of mass, where the effect on the secondary component plays determining role (see Vokrouhlický *et al.* (2004b) for detailed discussion). With that result, Vokrouhlický *et al.* (2004b) conclude the YE should be comfortably detected for this system during its close encounter in August 2016 provided accurate radar observations are acquired in September 2008.

Acknowledgements

This work was supported by the Grant Agency of the Czech Republic under the contract No. 205/02/0703. We thank Bill Bottke for several useful suggestions that helped to improve the final form of this paper.

References

Bottke, W.F., Vokrouhlický, D., Rubincam, D.P. & Brož, M. 2003, in: Bottke, W.F., Cellino, A., Paolicchi, P. & Binzel, R.P. (eds.), *Asteroids III* (The University of Arizona Press, Tucson), p. 395

Bottke, W.F., Vokrouhlický, D., Brož, M., Nesvorný, D. & Morbidelli, A. 2001, *Science* 294, 1693

Chesley, S.R., Ostro, S.J., Vokrouhlický, D., Čapek, D., Giorgini, J.D., Nolan, M.C., Margot, J-L., Hine, A.A., Benner, L.A.M. & Chamberlin, A.B. 2003, *Science* 302, 1739

Farinella, P. & Vokrouhlický, D. 1999, *Science* 283, 1507

Farinella, P., Vokrouhlický, D. & Hartmann, W.K. 1998, *Icarus* 132, 378

Giorgini, J.D., Ostro, S.J., Benner, L.A.M., Chodas, P.W., Chesley, S.R., Hudson, R.S., Nolan, M.C., Klemola, A.R. Standish, E.M., Jurgens, R.F., Rose, R., Chamberlin, A.B., Yeomans, D.K. & Margot, J.-L. 2002, *Science* 296, 132

Howell, E.S., Britt, D.T., Bell, J.F., Binzel, R.P. & Lebofsky, L.A. 1994, *Icarus* 111, 468

Hudson, R.S. & Ostro, S.J. 1995, *Science* 270, 84

Hudson, R.S., Ostro, S.J. & Scheeres, D.J. 2003, *Icarus* 161, 346

Kryszczyńska, A., Kwiatkowski, T., Breiter, S. & Michałowski, T. 1999, *Astron. Astrophys.* 345, 643

Landau, L.D. & Lifschitz, E.M. 1976, *Mechanics* (3rd Ed.) (Pergamon, Oxford)

Margot, J.-L., Nolan, M.C., Benner, L.A.M., Ostro, S.J., Jurgens, R.F., Giorgini, J.D., Slade, M.A. & Campbell, D.B. 2002, *Science* 296, 1445

Merline, W.J., Weidenschilling, S.J., Durda, D.D., Margot, J.-L., Pravec, P. & Storrs, A.D. 2003, in: Bottke, W.F., Cellino, A., Paolicchi, P. & Binzel, R.P. (eds.), *Asteroids III* (The University of Arizona Press, Tucson), p. 289

Milani, A., Nobili, A.-M. & Farinella, P. 1987, *Non-gravitational Perturbations and Satellite Geodesy* (A. Hilger, Bristol)

Morbidelli, A. & Vokrouhlický, D. 2003, *Icarus* 163, 120

Nesvorný, D. & Bottke, W.F. 2004, *Icarus* 170, 324

Ostro, S.J., Hudson, R.S., Rosema, K.D., Giorgini, J.D., Jurgens, R.F., Yeomans, D.K., Chodas, P.W., Winkler, R., Rose, R., Choate, D., Cormier, R.A., Kelley, D., Littlefair, R., Benner, L.A.M., Thomas, M.L. & Slade, M.A. 1999, *Icarus* 137, 122

Öpik, E.J. 1951, *Proc. Roy. Irish Acad.* 54, 165

Öpik, E.J. 1976, *Interplanetary Encounters* (Amsterdam: Elsevier Press)

Pravec, P., Harris, A.W., Scheirich, P., Kušnirák, P., Šarounová, L., Hergenrother, C.W., Mottola, S., Hicks, M.D., Masi, G., Krugly, Y.N., Shevchenko, V.G., Nolan, M.C., Howell, E.S., Kaasalainen, M., Galád, A., Brown, P., DeGraff, D.R., Lambert, J.V., Cooney, W.R. & Foglia, S. 2004, *Icarus*, in press

Spencer, J.R., Lebofsky, L.A. & Sykes, M.V. 1989 *Icarus* 78, 337

Spitale, J.N. & Greenberg, R. 2000, *Icarus* 149, 222

Tsiganis, K., Varvoglis, H. & Morbidelli, A. 2003, *Icarus* 166, 131

Vokrouhlický, D. 1998, *Astron. Astrophys.* 335, 1093

Vokrouhlický, D. 1999, *Astron. Astrophys.* 344, 362

Vokrouhlický, D. & Farinella, P. 1999, *Astron. J.* 118, 3049

Vokrouhlický, D. & Milani, A. 2000, *Astron. Astrophys.* 362, 746

Vokrouhlický, D., Milani, A. & Chesley, S.R. 2000, *Icarus* 148, 118

Vokrouhlický, D., Brož, M., Farinella, P. & Knežević, Z. 2001, *Icarus* 150, 78

Vokrouhlický, D., Brož, M., Morbidelli, A., Bottke, W.F., Nesvorný, D., Lazzaro, D. & Rivkin, A.S. 2002, Yarkovsky footprints in the Eos family, abstract 12-01, ACM 2002 meeting in Berlin, p. 115

Vokrouhlický, D., Čapek, D., Chesley, S.R. & Ostro, S.J. 2004a, *Icarus*, in press

Vokrouhlický, D., Čapek, D., Chesley, S.R. & Ostro, S.J. 2004b, *Icarus*, submitted

Vokrouhlický, D., Čapek, D., Kaasalainen, M. & Ostro, S.J. 2004c, *Astron. Astrophys.* 414, L21

Wechsler, A.E., Glaser, P.E., Little, A.D. & Fountain, J.A. 1972, in: Lucas, J.W. (ed.), *Thermal Characteristics of the Moon* (The MIT Press), p. 215

Wesselink A.J. 1948, *Bull. Astron. Inst. Nether.* 10, 351

Yomogida, K. & Matsui, T. 1983, *J. Geophys. Res.* 88, 9513

Dynamics of Populations of Planetary Systems
Proceedings IAU Colloquium No. 197, 2005
Z. Knežević and A. Milani, eds.

© 2005 International Astronomical Union
DOI: 10.1017/S1743921304008634

The population of asteroids in the 2:1 mean motion resonance with Jupiter revised

Miroslav Brož[1], D. Vokrouhlický[1], F. Roig[2], D. Nesvorný[3], W.F. Bottke[3] and A. Morbidelli[4]

[1]Institute of Astronomy, Charles University, V Holešovickách 2, 18000 Prague, Czech Republic
email: mira@sirrah.troja.mff.cuni.cz, vokrouhl@mbox.cesnet.cz

[2]Observatório Nacional, Rua Gal. José Cristino 77, Rio de Janeiro, 20921-400 RJ, Brasil
email: froig@on.br

[3]Department of Space Studies, Southwest Research Institute, 1050 Walnut St., Suite 400, Boulder, CO 80302, USA
email: davidn@boulder.swri.edu, bottke@boulder.swri.edu

[4]Observatoire de la Côte d'Azur, Dept. Cassiopee, BP 4224, 06304 Nice Cedex 4, France
email: alessandro.morbidelli@obs-nice.fr

Abstract. We study the population of asteroids inside the 2/1 mean motion resonance with Jupiter, in the so called Hecuba gap. Origin of these bodies is not well understood: (i) the long-lived (stable) population may be primordial, but this contradicts its steep size distribution, while (ii) the short-lived (unstable) population requires an efficient sustaining mechanism. Our working hypothesis is that the unstable asteroids are continuously resupplied from outside the resonance by the Yarkovsky effect. As a first step toward comparison of such model with observations, we report here an update of the observed population of asteroids residing in the 2/1 Jovian resonance, mainly because the number of cataloged orbits increased substantially during the last few years. We found there are 153 numbered and multi-opposition resonant asteroids in total and we classified them into the three sub-populations according to their dynamical lifetime. Our work also allowed us to derive several important parameters such as asteroid locations inside the resonance or size distribution of the sub-populations. As a particular novelty, we identified 6 asteroids located inside the high-eccentricity quasi-regular stable island, which previously seemed empty.

Keywords. Minor planets, asteroids

1. Introduction and current state of art

The structure of the Jovian mean motion resonances (MMRs) has been thoroughly studied for a long time. In the 1980's, there seemed to be a gap, almost void of any asteroids, in the 2/1 MMR, and that was in contrast with the 3/2 MMR, where numerous asteroids (so called Hildas) were observed. The 3-body models of motion in the 2/1 MMR (Murray 1986; Henrard & Lemaître 1987; Lemaître & Henrard 1990) failed to find any large chaotic regions inside, which could increase the eccentricity of the resonant objects and drive them to planet-crossing orbits, a preferred theory for the depletion of resonances since the work of Wisdom (1982). However, the early N-body numerical simulations (Wisdom 1987) showed the mechanism should work. Improved 3-body models (Morbidelli & Moons 1993), with main secular perturbations of Jupiter's orbit, allowed to calculate borders of secular resonances embedded in the 2/1 MMR, but the key point was to use at least a 4-body model (Ferraz-Mello 1994, Henrard *et al.* 1995), because only then low- and high-eccentricity chaotic regions, caused by overlapping secondary or secular resonances, are interconnected and allow large scale diffusion. The extensive N-body simulations (Nesvorný & Ferraz-Mello 1997) and semi-analytical analysis (Moons *et al.* 1998) allowed to map the structure of the 2/1 MMR in detail. Two macroscopic islands of stability, separated by ν_{16} secular resonance, were denoted A and B (see also Fig. 4). In the 1990's, there were already 10 resonant asteroids discovered.

In general, observations support the latest theoretical results, confirming that the 2/1 Jovian MMR harbors a small population of asteroids. Recent numerical analysis (Roig *et al.* 2002) allowed to classify these objects in three groups: (i) *Zhongguos* residing on stable orbits, whose dynamical lifetimes are comparable to the age of the Solar System, (ii) *Griquas* residing on orbits, whose dynamical lifetimes are several hundreds of My, and (iii) *unstable asteroids*, with a typical lifetime of $\simeq 10\,$My. Sub-populations (i) and (iii) are distinct not only by their very different lifetime, but also by their location in the resonance. Until now, no consistent scenario could make them linked together or to a common source.

Our goal is to prove that majority of the unstable asteroids have been pushed to the resonance from the main asteroid belt by the Yarkovsky effect. Before justifying any such hypothesis, we find useful to revise the 2/1 MMR asteroid population anew, because the automated search programs significantly boosted the number of known asteroids and smart identification procedures allowed to efficiently link past observations with the new discoveries, making thus the orbits more accurate.

In this paper we report results of our new search of the 2/1 resonant asteroids. The two main achievements are: (i) the population has grown to 153 objects with multi-opposition orbits (thus nearly tripled during the last 3 years), and (ii) we found the first asteroids in the previously empty island A. Both results have interesting implications: more numerous populations allow to better characterize their size distributions — an information directly constraining their origin. In the same way, the numbers of asteroids in the stable islands A and B constrain mechanisms by which they have been injected in the resonance or later destabilized.

2. The resonant population revised

Selection procedure.– The primary sample of asteroids we used are the numbered and multi-opposition bodies from the `AstOrb` database of the Lowell observatory (`ftp://ftp.lowell.edu/pub/elgb/` as of April 2004) whose osculating semimajor axis a and eccentricity e satisfy $e > \frac{0.45}{0.14}(3.24 - [a]_{\mathrm{AU}})$ and $e > \frac{0.5}{0.12}([a]_{\mathrm{AU}} - 3.34)$. This condition delimits a broad region near the 2/1 MMR (see Fig. 1 in Roig *et al.* 2002) containing ≈ 4000 asteroids. We numerically integrated their orbits for $10\,$ky using SWIFT-MVS2-FP integrator.† Our simulation included 4 giant planets. We applied a barycentric correction to the initial conditions of the planets' and asteroids' orbits (to partially account for the indirect perturbation by terrestrial planets). The second-order symplectic integrator allows a longer time-step than SWIFT-MVS, in our case 91.3125 days.

For each of the orbits we computed the evolution of the critical angle $\sigma_{\mathrm{J}2/1} = 2\lambda_{\mathrm{J}} - \lambda - \varpi$ that characterizes motion in the 2/1 Jovian MMR (λ_{J} and λ are the longitude in orbit for Jupiter and for the asteroid, and ϖ is the longitude of pericenter of the asteroid). The time-span of the integration was $10\,$ky and the output sampling was $10\,$y. We computed two diagnostic parameters (Fig. 1): (i) the ratio $r_\sigma = (n_+ - n_-)/(n_+ + n_-)$, where n_+ (or n_-) is the number of data points for which $\sigma_{\mathrm{J}2/1}(t)$ increased (or decreased) in two successive steps, and (ii) the difference $\Delta\sigma$ between the maximum and minimum value

---

† This is a modification of SWIFT package (Levison & Duncan 1994) with the second-order symplectic integrator included (see Laskar & Robutel 2001). Our program also contains on-line digital filters adapted to compute both synthetic proper orbital elements for the non-resonant motion and pseudo-proper elements for the resonant motion.

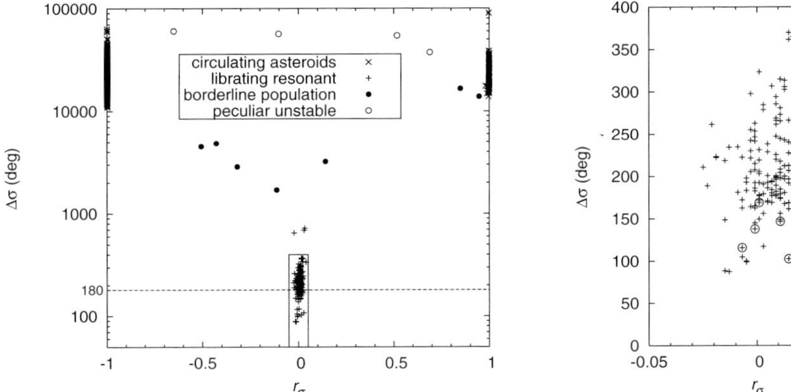

Figure 1. *Left:* The behavior of the critical angle $\sigma_{J2/1}$ of the 2/1 MMR with Jupiter for asteroids in the broad region around and in the resonance. The total change $\Delta\sigma$ of $\sigma_{J2/1}$ (i.e. the maximum minus minimum value) vs. the ratio r_σ (defined in the text) is plotted; both variables were computed from a 10 ky numerical integration with an output time-step 10 y. Bodies for which $\sigma_{J2/1}$ steadily circulates have either $r_\sigma = 1$ or $r_\sigma = -1$; the four asteroids with $\Delta\sigma > 30,000°$ and $|r_\sigma| < 0.98$ are either highly unstable or exhibit one abrupt change of $\sigma_{J2/1}(t)$ evolution. Asteroids with $r_\sigma \simeq 0$ and $\Delta\sigma \lesssim 360°$ reside inside the resonance. *Right:* A zoom in the resonant population. Asteroids highlighted by circles were later identified to be located in the stable island A (see the text).

of $\sigma_{J2/1}$. Regularly librating resonant asteroids appear closely clustered near the point $(r_\sigma, \Delta\sigma) = (0, 180°)$ (in practice we used a $|r_\sigma| < 0.05$ threshold). Asteroids outside the resonance, for which the critical angle circulates, have all $r_\sigma \doteq \pm 1$ and a value of $\Delta\sigma$ depending on the mean rate of their circulation. Since each of the parameters r_σ, $\Delta\sigma$ should allow by itself to detect the resonant asteroids, we can expect their values to be correlated – Fig. 1 confirms this conclusion. Several intermediate cases are: (i) borderline asteroids for which the critical angle alternates between periods of libration and circulation (full circles), and (ii) high-eccentricity asteroids with very unstable orbits or with peculiar orbits for which the sense of circulation of $\sigma_{J2/1}$ changes abruptly at some instant (open circles). Bodies with orbits for which $\sigma_{J2/1}$ librates, or temporarily librates, but which are non-resonant were excluded from Fig. 1. This is the case of motion near the pericentric and apocentric branches of periodic orbits in the three body problem. These orbits would have $|r_\sigma| \in (0.05, 0.98)$ in our plot and also $N = \sqrt{a}\,(2 - \sqrt{1 - e^2}\cos I) \leqslant$ 0.8 (a is the osculating semimajor axis, e the eccentricity and I the inclination; see e.g. Morbidelli & Moons 1993). There are many such pericentric alternators with $a \leqslant$ 3.275 AU and $e \leqslant 0.15$, but we do not study them here.

The procedure described above led to us to identify 153 asteroids residing inside the 2/1 MMR with Jupiter, making the population significantly grown since the last analysis (Roig *et al.* 2002). The complete list of the resonant asteroids, which is too long to be published here, is available at `http://sirrah.troja.mff.cuni.cz/yarko-site/`.

Resonant asteroids: dynamical lifetime and classification into sub-populations.– Motion inside MMRs is chaotic enough that any single numerical integration of a chosen orbit may not necessarily represent the true future motion. A safer procedure is to consider (for each asteroid) a multitude of orbits which are close enough to be all compatible with the observations.† Hence, unlike the previous studies, we used

---

† In practice this means to consider orbits with slightly modified initial elements, well inside the uncertainty region of the orbit determination. We call such fictitious orbits, but otherwise statistically equivalent, "close clones".

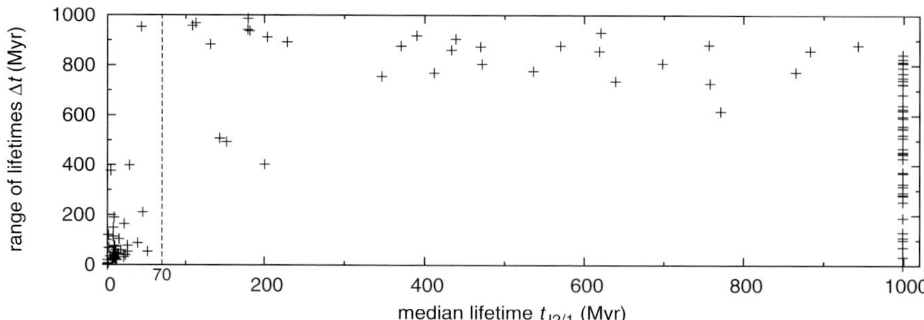

Figure 2. Median value $t_{J2/1}$ of lifetime vs. difference Δt between the maximum and minimum lifetime values of the close clones for asteroids observed inside the 2/1 MMR with Jupiter. Asteroids with $t_{J2/1} \leqslant 70\,\mathrm{My}$ have usually small Δt, indicating that all close clone orbits are unstable with a comparable timescale. Conversely, asteroids with $t_{J2/1} \geqslant 70\,\mathrm{My}$ have large Δt, indicating some clones may be long-lived, while others are short-lived. This seems to distinguish Griqua-like orbits from the unstable orbits. Most Zhongguos (with $t_{J2/1} = 1\,\mathrm{Gy}$) have again small or even zero value of Δt, since nearly all clones have the maximum lifetime.

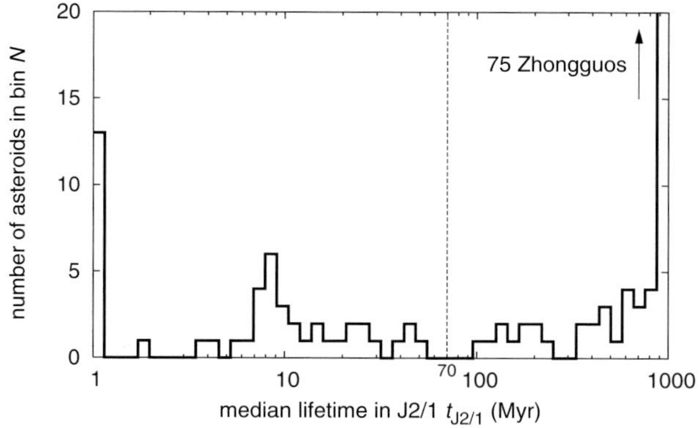

Figure 3. Distribution of the median lifetime values $t_{J2/1}$ for asteroids residing the 2/1 MMR. Note the log-binning and scale at the abscissa. Bodies with $t_{J2/1} \leqslant 70\,\mathrm{My}$ seem to form a distinct population with unstable orbits; Zhongguos on stable orbits "pile up" at $t_{J2/1} = 1\,\mathrm{Gy}$, limited in fact by the time-span of our integration. Griquas, with $t_{J2/1} \in (70\,\mathrm{My}, 1\,\mathrm{Gy})$, seem to be adjacent to the Zhongguo group.

the close-clones technique to determine dynamical lifetimes of the resonant asteroids: for each body we generated 12 clones by changing the orbital semimajor axis and eccentricity by multiples of $\delta a = 10^{-9}\,\mathrm{AU}$ and $\delta e = 10^{-9}$; orbital inclination and angular elements of clones were the same as of the original orbit. We numerically integrated resulting 13×153 orbits using SWIFT-MVS2-FP integrator upto $1\,\mathrm{Gy}$ or until the filtered mean semimajor axes (i.e. with fast oscillations removed) showed a significant departure (larger than $0.05\,\mathrm{AU}$) from the centre of the resonance at $3.27\,\mathrm{AU}$. This procedure was found reliable enough to characterize the residence time inside the resonance. Since all close clones are statistically equivalent realizations of the orbit, the characteristic dynamical lifetime $t_{J2/1}$ of the corresponding asteroid was determined as a median value of the individual lifetimes of the clones.

Figures 2 and 3 summarize the result. The first, showing aside the median lifetime $t_{J2/1}$ also the range Δt of the individual close-clone lifetime values, suggests there is a transition at about $t_{J2/1} \simeq 70\,\mathrm{My}$: (i) asteroids with smaller values of $t_{J2/1}$ reside on unstable orbits so that almost none of the close clones happen to have a long dynamical lifetime; (ii) asteroids with larger values of $t_{J2/1}$ reside on marginally stable or

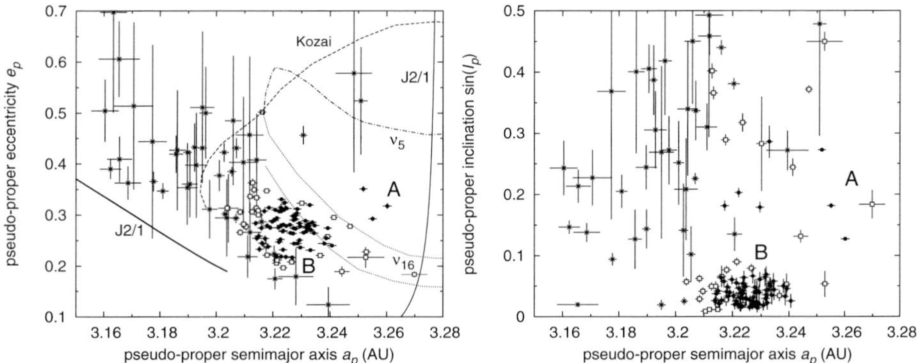

Figure 4. Distribution of the resonant asteroids in the space of pseudo-proper elements a_p, e_p, $\sin I_p$. Zhongguos are denoted by filled circles, Griquas by open squares and the unstable asteroids by crosses. The error bars denote standard deviation of pseudo-proper elements computed over 1 My interval of time. The thin and thick solid lines in the (a_p, e_p) projection (left) are the libration centre and separatrix of the 2/1 resonance (both shown for $I = 0°$). Borders of chaotic zones associated with embedded secular resonances — ν_5, ν_{16} and the Kozai resonance — are the dashed and dashed-dotted lines (all shown for $I = 10°$; adapted from Moons *et al.* 1998). A majority of the stable asteroids is clustered in the island B, while a few of them (see Table 1) are located in the island A, characterized by a higher mean eccentricity and inclination. All unstable asteroids are located in the chaotic zone where various secular resonances overlap. Griquas are a border-line population mostly at the edge of the B-region. In fact, 2-D projections shown here always lack clarity in showing 3-D structures; for that reason we posted a 3-D animation of the resonance structure with positions of the embedded asteroids at our web-site `http://sirrah.troja.mff.cuni.cz/yarko-site/`.

stable orbits, characterized with some (or all) clones having very long dynamical lifetimes (eventually $\geqslant 1$ Gy). Additionally, Fig. 3 suggests bodies with $t_{J2/1} < 70$ My are perhaps composed of two distinct groups: (i) very short-lived with $t_{J2/1} \leqslant 1$ My, and (ii) short-lived with $t_{J2/1} \geqslant 2$ My. The very short-lived group ($\simeq 25\,\%$ of the unstable asteroids) might be populated from a different source than the adjacent main belt, such as the Jupiter family comets or near-Earth asteroids, but this issue needs to be studied with care before drawing any conclusions. The long-lived asteroids ($t_{J2/1} \geqslant 70$ My) are either marginally stable Griquas or stable Zhongguos.† The respective numbers of asteroids are: 47 unstable asteroids, 31 Griquas and 75 Zhongguos.

Resonant asteroids: location inside the resonance.– Our numerical integration allowed us to characterize also the location of asteroids inside the 2/1 MMR. This was done by computing so called pseudo-proper (resonant) orbital elements (e.g. Roig *et al.* (2002)) — semimajor axis a_p, eccentricity e_p and inclination I_p. They are defined as osculating orbital elements at the time when the orbit intersects a surface of section defined by $\sigma_{J2/1} = 0°$ and $\dot{\sigma}_{J2/1} < 0$ and $\varpi - \varpi_J = 0°$ and $\Omega - \Omega_J = 0°$, where $(\Omega, \varpi; \Omega_J, \varpi_J)$ are the longitude of node and pericenter of the asteroid and of Jupiter, respectively, and $\dot{\sigma}_{J2/1}$ is the time derivative of $\sigma_{J2/1}$. In practice, though, we used a weakened condition $\sigma_{J2/1} \leqslant 5°$ and $\Delta\sigma < 0$ and $|\varpi - \varpi_J| \leqslant 5°$ (where $\Delta\sigma$ is a difference between the two successive values of $\sigma_{J2/1}$), in accord with Roig *et al.* (2002). The pseudo-proper resonant elements are near constants of motion in the averaged three-body problem with Jupiter on a fixed circular orbit; in reality the motion of both the asteroids and planets is more complicated and this may cause variations of (a_p, e_p, I_p). We thus computed mean values of (a_p, e_p, I_p)

† In fact, our 1 Gy simulation is still inappropriate to quantify lifetimes of Zhongguos, it only indicates they are longer than 1 Gy. Griquas are separated from Zhongguos only conventionally and up to this time-span they seem to form a shorter-lived "tail" of the Zhongguo population.

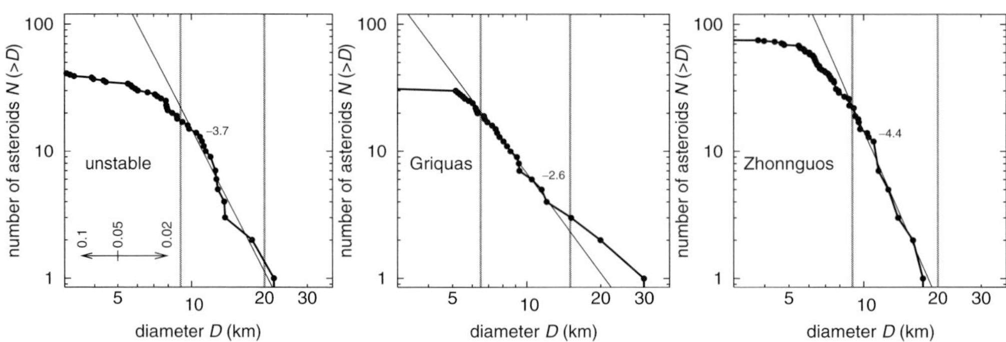

Figure 5. Cumulative size-frequency distributions of highly unstable asteroids (left), Griquas (middle) and Zhongguos (right). The solid lines indicate the best fit power-law approximations within the size interval delimited by the vertical grey lines. There are statistically significant differences in steepness of the size distributions of the three sub-populations. We assumed the albedo of all asteroids to be 0.05. Arrows in the left-most plot indicate a horizontal displacement of the distributions, if we assume different values of the albedo — 0.1 and 0.02.

and standard deviations $(\sigma_a, \sigma_e, \sigma_I)$ during the first 1 My of our integration (only nominal orbits were used for that purpose). Results are shown in Fig. 4, where we give positions of the resonant asteroids projected onto the planes (a_p, e_p) and $(a_p, \sin I_p)$. We note there is a strong correlation between the lifetimes of objects and their positions in the resonance: (i) the unstable orbits are located inside the chaotic regions where secular resonances overlap, (ii) the stable orbits (Zhongguo-like) are inside the stable islands B and A, and (iii) the marginally stable (Griqua-like) are right at the outskirts of the stable regions.

Resonant asteroids: size distribution.– Having been able to classify resonant asteroids into the three sub-groups we are now ready to study some of their parameters indicative of their origin or evolution. An important information is a size-frequency distribution (SFD) shown in Fig. 5. We assumed the value of albedo to be 0.05, appropriate for the prevalent C- and D-type asteroids in the outer part of the main asteroid belt; higher or smaller mean albedo values would displace the distributions toward smaller or larger sizes as shown in the left panel. The largest members of each of the sub-populations are of about the same size ($\simeq 20\,\mathrm{km}$). We fitted SFDs with a power-law $N(>D) \sim D^\alpha$, usually down to the diameter $\simeq 9\,\mathrm{km}$, that does not seem to be significantly affected by the observational incompleteness. We found the following values of the SFD exponent α: (i) -4.4 ± 0.3 for Zhongguos, (ii) -2.6 ± 0.5 for Griquas, and (iii) -3.7 ± 0.2 for the unstable asteroids.† In spite of relatively large standard deviations of the exponents, the SFDs differ from each other with a high statistical significance. The SFDs of Zhongguos and the unstable asteroids are also significantly steeper than -2.5, what one would expect for a collisionally evolved isolated system (e.g. Dohnanyi 1969). Even if generalizations of the original Dohnanyi's model are taken into account, such as a size-dependence of the disruption Q-factor (e.g. O'Brien & Greenberg 2003), the SFD of the unstable asteroids is hard to reconcile with a system evolved solely by collisions. It is also interesting to compare the SFDs of the resonant asteroids, the unstable asteroids in particular, with that of a large population of asteroids adjacent to the 2/1 MMR. Within a comparable interval of sizes ($9 - 50\,\mathrm{km}$, say) we found that the cumulative SFD of the main belt population is well approximated by a power-law with an exponent -2.60 ± 0.05.

† We warn the results for the long-lived populations – Zhongguos and Griquas – are partly ambiguous, because these populations are not well separated. The current division at $t_{\mathrm{J2/1}} = 1\,\mathrm{Gy}$ was chosen a little arbitrarily. If the whole long-lived population (i.e. with $t_{\mathrm{J2/1}} \geqslant 70\,\mathrm{My}$) is considered together its SFD has an exponent -3.5 ± 0.2.

Table 1. Parameters of Zhongguo- and Griqua-like asteroids found to reside in the stable island A of the 2/1 MMR with Jupiter. a_p, e_p and I_p are pseudo-proper semimajor axis, eccentricity and inclination, σ_a, σ_e and σ_I their standard deviations during a 1 My numerical integration, $t_{J2/1}$ median lifetime, H absolute magnitude and D diameter, calculated from H using the value of albedo 0.05. The last column is the classification: G – Griquas, Z – Zhongguos. Asteroid (4177) Kohman is a border-line case (see the text for discussion).

No.	Name	a_p [AU]	e_p	I_p [deg]	σ_a [AU]	σ_e	σ_I [deg]	$t_{J2/1}$ [My]	H [mag]	D [km]	Class.
78801	2003 AK$_{88}$	3.260	0.318	7.309	0.002	0.006	0.14	1000	15.20	5.50	Z
	1999 VU$_{218}$	3.241	0.295	14.125	0.001	0.002	0.82	771	15.25	5.37	G
	2001 FY$_{84}$	3.253	0.217	26.727	0.007	0.020	0.89	152	14.06	9.29	G
	2003 SA$_{197}$	3.252	0.351	15.807	0.001	0.006	0.09	1000	14.63	7.15	Z
	2003 YN$_{94}$	3.255	0.293	10.451	0.002	0.005	0.24	1000	15.20	5.50	Z
	2004 FG$_{32}$	3.247	0.278	21.816	0.001	0.004	0.37	535	14.53	7.48	G
4177	Kohman	3.233	0.320	16.598	0.001	0.001	1.52	1000	12.70	17.38	Z

This is interesting because its difference from the SFD exponent of the unstable asteroids is approximately -1.† This supports models in which the unstable asteroids are resupplied with a size-selective process which strength is inversely proportional to size. The Yarkovsky effect (e.g. Vokrouhlický *et al.*, this volume) meets this constraint (see also Morbidelli & Vokrouhlický 2003). These models, however, cannot be extended to Zhongguos, because of their anomalously steep SFD.

Resonant asteroids: objects inside the stable island A.– Among the newly identified asteroids inside the 2/1 MMR with Jupiter we found 6 cases that lie above the separatrix of the ν_{16} secular resonance (see Table 1 and Fig. 4). These are the first known objects in the previously void stable island A. In comparison to B-Zhongguos, A-Zhongguos (and also A-Griquas) have typically smaller ($< 90°$) amplitude of $\sigma_{J2/1}(t)$ (see Fig. 1). There is also a clear secular term with an approximate amplitude of $20°$ and period $\simeq 9$ ky. The sizes of objects, excluding Kohman, range from 5.5 km to 7.5 km, if we adopt the nominal geometric albedo 0.05. The ratio of the number of A-Zhongguos (excluding A-Griquas) and the number of B-Zhongguos is equal to $3/71 \doteq 0.04$. However, this ratio has only a low statistical significance, because the number of A-Zhongguos, discovered upto now, is too small and new discoveries will change the ratio a lot.

Models that would attempt to explain the origin and evolution of the stable population in the 2/1 MMR should consider the differences between island A and island B population as important constraints. For instance, Ferraz-Mello *et al.* (1998) discussed a possibility, that Zhongguos are relics of a primordial population. They proved the island A would become significantly more (but not completely) depleted if the mutual position of Jupiter and Saturn were such that the period of Great Inequality in Jupiter's motion would be shorter. (This is a plausible scenario since the giant planets likely migrated during the early evolution of the Solar system.) On the other hand, the steep SFD of Zhongguos (Fig. 5) would indicate a recent collisional origin of a majority of this population. However, the mean values of both eccentricity and inclination for A-Zhongguos and B-Zhongguos differ so much (the ejection velocities would be of order 1 km/s), that a possibility of a collisional break-up taking part in the island B and populating with fragments the island A is very unlikely.

† In fact, when we discard the extremely unstable asteroids with $t_{J1/2} \leqslant 1$ My (see Fig. 3), which may have anomalous origin, the size distribution of the remaining unstable asteroids becomes slightly shallower and the difference from the main belt population is $\simeq -0.8$.

We finally note that the asteroid (4177) Kohman, previously classified as Griqua-like by Roig *et al.* (2002), is a border-line case residing very close to the separatrix of the ν_{16} resonance. Indeed, we found the critical angle of this resonance ($\sigma_{16} = \Omega - \Omega_S$) exhibits periods of librations and circulations in Kohman's case.[†] With its large size (17 km in diameter) this body would by far dominate the island A population.

3. Conclusions

We revised the currently observed population of asteroids residing in the 2/1 mean motion resonance with Jupiter. We identified 153 resonant asteroids, and among them six new located in the stable island A. The properties of the resonant population will serve as important constraints for models of its origin and evolution (e.g. for the model involving the Yarkovsky effect to push main belt asteroids onto unstable resonant orbits). To make such models even more constrained, the following further observations would be particularly useful: (i) accurate astrometry of faint asteroids to complete the known population down to smaller sizes, with the goal to reduce uncertainty in determination of the power-law representation of the size-frequency distribution; (ii) spectral or multi-color observations to investigate similarity or diversity of the resonant sub-populations and of possible sources, as far as the surface mineralogy is concerned; (iii) lightcurve observations of the unstable asteroids with the goal to obtain (or at least constrain) their pole positions.[‡] However, spectral and lightcurve observations of a large number of small main-belt asteroids will probably not be available until we have data from space-born projects (such as GAIA).

Acknowledgements

We thank the referee, Sylvio Ferraz-Mello, for valuable comments and discussion. The work of MB and DV was supported by the Grant Agency of the Czech Republic.

References

Dohnanyi, J.W. 1969, *J. Geophys. Res.* 74, 2531

Ferraz-Mello, S. 1994, in *Asteroids, Comets, Meteors 1993*, eds. A. Milani *et al.*, Kluwer Academic Publishers, Dordrecht, 1994, p. 175

Ferraz-Mello, S., Michtchenko, T.A. & Roig, F. 1998, *Astron. J.* 116, 1491

Henrard, J. & Lemaître, A. 1987, *Icarus* 69, 266

Henrard, J., Watanabe, N. & Moons, M. 1995, *Icarus* 115, 336

Laskar, J. & Robutel, P. 2001, *Celest. Mech. Dyn. Astr.* 80, 39

Lemaître, A. & Henrard, J. 1990, *Icarus* 83, 391

Levison, H.F. & Duncan, M.J. 1994, *Icarus* 108, 18

Moons, M., Morbidelli, A. & Migliorini, F. 1998, *Icarus* 135, 458

Morbidelli, A. & Moons, M. 1993, *Icarus* 102, 316

Morbidelli, A. & Vokrouhlický, D. 2003, *Icarus* 163, 120

Murray, C.A. 1986, *Icarus* 65, 70

Nesvorný, D. & Ferraz-Mello, S. 1997, *Icarus* 130, 247

O'Brien, D.P. & Greenberg, R. 2003, *Icarus* 164, 334

Roig, F., Nesvorný, D. & Ferraz-Mello, S. 2002, *Mon. Not. R. Astron. Soc.* 335, 417

Wisdom, J. 1982, *Astron. J.* 87, 577

Wisdom, J. 1987, *Icarus* 72, 241

[†] The close proximity of ν_{16} resonance leads to a relatively larger value of σ_I for this body; similarly, the asteroid 1999 VU$_{218}$ is near the resonance, though in this case the critical angle circulates.

[‡] Note the Yarkovsky-driven origin of the unstable population would, for instance, predict most of these asteroids should have prograde rotation.

Dynamics of Populations of Planetary Systems
Proceedings IAU Colloquium No. 197, 2005
Z. Knežević and A. Milani, eds.

© 2005 International Astronomical Union
DOI: 10.1017/S1743921304008646

On the reliability of computation of maximum Lyapunov Characteristic Exponents for asteroids

Zoran Knežević and Slobodan Ninković

Astronomical Observatory, Volgina 7, 11160 Belgrade 74, Serbia and Montenegro
email: zoran@aob.bg.ac.yu; sninkovic@aob.bg.ac.yu

Abstract. We present an analysis of the reliability of computation of maximum Characteristic Exponents of Lyapunov from the numerical integrations of asteroid orbits over finite intervals of time. We used two complementary approaches - a comparison of the LCE estimates from the backward and forward integrations of orbits, and a comparison of the estimates coming from the integrations of the same initial conditions over different time spans. The main conclusion is that for a vast majority of asteroids ($> 80\%$) the results can be considered as reliable enough to reveal the very nature and basic properties of the motion.

Keywords. Asteroids, chaotic dynamics, maximum Lyapunov Characteristic Exponents

1. Introduction

Chaotic dynamics in the real systems can be quite complex, thus giving rise to very different diffusion times and macroscopic effects. In order to assess the stochastic motion, to identify and quantitatively describe the chaos, one usually computes some sort of indicator or measure, which reveals the very character of the motion, or distinguishes between phenomena related to the chaotic dynamics.

The most commonly used measure of chaos is the so-called maximum Lyapunov Characteristic Exponent (LCE). It measures the rate at which two initially nearby orbits diverge with respect to each other, in terms of some convenient metrics in the phase space of state vectors (Milani & Mazzini 1997). The stronger the chaos, the faster the divergence and the larger the LCE, so that a large, positive LCE typically means that the corresponding orbit is strongly chaotic.

By definition, Lyapunov exponents are all the real numbers obtained from a limit:

$$\chi(X_0, V_0) = \lim_{t \to \infty} \frac{\gamma(t)}{t} \; ; \qquad \gamma(t) = \log_e \frac{|V(t)|}{|V_0|} . \tag{1.1}$$

Given the dynamical system $\dot{X} = F(X)$ and its general solution $\Phi^t(X_0) = X(t)$ in the form of an integral flow, $V(t)$ is found as a solution (with initial condition V_0) of a variational equation (with respect to the orbit with initial condition X_0):

$$\dot{V} = \frac{\partial F}{\partial X}(X(t)) \, V \, , \tag{1.2}$$

associated with the dynamical system.

In practice, however, we are more often using the Lyapunov time $T_L = 1/\chi$, which is defined as a simple inverse of the LCE. Lyapunov time is easier to comprehend and interpret, in particular when discussing the dynamical mechanisms at work, or when we are to compare the chaotic behavior of different bodies. Obviously, T_L represents the time needed to increase the distance between the orbits exp(1) times.

Many alternative methods to identify and quantify chaos and to distinguish between phenomena related to chaotic dynamics - invariant tori, islands of libration, chaotic zones, etc., have been proposed over the last 15 years, like the frequency map analysis and the related sup-map analysis (Laskar 1990; 1993), the short-time measures like the local Lyapunov characteristic numbers (Froeschlé *et al.* 1993), or stretching numbers (Voglis & Contopoulos 1994), helicity and twist angles, azimuthal and rotation angles (Contopoulos *et al.* 1997), the so-called fast Lyapunov indicators (Froeschlé *et al.* 1998), etc. A comprehensive review of various methods can be found in Knežević (2000).

For the purpose of this paper, however, we stick with the LCE only. In Section 2 we describe the specific problem we are dealing with and the methods used to tackle it, in Section 3 we give and discuss the results, and in Section 4 we offer some conclusions.

2. Computation of maximum LCEs for asteroids

In the case of solar system bodies, like planets or asteroids, the dynamical system is well-known to be non integrable, so that an exact computation of LCE for $t \rightarrow \infty$ is not possible. The simplest way to get an approximate, temporary estimate of the value of maximum LCE is by monitoring the variation of $\gamma(t)$ with time, computed on-line, in the course of the numerical integration of the orbit itself; if this variation has an asymptotic character, the slope of the variation gives the estimated value of the maximum LCE.

In order to compute an estimate of the maximum LCE reliably, one has to integrate the orbit for at least 6-10 times T_L or even longer (Froeschlé *et al.* 1998); in the case of weak, slow chaos this can become quite a formidable task and a very time-consuming procedure. On the other hand, a single integration of a strongly chaotic orbit does not posses any predictive value; only a large ensemble of integrations with the same or similar initial conditions, integration methods and platforms can supply us with a valid indication on the nature and characteristics of the chaotic motion; again this is a time-consuming procedure, applicable only in some special cases.

For authors of asteroid catalogs which include hundreds of thousands of bodies, where LCEs are provided with other orbital information, the essential becomes to know *a priori* whether the effort of computing all these estimates is really meaningful, that is, whether the Lyapunov Characteristic Exponents can at all be reliably determined from the integrations covering only a limited time span. For users of these catalogs the question is whether they can believe and apply these estimates at face value, or, at least, what is the probability that the computed values are indeed indicative of the true nature of motion.

We used in our analysis the data obtained as part of the regular updates of the AstDyS database†, but also a large sample of integrations adapted to and carried out specifically for the purpose of this study.

We employed two different, but complementary approaches: (i) the comparison of the LCEs derived separately from two integrations, starting from the same initial conditions, using the same method, code and machines, and covering the same interval of time with the same step size, but once with the negative step (that is, propagating orbit into the past) and the other time with the positive step (propagating into the future); (ii) the comparison of the LCEs derived from the integrations starting again from the same initial initial conditions, using the same method, code and machines, but this time covering different intervals of time and integrating either forward or backward in time.

† http://hamilton.dm.unipi.it/astdys

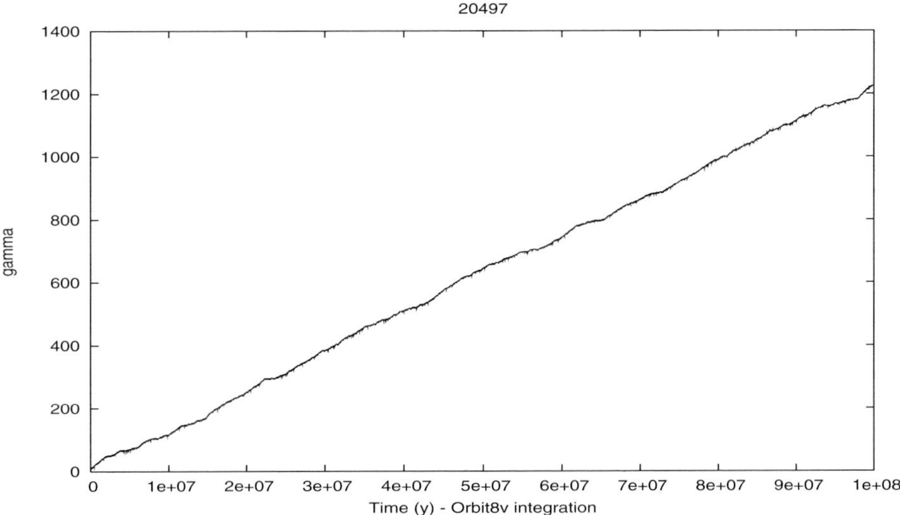

Figure 1. Asteroid 20497. A smooth variation of $\gamma(t)$. Maximum LCE can be estimated in a reliable manner. Orbit is moderately chaotic with $T_L \approx 83\,000$ yr.

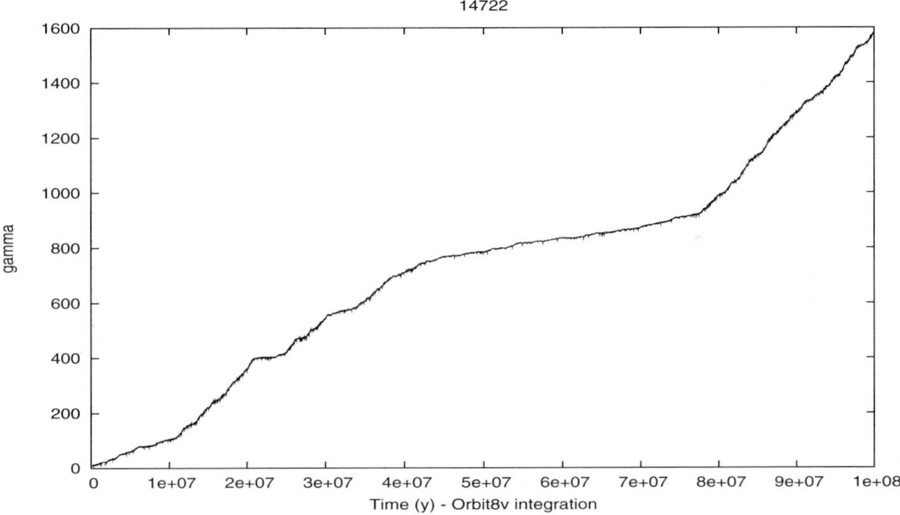

Figure 2. Maximum LCE estimate for asteroid 14722. Corresponding Lyapunov time is $T_L \approx 75\,000$ yr. Note abrupt changes of the slope of $\gamma(t)$, including episodes of strong chaos ($T_L \approx 26\,000$ yr). This orbit passes through different dynamical regimes in the course of time, and the estimate of the corresponding LCE in this case critically depends on the integration time span.

In the first set of experiments we in fact varied the configuration of the dynamical system and thus the circumstances of the determination of corresponding LCEs. We used integrations of asteroid orbits covering 1 Myr in the past and 1 Myr in the future. The LCEs were determined from the two integrations separately and then simply compared, assuming the values to be the same if the difference was less than 10^{-5} yr^{-1} (this values corresponds to $T_L > 100\,000$ yr, i.e. to a somewhat conservative requirement that for reliable computation of LCEs we need integrations covering at least $10\,T_L$).

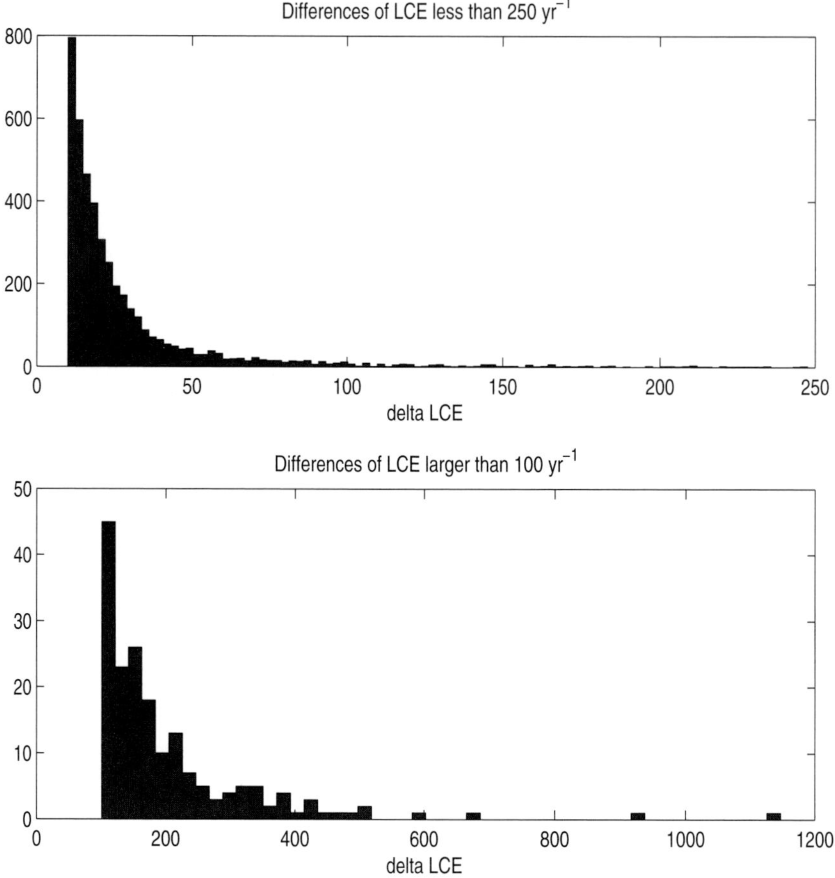

Figure 3. Number frequency distributions of differences of maximum LCEs in units of 10^{-6}. Above: asteroids with differences < 250 yr^{-1}. Below: the large differences tail of the distribution.

In the second set of experiments we were repeating computations over intervals of time spanning the two orders of magnitude. More precisely, we compared the outcomes of integrations covering 2, 10 and 100 Myr. The problem we are dealing with in this case is illustrated with the examples of the computation of $\gamma(t)$ given in Figs. 1 and 2.

3. Results

All the computations in this study were performed by using the public domain software packages OrbFit and Orbit9, available from the aforementioned AstDyS site. In particular, we made use of the Orbit9e integrator and of the software to compute synthetic proper elements, described in more detail in Knežević & Milani (2000).

Summarizing the results of the first set of experiments, let us repeat that we compared two estimates of the maximum LCEs obtained from the forward and backward 1 Myr integrations for a total of 23767 asteroids. Out of these, for 19371 (or 81.5%) the differences of LCEs were $\leqslant 10^{-5}$ yr^{-1}, that is below the adopted accuracy threshold for reliable LCE computation. Thus, for the vast majority of asteroids the results derived from two integrations can be considered the same and safely used as indicator of the true nature of motion.

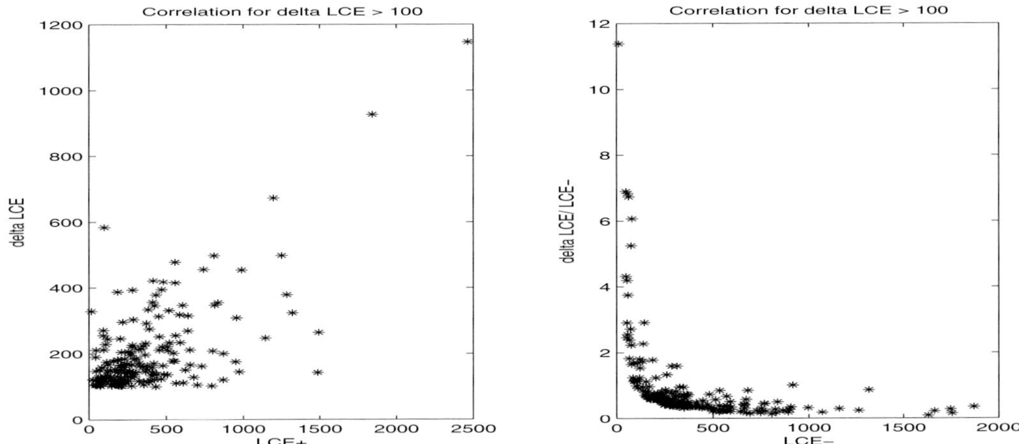

Figure 4. Differences of maximum LCE vs. one of the estimates for 183 objects with largest LCE differences (left). Relative difference vs. the other estimate for the same objects (right). All quantities are in units of 10^{-6}.

Of particular interest are, however, objects for which large differences of LCE estimates were found. There were 4396 asteroids with difference $> 10^{-5}$ yr^{-1} (the number frequency distribution of the differences of maximum LCEs up to 2.5×10^{-4} yr^{-1} is given in the upper panel of Fig. 3), out of which 183 with difference $> 10^{-4}$ yr^{-1} (see the distribution in lower panel of Fig. 3), and 40 with difference $> 2.5 \times 10^{-4}$ yr^{-1}. The median difference for all 4396 asteroids with LCE differences above the accuracy threshold is found to be $\approx 1.9 \times 10^{-5}$ yr^{-1}. Thus, even if the maximum LCEs for these bodies have been estimated with lower accuracy, there is still a good fraction of objects for which this estimate reliably reflects at least the very nature of the motion (e.g. strongly vs. moderately chaotic).

In order to identify the "troublemakers", in Fig. 4 we plotted the differences of the estimates against the estimates themselves, both in absolute and relative terms. We expected to find correlation in the sense - more chaotic the motion, less reliable the computation of the corresponding LCE; this sort of correlation appears quite obvious in the left panel, showing the absolute values of the differences vs. differences themselves. The result in the right hand side panel, showing the relative differences vs. differences themselves, indicates, however, that comparatively more important appear to be differences for the moderately chaotic orbits. For the asteroids with such differences computation of the maximum LCE is unreliable and the obtained values cannot be considered useful.

In Table 1 we give asteroids from our sample of 183 bodies with significant differences of LCE that have largest absolute and relative differences of the computed LCEs. In the first three columns we list asteroid number, maximum LCE as computed from the forward integration, and the corresponding LCE difference for 12 asteroids for which this difference was found to be larger then 4×10^{-4} yr^{-1}. In the remaining three columns we show again asteroid number, but with maximum LCE as computed from the backward integration, and the corresponding relative LCE difference for 9 asteroids for which this difference was larger then 3.

Repeating now more-or-less the same exercise for the second set of experiments, we begin with comparison of the LCEs computed from the integrations covering 2 and 10 Myr. Let us emphasize that asteroids for which we extend integrations from 2 to 10 Myr in the regular AstDyS database updates are those for which some sort of problem has been detected in the shorter integration; typically this might have been either the

Table 1. Asteroids from our sample of 183 bodies with largest absolute and relative differences of the computed LCEs

Asteroid	$LCE+$ $\times10^{-6}$ yr^{-1}	Abs. Diff $\times10^{-6}$ yr^{-1}	Asteroid	$LCE-$ $\times10^{-6}$ yr^{-1}	Rel. Diff
22541	1198	673	22269	29	4.2
29180	480	418	22272	18	6.7
37152	742	456	22378	68	6.0
42485	557	478	22383	65	11.4
42600	415	422	22467	10	6.8
43946	558	415	22502	10	6.9
49737	989	455	22543	37	3.7
53463	1844	927	22553	10	4.3
58949	2465	1147	22582	13	5.3
72205	811	498			
75246	1251	499			
80143	99	584			

poor accuracy of the resulting synthetic proper elements, or large positive LCE indicating strongly chaotic orbit. It is thus no surprise that out of a total of 5768 asteroids considered in the second set of experiments, in only 3663, or 63.5% of the cases, we found good agreement of the estimates of maximum LCE from the two integrations. In the remaining 2105 cases the difference was larger than 5×10^{-6} yr^{-1}, which was in this case the accuracy threshold for the shorter integration. Nevertheless, the number of significantly different estimates was not very high, just 113 objects with differences $> 10^{-4}$ yr^{-1} and 17 objects with differences $> 2.5\times10^{-4}$ yr^{-1} ; median for objects with difference above the accuracy threshold was $\approx 2.2 \times 10^{-5}$ yr^{-1}. Fig. 5, analogous with Fig. 3, shows the same number frequency distributions as in the previous case.

The most interesting in this context are, however, the results of comparison with integrations covering 100 Myr time span. Here we compared our estimates of LCE for a specially selected group of asteroids belonging to Veritas asteroid family; these asteroids exhibit very different dynamical behaviors, ranging from very stable to strongly chaotic (Knežević *et al.* 2002). We collected all the interesting results in Table 2. We mixed different integrations here, in order to check all possible outcomes and to partly compensate for a small number of experiments performed in this case. Thus we compare estimates of maximum LCE from short integrations, covering either 2 or 10 Myr, with 100 Myr forward or backward integrations. As one can easily see, even under these "extreme" circumstances the results in most cases agree quite well (note that in columns 4 and 5 all the printed figures are significant). In each of the several cases in which the agreement was found not to be that good, the comment "chs" indicates that the body changed the dynamical state during the interval of time covered by the integration, and that therefore we detected a significant change of slope of $\gamma(t)$ like in the example shown in Fig. 2.

4. Conclusions

We have conducted a comprehensive analysis of the reliability of computation of maximum Characteristic Exponents of Lyapunov from the numerical integrations of asteroid orbits covering finite intervals of time. We used two complementary approaches - a comparison of the results coming from the backward and forward integrations of orbits, and similar comparison of the results coming from the integrations of the same initial conditions, but over very different time spans.

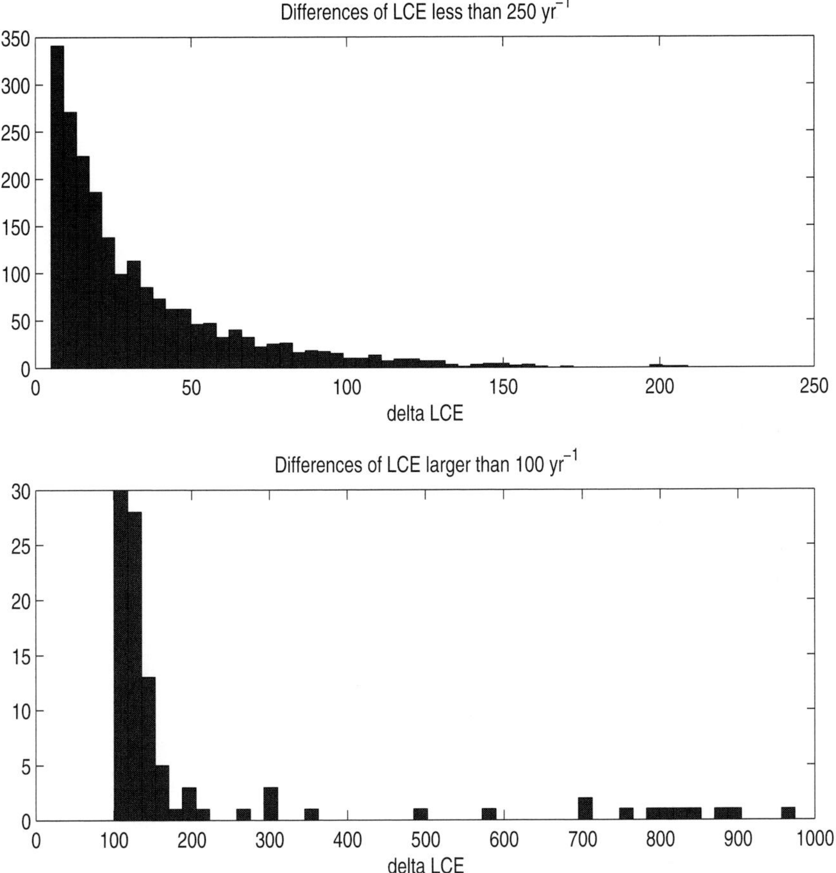

Figure 5. Number frequency distributions of differences of maximum LCEs in units of 10^{-6} obtained from integrations covering 2 and 10 Myr. Above: asteroids with differences $< 250 \ \mathrm{yr}^{-1}$. Below: the large differences tail of the distribution.

The main conclusion which can be drawn from this analysis is that for a vast majority of asteroids ($> 80\%$) the results of the computation, even if stochastic by definition, can be considered as reliable enough to reveal the very nature and basic properties of the motion. For chaotic bodies the LCE estimates are also in most cases ($> 60\%$) precise enough at least to clearly distinguish between different dynamical regimes (weak to moderate vs. strong chaos). Still, we recommend to always take the computed values of LCEs with some caution, in particular when dealing with outcomes of short integrations or with strongly chaotic motion. Even in the case the computed values indicate very stable motion, they should not be taken uncritically and at face value, but always bearing in mind their probabilistic definition and interpretation.

Acknowledgements

We would like to acknowledge the contribution of A. Milani, the principal author of the software packages OrbFit and Orbit9 which we used for all the computations. This research is supported by the Ministry of Science and Environmental Protection of Serbia through project No 1238.

Table 2. Asteroids in Veritas asteroid family exhibit very different dynamical behavior – from very stable to strongly chaotic. In the first column we give asteroid number, maximum LCE in the second column was obtained from the integration covering a time span indicated in the third column (negative time refers to the backward integration), LCEs as obtained from the forward and backward integrations are shown in the fourth and fifth columns, respectively; the comment is added in the final column ("chs" stands for the "changing state").

Asteroid	LCE $\times 10^{-6}$ yr^{-1}	ΔT Myr	LCE+100 $\times 10^{-6}$ yr^{-1}	LCE-100 $\times 10^{-6}$ yr^{-1}	Comment
490	98	10		112.8	
1086	5	-2	0.2		
2147	7	-2	8.1		
2428	5	-2	2.9		
2934	5	-2	3.7		
3090	5	-2	0.6		
3542	105	10		107.7	
5594	29	-2	22.6		
6374	5	-2	0.0		
7231	5	-2	0.0		
7612	30	-2	18.6		
7626	5	-2	4.5		
8726	107	10		98.5	
9715	5	-2	0.6		
10120	114	10		91.9	
10414	5	-2	0.9		
10793	37	-2	22.5		
13537	12	-2	11.1		
14722	39	-2	13.2		
19280	31	-2	20.1		
20497	9	-2	12.2		
21454	98	10		93.2	
24718	105	10		79.5	chs
32416	95	10		91.8	
38447	112	10		103.0	
39290	108	10		111.2	
42218	110	10		106.4	
43095	35	10		102.2	chs
45623	106	10		103.0	
46105	110	10		82.9	chs
51136	104	10		14.9	chs
54869	108	10		93.0	chs

References

Contopoulos, G., Voglis, N., Efthymiopoulos, C., Froeschl, C., Gonczi, R., Lega, E., Dvorak, R., & Lohinger, E. 1997, *Cel. Mech. Dyn. Astron.* 67, 293

Froeschlé, C., Froeschlé, Ch., & Lohinger, E. 1993, *Cel. Mech. Dyn. Astron.* 56, 307

Froeschlé, C., Gonczi, R., Lega, E., & Locatelli, U. 1998, *Cel. Mech. Dyn. Astron.* 69, 235

Knežević Z. 2000, in: N. Bokan (ed.) *Proceedings Symp. Contemp. Math.: 125 years of Faculty of Mathematics*, Fac. Math. Univ. Belgrade, p. 269

Knežević Z., & Milani A. 2000, *Cel. Mech. Dyn. Astron.* 78, 17

Knežević Z., Tsiganis K., & Varvoglis H. 2002, in *Asteroids, Comets, Meteors ACM2002.*, (European Space Agency SP-500), 335.

Laskar, J. 1990, *Icarus* 88, 266

Laskar, J. 1993, *Physica D* 67, 257

Milani, A. & Mazzini, G. 1997, *Sistemi dinamici* (Pisa University Editorial Service)

Voglis, N. & Contopoulos, G. J. 1994, *J. Phys. A: Math. Gen.* 27, 4899

Dynamics of Populations of Planetary Systems
Proceedings IAU Colloquium No. 197, 2005
Z. Knežević and A. Milani, eds.

© 2005 International Astronomical Union
DOI: 10.1017/S1743921304008658

Nekhoroshev stability estimates for different models of the Trojan asteroids

Christos Efthymiopoulos

Research Center for Astronomy and Applied Mathematics, Academy of Athens
Soranou Efessiou 4, 115 27 Athens, Greece
e-mail: cefthim@cc.uoa.gr

Abstract. Estimates of the region of Nekhoroshev stability of Jupiter's Trojan asteroids are obtained by a direct (i.e. without use of the normal form) construction of formal integrals near the Lagrangian elliptic equilibrium points. Formal integrals are constructed in the Hamiltonian model of the planar circular restricted three body problem (PCRTBP), and in a mapping model (Sándor *et al.* 2002) of the same problem for small orbital eccentricities of the asteroids. The analytical estimates are based on the calculation of the size of the remainder of the formal series by a computer program. An analysis is made of the accumulation of small divisors in the series. The most important divisors introduce competing Fourier terms with sizes growing at similar rates as the order of truncation increases. This makes impossible to improve the estimates by considering nearly resonant forms of the formal integrals for particular near-resonances. Improved estimates were obtained in a mapping model of the PCRTBP. The main source of improvement is the use of better variables (Delaunay). Our best estimate represents a maximum libration amplitude $D_p = 10.6^0$. This is a quite realistic value which demonstrates the usefulness of Nekhoroshev theory.

Keywords. Trojan asteroids; Nekhoroshev stability

---

1. Introduction

Stability estimates in nonlinear Hamiltonian dynamical systems are obtained by applications of either the KAM theorem (Kolmogorov 1954, Arnold 1963a,b, Moser 1962), or the Nekhoroshev theorem (Nekhoroshev 1977, Benettin et al. 1985, , Lochak 1992, Pöschel 1993). A connection between the two theorems is provided by the theorem of superexponential stability (Morbidelli and Giorgilli 1995a,b).

The question of obtaining analytical stability estimates for the orbits in specific subsystems of the solar system has raised considerable interest in recent years. In the case of KAM stability, the goal is to prove theoretically the existence of a KAM torus with fixed frequencies. Stability is guaranteed for all times for orbits with initial conditions on the torus. In the case of two degrees of freedom systems, KAM stability implies also the stability of all the orbits with initial conditions in an open domain inside the last KAM librational torus. This is no longer true for systems of three or more degrees of freedom, because of the a priori possibility of chaotic orbits to exhibit Arnold diffusion. Estimates of KAM stability refer to upper bounds with respect to either a) the distance from a local equilibrium solution, or b) the size of the effective perturbation (e.g. the mass parameter or the eccentricity) for which the existence of a torus is guaranteed. Examples of this approach, in Celestial Mechanics, were given by Robutel (1995), Celletti and Chierchia (1997, 1998), Locatelli (1998), and Locatelli and Giorgilli (2000).

Nekhoroshev stability, on the other hand, is applicable to all the orbits in open domains of initial conditions, independently of whether a particular orbit is regular, i.e., lies on a KAM torus, or chaotic. The Nekhoroshev formula predicts a finite time of stability for both types of orbits, regular or chaotic. This time is a lower limit, i.e., an underestimate of the real time of stability.

Nekhoroshev estimates applied to solar system dynamics refer to a) analytical upper bounds of the perturbation parameter so that the conditions for the application of Nekhoroshev theorem are fulfilled (e.g. Benettin et al. 1998), b) numerical verifications that there are open sets of the Arnold web being in a 'Nekhoroshev regime', i.e., where resonances do not overlap (Guzzo and Morbidelli 1997, Guzzo et al. 2000), and, finally c) estimates of the size of the stability region for times equal to the age of the solar system. A particular example of the latter approach are the estimates of the stability region of the Trojan asteroids in the neighborhood of Jupiter's triangular points given by Simó (1989), Celletti and Giorgilli (1991), Giorgilli and Skokos (1997) and Skokos and Dokoumentzidis (2001).

The present paper reports Nekhoroshev stability estimates for the Trojan asteroids obtained by calculating formal integrals in the neighborhood of the triangular points with a direct method due to Whittaker (1916), Cherry (1924) and Contopoulos (1960). The Hamiltonian model is the planar circular restricted three body problem. Then the discrete analog of the direct method is used to construct formal integrals in a mapping model (Sándor et al. 2002) which describes the dynamics of the PCRTBP for low orbital eccentricities of the asteroids. Nekhoroshev estimates are obtained for this mapping model which are expressed in terms of the proper elements of the orbits. Our best result corresponds to a libration amplitude $D_p = 10.6^o$ which is quite realistic (most asteroids have $D_p < 35^o$).

2. Nekhoroshev estimates in the Hamiltonian model

Following Giorgilli and Skokos (1997), the Hamiltonian of the planar, circular restricted three-body problem is written in heliocentric polar canonical coordinates $(\rho, \theta, p_\rho, p_\theta)$ as:

$$H = \frac{1}{2}\left(p_\rho^2 + \frac{p_\theta^2}{\rho^2}\right) - p_\theta - \mu\rho\cos\theta - \frac{1-\mu}{\rho} - \frac{\mu}{\sqrt{\rho^2 + 1 + 2\rho\cos\theta}} \tag{2.1}$$

The coordinates of L4 are $\rho = 1$, $\theta = 2\pi/3$, $p_\rho = 0$, $p_\theta = 1$, and the mass parameter $\mu = 0.00095387536$.

Following the canonical change of variables:

$$\rho = 1 + 1.00599x_1 + 0.329451x_2 \tag{2.2}$$

$$\theta = \frac{2\pi}{3} + 0.00125186x_1 + 0.0473548x_2 + 2.01417y_1 - 6.15034y_2$$

$$p_\rho = 1.00273y_1 - 0.0265079y_2$$

$$p_\theta = 1 + 1.00599x_1 + 0.329451x_2 \tag{2.3}$$

the Hamiltonian (2.1) can be expanded as

$$H(x_1, x_2, y_1, y_2) = H_2 + H_3 + \ldots \tag{2.4}$$

where the quadratic part H_2 is given by

$$H_2 = \frac{\omega_1}{2}(x_1^2 + y_1^2) + \frac{\omega_2}{2}(x_2^2 + y_2^2) \tag{2.5}$$

i.e., it is diagonal in the canonically conjugate pairs (x_i, y_i), $i = 1, 2$. The frequencies are $\omega_1 = 0.99675752552$ and $\omega_2 = -0.080463875837$. The time unit is equal to $T_J/2\pi$, with

$T_J = 11.84yr$. The period $T_1 = 2\pi/\omega_1$ corresponds to the period of oscillation of the quantity $\omega - M'$ in the linearized approximation around L4, where ω is the argument of the perihelion of the asteroid and M' is the mean anomaly of Jupiter. Since the precession of the perihelion is very slow, the frequency ω_1 is very close to 1, i.e., to the mean motion of Jupiter. On the other hand, the frequency ω_2 corresponds to the frequency of oscillation of the major semi-axis of the asteroid in the linearized approximation around L4, which is about twelve times smaller than the frequency ω_1.

A second formal integral $\Phi = \Phi_2 + \Phi_3 + \ldots$ of the hamiltonian (2.4) can be constructed by solving recursively the equations:

$$D_\omega \Phi_s = -\{\Phi_{s-1}, H_3\}, \qquad D_\omega \cdot = \{\cdot, H_2\} \tag{2.6}$$

where $\{\cdot, \cdot\}$ denotes Poisson brackets and the linear differential operator D_ω is defined as

$$D_\omega = \{\cdot, H_2\} \tag{2.7}$$

The construction of a formal integral starts with a particular choice for the second order term, i.e., $\Phi_{i,2} = \frac{1}{2}(y_i^2 + x_i^2)$, where $i = 1$ or $i = 2$. Then, Eq. (2.6) can be solved order by order and it ensures that the quantity $\Phi_i = \Phi_{i,2} + \Phi_{i,3} + \ldots$ has zero Poisson bracket with the Hamiltonian, i.e. it is a formal integral of motion. This 'direct' method of calculating the integrals step by step via Eq.(2.6) (Whittaker 1916, Cherry 1924, Contopoulos 1960) is as old as the Birkhoff - Gustavson method (Birkhoff 1927, Gustavson 1966) of determining the formal integrals via normal forms. Its extension to deal with resonance cases was given by Contopoulos (1963).

The series Φ is, in general, divergent (Siegel 1941). However, a proper truncation of it, say at order N, gives a polynomial function $\Phi^{(N)}$ which represents an approximate integral of the Hamiltonian H. The time derivative of the truncated series is given by

$$\frac{d\Phi^{(N)}}{dt} = R^{(N)} = \sum_{j=N+1}^{\infty} U_j \tag{2.8}$$

where

$$U_r = \sum_{k=2}^{N} \{\Phi_k, H_{r-k}\}, \qquad r > N \tag{2.9}$$

It can be shown that the series $R^{(N)}$, called the remainder of the integral, is convergent (Giorgilli 1988). The size of the remainder can be bounded from above by a quantity (Giorgilli 1988)

$$||R^{(N)}|| \leqslant \frac{AN!}{\prod_{s=2}^{N} a_s} \tag{2.10}$$

where the norm $||\cdot||$ is defined as the sum of the moduli of the polynomial coefficients, A is a positive constant, depending essentially on the size of the Hamiltonian perturbation $H_3 + \ldots$, and the sequence a_s, $s = 2, 3, \ldots$ refers to the smallest divisors that may appear in the formal series at order s. These are given by

$$a_s = \min\{|k \cdot \omega|, \ k \equiv (k_1, k_2), \ \omega \equiv (\omega_1, \omega_2), \ |k_1| + |k_2| \leqslant s, (|k_1| + |k_2|)mod2 = smod2\} \tag{2.11}$$

where the modulo 2 restriction comes from the fact that the order $|k_1| + |k_2|$ of any Fourier mode $\exp(ik \cdot \phi)$ in the formal series at order s has the same parity as s.

We say that the frequencies ω_i satisfy a diophantine condition, if there are positive constants γ, τ such that $|k \cdot \omega| \geqslant \gamma/|k|^\tau$ for all $k \equiv (k_1, k_2)$ with $k_1, k_2 \in \mathbb{Z}$, $|k| \equiv |k_1| + |k_2| \neq 0$ and $\tau \geqslant n - 1$ where n is the number of degrees of freedom. In the diophantine case, the sequence a_s satisfies the estimate $a_s \sim 1/s^\tau$. This will be called a 'diophantine sequence'. The product $\prod_{s=2}^{N} a_s$ in Eq.(2.10) is estimated as $\sim 1/N!^\tau$. Thus, at distance ρ from the equilibrium point, Eq.(2.10) yields the estimate

$$||R_N||_\rho \leqslant BN!^{\tau+1}\rho^N \tag{2.12}$$

where $||R_N||_\rho = ||R_N||\rho^N$ measures the size of the remainder terms at the distance ρ, and B is a positive constant. The optimal estimate is found by finding the optimal order of truncation N for which the r.h.s. of (2.12) has a minimum with respect to N. This minimum value is $O(\exp(-1/\rho^{\frac{1}{\tau+1}}))$. Thus, in view of Eq.(2.8), the time of stability is exponentially long in the inverse of the distance ρ, i.e.,

$$T \sim O(\exp(1/\rho^{\frac{1}{\tau+1}})) \tag{2.13}$$

If T is fixed, say the age of the solar system, Eq.(2.13) can be used to estimate the size of the Nekhoroshev stability region around the equilibrium points.

It should be stressed that the above rigorous estimates are rather pessimistic. A rigorous improvement of these estimates was given by Fassò et al. (1998), Guzzo et al. (1998) and Niederman (1998). On the other hand, a careful analysis of the accumulation of small divisors in the formal series (Efthymiopoulos et al. 2004) has shown that this cannot be in the form of products $\prod_{s=2}^{N} a_s$. Let d_s $s = 2, \ldots N$ denote the sequence of divisors appearing in the fastest growing terms of the remainder. Then, at most every second divisor can satisfy the equality $d_s = a_s$, i.e., be equal to the minimum divisor at the same order s. This leads to an improved estimate $R_{opt} \sim O(\exp(-1/\rho^{\frac{2}{\tau+1}}))$ which is in close agreement with estimates found by computer experiments (Contopoulos et al. 2003, Efthymiopoulos et al. 2004). But even this formula is a simplification, because the sequence a_s gives divisors equal to the diophantine limit γ/s^τ only at particular orders s determined by the continued fraction approximation of the frequency ratio ω_2/ω_1. A further complication is introduced by the 'inversion' and 'delay' effects (Efthymiopoulos et al. 2004).

On the other hand, it is possible to provide Nekhoroshev estimates by calculating precisely the size of the remainder with a computer program performing the algebraic manipulations. This allows one to determine the size of the region of stability in terms of a radius ρ given by $\rho = \min(\rho_1, \rho_2)$, where the radii ρ_1 and ρ_2 correspond to disks around the origin in the subspaces (x_1, y_1) and (x_2, y_2) respectively defined so that the variations of the corresponding formal integrals do not exceed an upper limit (Giorgilli and Skokos 1997). The final stability region is defined as the interior of a torus corresponding to the product of two circles of radius ρ. In the present calculations, the precise values of the radii ρ_1 and ρ_2 are specified by the same procedure as in Giorgilli and Skokos (1997). The radius ρ is given essentially by the formula

$$\rho \sim \left(\frac{1}{TCs||R_{s+1}^{(s)}||}\right)^{1/s} \tag{2.14}$$

where $T = 10^9$ (in periods of Jupiter) is the Nekhoroshev time, $||R_{s+1}^{(s)}||$ denotes the size of the first order of the remainder $R^{(s)}$, and C is a constant equal to the ratio of the total size of the remainder $||R^{(s)}||$ over $||R_{s+1}^{(s)}||$. The radius ρ is, thus, a function of the order of truncation s. The optimal order of truncation N_{opt} corresponds to the order $s = N_{opt}$ at which ρ, given by (2.14), is maximum.

By calculating the formal integrals as described above we found $N_{opt} = 34$ and $\rho = 0.0304$. This radius is marginally better than the radius found by computer

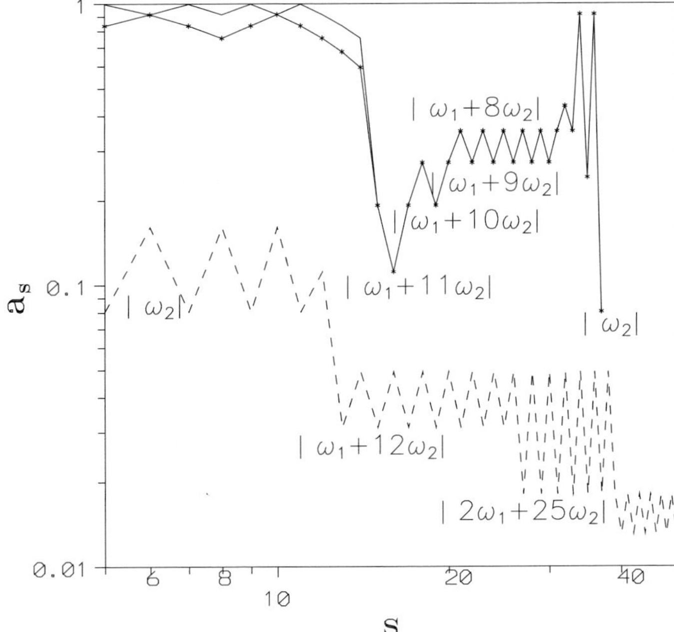

Figure 1. The sequence of small divisors a_s corresponding to the continued fraction expansion of the frequency ratio $a = \omega_2/\omega_1$, as a function of the order of truncation s (dashed line). The solid lines correspond to the sequence of divisors $|k \cdot \omega|$ for which the corresponding Fourier term $\exp(ik \cdot \phi)$ is the leading term of the remainder at order s, for the formal integrals Φ_1 (solid line with stars) and Φ_2 (solid line without stars).

implementation of the normal form method (Giorgilli and Skokos 1997). In physical units, it represents about 1/10 of the distance from L4 to Jupiter.

We examine now how do small divisors accumulate in the formal series so as to produce the observed asymptotic behavior of the series.

In the case of the Trojan problem, the frequency ratio $a = |\omega_2/\omega_1| = 0.0807256266212$ is expressed as a continued fraction expansion $a = [12, 2, 1, 1, 2, \ldots]$, with rational truncations $0/1, 1/12, 2/25, 3/37, 5/62, \ldots$. As shown in Figure 1, as the order of the expansion s increases, newborn small divisors $|k_1\omega_1 + k_2\omega_2|$ appear at the orders $s = k_1 + k_2$, where k_1/k_2 belongs to the sequence of rational truncations of a. Thus, newborn small divisors appear at the orders $1 = 1 + 0$, $13 = 1 + 12$, $27 = 2 + 25$, $40 = 3 + 37$, etc.

Nevertheless, a detailed analysis of the leading Fourier modes in the series shows that the divisors belonging to the diophantine sequence a_s are not the most important, in the sense that they are not the ones producing the terms of maximum size in the series. The solid lines in Figure 1 represent the sequences of divisors $|k \cdot \omega|$ corresponding to the dominant Fourier modes $\exp ik \cdot \phi$ in the remainder (where $\phi \equiv (\phi_1, \phi_2)$ are canonical angles), at successive orders of truncation s of the formal integrals $\Phi_1 = (x_1^2 + y_1^2)/2 + \ldots$, and $\Phi_2 = (x_2^2 + y_2^2)/2 + \ldots$. After a few transient steps, the two lines (for Φ_1 and Φ_2) coincide, beyond the order $s = 14$. This implies that, as s increases, the same Fourier modes become dominant at successive steps in the two series.

It can be observed that from order $s = 14$ to $s = 33$, there is a sequence of divisors with values of the same order of magnitude $\sim 10^{-1}$, namely $|\omega_1 + 11\omega_2| = 0.1116$, $|\omega_1 + 10\omega_2| = 0.1921$, $|\omega_1 + 9\omega_2| = 0.2726$, $|\omega_1 + 8\omega_2| = 0.3505$, which produce dominant Fourier terms in the series, namely the terms $\exp(i(\phi_1 + 11\phi_2))$, $\exp(i(\phi_1 + 10\phi_2))$, $\exp(i(\phi_1 + 9\phi_2))$, and $\exp(i(\phi_1 + 8\phi_2))$ respectively. At all orders s in the interval $14 \leqslant s \leqslant 33$, these terms have comparable size in the series. On the other hand, the size of the Fourier term corresponding to the diophantine divisor $a_{27} = |2\omega_1 + 25\omega_2| = 0.01808$ is much smaller than the size of the previous Fourier terms, at least up to order 40 (which is beyond the optimal order of truncation of the series). This is despite the fact that the divisor a_{27} is one order of magnitude smaller ($\sim 10^{-2}$) than the above listed divisors, which are of order 10^{-1}. This phenomenon is caused by the 'delay' mechanism (Efthymiopoulos et al. 2004). Namely, although the Fourier term $\exp(i(2\phi_1 + 25\phi_2))$ is the fastest growing term beyond order $s = 27$, when this term appears for the first time, at order $s = 27$, it has size much smaller than the size of other Fourier terms which are temporarily dominant at $s = 27$, and many iterations of the recurrent relation (2.6) are required before the size of the term $\exp(2\phi_1 + 25\phi_2)$ becomes dominant. This phenomenon appears even for good diophantine frequency ratios, because the intervals of orders s separating the appearance of successive new small denominators increases exponentially with s.

Another phenomenon observed near the order $s = 27$ is 'inversion'. Namely, a Fourier term with relatively large divisor becomes temporarily dominant while other terms, with smaller divisors, are temporarily less important. This is shown as two consecutive peaks of the solid lines at orders $s = 32$ and $s = 34$ (Figure 1).

The fact that many Fourier terms, with small divisors of similar size, become successively dominant in a short interval of values of s, implies that it is not possible to obtain better Nekhoroshev stability estimates by considering near-resonant forms of the formal integrals innstead of non-resonant formal integrals. In fact, a near-resonant construction deals with only one near-resonance at the time, eliminating the effect of the small divisors associated with this resonance. This can improve the estimates up to a particular order of truncation s, on the condition that there are no other near-resonances producing terms of considerable size up to order s. But figure 1 shows that this condition is not fulfilled in our case.

3. Nekhoroshev estimates with Delaunay variables in a mapping model

The most relevant variables for the description of the stability region around L4 and L5 are the Delaunay action-angle variables

$$x = \sqrt{\frac{\alpha}{\alpha'}} - 1, \quad \tau = \lambda - \lambda'$$

$$x_2 = \sqrt{\frac{\alpha}{\alpha'}}\left(\sqrt{1 - e^2} - 1\right), \varpi \tag{3.1}$$

where α, e are the semi-major axis and eccentricity of the asteroid, α' is the semi-major axis of Jupiter, τ is the critical argument, i.e., the difference between the mean longitude of the asteroid and of Jupiter, and ϖ is the longitude of the pericenter of the asteroid. These variables have the advantage that the pair of action-angle variables x, τ are immediately translated in the motion of the asteroid in configuration space. Namely, x gives the amplitude of librations perpendicularly to the circle of the 1:1 coorbital motion, while $\tau - \tau_0$ measures the synodic libration around the equilibrium values $\tau_0 = \pi/3$ (for L4) or $5\pi/3$ (for L5). If both librations are considered nearly harmonic, then, following Érdi (1988), the amplitude of librations is measured by the parameter D_p given by

$$\Delta\alpha = \alpha - \alpha' \simeq 2\alpha'x \simeq \sqrt{3\mu}\alpha'D_p\sin\phi, \quad \Delta\tau = \tau - \tau_0 \simeq D_p\cos\phi \tag{3.2}$$

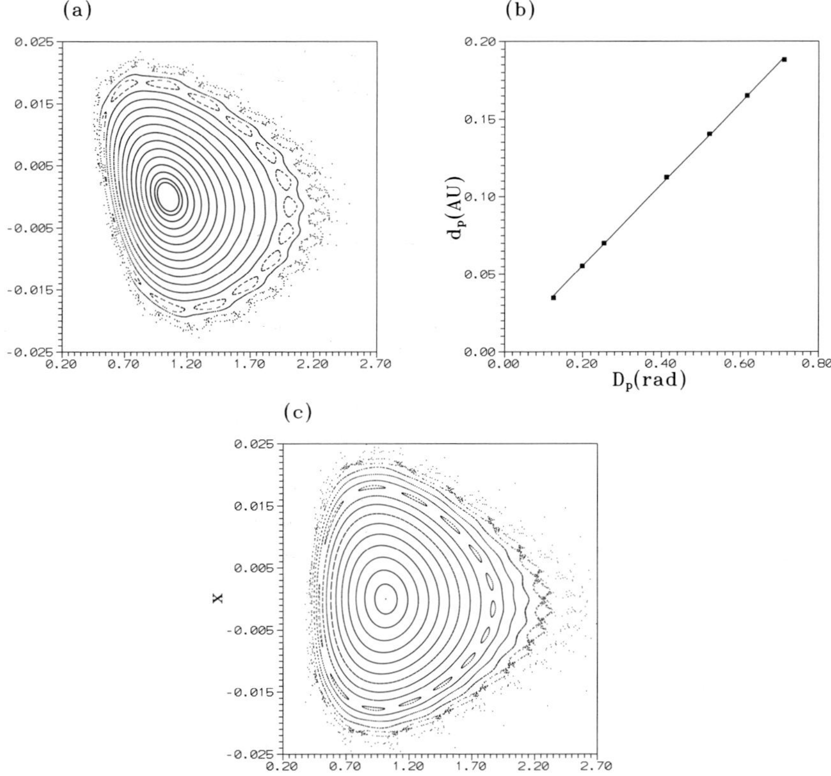

Figure 2. (a) The phase portrait of the mapping (3.3). (b) The relation d_p versus D_p (see text for definitions) as found by the invariant curves of the mapping (3.3). (c) The Poincaré surface of section for the equations of motion under the full Hamiltonian of the planar circular restricted three-body problem. The section is given by the variables (τ, x) when the surface $\varpi - \lambda' = 0$ is crossed by an orbit at the positive sense. The Jacobi constant is taken equal to $E = -1.49948$, which is very close to the Jacobi constant at $L4$, $E_{L4} = -1.49952$. This corresponds to eccentricities $e < 0.04$.

where ϕ is the phase of a libration, which is defined by the position of the guiding center of the motion of the asteroid along a tadpole-shaped orbit. According to Eq.(3.2) the parameter D_p is the amplitude of libration of the critical argument. For most asteroids the amplitude of libration D_p ranges from a few degrees up to about $D_p \simeq 35^o$ (Milani 1993, Érdi 1997, Levison et al. 1997), while in general D_p decreases as the eccentricity increases.

The usual analytic expansion of the Hamiltonian (2.1) cannot be transformed to an expansion in the variables (3.1) with analytically calculated coefficients. The most relevant expansion is the semi-analytic expansion of Beaugé and Roig (2001), which gives the averaged Hamiltonian over fast angles. A way to circumvent this problem is by using the 2D mapping model of Sándor et al. (2002) which accurately reproduces the dynamics at low orbital eccentricities of the asteroids. At the limit of zero eccentricity, the mapping reads

$$x_{n+1} = x_n + 2\pi\mu \sin\tau_n \left(1 - \frac{1}{(2 - 2\cos\tau_n)^{3/2}}\right)$$

$$\tau_{n+1} = \tau_n + 2\pi \left(\frac{1}{(1 + x_{n+1})^3} - 1\right) \tag{3.3}$$

Figure 2a shows the phase portrait of the mapping (3.3). The invariant KAM curves of this mapping correspond to librations around L_4. Each invariant curve defines approximate proper elements corresponding to the amplitudes of libration $D_p = \tau_{max} - \tau_{min}$, and $d_p = \alpha_{max} - \alpha_{min}$, where α is defined in terms of x by Eq.(3.1). Figure 2b shows D_p as a function of d_p, calculated from Figure 2a, by considering ten invariant curves of Figure 2a with initial conditions $x = 0$ and $\tau = \pi/3 + n\Delta\tau$ with $n = 1, 2, ..., 10$ and $\Delta\tau = \pi/60$. Notice that the points of Figure 2b are almost on a straight line with slope $\simeq (1/0.273)(rad/AU)$. The theoretical value given by Érdi (1988) is $(1/0.2783)(rad/AU)$.

Figure 2c shows the Poincaré surface of section for the exact Hamiltonian model of the CRTBP, at the Jacobi constant $E = -1.49948$, which is very close to the value at L4 $E_{L4} = -1.49952$. The maximum eccentricity, at the outer invariant curve is small ($e = 0.04$), thus the surface of section can be compared with the mapping phase portrait, which corresponds to $e = 0$. The mapping portrait is angularly deformed with respect to the hamiltonian portrait. This phenomenon is an artifact of the method used to produce the mapping. However, the extend of the stability region is the same in the two portraits, and the resonant chains of same multiplicity are at approximately the same distances from the center. The rotation number of the central point of figure 2c can be used to compare the sequences of divisors found for the mapping model (3.3) with the corresponding sequences for the hamiltonian model. The two rotation numbers are found equal up to three significant figures.

A formal integral Φ for the mapping (3.3) is calculated by a computer program, by implementing a direct method for mappings which is the discrete analog of the direct method for Hamiltonian systems. This method consists of solving recurently the homological equation (Bazzani and Marmi 1991)

$$\Phi(z', z'_*) = \Phi(z, z_*) \tag{3.4}$$

where

$$z' = e^{i\omega}[z + F_2(z, z_*) + F_3(z, z_*) + ...] \tag{3.5}$$

is the Taylor-expanded mapping (3.3) in complex-conjugate coordinates, after a trivial normalization which transforms the ellipses of the linearized mapping around the elliptic equilibrium into circles. Our calculations were extended to the order of expansion $N = 60$. Details of the algorithm, as well as an analytical treatment of the Nekhoroshev stability estimates for mappings are given in Efthymiopoulos (2004). It turns out that the size of the region of stability is determined by a formula very similar to Eq.(2.14). Precisely, we have

$$I_s = \left(\frac{s-1}{s+1}\right)\left(\frac{2(1-A)^{\frac{s+1}{2}}}{(s+1)B_{\rho_*}||U_{s+1}^{(s)}||T}\right)^{\frac{2}{s-1}} \tag{3.6}$$

where $||U_{s+1}^{(s+1)}||$ is the size of the first order of the remainder at the order of truncation s, B_{ρ_*} is a constant with similar meaning as the constant C in Eq.(2.14), and I_s is the value of the outermost level curve of the integral Φ for which stability is guaranteed for all times $t \leqslant T$. Thus, the stability estimates are given in terms of a curve which is a deformed circle, while the stability domain is the interior of this curve. The constant A measures the degree of deformation of the level curve $\Phi(z, z_*) = I_s$ from a perfect circle (see Efthymiopoulos 2004 for details).

The main advantage of this approach is that it allows to express the results directly in terms of the proper elements D_p or d_p. As in the case of formula (2.14), the estimate (3.6) is optimized with respect to s. We found the optimal order $s = N_{opt} = 38$, yielding the stability estimates

$$D_p \leqslant 10.6^o, \qquad d_p \leqslant 0.0512 \text{ AU} \tag{3.7}$$

This result represents an improvement over previously obtained Nekhoroshev estimates of the region of effective stability (Giorgilli and Skokos 1997, Skokos and Dokoumentzidis 2001). The region of stability where real asteroids are observed extends to $D_p \simeq 35^o$, meaning that the region given in Eq.(3.7), by analytical methods, has a size equal to about one third the real size of the observed region of stability. Previous estimates were giving a size smaller by a factor 10 for most asteroids, and up to a factor 30 in the worst case (Giorgilli and Skokos 1997). It should be stressed, however, that the mapping model used here is also a simplification of the Hamiltonian problem, which reproduces approximately the dynamics only at low proper eccentricities. Thus, in a strict sense, the two models are not comparable.

Finally, let us note that the planar circular restricted three body model represents a great simplification of the real problem of stability of the Trojans. In particular, the border of the stability region is shaped by the overlapping of secular resonances caused either by the elliptic motion of Jupiter or by the direct or indirect effects of other major planets (e.g. Tsiganis et al. 2002, Robutel 2004).

Obtaining analytical Nekhoroshev stability estimates by adding more degrees of freedom to the problem represents a challenge from many points of view. First, the number of terms of a formal series up to order s depends on the number of degrees of freedom n as $O(s^{n+1})$. In the 4 DOF case, relevant results should involve about 10^8 terms, which is at the limit of the present computational capacities. Second, some secular variations of the orbit of Jupiter can be introduced in the Hamiltonian only as time-dependent terms. It is unclear how these terms should be treated from the point of view of Nekhoroshev theory. Finally, Nekhoroshev theory itself is far from giving optimal estimates of the time (or size of the area) of stability, even in simple Hamiltonian models.

In conclusion, the optimization of the Nekhoroshev stability estimates in the case of the Trojan asteroids represents a mathematically, computationally and physically interesting question. However, even the presently obtained Nekhoroshev estimates are realistic, and demonstrate the usefulness of Nekhoroshev theory.

Acknowledgements

I am grateful to Prof. G. Contopoulos, as well as Drs. K. Tsiganis, F. Roig, and Z. Sándor, for many stimulating discussions on the problem of stability of Trojan asteroids.

References

Arnold, V.I. 1963a, *Russ. Math. Surveys* 18, 9
Arnold, V.I. 1963b, *Russ. Math. Surveys* 18, 85
Bazzani A., and Marmi, S. 1991, *Nuovo Cimento* 106B, 673
Beaugé, C., and Roig, F. 2001, *Icarus* 153, 391
Benettin, G., Galgani, L., and Giorgilli, A. 1985, *Cel. Mech.* 37, 1
Benettin, G., Fassò, F., and Guzzo, M 1998, *Reg. Ch. Dyn.* 3, 56
Birkhoff, G.D. 1927, *Dynamical Systems*, Amer. Math. Soc., Providence, R.I
Celletti, A., and Giorgilli, A. 1991, *Cel. Mech. Dyn. Astron.* 50, 31
Celletti, A., and Chierchia, L. 1997, *Comm. Math. Phys.* 186, 413
Celletti, A., and Chierchia, L. 1998, *Planet. Space Sci.* 46, 1433
Cherry, T.M. 1924, *Proc. Cambridge Phil. Soc.* 22, 325
Cherry, T.M. 1924, *Proc. Cambridge Phil. Soc.* 22, 510
Contopoulos, G. 1963, *Astron. J.* 68, 763

Contopoulos, G. 1960, *Z. Astroph.* 49, 273

Efthymiopoulos, C., Contopoulos, G., and Giorgilli, A. 2004, *J. Phys. A* 37, 10831

Efthymiopoulos, C 2004, *Cel. Mech. Dyn. Astron.* (in press)

Érdi, B. 1988, *Cel. Mech.* 43, 303

Érdi, B. 1997, *Cel. Mech. Dyn. Astron.* 65, 149

Fassò, F., Guzzo, M., and Benettin, G. 1998, Commun. Math. Phys. 197, 347

Giorgilli, A. 1988, *Ann. Inst. H. Poincaré* 48, 423

Giorgilli A., and Skokos Ch. 1997, *Astron. Astrophys.* 317, 254

Gustavson, F. 1966, *Astron. J.* 71, 670

Guzzo, M., and Morbidelli, A. 1997, *Cel. Mech. Dyn. Astron.* 66, 255

Guzzo, M., Fassò, F., and Benettin, G. 1998, *Math. Phys. Electronic. J.* 4, paper 1

Guzzo, M., Knežević, Z., and Milani, A., 2002 *Cel. Mech. Dyn. Astron.* 83, 121

Kolmogorov, A.N., 1954 *Dokl. Akad. Nauk. SSR* 98, 469

Levison, H., Shoemaker, E.M., and Shoemaker, C.S. 1997, *Nature* 385, 42

Locatelli, U., 1998 *Planet. Space Sci.* 46, 1453

Locatelli, U., and Giorgilli, A 2000, *Cel. Mech. Dyn. Astron.* 78, 47

Lochak, P. 1992, *Russ. Math. Surv.* 47, 57

Milani, A. 1993, *Cel. Mech. Dyn. Astron.* 57, 59

Morbidelli, A., and Giorgilli, A. 1995a, *J. Stat. Phys.* 78, 1607

Morbidelli, A., and Giorgilli, A. 1995b, *Physica D* 86, 514

Moser, J. 1962, *Nach. Akad. Wiss. Göttingen, Math. Phys. KL II* 1, 1

Nekhoroshev, N.N. 1977, *Russ. Math. Surv.* 32(6), 1

Niederman, L. 1998, *Nonlinearity* 11, 1465

Pöshel, J. 1993, *Math. Z.* 213, 187

Robutel, P. 1995, *Cel. Mech.* 62, 219

Robutel, P. 2004, presentation at the 6th Alexander von Humboldt Symposium, Bad Hofgastein, Austria

Sándor, Z., Érdi, B., and Murray, C. 2002, *Cel. Mech. Dyn. Astron.* 84, 355

Skokos, Ch., and Dokoumetzidis, A. 2001, *Astron. Astrophys.* 367, 729

Siegel, C.L. 1941, *Ann. Math.* 42, 806

Simó, C. 1989, *Memorias de la Real Academia de Ciencias y Artes de Barcelona* 48, 303

Tsiganis, K., Varvoglis, H., and Hadjidemetriou J.D. 2002, *Icarus* 155, 454

Whittaker, E.T. 1916, *Proc. Roy. Soc. Edinburgh* 37,95

Dynamics of Populations of Planetary Systems
Proceedings IAU Colloquium No. 197, 2005
Z. Knežević and A. Milani, eds.

© 2005 International Astronomical Union
DOI: 10.1017/S174392130400866X

The role of the resonant "stickiness" in the dynamical evolution of Jupiter family comets

A. Alvarez-Candal and F. Roig

Observatório Nacional, Rua Gal. José Cristino 77, 20921-400, Rio de Janeiro, RJ, Brazil.
email: alvarez@on.br

Abstract. We analyze the effect of the temporary capture of comet-like orbits in asteroid mean motion resonance by following the dynamical evolution of 2090 Jupiter-family comet-like orbits over 10^7 yr under the perturbation of the four major planets. The resonant capture may be related to the phenomenon known as "resonant stickiness" consisting in the temporary stabilization of very eccentric orbits near the separatrices of the mean motion resonances. We found that the population of orbits that were captured at least once during the simulation has a median lifetime larger than that of the complete sample.

Keywords. Celestial Mechanics, comets: general.

---

1. Introduction

The resonance stickiness arises when a very eccentric orbit becomes temporarily captured near the separatrix of a mean motion resonance. The "sticky" orbits are chaotic although practically stable, having dynamical lifetimes longer than we should expect from their short Lyapunov times. These orbits usually evolve confined in a limited region of the phase space (Murison, Lecar, and Franklin, 1994).

In the asteroid belt the resonant stickiness has been identified as being responsible for the "stable chaos" (Milani and Nobili, 1992; Milani, Nobili and Knežević, 1997) detected in some high-order mean motion resonances (Tsiganis, Varvoglis, and Hadjidemetriou, 2000; 2002). On the other hand, Malyshkin and Tremaine (1999) found that the rate of decay of the population of ecliptic comets, and in particular of the Jupiter-family comets (JFC), is mostly explained by considering stickiness phenomena of the orbits near resonant islands in the phase space.

In this contribution, we analyzed the dynamical evolution of a sample of JFC-like orbits, aiming to study the effect of the resonance stickiness in their dynamical lifetime.

1.1. *The numerical setup*

From a list of cometary nuclear magnitudes provided by G. Tancredi (pers. comm.), we selected 190 known JFC with orbital elements referred to JD 2452700.5. Then, we generated a set of JFC test orbits by cloning each real orbit ten times. The orbital elements of these clones were chosen at random in the intervals $\Delta a = \pm 0.1$ AU, $\Delta I = \pm 1.5°$, and $\Delta M = \pm 10°$ around the reference orbit (M denotes the mean anomaly). The eccentricity was cloned so as to preserve the value of the Tisserand parameter (with respect to Jupiter) of the reference orbit.

The dynamical evolution of this set of 2090 test particles (real JFC plus clones) was followed over 10^7 yr using the symplectic integrator SWIFT_RMVS3 (Levison and Duncan, 1994) and taking into account the perturbation of the four major planets. The integration time-step was 1.8 days, and the simulation of a particle was interrupted whenever: (i) it came closer than 0.01 Hill radius to some planet, (ii) it collided with the Sun, or (iii) its heliocentric distance became larger than 200 AU.

2. Results

Figure 1 shows the depletion of the entire set of test orbits over time. The median lifetime of the set is $\tau = 1.7 \times 10^5$ yr, which is about half the value estimated by Levison and Duncan (1994). This discrepancy may be due to the different model used by these authors, or to the much smaller sample of test orbits simulated by them.

A two-stages behavior is evident from Fig. 1. Before $\tau_s \sim 3 \times 10^5$ yr, the population decays following an exponential law $(N_1(t) \propto e^{(-4 \times 10^{-6} t)})$. After τ_s, the decay switches to a power law $(N_2(t) \propto t^{-1.104})$. This kind of behavior has been previously reported by Kerney (1983), Levison and Duncan (1994) and Malyshkin and Tremaine (1999). The exponential decay is expected from a random-walk behavior induced by close encounters with the planets and also from the existence of hyperbolic regions of the phase-space, devoided of stability islands. The switch to a power law decay indicates that some mechanism or mechanisms are delaying the depletion of orbits. This is usually interpreted as a consequence of the resonance stickiness, and the time τ_s is usually referred to as the "stickiness time".

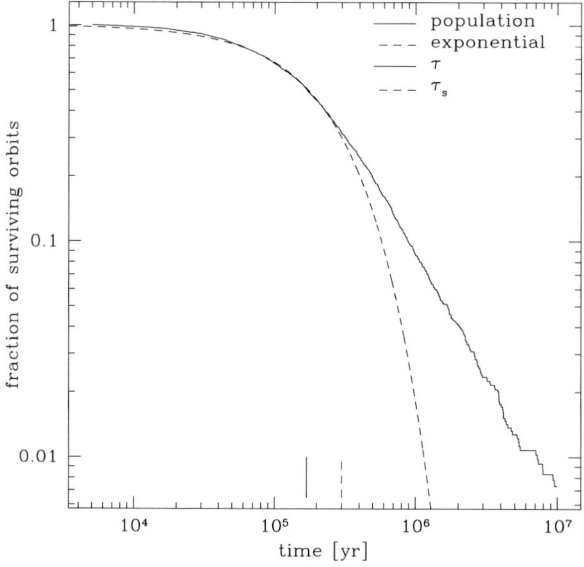

Figure 1. The decay of our set of test orbits, normalized to the total number of initial orbits.

To analyze the resonant capture of test orbits during the simulation, we first computed the mean orbital elements of the orbits by running averages of the osculating elements. We applied a running window 10,000 yr wide, and mean elements were determined every 500 yr. Then, we considered that an orbit suffered a resonant capture if, at any moment during the simulation, it had mean semi-major axis $\bar{a}$ satisfying the condition $|\bar{a} - a_{res}| < \Delta a_{res}$ over at least 10,000 yr. The nominal semi-major axes, a_{res}, of the considered resonances as well as their corresponding widths, Δa_{res}, are shown in Table 1. We found that 304 test particles from our sample became temporarily captured in any of these resonances. This number does not account for the test orbits that were already captured

Table 1. Nominal semi-major axes and corresponding widths of the analyzed resonances. The resonance width was defined as half the minimum distance to the adjacent resonances. The last column indicates the total number of captured orbits.

p:q	a_{res} [AU]	Δa_{res} [AU]	N	p:q	a_{res} [AU]	Δa_{res} [AU]	N
8:3	2.704602	0.06	24	7:5	4.155888	0.025	19
5:2	2.823509	0.06	21	4:3	4.293289	0.055	35
7:3	2.956410	0.035	33	9:7	4.398652	0.04	9
9:4	3.028965	0.035	52	5:4	4.482042	0.04	29
2:1	3.276392	0.125	36	6:5	4.605694	0.045	23
9:5	3.514802	0.035	28	7:6	4.693009	0.035	12
7:4	3.581436	0.03	26	8:7	4.757965	0.025	2
5:3	3.699844	0.055	48	9:8	4.808182	0.02	3
8:5	3.801917	0.05	34	10:9	4.848167	0.02	3
3:2	3.969067	0.07	37	1:1	5.200949	0.18	112
10:7	4.100290	0.03	18				

from the very beginning of the simulation (77 orbits). Moreover, in order to safely avoid any possible bias caused by the cloning process, we did not take into account for tests orbits that became captured in any of these resonances during the first 30,000 yr of the simulation either.

Analyzing this sub-population of captured orbits, we found that their residence times (defined as the time interval for which the orbit remains captured in the resonance) are generally shorter than 10^5 yr. However, in some particular cases, as for example the 7:4 resonance, it may be as long as 10^6 yr (Fig. 2). We also found that the sub-population of captured orbits has a median lifetime which is twice that of the entire population (Fig. 3).

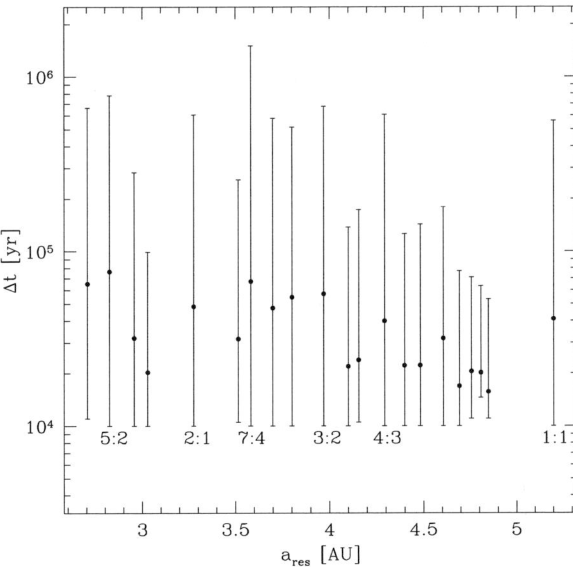

Figure 2. Mean residence times in the analyzed resonances. The minimum and maximum residence times are indicated by the error bars.

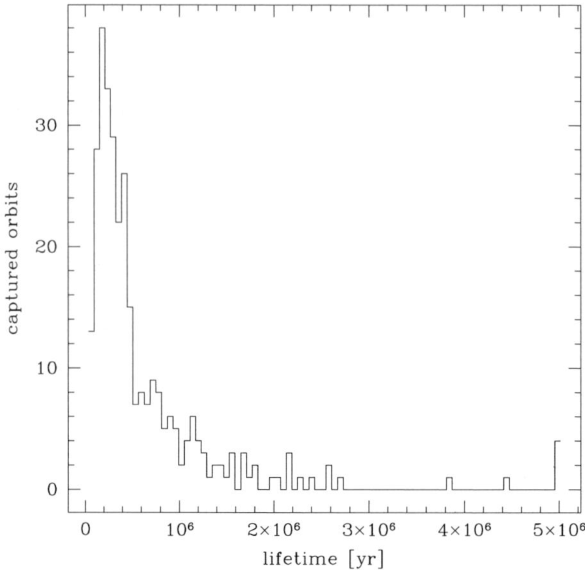

Figure 3. The dynamical lifetimes of the orbits that became captured in any resonance during the simulation, distributed in bins of $60,000$ yr. The median lifetime is $\tau \sim 3.6 \times 10^5$ yr.

3. Conclusions

According to our simulations, the population of resonant captured orbits has a median lifetime longer than the complete sample, which seems to support the hypothesis of the resonance stickiness as the mechanism which temporarily stabilizes cometary orbits in mean motion resonances. We must recall that we are talking here about "long-term" resonant captures, with residence times of the order of 10^4–10^5 yr. These must be distinguished from the "short-term" captures reported by Carusi et al. (1985) which last for only a few hundred years. The existence of these long-term captures may imply that a non negligible fraction of the known resonant asteroids evolving in chaotic orbits could actually be JFCs. This puts interesting constraints to the search of possible extinct cometary nuclei evolving on highly eccentric NEAs orbits or on asteroid-like orbits in the outer belt. However, these results are very preliminar and a better analysis of the data is required before drawing definite conclusions.

Acknowledgements

This work has been supported by CNPq and CAPES.

References

Carusi, A., Kresak, L., Perozzi, E., & Valsechi, G.B. 1985, *Long term evolution of short-period comets*, Adam Hilger, Bristol
Karney, C.F.F. 1983, *Physica D* 8, 360
Levison, H. & Duncan, M. 1994, *Icarus* 104, 18
Malyshkin, L. & Tremaine, S 1999, *Icarus* 141. 341
Milani, A. & Nobili, A. 1992, *Nature* 357, 569
Milani, A., Nobili, A., & Kneževic, Z 1997, *Icarus* 125, 13
Murison, M., Lecar, M., & Franklin, F. 1994, *Astron. J.* 108, 2323
Tsiganis, K., Varvoglis, H., & Hadjidemetriou, J. 2000, *Icarus* 146, 240
Tsiganis, K., Varvoglis, H., & Hadjidemetriou, J. 2002, *Icarus* 155, 454

Dynamics of Populations of Planetary Systems
Proceedings IAU Colloquium No. 197, 2005
Z. Knežević and A. Milani, eds.

© 2005 International Astronomical Union
DOI: 10.1017/S1743921304008671

Regimes of stability and scaling relations for the removal time in the asteroid belt: a simple kinetic model and numerical tests

Mihailo Čubrović

Department of Astronomy – Petnica Science Center, P. O. B. 6, 14104 Valjevo, Serbia and Montenegro
email: cygnus@EUnet.yu

Abstract. We report on our theoretical and numerical results concerning the transport mechanisms in the asteroid belt. We first derive a simple kinetic model of chaotic diffusion and show how it gives rise to some simple correlations (but not laws) between the removal time (the time for an asteroid to experience a qualitative change of dynamical behavior and enter a wide chaotic zone) and the Lyapunov time. The correlations are shown to arise in two different regimes, characterized by exponential and power-law scalings. We also show how is the so-called "stable chaos" (exponential regime) related to anomalous diffusion. Finally, we check our results numerically and discuss their possible applications in analyzing the motion of particular asteroids.

Keywords. Minor planets, asteroids; diffusion; celestial mechanics; methods: analytical

1. Introduction

Despite some important breakthroughs in the research of transport mechanisms in the Solar system in the past decade, we still lack a general *quantitative* theory of chaotic transport, which is especially notable for the so-called stable chaotic bodies. In this paper, we sketch a new kinetic approach, which, in our opinion, has a perspective of providing us such a theory sometime in the future.

A kinetic model of transport has already been proposed by Murray & Holman (1997). Although it is an important step forward, this model fails to include a number of important effects. We also wish to emphasize the role of phase space topology in the transport processes. This has only recently been understood in papers by Tsiganis, Varvoglis & Hadjidemetriou (2000, 2002a, 2002b). Still, the exact role of cantori and stability islands in various resonances remains unclear. This is one of the issues we intend to explore in this paper. We argue that, due to the inhomogenous nature of the phase space, a separate kinetic equation for each transport mechanism should be constructed; after that, one can combine them to obtain the description of long-time evolution. This is the basic idea of our approach, which leads to some interesting statistical consequences, such as anomalous diffusion and approximate scaling of removal times with Lyapunov times.

2. The kinetic scheme

In order to model the transport, we use the "building block approach" we have recently developed for Hamiltonian kinetics (Čubrović 2004). We use the Fractional Kinetic Equation (FKE), a natural generalization of the diffusion equation for self-similar and strongly inhomogenous media (e. g. Zaslavsky 2002):

$$\frac{\partial^\beta f(I,t)}{\partial t^\beta} = \frac{\partial^\alpha}{\partial |I|^\alpha}\left[\mathcal{D}(I)\,f(I,t)\right] \tag{2.1}$$

Thus, the evolution of the distribution function $f(I,t)$ is governed by the transport coefficient $\mathcal{D}$ (the generalization of the diffusion coefficient) and by the (non-integer, in general) order of the derivatives α ($0 < \alpha \leqslant 2$) and β ($0 < \beta \leqslant 1$). The quantity $\mu \equiv 2\beta/\alpha$ is called the transport exponent (for the second moment the following holds asymptotically: $\langle \Delta I^2 \rangle \propto t^\mu$). If $\mu \neq 1$, the transport is called anomalous (in contrast to normal transport or normal diffusion†).

We shall now very briefly describe each of the four building blocks; unfortunately, most expressions are cumbersome and complicated, so we limit ourselves in this paper to merely state the basic ideas and final results of the method. We use the planar MMR Hamiltonians $H_{2BR} = H^0_{2BR} + H'_{2BR}$ and $H_{3BR} = H^0_{3BR} + H'_{3BR}$ for two- and three-body resonances, taken from Murray, Holman & Potter (1998) and Nesvorný & Morbidelli (1998), respectively. Under H^0 we assume the action-only part of the Hamiltonian. H_{2BR} was modified to account for the purely secular terms; also, both Hamiltonians were modified to include the proper precessions of Jupiter and Saturn‡:

$$H'_{2BR} = \sum_{m=0,1;\,s=0...k_J-k}^{u_5,u_6} c_{msu_5u_6}\cos\left[m\left(k_J\lambda_J - k\lambda\right) + sp + \left(u_5g_5 + u_6g_6\right)t + u_5\beta_J + u_6\beta_S\right] \tag{2.2}$$

$$H'_{3BR} = \sum_{m=0,1;\,s}^{u_5,u_6} c_{msu_5u_6}\cos\left[m\left(k_J\lambda_J + k_S\lambda_S + k\lambda\right) + sp + \left(u_5g_5 + u_6g_6\right)t + u_5\beta_J + u_6\beta_S\right] \tag{2.3}$$

The notation is usual. In H'_{2BR}, we include all possible harmonics; in H'_{3BR}, we include only those given in Nesvorný & Morbidelli (1998). In what follows, we shall consider only the diffusion in eccentricity, i. e. P Delaunay variable. Inclusion of the inclination could be important but we postpone it for further work.

We estimate the transport coefficient as:

$$\mathcal{D} = \frac{T_{lib}^{(\alpha-\beta)}}{2}\sum_s s^\alpha P^{s\alpha}\left(\sum_{u_5,u_6} c'_{0su_5u_6}(\alpha) + jc'_{1su_5u_6}(\alpha)\right) \tag{2.4}$$

where T_{lib} denotes the libration period while $c'_{msu_5u_6}$ are coefficients dependent on the exponent α from (2.1), independent on angles and P, which were computed using the algorithm from Ellis & Murray (2000), for H_{2BR}, or taken from Nesvorný & Morbidelli (1998), for H_{3BR}. The indicator j can be equal to 0 or 1 (i. e. omission or inclusion of the resonant terms), depending on the building block (see bellow). Although we were able to compute also the higher-order corrections to this quasilinear result in some cases, we neglect them in what follows, in order to be able to solve the FKE analytically.

---

† From now on, we will refer to any transport in the phase space (i. e. evolution of the momenta of the action I) as to "diffusion"; for the "classical" diffusion, we shall use the term "normal diffusion".

‡ All our computations, analytical and numerical, are performed with the osculating elements, in order to gain as much simplification as possible. However, in order to avoid the non-diffusive oscillations of the osculating elements, one should use the proper elements instead; we plan to do this in the future.

The first class of building blocks we consider are the overlapping stochastic layers of subresonances. In this case, one expects a free, quasi-random walk continuous in both time and space, since no regular structures are preserved. Therefore, the FKE simplifies to the usual diffusion equation, i. e. we have $\alpha = 2$, $\beta = 1$ in (2.4); also, $j = 1$ (the resonant harmonics are actually the most important ones).

The above reasoning is only valid if the overlapping of subresonances is not much smaller than 1. Otherwise, the diffusion can only be forced by the secular terms. Also, long intervals between subsequent "jumps" induce the so-called "erratic time", i. e. β can be less than 1, its value being determined by the distribution of time intervals between "jumps" $p(\Delta t) = 1/\Delta t^{1+\beta}$. So, the transport coefficient (2.4) now has $j = 0$, $\alpha = 2$ and $\beta < 1$.

Our third class of building blocks are the resonant stability islands. To estimate β we use the same idea as in the previous case; after that, we compute α from β and μ, the transport exponent, which we deduce using the method developed in Afraimovich & Zaslavsky (1997). Namely, analytical and numerical studies strongly suggest a self-similar structure characterized by a power-law scaling of trapping times λ_T, island surfaces λ_S and number of islands λ_N at each level. The transport exponent is then equal to $\lambda_N \lambda_S / \lambda_T$. For some resonances and for some island chains, we computed the scaling exponents applying the renormalization of the resonant Hamiltonian as explained in Zaslavsky (2002); in the cases when we did not know how to do this, we used the relation between the transport exponent and the fractal dimension d_T of the trajectory in the (P, p) space (the space spanned by the action P and the conjugate angle p), which is actually the dimension of the Poincare section of the trajectory:

$$d_T = \frac{2\lambda_T}{\lambda_N \lambda_S} \tag{2.5}$$

The last remaining class of blocks are cantori. Here, we assume the scaling of gap area on subsequent levels with exponent λ_S and an analogous scaling in trapping probability with exponent λ_p, which determines the transport exponent as $2 \ln \lambda_p / \ln \lambda_S$; see also the reasoning from Shevchenko (1998). The scaling exponents were estimated analogously to the previous case.

For each building block, we construct a kinetic equation and solve it. We always put a reflecting barrier at zero eccentricity and an absorbing barrier at the Jupiter-crossing eccentricity. The solution in Fourier space (q, t) can be written approximately in the following general form:

$$f_i(q, t) = E_\beta(-|q^\alpha| \star \hat{\mathcal{D}}_i t^\beta) \tag{2.6}$$

where E_β stands for the Mittag-Leffler function and $\hat{\mathcal{D}}_i$ denotes the Fourier transform of $\mathcal{D}_i$. The index i denotes a particular building block. The key to obtaining the global picture is to perform a convolution of the solutions for all the building blocks. Furthermore, one must take into account that the object can start in different blocks and also that, sometimes, different ordering of the visited blocks is possible. Therefore, one has the following sum over all possible variations of blocks (we call it Equation of Global Evolution - EGE):

$$f(P, t) = \sum [p_1 f_1(P, t) \star p_2 f_2(P, t) \star \ldots \star p_i f_i(P, t) \star \ldots] \tag{2.7}$$

To calculate it, one has to know also the transition probabilities p_i, which is not possible to achieve solely by the means of analytic computations. That is why we turn again to semi-analytic results.

3. Removal times, Lyapunov times and $T_L - T_R$ correlations

The first task is to determine the relevant building blocks and transitional probabilities. We do that by considering the overviews of various resonances as given in Morbidelli & Moons (1993), Moons & Morbidelli (1995), Moons, Morbidelli & Migliorini (1998). For the resonances not included in these references, we turn again to the inspection of Poincare surfaces of section, integrating the resonant models (2.2) and (2.3). In this case, the probabilities are estimated as the relative measures of the corresponding trajectories on the surface of section.

The result of solving the EGE is again a Mittag-Leffler function:

$$f_{global}(q,t) = E_\gamma(-|q|^\delta t^\gamma) \tag{3.1}$$

The asymptotic behavior of this function, described e. g. in Zaslavsky (2002), has two different forms: the exponential one and the power-law one, depending on the coefficients γ and δ, which are determined by the probabilities p_i and transport coefficients and exponents of the building blocks. In the small γ limit, the behavior is exponential and the second momentum scales with τ_{cross}, where τ_{cross} is the timescale of crossing a single subresonance, which we interpret as the Lyapunov time†. When γ becomes large and the role of stickiness more or less negligible, one gets a power-law dependance on τ_{cross}, i. e. T_L. So, we have the expressions:

$$T_R \propto \exp\left(T_L^x\right) \Phi\left(\Lambda_0, \cos(\ln P_0), Q_0\right) \tag{3.2}$$

for the exponential or stable chaotic regime, and:

$$T_R \propto (T_L^y) \Phi\left(\Lambda_0, \cos(\ln P_0), Q_0\right) \tag{3.3}$$

for the power-law regime. The scalings are not exact because the fluctuational terms $\Phi\left(\Lambda_0, P_0, Q_0\right)$ appear. These terms are log-periodic in P_0 and can explain the log-normal tails of the T_R distribution, detected numerically e. g. in Tsiganis, Varvoglis & Hadjidemetriou (2000).

4. Results for particular resonances

We plan to do a systematic kinetic survey of all the relevant resonances in the asteroid belt. Up to now, we have only preliminary results for some resonances.

Table 1 sums up our results for all the resonances we have explored. For each resonance, we give our analytically calculated estimates for T_L and T_R. We always give a range of values, obtained for various initial conditions inside the resonance. If the "mixing" of the phase space is very prominent, we sometimes get a very wide range, which includes both normal and stable chaotic orbits. One should note that the "errorbars" in the plot are simply the intervals of computed values – they do not represent the numerical errors. We also indicate if the resonance has a resonant periodic orbit, which is, according to Tsiganis, Varvoglis & Hadjidemetriou (2002b), the key property for producing the *fast* chaos‡. Bulirsch-Stoer integrator with Jupiter and Saturn as perturbers was used for the integrations.

We have also tried to deduce the age of the Veritas family, whose most chaotic part lies inside the $5 - 2 - 2$ resonance. Our EGE gives an approximate age about 9 Myr while, assuming a constant diffusion coefficient (Knežević, personal communication), one gets

---

† One should bear in mind that this is just an approximation; strictly speaking, Lyapunov time is not equal, nor simply related to the subresonance crossing time.

‡ The existence of the periodic orbit for $13:6$ and $18:7$ resonances has not been checked thus far; however, we think this would be highly unlikely for such high-order resonances

Table 1. Analytical and numerical values of the Lyapunov time (in Kyr) and removal time (in Myr). Existence of the periodic orbit in the planar problem for the 2BR is also indicated; the data in this column were taken from Tsiganis, Varvoglis & Hadjidemetriou (2002b).

Resonance	T_L (Kyr)	T_L^{Num} (Kyr)	T_R (Myr)	T_R^{Num} (Myr)	Per. orbit ?
$2:1$	1.1–3.4	2.9–5.2	0.8–19.4	1.1–31.2	Yes
$3:2$	1.4–3.5	3.1–6.2	0.9–19.1	3.2–142.7	Yes
$3:1$	6.2–8.2	6.8–9.4	0.8–8.0	1.0–21.2	Yes
$5:3$	1.9–2.2	1.1–3.5	0.9–4.1	0.8–7.4	Yes
$5:2$	6.9–9.2	8.3–14.4	7.6–13.4	9.2–28.7	Yes
$7:4$	2.7–5.3	1.8–3.9	6.9–23.4	12.2–41.3	Yes
$8:5$	3.7–6.7	4.1–7.6	8.2–18.2	7.3–23.4	No
$7:3$	5.0–6.9	6.9–9.1	16.7–62.2	22.4–91.2	No
$9:5$	5.4–6.8	3.1–5.2	23.2–86.7	11.3–104.2	No
$11:7$	7.1–14.1	3.4–9.7	10.6–73.4	8.4–88.2	Yes
$11:6$	10.8–22.7	10.1–16.5	16.4–330.3	13.2–≈ 500	No
$12:7$	11.1–14.7	4.1–15.2	4.6–278.1	5.2–≈ 1000	No
$13:7$	65.1–84.6	43.3–76.1	23.2–≈ 1000	14.2–> 1000	No
$13:6$	55.1–77.6	14.6–24.5	41.3–≈ 1000	21.1–> 1000	No
$18:7$	≈ 500	300-600	≈ 20000	> 1000	No
$5-2-2$	11.4–13.7	8.4—10.9	≈ 10000	> 1000	No
$2+2-1$	90–160	130–240	≈ 20000	> 1000	No
$6+1-3$	130–150	130–170	≈ 40000	> 1000	No

about 8.3 Myr. The similarity is probably due to the young age of the family: were it older, the effects of non-linearity would prevail and our model would give an age estimate which is substantially different from that obtained in a linear approximation.

Figure 1 gives the results from table 1 plotted along the semimajor axis. It can be noted that the agreement is good within an order of magnitude, with some exceptions. Actually, one can see that the disagreement with the simulations is most significant exactly in the resonances with a periodic orbit, which might actually require a completely different treatment of transport.

In Figure 2, we plot the numerical T_L – T_R relation for the resonances $5:3$ and $12:7$, examples of normal and stable chaos, respectively. The largest discrepancies in the Figure 2a are probably for objects near the stability islands; in the Figure 2b, the fit fails completely. To check the assumption that this is due to the mixing of populations, we integrate a larger population of objects and divide them into two classes (the criterion being the prominence of anomalous diffusion, see later). For each class, we perform a separate fit with the corresponding T_L – T_R relation. Now most objects can be classified into one of the two scaling classes. In particular, this shows that the famous stable-chaotic object 522 Helga is probably not a remnant of some larger initial population but rather a member of one of the two populations existing in this resonance.

Finally, in Figure 4, we give the time evolution of the dispersion in P (i. e. $\langle \Delta P^2 \rangle$) for a set of clones of 522 Helga, using the procedure described in Tsiganis, Anastasiadis & Varvoglis (2000). Anomalous diffusion is clearly visible. This confirms the stable chaotic nature of this object and shows that we can use the anomalous character of diffusion as an indicator of stable chaos.

5. Conclusions and discussion

We have given a kinetic model of chaotic transport in the asteroid belt, based on the concept of convolution of various building blocks. Combining numerical and analytical results, we have shown how the removal time can be calculated and interrelated with the

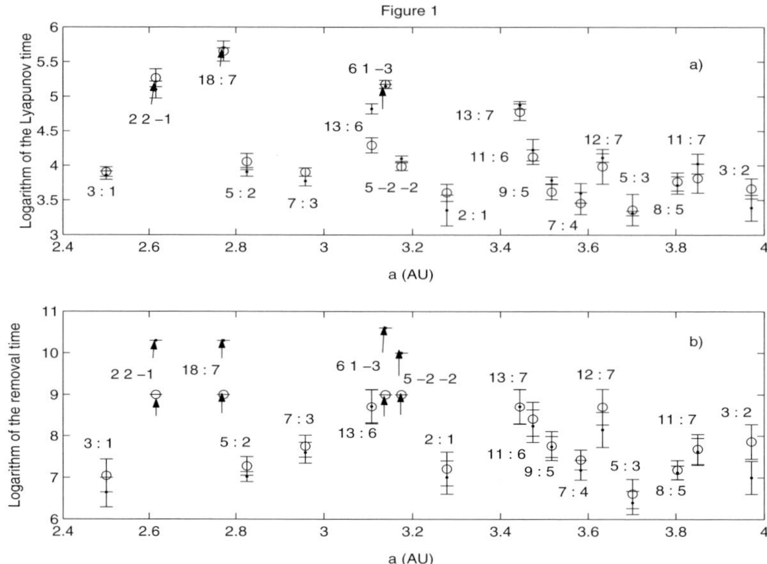

Figure 1. Analytical (points) and numerical (circles) values from table 1, for Lyapunov time (a) and removal time (b). Arrows correspond to extremely uncertain numerical values, usually the values close to or larger than the integration timespan (1 Gyr).

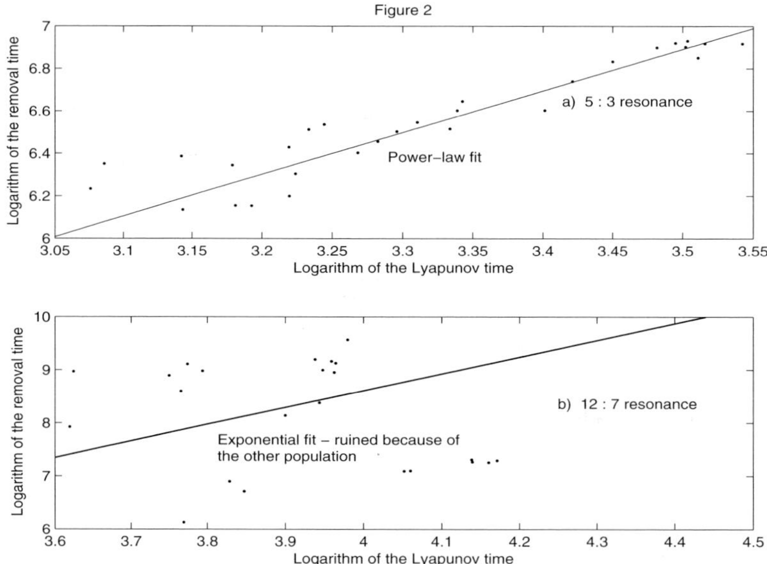

Figure 2. Plot of $T_L - T_R$ dependance for the resonances $5 : 3$ (a), fit with a power-law, and $12 : 7$ (b), fit with the exponential law. The straight line obtained by linear regression in the logarithmic scale is an obviously bad fit. See text for comments.

Lyapunov time. We have obtained two regimes for chaotic bodies, the power-law one and the exponential one. Due to the fractal structure of the phase space, however, asteroids from different regimes can be "mixed" in a small region of the phase space.

We would like to comment briefly on the controversial issue of the $T_L - T_R$ relation. First of all, the correlations we have found are *of statistical nature only* and should not

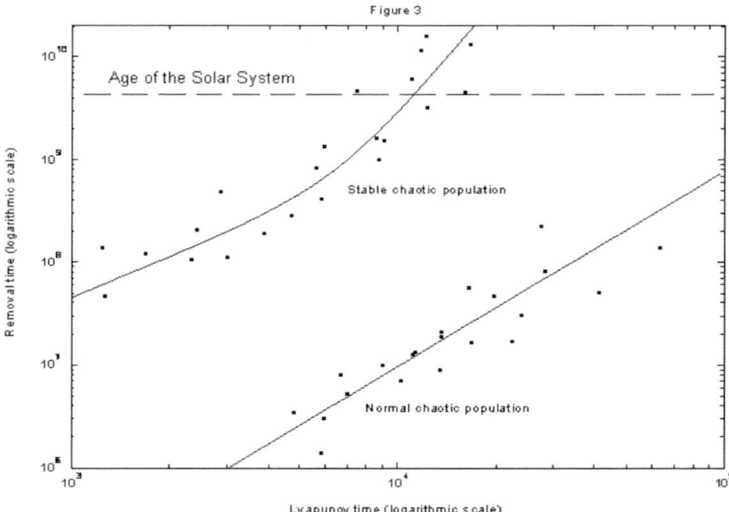

Figure 3. The same as in Figure 2b but with two separate plots for the exponential regime population, and for the power-law regime population. Obviously, we have a substantially better fit.

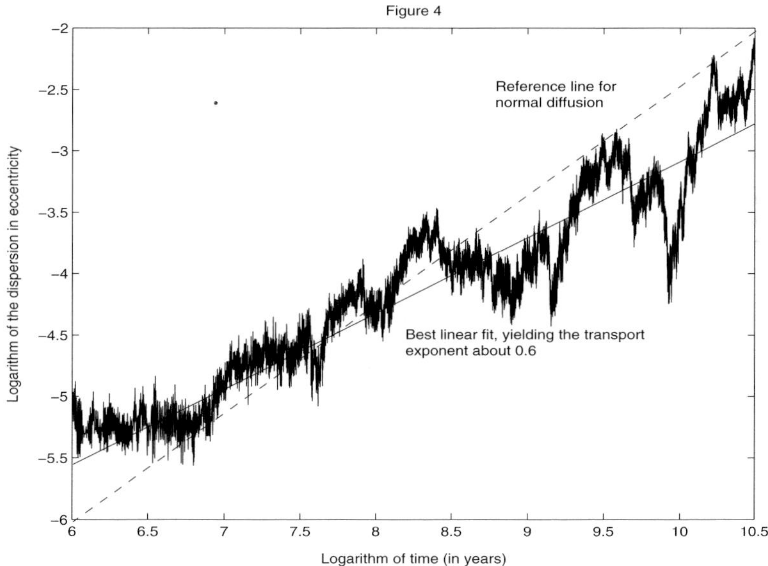

Figure 4. Diffusion in eccentricity for a set of clones of 522 Helga. Domination of anomalous transport is obvious. The reference line for normal diffusion ($\mu = 1$) is also plotted.

be regarded as "laws" in the sense of Murison, Lecar & Franklin (1994). Furthermore, due to their statistical nature, they cannot be used for any particular object, only for populations. Finally, it is clear that the scalings are *non-universal*, i. e. the scaling exponents are different for different resonances (possibly also in disconnected regions of a single resonance).

The exponential regime, characterized mainly by anomalous transport through various quasi-stable structures, corresponds to the stable chaotic regime, discussed e. g. in Tsiganis, Varvoglis & Hadjidemetriou (2000) and Tsiganis, Varvoglis & Hadjidemetriou (2002a). The reason that the exponential $T_L - T_R$ correlation was not noticed thus far are in part very large values of T_R in this regime, and in part the fact that stable chaotic objects are typically mixed with the objects in the normal chaotic regime. Also, it is interesting to note that the exponential scalings are of the same form as those predicted in Morbidelli & Froeschlé (1996) for the Nekhoroshev regime; therefore, it seems that the exponential stability can arise also due to stickyness, not necessarily as a consequence of the Nekhoroshev structure.

Finally, we hope that our research will stimulate further work in this field, since the results presented here are no more than just a sketch of possible general theory.

Acknowledgements

I am greatly indebted to Zoran Knežević for helpful discussions and for permission to cite his yet unpublished results. I am also grateful to George M. Zaslavsky, Harry Varvoglis and Alessandro Morbidelli for sending me copies of some of the references.

References

Afraimovich, V. & Zaslavsky, G. M. 1997, *Phys. Rev. E* 55, 5418
Čubrović M. 2004, in preparation.
Ellis, K. M. & Murray, C. D. 2000, *Icarus* 147, 129
Moons, M., Morbidelli, A. 1995, *Icarus* 114, 33
Moons, M., Morbidelli, A. & Migliorini, F. 1998, *Icarus* 135, 458
Morbidelli, A. & Froeschlé, C. 1996, *Cel. Mech. Dyn. Ast.* 63, 227
Morbidelli, A., Moons, M. 1993, *Icarus* 102, 316
Murison, M., Lecar, M. & Franklin, F. 1994, *Astron. J.* 108, 2323
Murray, N. & Holman, M. 1997, *Astron. J.* 114, 1246
Murray, N., Holman, M. & Potter, M. 1998, *Astron. J.* 116, 2583
Nesvorný, D. & Morbidelli, A. 1998, *Cel. Mech. Dyn. Astron.* 71, 243
Shevchenko, I. 1998, *Phys. Lett. A* 241, 53
Tsiganis, K., Anastasiadis, A. & Varvoglis, H. 2000, *Cel. Mech. Dyn. Ast.* 78, 337
Tsiganis, K., Varvoglis, H. & Hadjidemetriou, J. D. 2000, *Icarus* 146, 240
Tsiganis, K., Varvoglis, H. & Hadjidemetriou, J. D. 2002a, *Icarus* 155, 454
Tsiganis, K., Varvoglis, H. & Hadjidemetriou, J. D. 2002b, *Icarus* 159, 284
Zaslavsky, G. M. 2002, *Phys. Rep.* 371, 461

Part 3

ASTEROID ORBIT DETERMINATION AND IMPACTS

Dynamics of Populations of Planetary Systems
Proceedings IAU Colloquium No. 197, 2005
Z. Knežević and A. Milani, eds.

© 2005 International Astronomical Union
DOI: 10.1017/S1743921304008683

Virtual asteroids and virtual impactors

Andrea Milani

Department of Mathematics, University of Pisa, Via Buonarroti 2, 56127 Pisa, Italy
email: milani@dm.unipi.it

Abstract. When a celestial body, e.g., an asteroid, has been observed only over a short time, its orbit is not well determined but may be anywhere in a *confidence region* where the astrometric residuals are acceptable. This region can be sampled by a swarm of *Virtual Asteroids (VA)* sharing the reality of the asteroid: one of them is real, but we do not know which one. The problem is how to sample the confidence region with a small number of VA, still being able to solve the main problems of asteroid recovery/identification and impact monitoring.

One class of methods uses *random sampling* of the confidence region to mimic with the VA population the probabilistic distributions of the orbits. This class includes the Monte Carlo and the Statistical Ranging methods. When it is critical to detect a very small probability (e.g., of a catastrophic impact) by computing a small number of VA orbits, and also when a large catalog of asteroids has to be handled, it is more efficient to sample the confidence region with a *geometric object,* such as a smooth manifold: it can be sampled uniformly, taking into account its dimension. Our group has developed in the last 6-7 years 1-dimensional sampling methods based upon a differentiable curve, the *Line Of Variations (LOV),* which can represent, in suitable cases, the spine of the confidence region. The LOV is sampled by uniformly spaced VA, thus interpolation between consecutive VA is possible. This is the basis for the current algorithms of *Impact Monitoring,* used in Pisa and at JPL. The LOV method is also used for recovery of lost asteroids and for identification of independent discoveries of the same object.

When the asteroid has moved on the sky while being observed by $< 1°$, the confidence region is wide in two directions and the LOV may be an inappropriate way of sampling it. We have recently developed 2-dimensional sampling methods based upon the concept of *Admissible Region,* a 2-dimensional manifold parameterized by a compact subset of the range/range-rate plane. This region is then sampled by *triangulation,* with each node used as a VA. This allows to define methods for asteroid identification/recovery and for impact monitoring starting from very poor data, such as the ones collected during a single night of observations.

Keywords. orbit determination, surveys, recovery, identification, impacts, multiple solutions

---

1. Population orbit determination

Modern orbit determination was born with the space age, thus its concepts, algorithms and ways of thinking were based on the problem of determining the orbit of a spacecraft by using the tracking data. The key hypothesis is not that the celestial body is artificial, but that it has a device answering the signals from Earth: this is what I call *collaborative orbit determination.* It is possible to connect a trasponder with a planet, e.g., the Viking landers on Mars allowed to solve for the orbit of the planet within tens of meters, and to determine the orbit of a planet by tracking an orbiting spacecraft, e.g., with BepiColombo the orbit of Mercury will be determined to an accuracy of the order of 10 cm.

The opposite case, *population orbit determination,* is the one in which the observations are the scarce resource. This can occur with natural objects (asteroids, transneptunians, comets) and with artificial ones (space debris). The total number of data points may be comparable to the spacecraft tracking case but, the population being large, the number of observations for each object is on average low.

As an example, we have now $\simeq 600,000$ independent detections of asteroids for a total of $\simeq 50,000,000$ scalar data points†. The problem is made much more difficult by the fact that the non-collaborative observations do not contain a signature identifying the observed body. Thus for of each couple of independent detections of asteroids we do not know whether they correspond to the same body: finding the independent detections of the same body and an orbit fitting all the data is the *asteroid identification problem.*

Additional difficulties arise when, given a population, e.g., of Near Earth Asteroids (NEA), we need to use the few available data for a very accurate prediction, as in the case in which we try to predict whether an impact on Earth in a given year is possible.

To solve these problems we need some new mathematics, and as it is often the case in mathematical research one way forward is an act of courage: if the problem is difficult, let us make it more difficult by generalizing it. If the population, e.g., of asteroids, is big, let us make it bigger: we shall replace each detected asteroid with a swarm of *Virtual Asteroids (VA)* sharing the reality of the asteroid among them, that is, one of them is a good approximation of the real object orbit, but we do not know which one, and the others simply do not correspond to any existing objects. Then we can use the augmented population of all VA for all asteroids to tackle in a more effective way both the identification and the impact prediction problem.

2. The quasi-linear algorithms

In collaborative orbit determination (also for the best observed asteroids, the numbered ones) there is a *nominal solution* X^*, selected according to the least squares principle, that is a point of minimum for the *cost function* Q, which is computed from the sum of squares of all the m observation residuals‡. The nominal solution, obtained by *differential corrections*, that is by iteration of the least squares algorithm, is surrounded in the space of orbital elements X by a *confidence region* where the *penalty* $\Delta Q = Q(X) - Q(X^*)$ has an acceptable value. Since the confidence region is small, it is well approximated by a *confidence ellipsoid* $Z(\sigma)$ of the form

$$m\Delta Q(X) \simeq (X - X^*) \cdot C\,(X - X^*) \leqslant \sigma^2$$

where C is the *normal matrix* and the parameter σ corresponds to some confidence level.

Under these *quasi-linear* conditions, even the most complex problems of population orbit determination would have a comparatively simple solution.

2.1. *Identifications*

Given two independent sets of observations, let $m_j, X_j, Q_j, C_j, j = 1, 2$ be the number of residuals, the nominal solution, the minimum of the cost function and the 6×6 normal matrix (at some common epoch) for each of them. Under the hypothesis that they belong to the same object, the increase in the cost function for both sets fitted together is

$$(m_1 + m_2)\Delta Q(X) = (m_1 + m_2)\,Q(X) - (m_1\,Q_1 + m_2\,Q_2),$$

which can be computed in the linear approximation as

$$(X - X_1) \cdot C_1\,(X - X_1) + (X - X_2) \cdot C_2\,(X - X_2) = (X - X_0) \cdot C_0\,(X - X_0) + K$$

† These numbers refer to the published data. The detections kept secret by some observers are believed to amount to additional hundreds of thousands. Two thirds of the published data points refer to the $\simeq 80,000$ numbered asteroids.

‡ Q is either the sum of squares divided by m, or more generally a quadratic function of the residuals, taking into account uneven precisions, correlations and biases (Carpino *et al.* 2003).

where X_0 is the best "compromise solution" and the minimum K of the combined quadratic form is the *orbit identification penalty*. This algorithm has a geometric interpretation in terms of intersections of the two families of confidence ellipsoids. Selecting the couples with low orbit identification penalty as candidate for identification, then checking them by differential correction with X_0 as first guess is a very effective procedure to find identifications when the quasi-linear conditions apply (Milani *et al.* 2000a).

2.2. *Close approaches*

The close approach predictions can be handled in quasi-linear conditions by computing the *covariance matrix* $\Gamma = C^{-1}$ and then propagating it to the *target plane* of some future encounter with a planet (a plane perpendicular to the geocentric velocity at closest approach). The confidence ellipsoid in the space of initial conditions projects onto a *confidence ellipse* on the target plane, an impact is possible if this ellipse touches the impact cross section of the planet; the probability of the impact can be estimated by a classical Gaussian formalism (Milani *et al.* 2002, Milani and Valsecchi, 1999).

3. Sampling the confidence region

When the observational data are few and limited to a very short time span the orbit determination is strongly nonlinear, that is Q cannot be approximated by a quadratic function and $Z(\sigma)$ has a shape very different from an ellipsoid. Then, although the quasi-linear algorithms can be computed, they do not provide reliable predictions†.

If observations are scarce and computers are powerful, the logical approach is to use computationally intensive methods to provide fully nonlinear predictions. Thus the confidence region is sampled by a number of VA and for each of them the orbit is propagated: to the time of a potential recovery observation, to the time close to another detection candidate for identification, to the time of a possible close approach. The problem is how to select the VA in such a way that they are representative of all the possible outcomes and still the number of orbits to be computed is compatible with the available resources.

One class of methods uses *random sampling* of the confidence region to mimic with the VA population the probability density of some observation error model (e.g., Gaussian). This class includes the *Monte Carlo (MC)* and the *Statistical Ranging (SR)* methods.

3.1. *Monte Carlo*

The MC method uses the probabilistic interpretation of the least squares principle. If the residuals with respect to the "true" orbit are random observation errors distributed according to some known probability density, e.g., they are Gaussian, then this distribution can be sampled providing a set of equally probable VA (Milani *et al.* 2002).

In the quasi-linear case, the distribution of X can be approximated by a multivariate Gaussian with the nominal solution as mean and Γ as covariance matrix; this is the version of the MC method most often used. However, in the strongly nonlinear case there is no simple analytic way to compute the probability density of X, and the distribution needs to be sampled in the residuals space; this is done in a *nonlinear* version of MC method. Anyway, in the space of the orbital elements X the VA will be distributed in a non-uniform way, more densely packed near the nominal solution.

The MC method has been successfully used to compute nonlinear predictions, including computations of collision probabilities (Chodas and Yeomans 1996, Chodas and

---

† This does not occur in collaborative orbit determination: the tracking is planned at the mission definition stage, in such a way that the amount of data is enough to ensure convergence of the quasi-linear algorithms.

Yeomans 1999). However, when it is critical to detect a very small probability (e.g., of a catastrophic impact) the MC method may not be efficient enough: to detect a possible impact it requires that one of the VA has an impact orbit. E.g., to detect an impact with a probability of 10^{-8} we would need to compute a number of orbits of the order of 10^8.

The MC method is not used to handle at once a large real population, as in the asteroid identification problem, because the resulting VA population would be unpractically large.

3.2. *Statistical Ranging*

The SR method has been developed to improve upon the MC method by exploiting our understanding of the properties of the confidence region, especially in the case of *Too Short Arcs (TSA)*, that is sets of observations insufficient to compute an orbit according to the least squares principle. The method can work with just two observations, when there are only 4 equations in the six unknown coordinates of the initial conditions and there is no way to define "the orbit" satisfying the observations. The basic idea is that the two-dimensional manifold of orbits exactly satisfying two observations can be parameterized by the unknown values r_1, r_2 of the corresponding distances from the observer. If the 2-vector (r_1, r_2) is assumed, then the position of the asteroid is known at two different times (Lambert's problem) and a Keplerian orbit can be computed.

The SR method (Virtanen *et al.* 2001, Virtanen *et al.* 2003) generates a large population of VA by sampling at random the (r_1, r_2) plane and computing the corresponding keplerian elements; the elements corresponding to very unlikely orbits (e.g., hyperbolic) are discarded. To take into account the observational errors, the two observations used are changed at random by sampling the (supposedly known) error model; thus the SR method can be considered an implementation of the nonlinear MC method. This population of VA samples the confidence region, defined as the set of orbits not contradicting the observations, even when its shape has nothing to do with an ellipsoid. This is why with few observations the SR methods is more reliable than the MC: indeed it can be used, even starting from a TSA, to compute a nonlinear confidence region for interesting predictions, such as the position in the sky for recovery observations (Granvik *et al.* 2003) and the target plane positions for detecting a possible impact (Muinonen *et al.* 2001).

However, both the SR and the MC methods are limited in the resolution of the VA sampling of the confidence region. By design they attempt sampling of the entire confidence region, which has dimension 6, thus to decrease by an order of magnitude the typical distance among VA in the elements space the number of VA needs to increase by a factor one million. The SR method itself, with its 2-dimensional space of undetermined parameters (r_1, r_2), suggests that, for a TSA, the confidence region is "flat", that is it has a 2-dimensional spine. The confidence region, in such a strongly nonlinear case, is a thin tubular neighborhood of a 2-dimensional submanifold in the elements space. Thus it should be possible to devise a method of sampling in which we can decrease the typical distance by a factor d with an increase in the number of VA of the order of $1/d^2$.

4. Sampling by strings: the Line Of Variations

If the orbit of an asteroid is propagated for a time interval much longer than the time span of the observations, then the confidence region for the elements at the new epoch has a shape very different from an ellipsoid, even if the one at the original epoch was well approximated by an ellipsoid†. Given a generic set of VA sampling the confidence

---

† By the quasi-linear algorithms, the normal matrix at the new epoch is obtained by using the state transition matrix as a coordinate change in the tangent space: it defines a very elongated confidence ellipsoid, too large, especially in the along track direction, to be a good approximation.

region at an epoch close to the observations, as time goes by the separation increases along track (because different VA have different semimajor axis); they thus become like a string of pearls, "the wampum of the night" (Dickinson 1859). This suggests that, at least when predictions have to be provided for a time remote from the ones of the available observations, a sampling following a suitable one dimensional curve could be more effective than random sampling.

The idea of a *Line Of Variations (LOV)* representing somehow the spine of the confidence region is recurrent in the history of Celestial Mechanics, the oldest reference apparently being Le Verrier 1844. A simpler version, the orbits with all the keplerian elements equal but for the mean anomaly, has been in use for long time; also the method of solving for only five orbital parameters, e.g., by assuming the eccentricity, was used.

Recently this idea has been revived (Milani 1999, Milani *et al.* 2004a) by giving a rigorous definition of the LOV as a differentiable curve. At a point X in the elements space (at an epoch near the observations) the orbit determination has a *weak direction* $V_1(X)$, corresponding to the long axis of the confidence ellipsoid computed with the normal matrix $C(X)$. The point X is on the LOV if the cost function Q restricted to the hyperplane normal to the weak direction $V_1(X)$ has a local minimum.

The advantage of this definition is that there are effective algorithms to compute VA on the LOV, starting from a nominal solution when it is available. It is also possible to compute a LOV solution starting from a rough preliminary orbit, even in cases in which the quasi-linear differential corrections fail to provide a nominal solution; thus it is possible to sample the LOV for more asteroids than those for which it is possible to compute a nominal solution!

The main problem is that this definition of the LOV depends upon the metric of the orbital elements space, it is not invariant with respect to changes of coordinates and units. However, when the observed arc is not too short, the direction of the LOV changes very little with the coordinates used (see Milani *et al.* 2004a, Figures 3 and 4). Thus, if the LOV is sampled uniformly by a moderate number of VA, years after the observations the "pearls" are still evenly spaced along track (provided there has not been a very close approach), and they sample in an effective way the confidence region at a later epoch.

4.1. *Recovery*

LOV sampling is effective in organizing the recovery of a *lost asteroid*, an object such that the confidence region for the next possible observation is very large, e.g., spanning many degrees on the sky. In such cases the nominal prediction may be far from the real position, and the confidence ellipsoid computed by the standard linear theory may fail to identify the region to be scanned. A set of VA, regularly spaced along the LOV, is prepared and the predictions are computed for each of them. The number of LOV points should be selected in such a way that the spacing in the prediction between two consecutive ones is comparable to the field of view of the telescope used for the search. This method has allowed recoveries of especially interesting asteroids, by scanning the LOV trace either on the sky or on archives plates (Boattini *et al.* 2001).

4.2. *Identification*

The quasi-linear algorithm to identify two independent detections of the same asteroid requires the propagation of both orbits with their normal matrices to a common epoch. If the two detections are years apart, the confidence regions are very different from the confidence ellipsoids. Thus the intersections of the confidence regions, where the compromise orbit should be found, cannot be reliably computed as the intersections of the confidence ellipses and the quasi-linear method often fails.

Sampling the LOV of both detections (at epochs near the respective observations) with a number of VA X_1^i and X_2^k, then propagating all to the common epoch and computing the identification penalty $K_{i,k}$ for each "virtual couple" allows to propose many additional identifications with respect to the ones proposed by the quasi-linear algorithm. Moreover, the use of LOV solutions when a nominal solution is not available increases the size of the catalogues of orbits for which an identification can be attempted. The combined effect of these two improvements has increased the number of new confirmed identification, with the same data, by an order of magnitude (Milani *et al.* 2004a, Table 4).

4.3. *Impact monitoring*

The goal of *impact monitoring* is to establish whether the confidence region for a given NEA contains some *Virtual Impactor (VI)*, a small subset of initial conditions leading to a collision with Earth †. The *Impact Probability (IP)* of a VI is roughly proportional to the volume of the VI in the elements space. To find an initial condition belonging to the VI, the *VI representative*, when the IP is minute, would require a very dense sampling.

The current impact monitoring systems, CLOMON2 (Universities of Pisa and Valladolid, at http://newton.dm.unipi.it/neodys and http://unicorn.eis.uva.es/neodys) and Sentry (JPL at http://neo.jpl.nasa.gov/risk/) have solved this problem by using two improvements. First, the confidence region is sampled along the LOV, thus a moderate $(1,000$ to $10,000)$ number of VA allows to reproduce the different dynamical behaviors (Milani *et al.* 1999). Then all the VA orbits are propagated for the next 80 to 100 years and all the close approaches to the Earth are recorded, together with the confidence ellipse on the target plane (according to the quasi-linear algorithm of Section 2). Second, by using the LOV continuous structure: when two VA contiguous in the natural ordering of the LOV have a close approach (at the same epoch) such that there must be a minimum of the approach distance in the segment between the two, this minimum is found by interpolation on the LOV and iterative methods (e.g., *regula falsi*; Milani *et al.* 2004b). If this minimum is less than one Earth radius, a VI representative has been found. If the minimum corresponds to a very deep approach, such that the target plane confidence ellipse touches the Earth, either a VI representative is obtained by another iterative method (Milani *et al.* 2000b) or the local linearity is checked to confirm the VI.

The impact monitoring systems have sometimes to handle more complex behaviors of the LOV trace on the target plane, such as *interrupted returns* (Milani *et al.* 2004b), which can be understood in terms of the theory of *resonant returns* (Valsecchi *et al.* 2003). Anyway the current monitoring systems are very effective in detecting minute VI, at least when the LOV sampling represents well the different possible dynamical conditions. After a VI has been detected, and announced by the impact monitoring systems, the asteroid is reobserved, and this can "destroy" the VI: the confidence region in the space of initial conditions becomes smaller, and the VI may not be a subset anymore.

On average about once per week newly discovered NEA are found to have VI and the case is solved by re-observation. Not only the mathematical but also the public relations problem has been solved: the media and the public opinion now understand that the appearance and then destruction of some VI indicates that the system (the surveys, the orbit computers, the impact monitoring and the follow up observers) is working well and is effectively decreasing the impact risk with respect to the background risk.

---

† The collision subset for a given epoch may be disconnected, when the same collision can be reached through different dynamic routes, e.g., resonant returns in different resonances. In this case a VI is a connected component of the collision subset.

5. Sampling by surfaces: the Admissible Region

A TSA is a set of data such that the positions on the celestial sphere do not significantly deviate from a great circle (the differences being of the order of the expected observation error). This condition typically applies when the available observations only span an arc of $\simeq 1°$ or less on the celestial sphere. Then the data provide only four independent observed quantities, the two angles (α, δ) at some mean time t, and two angular rates $(\dot{\alpha}, \dot{\delta})$, like an arrow on the sky; such a 4-dimensional observable is called an *attributable* (following Milani *et al.* 2001). An average apparent magnitude h may also be available.

Under these conditions, the two coordinates $(r, \dot{r})$ are not constrained at all by the observations. The only way to restrict the set of possible values in the $(r, \dot{r})$ half plane ($r > 0$) is to use a priori information and/or hypotheses independent from the observations. The range and range rate are constrained if we assume that the body belongs to the solar system, but it is not a satellite of the Earth. We can also set a lower limit to the size of the object by an upper bound for the absolute magnitude H: since $H - h$ is a function of r, this condition sets a minimum distance, excluding "shooting stars". The combination of the three conditions above defines a compact subset of the $(r, \dot{r})$ space, the *Admissible Region*; it can have either one or two connected components, the boundary is defined by explicitly computable algebraic curves (Milani *et al.* 2004c).

Thus it is possible to sample the finite Admissible Region with a finite number of points $(r_j, \dot{r}_j)$, $j = 1, \ldots, N$. Different sampling strategies can be used†, however sampling with a geometric object has significant advantages. For 1-dim objects the sampling is done by intervals, with optimal metric properties (equal spacing in the LOV parameter). For 2-dim objects such as the Admissible Region the sampling should be done by triangles, with some optimal metric property. The ideal triangulation would include only equilateral triangles, for a complicated shape region this is not possible but we can use the *Delaunay triangulation*, with the largest possible minimum angle among all triangles. Then the nodes of the triangulation are the selected points $(r_j, \dot{r}_j)$ (Milani *et al.* 2004c).

5.1. *Attributable elements*

Given an attributable $A = (\alpha, \delta, \dot{\alpha}, \dot{\delta})$, and given a set of points $(r_j, \dot{r}_j)$, $j = 1, \ldots, N$, this defines a set of VA, each one with initial conditions‡ uniquely determined by the six dimensional vector $(\alpha, \delta, \dot{\alpha}, \dot{\delta}, r_j, \dot{r}_j)$. These *attributable elements* can be converted into Cartesian position and velocity (topocentric, then heliocentric), then to any other coordinates, such as keplerian elements.

Such a set of VA can be used, e.g., to compute multiple ephemerides for a planned recovery, as in Figure 1. Each of the *virtual ephemerides* obtained in this way has an associated uncertainty, which is obtained by propagation of the covariance matrix defined by the (linear) least squares fit used to compute the attributable A (Milani et al., in preparation); this uncertainty is represented by ellipses in the plane of Figure 1. The predictions are attributables, the uncertainty is represented in full by ellipsoids in 4-dim space. The union of all the confidence ellipsoids surrounding the alternate ephemerides, from each triangulation node, is an approximation of the confidence region for the ephemerides.

5.2. *Preliminary orbits*

Today the surveys produce large sets of TSA, i.e. of attributables, rather than individual observations. Thus the problem of computing preliminary orbits (to be used as first guess for differential corrections) has to be reformulated. We would like to perform a *linkage*,

---

† Tholen and Whiteley (private communication) and Marsden (private communication) use a rectangular grid, then discard hyperbolic orbits.
‡ The epoch corresponding to the initial conditions is $t_j = t - r_j/c$, with c the speed of light, to take into account the light travel time.

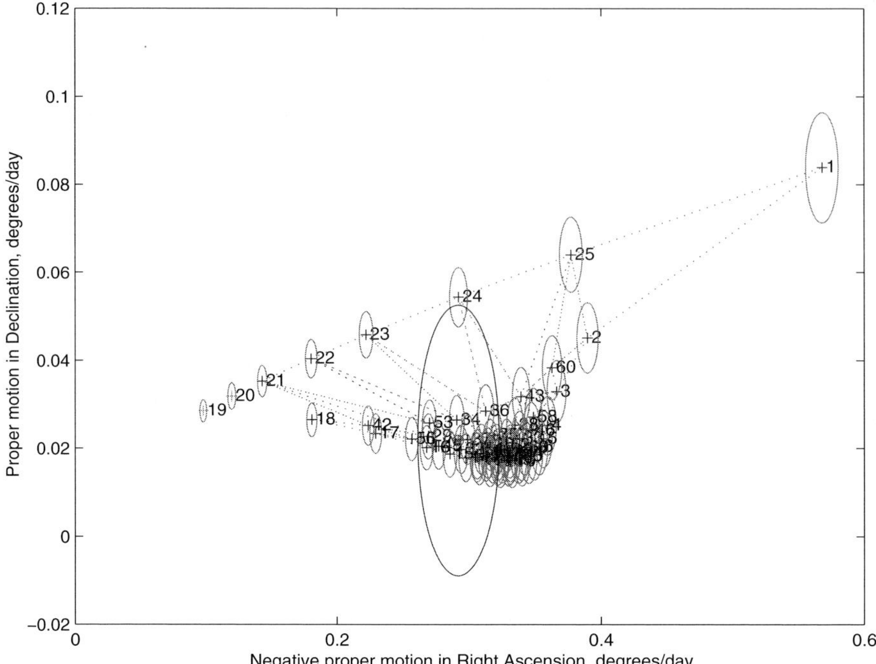

Figure 1. By using only the observations of the discovery night of the asteroid 2003 BH$_{84}$, recovery observations have been predicted for 11 days later in the plane of the proper motion (angular rates); for each VA the prediction is marked with the corresponding integer index, the sides of the triangles are dotted. The small ellipses give the uncertainty of the prediction, the larger ellipse gives the uncertainty of the attributable computed with the recovery data.

a special form of identification: to compute some orbit (either nominal solution or LOV) fitting together the data from two TSA, that is two attributables. The method of Gauss for a preliminary orbit requires 3 well separated observations, thus it is not applicable.

Given an attributable A_0 at time t_0 and the triangulation of the corresponding Admissible Region, for each VA we can compute for the time t_2 of another attributable A_2 the predictions $A_1(j)$ with uncertainty described by the normal matrix $C_1(j)$. As shown in Figure 1, the second attributable comes with its own uncertainty, represented by a normal matrix C_2. The same algorithm used for identification of orbits in a 6-dim space can be used to identify attributables in a 4-dim space: the formulas are the same of Section 2, only the size of the vectors and matrices are different. In this way we define a *attributable identification penalty* $K(j)$ for each VA, and we can select the values of the index j for which the penalty is low (if any). Moreover, for each one of these we can use the "compromise attributable", together with the values of $(r_j, \dot{r}_j)$, to define an orbit fitting both attributables with a moderate cost function: this is the new type of preliminary orbit (Milani at al., in preparation).

5.3. *Identifications*

If from the two attributables A_0 and A_2 it is possible to generate some of these preliminary orbits then they can be used as first guess in a differential corrections procedure, which may converge to a nominal solution or at least to a number of LOV solutions. Figure 2 shows this procedure: from the nodes with low values of $K(j)$, a number of LOV solutions have been computed. The LOV is very close to a straight line in the $(r, \dot{r})$ plane, the LOV solutions are marked with the index j. The "best" label indicates the nominal solution, the $+$ sign indicates the position of the real asteroid (as determined a posteriori).

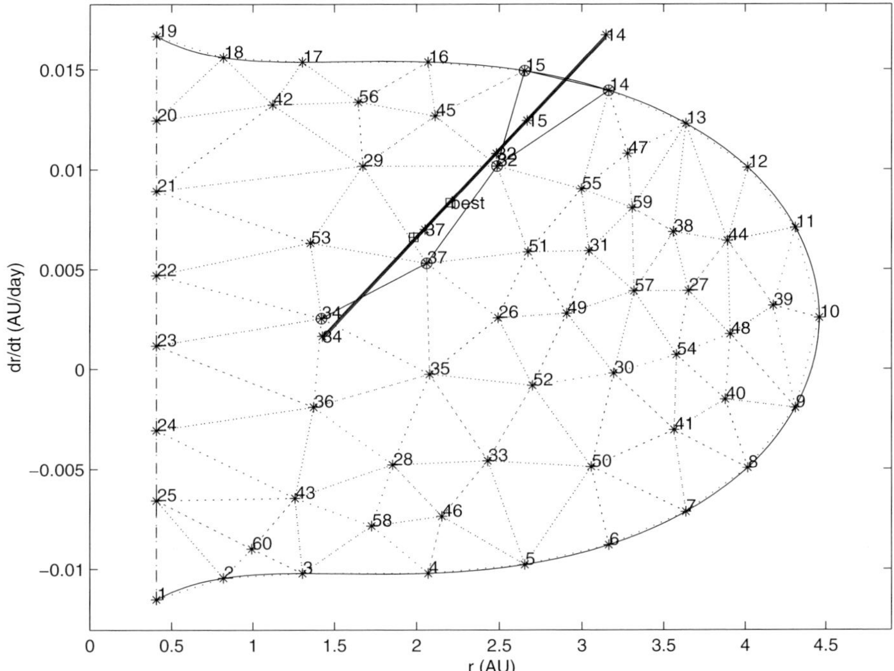

Figure 2. The Admissible region in the $(r, \dot{r})$ plane for the attributable from the discovery night of 2003 BH$_{84}$ with its the Delaunay triangulation. The sides marked with continuous lines join the nodes with (normalized) square root identification penalty $\sqrt{K(j)} < 9$ for identification with the recovery attributable of 11 days later.

If this procedure is successful, starting from a set of VA sampling a 2-dim set it provides another set of VA sampling a 1-dim set. Indeed, the orbit fitting two independent detections is "less undetermined", but still quasi-linear conditions do not apply. Then the procedure is repeated to try and find a third attributable. Each one of the second generation VA is propagated to the time of a third attributable candidate for *attribution*, the kind of identification in which an attributable is identified with an object for which an orbit is already available (Milani *et al.* 2001, Milani *et al.* 2004a). If the identification penalty is low, differential corrections can be attempted and they provide a nominal orbit fitting three attributables. Then these orbits are used again to search for an identification with a fourth attributable, and so on: the procedure is recursive, that is the algorithms and even the software are the same to fit together 3,4,5,... attributables.

These new algorithms have so far been tested only on simulations of future surveys. The tests on real data will be more demanding: the problem is to organize a computationally efficient procedure to handle the very large data sets expected from the next generation of asteroid surveys. However, the simulations results are very encouraging: they indicate that the algorithms described above can be used to solve the identification problem for large observational data sets, given enough computing power.

5.4. *Impact monitoring*

When a NEA has been observed over a very short arc, like $1°$, the impact monitoring techniques based upon LOV sampling may not be effective. The confidence region being like a flat disk, the LOV definition is strongly dependent upon the coordinates and the units used (see Milani *et al.* 2004a, Figures 1 and 2). The LOV is just one chord of a such disk, thus LOV sampling is not representative of the entire confidence region. The use of geometric 2-dim sampling should solve this problem; this is the subject of current research (see Tommei, these proceedings).

6. Future applications

As computer technology progresses, the balance between the cost of new data and the cost of computations will tip more and more in favor of computationally intensive methods. This does not imply that brute force methods will dominate: the need for sophisticated analytical tools from celestial mechanics will remain. Thanks to the computing power available and to the new methods such as geometric sampling, we will tackle more and more complex problems such as orbit determination for large populations.

Among the next targets of research in this field, the detection of VI for asteroids without a least squares orbit (TSA), the algorithms to process the data of the next generation surveys (expected to discover $\simeq$ 100 times more asteroids that the current ones), a catalog of space debris complete to very small sizes, and more yet to be invented.

Acknowledgements

In this paper I have also reported on my own research, for which I would like to acknowledge the essential contribution of many coworkers, in alphabetical order: O. Arratia, A. Boattini, M. Carpino, S.R. Chesley, P. W. Chodas, M. de' Michieli Vitturi, G. Gronchi, Z. Knežević, A. La Spina, M.E. Sansaturio, G. Tommei, G.B. Valsecchi. I also thank J. Virtanen for her referee report, containing useful suggestions.

References

Boattini, A., D'Abramo, G., Forti, G. and Gal, R. 2001, *Astron. Astrophys.* 375, 293

Carpino, M., Milani, A. and Chesley, S.R. 2003, *Icarus*, 166, 248

Chodas, P. W. and Yeomans, D. K. 1996, in: *IAU Colloq. 156: The Collision of Comet Shoemaker-Levy 9 and Jupiter*, p. 1

Chodas, P. W. and Yeomans, D. K. 1999, Paper AAS 99-462, AAS/AIAA Astrodynamics Specialists Conference, Girdwood, Alaska.

Dickinson, E. 1859, Poem 128.

Granvik, M., Virtanen, J., Muinonen, K., Bowell, E., Koehn, B. and Tancredi, G. 2003, *Earth, Moon, Planets*, 92, 73

Le Verrier, U.-J. 1844, *Compt. Rend. Acad. Sci.* 19, 982

Milani, A. 1999. *Icarus*, 137, 269

Milani, A. and Valsecchi, G.B. 1999, *Icarus* 140, 408

Milani, A., Chesley, S.R. and Valsecchi, G.B. 1999, *Astron. Astrophys.* 346, L65

Milani, A., La Spina, A., Sansaturio, M. E. and Chesley, S. R. 2000a, *Icarus*, 144, 39

Milani, A., Chesley, S.R., Boattini, A. and Valsecchi, G.B. 2000b, *Icarus* 145, 12

Milani, A., Sansaturio, M.E. and Chesley, S.R. 2001, *Icarus* 151, 150

Milani, A., Chesley, S.R., Chodas, P.W. and Valsecchi, G.B. 2002, in: W.F. Bottke Jr. et al. (eds.), *Asteroids III*, Univ. Arizona Press, p. 55

Milani, A., Sansaturio, M.E., Tommei, G., Arratia, O. and Chesley, S.R. 2004a, *Astron. Astrophys.*, in press

Milani, A., Chesley, S.R., Sansaturio, M.E., Tommei, G. and Valsecchi, G. 2004b, *Icarus*, in press

Milani, A., Gronchi, G.F., de' Michieli Vitturi, M. and Knežević, Z. 2004c, *Cel. Mech. Dyn. Astron.*, in press

Muinonen, K., Virtanen, J. and Bowell, E. 2001, *Cel. Mech. Dyn. Astron.*, 81, 93

Valsecchi, G.B., Milani, A., Gronchi, G.-F. and Chesley, S.R. 2003, *Astron. Astrophys.* 408, 1179

Virtanen, J., Muinonen, K. and Bowell, E. 2001, *Icarus*, 154, 412

Virtanen, J., Muinonen, K., Tancredi, G. and Bowell, E. 2003, *Icarus*, 161, 419

Dynamics of Populations of Planetary Systems
Proceedings IAU Colloquium No. 197, 2005
Z. Knežević and A. Milani, eds.

© 2005 International Astronomical Union
DOI: 10.1017/S1743921304008695

Asteroid population models

Alessandro Morbidelli

Observatoire de la Côte d'Azur, Nice, France
email: Alessandro.MORBIDELLI@obs-nice.fr

Despite of the large number of detections of new asteroids, both in the main belt and in the near-Earth space, our observational knowledge of the asteroid populations is not quite complete. Our catalogs are affected by observational biases. Thus, a modeling work is required in order to infer the real distributions of the asteroid populations.

In collaboration with scientists of the Southwest Research Institute in Boulder, the Universities of Hawaii and Prague and the Nice observatory, I have been heavily involved in the development of models of the de-biased distributions of main belt objects and NEAs. In this short paper, I will briefly summarize the principles on which our models are based and their main results, with references to the key papers that discuss these models in detail. I will end on a model of the Yarkovsky-driven origin of NEAs from the main belt, which links the two models together and enlightens their mutual compatibility.

The NEA population is made of transient objects, with quite short individual dynamical lifetimes (a few million years only; Gladman *et al.* 1997). But the NEA population has remained roughly constant over the last ~ 3 Gy (Grieve and Shoemaker, 1994). This argues that the population is maintained in a sort of steady state by the continuous influx of new objects from the main belt. From numerical simulations, we determined the steady state orbital distributions of the NEAs coming from the main source resonances in the belt (Bottke *et al.* 2000). The problem was then to compute the relative contributions of these various sources to the overall NEA population. We did this by fitting our combined model distribution (i.e. the linear combination of the distributions related to each of the sources) to the observed distribution. This required an accurate evaluation of observational biases. For this reason, we limited our observational sample to the NEAs detected by the Spacewatch survey, whose bias had been previously quantified (Jedicke, 1996). Our model accounts for the existence of ~ 1200 NEAs with absolute magnitude $H < 18$. In order to keep the NEA population in steady state, about 55 new asteroids with $H < 18$ should become NEAs per My from the ν_6 resonance, and 100 from the 3:1 resonance. (Bottke *et al.* 2002). Knowing the albedo distribution of the main belt asteroids close/in the NEA sources, we also derived an estimate of the albedo distribution of the NEAs. This allowed us to convert the magnitude distribution into a size distribution and compute the impact probability of NEAs with the Earth, as a function of impact energy. We estimated the existence about ~ 1000 NEAs larger than 1 km in diameter, with an exponent of the size distribution equal to -1.75. The average impact rate for energies larger than 1,000 MT should be of 1 collision every 55,000 y (Morbidelli *et al.* 2002a). We are currently in the process of revising our model, with an improved calibration of the relative contributions of the sources, done using the much larger sample of NEAs detected by the LINEAR survey. The results, however, do not seem to change significantly.

For the main belt population, we used a completely different strategy (Morbidelli and Vokrouhlicky, 2003). First, we adopted the cumulative luminosity function of the main belt provided by the SDSS survey (Ivezic *et al.* 2001), and normalized the counts at the bright end using the known object catalog. This gave a total population of 1,300,000 asteroids with $H < 18$. It implied that the observed population was complete up to $H \sim 14$. For $H > 14$ we generated synthetic objects in order to complete the observed population. Two problems needed to be solved, though. The first problem was to decide, as a function of magnitude, which fraction of the synthetic objects should belong to one of the various dynamical families, and which to the background population. We did this using the relative de-biasing technique for families and local background populations, developed in Morbidelli *et al.* (2003). The second problem was to decide the orbital distribution of the synthetic objects, separately for family and background members. The analysis of families showed that the orbital distribution of the observed members broadens with increasing magnitude up to $H \sim 14$–15, and then it is roughly invariant. Given that all our synthetic objects have $H > 14$, we distributed them according to these invariant distributions. For the background objects, we divided the main belt in semi-major axis zones, and assumed that the observed e, i distributions are representative of the real ones. Then we placed our synthetic background objects on orbits drawn from these distributions. In our resulting main belt model, about 60% of the asteroids with $H < 18$ are members of the background population. Once magnitudes are converted into sizes, the exponent of the size distribution for objects of about 1 km in diameter (or $H \sim 18$) is -1.3.

It has been generally believed that the collisional activity in the main belt, which continuously breaks-up large asteroids, is the main mechanism that supplies bodies to the NEA source resonances. The size distribution of the NEAs, however, is only moderately steeper than that on the main belt. It should be much steeper if the NEAs were fresh collisional debris. This suggests that Yarkovsky thermal drag, rather than collisional injection, plays the dominant role in delivering material to the NEA source resonances (Morbidelli *et al.* 2002b). In Morbidelli and Vokrouhlicky (2003), starting from our Main Belt model, assuming current wisdom lifetimes for asteroid collisional disruption and reorientation, and using state-of-the-art estimates for the magnitude of the Yarkovsky and of the YORP effects, we showed that the populations of objects delivered per million year to the ν_6 and 3:1 resonances are in agreement – for what concerns total number and size distribution– with those deduced from our NEA model. This shows that (i) our main belt model is compatible with our NEA model and (ii) the Yarkovsky is the major mechanism by which main belt asteroids become NEAs.

References

Bottke W. F., Jedicke R., Morbidelli A., Petit J. M. & Gladman B. 2000, *Science*, 288, 2190

Bottke W. F., Morbidelli A., Jedicke R., Petit J. M., Levison H. F., Michel P. & Metcalfe T.S. 2001, *Icarus*, 156, 399

Gladman B., Migliorini F., Morbidelli A., Zappalà V., Michel P., Cellino A., Froeschlé Ch., Levison H., Bailey M. & Duncan M. 1997, *Science*, 277, 197

Grieve R. A. & Shoemaker E. M. 1994, in: T. Gehrels & M. S. Matthews (eds.), *Hazards Due to Comets and Asteroids*, University of Arizona Press, p. 417

Ivezić Z. *et al.* 2001, *Astron. J.*, 122, 2749

Jedicke R. 1996, *Astron. J.*, 111, 970

Morbidelli A., Jedicke R., Bottke W. F., Michel P. & Tedesco E. F. 2002a, *Icarus*, 158, 329

Morbidelli A., Bottke W. F., Froeschlé C. & Michel P. 2002b, in: Bottke *et al.* (eds.), Asteroids III, University of Arizona Press, p. 409

Morbidelli A., Nesvorný D., Bottke W. F., Michel P., Vokrouhlický D. & Tanga P. 2003, *Icarus*, 162, 328

Morbidelli A. & Vokrouhlický D. 2003, *Icarus*, 163, 120

Dynamics of Populations of Planetary Systems
Proceedings IAU Colloquium No. 197, 2005
Z. Knežević and A. Milani, eds.

© 2005 International Astronomical Union
DOI: 10.1017/S1743921304008701

Linking Very Large Telescope asteroid observations

M. Granvik,[1] K. Muinonen,[1] J. Virtanen,[1] M. Delbó,[2] L. Saba,[2]
G. De Sanctis,[2] R. Morbidelli,[2] A. Cellino,[2] and E. Tedesco[3]

[1]Observatory, P.O. Box 14, FIN-00014 Univ. Helsinki, Finland
email: mikael.granvik@astro.helsinki.fi

[2]Turin Astronomical Observatory, via Osservatorio 20, I-10025 Pino Torinese (TO), Italy

[3]Space Science Center, Univ. New Hampshire, 39 College Road, Durham, NH 03824, USA

Abstract. A novel method for the preliminary identification of asteroids at discovery and a few days thereafter is being developed in Helsinki. Having two different sets of asteroid observations, the goal is to identify all possible pairs of objects between the sets. An arbitrary asteroid can either remain unidentified, or be preliminary linked to one or more asteroids. In the case of ambiguity, the final decision must usually be based on additional observations. We use a multistep approach, during which possible pairs of objects are first selected by comparing ephemerides that have been generated for three common epochs. The method has been successfully tested using both Very Large Telescope observations, and simulated observations of near-Earth and main-belt objects. Identification results of simulated observations indicate that the observing strategy promoted by the Minor Planet Center might not be the best one, at least for the purposes of identification. The ultimate goal is to produce a real-time asteroid identification tool for ESA's astrometric space observatory Gaia, the Lowell Observatory Near-Earth-Object Search, the Near-Earth Space Surveillance mission, and the Nordic Near-Earth Object Network. The tool could also benefit large-scale surveys done with the Large Synoptic Survey Telescope, and the Discovery Channel Telescope.

Keywords. Celestial mechanics, methods: numerical, methods: statistical, asteroids, surveys

1. Introduction

Possible linkages between asteroid observations are usually sought by comparing the observed positions with ephemerides, or by comparing orbital elements inverted from two separate sets of observations. The first approach is used daily by, for example, the Minor Planet Center (MPC) for several purposes. First, it is used when identifying reasonably well-known asteroids in a batch of new observations. The asteroids do not need to be numbered, because in most cases the mean anomaly M contains the most significant error, and a fit to the observations can thus be obtained by varying M. Second, the first approach is used when searching for linkages between single-night observations by comparing the computed positions and motions with observed positions and motions. Due to too few observations, and/or too short observational arcs, a rigorous orbital inversion is not possible when using deterministic approaches, and therefore the computed positions and motions are obtained by using Väisälä-type orbits. As the Väisälä method makes assumptions on the time of perihelion and the geocentric distance, this method can be reliable only for a couple of days (if at all). The Väisälä method has, however, proven to be useful in many cases, and is therefore also used by, for instance, the Lowell Observatory Near-Earth-Object Search (LONEOS). Milani *et al.* (2001) used the first approach in a method which they called *attribution*. The attribution method allows linking of short-arc observations, or, more accurately, a representation of those observations, with a least-squares orbit of an unnumbered asteroid over longer time spans than is possible with the M-variation technique. The same group use the second approach when tentatively identifying two reasonably well-determined orbits, for which the least-squares approximation is valid, over a long time span (Milani *et al.* 2000).

In the present method a third possibility is used: namely, the comparison of ephemeris clouds obtained with statistical orbital ranging (Virtanen *et al.* 2001, Muinonen & Bowell 1993). In contrast to the methods briefly described above, the present method is particularly suitable for analysing exiguous single-night data, such as a set of Very Large Telescope (VLT) observations analysed in this paper. As is the case with all Monte Carlo (MC) inversion methods, statistical ranging outputs a relatively large amount of data. In fact, the processing and interpretation of the output data turns out to be the bottleneck in most applications relying on these methods. Particularly in the case of asteroid identification, where a large number of objects have to be processed as fast as possible, the key element of the whole process is data mining of the statistical ranging output. In the current identification method, data mining is efficiently carried out by using the so-called address comparison technique.

The paper is organised as follows. Section 2 briefly describes the statistical ranging method, which is the inversion method used throughout this paper. The overall identification scheme is presented in Section 3, and the most important building block is described in Section 4. In Section 5, results produced by the identification method are presented and discussed. Finally, in Section 6, the key findings are summarised.

2. Statistical orbital ranging

The probability density of orbital elements is examined using MC selection of orbits in orbital element space in the following way:

- Two observations are chosen (usually the first and the last), and angular deviations mimicking the observational errors in R.A. and Dec. are introduced.
- Topocentric ranges (distances) are assumed corresponding to the observation dates. In other words, two positions equalling six constants of integration are known.
- A trial orbit is first computed using the p-iteration method and is then compared to all observations. If the trial orbit fits the observations to predefined accuracy (defined as a $\Delta\chi^2$-threshold and maximum sky-plane residuals), it is added to the sample of possible orbits.

In the basic version of statistical ranging, the initial topocentric range intervals are determined manually using an educated guess, whereas in the automated version the topocentric range intervals are further improved using the 3-σ cutoff values of the range probability density. By increasing the number of generated sample orbits ($10 \rightarrow 200 \rightarrow n$), an unbiased phase-space region of possible orbits is found. Each sample orbit is assigned a weight, which describes how well it explains the observations. Ignoring the weights, the distribution merely shows the extent of different orbital solutions in the orbital element space assuming predefined observational errors. Ephemerides are generated by transforming the orbital elements of every sample orbit to a position on the celestial sphere at a given epoch, the result thus being an ephemeris cloud.

3. The identification scheme

It is, in principle, possible to use a straightforward trial-and-error-scheme while searching for identifications using statistical ranging. The idea is to *try* to perform the inversion using statistical ranging by using all observations corresponding to two objects. If the inversion succeeds—that is, at least one orbit can be found, it shows that the objects can be tied together using the same orbit within the assumed observational errors.

The direct identification approach is, however, not a particularly efficient technique. Assume, for instance, two observation sets containing 1000 objects each. Simplistically, the number of pairs to be checked is 1 million, while at most only 0.1% are correct identifications. We use a novel method termed ephemerides address comparison (at common epochs) to efficiently reduce the huge initial number of object pairs (see Section 4). The remaining, reasonably probable, pairs are examined by using the direct approach after an intermediate step. The intermediate step is similar to the last step, but it only uses the two first and the two last observations of the combined observation set.

4. Ephemerides address comparison at common epochs

The idea is to generate ephemeris clouds (R.A. and Dec.) for all objects in both sets for three common epochs, and then find out whether any objects in different sets have similar ephemerides at all three epochs, which would indicate a possible identification (Granvik *et al.* 2004, *in preparation*). The choice of epochs can be optimised, but the use of the observational mid-epoch as the first epoch is a good first approximation. The choice can be justified based on the knowledge that the ephemeris uncertainty grows with increasing time since last observation (Muinonen *et al.* 1994). The second and third ephemerides are produced by propagating orbits from the first epoch 12 and 24 hours forward in time, respectively, and transforming the corresponding orbital elements to ephemeris clouds as described in Section 2. The search for similarities among the two ephemerides is carried out efficiently using the address-comparison technique, which is presented in Section 4.1. Similar ephemerides at several epochs indicate a tentative linkage, which requires further investigation using either a statistical-ranging inversion (required for single-night linkages due to short observational arcs), or differential correction of a least-squares orbit.

4.1. *Address comparison*

When searching through the bins of the discretised ephemeris clouds to find overlapping ephemerides, most time is spent checking empty bins, which is inefficient. Instead of using the whole map, or multidimensional array, one can write an address to each bin and just compare the addresses that are occupied with orbits (Muinonen *et al.* 2004). In practice, the address is an integer $i \in \mathbb{N}_+$, transformed from an array of elements $p \in \mathbb{R}^n$ (here, $n = 6$) using a transformation algorithm f, i.e., $i = f(p, ...)$. The transformation algorithm f essentially does the same as a basic binning algorithm, but instead of returning the coordinates of a bin in multidimensional space (a bi-product of the algorithm), it transforms the coordinates to a single integer i. The integer is the individual ID-number of a bin in the original multidimensional bin-network. Besides p, the essential input values for the transformation algorithm f are the boundary values of the multidimensional space and the bin sizes. At present we take into account the whole sky, so the only essential input value is the bin size, which is currently one arcmin for both R.A. and Dec.

Because the observations of an object are inverted to a sample of orbits, and every orbit in the sample is transformed to three ephemerides and further to a value i, each object will get a one-dimensional array containing the i-values. Potential identifications are sought by comparing the i-arrays of objects in the first set with the i-arrays of objects in the second set. The search can move to the next candidate pair as soon as a single pair of equal addresses, or integers, is found.

When dealing with an array of integers, the search algorithm can be optimised more easily than when searching a multidimensional array. By sorting the i-values in ascending order, a *binary search* algorithm can be used for the search of similar elements, which significantly accelerates the comparison algorithm.

5. Application to VLT observations

Observations with the VLT and the Canada-France-Hawaii Telescope were obtained in January 2004 to provide ground-based follow-up for the Spitzer First Look Survey Ecliptic Plane Component (FLS/EPC). See Meadows *et al.* (2004) for a description of, and preliminary results from, this Spitzer program. The results discussed here, using only the VLT data, represent the first step in applying this method to obtaining orbits for unknown asteroids observed in the Spitzer FLS/EPC. The requirement for the Spitzer observations to be made at a solar elongation of $115°$ considerably complicates the identifications because asteroids along the line of sight are close to their turning points and are consequently moving slowly and along curved paths. The observational set contains 532 detections of asteroids at $V \lesssim 26^{\mathrm{m}}$ unevenly spread over five nights (Fig. 1). An estimated, rather pessimistic, accuracy of $\sigma = 0.5''$ was used in the identification procedure.

The observations were provided as single observations, not pairs of observations per object as is usual, and the first objective was therefore to create tentative objects within each nightly set. This was accomplished by assuming linear motion within the few hours of observation each night. A reasonable upper limit for the coordinate motion was found by requiring that the object can be found within the observed area on two consecutive nights. If the true motion for an object is higher than the given limit, it could not be linked, and can therefore be omitted from the set of objects to be scanned for linkages. The maximum motion that a linkable object could have is thus $1.5°/\mathrm{d} = 0.0625''/\mathrm{s} \approx 0.1''/\mathrm{s}$, deriving from the width of the observational area. Given the upper limit of motion, all possible pairs of observations were generated for each nightly set. Additional observations for each observation pair were also searched for by assuming linear motion and a maximum deviation of 6-σ from the nominal position. Up to two additional observations could be found, resulting in a maximum of four observations per object per night.

Linkages between nights were searched using a cumulative strategy: the first-night objects were linked with the second-night objects, and the preliminary identifications as well as all individual objects from both the first and the second nights were linked with the third-night objects, and so on. It turned out that ambiguous linkages occurred even if there were three nights of observations for two objects (see Section 5.1). Identifications containing four or five nights of observations were, however, self-consistent; i.e., there was no overlapping of observations between different identifications.

Including ambiguous linkages, a total of 73 preliminary linkages consisting of different combinations of 429 detections were found by using the new method. The remaining 103 detections have not yet been analysed in detail, but it seems plausible, from the experience gained with simulated observations, that most of them were either too fast-moving or too faint to be detected on two separate nights.

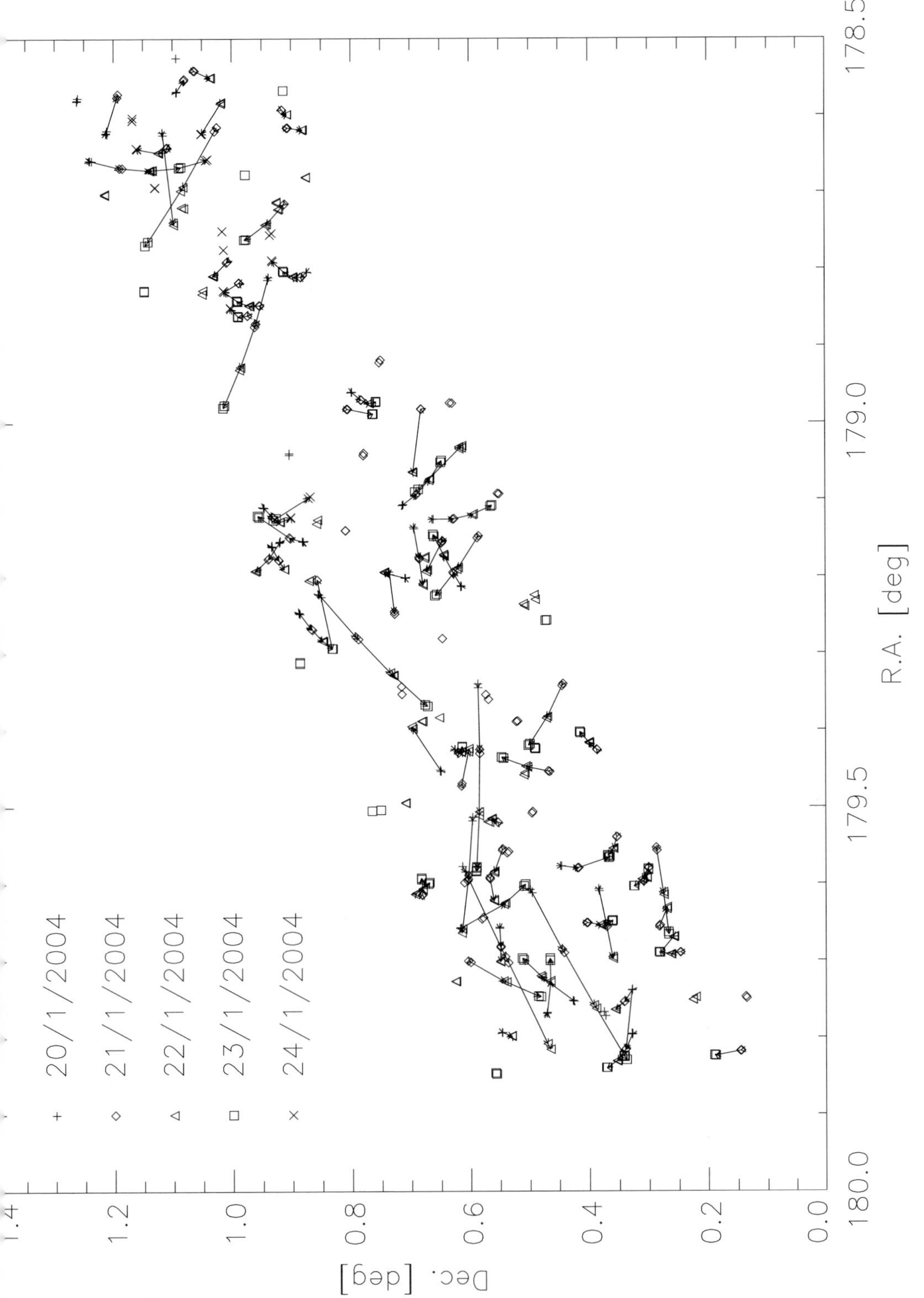

Figure 1: The figure shows 532 asteroid detections at $V \lesssim 26^{m}$ spread over five nights obtained with the VLT in January 2004. 76 preliminary inter-night identifications consisting of different combinations of 429 detections are shown with arrows pointing in the direction of motion.

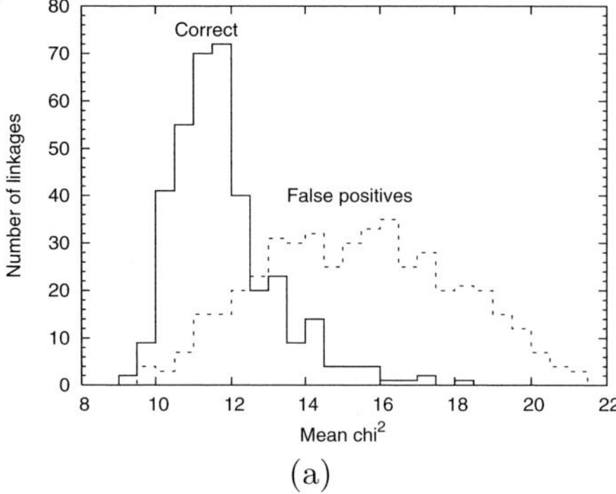
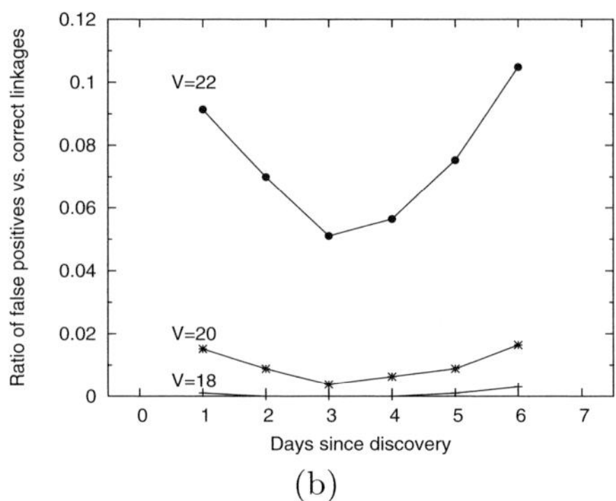

Figure 2: Results relating to the false positives problem obtained from simulated, opposition-centered NEO and MBO observations. (2a) The mean χ^2-values of sample orbits are similar for both correct linkages and false positives, which makes them useless when deciding individually between multiple tentative linkages. (2b) Assuming the discovery night and the night one week after discovery being fixed, the figure shows the ratio of false positives vs correct linkages as a function of the epoch of the middle observations after all three nights have been processed with the identification software. It is evident, that the choice of observation strategy becomes more important as the limiting magnitude increases.

5.1. *Minimising false positives*

To examine more closely the problem of *false positives*, that is, the tentative linkages for which orbits can be found even though the observations actually belong to different objects, we generated simulated observations of NEOs and MBOs according to the following scheme. Two nightly, geocentric snapshots of the same region of the sky for three different limiting magnitudes (18, 20, and 22) were generated during eight sequential nights, resulting in 1100 objects on average per limiting magnitude. The direction of opposition at the discovery epoch was chosen to be the center of the observation window. The time interval between two nightly snapshots was approximately one hour, which is roughly the same as that of current search programs. Random Gaussian noise ($\sigma = 0.5''$) was also added to the observations.

The first issue was to find out whether there is a difference in the goodness of fit between correct and erroneous linkages. According to our results so far, it seems like there would not necessarily exist a statistical measure to decide individually (that is, one linkage at a time) between the two when the observational arcs are short and the numbers of observations are small in both cases. As seen in Figure 2a, the mean χ^2-values of sample orbits are similar for both correct linkages and false positives. These results were obtained for the worst scenario case at limiting magnitude 22 and a time span of one week between the observational sets.

Examination of the effect of observing strategy on the number of erroneous linkages was the next logical step. Assuming that the observations at discovery and a week from discovery are fixed in time, we asked ourselves: What would be the optimum epoch for the middle observations so as to minimise the number of false linkages when all three nights have been linked? In this study we used the cumulative strategy, with the exception that only preliminary linkages between the first and second sets were processed with the third set. According to the results seen in Figure 2b, the observing strategy promoted by the

MPC (observations on the first, second, and eighth night) would not be the best choice for identification purposes. It seems like the strategy used by automatic surveys (equally distributed nights) would get better than the MPC strategy as the limiting magnitude increases. It should be stressed that the survey strategy is efficient only if large regions of the sky can be covered within a reasonable time. Otherwise a substantial number of objects will escape detection due to fast motion, which would not be the case when using the MPC strategy. It should also be noted that a single set (per limiting magnitude) of simulated observations obtained with a statistical tool is not statistically significant. It is, however, possible to draw preliminary conclusions of the results, as the outcome at all three limiting magnitudes point in the same direction.

6. Conclusions

The present method, and particularly the ephemerides address comparison, has been succesfully tested with both VLT observations and simulated observations of near-Earth and main-belt objects.

A relevant issue, which was brought up by A. Milani, is to find out how the whole identification method and, particularly, the address comparison technique, scales when the number of objects is increased by two, or even three, orders of magnitude. Making the address comparisons between two sets containing $1,000$ objects each takes a few minutes or less with current single-processor workstations. Assuming quadratical scaling, only the address comparison of sets containing $100,000$ objects each would thus require three weeks of cpu-time on a workstation. Fortunately, the present method is easy to parallelise on several levels starting from orbital inversion and address comparison all the way to target (that is, asteroid) management. The scaling problems should therefore not be insuperable.

The ultimate goal of this research is to produce a real-time asteroid identification tool for ESA's astrometric space observatory Gaia, LONEOS, the Near-Earth Space Surveillance mission, and the Nordic Near-Earth Object Network. The tool should also prove useful to large-scale surveys connected to telescopes such as the Large Synoptic Survey Telescope and the Discovery Channel Telescope.

Acknowledgements

The authors wish to thank Ted Bowell and the anonymous reviewer for many valuable comments on the manuscript. The research is supported by the Univ. Helsinki three-year research grant and by the Academy of Finland. MG is grateful to Magnus Ehrnrooth Foundation and Univ. Helsinki Chancellor for supporting his visit to the Lowell Observatory.

References

Meadows, V.S., Bhattacharya, B., Reach, W.T., Grillmair, C., Noriega-Crespo, A., Ryan, E.L., Tyler, S.R., Rebull, L.M., Giorgini, J.D. & Elliot, J.L. 2004, *Astrophys. J. Suppl. Ser.* 154, 469

Milani, A., La Spina, A., Sansaturio, M.E. & Chesley, S.R. 2000, *Icarus* 144, 39

Milani, A., Sansaturio, M.E. & Chesley, S.R. 2001, *Icarus* 151, 150

Muinonen, K. & Bowell, E. 1993, *Icarus* 104, 255

Muinonen, K., Bowell, E. & Wasserman, L.H. 1994, *Planet. Space Sci.* 42, 307

Muinonen, K., Virtanen, J., Granvik, M. & Laakso, T. 2004, in *Three-Dimensional Universe with Gaia*, ESA Special Publications SP-576
Virtanen, J., Muinonen, K. & Bowell, E. 2001, *Icarus* 154, 412

Dynamics of Populations of Planetary Systems
Proceedings IAU Colloquium No. 197, 2005
Z. Knežević and A. Milani, eds.

© 2005 International Astronomical Union
DOI: 10.1017/S1743921304008713

Collision orbits and phase transition for 2004 AS$_1$ at discovery

Jenni Virtanen, K. Muinonen,
M. Granvik and T. Laakso

Observatory, University of Helsinki, P.O.Box 14, FIN-00014 Helsinki, Finland
email: jenni.virtanen@helsinki.fi

Abstract. We evaluate asteroid orbital uncertainties from the discovery night onwards using 6D orbit computation tools based on statistical techniques. In particular, we outline a new nonlinear Monte Carlo technique of phase-space sampling that helps us in assessing the nonlinear phase transition from extended orbital-element distributions to well-constrained ones as the observational arc and number of observations grows. We apply the statistical techniques for near-Earth asteroid 2004 AS$_1$ to examine the time evolution of the orbital uncertainties and to assess the asteroid impact risk immediately after discovery. We start with the technique of statistical ranging for exiguous data, continue with the phase-space sampling technique for moderate data, and conclude with the standard least-squares fit for extensive data.

Keywords. Celestial mechanics, methods: statistical, minor planets, asteroids.

1. Introduction

For more than a decade, impacts of asteroids have been recognised as a significant threat to humankind, comparable to other major natural disasters. Due to the active monitoring of the impact risk, several new objects having non-zero probabilities for an Earth impact within the next century are identified each year. The realization of this cosmic threat has motivated the development of several theoretical and computational tools for relatively short-term collision probability assessment (over timescales of decades). The backbone of collision probability computations is the assessment of orbital uncertainty of the given object. The orbital uncertainty evolves rapidly after an object has been discovered, typically decreasing when new observations roll in and increasing when follow-up observations have failed. Thus also the asteroid collision probability evolves when the observational arc and the number of observations grows and, in fact, for most of the discovered impactor candidates the probabilities for an Earth impact have been shown to vanish after days or weeks of observations after the discovery.

In Muinonen and Virtanen (2002), a nonlinear collapse was seen in the orbital uncertainties as a function of the improving accuracy of astrometric observations. This phase transition effect can also be recognized in the time evolution of the orbital uncertainty: there is a threshold for the length of the observational arc (and the number of observations) over which the orbital-element distributions nonlinearly evolve from extended to well-constrained ones.

Most of the theoretical and computational work for collision probability assessment, such as that by Yeomans and Chodas (1994) and Chodas and Yeomans (1996, 1999), is however bounded by the use of linear approximation which has been shown to fail for short observational arcs. Milani and Valsecchi (1999) and Milani et al. (1999; 2000a, b, c) have made use of both linear and semilinear methods for the propagation of orbital uncertainty

239

(Milani 1999). They illustrate the probability of a collision with virtual impactors (for a review paper, see Milani 2004, this Proceedings).

We attack the problem of impact risk assessment at the very moment of discovery. The ill-posedness of this problem stems from the inverse problem of computing the orbit from a small number of observations covering a short observational arc. With statistical orbit computation techniques, the probability density of the orbital parameters, and in turn the collision probability, can be evaluated rigorously even for exiguous data, although the strict probabilistic interpretation of inversion results may be in doubt. Information on the orbital characteristics of the object, such as the minimum encounter distance with the planets in the near future, can nevertheless be extracted. To stress this ambivalence, in this paper we use the term "collision probability" only in connection to purely theoretical work, for practical purposes we replace it with the term "collision risk".

We provide a continuum of statistical orbit computation techniques that can be applied sequentially accounting for the phase transition effect. In particular, we outline an innovative 6D technique for asteroids with moderate time arcs and/or moderate numbers of observations. The technique of sampling in phase-space volume of variation makes use of local linear approximations by varying one or more orbital elements and fitting the remaining ones. The use of the curve (or line) of variation in evaluation of orbital uncertainties has already been brought up by Bowell *et al.* (1993) and Milani (1999), and has thereafter been used extensively by Milani in several application, such as ephemeris prediction, asteroid linkage, and collision probability assessment. We generalize the concept from a line to a volume of variation, and introduce Monte Carlo sampling in the phase-space volume for a fully nonlinear treatment.

We apply the orbit computation scheme to the case of 2004 AS1, an NEO discovered in early 2004 having a significant risk for a collision with the Earth some 48 hours following the discovery moment. While the risk of the immediate impact was removed with observations from the second night, the extreme nature of the first prediction calls for a detailed analysis.

2. Orbit computation

2.1. *Inverse problem*

We denote the osculating orbital elements of an asteroid at a given epoch t_0 by $\boldsymbol{P} = (a, e, i, \Omega, \omega, M_0)^T$ (T is transpose). The elements are, respectively, the semimajor axis, eccentricity, inclination, longitude of ascending node, argument of perihelion, and mean anomaly. The three angular elements i, Ω, and ω are currently referred to the ecliptic at equinox J2000.0.

The orbital-element probability density function (p.d.f.) p_{p} is proportional to the a priori (p_{pr}) and observational error (p_ϵ) p.d.f.'s, the latter being evaluated for the sky-plane ("O-C") residuals $\Delta\psi(\boldsymbol{P})$ (Muinonen and Bowell 1993),

$$p_{\mathrm{p}}(\boldsymbol{P}) \propto p_{\mathrm{pr}}(\boldsymbol{P})p_\epsilon(\Delta\psi(\boldsymbol{P})), \tag{2.1}$$

where p_ϵ can usually be assumed to be Gaussian. For the mathematical form of p_{p} to be invariant in transformations from one orbital element set to another (e.g., from Keplerian to equinoctial or Cartesian), we regularize the statistical analysis by Jeffreys' noninformative a priori p.d.f. (Jeffreys 1946, Muinonen *et al.* 2001),

$$\begin{aligned} p_{\mathrm{pr}}(\boldsymbol{P}) &\propto \sqrt{\det \Sigma^{-1}(\boldsymbol{P})}, \\ \Sigma^{-1}(\boldsymbol{P}) &= \Phi(\boldsymbol{P})^T \Lambda^{-1} \Phi(\boldsymbol{P}), \end{aligned} \tag{2.2}$$

where Σ^{-1} is the information matrix (or the inverse covariance matrix) evaluated for the orbital elements $\boldsymbol{P}$, Φ contains the partial derivatives of right ascension (R.A.) and declination (Dec.) with respect to the orbital elements, and Λ is the covariance matrix for the observational errors. By the choice of the a priori p.d.f., the transformation of rigorous p.d.f.'s becomes analogous to that of Gaussian p.d.f.'s.

The final a posteriori orbital-element p.d.f. is

$$p_{\mathrm{p}}(\boldsymbol{P}) \propto \sqrt{\det \Sigma^{-1}(\boldsymbol{P})} \, \exp\left[-\frac{1}{2}\chi^2(\boldsymbol{P})\right],$$

$$\chi^2(\boldsymbol{P}) = \Delta\boldsymbol{\psi}^T(\boldsymbol{P})\Lambda^{-1}\Delta\boldsymbol{\psi}(\boldsymbol{P}). \tag{2.3}$$

As a consequence of securing the invariance in orbital-element transformations, e.g., ephemeris uncertainties and collision probabilities based on the orbital-element p.d.f. are independent of the choice of the orbital element set. Note that assuming constant p_{pr} is acceptable, when the exponent part of Eq. (2.3) confines the p.d.f. into a phase space regime, where the determinant part reduces to a constant (typical for well-constrained p.d.f.'s, i.e., long observational arcs). Note also that a predominating role of the a priori p.d.f. casts doubts upon the reliability of the probabilistic interpretation.

2.2. *Volume-of-variation sampling*

The a posteriori p.d.f. in Eq. (2.3) allows the derivation of a local linear approximation in the orbital-element phase space (Muinonen *et al.* 2004). We can select one or more elements as "the elements to be varied" systematically and derive a linear approximation for "the remaining elements to be fitted". For simplicity, we illustrate the local linear approximations below in the case of a single mapping element and note that the formulation is analogous for more numerous mapping elements.

We rewrite the a posteriori p.d.f. in Eq. (2.3) explicitly in terms of the mapping element P_{m} and the five remaining elements $\boldsymbol{P}'$ (here and below, the prime denotes five-dimensional quantities),

$$p_{\mathrm{p}}(P_{\mathrm{m}}, \boldsymbol{P}') \propto \sqrt{\det \Sigma^{-1}(P_{\mathrm{m}}, \boldsymbol{P}')} \cdot \exp\left[-\frac{1}{2}\chi^2(P_{\mathrm{m}}, \boldsymbol{P}')\right],$$

$$\chi^2(P_{\mathrm{m}}, \boldsymbol{P}') = \Delta\boldsymbol{\psi}^T(P_{\mathrm{m}}, \boldsymbol{P}')\Lambda^{-1}\Delta\boldsymbol{\psi}(P_{\mathrm{m}}, \boldsymbol{P}'). \tag{2.4}$$

For a given P_{m}, we define the local linear approximation as follows,

$$p_{\mathrm{p}}(P_{\mathrm{m}}, \boldsymbol{P}') \propto \sqrt{\det \Sigma^{-1}(P_{\mathrm{m}}, \boldsymbol{P}'_{\mathrm{ls}})} \cdot \exp\left[-\frac{1}{2}\chi^2(P_{\mathrm{m}}, \boldsymbol{P}'_{\mathrm{ls}})\right] \cdot$$

$$\exp\left[-\frac{1}{2}\Delta\boldsymbol{P}'^T\Sigma'^{-1}(P_{\mathrm{m}}, \boldsymbol{P}'_{\mathrm{ls}})\Delta\boldsymbol{P}'\right],$$

$$\Delta\boldsymbol{P}' = \boldsymbol{P}' - \boldsymbol{P}'_{\mathrm{ls}}(P_{\mathrm{m}}), \tag{2.5}$$

where $\boldsymbol{P}'_{\mathrm{ls}} = \boldsymbol{P}'_{\mathrm{ls}}(P_{\mathrm{m}})$ is the local least-squares solution for the elements $\boldsymbol{P}'$. The local 5×5 covariance matrix Σ' is recomputed at the local least-squares solution P' and defines a hyperellipsoid centered at these elements. Note that both Σ^{-1} and Σ'^{-1} enter the local linear approximations above. The sequence of orbital elements $P_{\mathrm{m}}, \boldsymbol{P}'_{\mathrm{ls}}(P_{\mathrm{m}})$ defines the line of variation in the orbital-element phase space. The local covariance matrix Σ' defines a hyperellipsoid centered at the local least-squares orbital elements,

$$\Delta\chi^2(\boldsymbol{P}') = \Delta\boldsymbol{P}'^T\Sigma'^{-1}\Delta\boldsymbol{P}' \;\; = \;\; \Delta\chi_0^2, \tag{2.6}$$

where $\Delta\chi_0^2$ is a constant. The boundaries of, for example, the commonly used 68.3 % or 95.4 % -probability hyperellipsoids are $\Delta\chi_0^2 \approx 5.89$ or $\Delta\chi_0^2 \approx 11.3$, respectively.

It is convenient to express the differences $\Delta \boldsymbol{P}'$ in terms of the standard deviations $\sigma_j' = \sqrt{\Sigma_{jj}'}$ $(j = 1, \ldots, 5)$ and to utilize the dimensionless correlation matrix C'; with the help of the diagonal standard deviation matrix S',

$$\Delta \boldsymbol{Q}' = S'^{-1} \Delta \boldsymbol{P}',$$
$$C' = S'^{-1} \Sigma' S'^{-1},$$
$$S_{jk}' = \sigma_j' \, \delta_{jk}, \quad j, k = 1, \ldots, 6, \tag{2.7}$$

where δ_{jk} is the Kronecker symbol. The hyperellipsoid is thus defined by

$$\Delta \boldsymbol{Q}'^T C'^{-1} \Delta \boldsymbol{Q}' = \Delta \chi_0^2, \tag{2.8}$$

where all the parameters are dimensionless. The eigenvalues λ_j' $(j = 1, \ldots, 5)$ for the correlation matrix C' are normalized variances along the principal axes of the hyperellipsoid, the directions of the axes being given by the orthonormal eigenvectors $\boldsymbol{X}_j'$,

$$C' \boldsymbol{X}_j' = \lambda_j' \boldsymbol{X}_j', \quad j = 1, \ldots, 5. \tag{2.9}$$

Since C' is a real and symmetric matrix, the eigenproblem is readily solved via Jacobi transformations.

It is our goal to draw sample orbits from the rigorous orbital-element p.d.f. with the help of the local linear approximations. First, we specify the variation interval for the mapping element with the help of the covariance matrix Σ derived in the global linear approximation and emphasize that the variation interval must be subject to iteration. For example, one may utilize the one-dimensional 3σ variation interval as given by the linear approximation so that

$$P_{\mathrm{m}} \in [P_{\mathrm{m,ls}} - 3\sigma_{\mathrm{m}}, P_{\mathrm{m,ls}} + 3\sigma_{\mathrm{m}}], \tag{2.10}$$

where $P_{\mathrm{m,ls}}$ is the global least-squares value for the mapping element. Second, the remaining elements are sampled with the help of the local intervals of variation so that

$$\boldsymbol{P}' = \boldsymbol{P}_{\mathrm{ls}}'(P_{\mathrm{m}}) + \sum_{j=1}^{5} (1 - 2r_j) \cdot \sqrt{\Delta \tilde{\chi}^2 \lambda_j'(P_{\mathrm{m}})} S'(P_{\mathrm{m,ls}}) \boldsymbol{X}_j'(P_{\mathrm{m}}), \tag{2.11}$$

where $r_j \in (0, 1)$ $(j = 1, \ldots, 5)$ are independent uniform random deviates and $\Delta \tilde{\chi}^2$ is a scaling parameter to be iterated so that the entire orbit solution space is covered and the final results have converged. Initially, one may start with $\Delta \tilde{\chi}^2 = 11.3$ and slowly increase its value. $S'(P_{\mathrm{m,ls}})$ designates the single standard deviation matrix used throughout the interval of the mapping parameter, which allows a straightforward debiasing of the sample orbits at the end of the computation.

In Eq. (2.11), we sample the local phase-space volume using the principal-axis directions following from the local linear approximation, after diagonalization by the solution of the eigenproblem in the units specified by the S' matrix. In the present context, the shape of the local sampling volume is that of a five-dimensional rectangular parallelepiped.

In practical computations, we need to discretize the interval of the mapping element and, after solving the five-dimensional local least-squares problem, interpolate the interval parameters for Eq. (2.11). Once the entire variation-interval map is available across the interval of the mapping parameter, trial orbits are generated in a straightforward way. First, a value for the mapping orbital element is obtained from uniform sampling over the mapping interval. Second, the remaining five elements are generated by interpolating their variation intervals based on the precomputed map. Third, the trial orbit qualifies for a sample orbit if it produces an acceptable fit to the observations. Each sample orbit is accompanied by the weight factor

$$w(P_\mathrm{m}, \boldsymbol{P}') \propto \sqrt{\det \Sigma^{-1}(P_\mathrm{m}, \boldsymbol{P}')} \cdot \exp\left[-\frac{1}{2}\chi^2(P_\mathrm{m}, \boldsymbol{P}')\right] \cdot$$

$$\sqrt{\frac{\lambda_1(P_\mathrm{m}) \cdot \ldots \cdot \lambda_5(P_\mathrm{m})}{\lambda_1(P_\mathrm{m,ls}) \cdot \ldots \cdot \lambda_5(P_\mathrm{m,ls})}}. \tag{2.12}$$

2.3. *Statistical orbital ranging*

For initial orbit computation, we make use of Ranging (Virtanen *et al.* 2001, Muinonen *et al.* 2001). In Ranging, two observation dates (here A and B) are chosen from the complete observation set. The corresponding topocentric distances (or ranges ρ_A and ρ_B), as well as the R.A. (α_A and α_B) and Dec. (δ_A and δ_B) angles are MC sampled using intervals subject to iteration, resulting in altogether 12 interval boundary parameters. Explicitly,

$$\begin{cases} \rho_\mathrm{A} & \in \quad [\rho_\mathrm{A}^-, \rho_\mathrm{A}^+], \\ \alpha_\mathrm{A} & \in \quad [\alpha_\mathrm{A}^-, \alpha_\mathrm{A}^+], \\ \delta_\mathrm{A} & \in \quad [\delta_\mathrm{A}^-, \delta_\mathrm{A}^+], \end{cases} \qquad \begin{cases} \rho_\mathrm{B} & \in \quad [\rho_\mathrm{A} + \rho_\mathrm{B}^-, \rho_\mathrm{A} + \rho_\mathrm{B}^+], \\ \alpha_\mathrm{B} & \in \quad [\alpha_\mathrm{A}^-, \alpha_\mathrm{A}^+], \\ \delta_\mathrm{B} & \in \quad [\delta_\mathrm{A}^-, \delta_\mathrm{A}^+]. \end{cases} \tag{2.13}$$

Note that it is computationally efficient to generate ρ_B based on ρ_A generated at an earlier stage. The boundary values $\rho_\mathrm{B}^\pm$ must be carefully chosen so as to secure the coverage of the entire relevant interval in ρ_B. Once the two sets of spherical coordinates have been generated, the two corresponding Cartesian positions $(X_\mathrm{A}, Y_\mathrm{A}, Z_\mathrm{A})^T$ and $(X_\mathrm{B}, Y_\mathrm{B}, Z_\mathrm{B})^T$ lead to an unambiguous set of orbital elements $\boldsymbol{P}$, based on well-established techniques in celestial mechanics.

The set of trial orbital elements $\boldsymbol{P}$ is included in the set of sample orbital elements if and only if it produces an acceptable fit to the entire set of observations, that is, with the help of Eq. (2.3),

$$\exp\left[-\frac{1}{2}(\chi^2(\boldsymbol{P}) - \chi^2(\boldsymbol{P}_\mathrm{ref})) + \right.$$

$$\left. \ln \sqrt{\det \Sigma^{-1}(\boldsymbol{P})} - \ln \sqrt{\det \Sigma^{-1}(\boldsymbol{P}_\mathrm{ref})}\right] \geqslant c_\mathrm{min}, \tag{2.14}$$

where c_min is the level of acceptance and $\boldsymbol{P}_\mathrm{ref}$ refers to the best-fit orbital solution available, constantly updated during the iterative computation. The acceptance criterion thus becomes analogous to the $\Delta\chi^2$ criterion for Gaussian p.d.f.'s (Eq. 2.6).

In order to establish the uniform sampling of orbital elements $\boldsymbol{P}$, each set of sample orbital elements is weighted by the Jacobian of the transformation from "the phase space" of the two spherical positions $(\rho_\mathrm{A}, \alpha_\mathrm{A}, \delta_\mathrm{A})^T$ and $(\rho_\mathrm{B}, \alpha_\mathrm{B}, \delta_\mathrm{B})^T$ to the phase space of the orbital elements,

$$w(\boldsymbol{P}) \propto \det\left[\frac{\partial(\rho_\mathrm{A}, \alpha_\mathrm{A}, \delta_\mathrm{A}; \rho_\mathrm{B}, \alpha_\mathrm{B}, \delta_\mathrm{B})}{\partial \boldsymbol{P}}\right]^{-1}. \tag{2.15}$$

Ranging is repeated to obtain a large number of sample orbits, simultaneously iterating the 12 interval boundary parameters and updating the reference orbital elements in order to secure the coverage of the full orbit solution space.

2.4. *Collision probability using 6D orbital-element p.d.f.'s*

Muinonen (1999) and Muinonen *et al.* (2001) described a set of techniques for the computation of the collision probability for Earth-crossing asteroids. Their formulation of the collision probability computation relies on the Bayesian a posteriori probability density function p_{p} of the orbital elements $\boldsymbol{P}$, from which the collision probability follows through the integral:

$$P_{\mathrm{c}}(\tau) = \int_{C(\tau)} d\boldsymbol{P}\, p_{\mathrm{p}}(\boldsymbol{P}, t_0), \qquad (2.16)$$

where the phase space domain $C(\tau)$ contains the elements leading to a collision on the interval τ.

In particular, Muinonen *et al.* introduced new techniques for computing the collision probability for the case of short observational arcs and/or small numbers of observations. The orbital-element p.d.f. can in this case be very complicated, and it must be assessed rigorously using Monte Carlo simulations for large numbers of sample orbits. Once the extent of the orbital-element probability density has been mapped by generating a large, unbiased set of sample orbits, these orbits can be propagated through the assumed impact interval τ to derive the weighted fraction of collision orbits, i.e., the collision probability.

Even if none of the sample orbits lead to collision within the impact interval, the probability for a collision within the interval is not necessarily zero. This can simply mean that the sampling of the phase space for acceptable orbital solutions has not been dense enough. Either because, at an epoch close to the observations this 6D phase-space region is extremely large, as it is for short observational arcs, and it cannot be efficiently mapped keeping the sample size computationally reasonable. Or, if the time scale of propagation is very long, repeated close-approaches introduce chaotic effects causing the extent of an initially small phase-space volume (computed close to the epoch of the observations) to widen so that the sampling of the impact parameter space is, again, too sparse. In both cases, the result is that the impacting orbit solutions are missed, because they are typically confined to very narrow regions in the phase space.

The above limitations of the direct propagation of the orbital-element p.d.f. through the impact interval can be solved by changing from discretized p.d.f.'s to continuous functions using intervals of collision orbits (Muinonen *et al.*). In this method, intervals of acceptable orbits are computed in the vicinity of each sample orbital element set, and within these intervals a sub-interval is mapped including orbits that result in collision in the given time interval. The limitation of this method is that the computational load can easily become excessive when the unbiased orbital intervals are searched.

For robust and rapid evaluation of collision probability, Muinonen et al. provided upper bounds in terms of maximum likelihood (ML) collision orbit. This can be determined once collision orbits have been found in the neighborhood of the orbit maximizing the orbital-element p.d.f., i.e., around the maximum likelihood orbit. The ML collision orbit is then the collision orbit that lies probabilistically closest to the ML orbit. However, the upper bound values have been shown to give only loose bounds to the collision probability, so more rigorous means should be used whenever the more severe computational load can be handled.

3. Results and discussion

The orbital-element p.d.f. is here assessed using the two techniques described in the previous section. Ranging has proven to be a practical way to assess the probability density for objects with very short observational arcs, while the new VOV sampling has turned out useful for the moderately observed ones. For well-observed objects, we use the standard least-squares covariances to estimate the widths of the orbital-element p.d.f.'s.

For collision risk assessment, we applied the most simple approach described in the previous section: the direct propagation of the orbital-element p.d.f. through the predicted impact interval. In what follows, we will see that this turned out to be the first time that the collision risk could indeed be estimated directly using sample collision orbits. During the integrations, we also monitor for (temporary) satellite captures and Earth swing-by's to hyperbolic orbits as well as impacts to the Moon.

3.1. *2004 AS$_1$*

The Apollo asteroid 2004 AS$_1$ grabbed the attention of asteroid scientists for several days in January 2004 when, based on single-night data from only one observatory, this asteroid appeared to be on a collision course with the Earth with the timing of the predicted impact only about 48 hours after the discovery. Although the risk of immediate impact was ruled out as soon as the follow-up observations from the second night came in, the extreme nature of the first prediction called for a detailed analysis. First, we present the results for orbit computations, then we return to the analysis of the actual collision probability.

3.1.1. *Phase transition*

To locate the phase transition regime, we carried through the following orbit computation scheme using methods described in Section 2. As already implied by the preliminary study in Muinonen and Virtanen (2002) in the case of improving observational accuracy, the occurence of this nonlinear collapse in the orbital distributions is most likely highly case-sensitive. Thus, pinpointing its exact location for each object requires breaking down of the object's observational data night by night and applying the different statistical methods to the data in a sequential manner.

For 2004 AS$_1$, we considered the observational data for the first month after discovery which we divided into the following seven cases in the order of increasing time arc: 0.04 ($\sim$ 1 hr), 1.1, 1.5, 3.0, 5.0, 8.0, and 30.0 days. Ranging was applied to the five first cases, i.e., up to 5 days of observations, VOV-sampling from the third arc onwards (1.5 days) and least-squares fit from 1.1 days onwards. The resulting orbital-element p.d.f.'s are compared in Figure 1 where we plot for the semimajor axis the standard deviations of the marginal p.d.f.'s obtained with the different techniques.

The phase transition regime extents from the discovery night (1 hr arc) to the second night of observations, with the most abrupt collapse taking place between arcs of 1.1 and 1.5 days. Across this regime, the orbital uncertainties decrease by several orders of magnitudes for all the elements, but the improvement is most significant for the semimajor axis, close to a factor of 10^5. It is notable that for this NEO, for the semimajor axis, the uncertainties based on the linear technique, i.e., the least-squares covariances, are in good agreement with the ones from rigorous nonlinear techniques.

3.1.2. *Collision risk at discovery*

The discovery night observations of 2004 AS$_1$ consisted of four observations over 1 hour from a single observatory. The first published ephemerides by the Minor Planet Center (MPC) showed a rapid increase in the apparent brightness and indicated that this asteroid would indeed be passing within the Earth radius some 48 hours later. Although the orbit solution behind this prediction was quickly replaced by another equally acceptable solution at the MPC (resulting in ephemerides that no more passed through the planet), it was not before the new observations from the second night came in that the possibility for an immediate (and future) impact could be overruled. Further observations in the following months finally showed that the accuracy of the four initial observations was

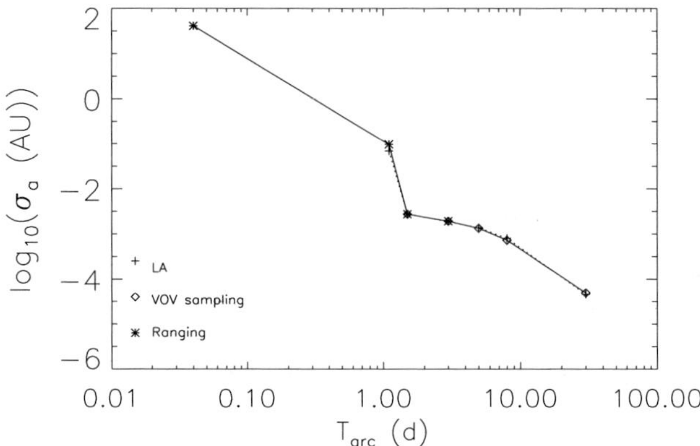

Figure 1: Phase transition in the accuracy of the semimajor axis as a function of the increasing observational arc for $2004\,\mathrm{AS}_1$ (obs. accuracy $\sim 0.5\,\mathrm{arcsec}$). Note the logarithmic scale on the vertical axis. LA stands for linear approximation, i.e., 6D covariances of the least-squares solutions.

rather poor (rms-values of 1.4 and 1.3 arcseconds for R.A. and Dec.) and that one observation was clearly an outlier (residuals of 2.4 and 1.9 arcseconds for R.A. and Dec.). The overall evolution of this case raised several questions: what was the effect of discordant observations in collision probability assessment? Can outliers be recognized from very short arcs? How did the collision probability estimates evolve as the observational arc increased?

We used the Ranging technique to map the 6D orbital element p.d.f. at discovery with 50,000 sample orbits. As the first orbit solution used by the MPC led to a collision, it seemed a priori evident that collision orbits would be among the best-fit orbit solutions. The collision risk was thus estimated by deriving the impact parameter distribution by propagating the 50,000 orbits through the iterated impact interval of January 14 to 17, 2004. We first made a conservative assumption of 1.0 arseconds for the observational noise. The collision risk is illustrated with the cumulative distribution function (c.d.f.) for the impact parameter in Figure 2, which confirms the drastic first prediction: for 1.0 arcsecond noise, the probability for an Earth impact was 12 %, several orders of magnitude higher than any of the previously estimated asteroid impact risks.

We then repeated the analysis (orbit and impact parameter computation) for different assumptions of the observational noise: 2.0, 1.5, 0.5, and 0.3 arcseconds. The evolution of collision risk as a function of the noise assumption is shown in Figure 3a: when the observations are assumed more accurate, the impact probability increases. The reason can be found by looking at the results from Ranging: assumption that the data was more accurate (less affected by noise) made the asteroid's plausible topocentric range interval to concentrate towards the Earth. This again shows that the best-fit orbits actually led to collisions within the predicted interval.

To study the evolution of collision risk as a function of time, we added one simulated observation to the data set: since the collision risk vanishes after the second night of observations, simulated observation was propagated between the first and second night. This study implies that a fifth observation 30 minutes after the last one in the original data set (increasing the time arc from 1 to 1.5 hr) could already have made the difference and removed the impact risk.

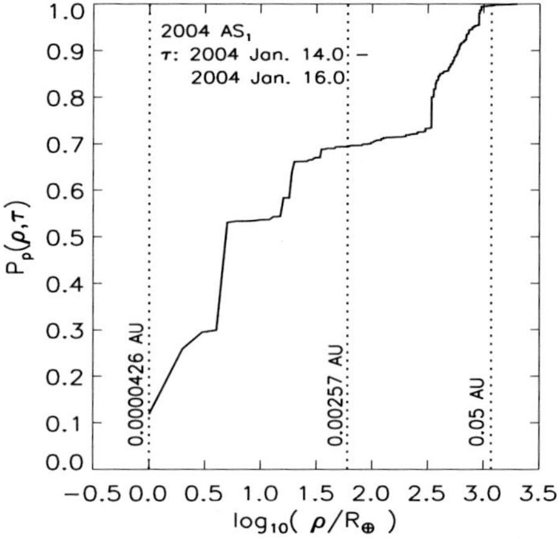 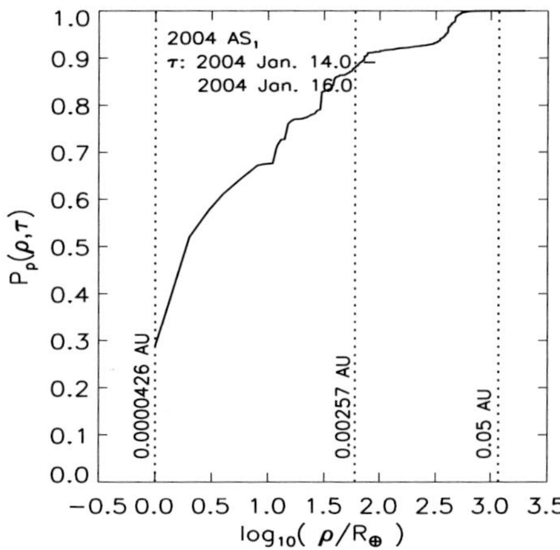

Figure 2: Collision risk with the Earth for 2004 AS$_1$ during January 14-17, 2004. Impact parameter c.d.f.'s have been computed with different assumption of the observational rms: 1.0 arcsec (left) and 0.5 arcsec (right). The estimates for the collision risk are 12% and 28%, respectively.

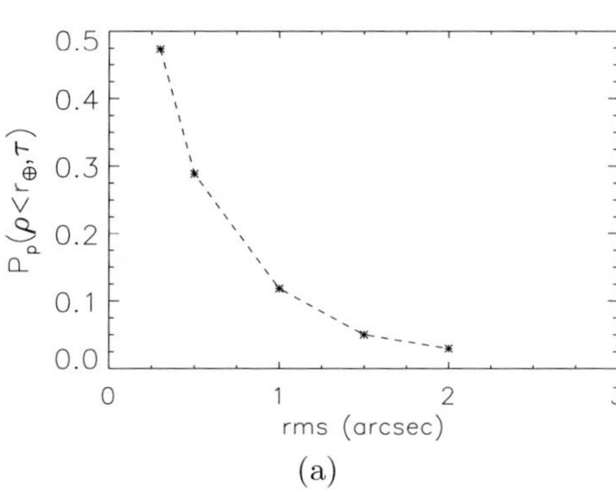

(a)

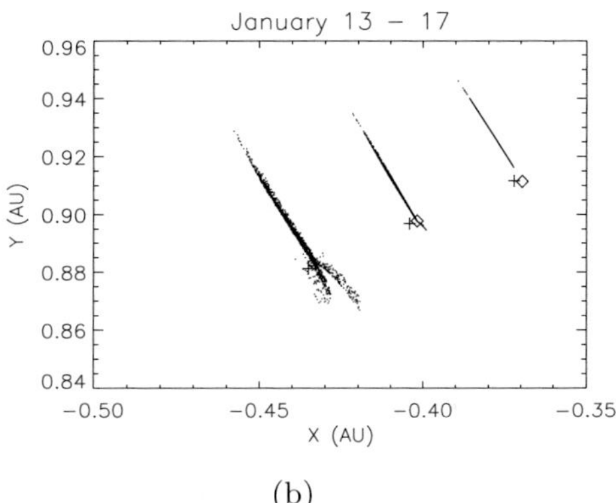

(b)

Figure 3: **a)** Evolution of the collision risk for 2004 AS$_1$ as a function of the assumed observational rms. **b)** Evolution of the positional uncertainty for 2004 AS$_1$ in the ecliptic plane during the assumed impact interval. Position distributions are shown for the following dates (from right to left): January 13th, 15th and 17th. Positions of Earth and Moon are shown with diamond and cross, respectively.

As to the effect of the discordant observations, first, the distributions of residuals from Ranging technique (cf. Fig. 7 in Bowell *et al.* 2002) did not reveal the outlier, although there were indications that the observations in general were not very accurate. Second, removing one observation at a time and repeating the above analysis for collision risk did not have major impact to the estimated values. Thus, the presence of discordant observations could not be deduced from the initial data.

Finally in Figure 3b, we show the highly nonlinear nature of the impact event with the evolution of the positional uncertainty over the impact interval. The close approaches to the Earth have scattered the position distribution of 2004 AS$_1$ significantly at the post-impact epoch of January 17, 2004 (left-most cloud of points).

4. Conclusions

We have applied our spectrum of statistical orbit computation techniques for $2004\,AS_1$ to study the evolution of orbital uncertainties as a function of time. We find that the nonlinear phase transition in the orbital p.d.f.'s for this asteroid takes place after 1.5 days of observations. We will continue to study the phase transition in more detail for different populations of minor planets, paying also attention to the case of high-precision astrometry obtained, e.g., from future space missions, such as the European Space Agency's GAIA space observatory. The succesful application of VOV technique and the consistency of the results with Ranging indicates that the new technique will be useful at the phase transition regime.

We confirm the extreme initial estimates for the collision risk of $2004\,AS_1$ for the time span of January 14-17. 2004. The risk of collision based solely on the discovery night data was indeed non-vanishing, estimates as high as 50% resulted from our analysis. However, we note that the strict probabilistic interpretation of both the inverse and prediction problem may no longer be valid for such exiguous data, and the probability values should be taken as loose estimates only. We investigate the use of regularizing a priori probability density further and present a more complete collision risk analysis for this case study in Virtanen *et al.* 2004.

Acknowledgements

The research is supported by the Finnish Graduate School in Astronomy and Space Physics and by the Academy of Finland.

References

Bowell, E., Wasserman, L.H. & Muinonen, K. 1993, *Bull. Am. Astron. Soc.* 25, 1118

Bowell, E., Virtanen, J., Muinonen, K. & Boattini, A. 2002, in: W.F. Bottke Jr., A. Cellino, P. Paolicchi & R.P. Binzel (eds.), *Asteroids III*, p. 27

Chodas, P.W. & Yeomans, D.K. 1996, in: K.S. Noll, H.A. Weaver & P.D. Feldman, (eds.), *The Collision of Comet Shoemaker-Levy 9 and Jupiter.* (Cambridge University Press)

Chodas, P.W. & Yeomans, D.K. 1999, in: 21st Annual AAS Guidance and Control Conference, Paper 99-002

Jeffreys, H. 1946, *Proceedings of the Royal Statistical Society of London, Series A* 186, 453

Milani A. 1999, *Icarus* 137, 269

Milani, A. & Valsecchi, G.B. 1999, *Icarus* 140, 408

Milani, A., Chesley, S.R. & Valsecchi, G.B. 1999, *Astron. Astrophys.* 346, L65

Milani, A., Chesley, S.R. & Valsecchi, G.B. 2000, *Planet. Space Sci., ACM'99 Special Issue*, Vol. 48, 945

Milani, A., Chesley, S.R., Boattini, A., Valsecchi, G.B. & Giovanni, B. 2000, *Icarus* 144, 12

Milani, A., LaSpina, A., Sansaturio, M.E. & Chesley, S.R. 2000, *Icarus* 144, 39

Muinonen, K. & Bowell, E. 1993, *Icarus* 104, 255

Muinonen, K. 1999, in: A.E. Roy & B.A. Steves (eds.), *The Dynamics of Small Bodies in the Solar System: A Major Key to Solar System Studies* (Kluwer), p. 127

Muinonen, K., Virtanen, J. & Bowell, E. 2001, *Cel. Mech. Dyn. Astron.* 81, 93

Muinonen, K. & Virtanen, J. 2002, in: S. Isobe & Y. Asakura (eds.), *International Workshop on Collaboration and Coordination Among NEO Observers and Orbital Computers*, p. 105

Muinonen, K., Virtanen, J., Granvik, M. & Laakso, T. 2004, *Mon. Not. R. Astron. Soc.*, to be submitted

Virtanen, J., Muinonen, K. & Bowell, E. 2001, *Icarus* 154, 412

Virtanen, J. & Muinonen, K. 2004, *Icarus*, submitted

Dynamics of Populations of Planetary Systems
Proceedings IAU Colloquium No. 197, 2005
Z. Knežević and A. Milani, eds.

© 2005 International Astronomical Union
DOI: 10.1017/S1743921304008725

The size of collision solutions in orbital elements space

G.B. Valsecchi[1], A. Rossi[2], A. Milani[3] and S.R. Chesley[4]

[1]INAF-IASF, via Fosso del Cavaliere 100, 00133 Roma, Italy
email: giovanni@rm.iasf.cnr.it
[2]ISTI-CNR, Area della Ricerca di Pisa, Via G. Moruzzi 1, 56124 Pisa, Italy
[3]Department of Mathematics, University of Pisa, Via Buonarroti 2, 56127 Pisa, Italy
[4]Jet Propulsion Laboratory, 4800 Oak Grove Drive, CA-91109 Pasadena, USA

Abstract. In the framework of the analytical theory of close encounters, and under suitable assumptions, we compute the size of the region in orbital elements space containing collisions solutions. In the linearized approximation in the semimajor axis/eccentricity plane the collision region is the interior of an ellipse. Examples are given from past cases of Near Earth Asteroids having the possibility of impacting our planet.

Keywords. Collisions, Near-Earth asteroids, planetary close encounters

1. Introduction

Impact monitoring programs like CLOMON2 and Sentry (Milani *et al.* 2004) routinely find collision possibilities of Near-Earth Asteroids (NEAs) with the Earth at specific dates, and characterize them with values of quantities like the stretching and the impact probability (Milani *et al.* 1999, Milani *et al.* 2004, Milani *et al.* 2002). These quantities depend, among other things, on the amount of observational data available for the given NEA, and on assumptions on the statistical features of the data; thus, they do not depend *only* on the orbital parameters of the NEA.

The question we address here is the following: leaving aside observations and statistics, and considering only the celestial mechanics side of the problem, can we meaningfully speak of the "size" of a collision solution and, if yes, how does it vary for different NEA collisions? We look for an answer in the framework of the recent extension (Valsecchi *et al.* 2003, Valsecchi 2004) of the analytical theory of close encounters (Öpik 1976).

2. Collision solutions

Let us start by considering the points in the space of orbital elements leading to an *exact* collision with a point-mass Earth. We consider the Earth to be on a circular orbit of radius 1 AU. The conditions on the orbital elements of the NEA for an exact collision at a given time t are:

$$\frac{a(1 - e^2)}{1 \pm e \cos \omega} = 1 \tag{2.1}$$

$$\Omega - \lambda_\oplus = \frac{\pi}{2} \mp \frac{\pi}{2} \tag{2.2}$$

$$\omega + f = \frac{\pi}{2} \mp \frac{\pi}{2}, \tag{2.3}$$

where the upper sign is for collisions at the ascending node and $\lambda_\oplus$ is the longitude of the Earth at the time t. Thus the collision solutions for a given date t lie on a 3-dimensional manifold in the 6-dimensional space of orbital elements.

Note that for a collision at another date equations (2.1)–(2.3) would be the same, with only $\lambda_\oplus$ changed. Thus, if we look for the set of collision solutions at an arbitrary date, equation (2.2) would not introduce any constraint on the orbital elements (a, e, i, ω, f), since Ω can be fixed at the suitable value as a function of $\lambda_\oplus$.

In the b-plane of Öpik's theory (Valsecchi *et al.* 2003), the exact collision condition is the point of coordinates $\xi = \zeta = 0$. The impact radius of the Earth on that plane is:

$$b_\oplus = \sqrt{r_\oplus^2 + 2cr_\oplus};$$

where $c = m/U^2$, m is the ratio of the mass of the Earth to that of the Sun,

$$U = \sqrt{3 - \frac{1}{a} - 2\sqrt{a(1 - e^2)}\cos i}$$

is the unperturbed geocentric speed of the NEA in units of the heliocentric velocity of the Earth, and $r_\oplus$ is the actual radius of the Earth; the unit of length used here is the AU. In the following, we use $b_\oplus$ to compute the region in elements space corresponding to a *physical* collision.

2.1. *The correspondence between orbital elements and b-plane coordinates*

If at $t = t_0$ the small body is near the Earth, at the point of geocentric cartesian coordinates (X_0, Y_0, Z_0), we have

$$\xi = X_0 \cos \phi - Z_0 \sin \phi$$
$$\zeta = (X_0 \sin \phi + Z_0 \cos \phi) \cos \theta - Y_0 \sin \theta,$$

where $\theta = \theta(a, e, i)$ and $\phi = \phi(a, e, i)$ are the usual angles of Öpik's theory, given by:

$$\cos \theta = \frac{\sqrt{a(1 - e^2)}\cos i - 1}{\sqrt{3 - \frac{1}{a} - 2\sqrt{a(1 - e^2)}\cos i}}$$

$$\sin \theta = \frac{\sqrt{2 - \frac{1}{a} - a(1 - e^2)\cos^2 i}}{\sqrt{3 - \frac{1}{a} - 2\sqrt{a(1 - e^2)}\cos i}}$$

$$\sin \phi = \pm\frac{\sqrt{2 - \frac{1}{a} - a(1 - e^2)}}{\sqrt{2 - \frac{1}{a} - a(1 - e^2)\cos^2 i}}$$

$$\cos \phi = \pm\frac{\sqrt{a(1 - e^2)}\sin i}{\sqrt{2 - \frac{1}{a} - a(1 - e^2)\cos^2 i}} .$$

In the expression for $\sin \phi$ the upper sign applies to encounters in the post-perihelion branch of the orbit, and in that for $\cos \phi$ to encounters close to the ascending node.

By using equations (2.1)–(2.3) we can express X_0, Y_0, Z_0 as functions of the orbital parameters (to first order in X_0, Y_0, Z_0):

$$X_0 = \frac{a(1 - e^2)}{1 + e \cos f_0} - 1$$

$$Y_0 = \Omega + \frac{\pi}{2} \mp \left\{\frac{\pi}{2} - \arctan[\cos i \tan(\omega + f_0)]\right\} - \lambda_\oplus$$

$$Z_0 = \sin i \sin(\omega + f_0),$$

where the upper sign is for collisions at the ascending node.

Let us now consider a NEA that, at a generic epoch t^*, has orbital parameters a, e, i, ω, Ω, f, and at $t = t_0$, when its true anomaly is f_0, has a collision with the Earth, located (at $t = t_0$) at longitude $\lambda_\oplus$. Note that

$$t_0 = t^* + 2h\pi a^{3/2},$$

where h is the non-integer number of heliocentric revolutions made by the NEA between t^* and t_0.

To understand what happens to ξ and ζ, when we apply small changes to the orbital elements, we can compute

$$d\xi = \sum_{i=1,6} \frac{\partial \xi}{\partial \mathcal{E}_i} d\mathcal{E}_i$$

$$d\zeta = \sum_{i=1,6} \frac{\partial \zeta}{\partial \mathcal{E}_i} d\mathcal{E}_i,$$

where $\mathcal{E}_1 = a, \mathcal{E}_2 = e, \mathcal{E}_3 = i, \mathcal{E}_4 = \omega, \mathcal{E}_5 = \Omega, \mathcal{E}_6 = f$.

2.2. *The derivatives of ξ, ζ with respect to the elements*

At the collision, the derivatives $\partial \xi / \partial \mathcal{E}_i$, $\partial \zeta / \partial \mathcal{E}_i$ have the form:

$$\frac{\partial \xi}{\partial \mathcal{E}_i} = \frac{\partial X_0}{\partial \mathcal{E}_i} \cos \phi - \frac{\partial Z_0}{\partial \mathcal{E}_i} \sin \phi$$

$$\frac{\partial \zeta}{\partial \mathcal{E}_i} = \frac{\partial X_0}{\partial \mathcal{E}_i} \cos \theta \sin \phi + \frac{\partial Z_0}{\partial \mathcal{E}_i} \cos \theta \cos \phi - \frac{\partial Y_0}{\partial \mathcal{E}_i} \sin \theta,$$

where we have dropped the terms with either X_0, or Y_0, or Z_0 as factor (since $X_0 = Y_0 = Z_0 = 0$). Note that, among the derivatives, only $\partial \xi / \partial a$ and $\partial \zeta / \partial a$ depend on the elapsed time $t_0 - t^*$; their expressions contain h, the non-integer number of heliocentric revolutions made by the NEA between t^* and t_0.

We can then write:

$$\begin{pmatrix} \delta \xi \\ \delta \zeta \end{pmatrix} = \begin{pmatrix} \frac{\partial \xi}{\partial a} & \frac{\partial \xi}{\partial e} & \frac{\partial \xi}{\partial i} & \frac{\partial \xi}{\partial \omega} & \frac{\partial \xi}{\partial \Omega} & \frac{\partial \xi}{\partial f} \\ \frac{\partial \zeta}{\partial a} & \frac{\partial \zeta}{\partial e} & \frac{\partial \zeta}{\partial i} & \frac{\partial \zeta}{\partial \omega} & \frac{\partial \zeta}{\partial \Omega} & \frac{\partial \zeta}{\partial f} \end{pmatrix} \begin{pmatrix} \delta a \\ \delta e \\ \delta i \\ \delta \omega \\ \delta \Omega \\ \delta f \end{pmatrix}. \tag{2.4}$$

Actually, since an explicit computation shows that $\partial \xi / \partial i = \partial \xi / \partial \Omega = \partial \zeta / \partial i = 0$, we have that:

$$\begin{pmatrix} \delta \xi \\ \delta \zeta \end{pmatrix} = \begin{pmatrix} \frac{\partial \xi}{\partial a} & \frac{\partial \xi}{\partial e} & \frac{\partial \xi}{\partial \omega} & 0 & \frac{\partial \xi}{\partial f} \\ \frac{\partial \zeta}{\partial a} & \frac{\partial \zeta}{\partial e} & \frac{\partial \zeta}{\partial \omega} & \frac{\partial \zeta}{\partial \Omega} & \frac{\partial \zeta}{\partial f} \end{pmatrix} \begin{pmatrix} \delta a \\ \delta e \\ \delta \omega \\ \delta \Omega \\ \delta f \end{pmatrix}.$$

Note that, to first order, the changes in (ξ, ζ) do not depend upon the inclination i; however, the partial derivatives do depend upon i, as also evidenced by the examples given in the next Section.

3. Collision solution size in a, e

As we have seen, the 2-dimensional vector giving a small displacement on the b-plane as function of the orbital elements is given by the product of a 2 rows by 6 columns matrix times a 6-dimensional vector in elements space (see equation 2.4). This corresponds to the fact that each point on the b-plane, if the time of close approach is variable, has as preimage a 4-dimensional manifold. However, if we fix four elements and leave only two as variables, the inversion becomes possible. To select the two elements we want to use as parameters, we have to take into account that the coordinates (i, Ω) play a different role: the partial derivatives with respect to i are zero, and Ω is only a function of $\lambda_\oplus$, that is of the close approach time t_0.

If, for example, we set $\delta\omega = \delta\Omega = \delta f = \delta i = 0$, we can find the size of a collision solution in the a-e plane. In this case, in fact:

$$\begin{pmatrix} \delta\xi \\ \delta\zeta \end{pmatrix} = \begin{pmatrix} \frac{\partial\xi}{\partial a} & \frac{\partial\xi}{\partial e} \\ \frac{\partial\zeta}{\partial a} & \frac{\partial\zeta}{\partial e} \end{pmatrix} \begin{pmatrix} \delta a \\ \delta e \end{pmatrix};$$

this can be inverted to give:

$$\begin{pmatrix} \delta a \\ \delta e \end{pmatrix} = \frac{1}{\begin{vmatrix} \frac{\partial\xi}{\partial a} & \frac{\partial\xi}{\partial e} \\ \frac{\partial\zeta}{\partial a} & \frac{\partial\zeta}{\partial e} \end{vmatrix}} \begin{pmatrix} \frac{\partial\zeta}{\partial e} & -\frac{\partial\xi}{\partial e} \\ -\frac{\partial\zeta}{\partial a} & \frac{\partial\xi}{\partial a} \end{pmatrix} \begin{pmatrix} \delta\xi \\ \delta\zeta \end{pmatrix}.$$

Using the above expressions we can find the values of δa and δe corresponding to a circle of radius $b_\oplus$ in the b-plane. The shape of the collision region, in this linear approximation, is an ellipse; it does not need to have the major and minor axes aligned with the coordinate axes.

3.1. *Application: the 2019 collision of 2002 NT$_7$ and the 2008 collision of 2003 EE$_{16}$*

We now compute the sizes in the a-e plane of the collision solutions of two NEAs that have been of some interest in the recent past.

In the summer of 2002 the confidence region of 2002 NT$_7$ in orbital elements space contained a collision taking place about $h = 7.3$ revolutions of the asteroid later, in early 2019. The value of the Palermo scale (Chesley *et al.* 2002) for this collision became, for a short time, larger than 0, before coming down quickly with the accumulation of new observations.

Substituting the numeric values of the orbital elements of 2002 NT$_7$ ($a = 1.74$, $e = 0.53$, $i = 42°\!.3$) in the expressions for δa and δe we get:

$$\delta a = \frac{-0.19\,\delta\xi - 1.7\,\delta\zeta}{3.4h}$$

$$\delta e = \frac{0.09\,\delta\xi + 0.79\,\delta\zeta}{-4.9h} - 0.62\,\delta\xi.$$

The left panel of Fig. 1 shows this collision in the a-e plane; to highlight the dependence on h (the non-integer number of heliocentric revolutions made by 2002 NT$_7$ before the collision), in the right panel we show the size of two collisions characterized by $h = 1$ and $h = 50$. The shrinking of the region along a, as h grows, is quite noticeable, as is the change in the overall slope of the collision region.

In March 2003 the confidence region of 2003 EE$_{16}$ in orbital elements space contained a collision taking place about $h = 2.9$ revolutions of the asteroid later, in early 2008. This collision solution did not go above the 0 of the Palermo scale at any time. Proceeding as

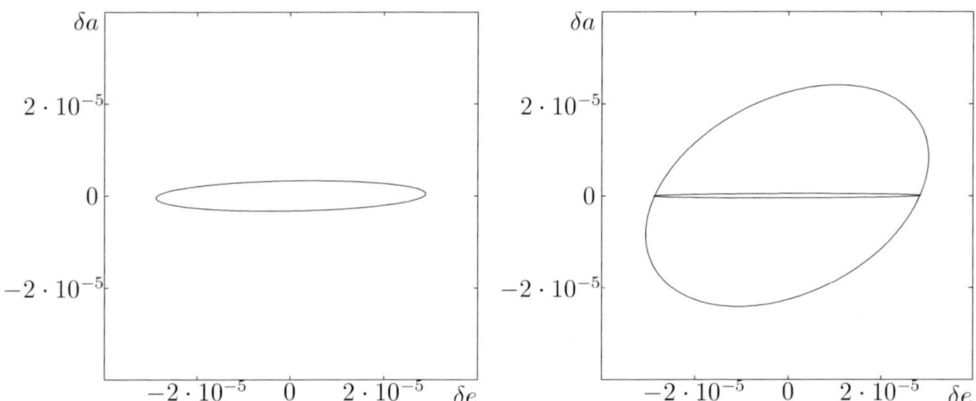

Figure 1. Left: the size of the collision solution of 2002 NT$_7$ in $\delta e, \delta a$ space for $h = 7.3$. Right: the size of collision solutions, of the same NEA, for $h = 1$ (larger ellipse) and $h = 50$.

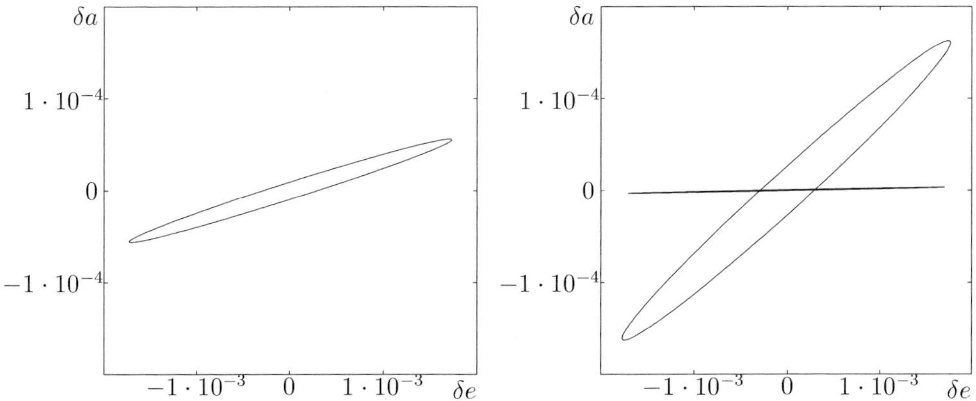

Figure 2. Left: the size of the collision solution of 2003 EE$_{16}$ in $\delta e, \delta a$ space for $h = 2.9$. Right: the size of collision solutions, of the same NEA, for $h = 1$ (larger ellipse) and $h = 50$. Note that the scale for δa is much smaller than for δe, that is, the ellipse is much wider in the δe direction.

before, for $a = 1.42$, $e = 0.62$, $i = 0°\!.6$ (the orbital elements of 2003 EE$_{16}$), we get:

$$\delta a = \frac{-0.19\,\delta\xi - 0.03\,\delta\zeta}{0.059h}$$

$$\delta e = \frac{0.09\,\delta\xi + 0.02\,\delta\zeta}{-0.071h} - 35\,\delta\xi.$$

Figure 2 shows the same information as Fig. 1 for 2003 EE$_{16}$. The most important difference between the two cases is the scale of the Figures: the collision solution of 2003 EE$_{16}$ is more than an order of magnitude larger in both coordinates. Of course this is due to the much lower inclination of 2003 EE$_{16}$.

4. Conclusions

We have shown that, under suitable assumptions, it is possible to deduce analytical expressions giving the size of the region in orbital elements space containing orbits colliding with the Earth at a given date. When computed in the a-e plane, the size depends on the number of heliocentric revolutions h made by the NEA before hitting the Earth, and is smaller for larger h.

Applying the above mentioned expressions to two NEAs, we find that the size of the collision region in the *a-e* plane differs significantly in the two cases.

We plan to continue this research, with the goal of providing an easy to compute first approximation of the location of a Virtual Impactor. This could be a useful tool for the impact monitoring systems.

Acknowledgements

We are grateful to G.F. Gronchi and G. Tommei for useful discussions on the subject of this paper.

References

Chesley, S.R., Chodas, P.W., Milani, A., Valsecchi, G.B., and Yeomans, D.K. 2002, *Icarus* 159, 423

Milani, A., Chesley, S.R., and Valsecchi, G.B. 1999, *Astron. Astrophys.* 346, L65

Milani, A., Chesley, S.R., and Valsecchi, G.B. 2000, *Planet. & Space Sci.* 48, 945

Milani, A., Chesley, S.R., Chodas, P.W., and Valsecchi, G.B. 2002, in: W.F. Bottke Jr., A. Cellino, P. Paolicchi and R.P. Binzel (eds.), *Asteroids III*, Univ. Arizona Press, p. 55

Milani, A., Chesley, S.R., Sansaturio, M.E., Tommei, G. and Valsecchi, G.B. 2004, *Icarus*, in press

Opik, E.J. 1976, *Interplanetary Encounters*, Elsevier

Valsecchi, G.B., Milani, A., Gronchi, G.-F., and Chesley, S.R. 2003, *Astron. Astrophys.* 408, 1179

Valsecchi, G.B. 2004, in: J. Souchay, S. Thoms, and G.Hakuba (eds.), *Dynamics of non point-like celestial bodies and of the rings*, Springer, in press

Dynamics of Populations of Planetary Systems
Proceedings IAU Colloquium No. 197, 2004
Z. Knežević and A. Milani, eds.

© 2004 International Astronomical Union
DOI: 10.1017/S1743921304008737

Very short arc orbit determination: the case of asteroid 2004 FU$_{162}$

Steven R. Chesley

Jet Propulsion Laboratory, M/S 301-150, Pasadena, CA 91109, USA
email: Steven.R.Chesley@jpl.nasa.gov

Abstract. We show the utility of the *Systematic Ranging* technique by analyzing the orbit determination of asteroid 2004 FU$_{162}$, which passed approximately 6400 km from the surface of the Earth on March 31, 2004. The limited observations introduce strong nonlinearities that must be accounted for when estimating the actual encounter distance.

Keywords. Orbit determination, ranging, impacts

On March 31, 2004, an interesting object, eventually designated 2004 FU$_{162}$, was detected by the telescopes of the LINEAR survey (Stokes *et al.* 2000). The position of 2004 FU$_{162}$ was measured four times over a 44 minute time interval, revealing a proper motion of several degrees per day. Thus 2004 FU$_{162}$ was recognized as a probable near-Earth asteroid, and the next day the measurements were flagged and forwarded to the Minor Planet Center. There the object was briefly posted to the World Wide Web for confirmation, but it was soon realized that it had already passed into the daylight sky and was no longer observable. However, it was already evident that this 5–10 meter object had passed very near the Earth, in fact far closer than any previous asteroid discovered beyond our atmosphere. However, 2004 FU$_{162}$ was never reobserved and searches for earlier observations proved fruitless.

In this paper we will consider the orbit determination challenges posed by the paucity of observational information available for 2004 FU$_{162}$. The few observations and very short data arc lead to substantial nonlinearity, making conventional covariance analyses inappropriate. In this case, although the least squares orbit determination does provide a nominal orbit, the confidence limits on the nominal solution are dominated by nonlinearities. Because of this, estimating the statistics of the close approach distance is not straightforward.

The techniques used for this analysis are thoroughly explained elsewhere (Chesley 2004). For brevity, only a summary will be included here. The method, here called *Systematic Ranging*, was originally proposed in an as yet unpublished manuscript by Tholen & Whitely (2002). Their basic idea exploits the fact that the range r and range rate $\dot{r}$ of a just-discovered object are poorly constrained, while the sky position and motion is directly measured. This suggests fixing r and $\dot{r}$ at realistic values and taking the best-fitting orbit given those constraints. In this way one systematically explores a raster in $(r, \dot{r})$-space, finding the best-fitting orbit and conditional uncertainty at each point. The RMS of the fit residuals for each raster point gives an indication of the quality of the fit, and standard chi-square probability theory can be used to derive confidence regions and even probability densities in the $(r, \dot{r})$-space.

The term Systematic Ranging is intended to contrast with an alternate procedure, dubbed Statistical Ranging (Virtanen *et al.* 2001), which is based on the Monte Carlo technique. The key difference between the two approaches is that with Statistical Ranging the $(r, \dot{r})$ values are selected randomly. Also, the technique randomly samples one pair

Table 1. 2004 FU_{162} Nominal Orbital Elements (J2000.0).

Epoch	2004 March 31.0 TDB
Semimajor axis	1.005 ± 0.010 AU
Eccentricity	0.342 ± 0.029
Long. of Ascending Node	$11°.1235 \pm 0°.0053$
Arg. of perihelion	$289°.24 \pm 0°.27$
Inclination	$2°.40 \pm 0°.24$
Mean anomaly	$289°.8 \pm 1°.1$
Absolute magnitude	28.7 ± 0.3

NOTE: Formal uncertainties are not particularly meaningful in the presence of strong nonlinearities. They are listed here as a general reference only.

observations and their observational errors, rather than obtaining the best-fitting sky motion that is consistent with all of the observations, given the the selected $(r, \dot{r})$.

The astrometry for 2004 FU_{162} was published on MPEC 2004-Q22 (2004); the following results assume 1 arcsec uncorrelated gaussian noise. The range of possible orbits (Fig. 1) permitted by the observations is really quite narrow considering the short interval of observations. With greater than 99% confidence, 2004 FU_{162} is an Earth-crossing asteroid with modest inclination and a diameter somewhere between a few meters and a few tens of meters. The best-fitting orbit, which does converge under general differential corrections, has an RMS of $0''.149$ and is detailed in Table 1.

The particulars of the close approach of 2004 FU_{162} are depicted in Fig. 2. Here we can see that the nominal close approach distance was just $2R_\oplus$ from the geocenter, or a scant 6400 km from the surface. The time of the encounter—give or take about an hour—is 2004 March 31.65, around 8 hours after the last observation. Close approach distances ranging from impact to $10R_\oplus$ can be found within the 99% confidence region. The probability of impact, derived using the fully nonlinear Systematic Ranging approach, is 0.33%†.

It is interesting to note that the nominal deflection of the asteroid due to perturbation of the Earth was substantial, approximately $20°$. The pre-encounter elements are listed in Table 1; the post-encounter elements are notably altered‡, indeed the category of the object was changed from Apollo to Aten by the approach.

Figure 3 reveals how the encounter distances for 2004 FU_{162} are distributed. The median distance is $2.8R_\oplus$ and the 99th percentile distance is $9R_\oplus$. The probability that 2004 FU_{162} passed closer than $7.8R_\oplus$, the record-setting flyby distance set by 2004 FH a few weeks earlier, is 97.5%.

Acknowledgements

Helpful discussions with Paul Chodas are gratefully acknowledged. This research was conducted at the Jet Propulsion Laboratory, California Institute of Technology, under a contract with the National Aeronautics and Space Administration.

References

Chesley, S.R. 2004, "Estimating the Near-term Impact Probability from Newly-discovered Asteroids" *Icarus*, submitted

Minor Planet Electronic Circ., 2004-Q22, 2004, Minor Planet Center, Cambridge, Mass.

† There was a well-publicized bolide impact on the east coast of Australia around 2004 March 31.4, but the timing and circumstances of that event appear to be wholly inconsistent with the close approach of 2004 FU_{162}.

‡ Post-encounter semimajor axis, eccentricity and inclination are 0.827 AU, 0.392 and $4.16°$, respectively. The longitude of perihelion changed from $300°$ to $331°$.

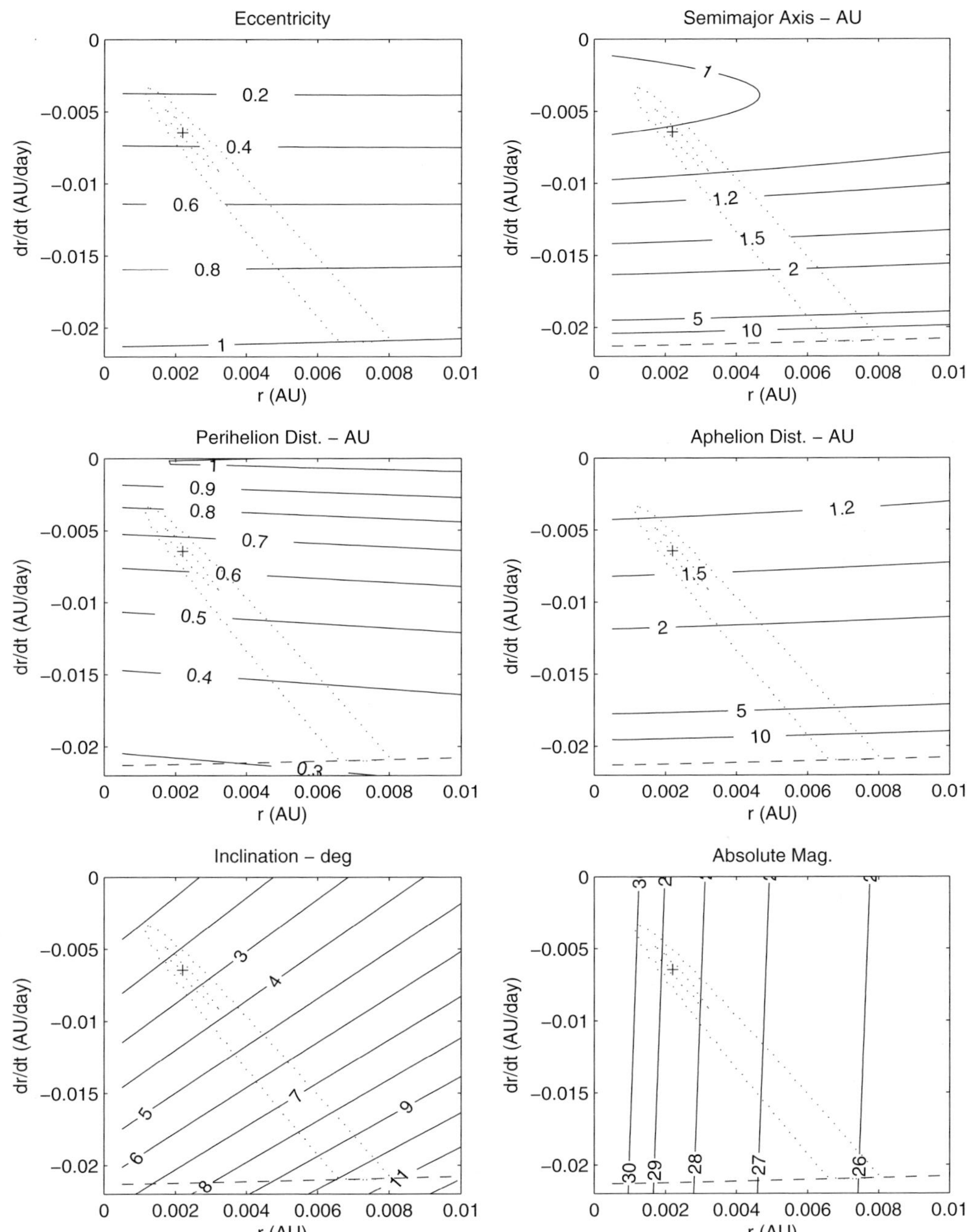

Figure 1. 2004 FU$_{162}$ orbital distribution in the $(r, \dot{r})$-plane. Dashed curves replicate the hyperbolic limit shown at upper left. Dotted curves depict the 50% and 99% confidence regions, assuming $1''$ astrometric noise. The best-fitting solution is marked with a +-sign.

Tholen, D,J., Whiteley, R.J. 2002, "Short Arc Orbit Solutions and Their Application to Near Earth Object," *Icarus*, submitted

Stokes, G.H., Evans, J.B., Viggh, E.M., Shelly, F.C., Pearce, E.C. 2000, *Icarus* 148, 21

Virtanen, J., Muinonen, K., Bowell, E. 2001, *Icarus* 154, 412

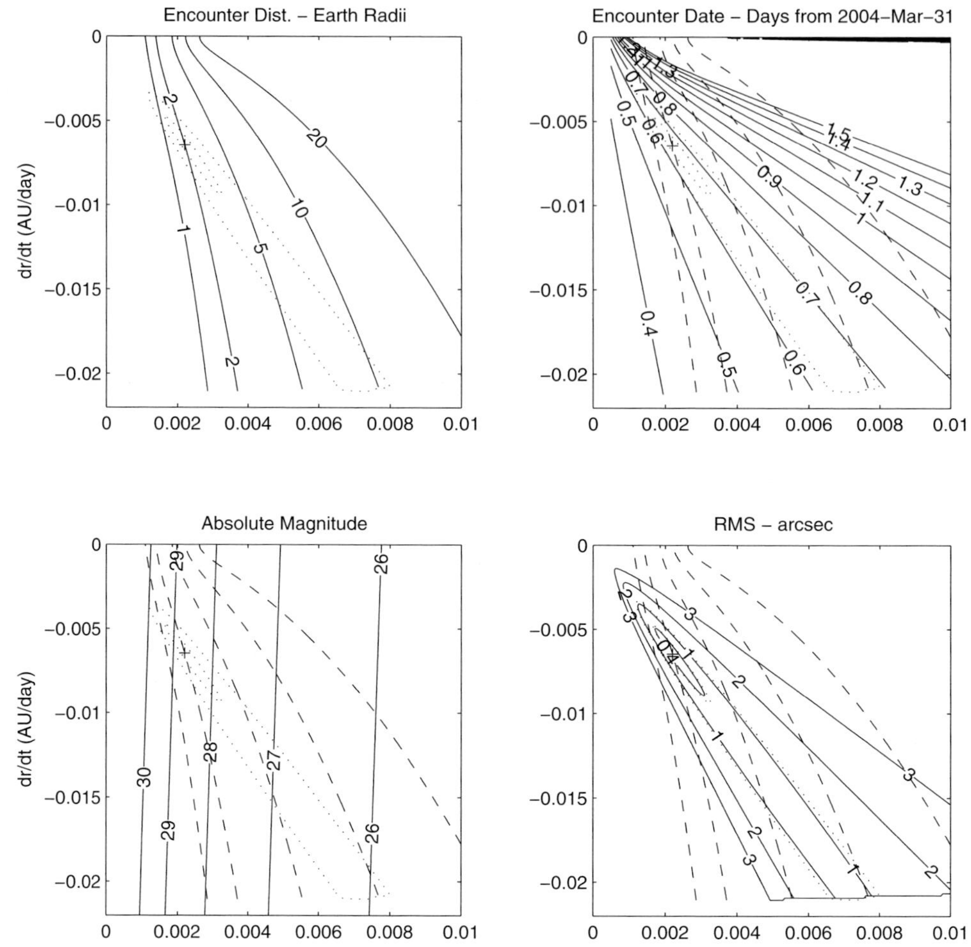

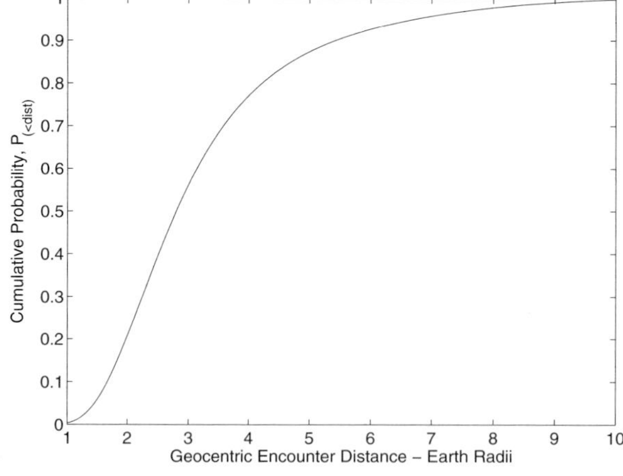

Figure 2. 2004 FU$_{162}$ close approach data. Dashed curves replicate the encounter distance curves shown at upper left. Dotted curves depict the 50% and 99% confidence regions, assuming $1''$ astrometric noise. The best-fitting solution is marked with a +-sign.

Figure 3. Probabilistic distribution of encounter distances for 2004 FU$_{162}$.

Dynamics of Population of Planetary Systems
Proceedings IAU Colloquium No. 197, 2005
Z. Knežević and A. Milani, eds.

© 2005 International Astronomical Union
DOI: 10.1017/S1743921304008749

Nonlinear impact monitoring: 2-dimensional sampling

Giacomo Tommei[1]

[1]Department of Mathematics, University of Pisa, Via Buonarroti 2, 56126 Pisa, Italy
email: tommei@mail.dm.unipi.it

Abstract. When a new Near Earth Asteroid is discovered, it is important to know if there is the possibility of an impact with the Earth in the near future. In these last years second generation software for impact monitoring (CLOMON2 and SENTRY) have been developed and the performances have been significantly increased in comparison to the earlier, simpler and solitary system CLOMON. The two systems use the Line Of Variations (LOV) approach: they sample the LOV, an 1-dimensional subspace, to perform the sampling of the 6-dimensional confidence region. This approach is very useful when the confidence region is elongated and thin, that is an eigenvalue of the covariance matrix is much bigger than the others. When the observed arc is short ($1°$ or less), usually for asteroids observed for few nights, the confidence region is like a flat disk and we propose to use a 2-dimensional sampling. We triangulate the admissible region in the $(r, \dot{r})$ plane, using the nodes of triangulations as Virtual Asteroids (VAs). After orbit propagation we project the VAs and the triangulation on the Target Plane (TP) of a given epoch to study the existence of a Virtual Impactor (VI) and complex dynamical behaviors such as folds.

Keywords. Celestial mechanics, asteroids, methods: numerical

1. Introduction

When an asteroid has just been discovered, its orbit is weakly constrained by the available observations spanning a short arc. In many cases a nominal orbital solution exists, but other sets of orbital elements, that correspond to RMS of the residuals not significantly above the minimum, are acceptable as solutions. This situation can be described by defining a *confidence region* $Z(\sigma)$ in the orbital elements space such that initial orbital elements belong to $Z(\sigma)$ if the *penalty* (increase in the target function) does not exceed some threshold depending upon the parameter σ.

In many applications we need to consider the set of the orbits with initial conditions in the confidence region as a whole. In impact monitoring we may need to predict some future close approach or even an impact, and this for all *possible* orbits, that is for every solution not resulting in too large residuals. Since the dynamic model for asteroid orbits is not integrable there is no way to compute all the solutions for some time span in the future. We can only compute a finite number of orbits by numerical integration. There are several methods to perform a sampling of the confidence region. In this paper we explore the geometrical sampling, reviewing the Line Of Variations (LOV) approach introduced by Milani (1999) (Section 2) and proposing a new 2-dimensional sampling for cases in which the LOV approach is not satisfactory (Section 3). In the last Section we show as this new sampling could be used to study complex dynamical behaviors such as folds.

2. The LOV approach in impact monitoring

The LOV is a 1-dimensional segment of a curved line in the 6-dimensional space of initial conditions. There is a number of different ways to define and compute the LOV, but the general idea is that a segment of this line is a kind of spine of the confidence region. Both CLOMON2 and SENTRY, the two independent systems of impact monitoring, use the LOV approach. The LOV is sampled at regular intervals obtaining an ordered set of Virtual Asteroids (VAs). In this way it is possible to interpolate between two consecutive VAs to prove the existence of a closest approach around a given encounter date. Using iterative algorithms it is possible to find the minimum approach distance and eventually a Virtual Impactor (VI) (Milani *et al.* (2004b)).

In Milani (1999) the LOV was defined as the solution of an ordinary differential equation of the first order, with the nominal least squares solution as initial condition, and the vector field defined by the weak direction of the covariance matrix, the eigenvector relative to the biggest eigenvalue of the matrix. Such differential equation is unstable, in particular when there is a largely dominant weak direction, the case when the LOV is most useful. Thus a corrective step was introduced, based upon differential corrections constrained on the hyperplane perpendicular to the weak direction. Milani *et al.* (2004c) have formulated a new definition which does not require the presence of a nominal solution. The LOV, in fact, is defined as a set of points in which the gradient of the cost function is in the weak direction. All the LOVs used in this paper to produce numerical results are obtained with this *constrained differential correction algorithm*.

There is an important fact to remark: the eigenvalues of covariance matrix are not invariant under a general coordinate change (only for an orthonormal one), so the LOV depends upon the coordinates used for initial orbit determination. Then it is possible to compute several LOVs and the choice of one of these to use in impact monitoring is not always a simple issue.

The next two figures show different LOVs for two Near Earth Asteroids (NEAs) in the plane $(r, \dot{r})$ with $\dot{r}$ scaled by the Gaussian constant. For each coordinates system (Cartesian, Attributable, Equinoctial, Cometary) we trace both the LOV (marked with 1), sampled by 41 VAs for $\sigma \in [-1, 1]$, and the *second* LOV (marked with 2), obtained using the eigenvector relative to the second largest eigenvalue of the covariance matrix.

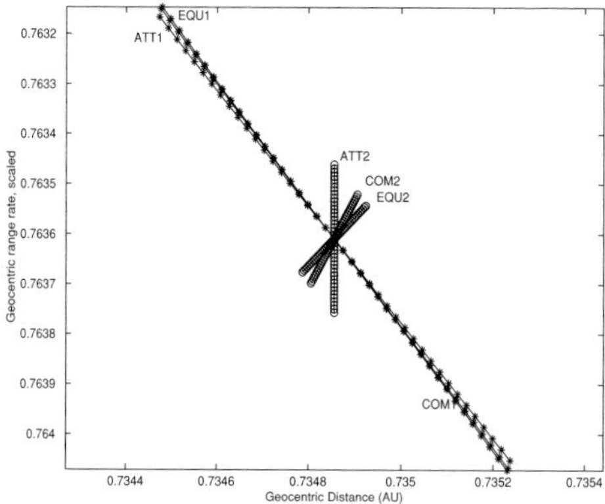

Figure 1. Milani *et al.* (2004c). LOVs in different coordinates for the asteroid 2002 NT$_7$ with the first 113 observations. The Cartesian and Attributable LOVs coincide so only the Attributable one is depicted.

Figure 1 shows the comparison of LOVs for asteroid 2002 NT$_7$ when the 113 available observations were spanning 15 days forming an arc of about $9°$. In this case the LOVs are very similar and the choice of one instead of another to use in impact monitoring does not produce significant changes in results. If we imagine the ellipse of confidence in this plane, having as axes LOV and second LOV, we note that it is very elongated and thin.

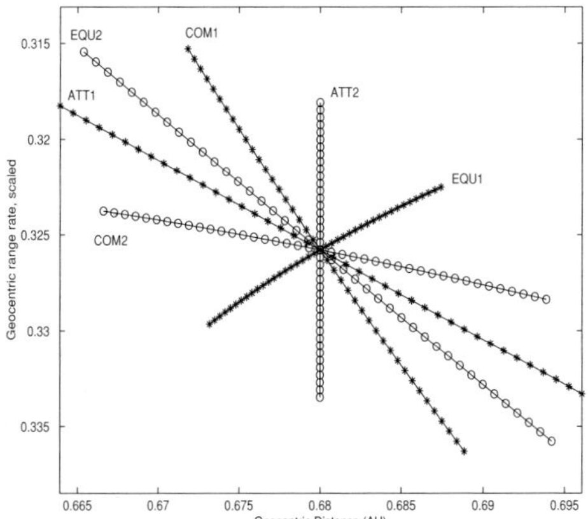

Figure 2. Milani *et al.* (2004c). LOVs in different coordinates for the asteroid 2004 FU$_4$ with the first 17 observations. The Cartesian and Attributable LOV coincide so only the Attributable one is depicted.

Figure 2 shows the LOVs for asteroid 2004 FU$_4$ when the 17 available observations were spanning 3 days forming an arc of about 1°. The dependence of the LOV on the coordinates is very strong here. Note that the LOV of the Attributable or Cartesian elements is near the second LOV of the Equinoctial elements. In this case the choice of a LOV as sampling of the confidence region to use in impact monitoring is not simple. In fact, with these data, using the LOV computed in Cartesian elements CLOMON2 did not find a VI in 2010, while using Equinoctial, or SENTRY, using Cometary, did find it. The ellipse of confidence in this figure is like a flat disk, so the idea is to switch to a new kind of sampling, to perform a better analysis of the geometric structure of this 2-dimensional subset.

3. 2-dimensional sampling

This new approach does not require the presence of a preliminary orbit, but only the observations. Starting from these it is possible to create an *attributable* which consists of a reference time (the mean of observations times), two average angular coordinates and two corresponding angular rates for the reference time.

Following the work of Milani *et al.* (2004a), we define an *admissible region* and we triangulate it using the Delaunay triangulation. Figure 3 shows the triangulated admissible region for the asteroid 2004 FU$_4$ with the first 17 observations, the case analyzed in the previous Section. Each node of the triangulation is used as VA and it is propagated to a target time recording all the close approaches to the Earth. Then, after a partition in showers, depending upon the time of close approaches, each VA is projected on the usual Target Plane (TP) of Öpik theory. Figure 4 shows the TP of first close approach of 2004 FU$_4$ after the discovery, with the triangulation obtained joining the nodes which belong to the same triangle in the original triangulation.

After some close approaches the geometry on the TP becomes very complicated. Figure 5 shows the TP of 2010 for 2004 FU$_4$ with the same data. The triangulation

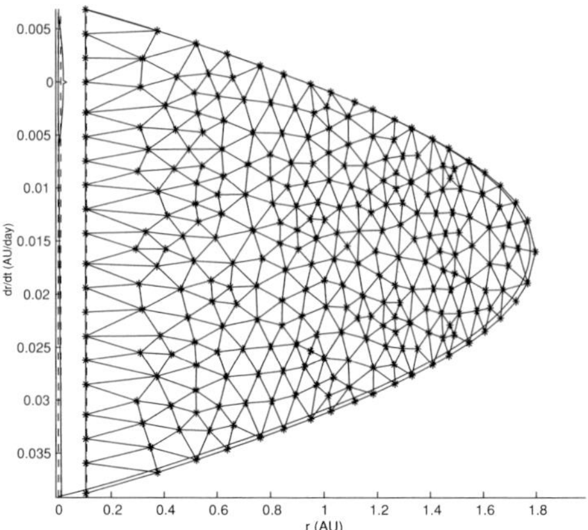

Figure 3. Triangulation of the admissible region in the plane $(r, \dot{r})$ for the asteroid 2004 FU$_4$ with the first 17 observations.

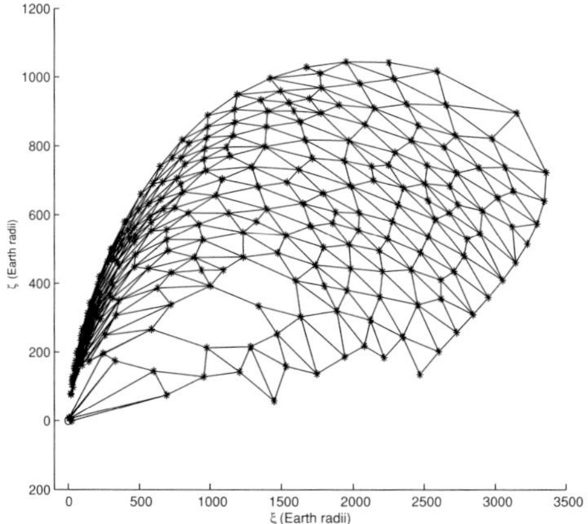

Figure 4. Triangulation on the TP of the first close approach in 2004 for the asteroid 2004 FU$_4$ with the first 17 observations.

appears very complicated but the remarkable fact is that the Earth is in at least one triangle, so the presence of a VI could not be excluded. In this figure it is also possible to note folds, complex behaviors derived from returns to close approaches that will be explained in the next Section.

4. An example of complex dynamical behavior: the folds

A return to a close approach, after a previous encounter (after the time of the initial conditions), is mathematically represented as a map between the TP of the first encounter and the TP of the second one. The behavior of differentiable maps $\mathbf{R}^2 \to \mathbf{R}^2$ has been studied by Witney (1950). He proved that generically (for a set of maps dense in the function space of all differentiable maps) the points of the first plane which are singular (that is, at which the differential of the map is not invertible) form smooth curves on the first plane and belong to only two geometrically distinct types: the folds and the cusps. A fold is such that the image on the second plane of the curve of singular points is a locally regular curve, while a cusp of the 2-dimensional map corresponds to a cusp (reversal of the tangent vector) on the image on the second TP of the curve of the singular

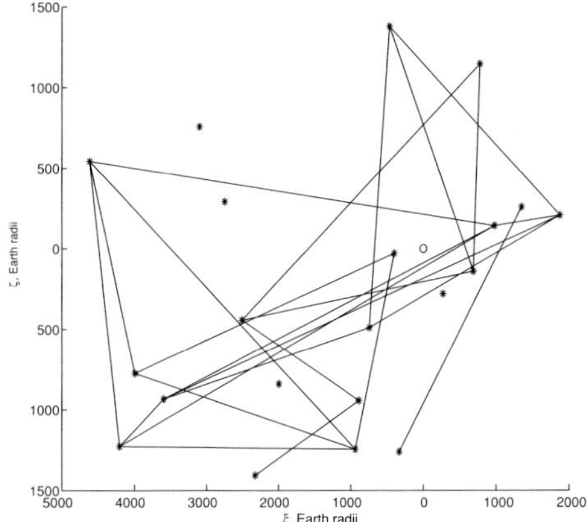

Figure 5. Triangulation on the TP of 2010 for the asteroid 2004 FU$_4$ with the first 17 observations.

points. The fold points are such that the map can be locally transformed, by suitable differentiable coordinate changes (in both planes), into the normal form

$$\xi'' = \xi \quad \zeta'' = \zeta^2.$$

In a intuitive way, the first TP folds along the $\zeta = 0$ fold line and covers twice the $\zeta'' \geqslant 0$ half plane. For the purpose of searching for VIs it is therefore essential to know if the fold is in the direction from the fold line towards the Earth or away from it. The role of cusps in impact monitoring will be the subject of future research.

Figure 6 shows an interrupted return, an example of fold obtained by a 1-dimensional sampling, using the LOV. The sequence of VAs approaches the Earth (the origin of the reference system) and then turns before reaching it.

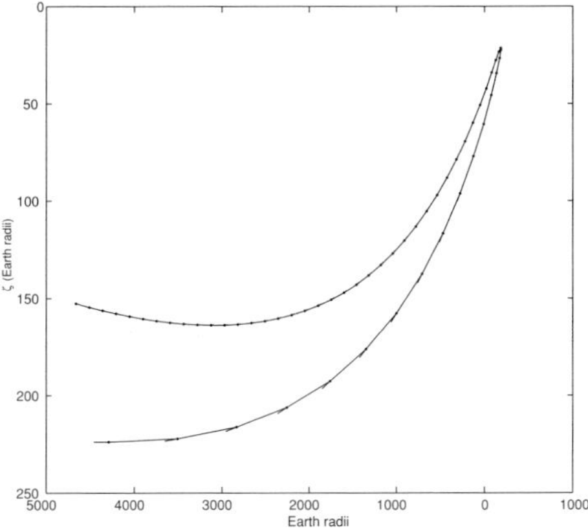

Figure 6. Interrupted return on the TP of 2025 for the asteroid 1994 UG.

In Figure 7 it is possible to appreciate the real 2-dimensional nature of a fold: a triangle folds and covers other triangles, but in such way that the Earth (depicted with a circle, not to scale) is protected by the fold and there is no VI.

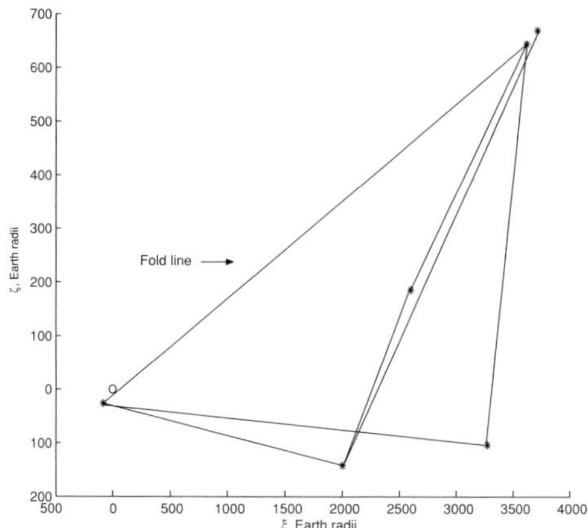

Figure 7. Example of fold on the TP of 2005 for the asteroid 2004 FU$_4$ (first 7 observations) found using a 2-dimensional sampling.

5. Conclusions and future work

The LOV approach works well when there is a largely dominant weak direction and it is possible to use the geometric properties of a string. When the confidence region has a 2-dimensional structure, it is useful to switch to the new approach explained in the paper. In practice this happens when the arc is very short (1° for example, as for 2004 FU4). The goal of this kind of geometrical sampling is to be able to analyze asteroids with few nights of observations. We would like to know if they could have either direct impacts (impacts before the next close approach) or VIs as returns after some close encounters. This new approach could also be useful to handle complex behaviors, such as folds, which could give us information about the geometry on the TP.

Next step will be to focus on the part of the triangulation which has acceptable residuals, using differential corrections in the 4-dimensional space orthogonal to the 2-dimensional triangulated manifold. Then, the idea is to create recursive algorithms to improve the triangulation obtained, defining criteria to detect VIs, to extrapolate an estimate of impact probability.

6. Acknowledgments

I would like to acknowledge A. Milani, who proposed the problem, for his suggestions, G.F. Gronchi, G.B. Valsecchi and S.R. Chesley for useful discussions.

References

Milani, A. 1999, *Icarus, 137, pp. 269-292*

Milani, A., Gronchi, G.F., De'Michieli Vitturi, M. & Knežević, Z. 2004a, *Cel. Mech. Dyn. Astron.*, in press

Milani, A., Chesley, S.R., Sansaturio, M.E., Tommei, G. & Valsecchi, G.B. 2004b, *Icarus*, in press

Milani, A., Sansaturio, M.E., Tommei, G., Arratia, O. & Chesley, S.R. 2004c, *Astron. Astrophys.*, in press

Dynamics of Populations of Planetary Systems
Proceedings IAU Colloquium No. 197, 2005
Z. Knežević and A. Milani, eds.

© 2005 International Astronomical Union
DOI: 10.1017/S1743921304008750

Searching for gravity assisted trajectories to accessible near-Earth asteroids

Ştefan Berinde

Babes-Bolyai University, Cluj-Napoca, Romania
email: sberinde@math.ubbcluj.ro

Abstract. This paper explores the advantages of using gravity assisted trajectories to perform flyby and rendezvous missions to accessible near-Earth asteroids (NEAs), in terms of the total velocity budget required for the mission. Combining the Opik's formalism of close encounters with the Monte Carlo sampling technique and Lambert trajectories, we give a general picture of the accessibility regions for NEAs in phase space of orbital elements, without considering the phasing requirements between bodies.

Keywords. Gravity assisted trajectories, flyby missions, near-Earth asteroids

---

1. Introduction

Given the increased interest for in-situ exploration of near-Earth asteroids (NEAs) in recent years, several papers explored already the possibility of direct transfer orbits for flyby and rendezvous missions (Perozzi, Rossi & Valsecchi 2001, Christou 2003). As we will highlight in this paper, the accessibility region in phase space of orbital elements is rather limited for such missions due to the high velocity budget involved.

Without considering the phasing requirements (proper alignment of the bodies on their orbits), which is just a matter of timing, this paper gives a general overview on the advantages of using gravity assisted trajectories to NEAs. The key parameters used to describe the accessibility regions are the nodal distances, eccentricity and inclination of the asteroid's orbit. Figure 1 shows the distribution of discovered NEAs in terms of these parameters. A strong observational bias near the Earth's orbit is evident.

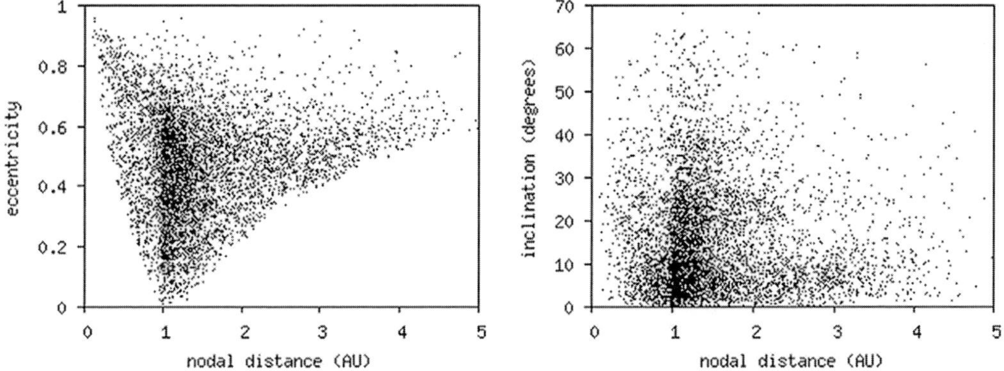

Figure 1. Distribution of discovered NEAs in *nodal distance-eccentricity* plane (left panel) and in *nodal distance-inclination* plane (right panel). Both nodal distances are considered for each asteroid.

2. Preliminaries

Throughout this paper we adopt the following measuring units for distance and time, such that the radius of the Earth's orbit (supposed to be circular) is 1 and its heliocentric velocity is also 1. It means that the heliocentric gravitational parameter $\mu_\odot = 1$.

We consider the initial state of an interplanetary spacecraft as being its parking LEO orbit (at about 200 km). From here, we will try to evaluate the velocity budget dv required to place the spacecraft in a flyby or rendezvous orbit with an asteroid. Let dv_0 be the excess velocity applied in LEO orbit. The unperturbed geocentric relative velocity u_0 (that is, velocity at infinity) is given by

$$u_0 = \sqrt{(v_{leo} + dv_0)^2 - v_{par}^2},\qquad(2.1)$$

where $v_{leo} = 0.26$ is the circular velocity in LEO orbit and $v_{par} = 0.37$ is the corresponding parabolic velocity. For escaping the Earth, we must have $dv_0 > 0.11$. We also note that $u_0 > dv_0$ only for $dv_0 > 0.13$. For practical reasons we can suppose that $dv_0 < 0.2$, since the total dv budget for NEAs missions is rather limited (6 km/s for the NEAR mission; see Farquhar *et al.* 1995).

Also in this section we will reproduce the formulas for the components of the heliocentric velocity vector of a body in an elliptic orbit, at one of its nodes. This will prove useful later when computing the excess velocity dv_{rend} required to perform a rendezvous mission at the nodes. Let be a rectangular reference frame oriented such that the y-axis points along the radial distance of the body towards the Sun, the z-axis is perpendicular on the ecliptic plane towards north pole and the x-axis completes the right-handed frame. Because we are at the nodes the x-axis is located on the ecliptic plane and we must have $v_x = \sqrt{v_x^2 + v_z^2}\cos i$ and $v_z = \pm\sqrt{v_x^2 + v_z^2}\sin i$. Writing the component v_y (the radial velocity) in terms of the keplerian orbital elements a, e, i, we get the final expressions

$$\begin{cases} v_x = \dfrac{\sqrt{a}}{d}\sqrt{1 - e^2}\cos i \\[2em] v_y = \pm\dfrac{\sqrt{a}}{d}\sqrt{e^2 - \left(1 - \dfrac{d}{a}\right)^2} \\[2em] v_z = \pm\dfrac{\sqrt{a}}{d}\sqrt{1 - e^2}\sin i, \end{cases}\qquad(2.2)$$

where d is the nodal distance. The signs of the right terms depend on the type of the node, if it is ascending or descending (for the component v_z) and if it is placed before or after the perihelion (for the component v_y).

3. Direct orbits

Following the basic optimal guidelines about orbital transfer in space (Marinescu 1982), a rendezvous mission with an asteroid includes, in general, two steps: a Hohmann transfer orbit from Earth to the farthest node of the asteroid's orbit, followed by a velocity impulse in order to match its orbital velocity.

A nodal distance d can be reached at aphelion on an orbit with the initial geocentric relative velocity u_0, along the Earth's orbit, given by

$$u_0 = \sqrt{\dfrac{2d}{(1 + d)} - 1}.\qquad(3.1)$$

The velocity at aphelion will be

$$v_{aph} = \sqrt{\frac{2}{d(1+d)}}. \tag{3.2}$$

In the previously introduced reference frame, this velocity is oriented along the x-axis. Using now the set of formulas (2.2), we obtain the required rendezvous velocity impulse at the node

$$dv^2_{rend} = \frac{2(2+d)}{d(1+d)} - \frac{1}{a} - \frac{2}{d}\sqrt{\frac{2a(1-e^2)}{d(1+d)}}\,\cos i, \tag{3.3}$$

where a, e, i, ω are the well known orbital elements of the asteroid. Note that we must have

$$d = \frac{a(1-e^2)}{1 \pm e\cos\omega}, \tag{3.4}$$

where the sign depends on the node's type.

From (2.1) and (3.1) we can get dv_0 in terms of d. Then $dv = dv_0 + dv_{rend}$ is the total velocity budget to accomplish the mission. Based on this approach, figure 2 gives the accessibility regions for such missions in the d-e plane. The required rendezvous velocity is minimal for $\omega = 0$ and $\omega = \pi$ and maximal for $\omega = \pi/2$ and $\omega = 3\pi/2$. It depends also strongly on the orbital inclination i, such that only orbits very close to the ecliptic are matched within the limit $dv < 0.2$.

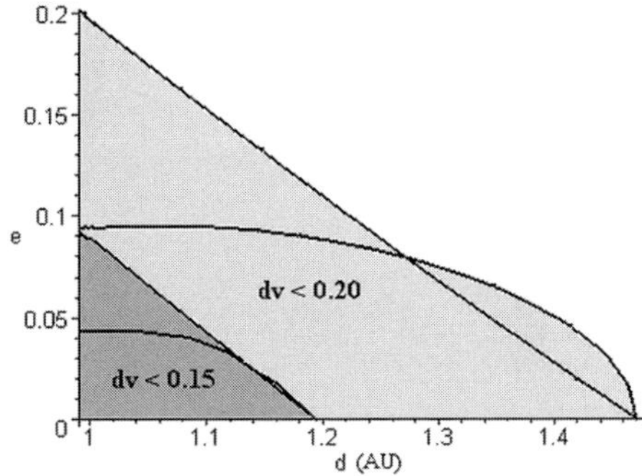

Figure 2. Accessibility regions in d-e plane for a rendezvous mission with an asteroid (see text for details). Four regions are depicted, depending on the total dv budget and on the perihelion argument of the orbit ($\omega = 0$ for straight borders and $\omega = \pi/2$ for curved borders). We set here $i = 0$.

As a matter of fact, if the asteroid's orbit intersect the Earth's one ($d = 1$, $u_0 = 0$), we get the well known formula (Carusi *et al.* 1990)

$$dv^2_{rend} = 3 - \frac{1}{a} - 2\sqrt{a(1-e^2)}\,\cos i = 3 - T, \tag{3.5}$$

where dv_{rend} has now the meaning of the unperturbed geocentric relative velocity and T is the Tisserand parameter.

It follows that the accessibility region of NEAs for rendezvous missions on direct orbits is very limited. If we consider only the flyby missions opportunities at the nodes, all NEAs are reachable since they have (by definition) at least one of their nodal distances between 1.0 and 1.3 AU.

4. VEGA and VEEGA gravity assisted orbits

In this section we want to explore the advantages of the gravity assisted orbits to the NEAs. It is practically impossible to give a general picture about the entire spectrum of gravity assisted orbits in the inner solar system. But, if we neglect the possibility of intermediary deep-space maneuvers involving significant velocity impulses (which was not the case for the NEAR mission; Farquhar *et al.* 1995), the only ways to increase orbital energy in a reasonable amount of time are VEGA (Venus-Earth Gravity Assist) and VEEGA (Venus-Earth-Earth Gravity Assist) transfer orbits (Longuski & Williams 1991).

Next we will work in the frame of Opik's geometric formalism (Carusi *et al.* 1990). This gives us the opportunity to develop simple algebraic formulas for various orbital quantities from the beginning of the mission (LEO orbit) till the end (asteroid flyby or rendezvous). We recall some of the principles of this approximation: planetary orbits are considered circular and coplanar, the spacecraft has a planetocentric hyperbolic orbit near a planet and a keplerian heliocentric orbit in interplanetary space and a planetary encounter is considered an instantaneous event when compared with the interplanetary travel time.

To perform an interplanetary journey Earth-Venus-Earth, the alignment of these two planets and the initial conditions are very sensitive to each other. Since we do not take into consideration here the phasing requirements, we focus on the initial conditions for VEGA orbital transfers.

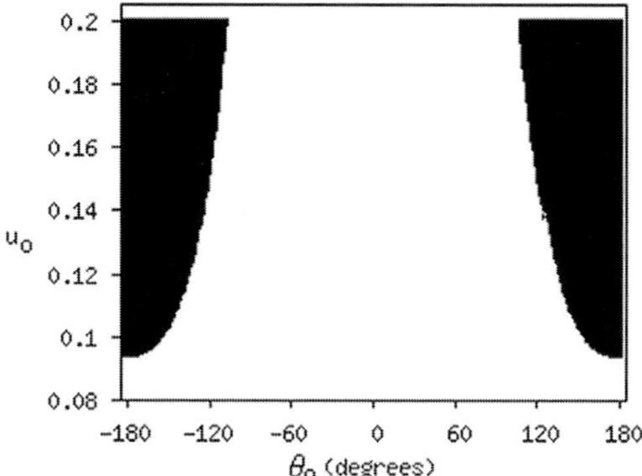

Figure 3. All possible starting conditions from Earth (θ_0, u_0) for VEGA transfer orbits.

Let θ_0 denote the angle between the Earth's velocity vector and the initial geocentric relative velocity vector of the spacecraft (with magnitude u_0). We consider this angle to be negative if the relative velocity vector points inside the Earth's orbit. The heliocentric starting conditions for a coplanar transfer are fully described by the pair (θ_0, u_0). Figure 3 shows all the pairs for which a VEGA orbit exists. These solutions were computed by Monte Carlo sampling technique, solving the Lambert problem twice and patching the conic sections at Venus encounter Battin (1987). This encounter has the effect of deflecting the spacecraft relative velocity vector u by the amount

$$\sin\frac{\gamma}{2} = \left(1 + \frac{\rho}{m_p}u^2\right)^{-1}, \tag{4.1}$$

where ρ is the minimum encounter distance and m_p the planetary mass. The angular deflection γ must not exceed γ_{max}, obtained from (4.1) when ρ equals the Venus planetary radius.

If we are interested in maximizing the relative velocity of the spacecraft u when it encounters the Earth again, for each value of u_0 we can find the right value of θ_0 for which the maximum occurs. Figure 4 presents the relation between u and u_0. An empirical linear dependence among these parameters is also determined. In these conditions, the spacecraft will encounter Earth at an angle $\theta \in (94°, 99°)$, defined analogously to θ_0.

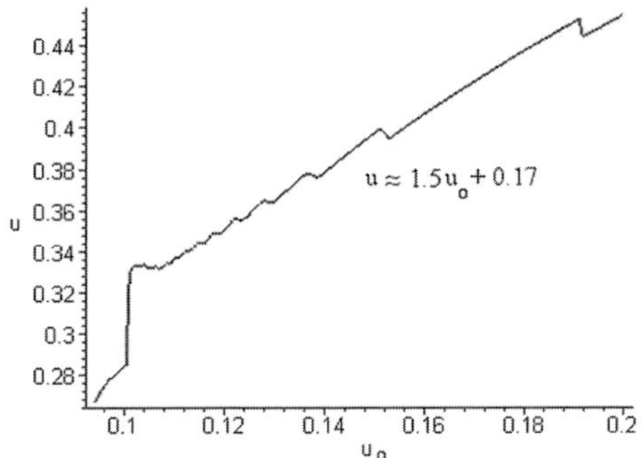

Figure 4. Maximum relative velocity u at Earth encounter for VEGA transfer orbits. An empirical linear dependence can be inferred between the initial and final relative velocities: $u \approx 1.5u_0 + 0.17$, valid for $u_0 > 0.1$.

If we define γ_{max} as the maximum angular deflection at Earth encounter (when the geocentric minimum distance ρ equals the radius of the LEO orbits), we get an approximate expression for it, $\sin(\gamma_{max}/2) \approx (1 + 14.5u^2)^{-1}$. This value imposes a threshold limit for the final orbital energy acquired in VEGA transfer orbits. But, since the relative velocity u and angle θ are rather large after this encounter, an additional Earth encounter allows a further variation of the semimajor axis of the spacecraft's orbit. In other words, the spacecraft must enter in a mean motion resonance with the Earth, on a VEEGA transfer orbit. If we set up a resonance ratio p/q and a corresponding semimajor axis of the post-encounter orbit $a = (q/p)^{2/3}$, there is a precise deflection angle γ_{res} for which the resonance is matched. Its value is given by $\gamma_{res} = \theta - \theta'$, where θ' is computed from

$$a = \frac{1}{1 - 2u\cos\theta' - u^2}, \tag{4.2}$$

a well known expression for the semimajor axis in the Opik's formalism. We found that the optimum low-order mean motion resonances are $1/2$ and $2/5$. The corresponding deflection angles $\gamma_{1/2}$ and $\gamma_{2/5}$ are computed in figure 5, for each value of the initial relative velocity u_0.

Besides the post-encounter semimajor axis a of the spacecraft's orbit, its eccentricity e can be also computed from

$$e = \sqrt{1 - \frac{1}{4a}\left(3 - \frac{1}{a} - u^2\right)^2}, \tag{4.3}$$

and so we can get the aphelion distance d for the final VEGA and VEEGA transfer orbits. Figure 6 gives these values computed with the condition $\theta' = \theta - \gamma_{max}$ (for VEGA orbits) and with the condition $\theta' = \theta - \gamma_{1/2} - \gamma_{max}/2$ for VEEGA orbits. For the last type of orbits, the second angular deflection was limited at $\gamma_{max}/2$, since a higher value will send the spacecraft beyond Jupiter.

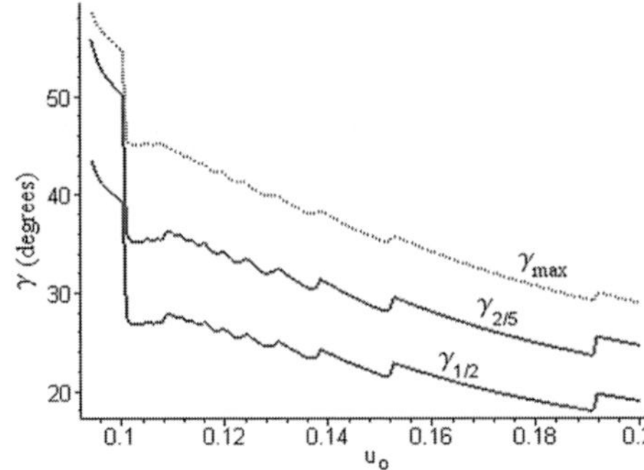

Figure 5. Angular deflection of the velocity vector at Earth encounter for a given initial relative velocity u_0. γ_{max} - maximum deflection, $\gamma_{2/5}$ - required deflection to set up an orbit in 2/5 mean motion resonance with the Earth, $\gamma_{1/2}$ - required deflection to set up an orbit in 1/2 mean motion resonance with the Earth.

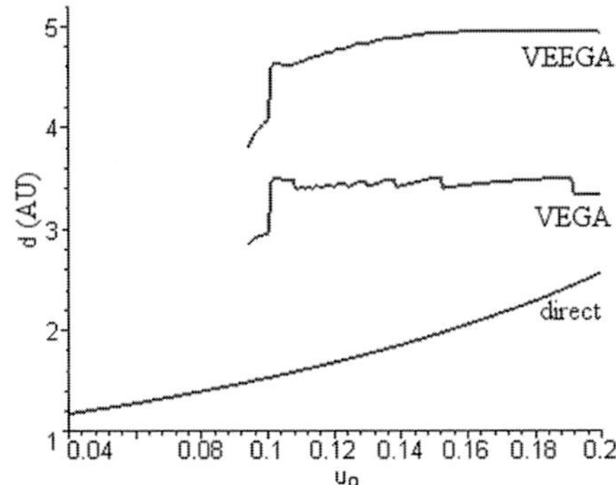

Figure 6. Maximum aphelion distance d, function of the initial relative velocity u_0, for different transfer orbits: direct orbits, VEGA orbits (computed for $\gamma = \gamma_{max}$) and VEEGA orbits (computed for $\gamma = \gamma_{1/2} + \gamma_{max}/2$).

As we can see from Figures 1 and 6, with a low initial velocity u_0 a spacecraft can reach both orbital nodes of the majority of the NEAs. The existence of such an orbit depends only on the phasing requirements. This permits a flyby at the farthest orbital node of an asteroid's orbit, where a rendezvous maneuver requires less additional velocity.

References

Battin, R.H. 1987, *An introduction to the mathematics and methods of astrodynamics*, AIAA Education Series, New-York

Carusi, A., Valsecchi, G.B. & Greenberg R. 1990, *Cel. Mech. Dyn. Astron.* 49, 111

Christou, A. 2003, *Planet. Space Sci.* 51, 221

Farquhar, R.W., Dunham, D.W. & McAdams J.V. 1995, *J. Astronautical Sci.* 43, 353

Longuski, J.M. & Williams S.N. 1991, *Cel. Mech. Dyn. Astron.* 52, 207

Marinescu, A. 1982, *Optimal problems in the dynamics of space flight* (in romanian), Editura Academiei, Bucuresti

Perozzi, E., Rossi, A. & Valsecchi G.B. 2001, *Planet. Space Sci.* 49, 3

Dynamics of Populations of Planetary Systems
Proceedings IAU Colloquium No. 197, 2005
Z. Knežević and A. Milani, eds.

© 2005 International Astronomical Union
DOI: 10.1017/S1743921304008762

KLENOT - Near Earth and other unusual objects observations

Michal Kočer, Jana Tichá and Miloš Tichý

Kleť Observatory, Zátkovo nábř. 4, CZ-37001 České Budějovice, Czech Republic
email: klet@klet.cz

Abstract. The KLENOT project is a project of the Kleť Observatory, Czech Republic, devoted to astrometric observations of Near-Earth objects, distant objects and comets. The improved effort of the large NEO surveys resulting in an increasing number of newly discovered NEOs calls for continuous follow-up astrometry to secure an accurate orbit determination of discovered bodies first in discovery opposition and then during next apparitions. Considering this urgent need of astrometric follow-up, the fact that many of these targets are fainter then magnitude 20.0 V and our results and experience in minor planet and comet CCD astrometry done at Kleť since 1993, we decided to bring into operation a new 1-m class facility working on a permanent basis - the KLENOT telescope. The regular observing of the telescope started in March 2002 (the MPC code 246). Beside methods and techniques we use for follow-up astrometry we present most important results of the project.

Keywords. Minor planets, asteroids

1. Introduction

The discovery of near-Earth objects (NEOs) is an important contribution, but it would mean little without follow-up observations used to refine orbits and to secure an identification of a possible impact in the future. This need for astrometric follow-up of NEOs increases with increasing capabilities of the existing surveys and a growing interest towards smaller objects. This is why the KLENOT (**KLE**ť observatory **N**ear-Earth and **O**ther unusual objects observations **T**eam and Telescope) project has been established to provide follow-up astrometry of NEOs up to magnitude limit $m_V = 22.0^m$.

2. The KLENOT telescope

The KLENOT telescope was built using an existing dome and infrastructure of the Kleť Observatory. The original mounting was upgraded and the optoelectronical control system was added. A new control and computer room was built.

Optical system: The KLENOT telescope has a 1.06-m primary mirror and a primary focus corrector. The main f/3.0 mirror was fabricated by Carl Zeiss Jena using a Sital glass (Zerodur type). The primary focus corrector was computed by Sincon, Turnov, Czech Republic, and was fabricated by the Optical Facility of Charles University, Prague, Czech Republic, led by Jindřich Walter. The corrector contains four spherical lenses. The optical configuration is f/2.7 at the folded prime focus where the CCD camera is located.

CCD camera: The CCD camera used for The KLENOT Project is CCD camera Photometrics Series 300. The CCD chip sensor is SITe SI003B with format 1024×1024 pixels, pixel size 24 × 24 microns, and is back illuminated with high quantum efficiency; Q.E. > 80% in range 5500 − 8000 Å. Imaging array size is 24.6 × 24.6 millimeters. The field of view of the KLENOT Telescope is 33 × 33 arcminutes using the CCD camera mentioned above. The image scale is 1.9 arcseconds per pixel. The limiting magnitude is $m_V = 21.5^m$ for 120-sec exposure time in standard weather condition.

3. The KLENOT Project

Confirmatory observations of newly discovered fainter NEO candidates: Some of new search facilities produce discoveries fainter than $m_V = 20^m$ which need a larger telescope for confirmation and early follow-up. A 1-m class telescope is also very suitable for confirmation of very fast moving objects and our larger FOV enables to search for NEO candidates having a larger ephemeris uncertainty.

Follow-up astrometry of poorly observed NEOs: It is necessary to observe newly discovered NEOs in a longer arc during the discovery opposition when they get fainter. Special attention is given to "Virtual Impactors" and PHAs, target of future space missions or radar observations. On the other hand, it is necessary to find and use an optimal observing strategy to maximize orbit improvement of each asteroid.

Recoveries of NEOs in the second opposition: For the determination of reliable orbits it is required to observe asteroids in more then one opposition. If the observed arc in a discovery apparition is long enough, the chance for a recovery in the next apparition is good. If the observed arc at single opposition is not so good, we plan to search along the line of variation. For this purpose a larger field of view is an advantage.

Follow-up astrometry of other unusual objects: We plan to make follow-up astrometry of other unusual objects, i.e. Centaurs and transneptunian objects, both in discovery opposition and next apparitions. To obtain positions of brighter transneptunians, we propose to use longer exposures with magnitude limit about $m_V = 22^m$. Considering the problem with securing adequate data for orbit computation of these objects, follow-up astrometry, at least of some of them, will be useful.

Cometary features: The majority of new ground-based discoveries of comets comes from large surveys devoted, predominantly, on Near Earth Asteroids. The first step in distinguishing these newly discovered members of the population of cometary bodies consists in confirmatory astrometric observations along with detection of their cometary features. A timely recognition of cometary features of a particular body having an unusual orbit can help in planning further observing campaigns.

Search for new asteroids: Our primary goal is astrometric follow-up of NEOs and other unusual objects. Moreover, all CCD images are processed not only for target objects, but also tested for possible new object(s). Obtained images will be processed with special reference to fast moving objects and slow moving objects.

Follow-up of Gamma-ray burst (GRB) optical counterparts: A part of observing time is devoted to GRB optical follow-up observations as a target of opportunity.

4. Technology

A special software package has been developed for the KLENOT Project at Klet using a combination of programs running on Windows and Linux platforms. The system consists of observation planning tools, data-acquisition, camera control and data processing tools. So, all the software mentioned bellow has been developed by KLENOT team.

Observation planning: The SQL database holds information on minor planets updated on daily bases from text-based databases; the MPC Orbit Database (MPCORB), maintained by the Minor Planet Center, and from the Asteroid Orbital Elements Database (ASTORB), created and maintained by E. Bowell at the Lowell Observatory. The asteroids listed on Spaceguard system Priority List and objects listed as a Virtual Impactors by SENTRY (JPL) or by CLOMON (NEODyS) are flagged in the SQL database as well. Besides the information on asteroids the database holds orbital elements and other useful data of all solar system objects discovered at Klet (database K_KLET) and also information on comets (database COMETS) created and updated from several sources by Klet. Also positions, times and observed objects on all of the processed plates and CCD images are stored in the database (CCD). For observation planning a web-based tool called **ephem** is used. The tool allows an observer to get ephemerides for one or more minor planets in a specified field in the sky at given time. The objects in the output list can be reduced to objects of given magnitude and/or type; i.e. to Near-Earth asteroids (NEAs), potentially hazardous asteroids (PHAs), Virtual Impactors (VI), Klet discoveries, critical list objects, unusual or distant minor planets, Trojans, Spaceguard Priority List objects and comets. The output list includes, beside designation, position in the sky, magnitude,and other usual ephemeris data, also information on object type, ephemeris uncertainty, date of last observation, length of orbital arc used in orbit computation. The output list is ordered by right ascension. Another program **KAC** – *Klet' Atlas Coeli* — shows stars and solar system objects with the line showing their daily motion on the sky on selected region in the sky. The size of the region corresponds to the FOV of the telescope used so it is also used to check the telescope position during observation. As a source of positions and magnitudes of stars the USNO-A2.0 star catalog is used.

Data acquisition and camera control: The CCD camera is controled by V++ scripts. For exposure control and data-acquisition a set of scripts in VPascal (build-in programming language in V++) has been written. The scripts store a sequence of several CCD frames (images) in one file in TIFF format. In the header of the sequence file information about the number of frames in the sequence, time, exposure time, equipment used and other information are included.

Data processing: The program **Astrometry** has been developed for the reduction of CCD images and identification of stars with USNO-A2.0 catalogue. Images taken by CCD camera are reduced and all objects with realized condition for signal to noise ratio are found on the image. These object are then identified with stars from the star catalogue. Equatorial coordinates of objects are then determined, at the same time stars with residuals greater than 1 arcsecond are excluded automatically and/or manually and magnitude of the objects is determined. The user then selects which object on the image the output should be made from.The output is directly in the MPC format. The time of observation and other information needed for the output are derived from the data stored in the header of the image file. Information about the processed CCD image (time, filename, frame number, equatorial coordinates of the center of the frame, telescope used, exposure time, position of objects on frame, etc.) are added into the SQL database of processed CCD images.

The measured astrometric position is checked before providing them to the community. Computing of residuals, i.e. differences between calculated and observed astrometric positions, is used for such proof (program **residua**). The calculation of residuals is based on osculating elements of the object near the current epoch, so they are acceptable mainly for the evaluation of actual observation data. Alongside the Δ–T variation of the mean anomaly is determined. Checking of both residuals and the Δ–T variation in mean anomaly helps in object identification. The next utility of the KLENOT's software package (program **orbit**) allows us a computation of preliminary orbital elements from observations for one or two nights on the assumption that an object is in perihelion, has been also developed. From observations over several nights the preliminary orbit of a new minor planet should be determined using Lagrange-Gauss method improved by variation of geocentric distances and variation of elements. Orbital elements of new discovered objects are stored in K_KLET database. We are working now on an improvement of this software for orbit determination taking into account perturbating mutual gravity within N-body system. For exact determination of planetary position the system uses The Planetary and Lunar Ephemerides DE405 provided by JPL.

‖	asteroids			NEA families		
year ‖	all	NEA	PHA ‖	Apollo	Amor	Aten
2002 ‖	6324	2305	625 ‖	2305	963	205
2003 ‖	12626	3603	831 ‖	3603	1953	304
2004 ‖	3285	1183	224 ‖	1183	528	90

Table 1. Total number of observations per year, total number of observation of NEAs and of the major NEA families obtained by KLENOT in the given year (by 2004 Aug.14)

	NEA		‖	PHA		
year ‖	all	follow-up	early follow-up ‖	all	follow-up	early follow-up
2002 ‖	377	213	145 ‖	75	47	23
2003 ‖	438	242	215 ‖	84	48	25
2004 ‖	298	151	95 ‖	56	24	10

Table 2. Contribution of KLENOT project to observation of newly discovered NEAs and PHAs in given years (by 2004Aug.14). **Legend: all** — total number of world-wide discovered bodies; **early follow-up** — total number of bodies observed by KLENOT in early follow-up, i.e. the KLENOT's observation was published in discovery circular of given body; **follow-up** — total number of bodies observed by KLENOT in a discovery opposition

5. Results

The regular observations of the KLENOT project started in March 2002. By 2004 Aug. 14 22.235 astrometric positions of Solar System Objects have been obtained; 7091 of them have been observations of NEAs (599 Atens, 3647 Apollos, 2845 Amors) and 755 have been observations of comets. Table 1 shows the the total number of observations and number of observation of NEAs/PHAs obtained by KLENOT and Table 2 shows the yearly contribution of KLENOT project to observation of newly discovered NEAs and PHAs in early follow-up as well as in later follow-up observation. Other important results of the KLENOT project are recoveries of 23 NEAs, recoveries of comets C/2003 A1, C/2003 A2, 100P and a discovery of Apollo-type asteroid 2003 LK and Aten-type asteroid 2003 UT55. Results obtained by the KLENOT team and telescope during its operation period show that this facility dedicated for permanent follow-up astrometry can significantly help in tracking NEAs worldwide.

Acknowledgements

This work has been sponsored by The Grant Agency of the Czech Republic Reg. number 205/98/0266, The 2000 NEO Shoemaker Grant of The Planetary Society, and The Grant Agency of the Czech Republic Reg. No. 205/02/P114.

References

Tichá, J., Tichý, M., and Moravec, Z. 2000a, *Planet. Space Sci.* 48, 787
Tichá, J., Tichý, M., and Moravec, Z. 2000b, *Planet. Space Sci.*, 48, 955
Tichá, J., Tichý, M., and Moravec, Z. 2000c, in: A. Fitzsimmons, D. Jewitt, R. M. West (eds.), *Minor Bodies in the Outer Solar System*, Springer, p 165
Tichá, J., Tichý, M., and Kočer, M. 2002, *ESA SP-500: ACM 2002*, 793
Spahr, T.B. 2002, Minor Planet Electronic Circular, 2002-L05.
Sekanina, Z., Chodas, P. W., Tichý, M., Tichá, J., and Kočer, M. 2003, *Astrophys. J.*, 591, L67

Part 4

COMETS AND TRANS-NEPTUNIAN OBJECTS

Dynamics of Populations of Planetary Systems
Proceedings IAU Colloquium No. 197, 2005
Z. Knežević and A. Milani, eds.

© 2005 International Astronomical Union
DOI: 10.1017/S1743921304008774

Transport of comets to the Inner Solar System

Hans Rickman

Astronomiska observatoriet, Box 515, SE–75120 Uppsala, Sweden
email: hans@astro.uu.se

Abstract. Recent years have seen a revolution in the possibility to understand cometary capture, i.e., the origin of the cometary population that moves in orbits confined to the inner Solar System. This is due to the discovery of the major source populations: the Edgeworth-Kuiper belt, and the scattered disk. We review the current understanding of the links between the distant sources, including the Oort cloud, and the observed, short-period population, and the problems that remain. Some highlights of present research in this field will serve to illustrate recent progress and major issues that are currently arising.

Keywords. Comets, cometary capture, Jupiter family, Oort cloud, Edgeworth-Kuiper belt, scattered disk

---

1. Introduction

One of the first facts that became known about cometary orbits is that the periods of revolution vary over a wide range. One often refers to long-period and short-period comets, but this distinction is mostly of historical interest, and the terms will not be used in a strict sense in this paper. Nonetheless, the fact that some comets are confined to the innermost part of the Solar System, while others make only temporary visits from very remote regions, is remarkable.

It was also realized long ago (Lexell 1778, 1780) that comets may pass close to Jupiter and receive perturbations that change their orbits drastically. Such instability offers a way for comets to evolve dynamically between long and short periods and thus to be "captured", at least temporarily, into the inner Solar System. The next important finding was that, in a physical sense, comets do not live forever. Especially famous are the 19th century cases of comets 3D/Biela, which disappeared after a major splitting, and 5D/Brorsen, which ceased to be observed for no obvious reason. This adds a new dimension to the capture issue in that there may be a sink of comets – especially those of short periods that return more frequently to the vicinity of the Sun and may hence be more vulnerable to destruction.

Another aspect is that comets are generally observable only when they penetrate deep within Jupiter's orbit and H_2O ice can sublimate efficiently from their nuclei. This means that the perihelion distance plays a crucial role for cometary discovery bias. Furthermore, all observable comets are ephemeral, since they can not live longer than the time it takes to exhaust their inventory of H_2O ice. The problem of comet evolution is very complex, involving both physical and dynamical effects as well as their mutual interactions – see (Jewitt 2004) for a recent review. At this point, suffice it to say that not only changes of orbital period are of interest. The delivery of comets into observable orbits by decrease of the perihelion distance is even more fundamental to our understanding of cometary populations.

This paper aims to describe the current understanding of how comets are transferred from distant source populations into observable orbits, especially those with short periods. To this end, the basic concepts will first be introduced by means of a historical background. The rest of the paper will discuss the most important current problems and some recent work on a few major issues.

2. Conceptual History

While, in principle, comets might be thought of as a temporary phenomenon that we happen to observe because we live at a privileged time, it is evidently necessary to look for a steady-state situation, where the losses are balanced by infeed from a source population. And even though there is no guarantee that the present cometary population is indistinguishable from that at other times, e.g. $\sim 10^7$ or 10^8 years ago, the latter pursuit has always proved successful. Typically, a source population and associated infeed mechanism has first been suggested on theoretical grounds, and later on the suggestion has been verified by observations – either partially or fully.

The first major structure to be recognized in the observed cometary population was the Jupiter family. Its primary characteristics – low inclinations and aphelia near the orbit of Jupiter – at once suggested gravitational captures by the giant planet during close encounters as the explanation. However, the source population remained elusive for a long time. The reason is obvious from the Tisserand criterion, which is derived from the Jacobi integral of the circular restricted three-body problem (see e.g. Danby 1962):

$$T = \frac{a_J}{a} + 2\sqrt{\frac{a}{a_J}(1 - e^2)} \cos i \tag{2.1}$$

where a and a_J are the semimajor axes of the comet and Jupiter, and e and i are the comet's eccentricity and inclination, respectively.

The comets of the Jupiter family have $\cos i \simeq 1$ and values of T between 2 and 3 (see Kresák 1972; Carusi *et al.* 1987). Since T is quasi-conserved for perturbations due to close encounters with Jupiter, the only possible connection with long-period orbits occurs for inclinations that remain low and perihelion distances that increase toward or just beyond a_J (see Fig. 1). Therefore, until improvements in observational techniques allowed the discovery of comets with perihelia near Jupiter's orbit, the source population had to remain hidden from view.

Only with the discovery of the 'Oort peak', i.e., a conspicious pile-up of original, long-period cometary orbits with $0 \lesssim 1/a_{\mathrm{orig}} \lesssim 1 \cdot 10^{-4}$ AU^{-1} creating a sharp peak in the distribution of binding energies, and the suggestion that this reveals a distant reservoir of comets (Oort 1950), was there an independent basis for identifying a particular source for the Jupiter family. Such a giant cloud, referred to as the 'Oort cloud', would be subject to external perturbations from passing stars and – as has later been realized – the tidal effects of the whole Galactic disk and bulge. The energy $H \propto a^{-1}$ is not very sensitive to those perturbations, but the angular momentum L may vary dramatically and reach values near zero, so that the perihelion distance $q \simeq L^2/2GM_\odot$ (for $e \simeq 1$) drops to observable values. On the way, comets would pass the region with $a_J/a \simeq 0$ and $q \simeq a_J$ in Fig. 1 and might thus offer a suitable source for the Jupiter family.

Orbital integrations including repeated encounters with Jupiter for a large sample of fictitious starting orbits (Everhart 1972) appeared to confirm this expectation, showing that captures into short-period orbits happened most easily for $\cos i_0 \simeq 1$ and $q_0 \simeq a_J$. However, problems remained. One was related to the efficiency of Everhart's capture process (Joss 1973), which relates the expected flux of Oort cloud comets in the 'capture

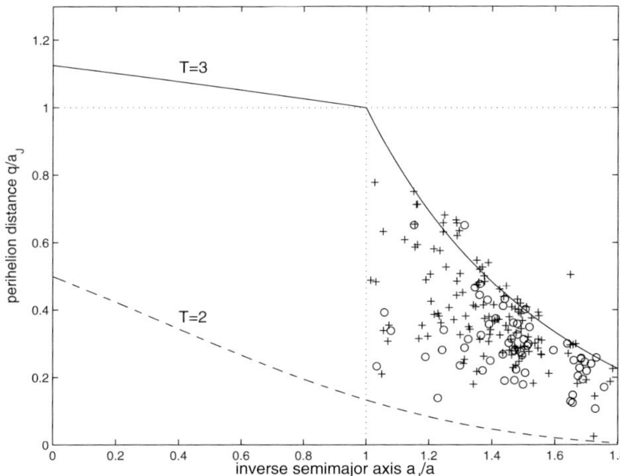

Figure 1. Perihelion distance is plotted vs inverse semimajor axis, both in units of Jupiter's semimajor axis. The solid and dashed curves show the relation given by the Tisserand criterion for zero inclination with two values of the Tisserand parameter T, as indicated. The symbols show the locations of all known comets with orbital periods shorter than Jupiter according to recent statistics. Circles denote comets discovered before 1950, and plus signs show more recent discoveries. The isolated symbol high above the $T = 3$ line is the peculiar comet 133P/Elst-Pizarro.

region' to the steady-state number of Jupiter family comets. This is a complex topic, clouded by uncertainties over the size and brightness distributions as well as fadings and observable lifetimes. A more straightforward argument came from the need to explain why the Jupiter family with its low inclinations is such an outstanding feature.

Simulations by Duncan *et al.* (1988) and Quinn *et al.* (1990) showed a general trend for the inclinations to stay within the range of the starting orbits, and the question hence arose: if the source population is isotropic with a flat distribution of $\cos i$, why is the Jupiter family so strongly concentrated to $\cos i \simeq 1$? High-inclination comets would run a larger risk of hyperbolic ejection before capture but, once captured, they would stay longer before being expelled again.

This issue is also somewhat confused by uncertainties over cometary lifetimes, because the lack of high-inclination, short-period comets might be partly explained, if such comets fade beyond detectability or disappear before they have time to be captured. But a consensus has grown around the inadequacy of such an explanation, and the need for a flattened, low-inclination source population has become generally accepted.

A very important concept was introduced by Kazimirchak-Polonskaya (1972) in a paper entitled 'The Major Planets as Powerful Transformers of Cometary Orbits', where she used a set of orbital integrations largely based on observed Jupiter family comets to suggest that comets could be captured via a multi-stage process involving all the giant planets. Each planet would be responsible for one part of the process, decreasing the orbital period and shifting the perihelion inward to the next planet while keeping the inclination low, as illustrated in Fig. 2. Further progress in understanding the dynamics of transformations between orbits nearly tangent to that of the planet at perihelion and aphelion was achieved by Carusi *et al.* (see Carusi and Valsecchi 1985).

This scenario invites a discussion of other sources for the Jupiter family, but even when restricting attention to the Oort cloud, it is seen that Jupiter does not need to capture comets single-handedly. In fact, each of the giant planets is able to start the process, if the

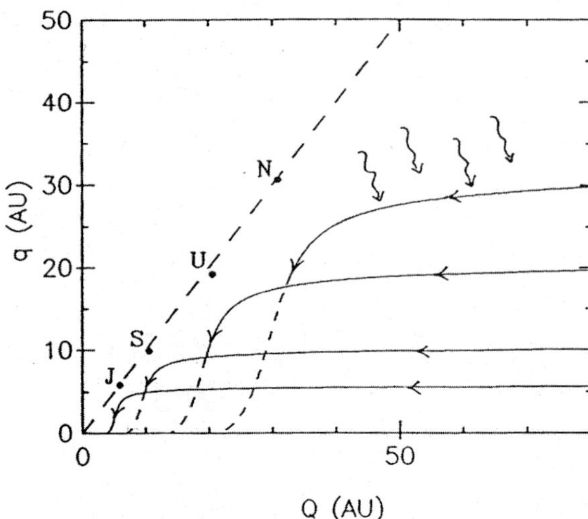

Figure 2. Low-inclination evolutionary tracks in the plane of aphelion distance (Q) vs perihelion distance (q) for multi-planet cometary capture. The solid curves indicate tracks of constant T as defined for each giant planet separately, and the dashed parts at the lower left are those where the planet in question does not dominate the evolution. From Rickman (1992).

perihelion of the comet falls in the vicinity of the planetary orbit. To this comes another feature, i.e., that the Oort cloud should have a broad energy distribution, where the Oort peak only represents the outer part for which the external perturbations are strong enough to bring the perihelia across the range of the giant planets in one revolution. Inside this there should be an inner core, for which the perihelia evolve in small steps, so that perturbations of orbital energy by the giant planets intervene before the comets become observable. As a result, hyperbolic ejections may occur but also captures of the multi-planet type illustrated in Fig. 2.

Bailey (1986) showed that this is likely to be important, but the intricate interplay of weak perturbations by the exterior giant planets and by Galactic tides for the inner core remains to be elucidated. Another idea, sketched by the wiggly arrows in Fig. 2, is that a source of comets is located just beyond the orbit of Neptune.

Whipple (1972) made interesting remarks about such a transneptunian comet belt but could not identify it as a source for the Jupiter family, since the dynamical mechanisms for a leakage from such a belt into Neptune-crossing orbits were not known. Fernández (1980) indeed proposed a transneptunian source for the Jupiter family as a remedy for the failure of jovian captures to produce the family directly from the Oort cloud. Finally, Duncan *et al.* (1988) used the argument of the inclination distribution to add further weight to the requirement of a flattened transneptunian source of comets, and they called this source the 'Kuiper belt' in remembrance of Kuiper's (1951) suggestion of a remnant, icy planetesimal population beyond Neptune's orbit.

The discovery of minor planet 1992 QB$_1$ (Luu and Jewitt 1993) marked the beginning of a new era in Solar System astronomy, where the exploration of the transneptunian population is one of the major issues. The number of discovered transneptunian objects is rapidly approaching 1000 (see `http://cfa-www.harvard.edu/iau/lists/TNOs.html`), and the amount of information on their physical properties is also growing quickly. We are now able to distinguish several major components of this population, all of which may contribute to the flux of captures into the Jupiter family. Fig. 3 shows how the transneptunians are distributed over semimajor axes and perihelion distances.

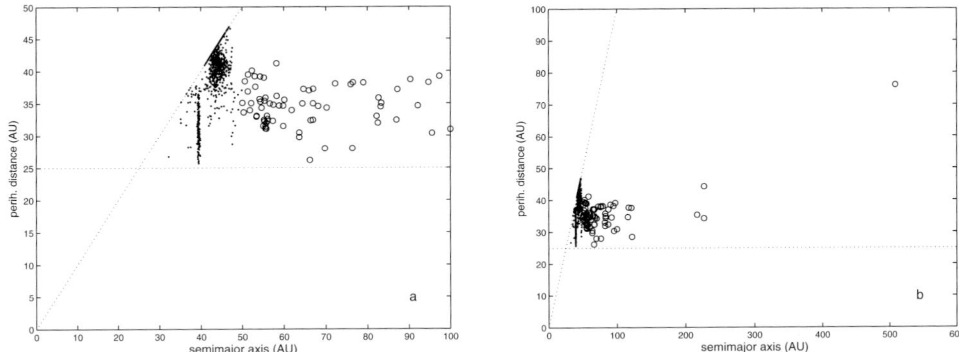

Figure 3. Perihelion distances vs. semimajor axes for the transneptunian population, according to recent data from the IAU Minor Planet Center web page. Only objects with $q > 25$ AU have been included. Dots denote objects with semimajor axes $a < 50$ AU, and circles show more distant orbits reaching further out. The latter roughly correspond to the scattered, and extended scattered disks. (a) Plot limited to $a < 100$ AU. (b) Plot showing all discovered objects.

The Plutinos are locked in a stable 2:3 resonance with Neptune and are not directly subject to captures. However, upon collisions with other transneptunian objects, fragments of Plutinos may approach Neptune and be transferred into the inner Solar System. The classical Kuiper belt, which is sometimes called the Edgeworth-Kuiper belt (see Edgeworth 1949), is a planetesimal population that derives from the cold disk resulting from the preplanetary accretion phase. For as yet unknown reasons this has been heated to significant eccentricities and inclinations, and the population is collisionally evolved (for a review, see Morbidelli and Brown 2004). Following collisions, small objects may enter into resonances and slowly reach eccentricities large enough to become planet-crossing, after which captures may occur.

The scattered disk is a distinct population of objects that originate from Neptune's accretion and have been expelled into orbits that often reach far outside all planets. They are not always able to approach Neptune in their current orbits, since long-term interactions with the planets may temporarily displace their perihelia somewhat (see e.g. Gladman *et al.* 2002). This was hypothesized by Fernández (1985). Subsequently, Duncan and Levison (1997) were able to demonstrate both that the scattered disk should exist (i.e., not all original objects have been removed) before the first object was discovered, and that it offers a viable source for the Jupiter family due to the continued approaches to Neptune. Duncan *et al.* (2004) estimate that the scattered disk predominates over the Edgeworth-Kuiper belt as such a source, based on available data on population sizes and dynamical time scales (see Section 3.2).

3. Current Problems

The orbital distribution of the Jupiter family can be adequately explained as a result of gravitational captures from either part of a low-inclination, transneptunian population via a process where all the giant planets participate. A detailed description of how this mechanism works is best offered by the team behind the simulations of such a capture (Duncan *et al.* 2004). The only additional assumption in order to produce an essentially perfect fit involves the observable lifetimes of Jupiter family comets. Newly captured comets tend to have very low inclinations, but repeated encounters with Jupiter thereafter scatter the inclinations over a range wider than observed. Fitting the observed inclination range is achieved by assuming an active lifetime between 3 000 and 25 000 years (Levison and Duncan 1997).

If the rest of the dynamical residence time in the Jupiter family is spent in a dormant state, then for each active comet there should be from two to six dormant ones. The statistics of observed asteroids in Jupiter family orbits appear to be consistent with this estimate (Rickman *et al.* 2001a), but no firm conclusions can be drawn at this stage.

3.1. *Population Characteristics*

A daunting, yet urgent task that has not yet been realized is to specify exactly what constraints can be placed on the abundance of the source. This means to specify the necessary rate of captures, based on the number of comets in the Jupiter family and their lifetimes. Therefore we need to characterize this population in terms of the distribution of brightness or nuclear size as well as that of perihelion distance. This is a very active area of research due to the compilation of large data bases on nuclear magnitudes (Fernández *et al.* 1999; Tancredi *et al.* 2000) or special observational programmes using front-line telescopes (Lamy *et al.* 2003; Meech *et al.* 2004). More research is needed in order to reach a coherent picture as to the power-law index for nuclear radius, and thus to define accurately the requirement on the capture rate.

The other major problem is to estimate the population sizes of the transneptunian sources and the associated infeed rates into Neptune-crossing orbits. Obviously, then, the observations so far refer to large objects with diameters $D \gtrsim 100$ km, and an estimate of the number of transneptunians with sizes similar to the observed Jupiter family comets must involve both a number $N(D \gtrsim 100$ km$)$ derived from observations, and a slope of the size distribution – assumed to be a power law, connecting with the number $N(D \lesssim 10$ km$)$. The latter is not very well constrained by observations and hence represents a significant source of uncertainty.

Jewitt *et al.* (1998) found the number of Edgeworth-Kuiper belt objects with $D \gtrsim 100$ km to be $\sim 1 \cdot 10^5$. The power-law slope of the differential distribution of radii appears to be $\alpha \simeq 4$, yielding an estimate of $\sim 1 \cdot 10^{10}$ for the number with diameters from 2 to 20 km. The scattered disk is less well constrained by observations, but Trujillo *et al.* (2000) found $N(D \gtrsim 100$ km$)$ to be $\sim (2-5) \cdot 10^4$ for $\alpha = 4$ and half of that number for $\alpha = 3$. If one prefers to use $\alpha = 4$ because this fits with the estimate for the Edgeworth-Kuiper belt, one gets $\sim 4 \cdot 10^9$ objects with $2 < D < 20$ km. Caution is however called for, since it is not certain that the scattered disk is collisionally evolved to the same degree as the Edgeworth-Kuiper belt. Using $\alpha = 3$ would yield 100 times fewer small-size objects.

For the scattered disk one may use the results by Duncan and Levison (1997) to relate the number of objects to that of the scattered-disk contribution to the Jupiter family with $q < 2.5$ AU. Thus, the ratio between those numbers was found to be $R = 1.3 \cdot 10^6$. If we take the analysis of the Jupiter family population and size distribution by Fernández *et al.* (1999), we get $N_{JF} \simeq 400$ for $D \gtrsim 2$ km and $q < 2.5$ AU. Taking the above R at face value, we would get $N_{SD} \simeq 5 \cdot 10^8$ for $D > 2$ km, which might appear to be on the low side. However, on the one hand there are uncertainties over the numbers and slope factors involved, and on the other hand it is likely that a typical Jupiter family comet has had its nucleus eroded by sublimation, so that the current radius may be about half of the original one. Allowing for this would increase the estimate of N_{SD} to $\sim 4 \cdot 10^9$, in perfect agreement with the above $\alpha = 4$ estimate.

3.2. *Capture Efficiencies*

The time scales of infeed into Neptune-crossing orbits for the Edgeworth-Kuiper belt and the scattered disk are worthy of further investigation, in particular as regards the values that prevail at the present time. This is particularly important, if the underlying history of the populations is considered in a more realistic way than previously. The scattered disk is the favoured source of Jupiter family comets, because its infeed time scale is believed to be much shorter than for the Edgeworth-Kuiper belt; thus the ratio R should be much larger for the latter source. However, the real infeed rate for the Edgeworth-Kuiper belt is difficult to determine, since it is set by resonant interactions that increase the eccentricities, and collisional evolution that produces an infeed into the resonances. The rate of this evolution depends on the population size and other uncertain parameters. The infeed rate for the scattered disk must be much smaller today than it was, when the Solar System was young, since the current population is made up of long-term survivors, protected by temporary displacement of the perihelia beyond Neptune's reach. Hence the ratio R must also have grown substantially, and the question remains, if the above value from Duncan and Levison (1997) is the one prevailing today. In conclusion, even if current evidence favours the scattered disk, the question of the main source for the Jupiter family is not yet definitively answered.

When estimating the infeed rate from the scattered disk, it is also important to consider all relevant perturbations. The effects of random passages of field stars, leading typically to changes of q up to several AU (Rickman *et al.* 2004), have not previously been included but may apparently influence the variation $R(t)$. But stellar perturbations may in fact have played a much more significant role in shaping the transneptunian population. Fernández (1997) drew attention to the possible importance of a dense stellar environment around the early Solar System for the formation of the Oort cloud, and Fernández and Brunini (2000) further explored the idea of the Sun's natal stellar cluster steering the early ejection of comets from the planetary accretion zones, creating a denser inner core of the Oort cloud than would otherwise have resulted. Ida *et al.* (2000) suggested that a close stellar encounter might have stirred up the early Edgeworth-Kuiper belt with important implications for its structure and collisional evolution.

3.3. *The Extended Scattered Disk*

However, the most important impulse to such research came with the discovery of minor planet 2003 VB_{12}, which is represented by the isolated data point at the upper right in Fig. 3b, and has a diameter $\simeq 3/4$ that of Pluto. Brown *et al.* (2004) described the discovery and its implications for the existence of a massive population of very remote objects, which may be identified with the "extended scattered disk" (Gladman *et al.* 2002) but turns out to be more significant than had earlier been expected. They also suggested that a slow and close encounter with a cluster star would offer the most likely mode of formation, by extracting the perihelia of scattered disk objects to large distances. This suggestion was supported by actual simulations (Morbidelli and Levison 2004), while Rickman *et al.* (2004) by their simulations found that random field stars passing during the age of the Solar System would also lead to a large, extracted population including orbits like that of 2003 VB_{12} via perturbations on the scattered disk (see Fig. 4).

The open questions now mainly concern the timing of the decisive event, and whether it had to occur early enough to strongly suggest a cluster star. A very early event might be required, if the mass of the extracted population demands that the scattered disk must have been very massive (e.g., ~ 10 Earth masses) by the time of the stellar encounter. The answer likely has to await relevant simulations of the evolution of the scattered disk in a realistic stellar environment as well as better knowledge of the population characteristics following further discoveries.

In any case, whatever the nature of the stellar passage, it also led to a massive injection of scattered-disk objects into the inner Solar System (see Fig. 4). Thus a random field star

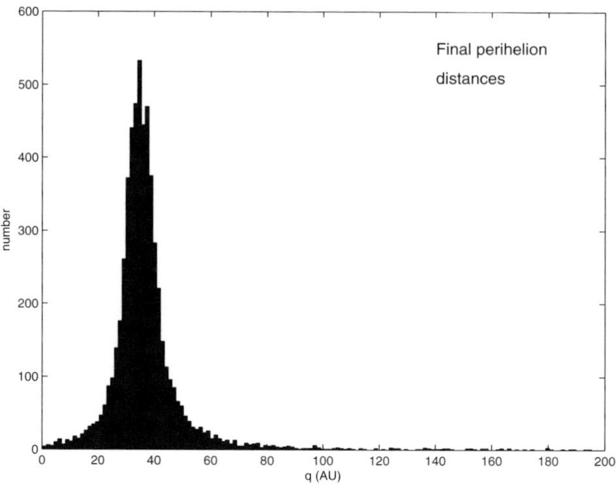

Figure 4. Distribution of perihelion distances acquired through passages of random field stars during 4 Gyr, according to a Monte Carlo simulation. Heliocentric impulses for a sample of 10 000 comets have been crudely estimated by the classical impulse approximation, and only those with an absolute value exceeding 0.00363 AU/yr have been considered as significant. The starting orbit in each case had a perihelion distance of 35 AU and an aphelion distance of 1000 AU. From Rickman *et al.* (2004).

passing ~ 4 Gyr ago may have triggered the late heavy bombardment, as mentioned by Rickman *et al.* (2004), along the lines first discussed by Mottman (1977). This illustrates very clearly that the cometary capture problem is not only relevant for explaining the observed distribution of cometary orbits; it also has implications for the cratering history of the terrestrial planets and, very likely, their history of volatile delivery.

Note that the extended scattered disk shares an interesting property with the inner core of the Oort cloud, namely, that on rare occasions it may be the source of strong, episodic infeed of cometary objects into the inner Solar System as a result of close stellar encounters. The term "cometary showers" (Hills 1981) has been coined to describe such events for the inner core, and the typical time scale for those is $\sim 10^7 - 10^8$ yr. Possibly, even more intense showers might arise from the extended and, to some extent, the normal scattered disk on a time scale of 10^9 yr due to even closer stellar encounters.

3.4. *Halley-Type Comets*

Comets with short orbital periods – typically, $P < 200$ years – and Tisserand parameters $T < 2$ are usually called Halley-Type Comets (HTCs; see Levison 1996). Their origin presents a separate problem, since captures from the Edgeworth-Kuiper belt or the scattered disk of the type discussed above will not produce HTCs. Many recent papers have dealt with that problem, but important issues remain to be resolved.

The Oort cloud provides an obvious source of HTCs, since it yields an infeed of Jupiter-crossing comets over the whole range of inclinations and values of T covered by the Halley-types. But other sources have recently been discussed as well, and once more we are faced with the problem to distinguish the different contributions. First, the Oort cloud flux extends over the entire planetary system with perihelia from 0 to $\gtrsim 30$ AU from the Sun, and while the simulations by Emel'yanenko and Bailey (1998) showed a predominance of original perihelia with $q_o < 2$ AU among comets captured into HTCs, the question remains how large the contribution from initial perihelia in the outer planetary system may be.

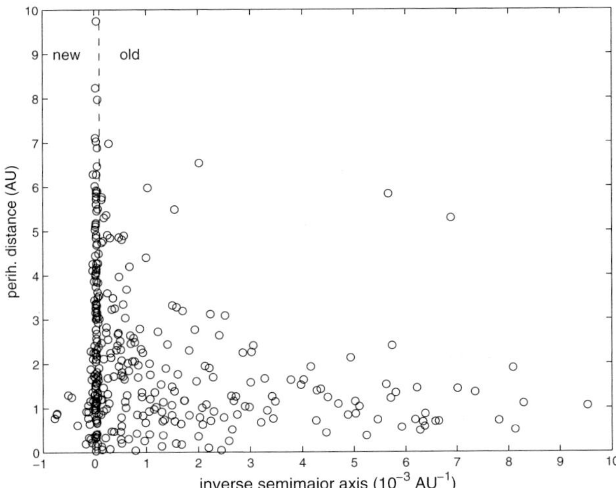

Figure 5. Perihelion distance vs semimajor axis of original orbits of long-period comets, according to a recent sample of observed comets. All quality classes of those orbits have been included. The data is from Marsden and Williams (2003).

The capture probability is then much smaller, but the flux of Oort cloud comets is larger outside the orbits of Jupiter and Saturn, where the inner core is also contributing (Hills 1981; Bailey 1990). An interesting feature of the evolution of such comets into HTCs is that the large changes of q are mostly driven by secular resonances (Bailey and Emel'yanenko 1996), which means that T is not conserved since Jupiter's eccentricity plays an essential role. Evaluation of the Oort cloud flux in the outer planetary system can only be done preliminarily as yet, based on models of the orbital energy distribution of the cloud and the effects of Galactic tides and stellar perturbations. Improvements may be expected from new developments in modelling the Galactic tides (e.g. Fouchard 2004) and the long-term effects of planetary perturbations (e.g. Rickman *et al.* 2001b).

The inclination distribution of the HTCs has been identified as a problem, if the isotropic Oort cloud is the only source, since there is a majority of prograde orbits among the HTCs. Prograde orbits are more strongly perturbed on the average than retrograde ones, but as a consequence the retrograde comets stay longer after capture and, unless physical evolution makes the comets unobservable after some limited time, there should be an excess of retrograde HTCs from the Oort cloud (Levison *et al.* 2001). These authors suggested a flattened inner Oort cloud with a preference for low inclinations as a supplement to the isotropic, outer Oort cloud, in order to solve this problem. But more work in needed, in particular on clarifying the physical evolution of HTCs and observability conditions, before we can really understand the inclination distribution. A viable idea (Levison *et al.* 2004) is that the scattered disk is feeding the HTC population with comets in preferentially prograde orbits via gravitational scattering into semimajor axes $a \gtrsim 10\,000$ AU, decrease of q by Galactic tides, and capture into HTC orbits by the giant planets.

Finally, observable lifetimes are a major issue when discussing cometary capture, especially for HTCs. The well-known "fading problem" (Weissman 1980; Bailey 1984) means that there is a lack of returning, long-period comets with respect to what would be inferred from the flux of new Oort cloud comets on the basis of pure dynamics. Fig. 5 shows statistics of original orbits of long-period comets, and it is clear that the concentration to the interval with $1/a < 1 \cdot 10^{-4}$ AU^{-1} is even sharper for large-q comets than for the classically observed, low-q comets. This must mean that there is a rapid fading or a particular brightness of new comets, or that cometary nuclei are quickly disrupted during their first passages near the Sun. Emel'yanenko and Bailey (1998) found that only $\sim 1\%$ of the Oort cloud comets with $q < 2$ AU can survive the capture into HTCs as observable objects, which is yet another indication of the same, elusive phenomenon. Levison *et al.* (2002) compared the number of thus 'unobserved' HTCs and long-period comets with the number of discovered asteroidal objects in such orbits and came to the conclusion that nearly all the Oort cloud comets must become disrupted.

However, more work is needed before we fully understand the loss of comets. One particular problem with the disruption idea, in addition to the fact that disruptions are not ubiquitously observed, is the indication from Fig. 5 that large-q comets should be preferentially affected, contrary to the intuitive picture of disruptions as caused by effects of solar heating. Another problem is that Jupiter family comets do not appear to frequently suffer disruption during their capture, in stark contrast to the HTCs, even though the objects seem to have formed under similar conditions, and other major differences of physical or chemical constitution are not known.

4. Concluding Remarks

The transport of comets from distant sources into the inner Solar System and the origin of the observed short-period cometary populations present a complex problem involving both cometary physics and dynamics. This problem is currently pursued very intensely for several reasons that go beyond the mere explanation of observed cometary populations. One is that current and future, ambitious space missions to Jupiter-family comets are largely motivated by a wish to use them as cosmogonical probes, and this requires that we understand where they were formed and how they have evolved. Another reason is that the delivery of comets into Earth-crossing orbits may have played an important role in volatile delivery to our planet and may also have made a significant contribution to the terrestrial cratering history, possibly including the late heavy bombardment.

It is fair to say that the dynamics of such transport is generally well understood, thanks to work involving numerical simulations which was carried out during the last decade or two. The main bottleneck for understanding the problem at present lies in characterizing the populations involved. Thus the Jupiter family and HTC populations are fairly well observed, but uncertainties remain over discovery biases related to perihelion distance as well as the nuclear size distribution. The transneptunian sources including the Oort cloud are still unobserved as far as comet-sized objects are concerned, and we have to rely on uncertain extrapolations from the observed data on $D \gtrsim 100$ km objects. A particular problem is presented by the scattered disk, whose orbital distribution and infeed rate into Neptune-crossing orbits must have varied during the history of the Solar System, and for which even the current orbital distribution is not very well constrained so far.

Uncertainties also remain over some aspects of the dynamics of the distant sources, including the interplay of Galactic tides and stellar perturbations in shaping the Oort cloud and regulating its infeed of comets into the planetary system. Moreover, we need to better understand the formation and evolution of all distant sources in order to answer very important questions about the transport of comets in the early Solar System.

Future progress may be expected both from deeper surveys of the sky (e.g., Pan-STARRS) that will likely multiply the number of known objects by a large factor, from continued, in-depth study of cometary nuclei and transneptunian objects, and from more ambitious or more realistic simulations of the external perturbations on the distant sources.

References

Bailey, M.E. 1984, *Mon. Not. R. Astron. Soc.* 211, 347

Bailey, M.E. 1986, *Nature* 324, 350

Bailey, M.E. 1990, in: C.-L. Lagerkvist *et al.* (eds.), *Asteroids, Comets, Meteors III* (Uppsala Univ.), p. 221

Bailey, M.E. & Emel'yanenko, V.V. 1996, *Mon. Not. R. Astron. Soc.* 278, 1087

Brown, M.E., Trujillo, C. & Rabinowitz, D. 2004, *Astrophys. J. Letters*, subm.

Carusi, A. & Valsecchi, G.B. 1985, in: A. Carusi & G.B. Valsecchi (eds.), *Dynamics of Comets: Their Origin and Evolution* (Dordrecht/Boston/Lancaster: Reidel), p. 261

Carusi, A., Kresák, L', Perozzi, E. & Valsecchi, G.B. 1987, *Astron. Astrophys.* 187, 199

Danby, J.M.A. 1962, *Fundamentals of Celestial Mechanics* (New York: MacMillan)

Duncan, M. & Levison, H. 1997, *Science* 276, 1670

Duncan, M., Quinn, T. & Tremaine, S.D. 1988, *Astrophys. J. Letters* 328, L69

Duncan, M., Levison, H. & Dones, L. 2004, in: M.C. Festou, H.U. Keller & H.A. Weaver (eds.), *Comets II* (Tucson: Univ. Arizona), in press

Edgeworth, K. 1949, *Mon. Not. R. Astron. Soc.* 109, 600

Emel'yanenko, V.V. & Bailey, M.E. 1998, *Mon. Not. R. Astron. Soc.* 298, 212

Everhart, E. 1972, *Astrophys. Letters* 10, 131

Fernández, J.A. 1980, *Mon. Not. R. Astron. Soc.* 192, 481

Fernández, J.A. 1985, in: A. Carusi & G.B. Valsecchi (eds.), *Dynamics of Comets: Their Origin and Evolution* (Dordrecht/Boston/Lancaster: Reidel), p. 45

Fernández, J.A. 1997, *Icarus* 129, 106

Fernández, J.A. & Brunini, A. 2000, *Icarus* 145, 580

Fernández, J.A., Tancredi, G., Rickman, H. & Licandro, J. 1999, *Astron. Astrophys.* 352, 327

Fouchard, M. 2004, *Mon. Not. R. Astron. Soc.* 349, 347

Gladman, B., Holman, M.J., Grav, T., Kavelaars, J.J., Nicholson, P., Aksnes, K. & Petit, J.-M. 2002, *Icarus* 157, 269

Hills, J.G. 1981, *Astron. J.* 86, 1730

Ida, S., Larwood, J. & Burkert, A. 2000, *Astrophys. J.* 528, 351

Jewitt, D.C. 2004, in: M.C. Festou, H.U. Keller & H.A. Weaver (eds.), *Comets II* (Tucson: Univ. Arizona), in press

Jewitt, D., Luu, J. & Trujillo, C. 1998, *Astron. J.* 115, 2125

Joss, P.C. 1973, *Astron. Astrophys.* 25, 271

Kazimirchak-Polonskaya, E.I. 1972, in: G.A. Chebotarev, E.I. Kazimirchak-Polonskaya & B.G. Marsden (eds.), *The Motion, Evolution of Orbits, and Origin of Comets* (Dordrecht: Reidel), p. 373

Kresák, L'. 1972, *Bull. Astron. Inst. Czech.* 23, 1

Kuiper, G.P. 1951, in: J.A. Hynek (ed.), *Astrophysics: A Topical Symposium* (New York: McGraw-Hill), p. 357

Lamy, P., Toth, I., Fernández, Y.R. & Weaver, H.A. 2004, in: M.C. Festou, H.U. Keller & H.A. Weaver (eds.), *Comets II* (Tucson: Univ. Arizona), in press

Lexell, A.J. 1778, *Acta Acad. Sci. Petropol* 1, 332

Lexell, A.J. 1780, *Acta Acad. Sci. Petropol* 2, 328

Levison, H. 1996, in: T.W. Rettig & J.M. Hahn (eds.), *Completing the Inventory of the Solar System*, (San Fransisco: ASP), p. 173

Levison, H. & Duncan, M. 1997, *Icarus* 127, 13

Levison, H.F., Dones, L. & Duncan, M.J. 2001, *Astron. J.* 121, 2253

Levison, H.F., Morbidelli, A., Dones, L., Jedicke, R., Wiegert, P.A. & Bottke, W.F. 2002, *Science* 296, 2212

Levison, H.F., Duncan, M.J., Dones, L. & Gladman, B.J. 2004, *Icarus*, subm.

Luu, J. & Jewitt, D. 1993, *Nature* 362, 730

Marsden, B.G. & Williams, G.V. 2003, *Catalogue of Cometary Orbits 2003* (IAU: CBAT & MPC)

Meech, K.J., Hainaut, O.R. & Marsden, B.G. 2004, *Icarus* 170, 463

Morbidelli, A. & Brown, M.E. 2004, in: M.C. Festou, H.U. Keller & H.A. Weaver (eds.), *Comets II* (Tucson: Univ. Arizona), in press

Morbidelli, A. & Levison, H.F. 2004, *Astron. J.*, in press

Mottman, J. 1977, *Icarus* 31, 412

Oort, J.H. 1950, *Bull. Astron. Inst. Netherlands* 11, 91

Quinn, T., Tremaine, S. & Duncan, M. 1990, *Astrophys. J.* 355, 667

Rickman, H. 1992, in: D. Benest & Cl. Froeschlé (eds.), *Interrelations between Physics and Dynamics for Minor Bodies in the Solar System* (Gif-sur-Yvette: Ed. Frontières), p. 197

Rickman, H., Fernández, J.A., Tancredi, G. & Licandro, J. 2001, in: M.Ya. Marov & H. Rickman (eds.), *Collisional Processes in the Solar System* (Dordrecht/Boston/London: Kluwer), p. 131

Rickman, H., Valsecchi, G.B. & Froeschlé, Cl. 2001b, *Mon. Not. R. Astron. Soc.* 325, 1303

Rickman, H., Froeschlé, Cl., Froeschlé, Ch. & Valsecchi, G.B. 2004, *Astron. Astrophys.*, subm.

Tancredi, G., Fernández, J.A., Rickman, H. & Licandro, J. 2000, *Astron. Astrophys. Suppl.* 146, 73

Trujillo, C.A., Jewitt, D.C. & Luu, J.X. 2000, *Astrophys. J. Letters* 529, L103

Weissman, P.R., 1980, *Astron. Astrophys.* 85, 191

Whipple, F.L. 1972, in: G.A. Chebotarev, E.I. Kazimirchak-Polonskaya & B.G. Marsden (eds.), *The Motion, Evolution of Orbits, and Origin of Comets* (Dordrecht: Reidel), p. 401

Dynamics of Populations of Planetary Systems
Proceedings IAU Colloquium No. 197, 2005
Z. Knežević and A. Milani, eds.

© 2005 International Astronomical Union
DOI: 10.1017/S1743921304008786

Nongravitational Accelerations on Comets

Steven R. Chesley[1] and Donald K. Yeomans

Jet Propulsion Laboratory, M/S 301-150, Pasadena, CA 91109, USA
[1]email: Steven.R.Chesley@jpl.nasa.gov

Abstract.
The orbital motion of comets is difficult to characterize accurately due to the rocket-like outgassing of material from the cometary nucleus. The resulting nongravitational accelerations often appear to be fundamentally stochastic in nature and thus pose severe modeling challenges in orbit determination, especially when the comet has been observed for many revolutions. Even so, new techniques have arisen in recent years that give new insight, not only into the motion of the comets, but also into their physical characteristics and spin states. These approaches include modeling of spin axis precession over many decades and the consideration of the seasonal variation in the thrust from discrete jets acting on a rotating nucleus. Such advances have been enabled, in part, by the increasing efforts and capabilities of comet observers worldwide as more and more comets with longer and longer observing arcs become available for study. In this review we specifically consider the application of the Rotating Jet Model to several space mission targets, indicating how this model can often be used to infer the orientation of a comet's spin axis.

Keywords. Comets, orbit determination, nongravitational accelerations

---

1. Introduction

A detailed historical introduction to the nongravitational accelerations affecting the motions of comets has recently been provided by Yeomans *et al.* (2004a), and so here we will provide only a short introduction to this topic for completeness.

Even though it was only the second comet predicted to return to perihelion, Comet 2P/Encke exhibited anomalous orbital behavior already at its first predicted return in 1822, behavior that was inconsistent with a trajectory governed solely by the gravitational influences of the sun and planets. It was Encke himself who noted that his eponymous comet arrived at perihelion a few hours earlier than his predictions. He proposed an interplanetary resisting medium to explain the phenomenon (Encke 1823), a medium that he envisaged as an extension of the solar atmosphere or the remains of cometary and planetary atmospheres. This model allowed Encke to accurately predict the perihelion passages of his comet between 1825 and 1858. Variations on Encke's resisting model were utilized for much of the nineteenth century but in the first half of the twentieth century, problems arose when the motions of comets 14P/Wolf and 6P/d'Arrest were consistent with an increasing, rather than a decreasing, orbital period (Kamieński 1933, Recht 1940). Clearly a mechanism was required whereby orbital energy could be added to, as well as subtracted from, a comet's orbital motion.

While little recognized at the time, Bessel (1836) had pointed out that a comet expelling material in a radial sunward direction would suffer a recoil acceleration, and if the expulsion of material took place asymmetrically with respect to perihelion, there would be a decrease or increase in the comet's orbital period depending upon whether the comet expelled more material before or after perihelion. Although Bessel did not identify the expulsion of material with the vaporization of ices, his basic concept of cometary nongravitational accelerations would ultimately prove to be correct. The development of Whipple's (1950, 1951) icy conglomerate model for the cometary nucleus finally put the rocket-like effect of an outgassing cometary nucleus on a firm physical footing and laid the ground work for more sophisticated models that could account for the cometary nongravitational accelerations. The so-called standard model for cometary nongravitational accelerations, which has the comet's outgassing reaching a peak at perihelion, was introduced by Marsden *et al.* (1973). A modification of this model by Yeomans & Chodas (1989) allowed the comet's outgassing to peak on either side of perihelion. Since these two models will be discussed in some detail in Sec. 2, we will move on to summarize some alternate nongravitational acceleration models that have been employed to represent the long-term motion of active periodic comets.

Partly to account for the time dependence found for some cometary nongravitational effects, Whipple & Sekanina (1979) and Sekanina (1981, 1984) introduced a linear precession model for a spherically symmetric nucleus. In this case, the outgassing acceleration does not act upon the nucleus center-of-mass and hence a torque is introduced causing the nucleus spin axis to precess with time. This introduces a time-varying nongravitational effect from apparition to apparition in a natural manner. Within this model, Sitarski (1995, 1996) introduced a time shift that allowed the outgassing peak to be reached before or after perihelion. Sekanina's (1984) forced precession model for the cometary nucleus, where he derived formulae for changes of the spin-axis orientation, was extended by Królikowska *et al.* (1998) for a nonspherical nucleus. In models of this type, solutions are made for the corrections to the six orbital elements and several additional parameters that include the obliquity of the nucleus, the magnitude of the nongravitational acceleration at one AU and the three angles that describe the direction of the nongravitational acceleration vector in orbital coordinates.

More recently, a Rotating Jet Model (RJM) has been introduced to model discrete outgassing jets on the surface of a cometary nucleus. That is, the nongravitational thrusting acting upon the nucleus takes place only when these active surface areas are exposed to sunlight and this depends upon the orientation of the nucleus spin pole. In a series of papers, Sekanina (1988a, 1988b, 1993) discussed the rotating jet model and used it to interpret the observed sunward fan-like coma of Comet 2P/Encke as an effect of northern and southern localized jets upon the surface of this comet's nucleus. The rotation averaged orbital components of the nongravitational acceleration for a nucleus with active jets was adapted for orbital computations by Szutowicz (2000) and independently by Chesley (2002). This model is described further in Sec. 3, and used to model the motions of several comets in Sec. 4.

2. The Extended Standard Model

The notion that comet nongravitational accelerations were caused by sublimating ices eventually led to the development of a model that proved effective in describing the variation of outgassing activity with heliocentric distance. That model, introduced by Marsden *et al.* (1973), is explicitly given by

$$\mathbf{a} = g(r)\,(A_1\,\hat{\mathbf{e}}_R + A_2\,\hat{\mathbf{e}}_T + A_3\,\hat{\mathbf{e}}_N). \tag{2.1}$$

Here the so-called $g(r)$ function reflects the sublimation rate of water ice as a function of heliocentric distance; $g(r)$ decays roughly as r^{-2} out to around 2 AU, and then much more steeply ($\sim r^{-23.5}$) beyond about 3 AU where the sublimation of water ice is substantially halted. This function remains a central feature of virtually all nongravitational acceleration models, even as these models have become increasingly complex.

The A_i in Eq. 2.1 are constant parameters that give the nongravitational acceleration the comet would experience when it is at 1 AU in each of the three coordinate directions. The accelerations are referred to the classical heliocentric Radial-Transverse-Normal (RTN) reference frame, denoted by $\hat{\mathbf{e}}_R$, $\hat{\mathbf{e}}_T$ and $\hat{\mathbf{e}}_N$, respectively. Marsden *et al.* (1973) reported that it was generally sufficient to neglect out-of-plane accelerations, and so, since its introduction, this model has been usually restricted to in-plane accelerations, i.e., $A_3 = 0$. However, the majority of the comets that we have investigated as a part of potential or actual flight mission studies have revealed substantial out-of-plane accelerations, and so, while the practice of neglecting A_3 has become commonplace, we have found that this practice is often not well-founded.

The standard formulation places the peak acceleration at the point of perihelion passage, while there are numerous comets known to exhibit marked asymmetries with respect to perihelion in their outgassing activity. To model this effect, Yeomans & Chodas (1989) applied a time offset ΔT to $g(r)$, essentially shifting the peak outgassing activity some number of days before or after perihelion. In other words, they replaced $g(r)$ by $g(r')$, where $r' = r(t') = r(t + \Delta T)$. In doing so Yeomans & Chodas found markedly improved orbital fits and predictions for a few comets, most notably 6P/d'Arrest, and moderate improvements for several other comets. Interestingly, they reported little improvement for two comets, 67P/Churyumov-Gerasimenko and 26P/Grigg-Skjellerup, that were known to exhibit notable perihelion asymmetries in their brightness.

Combining the foregoing formulations, we obtain the Extended Standard Model (ESM), a four-parameter (A_1, A_2, A_3 and ΔT) model that is the most versatile model available. Its utility lies in its simplicity. There is no requirement for *a priori* knowledge of the comet's physical characteristics or spin state, and no allowance for the possibility that nongravitational effects could vary from revolution to revolution. The success of this approach is due to the fact that most comets do not behave erratically over intervals up to a few orbital periods. Thus it works remarkably well across a wide cross-section of comets and it can be applied with a minimum of preliminary analysis. Still, the model assumes uniform behavior from revolution to revolution and does not specifically model more complex factors such as seasonal fluctuations in outgassing activity. For these, more refined approaches are necessary.

3. Rotating Jet Model

The *instantaneous* acceleration of a cometary jet depends upon the intrinsic jet strength A_J, the heliocentric distance r and the solar zenith angle z at the jet source according to

$$\mathbf{a}_J = -A_J\, g(r) \cos z\, \hat{\mathbf{e}}_J \tag{3.1}$$

when $z \geqslant 0$. For $z < 0$, i.e., when the source is in darkness, we have $\mathbf{a}_J = 0$. Here $\hat{\mathbf{e}}_J$ denotes the unit vector of the instantaneous jet direction. This formulation allows one to interpret A_J as a normalized acceleration; it is the magnitude of the acceleration when the sun is at the jet's zenith and the comet is at 1 AU from the sun.

As explained in detail in Appendix A, the *mean* acceleration of a cometary jet, averaged over a single comet rotation, is given by

$$\bar{\mathbf{a}}_J = g(r)A_J\,(J_S\,\hat{\mathbf{e}}_S + J_P\,\hat{\mathbf{e}}_P), \tag{3.2}$$

where the unit vector $\hat{\mathbf{e}}_P$ denotes the direction of the rotation axis and the unit vector $\hat{\mathbf{e}}_S$ denotes the most sunward direction in the the comet's equatorial plane. (The averaging is independent of the rotation direction, and so $\hat{\mathbf{e}}_P$ could refer to either the north or south pole.) The parameters J_S and J_P account for the daily insolation experienced by the particular jet, assuming a spherical figure. They depend upon the "season" (subsolar latitude) and the thrust angle η that the jet forms with the rotation pole (i.e., the jet's colatitude). When the jet is in the polar night regime $J_S = J_P = 0$ and there is no acceleration. The equations for J_S and J_P in the diurnal and polar day regimes are given in Appendix A, which also describes an extension to account for diurnal lag in the jet's activity.

The RJM is particularly attractive because it is a much higher fidelity model of a comet's outgassing activity than the ESM. Moreover, multiple jets can easily be accounted for by superposition. The RJM fully accounts for the fact that at various points along the orbit some source regions are continually shutdown during polar night, others are continually active during polar day, and still others are active only part of the day during the diurnal regime. This varying activity leads to nongravitational accelerations that may not be well modeled by the ESM. Further, with the RJM approach, only certain pole orientations can produce the observed accelerations, which allows the spin axis to be estimated in some cases. Of course, one problem is that some *a priori* knowledge of the comet's spin axis and the location of jets is generally required. However, as we show below, even without this information, the RJM can be used with considerable success to derive these quantities.

4. Application to selected actual or potential space mission targets

We investigate the nongravitational accelerations acting on several selected space mission targets: past, present and future. The six selected comets and the respective observational data sets that we have used are detailed in Table 1. These comets (and their associated space missions) include, 19P/Borrelly (Deep Space 1), 81P/Wild 2 (Stardust), 9P/Tempel 1 (Deep Impact), 67P/Churyumov-Gerasimenko (Rosetta), along with 2P/Encke and 6P/d'Arrest (erstwhile targets of the ill-fated CONTOUR mission).

In Table 2 we list for each comet the best-fitting values for the ESM parameters, along with their formal uncertainties. Here the formal uncertainties are most useful to judge the statistical significance of a result, rather than as an indication of the true value of the parameter, since in some cases the values can vary significantly from apparition to apparition. We also list in Table 2 the pole that can be inferred from the estimated ESM parameters according to the technique described below for Borrelly. Finally, the table includes the normalized (i.e., dimensionless) Root Mean Square (RMS) of the fit.

Table 3 lists the best-fitting RJM parameters. In every case we have assumed two jets, one acting on each hemisphere, with fixed thrust angles. When jet locations are known from other sources we use the published values, otherwise we use a generic configuration with one near-polar jet on the northern hemisphere ($\eta = 10°$) and a mid-latitude jet on the southern hemisphere ($\eta = 135°$). Although the thrust angles were not varied, a systematic search for the best-fitting pole direction was conducted by exploring a 5°-raster in the R.A.-Dec. space of possible pole orientations. The best-fitting pole thus obtained is tabulated for each case, along with the associated jet strengths and the normalized RMS. In all of the RJM fits, we actually fit only the comet orbital elements and the jet strengths A_{J_i}; the thrust angles and pole orientations were held fixed during the fits.

4.1. Comet 19P/Borrelly

Comet 19P/Borrelly was the target of a September 2001 flyby of NASA's Deep Space 1 spacecraft. About two weeks before the encounter, the spacecraft imaged the comet and

Table 1. COMET OBSERVATIONAL DATA USED IN FITS

	No. Obs.	First Obs.	Last Obs.	No. App.
19P/Borrelly	1671	1980-Jul-21.3	2002-Jun-11.9	4
81P/Wild 2	1853	1988-Sep-09.5	2003-Dec-30.9	3
2P/Encke	1538	1989-Jun-01.6	2004-Jul-17.4	5
6P/d'Arrest	387	1982-Apr-23.3	2001-May-25.3	4
9P/Tempel 1	706	1967-Jun-08.4	2003-Dec-26.9	8
67P/C.-G.	1005	1988-Jul-06.4	2004-Jun-20.0	3

Table 2. RESULTS FOR EXTENDED STANDARD MODEL

	$A_1 \times 10^{10}$ (AU/d^2)	$A_2 \times 10^{10}$ (AU/d^2)	$A_3 \times 10^{10}$ (AU/d^2)	ΔT (days)	Inferred pole R.A.	Dec.	RMS
19P/Borrelly	18.2 ± 0.5	-0.6 ± 0.1	6.0 ± 0.5	-53 ± 2	$205°$	$-5°$	0.618
81P/Wild 2	15.8 ± 0.5	1.9 ± 0.3	-8.3 ± 0.4	-23 ± 6	$342°$	$20°$	0.704
2P/Encke	-0.02 ± 0.1	-0.102 ± 0.006	-9.2 ± 0.8	-1.7 ± 0.2	$260°$	$55°$	0.631
6P/d'Arrest	37.7 ± 0.6	-1.4 ± 0.3	-3.8 ± 1.0	101 ± 4	$27°$	$-14°$	0.787
67P/C.-G.	14.9 ± 0.7	-1.3 ± 0.2	5.9 ± 0.3	47 ± 4	$274°$	$-50°$	0.740
9P/Tempel 1	1.0 ± 0.3	0.06 ± 0.06	-0.4 ± 0.2	48 ± 25	N/A[†]		0.843

[†]For 9P/Tempel 1 the low significance of A_3 and ΔT indicate that a pole should not be inferred.

Table 3. RESULTS FOR ROTATING JET MODEL

	$A_{J_1} \times 10^{10}$ (AU/d^2)	$A_{J_2} \times 10^{10}$ (AU/d^2)	η_1	η_2	Best-fit pole R.A.	Dec.	RMS
19P/Borrelly	28 ± 1	22 ± 3	$10°$	$135°$	$200°$	$-10°$	0.623
81P/Wild 2	29 ± 1	98 ± 2	$8°$	$115°$	$320°$	$15°$	0.711
2P/Encke	7.9 ± 0.4	6.5 ± 0.5	$35°$	$165°$	$225°$	$40°$	0.628
6P/d'Arrest	69 ± 4	137 ± 3	$10°$	$135°$	$25°$	$-20°$	0.779
67P/C.-G., pole A	25 ± 1	47 ± 2	$10°$	$135°$	$275°$	$-55°$	0.734
67P/C.-G., pole B	90 ± 4	47 ± 2	$10°$	$135°$	$90°$	$75°$	0.739

NOTE: For 9P/Tempel 1 the rotating jet model does not reveal a preferred pole orientation.

the associated optical navigation measurements were made available for improving the comet's orbit. It was only with great difficulty that these observations from the spacecraft were able to be absorbed into the fit. Even after carefully weighting and debiasing the ground-based astrometry, an acceptable fit with the two-parameter standard model (i.e., A_1 & A_2 only) still required a very short fit span, just over a single comet revolution, in order to accommodate the spacecraft observations (Chesley *et al.* 2001). The short data arc did allow us to deliver a comet ephemeris that was accurate to within about 50 km, as judged by the ground truth revealed from the successful flyby imaging. Even so, there was clearly room for improvement in modeling the Borrelly trajectory.

As a result of this experience, we extended our orbit determination software to estimate A_3 and ΔT, in addition to the conventional A_1 and A_2 parameters. We also developed, implemented and tested the rotating jet model described above. The results with the improved models were very favorable for Borrelly, as indicated by dramatically better fits and predictions.

The spacecraft encounter revealed the presence of a very strong polar jet (Soderblom *et al.* 2002) that had already been postulated by several astronomers (e.g., Farnham & Cochran 2002 and Schleicher *et al.* 2003). Because of the presence of a single dominant jet that seems to be responsible for most of the nongravitational acceleration, the values of the four ESM parameters can actually be used to orient the jet, and hence obtain a pole

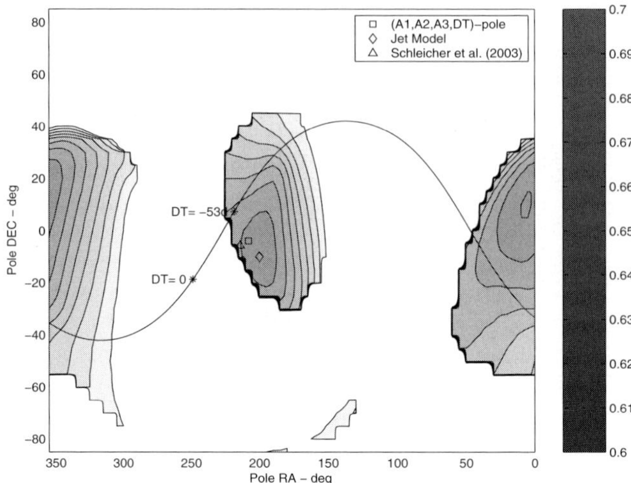

Figure 1. Contours of normalized fit RMS for 19P/Borrelly, revealing rotation poles favored by the rotating jet model. In this and subsequent plots, the blank regions of the plot correspond to nonphysical (negative thrust) jets and the solid sine wave is the locus of the comet's orbit plane. The sunward direction at $\Delta T = 0$ days (perihelion) and the best-fit $\Delta T = -53$ days (pre-perihelion) are marked. The poles derived from the RJM and ESM compare favorably with that of Schleicher *et al.* (2003) and several other published estimates (which are not plotted for the sake of clarity).

estimate. The idea is that the jet direction is given in the RTN frame by the acceleration vector (A_1, A_2, A_3) at time ΔT relative to perihelion. Simply rotating this acceleration into the equatorial frame yields the right ascension and declination of the polar jet. As indicated by Fig. 1, our pole estimate based on this method, $(\alpha, \delta) = (205°, -5°)$, compares well with the various estimates obtained by direct observation of the comet's coma.

Because Borrelly can be regarded as a ground truth for cometary jet activity and pole orientation, we use it as a test case to see whether these quantities can be inferred from the astrometry alone. To this end, we explore the entire space of possible pole orientations while assuming two jets, one at a high northern latitude of $+80°$ and another at a mid-southern latitude of $-45°$. For each possible pole on a $5°$ grid we obtain the best-fitting jet strengths A_{J_1} and A_{J_2}. We discard as nonphysical those cases where either jet has a statistically significant negative thrust; these regions are revealed by an absence of contours in Fig. 1 and similar plots. As indicated by Fig. 1, this experiment reveals two favored pole positions. One, known *a priori* to be correct, has the northern polar jet facing sunward 50–60 days before perihelion, and the other, just slightly favored by the fits, has the $180°$ opposite pole so that the mid-latitude southern jet is under continuous illumination at that time. To a large extent this is merely an ambiguity in the sense of rotation; the orbit fitting is simply striving to have a maximal thrust at 50–60 days pre-perihelion, a property also replicated by the ESM. The conclusion from this test is that both models have the potential to, at least in some cases, indicate the pole orientation of a subject comet.

4.2. *Comet 81P/Wild 2*

In January 2004 the Stardust spacecraft successfully imaged Comet 81P/Wild 2 during a close flyby, producing the highest resolution images of a comet's surface obtained to

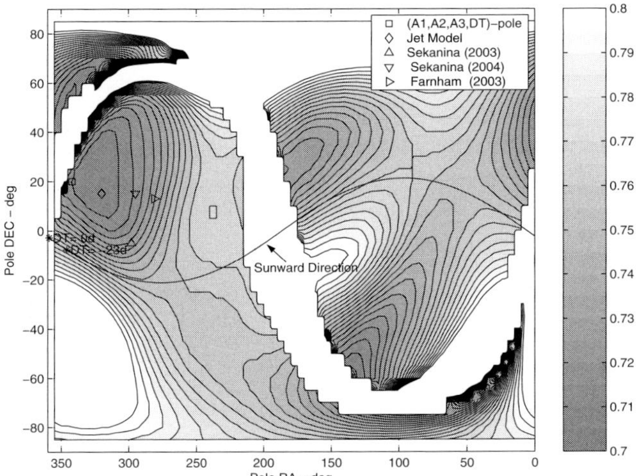

Figure 2. Contours of normalized fit RMS for 81P/Wild 2, revealing rotation poles favored by the rotating jet model. The solid sine wave defines the comet's orbit plane. The sunward direction at $\Delta T = 0$ days (perihelion) and the best-fit $\Delta T = -23$ days (pre-perihelion) are marked.

date. Sekanina *et al.* (2004) report that the flyby imaging was sufficient to reveal a spin direction around $(\alpha, \delta) = (295°, +15°)$, a result substantially consistent with earlier estimates from ground-based observations of jet activity (e.g., Farnham 2003, Sekanina 2003).

Following the RJM approach used above, we obtain the contours depicted in Fig. 2. As is clear from the figure, the poles derived from the RJM and ESM compare favorably with other published estimates obtained by direct observation. We note that Wild 2 is somewhat unusual among the comets studied so far because the 180° complement of the preferred pole positions is ruled out by the RJM. Of course, this does not reveal an independent determination of the *direction* of spin. Rather, recalling that for our tests the northen hemisphere held a near-polar jet, while the southern hemisphere had a mid-latitude jet, we conclude that the observed pre-perihelion increase in activity must be associated with a polar jet and not a mid latitude jet. Thus the seasonal effects of jetting are substantial and yield important constraints, not only on the pole position, but also on the thrust angles of the jets.

4.3. *Comet 2P/Encke*

Comet 2P/Encke was a primary target of the now-lost CONTOUR mission and was, as mentioned earlier, the first comet recognized to be affected by nongravitational accelerations. In a comprehensive study based on light curves and on the appearance of fans and jets in the morphology of Encke's coma, Sekanina (1988a, 1988b) was able to deduce two active jets, at latitudes $+55°$ and $-75°$, and he was also able to infer a precession in Encke's pole direction over the years from 1868 until 1924, beyond which point Sekanina (1988b) reports that he is unable to usefully constrain the pole position with his technique.

Using his two reported jet locations (which we note are roughly symmetric with those used above for Borrelly), we find the preferred polar orientations depicted in Fig. 3. The epoch of the pole estimate given by the contours in Fig. 3 is the midpoint of the Table 1 fit span, roughly year 1997.0. This suggests that the pole has been evolving steadily along

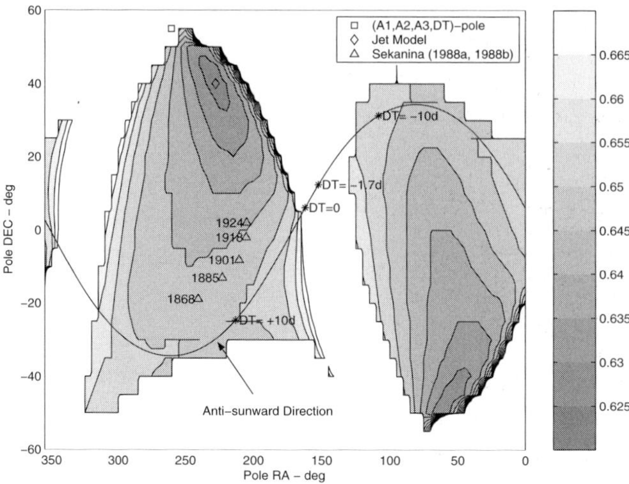

Figure 3. Contours of normalized fit RMS for 2P/Encke, revealing rotation poles favored by the rotating jet model. The solid sine wave denotes the comet's orbit plane. The anti-sunward direction at $\Delta T = 0$ days (perihelion), $\Delta T = \pm 10$ days and the best-fit $\Delta T = -1.7$ days are marked. (An anti-sunward north pole position faces Encke's south polar jet sunward.) The poles derived from the RJM and ESM suggest precession along a curved path at a more or less constant precession rate since 1868.

a roughly circular path since 1868, and fits with alternate observational arcs do indeed seem to confirm this trend. It is important to emphasize that the statistical significance of these conclusions is moderate, but not overwhelming. Even so, it is noteworthy that the inferred circular precession cycle contains the anti-perihelion vector $(\alpha, \delta) = (342°, -6°)$ near its center. Neishtadt *et al.* (2002) report that, for some comet configurations, a precession cycle centered on the perihelion direction can be a stable mode of spin state evolution, according to their modeling of discrete source outgassing.

4.4. *Comet 6P/d'Arrest*

Comet 6P/d'Arrest was a potential target for the CONTOUR mission and remains well situated for investigation by future space missions. In their original introduction of the use of ΔT, Yeomans & Chodas (1989) found this comet to have a statistically significant $\Delta T = +40$ days (with data from 1963 to 1988), in agreement with contemporaneous lightcurves. Fits of ΔT to progressively more recent data sets indicate that ΔT has been increasing steadily since about 1975, reaching a value of 101 ± 4 for the data arc considered here (Fig. 4). The most likely interpretation of this drift in DT is precession of the spin axis.

Comet 6P/d'Arrest is the first comet considered here for which we know of no independent pole information; however, we follow the approach used successfully in the three comets considered above in order to obtain a pole orientation for 6P/d'Arrest. Figure 5 indicates that the ESM-derived pole is similar to that favored by the RJM. In particular, pole positions in the region around our best-fit $(\alpha, \delta) = (25°, -25°)$ appear to be significantly favored by the RJM over those oriented in the opposite direction. This is a strong indication that the increase in post-perihelion activity is due to the action of a mid-latitude jet, rather than a near-polar jet. For this case, the right ascension of the pole is fairly well constrained, with uncertainty around $\pm 10°$, while the declination (and obliquity) are only determined to within roughly $\pm 30°$.

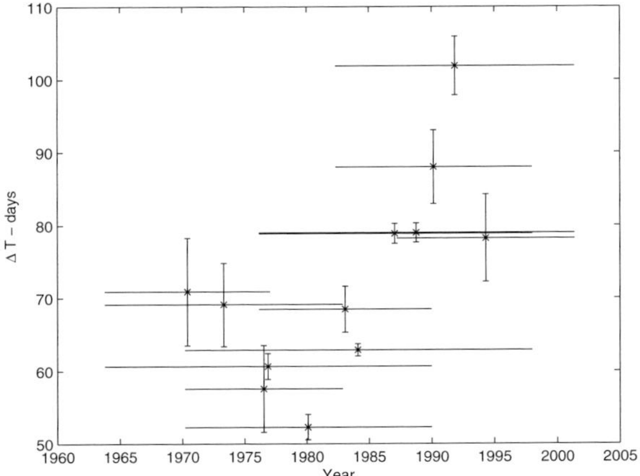

Figure 4. Best-fitting ΔT as a function of observational data set for 6P/d'Arrest with the Extended Standard Model. Horizontal lines depict the extent of observations used in fit; asterisks with error bars reflect the associated ΔT with formal uncertainty.

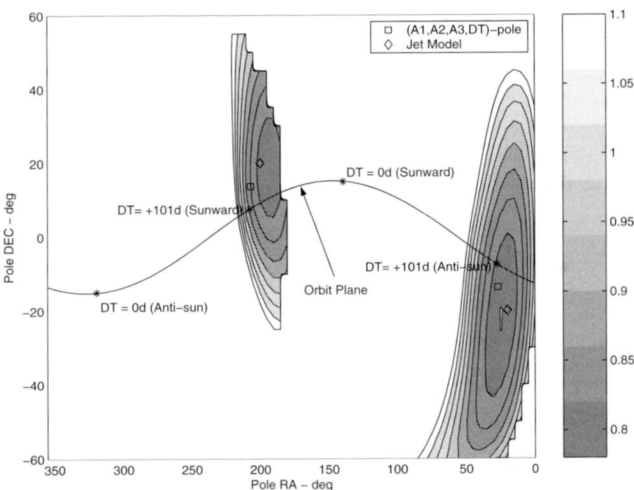

Figure 5. Contours of normalized fit RMS for 6P/d'Arrest, revealing rotation poles favored by the rotating jet model. The solid sine wave defines the comet's orbit plane. The sunward and anti-sunward directions at $\Delta T = 0$ days (perihelion) and the best-fit $\Delta T = +101$ days (post-perihelion) are marked.

4.5. *Comet 67P/Churyumov-Gerasimenko*

The recent re-targeting of the European Space Agency's Rosetta mission to 67P/Churyumov-Gerasimenko has made this comet the focus of considerable observational scrutiny. Several sources (e.g., Weiler *et al.* 2004) have reported substantially higher water and dust production rates post-perihelion, peaking 30–40 days after perihelion passage. Confirming these observations, it has been reported by Królikowska (2003) that both A_3 and ΔT are significant when using the ESM, a result with which we concur (Table 2).

We again explore the complete range of possible pole positions in the manner described earlier, and we find two favored pole orientations, which we label Pole A and Pole B. (See Table 3 and Fig. 6.) These are separated by $160°$, so they are two distinct solutions,

Chesley and Yeomans

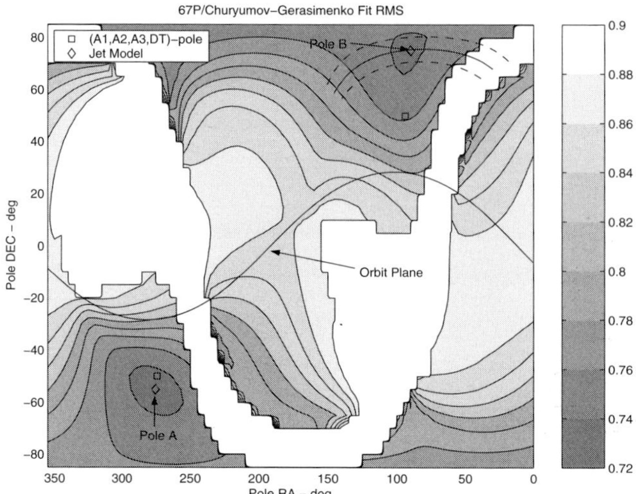

Figure 6. Contours of normalized fit RMS for 67P/Churyumov-Gerasimenko, revealing rotation poles favored by the rotating jet model. The solid sine wave defines the comet's orbit plane. The sunward and anti-sunward directions at $\Delta T = 0$ days (perihelion) and the best-fit $\Delta T = +47$ days (post-perihelion) are marked. The solid and dashed curves at upper right reflect the constraint provided by the March 2003 images of Weiler *et al.* (2004) as described in the text.

not just an ambiguity in the sense of rotation. Pole A has obliquity $117°$ with the north polar jet facing sunward around the time of peak outgassing. Pole B has obliquity $43°$ and has the mid-southern latitude jet responsible for the post-perihelion increase in activity. Statistically, there does not appear to be a substantial preference for either pole solution.

The chief distinction between the two pole solutions is whether a polar jet or a mid-latitude jet is responsible for the post-perihelion activity. Direct observational evidence strongly indicates that the latter is correct. Specifically, Weiler *et al.* (2004) report the presence of two persistent structures in images from 2003 March 7 and 28, and they suggest that these features could be the edges of a cone-shaped fan formed from a single active area on the rotating nucleus. We agree with this interpretation, and using their published images we find that the cone centerline, i.e., the projected pole, is in PA $165° \pm 5°$ and the fan extends $\pm 40°$ to either side.

In the case of a comet observed near opposition, as was the case for Churyumov-Gerasmenko in March 2003, we suppose that a rotating jet forming a cone will project as a fan-shaped structure if the pole has a modest projection angle towards the observer. If the angle is too great then the observer will be looking into the cone, which would appear more elliptical than fan-shaped; if the angle is too small, or even away from the observer (and sun), then the jet would not be sufficiently active to form a substantial cone. Thus we apply projection angles from $0°$ to $45°$ (towards the observer) to the projected PA 165 centerline and compute the associated pole positions. These are plotted in Fig. 6 as the solid curve at upper right, where the smaller projection angles appear at smaller right ascensions. The $\pm 5°$ variations in PA are plotted as dashed curves.

We find it remarkable that the constraints supplied by the Weiler *et al.* images are centered on pole B, and that for projection angles around $20°$—very realistic values—the two independent methods are in perfect agreement. Moreover, pole B, the pole orientation that is confirmed by the Weiler *et al.* images, is the one for which the mid-southern latitude ($-45°$) jet is responsible for the post-perihelion increase in activity, a point fully consistent with the 40° cone angle present on the March 2003 images. In contrast, if pole A were correct we would see a very narrow fan or a single collimated jet, as was the case with Borrelly. Further, in March 2003, pole A projects in PA 143°, a value incompatible with the Weiler *et al.* images.

4.6. *Comet 9P/Tempel 1*

NASA's Deep Impact mission, due to be launched in 2004 December, is planned to impact comet 9P/Tempel 1 on 2005 July 4 with a 360 kg projectile at a speed of around 10.2 km/s. The nongravitational accelerations appear to be rather unusual, being quite small and extraordinarily steady (Yeomans *et al.* 2004b). There is, for example, no difficulty in obtaining a simultaneous fit of all 8 apparitions observed since 1967. When fitting with the ESM (Table 2), only A_1 is significantly different from zero (although fixing $\Delta T = 0$ does lend substantial significance to the estimate for A_2). Also, the low significance levels for A_3 and ΔT do not allow a useful constraint to be placed on the pole orientation. While the RJM does give good fits for Tempel 1, the fit RMS is remarkably insensitive to the assumed pole position, and hence a statistically significant estimate of the pole position is not revealed. Thus, in contrast to the other five comets investigated here, neither model is capable of revealing the pole orientation of Tempel 1. All of this leads us to a few conclusions and speculations regarding Tempel 1:

- The stability of the nongravitational accelerations considerably eases the orbital prediction problem compared to other comets, making Tempel 1 an excellent target for an impacting mission such as Deep Impact.
- The low level of nongravitational acceleration implies a limited active surface area or a dwindling supply of volatiles. This line of reasoning suggests that Tempel 1 may be verging on extinction, either through depletion or some kind of mantling process.
- The nongravitational accelerations appear to have a negligible seasonal component. A possible explanation for this is that the comet's surface is more or less homogeneous so that there are not isolated active regions that account for most of the outgassing activity. Moreover, Tempel 1 has a period of about 42 hours (Belton *et al.* 2004), long enough that the diurnal thermal wave should penetrate relatively deeply, releasing volatiles that would otherwise be locked in the interior. Through this mechanism, the subsolar point may be the primary source of outgassing, effectively diminishing the signal from seasonal effects and body-fixed source regions.

5. Concluding Remarks

With these case studies we have shown that cometary spin axis information can, in many cases, be derived through measuring and modeling nongravitational accelerations on comets, although no information on the sense of rotation can be divined from these methods.

Poles inferred from the ESM assume that the nongravitational accelerations are dominated by a single high-latitude jet, a scenario that appears realistic in several of our tests, but is unlikely to be true in general. Even so, this argument suggests that a measured secular drift in ΔT should generally be interpreted as a drift in the pole orientation of the comet.

Poles obtained with the RJM are much more credible due to the more realistic nature of the modeling. Still, in the absence of corroborating measurements, such estimates should be viewed with some measure of skepticism.

We also show that the widespread assumption that comets generally have negligible out-of-plane accelerations should be revisited. In fact, we are now at a fortuitous point of convergence between increasing observing capabilities and improved nongravitational modeling. This suggests that a comprehensive review of the comet orbits with the techniques outlined in this paper could indicate more generally the extent to which out-of-plane accelerations are important and just how pervasive clear seasonal signals are among the comet population.

Acknowledgements

This research was conducted at the Jet Propulsion Laboratory, California Institute of Technology, under a contract with the National Aeronautics and Space Administration.

Appendix A. Rotating Jet Model

We intend to average $\mathbf{a}_J$ (Eq. 3.1) over one comet rotation. Assuming for the moment that the thrusting is symmetric about the solar meridian crossing, the averaged acceleration will be restricted to the plane containing the spin axis $\hat{\mathbf{e}}_P$ and the sun-comet line $\hat{\mathbf{r}} = \mathbf{r}/r$. This plane is defined by the normal vector

$$\hat{\mathbf{e}}_Q = \frac{\hat{\mathbf{r}} \times \hat{\mathbf{e}}_P}{|\hat{\mathbf{r}} \times \hat{\mathbf{e}}_P|} \tag{A 1}$$

The spin axis unit vector can be computed from the right ascension α and declination δ of the (assumed North) pole

$$\hat{\mathbf{e}}_P = (\cos\alpha \cos\delta,\ \sin\alpha \cos\delta,\ \sin\delta). \tag{A 2}$$

Finally, we define the $\hat{\mathbf{e}}_S$ axis

$$\hat{\mathbf{e}}_S = \hat{\mathbf{e}}_Q \times \hat{\mathbf{e}}_P, \tag{A 3}$$

which is opposite the projection of $\mathbf{r}$ on the comet's equatorial plane. This completes the right-handed (S, Q, P)-frame, in which the jet direction is given by

$$\hat{\mathbf{e}}_J = \cos\theta \sin\eta\, \hat{\mathbf{e}}_S + \sin\theta \sin\eta\, \hat{\mathbf{e}}_Q + \cos\eta\, \hat{\mathbf{e}}_P. \tag{A 4}$$

Here the rotation angle θ of the jet is measured about the pole from $\hat{\mathbf{e}}_S$ and the *thrust angle* η is the angle from $\hat{\mathbf{e}}_P$ to $\hat{\mathbf{e}}_J$. (Under the assumption that the jet is outgassing in a direction that is normal to the surface of a spherical comet, η is the colatitude of the jet's location.)

The insolation term $(\cos z)$ of Eq. 3.1 can be cast in terms of the rotation angle θ and the thrust angle η according to

$$\cos z = \cos\eta \cos\gamma + \sin\eta \sin\gamma \cos\theta, \tag{A 5}$$

where $\cos\gamma = -\hat{\mathbf{r}} \cdot \hat{\mathbf{e}}_P$, γ being the colatitude of the subsolar point.

The jet is on the terminator when $\cos z = 0$, that is, when

$$\cos\theta^* = -\frac{\cos\eta \cos\gamma}{\sin\eta \sin\gamma}, \tag{A 6}$$

where $\theta^* > 0$ is the rotation angle that places the jet on the afternoon terminator. If $\cos\theta^* < -1$ then the sun is never below the horizon and we explicitly define $\theta^* \doteq \pi$. Similarly, if $\cos\theta^* > 1$ then the sun is never above the horizon and $\theta^* \doteq 0$. These situations represent, respectively, the polar day and polar night regimes. Intermediate values of $\cos\theta^*$ correspond to the day-night, or diurnal, regime. In each of these three regimes the jet is illuminated over the interval $-\theta^* < \theta < \theta^*$, and this is the interval across which we wish to integrate Eq. 3.1 when averaging, since elsewhere the acceleration is zero.

Combining Eqs. 3.1 and A 4, we recast the instantaneous acceleration as

$$\mathbf{a}_J = -A_J\, g(r)\, \cos z\, (\cos\theta \sin\eta\, \hat{\mathbf{e}}_S + \sin\theta \sin\eta\, \hat{\mathbf{e}}_Q + \cos\eta\, \hat{\mathbf{e}}_P) \tag{A 7}$$

and average over one rotation according to

$$\bar{\mathbf{a}}_J = \frac{1}{2\pi} \int_{-\theta^*}^{\theta^*} \mathbf{a}_J\, d\theta = A_J\, g(r)\, \mathbf{J}, \tag{A 8}$$

where

$$\mathbf{J} = J_S\, \hat{\mathbf{e}}_S + J_P\, \hat{\mathbf{e}}_P. \tag{A 9}$$

The terms J_S and J_P are zero for the polar night regime ($\theta^* = 0$) and take on simple forms for the polar day regime ($\theta^* = \pi$)

$$J_S = -\frac{1}{2}\sin^2\eta \sin\gamma, \quad J_P = -\cos^2\eta \cos\gamma. \tag{A 10}$$

For the diurnal regime, we use Eq. A 5 to compute these terms

$$
\begin{aligned}
J_S &= -\frac{\sin\eta}{2\pi} \int_{-\theta^*}^{\theta^*} \cos z \cos\theta\, d\theta \\
&= -\frac{\sin\eta}{\pi} \left[\cos\eta \cos\gamma \sin\theta^* + \frac{1}{2}\sin\eta \sin\gamma\, (\theta^* + \cos\theta^* \sin\theta^*) \right] \\
J_P &= -\frac{\cos\eta}{2\pi} \int_{-\theta^*}^{\theta^*} \cos z\, d\theta = -\frac{\cos\eta}{\pi} \left(\theta^* \cos\eta \cos\gamma + \sin\eta \sin\gamma \sin\theta^* \right).
\end{aligned}
\tag{A 11}
$$

Note that the Q-component of acceleration has been eliminated by the averaging process because of the assumption that outgassing peaks at the solar meridian crossing and that the outgassing profile is symmetric about this point. However, if the diurnal acceleration peak actually occurs at a point other than $\theta = 0$, which may be expected if, for example, surface thermal inertia causes a diurnal lag, then there will be some component of acceleration in the $\hat{\mathbf{e}}_Q$ direction. To model this possibility we simply rotate $\hat{\mathbf{e}}_S$ by a diurnal lag angle $\Delta\theta$ to obtain a new equatorial direction for the averaged acceleration

$$\hat{\mathbf{e}}_E = \cos\Delta\theta\, \hat{\mathbf{e}}_S + \sin\Delta\theta\, \hat{\mathbf{e}}_Q, \tag{A 12}$$

so that, with a non-zero diurnal lag $\Delta\theta$,

$$\mathbf{J} = J_S \cos\Delta\theta\, \hat{\mathbf{e}}_S + J_S \sin\Delta\theta\, \hat{\mathbf{e}}_Q + J_P\, \hat{\mathbf{e}}_P. \tag{A 13}$$

If $\hat{\mathbf{e}}_P$ represents the North pole then one should expect $\Delta\theta > 0$ due to thermal inertia. On the other hand, $\Delta\theta < 0$ would suggest that $\hat{\mathbf{e}}_P$ is actually the South pole.

Of course, we can have more than one jet acting on the comet, in which case we can simply sum the contributions from each, using the subscript i to distinguish the contributions of the various jets:

$$\bar{\mathbf{a}}_J = g(r) \sum_{i=1}^{n} A_{J_i}\, \mathbf{J}_i. \tag{A 14}$$

The partial derivatives of the acceleration with respect to $\mathbf{r}$, $\mathbf{v}$ and any estimable model parameters are required as a part of the orbit determination process. The basic equations for these derivatives are

$$\frac{\partial \bar{\mathbf{a}}_J}{\partial \mathbf{r}} = \frac{\partial g(r)}{\partial \mathbf{r}} \sum_{i=1}^{n} A_{J_i} \mathbf{J}_i + g(r) \sum_{i=1}^{n} A_{J_i} \frac{\partial \mathbf{J}_i}{\partial \mathbf{r}}$$

$$\frac{\partial \bar{\mathbf{a}}_J}{\partial A_{J_i}} = g(r)\mathbf{J}_i; \quad \frac{\partial \bar{\mathbf{a}}_J}{\partial(\alpha,\delta)} = g(r)\sum_{i=1}^{n} A_{J_i}\frac{\partial \mathbf{J}_i}{\partial(\alpha,\delta)}$$

$$\frac{\partial \bar{\mathbf{a}}_J}{\partial \eta_i} = g(r)A_{J_i}\frac{\partial \mathbf{J}_i}{\partial \eta_i} = g(r)A_{J_i}\left(\frac{\partial J_{S_i}}{\partial \eta_i}\hat{\mathbf{e}}_E + \frac{\partial J_{P_i}}{\partial \eta_i}\hat{\mathbf{e}}_P\right)$$

$$\frac{\partial \bar{\mathbf{a}}_J}{\partial \Delta\theta} = g(r)\sum_{i=1}^{n} A_{J_i}\frac{\partial \mathbf{J}_i}{\partial \Delta\theta} = g(r)\left(-\sin\Delta\theta\,\hat{\mathbf{e}}_S + \cos\Delta\theta\,\hat{\mathbf{e}}_Q\right)\sum_{i=1}^{n} A_{J_i}J_{S_i}$$

The derivatives $\partial \mathbf{J}_i/\partial \mathbf{r}$ and $\partial \mathbf{J}_i/\partial(\alpha,\delta)$ are fairly complex and we compute these through finite differences. Other terms can be computed analytically.

References

Belton, M.J.S., Meech, K.J., A'Hearn, M.F. & 10 coworkers 2004, *Space Sci. Rev.*, in press

Bessel, F.W. 1836, *Astron. Nach.* 13, 345

Chesley, S.R. 2002, *Bull. Am. Astr. Soc.* 34, 869

Chesley, S.R., Chodas, P.W., Keesey, M.S., Bhaskaran, S., Owen, W.M., Jr., Yeomans, D.K., Garradd, G.J., Monet, A.K.B. & Stone, R.C. 2001, *Bull. Am. Astr. Soc.* 33, 1116

Encke, J.F. 1823, in: *Berliner Astronomische jahrbuch fur das jahr 1826* (Berlin) p. 124

Farnham, T.L. 2003, *Bull. Am. Astr. Soc.* 35, 971

Farnham, T.L. & Cochran, A.L. 2002, *Icarus*, 160, 398

Kamieński, M. 1933, *Acta Astron., Series A* 3, 1

Królikowska, M. 2003, *Acta Astron.* 53, 195

Królikowska, M., Sitarski, G. & Szutowicz, S. 1998, *Astron. Astrophys.* 335, 757

Marsden, B.G., Sekanina, Z. & Yeomans, D.K. 1973, *Astron. J.* 78, 211

Neishtadt, A.I., Scheeres, D.J., Sidorenko, V.V. & Vasiliev, A.A. 2002, *Icarus* 157, 205

Recht, A.W. 1940, *Astron. J.* 48, 65

Schleicher, D.G., Woodney, L.M. & Millis, R.L. 2003, *Icarus* 162, 415

Sekanina, Z. 1981, *Annual Rev. Earth Planet. Sci.* 9, 113

Sekanina, Z. 1984, *Astron. J.* 89, 1573

Sekanina, Z. 1988a, *Astron. J.* 95, 911

Sekanina, Z. 1988b, *Astron. J.* 95, 1455

Sekanina, Z. 1993, *Astron. J.* 105, 702

Sekanina, Z. 2003, *J. Geophys. Research* 108, 8112

Sekanina, Z., Brownlee, D.E., Economou, T.E., Tuzzolino, A.J. & Green, S.F. 2004, *Science* 304, 1769

Sitarski, G. 1995, *Acta Astron.* 45, 763

Sitarski, G. 1996, *Acta Astron.* 46, 29

Soderblom, L.A., Becker, T.L., Bennett, G. & 19 coworkers 2002, *Science* 296, 1087

Szutowicz, S. 2000, *Astron. Astrophys.* 363, 323

Weiler, M., Rauer, H., Helbert, J. 2004, *Astron. Astrophys.* 414, 749

Whipple, F.L. 1950, *Astron. J.* 111, 375

Whipple, F.L. 1951, *Astrophys. J.* 113, 464

Whipple, F.L. & Sekanina, Z. 1979, *Astron. J.* 84, 1894

Yeomans, D.K. & Chodas, P.W. 1989, *Astron. J.* 98, 1083

Yeomans, D.K., Chodas, P.W., Sitarski, G., Szutowicz, S. & Królikowska, M. 2004a, in: M. Festou (ed.), *Comets II* (Univ. Arizona), in press

Yeomans, D.K., Giorgini, J.D. & Chesley, S.R. 2004b, *Space Sci. Rev.*, in press

Dynamics of Populations of Planetary Systems
Proceedings IAU Colloquium No. 197, 2005
Z. Knežević and A. Milani, eds.

© 2005 International Astronomical Union
DOI: 10.1017/S1743921304008798

Interaction of planetesimals with the giant planets and the shaping of the trans-Neptunian belt

Harold F. Levison[1] and Alessandro Morbidelli[2]

[1]Southwest Research Institute, 1050 Walnut St., Suite 400, Boulder, CO 80302, USA
email: hal@boulder.swri.edu

[2]Obs. de la Côte d'Azur, B.P. 4229, 06034 Nice Cedex 4, France

Abstract. The trans-Neptunian region is inhabited by multiple dynamical populations, each of which have a complicated structure. For the most part, these structures cannot have been sculpted by the giant planets, once on their current orbital configuration. Thus, they represent important clues to the conditions that existed in the distant past. We argue in this paper, that most of what we see is the result of the outward migration of Neptune. By combining results from various authors, we can reproduce most of the observed properties of the trans-Neptunian region. Several aspects are not yet totally clear, and some may not be totally correct. But, for the first time, we have a view — if not not a detailed model — of how the system formed.

Keywords. Kuiper Belt, solar system: formation, solar system: general, comets: general

---

1. Introduction

It has been over a decade since the first object beyond the orbit of Neptune was discovered (Jewitt & Luu 1993). Since that time over 850 so called trans-Neptunian objects (or TNOs) have been discovered, and over 400 of these have been observed often enough so that their orbits are reliable†. As these data were collected, it became clear that rather than being the dynamically cold, smooth, and basically uninteresting system that most researchers expected, the trans-Neptunian region is home to a complex series of overlapping dynamical structures, each with a complicated, and sometimes diffcult to understand, configuration.

This complexity has been a significant boon to the planetary science community. In addition to keeping us funded, the trans-Neptunian region represents an excellent site to study the early dynamical evolution of the planetary system. This is due to the fact that, for the most part, the TNOs that we see cannot have formed in the orbits in which we see them in. Indeed, they must have accreted on low-inclination, nearly circular orbits (Stern 1996). In addition, they cannot have been placed on their current orbits by the planets in their current configuration. However, something must have moved the TNOs from their primordial orbits to the ones we observe, and thus, the outer planetary system must have looked very different in the distant past.

The current TNO orbits are, therefore, a diagnostic to what happened to the outer Solar System during the epoch when the planets formed. And just like a forensic scientists at a crime seen can determine the perpetrator of the crime, we believe that the dynamical structure of the trans-Neptunian region will reveal to us the mysteries of the birth of the Solar System. In this paper, we will discuss the current status of our investigation. In §2

† For more information see `http://cfa-www.harvard.edu/cfa/ps/lists/TNOs.html` and `http://cfa-www.harvard.edu/iau/lists/Centaurs.html`.

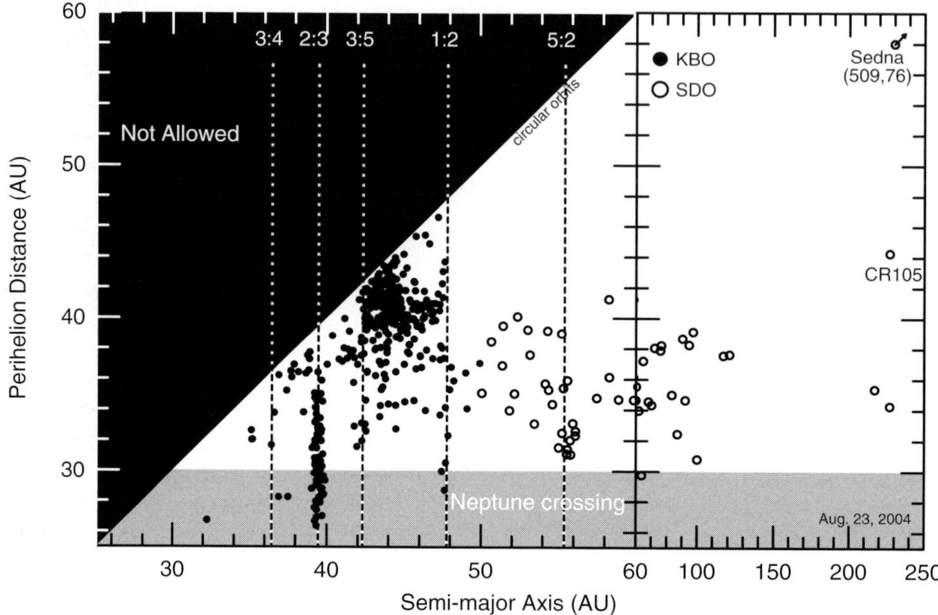

Figure 1. The semi-major axis – perihelion distance distribution for all the known TNOs with multi-opposition observations as of August 23, 2004. The filled and open dots represent the KBOs and SDOs, respectively, as determined my the Minor Planet Center. The dashed lines show a few of the mean motion resonances with Neptune. The black region is forbidden because perihelion distance cannot be larger than the semi-major axis. The gray region marked orbits that cross the orbit of Neptune.

we discuss the observed dynamical features and sub-populations of the trans-neptunian region. In §3 we briefly review planet migration. We discuss the models for the origin of these sub-populations in §4 – §9. We conclude in §10.

2. The Observed Structure of the Trans-Neptunian Region

Figure 1 shows the semi-major axis (a) – perihelion distance (q) distribution for all the TNOs that have been observed over multiple oppositions as of August 23, 2004. The figure shows the orbits of these objects as dots. The location of Neptune's important mean motion resonances are shown by vertical dashed lines. Objects in the gray region cross the orbit of Neptune. In general, they are stabilized because they are trapped in a mean motion resonance with Neptune.

The trans-Neptunian population is typically divided into two distinct structures, called the *Kuiper belt* and the *scattered disk*. Traditionally (i.e. four or five years ago), the scattered disk was thought to consist of objects that were placed on their current orbits by the giant planets in their current configuration (Duncan & Levison 1997). That is, they were thought to be objects that were scattered by Neptune onto orbits with large semi-major axes and large eccentricities. As a result, they should be on orbits that pass close to Neptune's orbit (i.e. with perihelion distance q less than $\sim 35\,\mathrm{AU}$). If this idea were true, they would represent a dynamically active population where objects are slowly

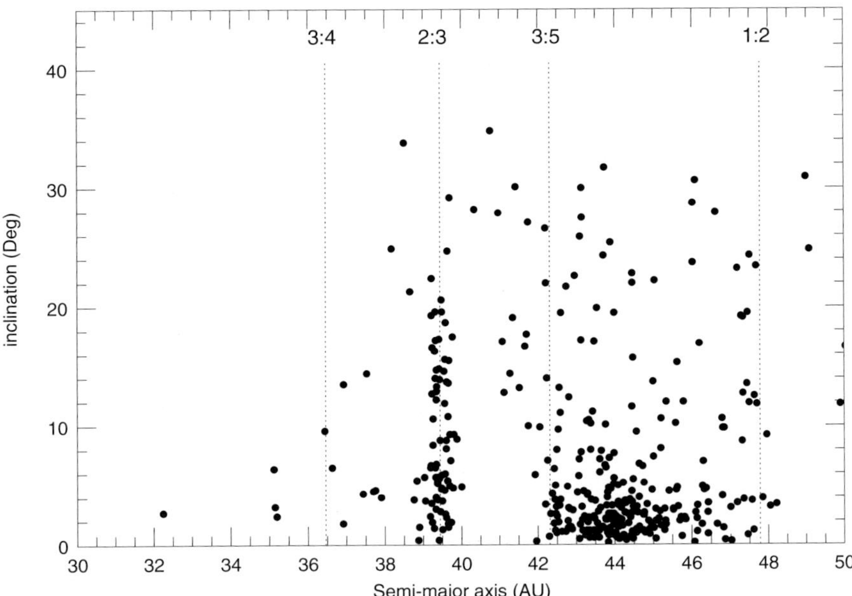

Figure 2. The semi-major axis – inclination distribution for all the known $a < 50\,\mathrm{AU}$ TNOs with multi-opposition observations as of August 23, 2004. The dashed lines show a few of the mean motion resonances with Neptune.

diffusing through orbital element space, and thus the bodies that belong in it would not provide us many direct clues to uncover the primordial architecture of the Solar System. As we describe more in details below, this definition is not strictly true, and so the scattered disk is the source of some important constraints on planet formation.

The *Kuiper belt* is basically the complement of the scattered disk in the $a > 30\,\mathrm{AU}$ region. As a result, it consists of objects which are essentially "frozen" in their current orbits (Duncan *et al.* 1995). As described above, all bodies in the Solar System must have been formed on orbits typical of an accretion disk (e.g. with very small eccentricities and inclinations; Stern 1996). Therefore, all the properties that make the current Kuiper belt very different from an accretion disk must have been acquired during a primordial phase of the Solar System.

In Figure 1, scattered disk objects (SDOs) and the Kuiper belt objects (KBOs) are represented by open and filled circles, respectively. However, to generate this figure we employed the classification scheme used by the Minor Planet Center, which simply calls any object with semi-major axis a smaller than 50 AU a KBO, while objects with $a > 50\,\mathrm{AU}$ are classified as SDOs. This scheme, of course, has the advantage that it does not require a detailed understanding of the behavior of TNOs. However, it has very little physical meaning. Having said this, the distinction between classes, as shown in the figure, is probably appropriate. The only major change that we would recommend is to consider the non-resonant objects with $40 \lesssim a \lesssim 50\,\mathrm{AU}$ and $q \lesssim 39\,\mathrm{AU}$ SDOs rather than KBOs.

Any theory of the early evolution of the outer Solar System must explain the following characteristics of the trans-Neptunian region:

(*a*) *The existence of resonant populations.* A glance at Figure 1 shows that the Kuiper belt is made of two distinct sub-populations: the *resonant population* and the *classical belt*. The former is made of the objects located in some major mean motion resonance with Neptune (essentially the 3:4, 2:3 and 1:2 resonances, but also the 2:5), while the classical belt objects are not in any noticeable resonant configuration. According to Trujillo *et al.* (2001) the resonant population accounts for only 10% of the total Kuiper belt population, when observational biases are corrected for.

(*b*) *The bi-modal inclination distribution of the classical belt.* Figure 2 shows that the bodies in the Kuiper belt can have very large inclinations. An analysis of observational biases by Brown (2001) suggests that the real inclination distribution of the non-resonant classical belt is bimodal (this is particularly noticeable in the region between the 2:3 and 1:2 mean motion resonance). About half of the objects have an inclination smaller than $4°$, while the remaining half have a very broad inclination distribution ranging up to $30°$ or more. We will refer to these two groups as the *cold* and the *hot* populations, respectively. Interestingly, these two populations seem to have different physical properties. The largest Kuiper belt objects are all in the hot population (Levison & Stern 2001; Trujillo & Brown 2003); moreover there is a statistically significant difference between the color distributions within the two populations (Tegler & Romanishin 2000; Doressoundiram *et al.* 2001; Trujillo & Brown 2002).

(*c*) *The outer edge of the classical belt.* It was generally expected that the number of the Kuiper belt objects should smoothly change with heliocentric distance. However the persisting lack of detection of objects beyond about 50 AU (Allen *et al.* 2001, 2002) cannot be explained by observational biases, but implies a statistically significant steep drop off in number density of large objects (Trujillo & Brown 2001). Figure 1 suggests that the edge of the classical belt coincides with the location of the 1:2 mean motion resonance with Neptune. In addition, there appears to be a correlation between eccentricity and semi-major axis in the classical Kuiper belt. There are many objects with $e \sim 0$ interior to 45 AU, but beyond $a = 45$ AU e on average increases with a.

(*d*) *Existence of an 'extended scattered disk' population.* Gladman *et al.* (2002) was the first to point out that there exists a population of SDOs (i.e. objects with $a > 50$ AU and highly eccentric orbits) that have large enough perihelion distances that they are decoupled from Neptune. The first obvious example of this population was 2000 CR$_{105}$ ($a = 230$ AU and $q = 44.17$ AU), however, Emel'yanenko *et al.* (2003) have shown that several others exist. The most extreme is (90377) Sedna, with a semi-major axis of 509 AU and a perihelion distance of 76 AU (Brown et al. 2004). We call these objects *extended scattered disk* objects because they are on orbits with semi-major axes similar to other scattered disk objects, but their perihelion distances are outside (or 'extended' beyond) the range for the normal scattered disk (Duncan & Levison, 1997, Gladman *et al.* 2001; Emel'yanenko *et al.* 2003). Their large eccentricities strongly suggest that they were gravitationally scattered onto their current orbits. However, this cannot have been done by the current planetary system.

(*e*) *The mass of the Kuiper belt.* The current mass of the Kuiper belt in the 40–50 AU region is estimated from detection statistics to be of order 0.1 Earth masses ($M_\oplus$) only (Jewitt *et al.* 1996, Chiang & Brown 1999, Trujillo *et al.* 2001, Gladman *et al.* 2001). However, models of the accretion process argue that there should have been several tens of Earth masses in this region, in order to produce the observed objects within a reasonable time interval of several 10^7 to 10^8 years (Stern, 1996; Stern & Colwell, 1997a; Kenyon and Luu, 1998, 1999a, 1999b). What happened to this mass is a critical issue. We refer to this as the *mass deficit problem.*

A large number of mechanisms have been proposed so far to explain some of the above mentioned properties of the Kuiper belt. For space limitation we debate here only those which in our opinion –at the light of our current observational knowledge of the Kuiper belt– played a role in the primordial sculpting of the trans-Neptunian population. A more exhaustive review can be found in Morbidelli and Brown (2003).

3. Giant planet migration

It is now well accepted that, after their formation, Uranus and Neptune underwent substantial outward migration due to the scattering of the planetesimals formed in their vicinity (Hahn & Malhotra 1999; Gomes *et al.* 2004). Planet formation is not expected to be 100% efficient. As the masses of Uranus and Neptune approached their full adult values, they must have gravitationally scattered most of the remaining nearby planetesimals both outward into the trans-Neptunian region and inward toward Jupiter and Saturn. Usually the objects scattered outward returned to the Uranus-Neptune region in the next orbit to be scattered again (either inward or outward). Jupiter and Saturn are so massive, however, that they can efficiently eject many of these planetesimals out of the planetary system.

Thus, the objects scattered outward usually return to Uranus and Neptune, while those scattered inward are effectively removed, thereby causing a net flux of objects toward Jupiter. This inward transport of mass required an outward migration of the orbits of Uranus and Neptune in order to conserve the angular momentum and energy of the system. For the same reason, Jupiter migrated inward.

We will argue in the following sections that much of the observed structure in the trans-Neptunian region is the direct result of the outward migration of Neptune. This is particularly true for the Kuiper belt. Indeed, we suggest that all of the observed KBOs were pushed onto their current orbits during this process.

4. Origin of the resonant populations

As we described in the last section, Neptune should have migrated outwards during the final stages of planet formation. Malhotra (1993, 1995) realized that, following Neptune's migration, the mean motion resonances with Neptune also migrated outwards, sweeping the primordial Kuiper belt until they reached their present position. From adiabatic theory (Henrard 1982), some of the Kuiper belt objects swept by a mean motion resonance would have been captured into resonance; they would have subsequently been push outward with the resonance in its migration, which would in turn has driven up their eccentricities. This model accounts for the existence of the large number of Kuiper belt objects in the 2:3 mean motion resonance with Neptune (and also in other resonances) and explains their large eccentricities (see Figure 3).

The mechanism of adiabatic capture into resonance requires that Neptune's migration happened very smoothly. If Neptune had encountered a significant number of large bodies (Lunar mass or more), its jerky migration would have jeopardized capture into resonances. Hahn & Malhotra (1999), who simulated Neptune's migration using a disk of Lunar to Martian-mass planetesimals, did not obtain any permanent capture.

Reproducing the observed range of eccentricities of the resonant bodies requires that Neptune migrated by at least 7 AU. Malhotra's simulations also showed that the bodies captured in the 2:3 resonance can acquire large inclinations, comparable to that of Pluto and other objects. The mechanisms that excite the inclination during the capture

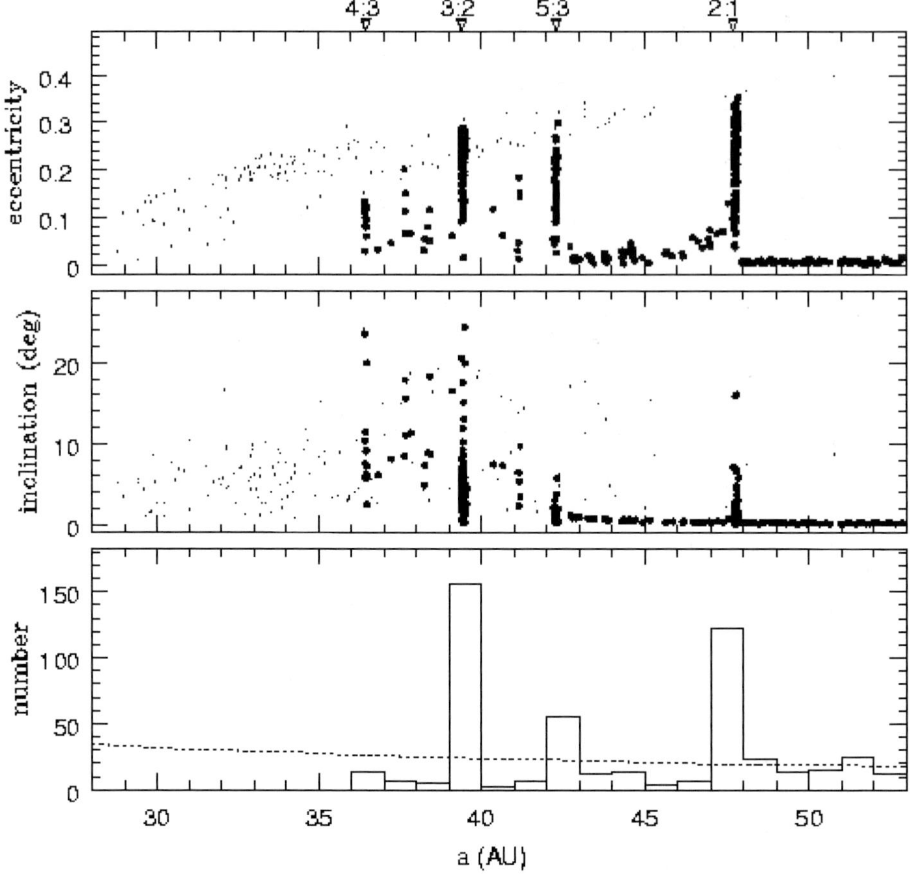

Figure 3. Top and middle panel: Final distribution of the Kuiper belt bodies according to the sweeping resonances scenario (courtesy of R. Malhotra). The simulation is done by numerical integrating, over a 200 Myr timespan, the evolution of 800 test particles on initial quasi–circular and coplanar orbits. The planets are forced to migrate (Jupiter: -0.2 AU;, Saturn: 0.8 AU; Uranus: 3 AU; Neptune: 7 AU) and reach their current orbits on an exponential timescale of 4 Myr. Large solid dots represent 'surviving' particles (i.e., those that have not suffered any planetary close encounters during the integration time); small dots represent the 'removed' particles at the time of their close encounter with a planet. Bottom panel: Histogram of semi-major axis distribution of the surviving bodies. The dashed curve denotes the initial distribution.

process have been investigated in detail by Gomes (2000), who concluded that, although large inclinations can be achieved, the resulting proportion between the number of high inclination vs. low inclination bodies and their distribution in the eccentricity vs. inclination plane do not reproduce well the observations. According to Gomes (2003) most high inclination Plutinos were captured during Neptune's migration from the scattered disk population, rather than from an originally cold Kuiper belt as in Malhotra scenario. We discuss Gomes's scenario in more detail in the next section.

5. Origin of the hot population

Gomes (2003) showed that the Neptune migration, in addition to the resonant populations, can also explain the origin of the hot classical Kuiper belt. Like Hahn & Malhotra

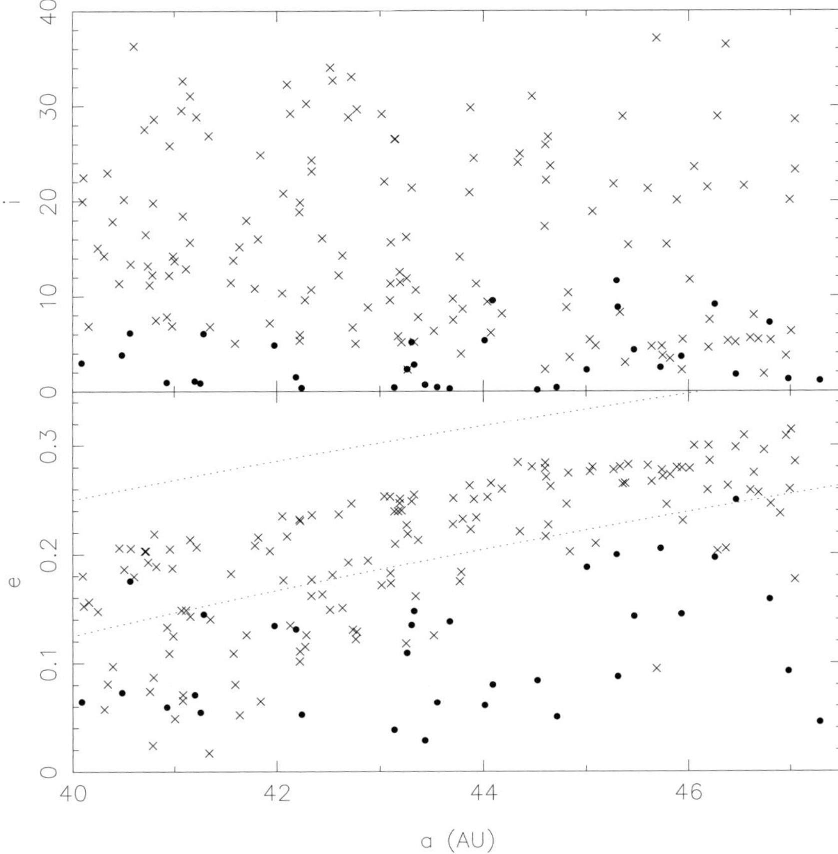

Figure 4. The orbital distribution in the classical belt according to Gomes' simulations. The dots denote a local population, which is only moderately dynamically excited as the mean motion resonances swept by. The crosses denote the bodies that were originally inside 30 AU and were trapped in the Kuiper belt via the mechanism described in the text. Note that the scattered population has a high inclination distribution, that is similar to the observed dynamically hot population. The dotted curves in the eccentricity vs. semi-major axis plot correspond to $q = 30$ AU and $q = 35$ AU. Courtesy of R. Gomes.

(1999), Gomes simulated Neptune's migration, starting from about 15 AU, by the interaction with a massive planetesimal disk with an inner edge slightly beyond Neptune's initial position. However, taking advantage of improved computer technology, he used 10,000 particles to simulate the disk population, with individual masses roughly equal to twice the Pluto's mass. This is in contrast to Hahn & Malhotra, who used only 1,000 particles, with Lunar to Martian masses.

He found that during its migration Neptune scattered planetesimals outward, forming a massive scattered disk. Some of the scattered bodies decoupled from the planet, by decreasing their eccentricity through the interaction with some secular or mean-motion resonance. If Neptune had not been migrating, the decoupled phases would have been transient, because the objects would have remained in the resonance and thus the process is reversible. However, Neptune's migration broke the reversibility because the resonances moved. Thus, some of the decoupled bodies managed to escape from the resonances, and remained permanently trapped in the Kuiper belt.

Since the particles that end up on stable Kuiper belt orbits had scattered off Neptune during their dynamical evolution, they tend to have an extended inclination distribution that is consistent with the observed hot population (see Figure 4). Recall that the inclination distribution of the Kuiper belt is bimodal. Gomes argued that the observations are well matched if it is assumed that the cold population is primordial and his mechanism produced the hot population. While this idea may be consistent with the observed inclination distribution, there are other reasons to believe that the cold population is not primordial, which we discuss in §7. However, we do think that Gomes's mechanism is the most likely cause for the dynamically hot population of the Kuiper belt.

Assuming that the bodies' color varied in the primordial disk with heliocentric distance, Gomes scenario also explains why the scattered objects, and hot classical belt objects, which mostly come from regions inside ~30 AU, appear to have similar color distributions, while the cold classical objects –the only ones that actually formed in the trans-Neptunian region– have a different color distribution. The Plutinos would be a mixture of the two populations. Similarly, assuming that the maximal size of the objects was a decreasing function of the heliocentric distance at which they formed, Gomes scenario also explains why the biggest Kuiper belt objects are all at large inclination.

Gomes's scenario also helps solve the mass depletion problem. In his simulations only ~0.2% of the bodies initially in the disk swept by Neptune remained in the Kuiper belt on stable high-i orbits at the end of Neptune's migration. This naturally explains the current low mass of the hot population. However, if the dynamically cold Kuiper belt is primordial, as Gomes believes, then the mass depletion problem still remains for this population.

6. Origin of the outer edge of the Kuiper belt

The existence of an outer edge to the Kuiper belt is, perhaps, its most intriguing characteristic. At least five mechanisms for its origin have been proposed, none of which has yet the general consensus of the community of the experts:

(*a*) A passing star tidally strips the Kuiper belt after the observed Kuiper belt objects formed (Ida *et al.* 2000; Kobayashi & Ida 2001; Melita *et al.* 2002).

(*b*) An edge formed prior to planetesimal formation due to aerodynamic drag (Youdin & Shu 2002).

(*c*) An edge formed during planet accretion due to size-dependent radial migration caused by gas drag (Weidenschilling 2003).

(*d*) Nearby early-type stars photo-evaporated the outer regions of the solar nebula before planetesimals could form (Hollenbach & Adams 2003).

(*e*) Magneto-hydrodynamic instabilities in the outer regions of the disk prevented the formation of planetesimals in these regions (Stone *et al.* 1998).

Unfortunately, space limitations do not allow us to discuss each of these in detail. However, a couple of points need to be made. Firstly, many of these mechanisms truncate the disk of planetesimals, from which the KBOs/SDOs form, without affecting the size of the solar nebula. Thus, observations of the sizes of extra-solar disks are not necessarily a direct constraint on where the edge of the Kuiper belt should lie.

In addition, more importantly to the current discussion, none of these mechanisms predict where the edge of the Kuiper belt should lie. In particular, none can explain why the outer edge of the Kuiper belt lies at the location of the 1:2 resonance. Either we must conclude that this is a coincidence, or we are still missing an important piece of the puzzle. We return to this issue again in the next section.

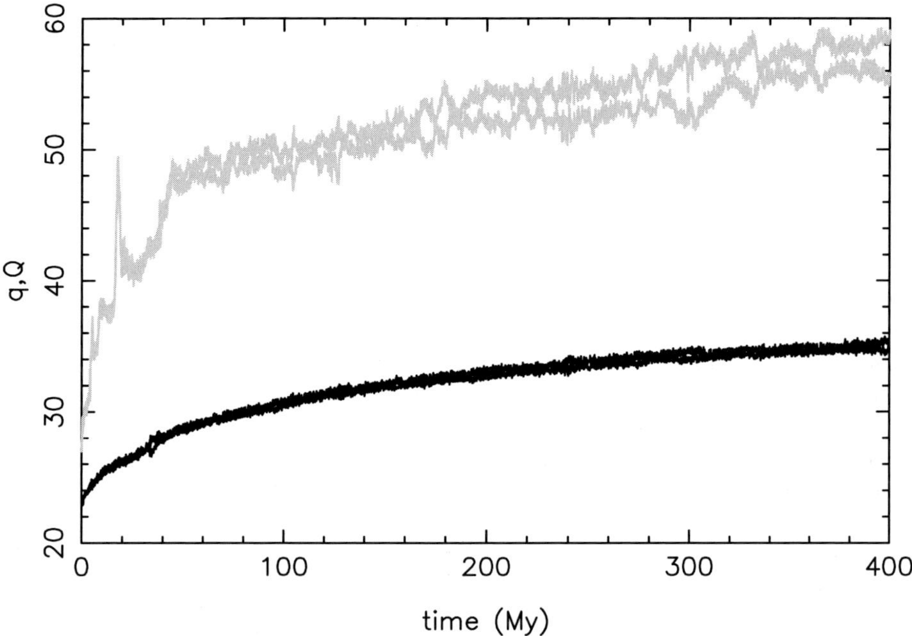

Figure 5. A self-consistent simulation of the Petit *et al.* (1999) scenario for the excitation and dynamical depletion of the Kuiper belt (from Gomes *et al.* 2004). Neptune is originally assumed at ~ 23 AU and an Earth-mass embryo at ~ 27 AU. Both planets are embedded in a 30 $M_\oplus$ disk, extending from 10 to 50 AU with a r^{-1} surface density profile (7.5 $M_\oplus$ between 40 and 50 AU). The black curve shows the evolution of Neptune's heliocentric distance, while the pair of gray curves refer to the perihelion and aphelion distances of the embryo. Notice that the embryo is never scattered by Neptune, unlike in Petit *et al.* simulations. It migrates through the disk faster than Neptune until the disk's outer edge. Neptune interacts with the entire mass of the disk, thanks to the dynamical excitation of the latter due to the presence of the embryo. Therefore, it migrates much further that it would if the embryo were not present, and reaches a final position well beyond 30 AU (it reaches 40 AU after 1 Gy).

7. The mass depletion of the cold population

As described in §5, the mechanism proposed by Gomes (2003) for the origin of the dynamically hot classical Kuiper belt naturally explains the mass depletion for this population. However, the population originally in the 40-50 AU region — which eventually became the cold population in Gomes scenario — would have been only moderately excited and not dynamically depleted during Neptune's migration, so that it should have preserved most of its primordial mass. This is still a significant problem in our understanding of the Kuiper belt, because of the observed low mass in this region.

Two general mechanisms have been proposed for the mass depletion: the dynamical ejection of most of the bodies from the Kuiper belt to the Neptune-crossing orbits and the collisional grinding of most of the mass of the Kuiper belt into dust.

All dynamical depletion mechanisms suffer from the same problem that is perhaps best illustrated by an example. The first such mechanism was proposed by Morbidelli & Valsecchi (1997) and Petit et al. (1999). In their scenario, a planetary embryo, with mass comparable to that of Mars or of the Earth, was scattered by Neptune onto a high-eccentricity orbit that crossed the Kuiper belt for $\sim 10^8$ y. The repeated passage of the embryo through the Kuiper belt excited the eccentricities of the Kuiper belt bodies, the vast majority of which became Neptune crosser and were subsequently dynamically eliminated by the planets' scattering action. In the Petit *et al.* (1999) integrations that supported this scenario, however, the Kuiper belt bodies were treated as test particles, and therefore their ejection to Neptune-crossing orbit did not alter the position of Neptune.

Gomes *et al.* (2004) have re-done Petit *et al.*-like simulations in the framework of a more self-consistent model, accounting for planetary migration. As we should have anticipated, the dynamical depletion of the Kuiper belt forces Neptune to migrate beyond 30 AU! The reason for this is that, thanks to the dynamical excitation of the distant disk provided by the embryo, Neptune interacts not only with the portion of the disk in its local neighborhood, but with the entire mass of the disk at the same time. As shown in Figure 5, even a low mass disk of 30 $M_\oplus$ between 10 and 50 AU (7.5 $M_\oplus$ in the Kuiper belt) drives Neptune well beyond 30 AU. Stopping Neptune's migration at $\sim$ 30 AU requires a disk mass of $\sim$ $15M_\oplus$ or less (depending on the initial Neptune's location). Such a mass and density profile would imply that there was only 3.75 $M_\oplus$ of material originally in the Kuiper belt between 40 and 50 AU, which is less than the mass required (10–30 $M_\oplus$) by the models of accretion of Kuiper belt bodies (Stern & Colwell 1997a; Kenyon & Luu 1999a, 1999b).

This result is quite general. We must conclude that Neptune never saw the missing mass of the Kuiper belt. Thus, any mechanism that depletes the Kuiper belt mass by handing it off to Neptune (for an other example see Nagasawa & Ida 2000) can be ruled out.

The collisional grinding scenario was proposed by Stern & Colwell (1997b) and Davis & Farinella (1997, 1998). A massive Kuiper belt with large eccentricities and inclinations would undergo a very intense collisional cascade. Consequently, most of the mass originally incorporated in bodies smaller than 50–100 km in size could be comminuted into dust, and then evacuated by radiation pressure and Poynting-Robertson drag. This would cause a substantial mass depletion, provided that the bodies larger than 50 km (which cannot be efficiently destroyed by collisions) initially represented only a small fraction of the total mass.

The collisional grinding scenario, however, has several apparent problems. First, it requires a peculiar size distribution, such that all of the missing mass was contained in small, easy to break, objects, while the number of large object was essentially identical to the current one.

Second, in order to reduce the mass of the Kuiper belt to less than an Earth mass over the age of the Solar System, Stern and Colwell (1997b) required large eccentricities and inclinations ($e \sim 0.25$ and/or $i \sim 7°$). This excitation is significantly larger than that characterizing the cold population.

Third, most of the binaries in the cold population would not survive the collisional grinding phase (Petit & Mousis 2004). In fact, the Kuiper belt binaries have large separations, so that it is easily shown that the impact on the satellite of a projective only 1% of the satellite's mass could unbind the binary. If the collisional activity was strong enough to cause an effective reduction of the overall mass of the Kuiper belt, these kind of collisions would have been extremely common. Thus, the existence of a large number of binaries is strong evidence against the collision grinding scenario.

So, at this point in the history of the field, we seemed to be faced with an unsolvable contradiction — we need to start with a large amount of mass within the 40 – 50 AU region in order to grow the Kuiper belt the we see, but we have no way to get rid of this mass and still be consistent with the observations. However, there is an implicit assumption that is being made in the above argue. In particular, we have been assuming that the cold population formed where we see it.

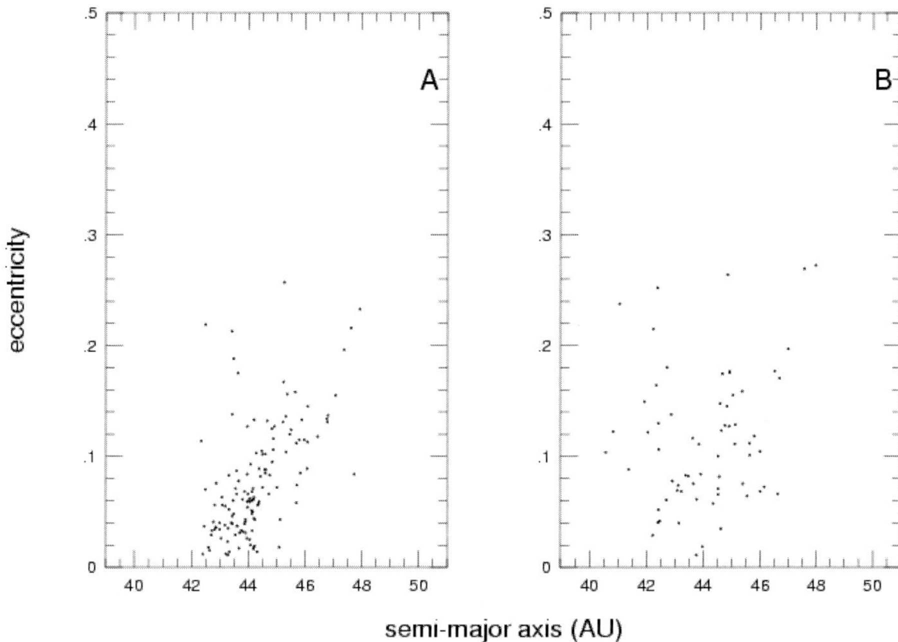

Figure 6. Left: the observed semi-major axis vs eccentricity distribution of the cold population. Only bodies with multi-opposition orbits and $i < 4°$ are taken into account. Right: the resulting orbital distribution in the scenario proposed by Levison & Morbidelli (2003).

8. Circumventing the mass depletion problem

A possible way out of this mass depletion problem has been proposed by Levison & Morbidelli (2003). In their preferred scenario, the primordial edge of the massive proto-planetary disk was somewhere around 30–35 AU and the *entire* Kuiper belt population — not only the hot component as in Gomes's scenario — formed within this limit and was transported to its current location during Neptune's migration.

In Levison & Morbidelli (2003)'s scenario, the transport process for the cold population was different from the one found by Gomes (2003) for the hot population. The bodies in the cold population were trapped in the 1:2 resonance with Neptune and transported outwards within the resonance, until they were progressively released due to the non-smoothness of the planetary migration. In the standard adiabatic migration scenario (Malhotra 1995), there would be a resulting correlation between the eccentricity and the semi-major axis of the released bodies. However Levison & Morbidelli found that this correlation is broken by a secular resonance embedded in the 1:2 mean motion resonance. Simulations of this process can match the observed (a, e) distribution of the cold population fairly well (see Figure 6), while the initially small inclinations are only very moderately perturbed.

In this scenario, the small mass of the current Kuiper belt population is simply due to the fact that, presumably, only a small fraction of the massive disk population was initially trapped in the 1:2 resonance and released on stable non-resonant orbits. The preservation of the binary objects is not a problem, because these objects were moved out

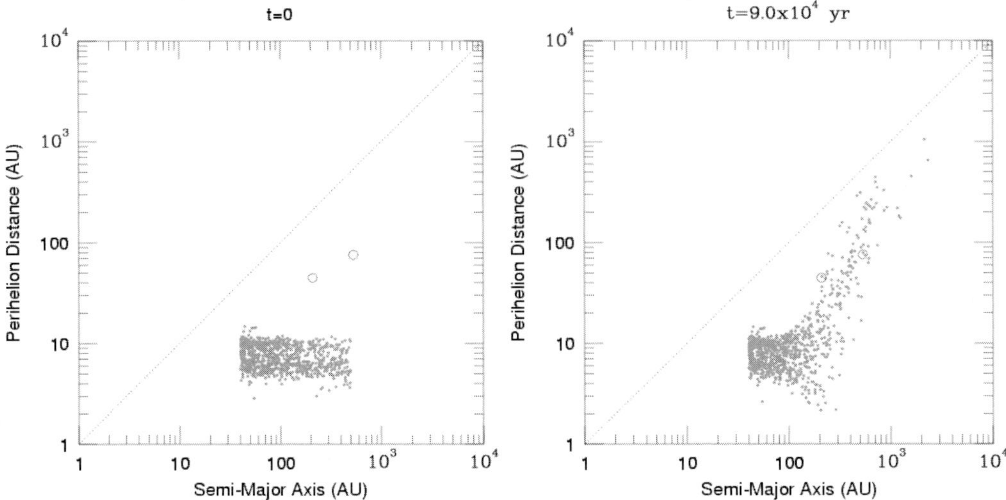

Figure 7. The semi-major axis – perihelion distance distribution of the scattered disk before and after Morbidelli & Levison (2004)'s nominal stellar passage with $q_\star = 800$ AU. The left panel is taken directly from the simulations in Dones *et al.* (2004). The right panel shows the effect of the passage. The pluses show the location of the simulated particles, while the filled circles show the real location of 2000 CR$_{105}$ and (90377) Sedna.

of the massive disk in which they formed by a gentle dynamical process. The final position of Neptune would simply reflect the primitive truncation of the protoplanetary disk (see §6). Conversely, this model opens again the problem of the origin of different physical properties of the cold and hot populations, because they would have both originated within 35 AU, although in somewhat different parts of the disk. This scenario also makes a simple prediction that will be confirmed or denied by future observations: the edge of the cold classical belt is exactly at the location of the 1:2 resonance.

9. Origin of the extended scattered disk

Explaining the origin of the orbits of 2000 CR$_{105}$ ($a \sim 230$ AU, $q \sim 45$ AU) and (90377) Sedna ($a = 509$ AU, $q = 76$ AU) is a major test for our understanding of the primordial evolution of the outer Solar System. Gladman *et al.* (2001) showed that 2000 CR$_{105}$ could not have been a normal member of the scattered disk that had its perihelion distance increased by chaotic diffusion. The same conclusion also clearly applies to (90377) Sedna. Thus, as with the Kuiper belt, these extended scattered disk objects are the result of dynamical processes that are no longer acting in the Solar System. Understanding how these objects could have formed will supply important clues to the origin of the Solar System.

In Morbidelli & Levison (2004), we explored five seemingly promising mechanisms for explaining the origin of the orbits of these peculiar objects: (i) the passage of Neptune through a high-eccentricity phase, (ii) the past existence of massive planetary embryos in the Kuiper belt or the scattered disk, (iii) the presence of a massive trans-Neptunian disk at early epochs that perturbed highly-inclined scattered disk objects, (iv) encounters with other stars that perturbed the orbits of some of the Solar System's trans-Neptunian planetesimals, and (v) the capture of extra-solar planetesimals from low mass stars or brown dwarfs encountering the Sun.

Of the five mechanisms listed above, only the two related to early stellar passages appear satisfactory. By satisfactory, we mean capable of producing the orbits of both

2000 CR$_{105}$ and Sedna at the same time, without generating a larger population of extended scattered disk objects with $q \sim 45$ AU but $a < 200$ AU. In our analysis, we put a lot of emphasis on the absence of detections of bodies with such orbital characteristics. Since observational biases (given an object's perihelion distance and absolute magnitude, and a survey's limiting magnitude of detection) sharply favor the discovery of objects with small semi-major axes, we think that it would be unlikely that the first two discovered bodies with $q > 44$ AU had $a > 200$ AU if the real semi-major axis distribution in the extended scattered disk were skewed toward smaller a.

We believe that the most likely scenario for the origin of these objects is that they were originally normal scattered disk objects that got placed on their current orbits by a close stellar encounter early in the history of the Solar System. Figure 7 shows an example of an encounter between a young scattered disk and an $1\,M_\odot$ star that passes at $800\,$AU on a slightly hyperbolic orbit. The initial condition for the scattered disk was taken from the simulations of Dones *et al.* (2004) and represent the scattered disk at $10^5\,$yrs. As the figure shows, there is excellent agreement between the model and the observations in that both the orbits of 2000 CR$_{105}$ and Sedna are reproduced while no small semi-major axis extended scattered disk objects are produced.

10. Concluding remarks

Over ten years of dedicated surveys have revealed unexpected and intriguing properties about the trans-Neptunian population. These characteristics include the existence of a large number of bodies trapped in mean motion resonances, the overall mass deficit, the large orbital eccentricities and inclinations, and the apparent existence of an outer edge at ~ 50 AU and of a correlation between inclinations, sizes and colors. Understanding how the Kuiper belt acquired all these properties would significantly constraint models of the formation of the outer planetary system and of its primordial evolution.

Up to now, a plethora of scenarios have been proposed by theoreticians. None of them can account for all the observations alone, and the solution of the Kuiper belt primordial sculpting problem probably requires a sapient combination of the proposed models. As we first suggest in Levison & Morbidelli (2003), we currently believe that the primordial planetesimal disk was truncated inside 40 AU and that the entire Kuiper belt was pushed outward by the migration of Neptune. If true, the Kuiper belt's complex structure is the result of a combination of the mechanisms presented in Malhotra (1995), Gomes (2003), and Levison & Morbidelli (2003).

However, Kuiper belt science is a rapidly evolving one. New observations change our view of the belt every year. Since the discovery of the first trans-Neptunian object 12 years ago, several review papers have been written, and all of them are already obsolete. No doubt that this will also be the fate of this chapter, but it can be hoped that the ideas presented here can continue to guide us in the direction of further understanding of what present observations of the Kuiper belt can tell us about the formation and evolution of the outer Solar System.

References

Allen, R.L., Bernstein, G.M. and Malhotra, R. 2001, *Astrophys. J.* 549, L241
Allen, R.L., Bernstein, G.M. and Malhotra, R. 2002, *Astron. J.* 124, 2949
Brown M. 2001, *Astron. J.* 121, 2804
Brown M., Trujillo C. and Rabinowitz D. 2004, *Astrophys. J. Letters*, in press.
Chiang, E.I. and Brown, M.E. 1999, *Astron. J.* 118, 1411
Davis D. R. and Farinella P. 1997. *Icarus* 125, 50

Davis D. R. and Farinella P. 1998, in: *Lunar Planet. Science Conf.* 29, 1437

Dones L., Levison, H.F., Duncan, M.J. and Weissman, P.R. 2004, *Icarus*, submitted.

Doressoundiram A., Barucci M.A., Romon J. and Veillet C. 2001, *Icarus* 154, 277

Duncan, M. J., Levison, H. F. and Budd, S. M. 1995, *Astron. J.* 110, 3073

Duncan, M. J. and Levison, H. F. 1997, *Science* 276, 1670

Emel'yanenko V. V., Asher D. J. and Bailey M. E. 2003, *Mon. Not. R. Astron. Soc.* 338, 443

Gladman, B., Kavelaars, J.J., Petit, J.M., Morbidelli, A., Holman, M.J. and Loredo, Y. 2001,
        Astron. J. 122, 1051

Gladman B., Holman M., Grav T., Kaavelars J.J., Nicholson P., Aksnes K. and Petit J.M. 2002,
        Icarus 157, 269

Gomes R. S. 2000, *Astron. J.* 120, 2695

Gomes R.S. 2003, *Icarus* 161, 404

Gomes R.S., Morbidelli A. and Levison H.F. 2004, *Icarus* 170, 492

Hahn J. M. and Malhotra R. 1999, *Astron. J.* 117, 3041

Henrard J. 1982, *Celest. Mech.* 27, 3

Hollenbach, D. and Adams, F.C. 2003, in: L. Caroff and D. Backman, (eds.) *Debris Disks and
        the Formation of Planets,* (San Francisco: ASP), in press

Ida S., Larwood J. and Burkert A. 2000, *Astrophys. J.* 528, 351

Jewitt, D. and Luu, J. 1993, *Nature* 362, 730

Jewitt, D., Luu, J. and Chen, J. 1996, *Astron. J.* 112, 1225

Kenyon, S.J. and Luu, J.X. 1998, *Astron. J.* 115, 2136

Kenyon, S.J. and Luu, J.X. 1999a, *Astron. J.* 118, 1101

Kenyon, S.J. and Luu, J.X. 1999b, *Astrophys. J* 526, 465

Kenyon S.J. and Bromley B.C. 2002, *Astron. J.,* 124, 1757

Kobayashi H. and Ida S. 2001, *Icarus* 153, 416

Levison H.F. and Stern S.A. 2001, *Astron. J.* 121, 1730

Levison H.F. and Morbidelli A. 2003, *Nature* 426, 419

Malhotra, R. 1993, *Nature* 365, 819

Malhotra, R. 1995, *Astron. J.* 110, 420

Malhotra, R. 1996, *Astron. J.* 111, 504

Melita M., Larwood J., Collander-Brown S, Fitzsimmons A., Williams I.P. and Brunini A. 2002,
        in: *Asteroids, Comets, Meteors,* ESA Spec. Publ. series, SP-500, p. 305

Morbidelli A. and Valsecchi G. B. 1997, *Icarus* 128, 464

Morbidelli A. and Brown M. 2003, in: M. Festou et al. (eds.) *Comets II,* Univ. Arizona Press,
        Tucson, in press

Morbidelli A. and Levison H.F. 2004, *Astron. J.,* in press.

Nagasawa M. and Ida S. 2000, *Astron. J.* 120, 3311

Petit J. M., Morbidelli A. and Valsecchi, G. B. 1999, *Icarus* 141, 367

Petit J. M. and Mousis O. 2004, *Icarus* 168, 409

Stern, S. A. 1996, *Astron. J.* 112, 1203

Stern, S. A. and Colwell, J. E. 1997a, *Astron. J.* 114, 841

Stern, S. A. and Colwell, J. E. 1997b, *Astrophys. J.* 490, 879

Stone, J.M., Gammie, C.F., Balbus, S.A. and Hawley, J.F. 1998, in: V. Mannings, A.P. Boss,
        and S.S. Russell (eds.) *Protostars and Planets IV,* p. 589

Tegler, S. C. and Romanishin, W. 2000, *Nature* 407, 979

Trujillo C.A. and Brown M.E. 2001, *Astrophys. J* 554, 95

Trujillo, C. A., Jewitt D. C. and Luu J. X. 2001, *Astron. J.* 122, 457

Trujillo C.A. and Brown M.E. 2002, *Astrophys. J.* 566, 125

Trujillo C.A., Brown M.E. 2003, *Earth, Moon, Planets* 92, 99

Weidenschilling S. 2003, in: M. Festou *et al.* (eds.), *Comets II,* Univ. Arizona Press, Tucson, in
        press

Youdin, A.N. and Shu, F.H. 2002, *Astron. J.* 580, 494

Dynamics of Populations of Planetary Systems
Proceedings IAU Colloquium No. 197, 2005
Z. Knežević and A. Milani, eds.

© 2005 International Astronomical Union
DOI: 10.1017/S1743921304008804

Transport of comets
to the outer planetary system

Arika Higuchi[1,2] **Eiichiro Kokubo**[1]
and Tadashi Mukai[2]

[1]Division of Theoretical Astronomy, National Astronomical Observatory of Japan,
2-21-1 Osawa Mitaka Tokyo, Japan
email: higuchia@th.nao.ac.jp
[2]Graduate School of Science and Technology, Kobe University
1-1 Rokkodai-cho Nada Kobe Hyogo, Japan

Abstract. In the late stage of planet formation, planetesimals are perturbed by large (proto) planets. There are four fates of planetesimals, (1) to collide with planets, (2) to escape from the planetary region, (3) to survive in the planetary region, and (4) to fall onto the central star. The ratios of these fates depend on initial orbital parameters. We performed numerical simulations of gravitational scattering of planetesimals by a planet. We obtained the escape rate of planetesimals and its dependence on the orbital parameters of the planetesimals and the planet. We also calculated the rate for increasing the semimajor axis to more than 3000AU. Using these results, we discuss the relative efficiency of the four giant planets of the solar system in the formation of the Oort cloud.

Keywords. Oort cloud, comets: general, solar system: formation

1. Introduction

The Oort cloud is the spherical comet reservoir surrounding the solar system, as first proposed by Oort (1950). He suggested that the Oort cloud is constructed from planetesimals through two dynamical stages: (1) transportation of planetesimals to the outer region of the solar system by planets, and (2) pulling up of their perihelion distances by the external forces of the galactic tide and passing stars. The first dynamical stage was studied by several authors (e.g., Safronov 1972, Weidenschilling 1975). For example, Weidenschilling (1975) studied the ejection rate of planetesimals by the four giant planets analytically. He concluded that Jupiter ejected planetesimals with very high rate because of its large mass and it can not form the Oort cloud. Simulations of both of the first and the second stages were recently done by Dones *et al.* (2004). They showed that Saturn is the planet most responsible for the Oort cloud formation. They deal with the solar system and treat the two dynamical stages together. However, they do not focus on the general properties of the elementary processes of the planet-planetesimal scattering.

In this paper, we investigate the first dynamical stage of the comet cloud formation by using numerical calculations. On this first stage, planetesimals have four fates under the strong gravitational influence of a planet: (1) collision with the planet, (2) escape from the planetary system, (3) survival as a planetesimal, and (4) fall onto the central star. We investigate the dependence of the rates of these fates upon the orbital parameters.

Table 1. The ranges of the initial parameters.

parameter	a	e	i [rad]	a_{p} [AU]	m_{p} [m_{J}]
range	$\frac{a_{\mathrm{p}}}{1+e} \lesssim a \lesssim \frac{a_{\mathrm{p}}}{1-e}$	0-0.9	0.01-0.1	1, 5, 10, 30	0.1-10

2. Method of Calculation

2.1. Model and Integration Method

We integrate the orbits of planetesimals numerically using the forth-order Hermite scheme (Makino & Aarseth 1992) with the hierarchical timestep (Makino 1991). We assume a planet around a central star on a circular orbit and a mass-less planetesimal orbiting the central star under the gravity of the planet. The equation of motion of the planetesimal is

$$\frac{d^2\mathbf{r}}{dt^2} = -Gm_\star \frac{\mathbf{r}}{r^3} - Gm_{\mathrm{p}}\left(\frac{\mathbf{r}-\mathbf{r}_{\mathrm{p}}}{|\mathbf{r}-\mathbf{r}_{\mathrm{p}}|^3} + \frac{\mathbf{r}_{\mathrm{p}}}{r_{\mathrm{p}}^3}\right), \tag{2.1}$$

where r is the heliocentric distance of the planetesimal, G is the gravitational constant, $m_\star$ is the mass of the central star, and m_{p} and r_{p} are the mass and the heliocentric distance of the planet. The last term on the r.h.s represents the indirect term.

2.2. Initial Conditions and Parameters

We assume an initial planetesimal disk which consists of mass-less planetesimals and a planet. All the planetesimals have the same eccentricity e and inclination i. The inner and outer edges, $a_{\min}$ and $a_{\max}$, of the disk are $a_{\min} = a_{\mathrm{p}}/(1 + e)$, $a_{\max} = a_{\mathrm{p}}/(1 - e)$, respectively. Here a_{p} is the semimajor axis of the planet. The argument of perihelion ω and the longitude of ascending node Ω are distributed randomly. The number density of planetesimals is proportional to a^{-1}. Our model contains 10^7 planetesimals per ring with the width of 1AU. The ranges of the initial parameters are shown in Table.1. We consider the parameter set $i = 0.05\mathrm{rad}, a_{\mathrm{p}} = 5\mathrm{AU}$, and $m_{\mathrm{p}} = m_{\mathrm{J}}$ as the standard case. We define $m_{\mathrm{J}} \equiv 0.001m_\star$ where we set $m_\star = m_\odot$ (solar mass). In all cases, we calculate orbits of planetesimals for 1 Kepler period (T_{K}). During the orbit integration, if the separation between a planet and a planetesimal becomes smaller than the radius of the planet R_{p}, or the heliocentric distance of a planetesimal becomes smaller than the radius of a central star $R_\star$, the planetesimal is counted as "collider" or "fall", respectively. Planetesimals not yet classified as colliders and falls are checked for their orbital elements at the final time T_{K}. If the perihelion distance of a planetesimal is smaller than $R_\star$, it is also counted as "fall". If the eccentricity of a planetesimal is larger than 1, it is counted as "escaper". Following to Dones *et al.* (2004), here we assume that planetesimals with $a > 3000\mathrm{AU}$ or $e > 1$ can go to the next dynamical stage. Thus a planetesimal with $a > 3000\mathrm{AU}$ and $e < 1$ is counted as "candidate", candidate for a member of the Oort cloud.

2.3. Definitions of Probability and Efficiency

We denote P_{esc} and P_{can} the probabilities of escape and candidate after a time span T_{K}. Using these probabilities P, we define the efficiencies K. Efficiencies K_{esc} and K_{can} mean the expected numbers of escaper and candidate per time and are defined as

$$K = \int_{a_{\min}}^{a_{\max}} PT_{\mathrm{K}}^{-1} n_{\mathrm{s}}(a) 2\pi a \, da, \tag{2.2}$$

where $n_{\mathrm{s}}(a)$ is a surface number density of planetesimals estimated as

$$n_{\mathrm{s}}(a) = n_0 a^\theta, \tag{2.3}$$

where n_0 is the surface number density at $a=1$AU, and θ is the power-index of the distribution.

3. Probabilities

3.1. *Escape from the Planetary System*

Figure 1 shows $P_{\rm esc}$ against a, with $e = 0.8$ and $e = 0.7$ for the standard case. The orbit-crossing regions with these parameters are from 2.78AU to 25AU and from 2.94AU to 16.7AU, respectively. Escapers appear almost over the entire orbit-crossing region. In this region, $P_{\rm esc}$ increases along a and suddenly drops around the end of the region. The orbit-crossing region increase with e. $P_{\rm esc}$ decreases with e at fixed a. There are no escapers

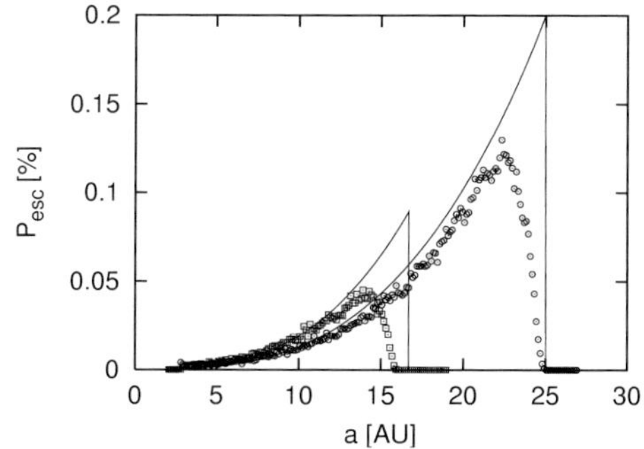
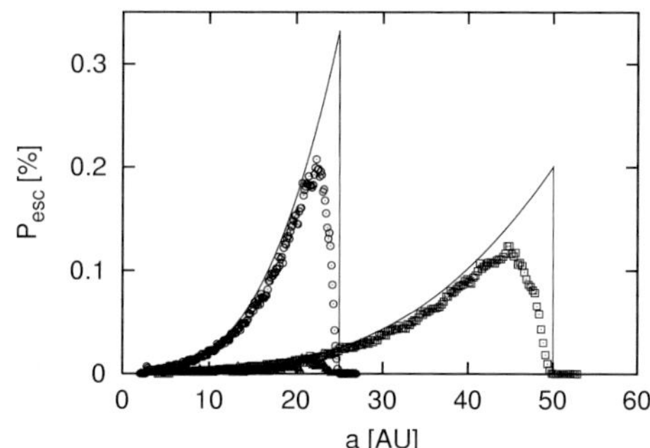

Figure 1. Probabilities of escape and the empirical fits are plotted against a. The circles and squares indicate $P_{\rm esc}$ with $e = 0.8$ and $e = 0.7$ for the standard case $(i, a_{\rm p}, m_{\rm p})=(0.05, 5, 1)$.

Figure 2. The same figure as figure 1. The symbols indicate $P_{\rm esc}$ for $e = 0.8$, $(i, a_{\rm p}, m_{\rm p})=$ (0.03, 5, 1)(circles), (0.05, 10, 1)(squares), and (0.05, 5, 0.3)(triangles), respectively.

for $e < 0.4$. This absence of escaper is explained by the relative velocity $v_{\rm r}$ between a planetesimal and a planet and the fly-by theory under the two-body approximation. Using the conservation of the Jacobi energy, $v_{\rm r}$ is written as $v_{\rm r}^2 = \{3 - 2[(1 - e^2)a/a_{\rm p}]^{1/2} \cos i - a_{\rm p}/a\}v_{\rm p}^2$, where $v_{\rm p}$ is the Kepler velocity of the planet. To escape from the planetary system, $v_{\rm r}$ need to satisfy $|\mathbf{v_{\rm p}} + \mathbf{v_{\rm r}}| > v_{\rm parabolic}$, where $v_{\rm parabolic}$ is the local escape velocity written as $\sqrt{2Gm_\star/a_{\rm p}}$. There is a minimum relative velocity $v_{\rm r}^{\rm min}$ to escape derived from the fly-by theory. A planetesimal gains the largest additional velocity of $v_{\rm r}$, if it is scattered toward the direction of the velocity vector of the planet. Here $v_{\rm r}^{\rm min}$ is given by

$$v_{\rm r}^{\rm min} = v_{\rm parabolic} - v_{\rm p}$$
$$= (\sqrt{2} - 1)v_{\rm p}. \tag{3.1}$$

To satisfy the condition $v_{\rm r} > v_{\rm r}^{\rm min}$, we need $e \gtrsim 0.4$. So planetesimals initially with $e \lesssim 0.4$ can not escape. Figure 2 is the same as figure 1 but with different i, $a_{\rm p}$, and $m_{\rm p}$. $P_{\rm esc}$ decreases with i. The orbit-crossing region increases with $a_{\rm p}$ and $P_{\rm esc}$ at fixed a decreases with $a_{\rm p}$ but they can be scaled by $a_{\rm p}$. This is because $v_{\rm r}$ is scaled by $a_{\rm p}$. $P_{\rm esc}$ increases with $m_{\rm p}$. This is because the cross section for strong scattering is proportional to $m_{\rm p}^2$. The main features of $P_{\rm esc}$ do not change when these parameters are changed. We plot the empirical fit with the approximated dependences, given by

$$P_{\mathrm{esc}}^{\mathrm{fit}} \sim 4 \left(\frac{a}{a_{\mathrm{p}}} \right)^3 (1 - e) \sin^{-1} i \, m_{\mathrm{p}}^2, \tag{3.2}$$

in figures 1 and 2. Here we set the unit mass as $m_\star$. The derivation of this expression is discussed in another paper.

3.2. Candidate for the Oort Cloud

Figure 3 shows P_{can} against a, with $e = 0.8$ for the standard case. Candidates appear almost over the orbit-crossing region. In this region, P_{can} increases along a and suddenly drops around the end of the region. This behavior is similar to that of P_{esc}. However, the dependence on a is different. Figure 3 also shows P_{can} with $e = 0.7$. Figure 4 is the same as figure 3 but with different i, a_{p}, and m_{p}. The dependences of P_{can} on i and m_{p} are the same as those of P_{esc}. The main features of P_{can} do not change when these parameters are changed. In figures 3 and 4, we plot the empirical fit given by

$$P_{\mathrm{can}}^{\mathrm{fit}} \sim 12 \left(\frac{a_{\mathrm{can}}}{a_{\mathrm{p}}} \right)^{-1} \left(\frac{a}{a_{\mathrm{p}}} \right)^5 (1 - e)^2 \sin^{-1} i \, m_{\mathrm{p}}^2. \tag{3.3}$$

This expression is valid under two conditions: (1) $m_{\mathrm{p}} \lesssim 3 m_{\mathrm{J}}$ and (2) $a_{\mathrm{can}}/a_{\mathrm{p}} \gtrsim 100$. The derivation of this expression is also discussed in another paper.

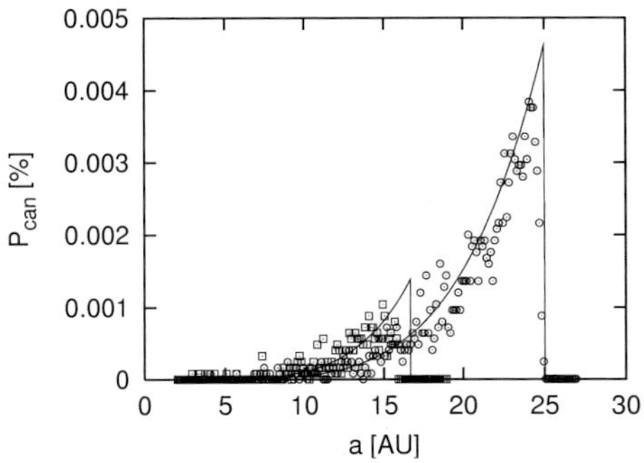
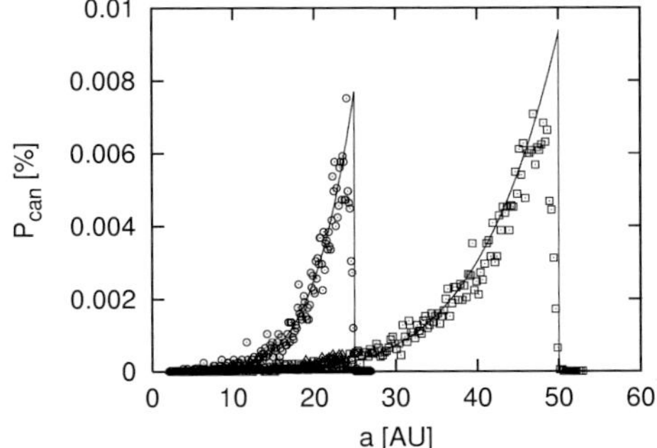

Figure 3. Probabilities of candidate and the empirical fits are plotted against a. Symbols and parameters are the same as figure 1.

Figure 4. The same figure as figure 2. Symbols and parameters are also the same as figure 2.

4. Efficiencies

Using the probabilities of P_{esc} and P_{can}, we obtain the efficiencies of escape K_{esc} and candidate K_{can}. Here we adopt the standard disk model of $\theta = -3/2$ (Hayashi 1981) and $n_0 = 1$ for simplicity. The integration is performed over the orbit-crossing region. Figures 5 and 6 show the dependence of K_{esc} and K_{can} on e. Both K_{esc} and K_{can} increase with e for $e > 0.4$. From the dependences of $P_{\mathrm{esc}}^{\mathrm{fit}}$, $P_{\mathrm{can}}^{\mathrm{fit}}$, (2.2), and the disk model we assumed, we obtain the dependences $K_{\mathrm{esc}}^{\mathrm{fit}} \propto \sin^{-1} i \, a_{\mathrm{p}}^{-1} \, m_{\mathrm{p}}^{-2}$ and $K_{\mathrm{esc}}^{\mathrm{fit}} \propto \sin^{-1} i \, a_{\mathrm{can}}^{-1} \, m_{\mathrm{p}}^{-2}$. For $e \gtrsim 0.5$ and $i < 0.07$, $K_{\mathrm{esc}}^{\mathrm{fit}}$ agrees K_{esc} within a factor of ~ 2 and $K_{\mathrm{can}}^{\mathrm{fit}}$ agrees K_{can} within a factor of ~ 4.

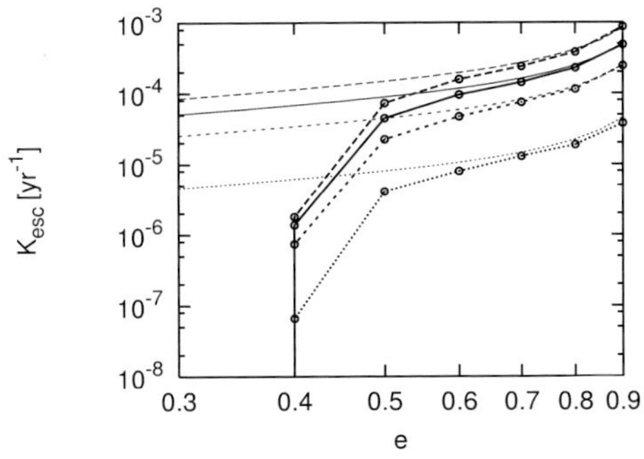 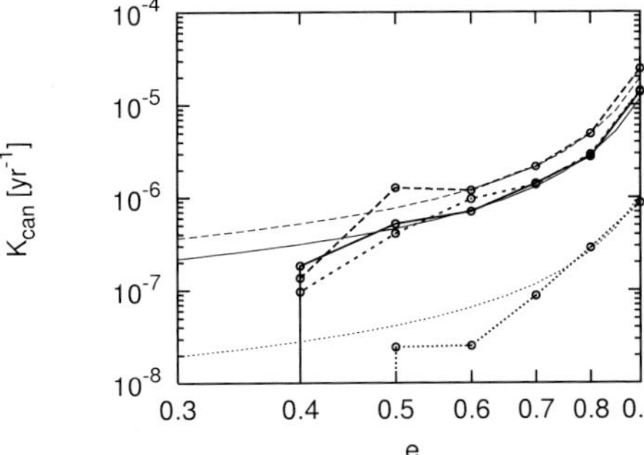

Figure 5. Efficiencies of escape(symbol) and the empirical fits are plotted against e. The solid, dash, short-dash, and dotted lines indicate $K_{\rm esc}$ for $(i, a_{\rm p}, m_{\rm p})= (0.05, 5, 1)$(standard case), $(0.03, 5, 1)$, $(0.05, 10, 1)$, and $(0.05, 5, 0.3)$, respectively.

Figure 6. Efficiencies of candidate and the empirical fits are plotted against e. Lines and parameters are the same as in figure 5.

5. Application to the Oort Cloud Formation

We apply the results to the Oort cloud formation. We assume that the planetesimal disk has the outer edge at 50AU and adopt the four giant planets with present $a_{\rm p}$ and $m_{\rm p}$.

5.1. *Standard Disk*

Figure 7 shows $K_{\rm esc}$ and $K_{\rm can}$ with $i = 0.05$rad and $n_{\rm s} = a^{-3/2}$. For Saturn, Uranus, and Neptune, $K_{\rm esc}$ and $K_{\rm can}$ decrease with e for $e > 0.8$, $e > 0.6$, and $e > 0.5$ respectively. This is because of the finite disk extent. Jupiter has the highest $K_{\rm esc}$ and $K_{\rm can}$. All planets have $K_{\rm esc}$ higher than $K_{\rm can}$. This means that Jupiter is the planet most responsible for transporting planetesimals to the Oort cloud, even if it produces many escapers. The roles of Uranus and Neptune are comparable and both are much less than that of the other two planets.

5.2. *Hot Disk*

We vary i of planetesimals around each planet. The massive planet effectively excites the eccentricity and inclination of nearby planetesimals. Here we adopt $i = (m_{\rm p}/3m_\star)^{1/3}$, the reduced Hill radius of the planet. Figure 8 shows $K_{\rm can}$ with $i = (m_{\rm p}/3m_\star)^{1/3}$ and $n_{\rm s} = a^{-3/2}$. In this case, the planetesimals around Jupiter have the highest i and this reduces $K_{\rm can}$. However, Jupiter still has the highest value and the relationships among the efficiencies of the four planets are not changed.

6. Summary and Discussion

We investigated the orbital evolution of planetesimals scattered by a planet using numerical calculations. We obtained the probabilities and the efficiencies in producing escapers and candidates and their empirical expressions. We applied these results to the Oort cloud formation. We consider two disk models: (1) standard disk and (2) hot disk. With both disk models, Jupiter is the most effective planet to produce escape and candidate for the Oort cloud. However, we have to consider more realistic initial condition, especially the orbital elements of planetesimals around a planet. They depend on the parameter of the planet, the existence of the gas disk, and the interaction among planetesimals. We will investigate this in the near future work.

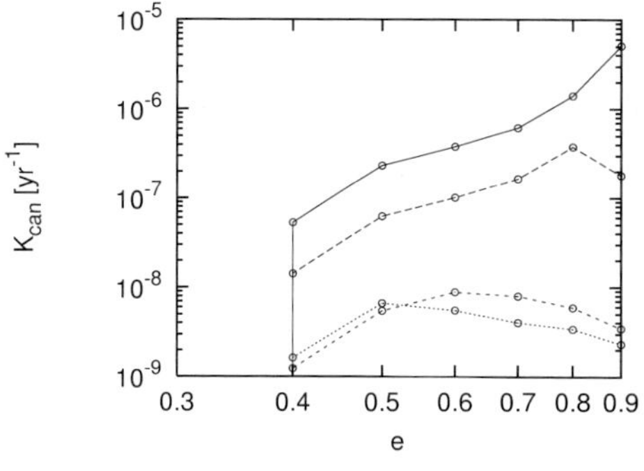

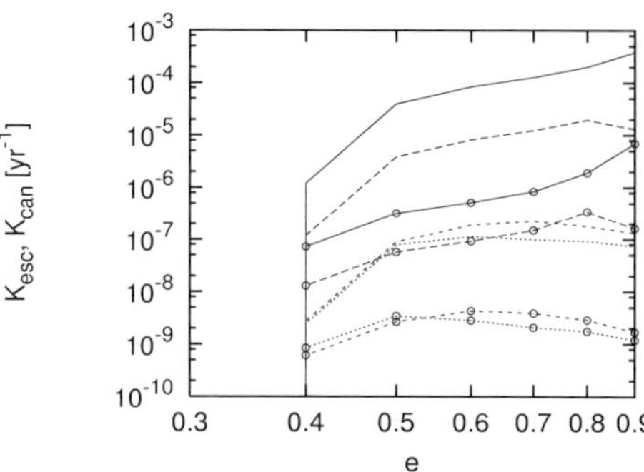

Figure 7. Efficiencies $K_{\rm esc}$ and $K_{\rm can}$(symbol) with the standard disk, i=0.05rad, $n_{\rm s} = a^{-3/2}$ are plotted against e. The solid, dash, short-dash, and dotted lines indicate Jupiter, Saturn, Uranus, and Neptune, respectively.

Figure 8. The same figure as 7 but with the hot disk, $i = (m_{\rm p}/3m_\odot)^{1/3}$, and $n_{\rm s} = a^{-3/2}$. Only $K_{\rm can}$ is plotted.

Our calculations deal with the elementary process of gravitational scattering of planetesimals by a planet. The parameter ranges we used are not restricted to the solar system. Thus, we can apply the results to the extrasolar planetary systems. Using these results, we can also estimate the formation rate of the comet clouds around the extrasolar planetary systems.

Acknowledgements

This work was supported by the 21st Century COE Program Origin and Evolution of Planetary Systems of the Ministry of Education, Culture, Sports, Science, and Technology, Japan, and JSPS Research Fellowship for Young Scientists.

References

Duncan, M., Quinn, T. & Tremaine, S. 1987, *Astron. J.* 94, 1330

Dones, L., Levison, H.F., Duncan, M. & Weissman, P. 2004, *Icarus*, in press

Makino, J 1991, *Publ. Astron. Soc. Japan* 43, 859

Makino, J. & Aarseth, S.J. 1992, *Publ. Astron. Soc. Japan* 44, 141

Hayashi, C 1981, *Prog. Theor. Phys. Suppl.* 70, 35

Oort, J.H 1950, *Bull. Astron. Inst. Neth.* 11, 91

Safronov, V.S. 1972, in: G. A. Chebotarev, E. I. Kazimirchak-Polonskaia & B. G. Marsden (eds.), *Proceedings IAU Symp.45: The Motion, Evolution of Orbits, and Origin of Comets*, (Dordrecht: Reidel), p. 329

Weidenschilling, S.J. 1975, *Astron. J.* 80, 145

Dynamics of Populations of Planetary Systems
Proceedings IAU Colloquium No. 197, 2005
Z. Knežević and A. Milani, eds.

© 2005 International Astronomical Union
DOI: 10.1017/S1743921304008816

Dynamical evolution of comets from high-eccentricity trans-Neptunian orbits to near-Earth space

Vacheslav V. Emel'yanenko

Department of Computational and Celestial Mechanics, South Ural University, Lenina 76 ,
Chelyabinsk 454080, Russia
email: vvemel@math.susu.ac.ru

Abstract. The contribution of high-eccentricity trans-Neptunian objects to the observed populations of both short-period and long-period comets is estimated. About 10^{10} objects with a radius $R > 0.7$ km in orbits with perihelion distances $28 < q < 35.5$ AU and semimajor axes $60 < a < 1000$ AU are the main source of Jupiter-family comets. If the population of high-eccentricity trans-Neptunian objects formed about 4.5 Gyr ago, the mean near-parabolic flux produced by these objects is on the order of 0.3-1.0 AU^{-1} yr^{-1} for comets with $R > 0.7$ km.

Keywords. celestial mechanics, comets: general, Kuiper Belt, Oort Cloud

1. Introduction

The population of trans-Neptunian objects (TNOs) in high-eccentricity orbits is widely accepted as a source of comets in the inner planetary region. There are two ways by which these TNOs can reach near-Earth space. Due to perturbations of Neptune they can move both inwards and outwards. The inner fraction of objects may directly become Neptune-crossing, Uranus-crossing, Saturn-crossing and Jupiter-crossing (Kazimirchak-Polonskaya 1972), finally appearing as members of the Jupiter-family (JF) comet system. The bodies diffusing into the outer region may reach large distances where galactic and stellar perturbations are able to inject some of them into the planetary region, creating a flux of near-parabolic bodies.

The basic idea about the origin of JF comets from the flat disc of objects located beyond Neptune can be traced to papers such as (Whipple 1964; Whipple 1972; Fernández 1980b) and others. The detailed study of the relationship between TNOs and JF comets was done in the papers (Levison & Duncan 1997; Duncan & Levison 1997). But now the structure of the trans-Neptunian region is known much better than at the time of those papers. Therefore, here we discuss the results of the investigation for the transfer of high-eccentricity TNOs to JF orbits, based on the present orbits of TNOs.

A scattering of bodies from the near-Neptune region to extremely large distances is usually associated with early stages of the Solar system formation (Oort 1950; Fernández 1980a). In combination with galactic and stellar perturbations, it creates the basis for the concept of the Oort cloud (with a dense inner core) (Hills 1981; Duncan, Quinn & Tremaine 1987) which is a source of observed near-parabolic comets. But it is evident now that high-eccentricity TNOs can penetrate to the Oort cloud region even at the present epoch. Therefore, we study the process of the dynamical evolution of objects from high-eccentricity trans-Neptunian orbits to near-parabolic orbits with perihelia in the inner planetary region.

323

Thus, the aim of this paper is to estimate the contribution of high-eccentricity TNOs to the observed populations of both short-period and long-period comets.

2. High-eccentricity trans-Neptunian objects as a source of Jupiter-family comets

The detailed investigation of the process by which high-eccentricity TNOs are transferred to short-period orbits was done by (Emel'yanenko, Asher & Bailey 2004). This work was based on 7 observed high-eccentricity Neptune-approaching orbits of multi-opposition TNOs. Each orbit was cloned 249 times. The resulted 1750 orbits were integrated for the age of the Solar system. The comparison of the results of integrations with the observed distribution of JF comets showed that the best fit took place by restricting to 1200 yr the maximum lifetime of comets after their first entry in the JF region defined by the following conditions: the perihelion distance $q < 1.5$ AU, the Tisserand parameter $T > 2$. With this restriction, the objects spend 460 yr in the JF region on average.

The calculations gave also that the relative fraction of objects captured per year from high-eccentricity orbits with semimajor axes $60 < a < 1000$ AU and perihelion distances $28 < q < 35.5$ AU to JF comets with $q < 1.5$ AU is 0.2×10^{-10}. Using this value, the fact that there exist about 90 JF comets with a nuclear radius of $\gtrsim 0.7$ km in the region $q < 1.5$ AU (Fernández, Tancredi, Rickman & Licandro 1999) and the above estimate for the lifetime of JF comets we obtained in the paper (Emel'yanenko, Asher & Bailey 2004) that there are $\sim 10^{10}$ TNOs of cometary size in orbits with $60 < a < 1000$ AU, $28 < q < 35.5$ AU.

3. High-eccentricity trans-Neptunian objects as a source of near-parabolic comets

Observed 'new' comets come from orbits with $a > 10^4$ AU. The structure of the trans-Neptunian population with such semimajor axes is not known. It is evident that this structure is different from that which can be obtained from the study of observed TNOs because the former has been built by a slow evolution of objects for the age of the Solar system. The present distribution of TNOs can give only a general indication about the original one. Therefore, we study the dynamical evolution of 667 objects with a quite broad distribution of initial orbits for 4.5 Gyr. Initial perihelion distances and semimajor axes are distributed uniformly in the range (25,45) AU and (50,300) AU, respectively; initial inclinations i are distributed between 0 and 40 degrees according to a sine law with a peak at 20 degrees. Later, different models of the original high-eccentricity trans-Neptunian population are tested by weighting initial orbits.

The integration of equations of motion is done by using the symplectic integrator (Emel'yanenko 2002; Emel'yanenko, Asher & Bailey 2003). We include the perturbations from the four outer planets, with the mass of the inner planets being added to that of the Sun. We take account of Galaxy in the way described earlier (Emel'yanenko 1999). Stellar perturbations are described by the impulse approximation, adopting the distribution of parameters for stars from (Heisler, Tremaine & Alcock 1987).

310 objects that reached the region $a > 1000$ AU were cloned 3 times by small changes of the mean anomaly. Then, in order to better study the near-parabolic population in the planetary region after the long-term evolution, all the objects reaching $q < 20$ AU later than 1.7 Gyr were cloned 19 times.

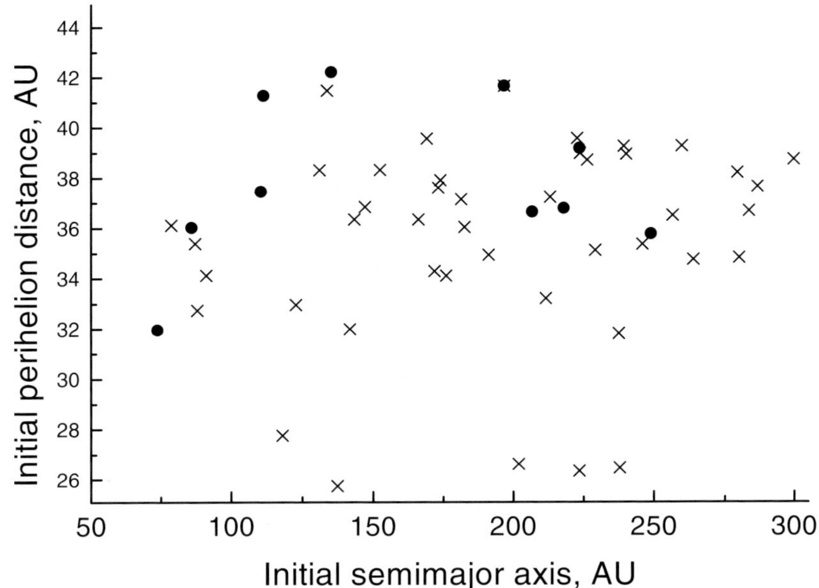

Figure 1. The distribution of initial a and q for objects in orbits with $28 < q < 35.5$ AU, $60 < a < 1000$ AU (circles) and $q < 20$ AU, $a > 1000$ AU (crosses).

In order to quantify the near-parabolic flux produced by TNOs we use the above estimate that there are $\sim 10^{10}$ objects with a radius $R > 0.7$ km in the near-Neptune region $60 < a < 1000$ AU and $28 < q < 35.5$ AU. In our integration of 667 objects, we have registered 10 objects remaining in this region after 4.5 Gyr. On the other hand, 667 initial objects produce on average 1.8 objects reaching the region $q < 5$ AU with $a > 10^4$ AU for the last billion years of our computations. Then we conclude that the mean near-parabolic flux produced by evolved TNOs is equal to $\nu = \frac{1.8 \times 10^{-9} \times 10^{10}}{10 \times 5} = 0.36$ AU^{-1} yr^{-1}.

Even this estimate based on very broad initial conditions shows that former TNOs give a substantial contribution to the near-parabolic flux. But this value depends heavily on the initial distribution of high-eccentricity TNOs. Figure 1 shows the distributions of initial perihelion distances and semimajor axes for objects that remained in the region $28 < q < 35.5$ AU, $60 < a < 1000$ AU after 4.5 Gyr and objects injected in the region $q < 20$ AU, $a > 10^3$ AU later than 3.5 Gyr. The majority of objects surviving in the near-Neptune region have initial $q > 36$ AU. If we just assume that high-eccentricity TNOs had mainly orbits with $30 < q < 36$ AU, $50 < a < 240$ AU about 4 Gyr ago, then there is only 1 object surviving in the near-Neptune region $60 < a < 1000$ AU and $28 < q < 35.5$ AU (Figure 1). At these initial conditions, we have registered on average 0.46 objects reaching the region $q < 5$ AU with $a > 10^4$ AU for the last billion years of our computations. Then the estimated value of $\nu = \frac{0.46 \times 10^{-9} \times 10^{10}}{1 \times 5} = 0.92$ AU^{-1} yr^{-1}. This is comparable with the observed magnitude. Although the various authors value for the observed near-parabolic flux has a rather wide spread, usually it is estimated as a few comets per year and AU for the size of $\gtrsim 0.7$ km (Bailey & Stagg 1988; Fernandez & Gallardo 1999; Wiegert & Tremaine 1999).

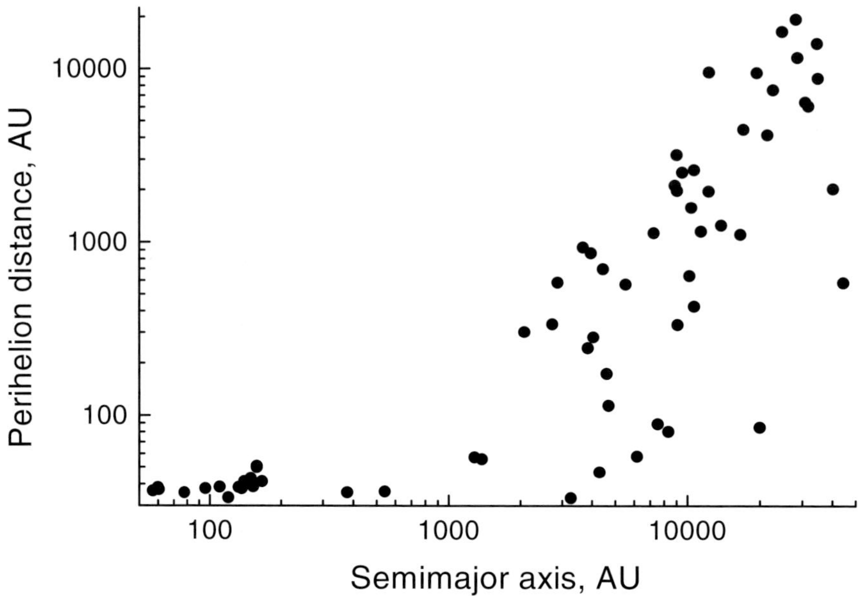

Figure 2. The distribution of a and q for objects with initial $30 < q < 36$ AU after 4.5 Gyr.

At the initial conditions $30 < q < 36$, $50 < a < 240$ AU, we have found 65 objects with $q > 30$ AU after 4.5 Gyr. Then the estimated number of bodies surviving in the Solar system beyond Neptune after 4.5 Gyr is equal to $\frac{65\times10^{10}}{1} = 6.5 \times 10^{11}$. The calculated orbital distributions for them are shown in Figures 2 and 3. If the upper limit for q is reduced to 34.5 AU, then our computations give on average 0.345 objects reaching the region $q < 5$ AU with $a > 10^4$ AU for the last billion years, 1 object surviving in the near-Neptune region $60 < a < 1000$ AU and $28 < q < 35.5$ AU, and 37 objects that have $q > 30$ AU after 4.5 Gyr. Then the estimated value of $\nu = \frac{0.345\times10^{-9}\times10^{10}}{1\times5} = 0.69$ AU^{-1} yr^{-1}, and the corresponding number of objects beyond Neptune is equal to $\frac{37\times10^{10}}{1} = 3.7 \times 10^{11}$. It is necessary to stress that these conditions do not conflict with the distribution of perihelia in the scattered disc model with migrating Neptune (Duncan & Levison 1997; Gomes 2003).

4. Conclusions

The presented computations show that high-eccentricity TNOs give a large contribution to both short-period and long-period comets. $\sim 10^{10}$ objects with $R > 0.7$ km in orbits with $28 < q < 35.5$ AU, $60 < a < 1000$ AU are the dominant source of JF comets. If the population of high-eccentricity TNOs formed about 4.5 Gyr ago, the mean near-parabolic flux produced by these objects is on the order of 0.3-1.0 AU^{-1} yr^{-1} for comets with $R > 0.7$ km.

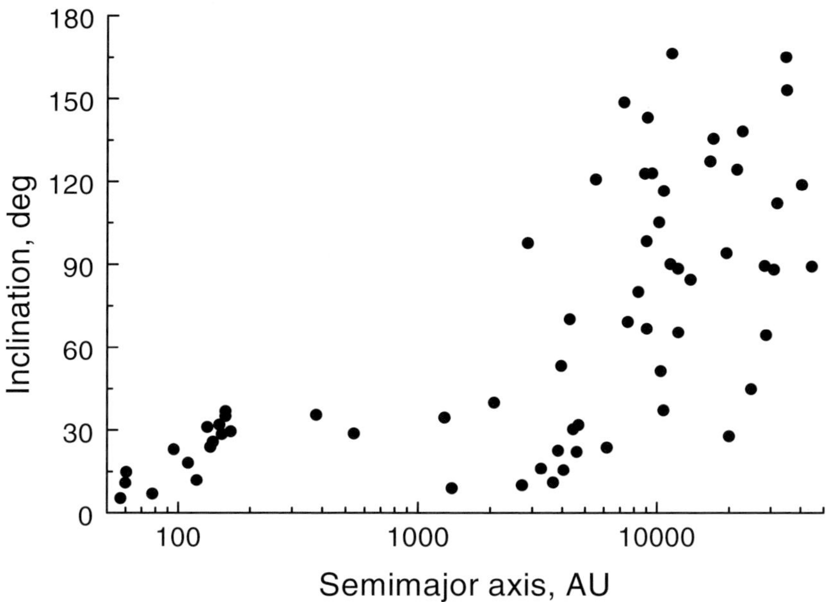

Figure 3. The distribution of a and i for objects with initial $30 < q < 36$ AU after 4.5 Gyr.

Acknowledgements

This work was supported by RFBR under grant number 04-02-96042 and INTAS under grant number 00-240. The author thanks an anonymous referee for a careful reading of the manuscript and helpful comments.

References

Bailey, M.E. & Stagg, C.R. 1988, *Mon. Not. R. Astron. Soc.* 218, 1

Duncan, M.J. & Levison, H.F. 1997, *Science* 276, 1670

Duncan, M., Quinn, T. & Tremaine, S. 1987, *Astron. J.* 94, 1330

Emel'yanenko, V.V. 1999, in: J. Svoren, E.M. Pittich & H. Rickman (eds.),*Evolution and Source Regions of Asteroids and Comets, Proc. IAU Coll. 173* (Tatranska Lomnica: Astron. Inst. Slovak Acad. Sci.), p. 339

Emel'yanenko, V.V. 2002, *Celest. Mech. Dyn. Astron.* 84, 331

Emel'yanenko, V.V., Asher, D.J. & Bailey, M.E. 2003, *Mon. Not. R. Astron. Soc.* 338, 443

Emel'yanenko, V.V., Asher, D.J. & Bailey, M.E. 2004, *Mon. Not. R. Astron. Soc.* 350, 161

Fernández, J.A. 1980a, *Icarus* 42, 406

Fernández, J.A. 1980b, *Mon. Not. R. Astron. Soc.* 192, 481

Fernández, J.A. & Gallardo, T. 1999, in: J. Svoren, E.M. Pittich & H. Rickman (eds.),*Evolution and Source Regions of Asteroids and Comets, Proc. IAU Coll. 173* (Tatranska Lomnica: Astron. Inst. Slovak Acad. Sci.), p. 327

Fernández, J.A., Tancredi, G., Rickman, H. & Licandro, J. 1999, *Astron. Astrophys.* 352, 327

Gomes, R.S. 2003, *Icarus* 161, 404

Heisler, J., Tremaine, S. & Alcock, C. 1987, *Icarus* 70, 269

Hills, J.G. 1981, *Astron. J.* 86, 1730

Kazimirchak-Polonskaya, E.I. 1972, in: G.A. Chebotarev, E.I.Kazimirchak-Polonskaya & B.G. Marsden (eds.), *Proc. IAU Symp. 45, The Motion, Evolution of Orbits, and Origin of Comets* (Dordrecht: Reidel) p. 373

Levison, H.F. & Duncan, M.J. 1997, *Icarus* 127, 13

Oort, J.H. 1950, *Bull. Astron. Inst. Netherlands* 11, 91

Whipple, F.L. 1964, *Proc. Nat. Acad. Sci. USA* 51, 711

Whipple, F.L. 1972, in: G.A. Chebotarev, E.I.Kazimirchak-Polonskaya & B.G. Marsden (eds.), *Proc. IAU Symp. 45, The Motion, Evolution of Orbits, and Origin of Comets* (Dordrecht: Reidel) p. 401

Wiegert, P. & Tremaine, S. 1999, *Icarus* 137, 84

Dynamics of Populations of Planetary Systems
Proceedings IAU Colloquium No. 197, 2005
Z. Knežević and A. Milani, eds.

© 2005 International Astronomical Union
DOI: 10.1017/S1743921304008828

The distributions of angular elements of new comets

L. Neslušan

Astronomical Institute, Slovak Academy of Sciences, Tatranská Lomnica, Slovakia
email: ne@ta3.sk

Abstract. We point out two important effects relevant to the information which can be obtained from the distributions of galactic angular elements of new comets. (i) The commonly used criterion for a selection of new comets from a catalog of long-period comets (reciprocal original semi-major axis $a < 1.0^{-4}\,\mathrm{AU}^{-1}$) is crude as already proved by Dybczyński in 2001. It is more relevant to regard as new the comets with previous perihelion distance $q > 15\,\mathrm{AU}$. (ii) The angular orbital elements of Oort-cloud comets referred to the galactic coordinate system undergo large changes at the observed (current) perihelion passage, therefore their values are chaotic enough. Thus the information contained in the distribution of angular elements is dimmed. We suggest constructing the distributions for the elements at other epoch, e.g. for those at the previous perihelion passage.

Keywords. comets: general, Oort Cloud

---

1. Introduction

The dynamically new comets are those comets in the zone of visibility (ZV), which were not significantly perturbed by planets at their previous perihelion passages. They are thus unique bodies carrying information about the Oort cloud (OC). Any model of the OC can be regarded as realistic only if we can predict, on its basis, all the characteristics of new comets.

For a long time, since Oort's era, the comets with $1/a < 10^{-4}\,\mathrm{AU}^{-1}$ were regarded as new. This criterion to select the new comets is, hereinafter, referred to as the *first criterion*.

In 2001, Dybczyński demonstrated that 41 of 85 comets, classified by the above criterion as new, were significantly perturbed by the planets at their previous perihelion passage. According his analysis, new comets are those of which the perihelion distance, q, at the previous perihelion passage is $q > 15\,\mathrm{AU}$, and current perihelion is situated in the ZV. This criterion is, hereinafter, referred to as the *q15 criterion*. Since there is no *previous* perihelion passage in the case of comets in hyperbolic orbits, these have to be ignored, when new comets are selected from a sample of orbits.

In this contribution, we show that the theoretically derived characteristics of new comets can be compared with their observed counterparts only if the comets to be considered new are correctly selected, that is, if the information is extracted from orbital parameters at an epoch different from the current perihelion passage.

2. Galactic angular elements of new comets

Due to observational selection, the observed distribution of q is too much biased to be used in a quantitative analysis. On the other hand, this problem is much less severe for the observed distribution of semi-major axis, a. In the following, we shall deal with the

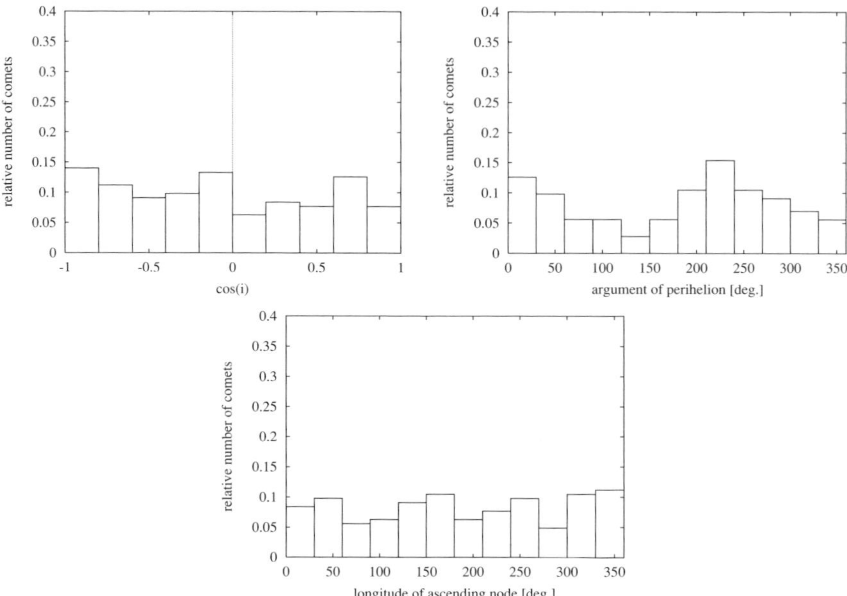

Figure 1. The distributions of dynamical galactic-coordinate-system $cos(i)$, ω, and Ω of new comets, selected from the catalog of the original orbits of long-period comets (Marsden & Williams 2003) by the *first* $(1/a < 10^{-4}\,\mathrm{AU}^{-1})$ criterion, for the current perihelion passage.

distributions of the angular orbital elements of new comets, which appears to be a non trivial undertaking.

In Fig. 1, there are demonstrated the distributions of angular elements of new comets for the current perihelion passage. The new comets are selected from the catalog of precise original orbits of 383 long-period comets (Marsden & Williams 2003) by the first $(1/a < 10^{-4}\,\mathrm{AU}^{-1})$ criterion. (The set of precise original $1/a$ published in the catalog is completed including the original angular elements calculated by Dybczyński (private communication) utilizing the method described in (Dybczyński 2001).) 143 of total 383 long-period (LP) comets should be classified as dynamically new according to this criterion. (The catalog contains 386 original orbits of LP comets. Three of them are however suspected to be fragments of split nuclei and discarded from the used data. For more details see (Neslušan & Jakubík 2004)). We can see that the distributions of cosine of galactic inclination, $cos(i)$, and galactic ascending node, Ω, (the first and third plots) are nearly flat. No structural patterns are apparent. In the distribution of galactic argument of perihelion, ω, (the second plot), two shallow peaks, at $\omega \approx 15^{o}$ and $\omega \approx 225^{o}$, occur. Nevertheless, also this element spans through the entire possible interval of values.

Similarly, in Fig. 2, there are the distributions of $cos(i)$, ω, and Ω of new comets for the current perihelion passage, this time selected from the catalog by the $q15$ criterion. Since the comets on hyperbolic orbits have to be ignored in the case of this criterion, the set of orbits in the used data is reduced to 349 LP-comet orbits. 53 of the total 349 LP comets are classified as new according to the $q15$ criterion. Concerning the behaviors of the distributions, the same facts as for the *first* criterion can be stated: the distributions of $cos(i)$ and Ω (the first and third plots in Fig. 2) are flat with some fluctuations. The peaks at $\omega \approx 15^{o}$ and $\omega \approx 225^{o}$ (the second plot in Fig. 2) are slightly higher, but the element still ranges through the entire interval of values.

From these almost structureless distributions we can scarcely draw information. We would however like to point out that such behaviors are a consequence of the following

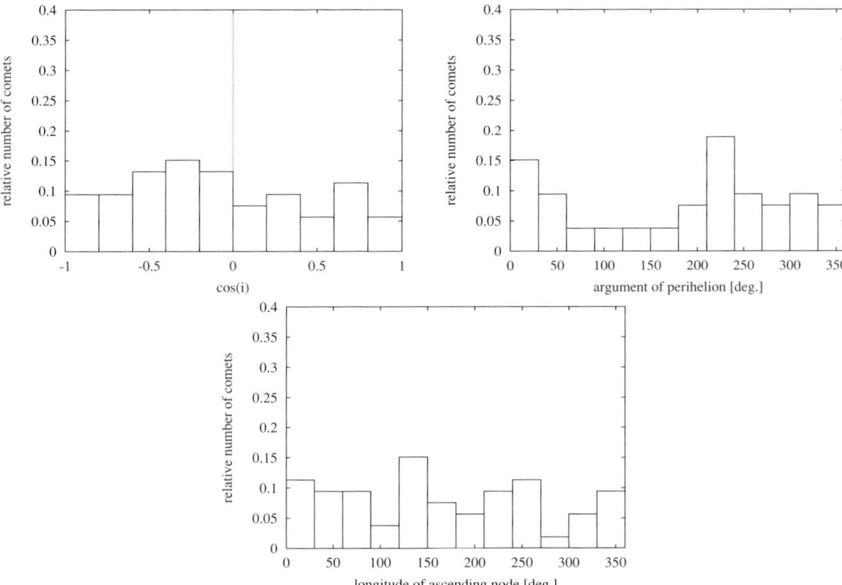

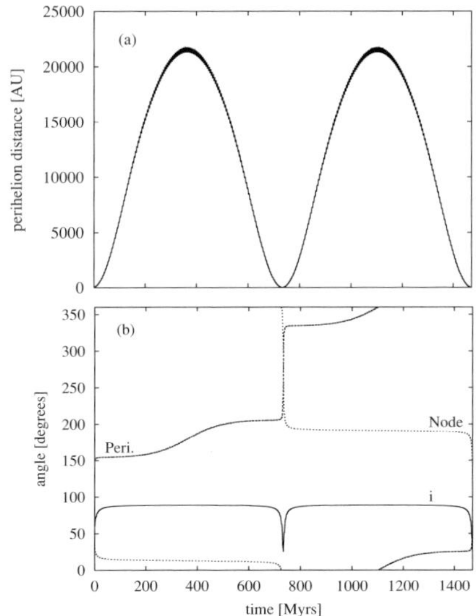

Figure 2. The distributions of dynamical galactic-coordinate-system $cos(i)$, ω, and Ω of new comets, selected from the catalog of the original orbits of long-period comets (Marsden & Williams 2003) by the $q15$ criterion, for the current perihelion passage.

Figure 3. The evolution of perihelion distance (plot a) and dynamical-galactic-coordinate-system angular orbital elements (plot b) of a typical outer-Oort-cloud comet under an action of the Galactic tide during about two libration cycles of perihelion distance.

specific circumstance. In Fig. 3 the behaviors of perihelion distance (plot a) and angular elements (plot b) are shown for the orbit of a typical OC comet perturbed by the Galactic tide, which has been proved to be the dominant perturber reducing the cometary perihelia to the ZV (Heisler & Tremaine 1986; Morris & Muller 1986; Bailey 1986; Duncan, Quinn & Tremaine 1987; Heisler, Tremaine & Alcock 1987). The libration of q and libration/circulation of angular elements, displayed in Fig. 3, was found earlier by Heisler

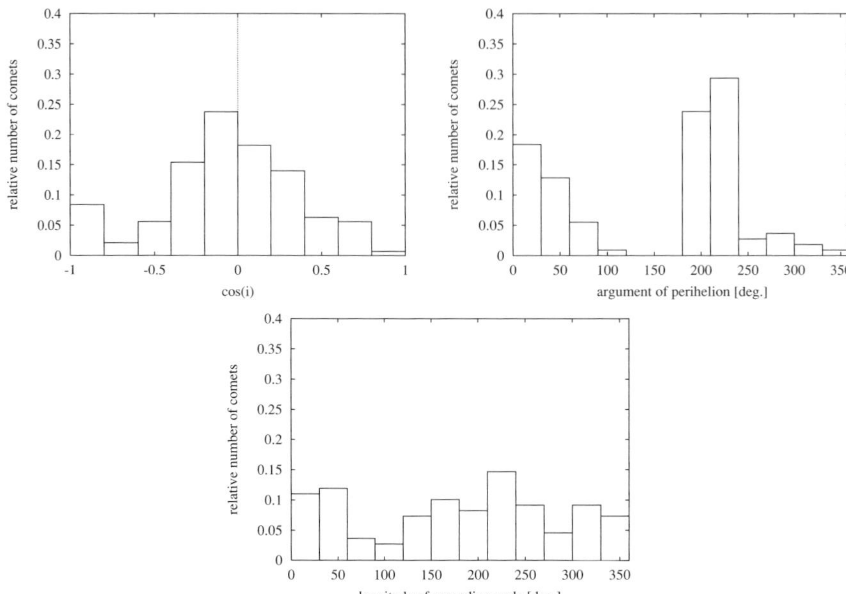

Figure 4. The distributions of dynamical galactic-coordinate-system $cos(i)$, ω, and Ω of new comets, selected from the catalog of the original orbits of long-period comets (Marsden & Williams 2003) by the *first* criterion, for the previous perihelion passage.

& Tremaine (1986) (for a more detailed demonstration of the effect see, e.g., Pretka & Dybczyński 1994). We can see that the new comets come to the ZV in the phase of their libration cycle of q, when q is in minimum. In this phase, the angular elements are macroscopically changing and, consequently, their instantaneous distributions appear chaotic enough. A possible structure is, thus, dimmed.

Because of the bad defined values of the angular elements in the period of the current, i.e. observed perihelion passage, it is necessary to construct their distribution for some other time. Since we anyway have to compute the values of elements for the previous perihelion passage in the case of $q15$ criterion, we suggest constructing the distributions just for the previous perihelion passage.

In Fig. 4, we can see the distributions of angular elements of new comets, selected from the catalog of original orbits of LP-comet original orbits by the *first* ($1/a < 10^{-4}\,\mathrm{AU}^{-1}$) criterion, for the previous perihelion passage. 109 of total 349 LP comets are classified as new according this criterion, by using the elliptical orbits at the previous perihelion passage. In the $cos(i)$ distribution (the first plot), a central peak, at $cos(i) \approx 0$, is apparent. A lower peak can be seen at $cos(i) \approx -0.9$: we can guess it is just a fluctuation. The peaks at $\omega \approx 15^o$ and $\omega \approx 225^o$ in the ω distribution (the second plot) are significantly higher than in the corresponding distributions for the current perihelion passage. There are only few comets having ω very different from these two values. The nearly flat behavior is conserved in the distribution of Ω, but the real behavior of this element is the least variable.

A single, high central peak also appears in the distribution of $cos(i)$ of the previous-perihelion-passage orbits constructed for new comets selected by the $q15$ criterion (the first plot in Fig. 5). In the distribution of ω (the second plot in Fig. 5), the cometary arguments of perihelia are clearly concentrated in the already noticed peaks at $\omega \approx 15^o$ and $\omega \approx 225^o$, when $q15$ criterion is applied. With one exception, there are no observed new comets having ω in the ranges from 90° to 180° and from 270° to 360°.

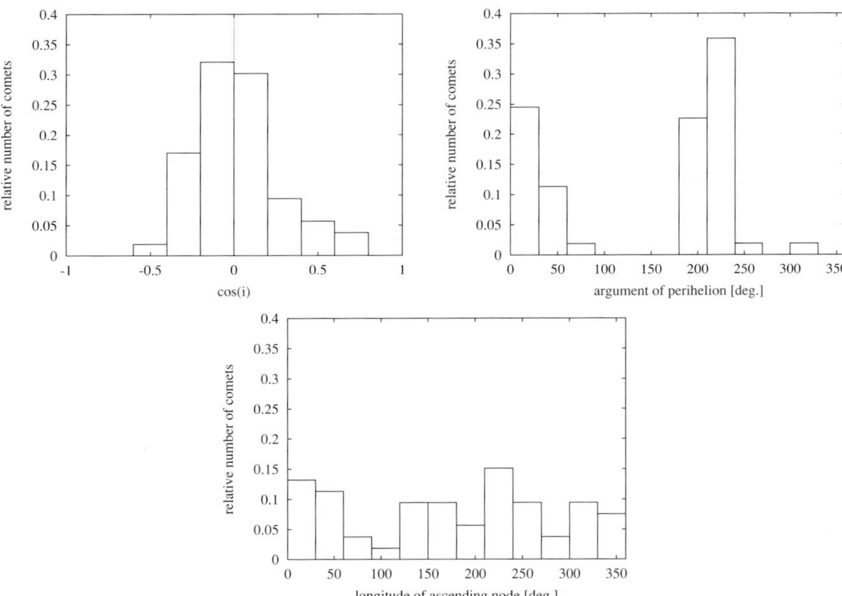

Figure 5. The distributions of dynamical galactic-coordinate-system $cos(i)$, ω, and Ω of new comets, selected from the catalog of the original orbits of long-period comets (Marsden & Williams 2003) by the $q15$ criterion, for the previous perihelion passage.

In conclusion, the measure and quality of information obtained from the distributions of angular elements of new comets depend on the particular choice of the criterion for a selection of new comets from a catalog of orbits of LP comets as well as on the choice of q-libration-cycle phase, in which the distributions are constructed. This is clearly documented by comparing the first plot in Fig. 1 with the first plot in Fig. 5 ($cos(i)$) and the second plot in Fig. 1 with the second plot in Fig. 5 (ω). In the case of Ω (the third plots in Figs. 1 and 5), the difference is small, obviously due to fact that the real distribution of this element is nearly flat.

It is necessary to note that our result has to be affected by the uncertainty of the original $1/a$. (The values of the other elements are not exact either, but their uncertainty is insignificant is the given context.) This is clearly indicated by the presence of hyperbolic $1/a$ in the used sample. The hyperbolicity has been interpreted as an error in the determination of $1/a$. As a result of the determination error, we may have classified, by the first criterion, some orbits with real $1/a \geqslant 10^{-4}\,\mathrm{AU}^{-1}$ as new. Similarly by using the $q15$ criterion, some comets in the ZV with the calculated previous $q > 15\,\mathrm{AU}$ could have their actual $1/a$ larger and, consequently, the actual current $q < 15\,\mathrm{AU}$. So, these comets could be classified as new, but they were not new in fact. To estimate a possible significance of this effect, we also repeated the construction of all distributions using only higher quality, class 1, orbits published in the used catalog (Marsden & Williams 2003). Though some small quantitative differences occurred, the same qualitative conclusions could be drawn from the corresponding series of graphs. (Unfortunately, the sample of class 1 orbits is less numerous than the entire sample, therefore a more reliable result is not warranted, despite of a higher quality.) The qualitative correctness, at least, of our conclusions is also supported by the appropriate shape of the $cos(i)$ and, especially, of ω distributions. The determination error is expected to cause some random fluctuations and, thus, a more dispersed behavior of whatever dependence. On contrary, we proceeded from the dispersed behaviors on Figs. 1ab and 2ab to much less dispersed behaviors on

Fig. 5ab. The absence of any comet in the intervals $\omega \in (90^o, 180^o)$ and $\omega \in (270^o, 360^o)$ (with one exception) in Fig. 5a can scarcely be any effect of the orbit determination error with its random character.

3. Summary

The best currently available criterion to choose new comets from a catalog is to require that their original-orbit perihelion distance q at the previous perihelion passage was $q > 15 \, \mathrm{AU}$.

The distributions of angular elements of new comets can provide a usable information only if they are constructed for the elements at an epoch different from the current (observed) perihelion passage. We suggest to construct them for the previous perihelion passage (as the orbital elements for the previous perihelion passage must be computed anyway).

Acknowledgements

This work was supported by VEGA – the Slovak Grant Agency for Science, grant No. 4012.

References

Bailey, M.E. 1986, *Nature* 324, 350

Duncan, M., Quinn, T. & Tremaine, S. 1987, *Astron. J.* 94, 1330

Dybczyński, P.A. 2001, *Astron. Astrophys.* 375, 643

Heisler, J. & Tremaine, S. 1986, *Icarus* 65, 13

Heisler, J., Tremaine, S. & Alcock, C. 1987, *Icarus* 70, 269

Morris, D.E. & Muller, R.A. 1986, *Icarus* 65, 1

Neslušan, L. & Jakubík, M. 2004, *Contrib. Astron. Obs. Skalnaté Pleso* 34, 87

Marsden, B.G. & Williams, G.V. 2003, *Catalogue of Cometary Orbits 2003*, 15-th edition, (Cambridge: Smithson. Astrophys. Obs.)

Pretka, H. & Dybczyński, P.A. 1994, in: K. Kurzyńska, F. Barlier, P.K. Seidelmann, & I. Wytrzyszczak (eds.), *Dynamics and Astrometry of Natural and Artificial Celestial Bodies*, (Poznań: Astron. Obs. of A. Mickiewicz Univ.), p. 299

Dynamics of Populations of Planetary Systems
Proceedings IAU Colloquium No. 197, 2005
Z. Knežević and A. Milani, eds.

© 2005 International Astronomical Union
DOI: 10.1017/S174392130400883X

Long term dynamical evolution of the Oort cloud comets: galactic and planetary perturbations

Piotr A. Dybczyński

Astronomical Observatory of the A. Mickiewicz Univ., Poznań, Poland
e-mail: dybol@amu.edu.pl

Abstract. The source of long-period comets can be numerically modeled by means of Monte Carlo simulation of the Oort cloud dynamics. Tracing a comet motion under the galactic perturbations over a long time interval requires taking into account of planetary perturbations. A method for the approximate treating of the planetary perturbations in such simulations is described. Some peculiarities found in planetary action on long-period simulated cometary sample are also discussed.

Keywords. Comets: general, Oort Cloud.

1. Introduction

The source of the observed long period comets may be modeled numerically with the Monte Carlo simulation of the long term dynamical evolution of the bodies in the Oort cloud under the influence of stellar and Galactic perturbations. In this paper we concentrate on the influence of the planetary perturbations during the cometary perihelion passages among the Solar System planets. It is very expensive (in CPU time) to account for planetary perturbations in an exact manner (e.g. by numerical integration of motion) because the simulation typically extends for $10^6 - 10^9$ years. In a highly simplified case one can assume either that each comet is removed from the observable part of the cloud after its first perihelion passage below the observability limit (OL), or that planetary perturbation might be completely ignored, thus allowing a comet to make several, even hundreds of observable perihelion passages without any change to its orbit.

As it was shown recently by Dybczyński (2004) both the above-mentioned approaches are far from the truth. A convenient solution has been proposed: the use of the Planetary System transparency coefficient P, defined as the probability that the planetary perturbations remove a particular comet from the observable part of the cloud during a single perihelion passage among planets.

Before we go into details let us recall the typical orbit evolution of an Oort cloud comet under the Galactic disk tide. Such an evolution for the case of $a = 20\,000\mathrm{AU}$ is presented in Fig. 1. In this kind of evolution the argument of perihelion ω , the inclination i and perihelion distance q evolve in a strictly synchronous manner: the minima of q and i coincide and ω crosses the 270° or 90° value at the same moment. The vertical lines in Fig. 1 represent changes in the heliocentric distance of a comet. Each vertical line marks one perihelion passage (due to the scale of the horizontal axis the descending and ascending branches overlap).

Planetary perturbations limit the number of consecutive perihelion passages of a particular comet and as a result they significantly decrease the new comet influx. Additionally,

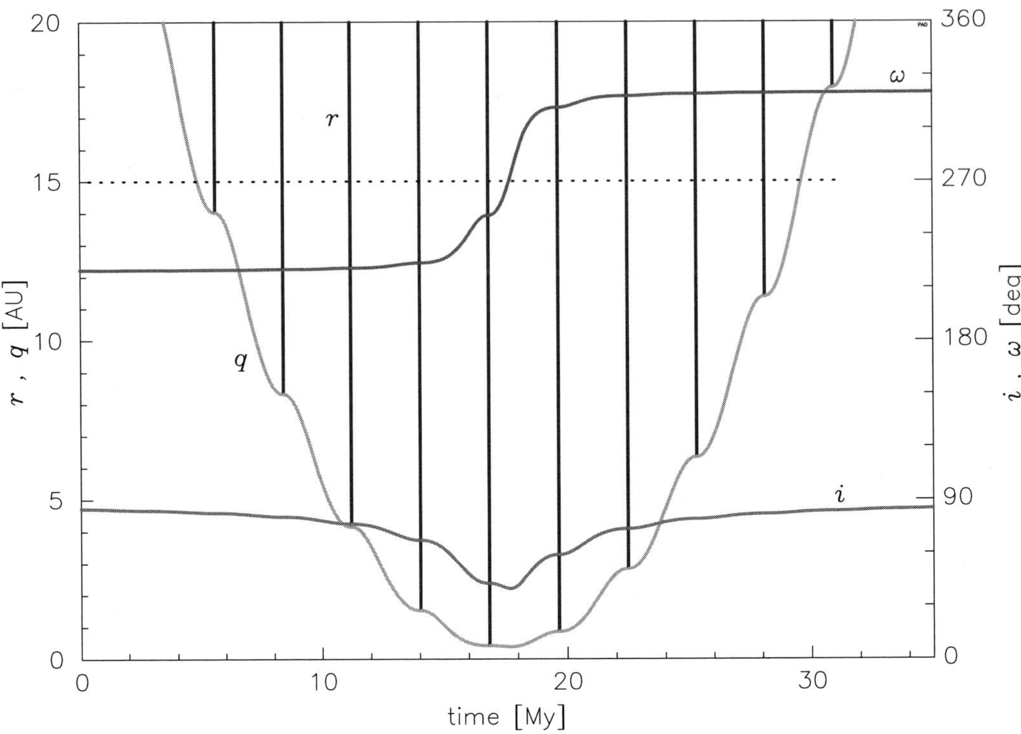

Figure 1. An example of an orbit evolution under the influence of the Galactic disk tide. In this plot the simultaneous long term evolution of the inclination i and the argument of perihelion ω (both in degrees, see right hand vertical axis) as well as perihelion distance q in AU (left hand vertical axis) is presented. Additionally, the changes in the cometary heliocentric distance r are plotted, which produces a series of vertical lines due to the relatively short time spans spent in the inner solar system (below 20 AU). Note that during a single minimum of the perihelion distance this comet crosses the adopted planetary perturbation limit at 15 AU (dotted line) 9 times. Here the semi-major axis $a = 20\,000$ AU.

due to the strictly synchronous changes in the orbital elements, their distributions depend on the strength of the planetary effect on a comet.

2. Planetary System transparency coefficient

Dybczyński (2004) estimated the value of the coefficient P on the basis of the observed, cloned and simulated samples of the Oort cloud comets. The method of estimation is based on the comparison of cometary orbits (especially the perihelion distances) one orbital revolution into the future, calculated twice, one with planetary perturbations completely omitted and the other one with perturbations fully taken into account. The dynamical model of the cometary motion among Solar System planets is based on the JPL DE406 planetary ephemerides (Standish 1998) and fully described in Dybczyński (2001). A special technique of comet cloning is proposed to obtain more valuable P estimations based on much more numerous cometary sample. An additional check is made by estimating the planetary system transparency coefficient from synthetic, simulated long-period observable comets. The are obtained are almost all in good agreement as shown in Table 1. The discrepancy between $P = 0.64$ obtained from the observed sample

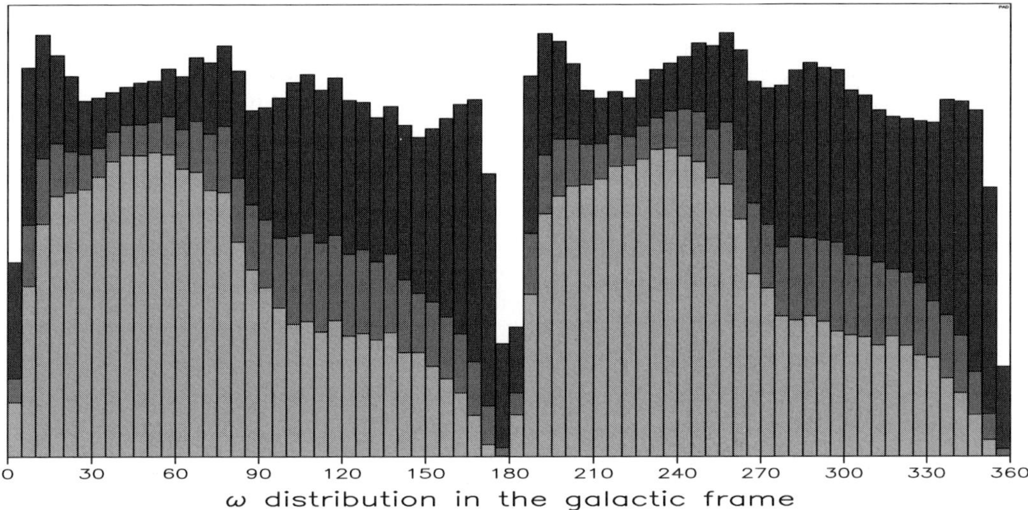

Figure 2. Three different (overlapped) distributions of the argument of perihelion of simulated observable comets with $a > 20\,000$AU for three different P values: $P = 0.05$ (dark-grey, background histogram), $P = 0.50$ (grey histogram) and $P = 0.95$ (light-grey, foreground histogram).

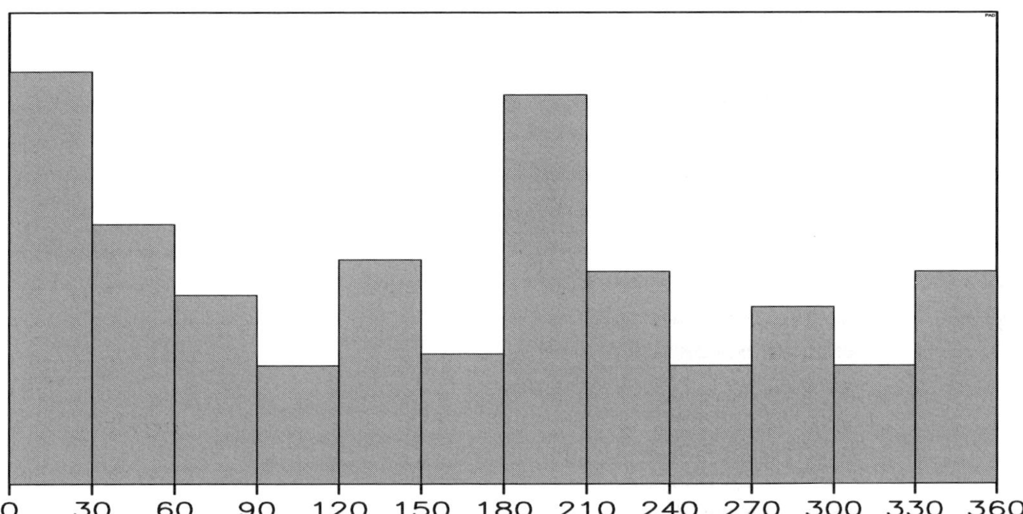

Figure 3. The distribution of the argument of perihelion with respect to the Galactic disk for 371 observed elliptic comets of classes 1 and 2, taken from the 15th edition of the Catalogue of Cometary Orbits(Marsden & Williams 2003) .

and $P = 0.48$ from the cloned and simulated ones is some bias resulting from the specific orientation of the elliptic cometary sample with respect to the Planetary System, as it is demonstrated and explained in detail in the above-mentioned paper.

The proposed approximation is a very convenient way of incorporating planetary perturbations into a dynamical model used in Oort cloud evolution simulations. In many cases it can be applied after the main simulation calculations, during the analysis stage of the project. Below we present two examples of the successful use of P in the Oort cloud modeling. The simulation code is highly simplified in both cases (Galactic center and stellar perturbations omitted) thus the results presented in the next two sections are not intended to be compared with observations, they only illustrate the improvements

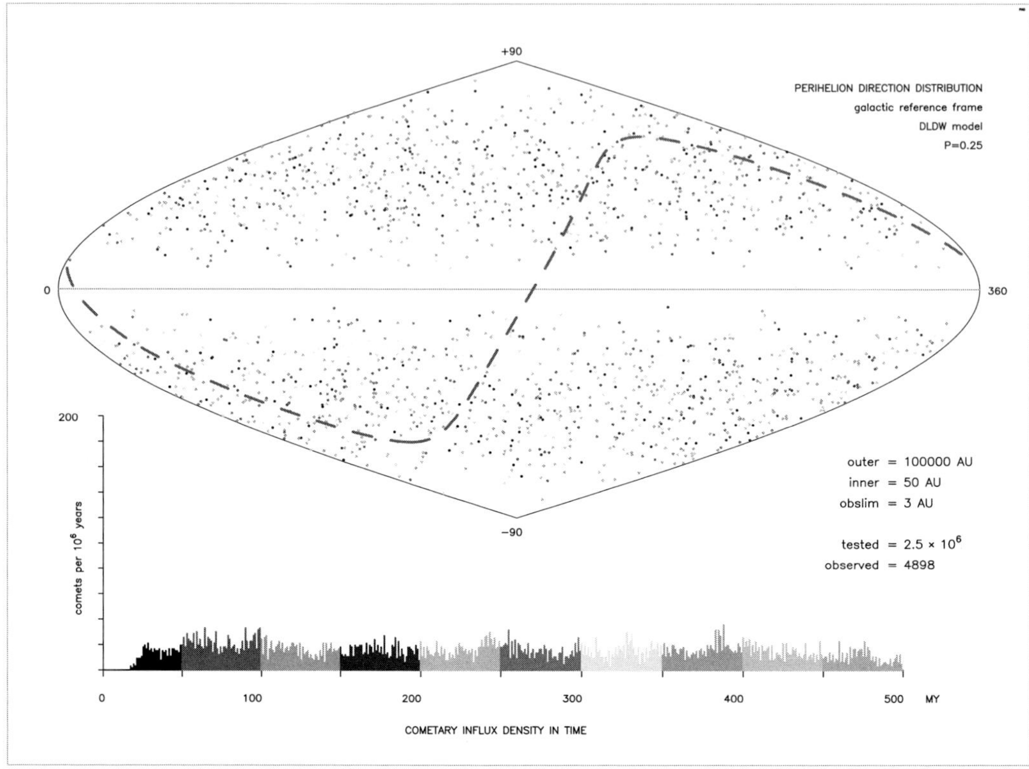

Figure 4. The perihelion direction distribution in galactic reference frame for comets with semi-major axis $a < 20\,000$ AU. The DLDW cloud model is flattened in its inner part towards the Solar System invariant plane (dashed curve in the main plot) which, combined with the orientation of the Galactic disk produces strong anisotropies. The overpopulation of these regions comes from ignoring planetary perturbations ($P = 0.0$).

Table 1. The Planetary System transparency coefficient P for the observability limit $OL = 3$ AU, estimated from different cometary samples by Dybczyński (2004).

Source of estimation	$a < 10\,000$AU	$a > 10\,000$AU
observed comets	0.23	0.64
cloned comets	0.24	0.48
simulated comets	—	0.49

possible to achieve when using the Planetary System transparency coefficient P. In all these calculations we used the DLDW model of the Oort cloud, based on the unpublished results obtained by Dones, Levison, Duncan and Weissman; for details see Dybczyński (2002).

3. Argument of perihelion distribution for simulated observable comets

Due to a synchronous variations of cometary orbital elements under the Galactic disk tidal perturbations the simulated distribution of the argument of perihelion with respect to the Galactic frame strongly depends on the Planetary System transparency coefficient P. In Fig. 2 we present this dependence for the simulated observable comets with $a >$

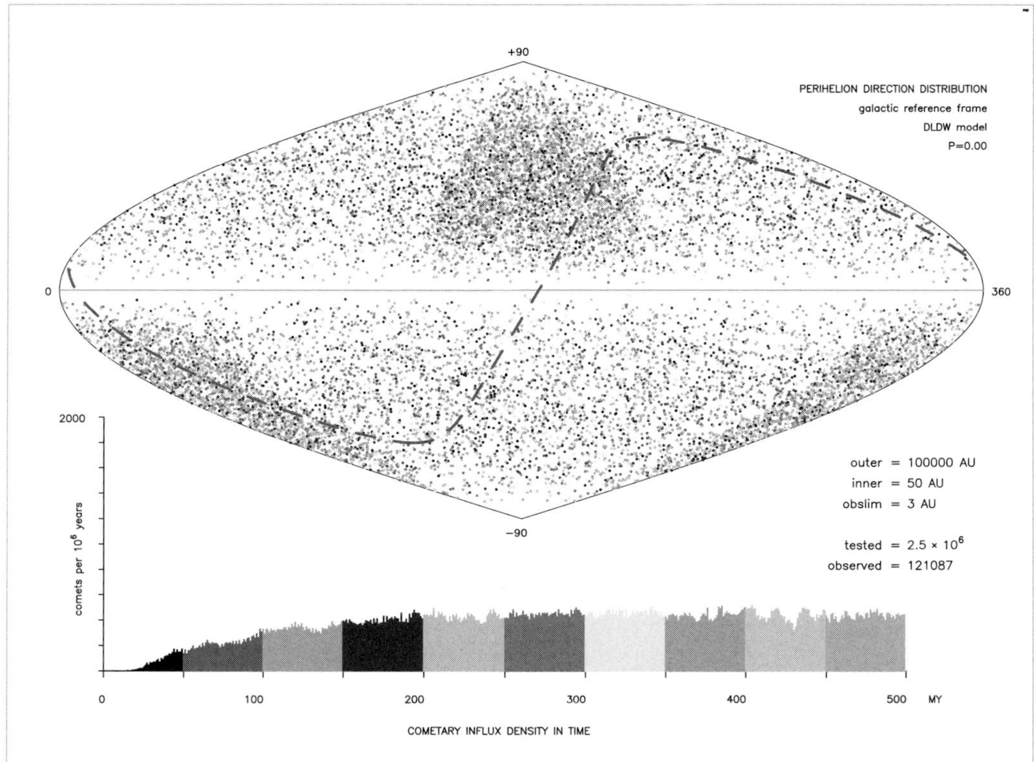

Figure 5. The same observable cometary sample as in Fig. 4 but after applying $P = 0.25$. Artificial anisotropies and concentrations disappeared.

20 000 AU. As it may be inferred from Fig. 1, the lower P value gives strong preferences for the low semi-major axis comets, resulting in much more uniform ω distribution (except regions close to 0° and 180°, which always remain empty in this simplified model). For higher P values those comets are generally removed from the observed flux and the distribution is dominated by comets observed during the first (and sometimes the second) perihelion passage. As a result the first and the third quarters of the ω distribution are relatively more populated. Such an asymmetry is seen in the really observed long-period comets population as it is shown in Fig. 3. Comparing this to the distributions presented in Fig. 2 confirms that using $P \simeq 0.5$ produces much more reliable simulation output than $P \simeq 0$, what correspond simply to ignoring the planetary perturbations.

4. Anisotropies for low semi-major-axes comets

Another example of a successful use of the Planetary System transparency coefficient is the analysis of the directional distribution of the simulated observable comets with low semi-major-axes. We restrict here our consideration to comets with the semi-major axes $a < 20\,000$ AU, what typically constitute more than 50% of the long-period cometary influx. If we plot perihelion directions of these comets on a celestial sphere, we obtain the distribution shown in Fig. 4. The mutual orientation of the flattened inner part of the cloud in the DLDW model and the Galactic disk generates here two privileged regions, highly overpopulated when omitting planetary perturbations ($P = 0$). If we apply $P = 0.25$, recommended for small semi-major axis comets, we obtain the distribution shown in Fig. 5. Note that due to the dramatic decrease in the cometary flux, the vertical scale of the flux density in time has to be ten times enlarged. It should be stressed here that the concentrations shown in Fig. 4 are completely absent in the observed samples of orbits thus applying the nonzero value of P brings the simulation output much closer to the reality. Additionally the overall number of the observed simulated comets decreased here from 121 087 down to 4 898 bodies. When scaled to the cloud population of 2×10^{12} comets it means that the influx of the new comets decreased from over 400/year down to ~ 20/year what is much closer to the estimations of the overall long-period comets influx for $OL = 3$ AU, see for example Wiegert & Tremaine (1999).

5. Conclusions

The Planetary System transparency coefficient is shown to be a useful tool for including planetary perturbations into a dynamical model used in numerical simulations of a long time evolution of the Oort cloud comets. We demonstrated that the asymmetric distribution of the galactic argument of perihelion of the observed long-period comets with high quality orbits can be successfully reproduced when $P \simeq 0.5$ is used. We also showed that the nonzero P value is necessary to remove some artificial directional concentrations of the simulated observable comets sample, especially for low semi-major axes comets from the flattened, inner part of the simulated Oort cloud. It has been also shown that the Solar System transparency coefficient P influences both the geometrical characteristics of the simulated comets as well as the long-period cometary influx obtained from simulations. The proposed method of including the planetary perturbations into the dynamical model of the Oort cloud dynamics used in numerical simulations proved to be powerful while still very simple. We incorporate this approximation into the extended simulation computer code for the long-term Oort cloud comets dynamical evolution under the influence of stellar and Galactic perturbations. This work is in progress.

Acknowledgements

The author is indebted to an anonymous referee for comments and suggestions which allowed to improve the quality of this manuscript. The research described in this paper was supported by KBN grant no. 2P03D01324.

References

Dybczyński, P. A. 2001, *Astron. Astrophys.* 375, 643

Dybczyński, P. A. 2002, *Astron. Astrophys.* 396, 283

Dybczyński, P. A. 2004, *Astron. Astrophys.* in press

Marsden, B. G. & Williams, G. V. 2003, *Catalogue of Cometary Orbits, 15th Edition* (Cambridge, Mass.: Smithsonian Astrophysical Observatory)

Standish, E. M. 1998, JPL Planetary and Lunar Ephemerides, *Interoffice Memorandum IOM 312.F – 98 – 048*, Jet Propulsion Laboratory

Wiegert, P. & Tremaine, S. 1999, *Icarus* 137, 84

Dynamics of Populations of Planetary Systems
Proceedings IAU Colloquium No. 197, 2005
Z. Knežević and A. Milani, eds.

© 2005 International Astronomical Union
DOI: 10.1017/S1743921304008841

Resonances near the orbit of 2003 VB_{12} (Sedna)

Matija Ćuk

Department of Astronomy, Cornell University, Ithaca, NY 14853, USA
email: cuk@astro.cornell.edu

Abstract. 2003 VB_{12} (Sedna) is as much distinguished by its considerable size as by its extremely unusual orbit, which has perihelion at about $q = 76$ AU with semi-major axis $a = 533$ AU (Brown *et al.* 2004, JPL Horizons†). Thus it is effectively decoupled from both Neptune and the Galactic tide (Fernandez 1997). Brown *et al.* (2004) and Morbidelli & Levison (2004) maintain that only scattering by a so-far-unobserved "Planet X" or by an errant star could produce such a high-perihelion orbit for a scattered-disk KBO. While a close encounter is plausible, given the Sun's likely birth in an open cluster, such an interaction would profoundly disturb the Oort cloud and would require fundamental revision to the present theories of its formation.

Although the planets cannot significantly affect VB_{12}'s orbit through close approaches, resonant perturbations could conceivably produce secular effects on it. To explore this possibility, we have numerically integrated test particles with $480 < a < 580$ AU and a fixed $q = 76$ AU. Including the four giant planets, but ignoring the Kuiper Belt and the inner Oort Cloud, as well as the Galactic tide, we find multiple resonances, some of which perturb significantly the test particles' eccentricity more strongly than the leading secular terms. We identify these resonances as variants of the very high-order ($n_N > 60\ n$) mean-motion commensurabilities between Neptune and VB_{12}. Although unprecedented, these extremely high-order resonances can be significant due to VB_{12}'s very high eccentricity ($e = 0.86$). Even powers of eccentricity beyond sixty are still on the order of 10^{-4}, which is comparable to the strength of low-order resonances involving near-circular orbits. We extrapolate the possible long-term drift rate and estimate the likelihood of such resonances producing an "inner Oort cloud" population consistent with VB_{12} over the age of the Solar System. Finally we discuss how planetary migration and the Kuiper-Belt's depletion might have affected VB_{12}'s putative resonance.

Keywords. Celestial mechanics, Kuiper Belt, minor planets, asteroids, Oort Cloud, solar system: formation

---

1. Introduction

Minor planet 2003 VB_{12} (unofficially known as "Sedna", hereafter "VB_{12}") is a unique Solar System object, and its discovery convinced many researchers that present theories of the Solar System's formation need some rethinking (Brown *et al.* 2004). After nine major planets, VB_{12} is most likely the largest known body that orbits the Sun; its very eccentric orbit has a perihelion that is too high for direct interactions with Neptune (Gladman *et al.* 2002), and an aphelion too low to be affected by the Galactic tide (Fernandez 1997). Given that its present orbit is extremely stable, it is unclear how it could have evolved into its present state.

Morbidelli & Levison (2004, ML04) discuss four possible solutions to the mystery of VB_{12}'s orbit: scattering by a more eccentric Neptune, scattering by an unknown distant major planet, secular interactions with an extended planetesimal disc and effects of a

† Orbital elements were obtained on August 13th 2004, through Jet Propulsion Laboratory's Horizons on-line ephemeris service, http://ssd.jpl.nasa.gov/

close stellar encounter. The authors show that, even if Neptune at some point had eccentricity as large as 0.4, VB_{12} could not attain such a high aphelion distance. ML04 find interaction with an unknown planet to be a more promising mechanism, although one that "raises more problems than it solves". They demonstrate that an Earth-mass planet would require billions of years to increase VB_{12}'s aphelion to almost 1000 AU, making it almost certainly both decoupled from Neptune and still extant. Neither scattering by Neptune nor in-situ formation are likely to produce such a body. Interactions with an extended planetesimal disk conserve the component of angular momentum normal to the disk, $H = \sqrt{\mu a(1 - e^2)}\cos i$, making it impossible for low-i objects like VB_{12} to have had a past eccentricity large enough to interact strongly with Neptune. ML04 find that a slow passage by a solar-mass star, with the perihelion distance of 800 AU, can explain the present characteristics of VB_{12}'s orbit, as well as that of another "extended scattered disk" object, 2000 CR_{105}.

Such an encounter is not unlikely given that most stars form in clusters; a similar encounter has been invoked by Ida *et al.* (2000) to explain the dynamical excitation of the "classical" Kuiper Belt. However, ML04 think that the existence of objects like 2000 CR_{105} and 2003 VB_{12} is the only compelling evidence for such a passage. A slow stellar passage would likely strip the Sun of the much of the existing Oort cloud, making its formation even more difficult (H. Levison 2004, personal communication). Therefore, before we can accept the necessity of a stellar passage early in the Solar System history, it is essential to consider all the alternatives.

In this paper, we will take a first look at the intermediate-term (3×10^7 yr) dynamics of objects with VB_{12}-like orbits, that is, with perihelia at 76 AU and semimajor axes in 480–580 AU range. In Section 2, we will present the results of a rough dynamical survey using a symplectic integrator. In Section 3, we will concern ourselves with features that are weakly dependent on mean motion, while the resonant features will be discussed in Section 4, with results summarized in Section 5.

2. Numerical Experiment

In order to study the secular dynamics of the inner Oort cloud, we have integrated orbits of 100 massless test particles for 3×10^7 yr period, starting at the epoch of midnight, September 24th 2003. The Sun and four giant planets were fully included into the integration, while Pluto, the Kuiper belt and any possible extended disk were ignored. The initial conditions were varied so that the range of average semimajor axes was 478–578 AU, with the step size of 1 AU. The perihelion distance was fixed at 76 AU for all particles, while the inclination and other angular variables were taken to be the same as those for VB_{12} (generated through JPL Horizons on August 13th 2004, for the above epoch). A home-made symplectic integrator based on the standard algorithm of Wisdom & Holman(1991) was used. The large dynamic range of mean motions made our integrations comparably inefficient: the timestep was dictated by Jupiter's orbital period, which is 1000 times shorter than those of our test particles. However, we could not avoid integrating directly all four giant planets, since we were interested in detecting any perturbations arising from frequencies associated with the planets' mean motions (some important perturbations, notably the near-resonance of Uranus and Neptune, have periods longer than 10^3 yr).

Since we are above all interested in perturbations that can change the perihelion distance of VB_{12}-like objects, in Fig. 1 we plot the change of pericenter distance in the course of the simulation, as a function of a particle's semimajor axis. Right away, despite the low resolution of our survey, we can see two major features of the plot: the

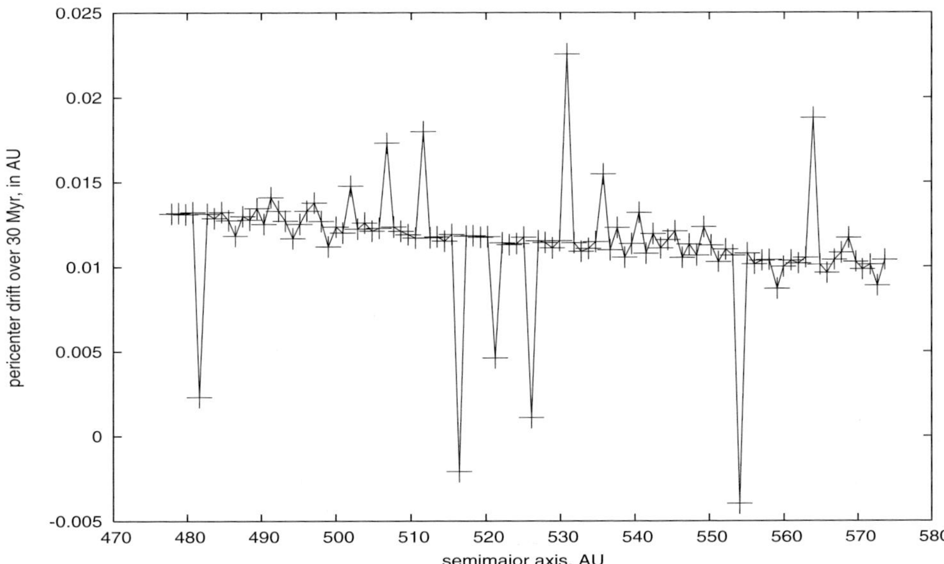

Figure 1. Change in perihelion distance (in AU) during the integration (3×10^7 yr) as the function of a test particle's semimajor axis. The change was computed as the difference between the average perihelia over the first and the last 5×10^5 yr, while the a shown is the average over the first of those intervals. The continuous background and discrete features are clearly visible. Varying intensities of the resonances are in large part due to rough sampling (1 AU).

constant background drift of the pericenter and multiple discrete features, within which the perihelion behaves significantly different from the background. The background drift is positive, amounting to about 10^{-2} AU over the whole integration, and it decreases noticeably for larger semimajor axes. Despite its secular appearance in our survey, this drift is just a consequence of a long-period oscillation in eccentricity, caused by the J_4 moment of the Solar System. Its direction is determined solely by the phase of the particle's argument of pericenter ω, and we will derive all of its important features in the next section.

Discrete features fill the whole range in a covered by the survey, with spacing of about 5 AU between neighboring ones. Their regularity and large number hint at each being a high-order mean-motion resonance with one of the giant planets, Neptune being most likely. Indeed, spacing between two neighbouring high-order resonances with Neptune of type $1 : k$ and $1 : (k+1)$ (where $k >> 1$) is expected to be

$$\frac{\Delta a}{a} = \frac{2\Delta n}{3n} \approx \frac{2n}{3n_N} = \frac{2}{3}\left(\frac{a}{a_N}\right)^{-3/2},$$

where a and n are the semimajor axis and mean motion of the particle, while a_N and n_N are those for Neptune. For $a = 530$ AU, the spacing will be $\Delta a = 4.8$ AU. Finally, we have plotted the resonant argument $\Psi = \lambda_N - k\lambda + (k-1)\varpi$ (where k is an integer) for the number of particles (an example is given in Fig. 2) and observed episodes of the libration of Ψ for all "excited" bodies, while Ψ circulated rapidly for all "non-excited" ones. In section 4 we will address the particulars of these resonances in more detail.

3. Secular Drift

Secular effects of the planets on the orbit of a "inner Oort cloud" body like VB$_{12}$ are relatively simple. We are interested in timescales much longer than the precession periods of the planet's perihelia and nodes, so the Solar System potential can be considered azimuthally symmetric. Stated this way, the problem of the secular evolution of VB$_{12}$'s orbit is very similar to that of a very eccentric artificial satellite orbit evolving in the field of an oblate planet. Brouwer (1959) has analytically studied that problem in some detail, and we will apply his results to the problem of the secular drift in VB$_{12}$'s eccentricity. More recently, Yokoyama *et al.* (2003) have independently derived the secular disturbing function to the fourth order in a/a' for a more general case of two well-separated bodies; when the inner body's eccentricity equals zero, their result becomes identical to that of Brouwer (1959).

The J_2 moment of the Solar system, defined here as $J_2 = (1/2) \sum_{k=1}^{4}(m_k/M)(a_k/a_4)^2$, where m_k and a_k are masses and distances of planets Jupiter through Neptune, and M is the Sun's mass, will cause only simple precession of a perturbed particle's orbit, with no effect on its e or i (see, e.g., Danby 1992). Therefore, the terms in the Hamiltonian involving $J_4 = (3/8) \sum_{k=1}^{4}(m_k/M)(a_k/a_4)^4$ need to be taken into account (the moment $J_3 = 0$ as the problem exhibits north-south symmetry). According to Brouwer (1959), the secular part of the disturbing potential arising from J_4, expressed in Delaunay elements, is:

$$U_4 = \frac{\mu^6 k_4}{L^6 G^7}\left(\frac{3}{8} - \frac{15}{4}\frac{H^2}{G^2} + \frac{35}{8}\frac{H^4}{G^4}\right)\left(\frac{5}{2} - \frac{3}{2}\frac{G^2}{L^2}\right)$$

$$+\left(-\frac{5}{6} + \frac{20}{3}\frac{H^2}{G^2} - \frac{35}{6}\frac{H^4}{G^4}\right)\left(\frac{3}{4} - \frac{3}{4}\frac{G^2}{L^2}\right)\cos(2g). \tag{3.1}$$

In Eq. 3.1, $\mu = GM$, $k_4 = (3/8)J_4$ while the Delaunay variables L, G, H and g have their usual meaning (Murray & Dermott 1999), except that the planetary radius (in our case replaced by Neptune's a) is used as the unit of length. The only part of (3.1) that is of interest to us is the one containing $\cos(2g)$, as the remaining terms can produce no change in e. Expressed in standard orbital elements, the term in question becomes:

$$U_4' = \frac{3}{8}\frac{\mu\, J_4\, a_N^4 e^2}{a^5(1 - e^2)^{7/2}}\left(\frac{15}{4}\sin^2 i - \frac{35}{8}\sin^4 i\right)\cos(2\omega). \tag{3.2}$$

Through the Lagrange equations (Danby 1992), we get the secular rate of change in eccentricity:

$$\dot{e} = \frac{15}{16}n\, J_4\left(\frac{a_N}{a}\right)^4\frac{e}{(1 - e^2)^3}\left(3\sin^2 i - \frac{7}{2}\sin^4 i\right)\sin(2\omega). \tag{3.3}$$

Using $n = 5.1 \times 10^{-4}$rad/yr, $J_2 = 2.4 \times 10^{-5}$, $a = 533$AU, $e = 0.857$, $i = 12°$, $\omega = 312°$, we obtain $\dot{e} = -7.5 \times 10^{-13}$yr^{-1}. The corresponding perihelion drift rate during our integration is $\Delta q = -a\dot{e}(3 \times 10^7 yr) = 1.2 \times 10^{-2} AU$, which agrees well with result plotted in Fig. 1. According to the above formulae, the magnitude of Δq should decrease with semimajor axis proportional to approximately $a^{-1.5}$, also in agreement with Fig. 1. This decrease is a smooth function of a and cannot be related to the discrete features in Fig. 1. Therefore those features must be caused by a different mechanism which cannot be explained by a purely secular theory.

It is interesting to note that secular evolution of a body perturbed by the system's J_4 moment is somewhat similar to that described by Kozai (1962). In the latter case, perturbee is orbiting inside the perturber's orbit, which need not be eccentric (classical application of Kozai's theory is the secular motion of a high-inclination asteroid perturbed

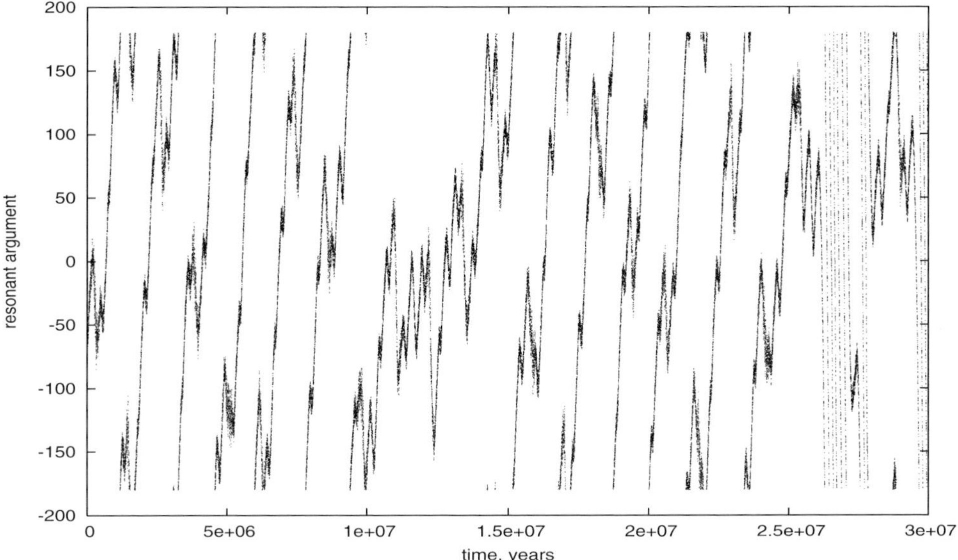

Figure 2. The resonant argument $\Psi = \lambda_N - 74\lambda + 73\varpi$ over the course of the integration for the particle associated with the resonant feature at 531 AU

by Jupiter). I both cases, e and i oscillate in counter-phase, and go through two cycles during one full precession period of ω. However, the terms in the disturbing potential which cause oscillations due to the Kozai mechanism and J_4 are not the same. Kozai mechanism relies on term $U_2 \sim a_1^2/a_2^3$ (subscripts 1 and 2 refer to inner and outer bodies), which becomes zero if $e_1 = 0$ (since ω_1 is then meaningless). Similarly, $U_3 \sim a_1^3/a_2^4$ is zero if either of the eccentricities is zero. Only $U_4 \sim a_1^4/a_2^5$ will not be zero if $e_1 = 0$ (which is an inevitable side effect of using the J_2/J_4 approximation). So it is somewhat incorrect to refer to oscillations in a particle's e and i caused by J_4 as "the Kozai behavior" in the usual sense, but the label "quasi-Kozai" would be more appropriate.

4. Resonances

To illustrate the resonant behavior, Fig. 2 plots the resonant argument $\Psi = \lambda_N - 74\lambda + 73\varpi$ for the test particle that produces the positive peak at $a = 531$ AU in Fig. 1. During most of the integration, Ψ shows intermittent episodes of slow circulation and libration. The transitions between different regimes toward the end of the integrations is correlated with abrupt (if small) changes in a and e. Based on the locations and spacing of these resonances, as well as the evidence from the resonant arguments, we have no doubt that they are very high-order mean-motion commensurabilities with Neptune.

The existence of meaningful resonances with orders as high as 73 (as that in Fig. 2) is somewhat surprising. Resonances with orders higher than a few are rare in the Solar System (Murray & Dermott 1999). This is mainly because every resonant term in the Hamiltonian containing a body's secular angles ϖ and Ω has to multiplied by a monomial in the same body's e and $\sin i$, respectively. The power of e or $\sin i$ that factors the resonant term in question is equal to the whole-number coefficient that multiplies the corresponding secular angle in the resonant argument. The total sum of all such whole-number coefficients factoring the secular angles in the resonant argument is, in turn, equal to the order of the resonance. Therefore, a term in the Hamiltonian associated with the mean-motion resonance of order n has to contain a factor $e^j \sin^k i$, where $j + k = n$. Since

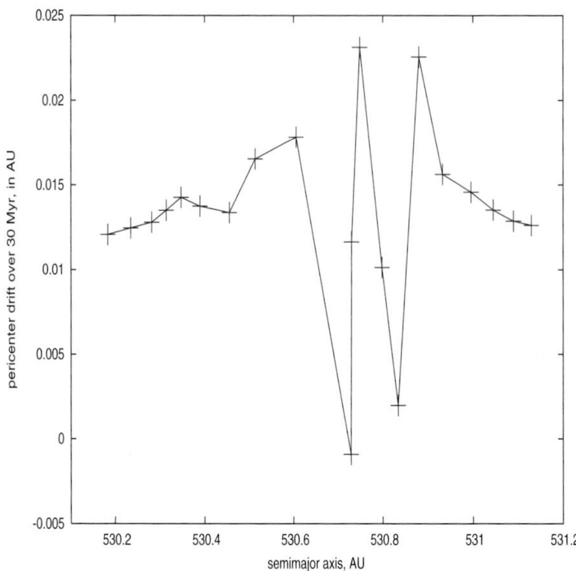

Figure 3. Same as Fig. 1, only for the region around 1:74 mean motion resonance with Neptune at 531 AU.

both e and $\sin i$ are small for most Solar System objects, the magnitudes of high-order resonant terms are usually vanishingly small. These relations describing resonant terms are commonly known as d'Alembert rules (Murray & Dermott 1999).

However, in the case of VB_{12}, eccentricity is by no means small, and even a high power of it like e^{73} is larger than 10^{-5}. This factor is quite comparable to the one a low-order resonance among bodies on low-eccentricity orbits (e.g., the major satellites of Saturn, Murray & Dermott 1999) would have. Therefore, even if a 73th-order resonance intuitively appears impossible, it cannot be ruled out on analytical grounds.

To probe in more detail the structure of one of these resonances, we also ran a higher-resolution probe with twenty test particles at semimajor axes within 1 AU surrounding the body featured in Fig. 2. Figure 3 is the same as Fig. 1, just restricted to the region around 531 AU, and shows the result of the higher-resolution probe. We can now see that the "wings" of the resonance are associated with a positive change in q, the magnitude of which increases toward the center, while the core itself appears more chaotic, without any preferred sign of the perturbation in q. In terms of the resonant argument, we have noticed a prevalence of Ψ's slow circulation among the particles affected by the "wings" of the resonance (most of the behavior shown in Fig. 2 is of this type), while the particles in the "core" show a chaotic mix of episodes of fast circulation and libration, the latter resulting in abrupt changes to the particles' a and e. Therefore we can tentatively conclude that while the bodies closest to the centers of the resonances usually have chaotically changing perihelia, it is likely that a uniform increase in q can be sustained for very long periods of time for some bodies.

How significant is this drift over the age of the Solar System? We will make a naive approximation that should give us some handle on the timescales needed for a significant change in the orbital elements of VB_{12} to occur. If we designate $C_e = de/dt$ at the present epoch ($e_0 = 0.857$), and assume that $a = const$ and $de/dt = (e/e_0)^{73}C_e$, then the time needed to change the perihelion to 45 AU ($e_1 = 0.916$) is given by:

$$T = \frac{1}{C_e} \int_{e_1}^{e_0} \left(\frac{e_0}{e}\right)^{73} de = -\frac{e_0}{72 C_e}\left[1 - \left(\frac{e_0}{e_1}\right)^{72}\right]. \tag{4.1}$$

Based on 4.1, we can estimate what C_e is required to change the eccentricity from e_1 to e_0 in $T = 4.5 \times 10^9$ yr:

$$C_e \simeq -\frac{e_0}{72 T} = -2.64 \times 10^{-12} \mathrm{yr}^{-1}, \tag{4.2}$$

which corresponds to $\Delta q = 4.2 \times 10^{-2}$ AU over the course of our integration. While this drift is several times larger than the deviation of resonant points from the background in Fig. 1, it is likely that a more careful sampling could find locations within the resonances where a sustained positive drift in Δq could be produced.

Given the extremely high orders involved, it appears unlikely that any one particular resonance would be much stronger than its immediate neighbors. However, some of them might be affected by secondary resonances; in particular, frequencies associated with multiples of the 4200-yr period of the "Lesser Inequality" (LI) of Uranus and Neptune might change the nature of local mean-motion resonances. Bodies with $a \simeq 544$ AU have orbital periods close to three times the LI; while we have not seen any primary resonances associated with this or any other harmonic of the LI, we still have to explore the possible secondary resonances associated with the LI. The frequency of the LI leaves a strong signal in Neptune's a, so it is conceivable that it could affect its mean-motion resonances, too.

5. Conclusions

In previous sections we have seen that high-order mean-motion resonances with Neptune can slowly but continuously change the perihelion of bodies on orbits similar to that of 2003 VB$_{12}$. Although of very high order (> 70), these resonances are still significant due to considerable eccentricity of 2003 VB$_{12}$ ($e \simeq 0.86$). Also, assuming that the strength of the resonance is a simple function of the perturbee's eccentricity, the strength of the observed resonances is of the right order of magnitude to produce the current orbit of 2003 VB$_{12}$ over the age of the Solar System.

While our present results are intriguing, they by no means prove that 2003 VB_{12} is in a mean-motion resonance with Neptune or that such a resonance produced its current orbit. Much more work, both theoretical and observational, is needed to test this hypothesis. More detailed surveys and longer integrations are needed to better understand the dynamics of these resonances. To fully explore all the possible paths 2003 VB$_{12}$ could have taken to its present dwelling place, simulations including migration of Neptune are also needed (especially so if the Lesser Inequality of Uranus and Neptune is also involved).

Further observations of 2003 VB$_{12}$ would be valuable not only to determine its precise mean motion, but also to put some constraints on its physical properties. If 2003 VB$_{12}$ is as large as currently thought, it implies a large mass for the "inner Oort Cloud", which then should be taken into account by all theories of Solar System formation. On the other hand, if the "observed" 2003 VB$_{12}$ refers to its coma or an extended atmosphere of some kind, the implied mass and the significance of similar bodies could be much less. In order to decide among the different models for the formation and migration of 2003 VB$_{12}$, it will be absolutely necessary to have some estimate of the total mass of all bodies on similar orbits. In any case, it is likely that "Sedna" and her yet unknown sisters will hold more surprises for observers and theorists alike.

Acknowledgements

The author would like to thank Joseph A. Burns for help and support since the beginning of this project, as well as for his comments on an early draft of this paper.

References

Brouwer, D. 1959, *Astron. J.* 64, 378

Brown, M.E., Trujillo, C. & Rabinowitz, D. 2004, *Astrophys. J. Letters*, in press

Danby, J.M.A. 1992, *Fundamentals of Celestial Mechanics* (2nd ed.), Richmond: Willman-Bell

Fernandez, J.A. 1997, *Icarus* 129, 106

Gladman, B.J., Holman, M., Grav, T., Kavelaars, J.J., Nicholson, P.D., Aksnes, J. & Petit, J.-M. 2002, *Icarus* 257, 269

Ida, S., Larwood, J. & Burkert, A. 2000, *Astrophys. J.* 528, 351

Kozai, Y. 1962, *Astron. J.* 67, 591

Morbidelli, A. & Levison, H.F. 2004, *Astron. J.*, in press

Murray, C. D. & Dermott, S.F. 1999, *Solar System Dynamics*, Cambridge Univ. Press

Wisdom, J. & Holman, M. 1991, *Astron. J.* 102, 1528

Yokoyama, T., Santos, M.T., Cardin, G. & Winter, O.C. 2003, *Astron. Astrophys.* 401, 762

Dynamics of Populations of Planetary Systems
Proceedings IAU Colloquium No. 197, 2004
Z. Knežević and A. Milani, eds.

© 2004 International Astronomical Union
DOI: 10.1017/S1743921304008853

Bifurcations of periodic orbits and potential stability regions in Kuiper belt dynamics

Thomas A. Kotoulas and George Voyatzis

Section of Astrophysics, Astronomy and Mechanics, Department of Physics, University of
Thessaloniki, GR-541 24 Thessaloniki, Greece
email: tkoto@skiathos.physics.auth.gr, voyatzis@auth.gr

Abstract. In the framework of the restricted three body problem, the resonant periodic orbits associated with the Kuiper belt dynamics are studied. Particularly, all the first, second and third order exterior mean motion resonances with Neptune located up to 50A.U. and the asymmetric resonances (beyond the 48 A.U.) are considered. We present the bifurcation points of families of periodic orbits of the planar circular problem from which families of periodic orbits are generated in the planar elliptic and in the 3D circular problem. Similarities and differences between the various resonant cases are noticed. The relation between the distribution of the bifurcation points and the population of small bodies at the particular resonances is discussed.

Keywords. Kuiper Belt, minor planets

1. Introduction

An interesting problem in Solar System science is the study of the dynamical evolution of small bodies beyond the orbit of Neptune, in the so called *Edgeworth-Kuiper (E-K) belt*. More than 700 trans-Neptunian objects (TNOs) have been observed up to now and are included in the associated list of the Minor Planet Center (http://cfa-www.harvard.edu/iau/ mpc.html). Most of the TNOs move in low eccentric orbits ($e \approx 0.1$) but a significant portion of them shows relatively high eccentric motion. The existence of high inclined orbits is also evident. A recent review about the structure of the Edgeworth-Kuiper belt was given by Morbidelli *et al.* 2003.

Many numerical and theoretical works were done to investigate the dynamics of TNOs in the framework of the restricted three body problem (RTBP) (e.g. Knežević *et al.* (1991), Levison and Duncan (1993), Holman and Wisdom (1993), see also references in Morbidelli (1999)). Even the simplest model, namely the planar circular restricted three-body problem, reveals interesting dynamical features (Malhotra (1996)). For models of few degrees of freedom, the existence and stability of periodic orbits determine critically the phase space structure. Therefore the regions round stable periodic orbits are possible places to host TNOs revolving in regular orbits. We studied the families of periodic orbits in many resonances in E-K belt and in the present work we discuss about their properties with respect to the bifurcation points and stability of the generating periodic orbits. In Section 2 we present the main features of the families of symmetric periodic orbits in the planar circular RTBP, while in Section 3 we present the bifurcation points of each family from the planar circular to the planar elliptic and to the 3D circular RTBP. The resonances of the type $n/n' = 1/q, q = 2, 3, ...$, the so-called asymmetric resonances, are discussed in Section 4.

2. Families of symmetric periodic orbits in the planar circular RTBP

We consider the rotating orthogonal frame of reference xOy where the Sun and the secondary body, particularly the Neptune, define the Ox-axis. For the mass of Neptune we have chosen the value $\mu=5.178\times10^{-5}$ and the orbital period of Neptune around the Sun is equal to 2π. In the above rotating frame, the system has families of symmetric periodic orbits ($y(0) = 0, \dot{x}(0) = 0$) classified in two different kinds:

- *Families of circular orbits (first kind)* : The periodic orbits correspond to nearly circular orbits for the small body. The ratio n/n' varies along each family.
- *Resonant Families (second kind)* : The periodic orbits correspond to elliptic orbits for the small body. The eccentricity increases along the family but the ratio n/n' is almost constant and rational $n/n' \approx p/q$, $p, q \in Z$.

For the exterior resonances, studied in this paper, it is $q > p$ and the difference $q - p$ defines the order of the resonance. The multiplicity of a periodic orbit is defined as the number of crosses of the orbit with the axis $y = 0$ in the same direction ($\dot{y} > 0$ or $\dot{y} < 0$) in a period. For each resonance there exist two different families (I and II) differing in phase. On family I the small body is initially at perihelion and on family II it is at aphelion. Both of them originate from the circular family. For the high order resonances (e.g. 1/3, 3/5, 1/4, 4/7) the continuation of the family of the circular orbits is possible for $\mu >0$ and the resonant families bifurcate from the circular family where $n/n' = p/q$. For the first order resonances (e.g. 1/2, 2/3, 3/4, 4/5) the above mentioned continuation is not possible (Hadjidemetriou and Ichtiaroglou (1984)). The resonant families join smoothly the circular family, which shows a gap at the position of the first order resonance for $\mu \neq 0$. A detailed study on the 1/2, 2/3 and 3/4 resonant families is given in Kotoulas and Hadjidemetriou (2002).

In all resonances, family I starts as unstable; a collision orbit with Neptune takes place at a specific value of eccentricity where the relation $a(1 - e) \approx 1$ holds. After the collision orbit the family I becomes stable and the multiplicity increases by one. In all resonances except for the asymmetric ones, family II is stable except for a small area in which a collision between the small body and Neptune occurs. Along both families the eccentricity increases starting from zero and we stopped following the families because of numerical difficulties at very high values where collision with Sun takes place.

3. Bifurcation points

3.1. *Bifurcations from planar circular RTBP to planar elliptic RTBP*

In the elliptic problem, where the eccentricity of secondary body is $e' \neq 0$, the families of periodic orbits bifurcate from the resonant families of periodic orbits of the circular problem ($e' = 0$) at those points where the period is a multiple of 2π. The eccentricity values that correspond to the bifurcation points from the planar circular to planar elliptic problem are shown in Table 1 and presented in the $a - e$ plane in Fig. 1a. Two pairs of resonant families bifurcate from each point and are continued for $e' > 0$. We call them as E_p and E_a. In family E_p Neptune is initially at perihelion and in family E_a it is initially at aphelion. In most of the resonant cases the family E_p is stable while the family E_a is unstable. In Table 1 the bifurcation points are presented and the symbols S and U denote stability and instability respectively of the families bifurcating from these points (the first symbol refers to family E_p and the second one to E_a).

For the particular value $e' = 0.00912$ (eccentricity of Neptune's orbit) the isolated periodic orbits correspond to eccentricity values which are slightly different of that obtained for $e' = 0$. The position of such periodic orbits in the $a - e$ plane are shown in Fig. 2

Table 1. Eccentricity values of bifurcation points (Bfn) from the planar circular to the planar elliptic RTBP. The symbols S and U denote the stability type. The first symbol refers to the generated family E_p and the second one to family E_a.

Resonance	a_{res}(A.U.)	Period	Bf1	Bf2	Bf3	Bf4
2/3	39.40	6π	0.469SU			
3/4	36.41	8π	0.329SU			
4/5	34.88	10π	0.253SU	0.871SU		
5/6	33.95	12π	0.205SU	0.749SU		
6/7	33.31	14π	0.172UU	0.649SU	0.960SU	
3/5	42.26	10π	0.427US	0.800SU		
5/7	37.62	14π	0.278SU	0.562SU	0.778US	0.936SU
7/9	35.54	18π	0.203SU	0.427SU	0.606US	0.766SU
4/7	43.65	14π	0.027UU	0.400UU	0.900SU	
5/8	41.12	16π	0.029SU	0.335US	0.800SU	
7/10	38.13	20π	0.025SU	0.249US	0.905SU	

Table 2. Eccentricity values of bifurcation points (Bfn) from planar to 3D circular RTBP. The character A or B denotes the symmetry type (see the text) of the orbits of the bifurcating family and the character S or U denotes the corresponding stability. The 3/5 resonant bifurcations indicated by an asterisk show exceptional characteristics (see the text).

Res.	Bf1	Bf2	Bf3	Bf4	Bf5	Bf6	Bf7	Bf8
2/3	0.421AS	0.450BU	0.968BU					
3/4	0.291AS	0.307BU	0.663AU	0.753AU	0.767BS			
4/5	0.222AS	0.233BU	0.624BU	0.729AS	0.825BU			
5/6	0.179AS	0.188BU	0.525BU	0.652AS	0.686BU			
6/7	0.150AS	0.157BU	0.578AS					
3/5	0.373AS	0.393*	0.705AS*	0.730AS	0.768BU	0.815BS*	0.820AS	
5/7	0.248BS	0.251AU	0.281AS	0.518BS	0.519AU	0.592BU	0.716BU	
7/9	0.175BS	0.179AU	0.228AU	0.388BS	0.394AU	0.470BU	0.560AU	0.678AU
4/7	0.051AS	0.064BU	0.359BS	0.369AU	0.727BU	0.808AS	0.860BU	
5/8	0.050AS	0.064BU	0.293BS	0.303AU	0.640BU	0.732AS	0.738BU	
7/10	0.049AS	0.064BU	0.210BS	0.222AU	0.517BU	0.565BS	0.597AU	

where the position of the observed TNO's, taken from the Minor Planet Center table, is also indicated. We may notice that the distribution of objects does not exceed the eccentricity values of the periodic orbits at $e \approx 0.4$. Especially in the case of 2/3 resonance, where the periodic orbits are located at $e \approx 0.47$, the distribution of objects shows an extensive spread up to $e \approx 0.35$. In Kotoulas and Voyatzis (2004) it is shown that for the planar elliptic problem and, particularly, for the resonances 2/3 and 3/4, regular orbits exist for eccentricity values almost up to the position of the unstable periodic orbits. The 4/7 resonance does not possess stable periodic orbits for low and moderate values of the eccentricity, but, as it is shown in Fig. 2, a large number of objects is observed at low eccentricities.

3.2. *Bifurcations from planar circular RTBP to 3D circular RTBP*

The three dimensional families of symmetric periodic orbits in the circular problem bifurcate from the vertical critical orbits of the corresponding planar circular problem (Hénon (1973)). The bifurcation points from the planar circular problem to the corresponding 3D one are presented in Fig.1b and in Table 2. The families consist of periodic orbits which are symmetric to xz-plane ($y(0) = \dot{x}(0) = \dot{z}(0) = 0$, type A) or to x-axis ($y(0) = \dot{x}(0) = z(0) = 0$, type B). For both symmetries, families with stable (S) or

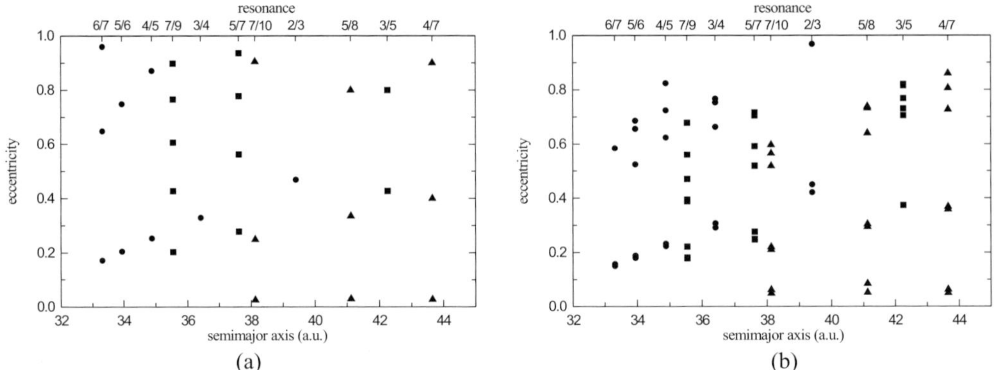

Figure 1. Bifurcation points a) from planar circular to planar elliptic problem b) from planar circular to 3D circular problem. Circles, squares and triangles refer to 1st, 2nd and 3rd order resonances, respectively.

unstable (U) periodic orbits exist. The multiplicity may change many times along the families. In Table 2 the type of symmetry and the stability of the bifurcating family is indicated. For the 1st order resonances the stability type does not change at least up to $i = 20^o$. For the 3/5 resonance some exceptions were found. From the bifurcation point at $e = 0.393$ two families are generated; one has symmetry of type A and is stable (S) and the other one has symmetry of type B and is unstable. Moreover, the bifurcating families at $e = 0.705$ and $e = 0.815$ start as stable but they become doubly unstable for $i \approx 9^o$ and $i \approx 25^o$, respectively. We should remark that for $e < 0.5$ all resonances have stable families which are of symmetry type A. Additionally, the second and third order resonances have families of stable orbits of symmetry type B too. A detailed study for the resonant cases 1/2, 2/3 and 3/4 is given in Kotoulas and Hadjidemetriou (2002).

4. Asymmetric resonances

We recall that the asymmetric resonances are of the form $n/n' = 1/q, q = 2, 3, ...$ and located beyond the 47 A.U. In all such resonances and in the framework of the planar circular RTBP, the family I presents the same characteristics as in the case of low or high-order resonances studied in the previous section. But along the family II a different structure is obtained. The family II of asymmetric resonances starts as stable; it becomes unstable at a certain value of eccentricity and then becomes again stable at a very high value of it. The critical points, where the stability changes, correspond to bifurcations to asymmetric periodic orbits (Voyatzis *et al.* (2004)). This scenario is valid for 1/2, 1/3, 1/4 and 1/6 mean motion resonances with Neptune. The eccentricity values of the above bifurcation points are given in Table 3. An exceptional case is the 1/5 resonance where a collision orbit is obtained along the family II. Asymmetric periodic orbits exist but their complete localization needs further investigation.

The bifurcation points from the planar circular to planar elliptic problem and the stability type of the generated families of periodic orbits are presented in Table 4 using the notation $\mathrm{Bf}_{el}n$. Checking the vertical stability of periodic orbits of the corresponding planar circular problem, we found that there exist vertically critical orbits along the families of the resonances 1/2 and 1/3. These points are presented in Table 4 too and using the notation $\mathrm{Bf}_{3D}n$.

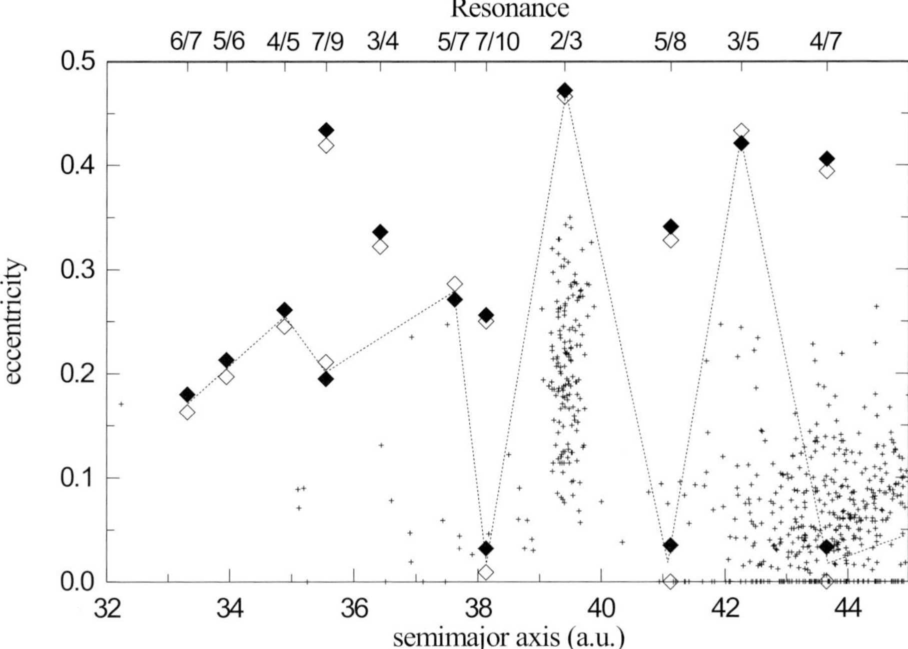

Figure 2. The distribution of TNOs on the $a - e$ plane and the position of the isolated periodic orbits of the planar elliptic problem for $e' = 0.00912$. Solid and empty symbols refer to the periodic orbits of the families E_p and E_a respectively.

Table 3. Eccentricity values of bifurcation points for asymmetric periodic orbits in the planar circular RTBP

Resonance	a_{res}(A.U.)	Bf1	Bf2
1/2	47.78	0.035	0.960
1/3	62.53	0.123	0.972
1/4	75.75	0.201	0.978
1/5	87.90	?	$\simeq 1.0$
1/6	99.26	0.322	0.984

Table 4. Eccentricity values of bifurcation points from the circular to the elliptic planar RTBP (Bf$_{el}n$) and from planar to 3D circular RTBP (Bf$_{3D}n$)

Resonance	a(A.U.)	Bf$_{el}$1	Bf$_{el}$2	Bf$_{el}$3	Bf$_{3D}$1	Bf$_{3D}$2
1/2	47.78	0.070UU	0.637UU		0.059AU	0.066BU
1/3	62.53	0.135UU	0.759UU	0.950UU	0.112AU	0.595AS
1/4	75.75	0.815UU				
1/5	87.90					
1/6	99.26	0.870UU				

5. Conclusions

We studied the resonant families of periodic orbits and presented the bifurcation points from the planar circular to planar elliptic and to three-dimensional circular one for all the first, second and third order E-K belt resonances. In most cases of the elliptic problem the bifurcating families appear in pairs, one with stable orbits and one with unstable orbits. The stability type does not change up to $e' = 0.1$. The 3D circular problem shows

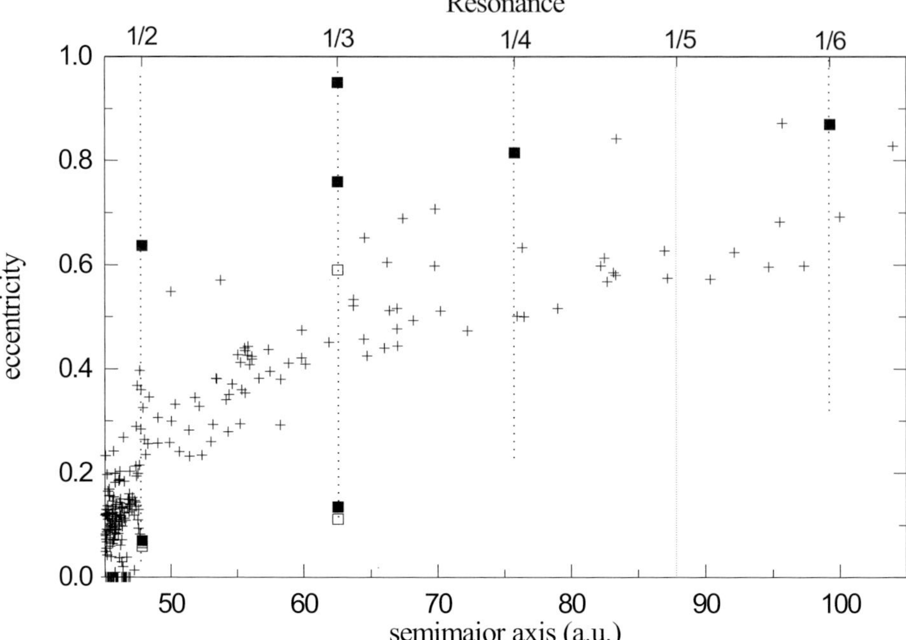

Figure 3. Bifurcation points and the distribution of TNOs at the region of the asymmetric resonances. Solid symbols refer to bifurcation points from the planar circular to the planar elliptic problem. Empty symbols refer to bifurcation points from the planar circular to the 3D circular problem. The dotted vertical lines indicate the eccentricity range where stable asymmetric periodic orbits can be found.

also both stable and unstable families. Generally, in this case, the families extend up to moderate or high inclination values preserving their stability type. The asymmetric resonances show bifurcation points from the planar circular to planar elliptic problem except the 1/5 one. Bifurcations from the planar circular to the three-dimensional circular problem were found only for the cases 1/2 and 1/3.

Acknowledgements

We would like to acknowledge the research program "EPEAEK II, PYTHAGORAS No.21878" of the Greek Ministry of Education and E.U., which supported the present work.

References

Hadjidemetriou, J.D. and Ichtiaroglou, S. 1984, *Astron. Astrophys.* 131, 20
Hénon, M. 1973, *Astron. Astrophys.* 28, 415
Holman M.J. and Wisdom J. 1993, *Astron. J.* 105, 1987
Kneževíc Z., Milani A., Farinella P., Froeschlé Ch. and Froeschlé C. 1991, *Icarus* 93, 316
Kotoulas T. and Hadjidemetriou, J.D. 2002, *Earth, Moon, Planets* 91, 63
Kotoulas T. and Voyatzis G. 2004, *Cel. Mech. Dyn. Astron.* 88, 343.
Levison, H.F. and Duncan, M.J. 1993, *Astrophys. J.* 406, L35
Malhotra, R. 1996, *Astron. J.* 111, 540
Morbidelli, A. 1999, *Cel. Mech. Dyn. Astr.* 72, 129
Morbidelli, A., Brown, M.E. and Levison, H.F 2003, *Earth, Moon, Planets* 92, 1
Voyatzis G., Kotoulas T. and Hadjidemetriou J.D. 2004, *Cel. Mech. Dyn. Astr.*, in press.

Part 5

METEORS, METEORITES AND DUST

Dynamics of Populations of Planetary Systems
Proceedings IAU Colloquium No. 197, 2005
Z. Knežević and A. Milani, eds.

© 2005 International Astronomical Union
DOI: 10.1017/S1743921304008865

The origin and evolution of stony meteorites

William F. Bottke[1], D. Durda[1], D. Nesvorný[1], R. Jedicke[2], A. Morbidelli[3], D. Vokrouhlický[4], and H. Levison[1]

[1]Southwest Research Institute, 1050 Walnut St., Suite 400, Boulder, CO 80302, USA
W. F. Bottke's email: bottke@boulder.swri.edu

[2]Institute for Astronomy, University of Hawaii, Honolulu, Hawaii 96822

[3]Obs. de la Côte d'Azur, B.P. 4229, 06034 Nice Cedex 4, France

[4]Institute of Astronomy, Charles University, V Holešovičkách 2, 180 00 Prague

Abstract. The origin of stony meteorites landing on Earth today is directly linked to the history of the main belt, which evolved both through collisional evolution and dynamical evolution/depletion. In this paper, we focus our attention on the main belt dynamical evolution scenario discussed in Petit *et al.* (2001). According to Petit *et al.*, during the planet formation epoch, the primordial main belt contained several Earth masses of material, enough to allow the asteroids to accrete on relatively short timescales.

After a few My, the accretion of planetary embryos in the main belt zone dynamically stirred the remaining planetesimals to high enough velocities to initiate fragmentation. After a short interval, perhaps as long as 10 My, the primordial main belt was dynamically depleted of $> 99\%$ of its material via the combined perturbations of the planetary embryos and a newly-formed Jupiter. The small percentage of objects that survived in the main belt zone became the asteroid belt. It has been shown that the wavy-shaped size-frequency distribution of the main belt is a "fossil" left over from this violent period (Bottke *et al.* 2004).

Using a collisional/dynamical model of this scenario, we tracked the evolution of stony meteoroids produced by catastrophic disruption events over the last several Gy. We show that most stony meteoroids are a byproduct of a collisional cascade derived from large and ancient asteroid families or smaller, more recent breakup events. The meteoroids are then delivered to Earth by drifting in semimajor axis via Yarkovsky thermal drag forces until they reach a resonance powerful enough to place them on an Earth-crossing orbit.

Keywords. Minor planets, asteroids; collisional physics, impact processes, origin, solar system

---

1. Introduction

Meteorites are hand-samples of asteroids (and possibly comets) that have survived passage through our atmosphere to reach the Earth's surface. Properly analyzed, the bulk composition, mineralogy, and petrology of meteorites can be used to constrain planet formation processes as well as the evolution of the solar nebula. For example, by identifying where meteorites with different chemical and isotopic signatures originated, we may be able to deduce the compositional and thermal gradients that existed in the solar nebula (e.g., Burbine *et al.* 2002).

One of the most challenging aspects of this problem is placing our meteorite samples into the appropriate geologic context. At present, we have limited information on which meteorites go with which parent bodies (or their immediate precursors). Determining the linkages between meteorites and their parent bodies is a critical goal of asteroid and meteorite studies.

Our first task is to identify the source regions for those meteorites landing on Earth today. Before a meteorite can reach Earth, it must first become part of the Earth-crossing object (ECO) population (perihelion distance $q \leqslant 1.017$ AU and aphelion distances $Q \geqslant 0.983$ AU). Bottke *et al.* (2000; 2002a) showed that nearly all of the observed ECOs (mainly diameter $D > 1$ km objects) originated in the main belt. Because most meteorites have physical properties similar to the observed main belt asteroids, and because meteoroids must follow the same dynamical pathways as km-sized ECOs (and near-Earth asteroids, or NEOs; $q \leqslant 1.3$ AU and $Q \geqslant 0.983$ AU) once they reach planet-crossing orbits, we infer that the majority of meteoroids reaching Earth are from the main belt (e.g., Vokrouhlický & Farinella 2000).

Our next task is to determine how meteoroids, the meter-sized precursors of the meteorites, are delivered to Earth. Numerical simulations have shown that meteorites travel from the main belt to Earth through a two-step process.

In the first step, main belt asteroids collide with one another at high velocities (~ 5 km s^{-1}; Bottke *et al.* 1994) and undergo fragmentation (see Asphaug *et al.* 2002 and Holsapple *et al.* 2002 for recent reviews). The orbits, spin states, shapes, and internal structures of the surviving bodies are determined from the kinetic energy and specific nature of the impact. The largest disruption events occurring in main belt over the last several Gy produced the observed asteroid families (e.g., Zappalá *et al.* 2002). The typical ejection velocities produced by these fragmentation events are $\lesssim 100$ m s^{-1} (Bottke *et al.* 2001; Michel *et al.* 2001; 2002; 2003); in most cases, this leaves the ejecta relatively close to the impact site (i.e., within a few 0.01 AU).

In the second step, fragments with diameter $0.0001 < D < 20$ km undergo slow but steady dynamical evolution via the Yarkovsky effect, a thermal radiation force that causes small objects to undergo semimajor axis drift as a function of their spin, orbit, and material properties. As described in the chapter by Vokrouhlický *et al.* (2004; this book), the drift rate for meter-size stones in the main belt is $\pm(0.01\text{-}0.001)$ AU My^{-1}, while the drift rate for iron meteoroids is ~ 10 times slower.

This process drives some of these objects into powerful resonances produced by the gravitational perturbations of the planets (e.g., the 3:1, ν_6 resonances). Numerical studies show that test objects placed in such resonance have their eccentricities pumped up to planet-crossing orbits (e.g., Wisdom 1983). Once on planet-crossing orbits, meteoroids have their dynamical evolution dominated both by resonances and gravitational close encounters with the planets. A small fraction of these bodies go on to strike the Earth ($\sim 1\%$; Gladman *et al.* 1997; Morbidelli & Gladman 1998), though most impact the Sun or are ejected from the inner solar system via a close encounter with Jupiter (Farinella *et al.* 1992; Gladman *et al.* 1997).

The missing component in this description is collisional evolution, namely the length of time meteoroids survive against collisional disruption. If collision disruption timescales are short, the immediate precursors of nearly all meteorites must be in the NEO/ECO populations, while if they are long, the immediate precursors are likely to be main belt asteroids. An important clue used to determine the immediate source region of most meteoroids is the cosmic-ray exposure (CRE) ages of meteorites. CRE ages measure the time interval spent in space between the meteoroid's formation as a $D < 3$ m body (following removal from a shielded location within a larger object) and its arrival at Earth (e.g., Morbidelli and Gladman 1998). The CRE ages of most stony meteorites are between $\sim 10\text{-}100$ My, while iron meteorites have CRE ages between $\sim 0.1\text{-}1.0$ Gy (Caffee *et al.* 1988; Marti & Graf 1992; Eugster 2003). Because these timescales are longer than the average dynamical lifetime of near-Earth asteroids (~ 4 My; Bottke *et al.* 2002a), we infer that many meteoroids, particularly iron meteoroids, had to obtain considerable damage from cosmic rays while drifting in the main belt as a meter-sized body. Thus, it is likely that most meteoroids originated in the main belt.

Given that there are more than 20,000 meteorites are in our existing collection (Grady 2000), and that the meteoroid delivery scenario described above allows nearly any main belt asteroid to produce meteoroids, one would expect the Earth to be on the receiving end of samples from thousands upon thousands of distinct parent bodies. Tests of these samples, however, suggests the opposite; our meteorite collection could represent as few as ~ 100 different asteroid parent bodies: ~ 27 chondritic, ~ 2 primitive achondritic, ~ 6 differentiated achondritic, ~ 4 stony-iron, ~ 10 iron groups, and ~ 50 ungrouped irons (Meibom & Clark 1999; Keil 2000; 2002; Burbine *et al.* 2002). If we ignore the stony-iron, iron, and differentiated meteorites for the moment, this number is reduced to ~ 30 objects. This value seems surprisingly small when one considers the large number of main belt asteroids that should have disrupted over the last 4.6 Gy.

What are our models missing? As far as we can tell, our numerical simulations are doing a reasonable job of tracking the dynamical evolution of meteoroids throughout the inner solar system. Our treatment of collisional evolution among the meteoroid population, however, is less advanced. At present, we have yet to realistically track how meteoroids evolve via comminution over several Gy of solar system history. This is a difficult problem, partly because main belt collisional models need accurate starting conditions and constraints, but also because the outcome is strongly linked to the main belt's dynamical history. For example, the dynamical events occurring during the planet formation era (e.g., wandering planetary embryos, the formation of the Jovian planets) almost certainly influenced the main belt population observed today.

Thus, to correctly model meteoroid evolution, we need to be able to:

(i) determine how the main belt population collisionally and dynamically evolved over its history,

(ii) model the disruption of individual parent bodies within such a model,

(iii) track how the fragment size distribution of each parent body evolves with time,

(iv) determine how the meteoroid population derived from each parent body is affected by sources (e.g., new meteoroids produced by a 'collisional cascade' among the parent body's fragment size distribution) and sinks (e.g., dynamical loss mechanisms such as Yarkovsky effect/resonances; collisional disruption),

(v) determine how many meteoroids from each disrupted parent body would have reached Earth over a window starting ~ 0.1 Ma (i.e., the maximum terrestrial residence time of stony meteorites in the Antarctic; Bland *et al.* 1998).

At present, we are still struggling to accurately model these processes. Despite this, our group has made enough recent progress to show some interesting preliminary results in this brief paper. Our focus will be on the origin and evolution of the stony meteorites; we save a discussion of iron meteorites for future work. We start with a discussion of how the main belt may have dynamical evolved over solar system history.

2. Dynamical Evolution of the Main Belt

One of the most important constraints on planet formation processes is the apparent mass depletion of the main belt. During the planet formation epoch, the primordial main belt was believed to have contained several Earth masses of material, enough to allow the asteroids to accrete on relatively short timescales (e.g., Weidenschilling 1977; Wetherill 1989). The present-day main belt, however, only contains ~ 0.0005 Earth masses of material (e.g., Britt *et al.* 2002). Though several scenarios have been proposed to explain this putative loss of mass (e.g., see review by Petit *et al.* 2002), the best available model at present is that described by Petit *et al.* (2001). We describe their results by dividing them into 3 distinct phases.

2.1. *Phase 1: Dynamical Excitation of the Primordial Main Belt by Embryos*

Planetesimals and planetary embryos accreted in the primordial main belt during the first few My of solar system history. According to planetary accretion models, runaway growth should have produced Moon- to Mars-size embryos throughout the inner solar system (< 4 AU) over a timescale of ~ 0.1-1 My (Wetherill 1989; Wetherill & Stewart 1989; Weidenschilling & Davis 2001). Models describing the evolution of planetary embryos and planetesimals in the inner solar system have been investigated by several groups (e.g., Wetherill 1992; Agnor *et al.* 1998; Chambers & Wetherill 1998; 2001; Petit *et al.* 2001).

Once formed, planetary embryos in the inner solar system perturbed both themselves and the surviving planetesimals. This was modeled by Petit *et al.* (2001), who tracked the evolution of 56 embryos started on circular, slightly inclined orbits ($0.1°$) between 0.5 and 4 AU using the MERCURY integration package (Chambers & Migliorini 1997), and then created several simulations where the recorded masses, positions, and velocities of the embryos were used to gravitationally perturb test bodies initially placed on circular, zero inclination orbits. The test bodies were designed to represent asteroids in the primordial main belt that failed to accrete with various planetary embryos during runaway growth. The dynamical evolution of the test bodies were tracked using the numerical integrator SWIFT-RMVS3 (Levison & Duncan 1994), with perturbations of the planetary embryos included using techniques described by Petit *et al.* (1999).

Petit *et al.* (2001) found that after 1-2 My, planetesimal e, i values were already high enough to initiate fragmentation, while at 10 My, the e, i values of some test bodies were as high as ~ 0.6 and $\sim 40°$, respectively. These perturbations cause collision velocities to steadily increase throughout Phase 1.

2.2. *Phase 2: Dynamical Depletion of the Main Belt by Embryos and Jupiter/Saturn*

In the core accretion model, planetary embryos were also forming near where Jupiter is found today (~ 5 AU). At some unknown time during Phase 1, these embryos merged and produced a ~ 2-20 Earth mass core, where the core's large size may have been a byproduct of water ice condensation beyond the so-called "snowline" (Wuchterl *et al.* 2000; Inaba *et al.* 2003). Numerical modeling results suggest that once this core was massive enough to accrete gas from the solar nebula, Jupiter obtained its full size within ~ 1 My (e.g., Pollack *et al.* 1996). It is believed that Jupiter (and presumably Saturn) formed $\lesssim 10$ My after the formation of the first solids (Pollack *et al.* 1996; Guillot & Hueso 2003).

The introduction of Jupiter and Saturn into the solar system, which marks the beginning of Phase 2, had a dramatic effect on the dynamical structure of the primordial main belt (Fig. 1). To simulate this, Petit *et al.* (2001) injected Jupiter and Saturn, with their present-day masses and orbital elements, into the evolving system of embryos described in Phase 1 at 10 My. They found that gravitational perturbations from Jupiter and Saturn, when combined with the mutual gravitational perturbations of the embryos, dynamically excited those bodies residing in the main belt. In most simulations, the embryos achieved high enough e, i values to push themselves out of the main belt. Eventually, as the embryos collided and merged, a system of terrestrial planets was formed.

Fig. 1 shows several snapshots of the evolution of the embryos in the Petit *et al.* simulations. In the end, they produced two terrestrial planets, one with 1.3 Earth masses ($a = 0.68$ AU, $e = 0.15$, and $i = 5°$) and another with 0.48 Earth masses ($a = 1.5$ AU, $e = 0.03$, and $i = 23°$). This result illustrates the success and failures of the current generation of late-stage planet formation models; it is possible to generate terrestrial planet systems reminiscent of our Solar System, but the planets are dynamically hotter than Earth and Venus. Despite these limitations, however, the terrestrial planets produced by Petit *et al.* (2001) are similar enough to our own system (e.g., no planet is crossing the main belt) that they can be used to investigate the evolution of the primordial main belt.

To determine what happened to the main belt after the formation of Jupiter, Petit *et al.* (2001) inserted 100 test bodies on circular, low inclination orbits between 1.0-2.0 AU and 1000 test bodies between 2.0-2.8 AU into the embryo/Jupiter/Saturn system described above (Fig. 1). The test bodies were tracked for 100 My, with their orbits modified by the combined perturbations of the embryos and Jupiter/Saturn. Fig. 1 shows the test bodies becoming highly excited, with most objects eliminated by striking the Sun or being thrown out of the inner solar system via a close encounter with Jupiter after a few My. For the 1000 bodies started between 2.0-2.8 AU, five survived to become embedded within the main belt zone. These results imply that the main belt may have dynamically lost $\gtrsim 99\%$ of its primordial material through dynamical excitation.

The end of Phase 2 is somewhat nebulous. We believe a good working definition is the time when the dynamically-excited population becomes depleted enough that it is no longer capable of producing a statistically significant number of disruptions in the main belt population. Using dynamical results from Morbidelli *et al.* (2001), we assume Phase 2 ends ~ 400 My after the end of accretion and the onset of fragmentation.

2.3. *Phase 3: Collisional Evolution in a Depleted Main Belt*

In Phase 3, the surviving main belt bodies were left in a dynamical state comparable to the current main belt population. Any loss of material in Phase 3 is produced by the Yarkovsky effect, which drives multi-km and smaller asteroids into mean motion and secular resonances capable of pumping their e values onto planet-crossing orbits (e.g., Bottke *et al.* 2000; 2002a), and comminution. Nearly all asteroids escaping the main belt become part of the NEO population, though their dynamical lifetime in the NEO region varies considerably (see chapter by Vokrouhlický *et al.*)

Though there much to like about the Petit *et al.* (2001) scenario, we caution that it does not yet include the effects of the putative Late Heavy Bombardment (LHB), an event ~ 3.9 Gy ago where the inner solar system was ravaged by numerous impactors (e.g., Hartmann *et al.* 2000). We refer the reader to recent work by A. Morbidelli, K. Tsiganis, R. Gomes, and H. Levison (e.g., Gomes *et al.* 2004; Tsiganis *et al.* 2004) for additional details on the nature of the LHB.

Using the Petit *et al.* (2001) dynamical evolution scenario, we are now ready to model the collisional evolution in the main belt over solar system history. Our code and results are described below.

3. Collisional and Dynamical Depletion Evolution Model (CoDDEM)

To model the evolution of the main belt as completely as possible, we integrated the dynamical results described above with a self-consistent 1-D collisional evolution code (Bottke *et al.* 2004a,b). The essentials of this code, called CoDDEM (for collisional and dynamical depletion model), are briefly described below.

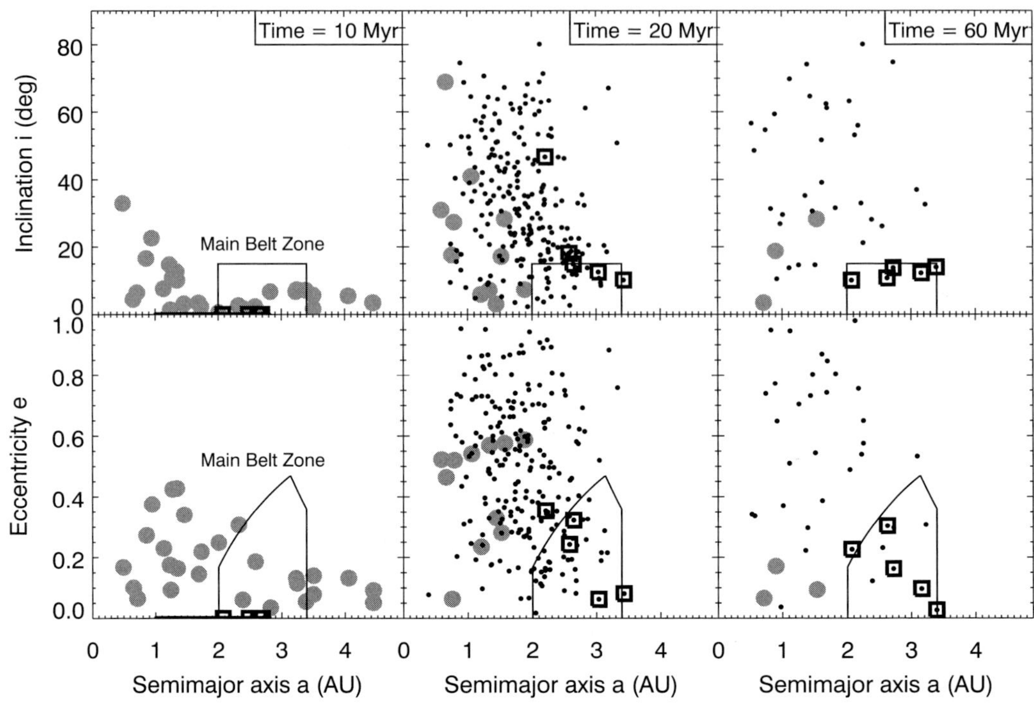

Figure 1. Three snapshots from a representative run in Petit *et al.* (2001), where the dynamical evolution of test bodies and planetary embryos were tracked for more than 100 My just after the formation of Jupiter (see Bottke *et al.* 2004b). The grey dots are planetary embryos, while the black dots are planetesimals. The black squares are centered on the 5 test bodies that will ultimately be trapped in the main belt zone. At $t = 20$ My, we see that nearly all of the test bodies have been pushed out of the main belt onto high e, i orbits. Perturbations from Jupiter have also eliminated or forced the merger of many embryos. By $t = 60$ My, the terrestrial planet system is nearly complete, with only a small remnant population left behind in the main belt zone. This system will eventually produce two terrestrial planets (see text).

We run CoDDEM by entering an initial main belt size-frequency distribution where the population (N) has been binned between 0.001 km $< D <$ 1000 km in logarithmic intervals $d\mathrm{Log}D = 0.1$. The particles in the bins are assumed to be spherical and are set to a bulk density of 2.7 g cm^{-3} (Britt *et al.* 2002; see Bottke *et al.* 2004a for details). CoDDEM then computes the time rate of change in the differential population N per unit volume of space over a size range between diameter D and $D + dD$ (Dohnanyi 1969; Williams & Wetherill 1994):

$$\frac{\partial N}{\partial t}(D, t) = -I_{\mathrm{DISRUPT}} + I_{\mathrm{FRAG}} - I_{\mathrm{DYN}}. \tag{3.1}$$

All these terms are defined below.

Here I_{DISRUPT} is the net number of bodies that leave the bin between D and $D + dD$ per unit time from catastrophic disruption. The collisional lifetime of a body is computed using standard methods (e.g., Öpik 1951; Wetherill 1967; Greenberg 1982; Farinella & Davis 1992; Bottke & Greenberg 1993), where the "intrinsic" collision probabilities and impact velocities V_{imp} between the target body and background population are computed directly from Phases 1-3 of the dynamical simulations described in Petit *et al.* (2001). For details, see Bottke *et al.* (2004b).

The projectile capable of disrupting D_{target} is defined as (d_{disrupt}):

$$d_{\text{disrupt}} = (2Q_{\text{D}}^*/V_{\text{imp}}^2)^{1/3} D_{\text{target}} \qquad (3.2)$$

Q_{D}^* is defined as the critical impact specific energy; the energy per unit target mass needed to disrupt the target and send 50% of its mass away at escape velocity. In CoDDEM, we assume that $D < 150$ km disruption events are barely-catastrophic (i.e., 50% of the target body's mass is ejected). This means we neglect both cratering events, which produce much less ejecta over time than barely-catastrophic disruption events (e.g., Dohnanyi 1969; Williams & Wetherill 1994) and highly-energetic catastrophic disruption events, which are unlikely to occur (e.g., Love & Ahrens 1997). For $D > 150$ km disruptions, we assume they are super-catastrophic (i.e., more than 50% of the mass is ejected). We base this on the observation that the largest bodies found in families formed by $D \gtrsim 150$ km disruption events frequently contain $< 50\%$ of the mass of the original parent body (e.g., Eos, Themis) (Tanga *et al.* 1999; Bottke *et al.* 2004a). More discussion of this issue is given below.

For the simulations described in this paper, we use the Q_{D}^* function defined for basaltic targets described by the hydrocode runs of Benz & Asphaug (1994). This selection was corroborated by Bottke *et al.* (2004a), who used a CoDDEM-like code to show that Q_{D}^* functions similar to those described by Benz & Asphaug (1994) provide main belt collisional evolution results that are consistent with the available constraints (see below).

To remove disrupted bodies from our size-frequency distribution, we treated breakup events as Poisson random events, with integer numbers of particles removed (or not removed) from a size bin within a timestep according to Poisson statistics (Press *et al.* 1989). This means that different seeds for the random number generator can produce different outcomes. For this reason, our CoDDEM results are treated in a statistical manner; to get a quantitative measure of how good a given set of input parameters reproduces observations, we perform numerous trials using different random seeds before comparing our results to observations.

We defined I_{FRAG} as the number of bodies entering a size interval per unit time that were produced by the fragmentation of larger bodies. To determine I_{FRAG}, and to keep things as simple as possible given our unknowns in this area, we assume the fragment size distribution (FSD) produced by each catastrophic disruption event was similar to those observed in asteroid families like Themis (super-catastrophic) or Flora (barely-catastrophic) (Tanga *et al.* 1999; see Bottke *et al.* 2004a for details of how we chose our FSD values). Themis-like FSDs were developed for $D > 150$ km disruption events. We assumed the largest remnant was 50% the diameter of the parent body, and the incremental power law index between the largest remnant and fragments 1/60th the diameter of the parent body was -3.5. Fragments smaller than 1/60th the diameter of the parent body follow a power law index of -2.0. Flora-style FSDs were developed for breakups among $D < 150$ km bodies. Here the diameter of the largest remnant is set to 80% the diameter of the parent body. We gave these FSDs incremental power law indices, from the large end to the small end, of -2.3, -4.0, and -2.0, with transition points at 1/3 and 1/40 the diameter of the parent body. Note that in both cases, we assume that some material is located below the smallest size used by CoDDEM ($D = 0.001$ km) in the form of small fragments or regolith. Accordingly, mass is roughly but not explicitly conserved.

We defined I_{DYN} to be the number of bodies lost via dynamical processes from the size interval per unit time. The loss mechanisms could be those described in Phase 2 above or by the Yarkovsky effect/resonances. For the latter, we assumed a size-dependent removal rate for main belt asteroids in each size interval (e.g., Farinella & Vokrouhlický 1999). Given the uncertainties in choosing these values (e.g., unknown spin rates, thermal conductivities, interaction between Yarkovsky and the so-called YORP mechanism that can change the spin vectors of asteroids; see Rubincam 2000), we decided base our Yarkovsky removal rate function for $D > 1$ km bodies on numerical results described in Morbidelli and Vokrouhlický (2003). Morbidelli and Vokrouhlický (2003) imposed a steady state on the inner main belt population and tracked their removal via Yarkovsky effect and resonances. Their model included factors such as collisional disruption, collisional spin axis reorientation events, and the effects of YORP. Accordingly, our removal rates decrease from 0.03% per My for $D = 1$ km bodies to 0.008% per My for $D = 10$ km bodies. This trend was continued down to 0.005% per My for $D = 20$ km bodies and 0.0002% for $D = 30$ km bodies. Beyond this point, we assume nothing escapes the main belt.

For $D < 1$ km bodies, our removal rate model is uncertain, with little trustworthy numerical work done on this issue to date. For this reason, we kept things simple and assumed sub-km asteroids ($0.001 < D < 1$ km) are removed at the same rate as $D \sim 1$ km asteroids (0.03% per My). Our main justification for this loss rate is that it provides a surprisingly good match to shape of the NEO population (see below). Vokrouhlický and Farinella (2000), however, show there are asteroidal thermal conductivity values that allow bodies ranging from cm- to km-sized objects to travel approximately the same distance in semimajor axis before disrupting. Given our uncertainty about asteroidal thermal conductivity values for sub-km-sized bodies, we believe our loss rate estimates are as good an estimate as any other currently available.

To calibrate our Yarkovsky effect/resonance removal rate functions, we ran numerous CoDDEM simulations and compared our model NEO population to estimates of the observed population between $0.001 < D \lesssim 10$ km (Stokes *et al.* 2003; Stuart & Binzel 2004). We found the above parameters produced a model NEO population that was a good match with observations.

4. Model Constraints

We tested our CoDDEM model results against a wide range of main belt constraints. One important constraint was the wavy-shaped main belt size distribution. To determine its value, we converted the absolute magnitude H distribution of the main belt described by Jedicke *et al.* (2002), who combined observations of bright main belt asteroids with renormalized results taken from the Sloan Digital Sky Survey (Ivezić *et al.* 2001), into a size distribution. This was accomplished using the relationship $D(\mathrm{km}) = \frac{1329}{\sqrt{p_v}} 10^{-H/5}$ (Fowler & Chilemi 1992), and a representative visual geometric albedo $p_v = 0.092$. This produces a main belt population with $1.36 \times 10^6, 680$, and 220 asteroids for $D > 1, 50$ and 100 km, respectively. These results are consistent with estimates from IRAS/color-albedo data (e.g., Farinella & Davis 1992; Tedesco *et al.* 2002), main belt population estimates from the ISO spacecraft (Tedesco & Desert 2002), and numerical simulations (Morbidelli & Vokrouhlický 2003).

A second set of constraints was provided by asteroids families, particularly those that are too large to be dispersed by the Yarkovsky effect over the age of the solar system (Nesvorný & Bottke 2004; Nesvorný *et al.* 2004). Using hydrocode simulations (Durda *et al.* 2004) to estimate the amount of material in families located below the observational detection limit, we computed that $\sim$ 20 families have been produced by the breakup of $D \gtrsim 100$ asteroids over the last $\sim$ 3.5 Gy (Bottke *et al.* 2004a).

CoDDEM adopts the same constraint; we assume that the size distribution bins centered on $D = 123.5, 155.5, 195.7, 246.4, 310.2$, and 390.5 km experienced 5, 5, 5, 1, 1, 1 breakups over the last 3.5 Gy, respectively.

A third constraint comes from studies of the cratering record on the lunar maria, which shows the NEO impact flux has been relatively constant over the last ~ 3 Gy (e.g., Grieve & Shoemaker 1994). This result implies the NEO population (and thus the main belt population) has been in a quasi-steady state over this time period.

Other constraints are described by Bottke *et al.* (2004a,b).

5. Initial Conditions and a Sample CoDDEM Run

The initial main belt population entered into CoDDEM is divided into two components that are tracked simultaneously: a small component of main belt asteroids that will survive the dynamical excitation event (DDE) described in Phase 2 ($N_{\rm rem}$) and a much larger component that will be excited and ejected from the main belt during the Phase 2 DDE ($N_{\rm dep}$). Thus, our initial population is $N = N_{\rm rem} + N_{\rm dep}$. We can use this procedure because we know in advance the dynamical fate of each population via the dynamics simulations described in Sec. 2. The two populations undergo comminution with themselves and with each other. At the end of Phase 2 (i.e., when $N_{\rm dep} = 0$), CoDDEM tracks the collisional and dynamical evolution of $N_{\rm rem}$ alone for the remaining simulation time.

The size and shape of our initial size distribution was determined by running different initial populations through CoDDEM-like codes and then testing the results against the constraints described in Sec. 4 (Durda *et al.* 1998; Bottke *et al.* 2004a,b). The size distribution that provided the best fit for $N_{\rm rem}$ followed the observed main belt for $D > 200$ km bodies, an incremental power law index of -4.5 for $110 < D < 200$ km bodies, and an incremental power law index of -1.2 for $D < 110$ km bodies (Bottke *et al.* 2004b). The initial shape of the $N_{\rm dep}$ population is always the same as $N_{\rm rem}$, but its size can be increased or decreased depending on when Jupiter reaches its full size (i.e., the length of Phase 1) (Bottke *et al.* 2004b). For the runs shown here, we set $N_{\rm dep} = 200 N_{\rm rem}$ and we assumed that Jupiter reached its full size 4 My after the onset of main belt fragmentation.

An example of one successful CoDDEM run is shown in Fig. 2. The $t = 0$ My timestep shows our initial conditions. The $t = 3$ My timestep shows a bump developing near $D \sim 2\text{-}3$ km for both $N_{\rm rem}$ and $N_{\rm dep}$; this is a by-product of the "V"-shaped $Q_{\rm D}^{*}$ function (Campo Bagatin *et al.* 1994; Durda *et al.* 1998; Davis *et al.* 2002). Self-gravity among $D > 200$ m objects makes them increasingly difficult to disrupt. This produces an "overabundance" of $D \sim 200$ m objects that induces a wave-like perturbation into the main belt size distribution. The $t = 30$ My timestep shows the $N_{\rm dep}$ population significantly depleted, with few members lasting beyond 100 My. Concurrently, impacts on $N_{\rm rem}$ from $N_{\rm dep}$ eventually produce a shape for its size distribution that approaches that of the observed main belt. [In this paper, when we say "observed main belt", we are usually including the portion of the main belt size distribution that has been inferred from debiased observational data (Jedicke *et al.* 2002; see also Bottke *et al.* 2004a).] This explains how the NEO/ECO population (and the impact flux on the Moon) could have been in quasi-steady state for several Gy. A good fit to the main belt population (and to the population of asteroid families) is achieved at 4.6 Gy. Our results show a good match to the estimated NEO population between $0.001 < D < 10$ km. This is not a surprise given that our Yarkovsky effect/resonance removal function was selected for this purpose.

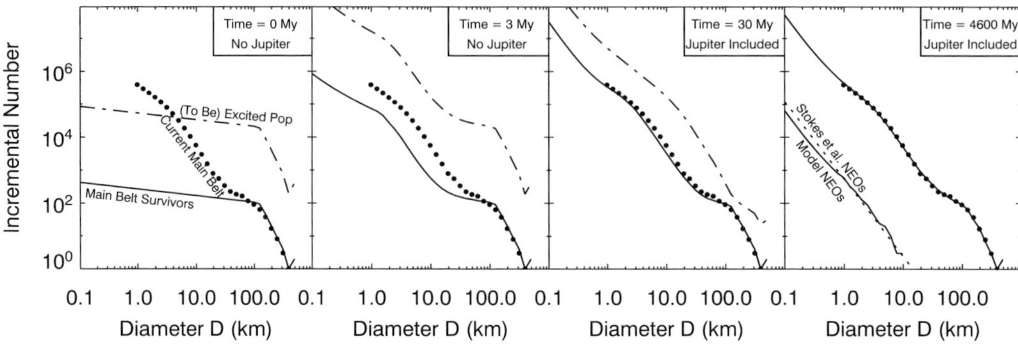

Figure 2. Four snapshots from a representative run where we track the collisional evolution of the main belt size distribution for 4.6 Gy. The starting population is divided into two components: the "main belt survivors", who will be left behind in the main belt zone, and the "(To be) excited population", who will be scattered out of the main belt via perturbations from embryos and Jupiter. Here the former population is 200 times smaller than the latter, with Jupiter entering the system at 4 My. The dots represent the observed main belt. The bump observed in most frames near $D \sim$ 2-3 km is driven by the shape of the asteroid disruption scaling law. The 30 My timestep shows this population significantly depleted, with few members lasting beyond 100 My. A good fit to the main belt and NEO populations is achieved at 4600 My. Note that the NEO population from Stokes *et al.* (2003) was determined by fitting a line to the $D > 1$ km bodies; our model NEO population actually provides a better fit to the available $D < 1$ km NEO data than Stokes *et al.*

However, if our Yarkovsky removal rate function were several times stronger or weaker, we would see a mismatch in not only in the observed number of NEOs but also in the number of main belt objects.

The interesting element here is that the small kink observed in the NEO population near $D \sim 0.7$ km is apparently real; note that our Yarkovsky/resonance removal rate function has no inflection points in that size range. A reasonable explanation for the kink is that it is produced by the Yarkovsky effect, which samples $D \lesssim 30$ km bodies from across the wave-shape main belt size distribution and delivers them to resonances. Thus, the kink provides additional evidence that most NEOs come from the main belt and that the Yarkovsky effect is the dominant physical process providing asteroids to resonances.

6. The Evolution of Asteroid Families and Stony Meteoroids

Now that we have a reasonably accurate model of how the main belt and NEO size distributions evolved over solar system history, we can return to the issue of stony meteoroid production, evolution, and delivery to the Earth. For the moment, we concentrate on what happens to main belt meteoroids produced by a single homogeneous and distinct parent body.

A cratering or catastrophic disruption event on our parent body will produce a fragment size distribution (FSD); we assume the event in question is large enough that the FSD includes both meter-sized bodies (meteoroids) as well as larger objects. Subsequent collisions onto bodies in the FSD act as a source for new meteoroids that are genetically the same as those created in the previous generation. This so-called collisional cascade guarantees that meteoroids from this parent body will be provided to the main belt population (and possibly to Earth) for some time after the initial impact event (e.g., Williams & Wetherill 1994). At the same time, dynamical processes and collisions onto the newly-created meteoroids act as a sink to eliminate them from the main belt. Thus, after the initial collision event, the sources and sinks drive the meteoroid population toward a quasi-steady state. If the sinks dominate the sources, the meteoroid population will undergo a slow (or not so slow) decay as collisions on the FSD deplete the reservoir of larger bodies capable of replacing them. The size and nature of this reservoir, therefore, determine the decay rate of the meteoroid population.

To simulate this process, we modified CoDDEM to track input FSDs evolving within the main belt population. We then tested how FSDs produced by parent bodies having the same diameter as the central body in each bin of our size distribution (e.g., $D = 31, 39, 47, 61, 77, 98, 123, 155, \ldots$) would react to collisions and dynamical depletion via the Yarkovsky effect/resonances over main belt history. We also assumed the collision probability between family members was the same as family members with the background population. This is an oversimplification of reality but reasonable given the approximations used in this work.

Our chief unknown was the nature of the FSD used in each simulation. Note that numerical hydrocode results and observations of asteroid families indicate that FSDs produced by cratering or disruption events can be very different from one another, with factors like impact energy, parent body size, impact angle, etc. playing critical roles in the outcome (e.g., Tanga *et al.* 1999; Michel 2001; 2002; 2003; Durda *et al.* 2004). For this reason, we decided to create a "toy" FSD for each body that was purposely exaggerated to test the extremes of the problem (and one that is inconsistent with the FSDs described previously). Our goal here was not to concentrate on absolute numbers delivered to Earth but instead to compute an approximate decay rate for the meteoroid population produced by each toy FSD. The criteria used to create our toy FSDs were as follows:

(i) the FSD had to have the same mass as the parent body when computed over the size intervals with $D > 0.001$ km,

(ii) the FSD had to follow a single power law slope, though that slope could be different for different parent body sizes

(iii) the number of meteoroids in the FSD had to be a factor of ~ 2 smaller than the number of meteoroids in the main belt population.

Accordingly, the breakup of our parent object would produce a FSD that would immediately dominate the main belt meteoroid population. This is not as absurd as it seems at face value; CoDDEM results predict the main belt contains $\sim 10^{12}$ meter-sized objects, which is only the mass equivalent of a single 10-20 km asteroid. Still, this estimate likely overestimates the number of meteoroids produced by small families. The reason we use it is that it places all our toy FSDs on the same initial footing. If smaller asteroid disruption events fail to dominate the meteoroid population under an exaggerated situation, they are similarly unlikely to dominate under more realistic conditions.

Fig. 3 shows the what happens when FSDs produced by $D = 30$ km and 100 km parent bodies are placed in the main belt ~ 3.1 Ga (1.5 Gy after solar system formation). We made sure this time took place after the LHB, whose effects on the main belt population may have been important. For the FSD derived from the 30 km body, we find that the initial meteoroid population (i.e., meter-sized bodies) drops by a factor of 100 and 10^5 within 130 My and 3.1 Gy, respectively (Fig. 4). Thus, FSDs produced by < 30 km

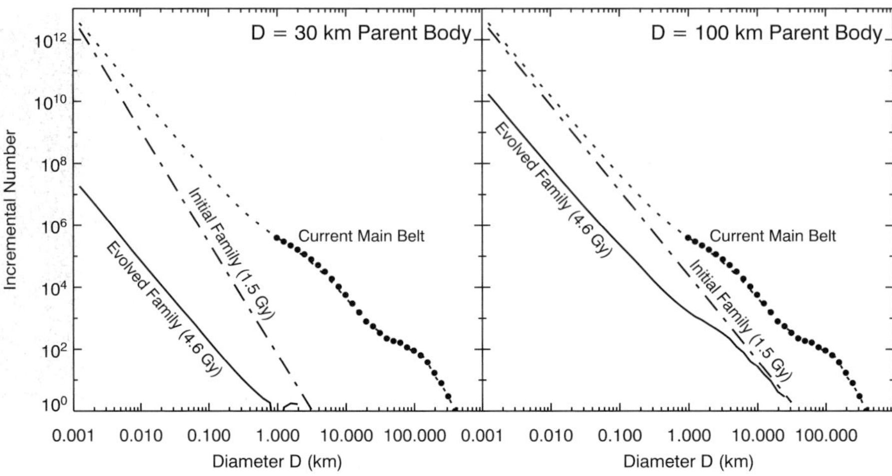

Figure 3. The collisional and dynamical evolution of two "toy" asteroid families produced by the disruption of a $D = 30$ and 100 km parent bodies. Both families were inserted into CoDDEM at 1.5 Gy after solar system formation. The meteoroid population is represented by the number of bodies in the $D \sim 0.001$ km size bin. The solid lines show the families at present (4.6 Gy). The smaller family has decayed significantly more than the larger family. Note the shallow slope of the $D = 100$ km family for $0.7 \lesssim D \lesssim 5$ km. This shape mimics the that of the background main belt population over the same size range. Morbidelli *et al.* (2003) found the same kinds of shallow slopes in many prominent asteroid families. This suggests these families are old enough to have experienced significant comminution.

parent bodies decay away so quickly that ancient ones are unlikely to deliver meaningful numbers of meteoroids to Earth today. For the 100 km parent body, the decay rate is significantly slower, with the meteoroid population only dropping by a factor of 100 over 2-3 Gy. This suggests that many meteoroids reaching Earth today come from prominent asteroid families with sizable FSDs, even if those families were created billions of years ago.

Another interesting aspect of the evolution of the 100 km parent body's FSD is that comminution produces a shallow slope between $0.7 \lesssim D \lesssim 5$ km. We note that shallow slopes in this size range have been observed among the FSD of many prominent asteroid families (Morbidelli *et al.* 2003). Our simulations suggest this slope is produced by collisional evolution and is a function of several factors: the "V"-shaped Q_D^* function used by CoDDEM (e.g., Benz & Asphaug 1999; Bottke *et al.* 2004a), the number of $D \gtrsim 10$ km objects in the FSD, and the bombardment of the FSD from the background main belt population. As the FSD evolves via collisions and dynamical depletion, smaller objects are eliminated faster than they can be replenished by disruption events among bigger objects. Conversely, $D > 10$ km objects in the FSD (and in the main belt) are hard to disrupt, partly because these bodies have relatively high Q_D^* values but also because the main belt population's wavy shape provides relatively few projectiles capable of disrupting $D > 10$ km targets. This means that objects moderately smaller than $D \sim 10$ km are not replenished very often. The net result is that (i) the smallest objects in the FSD become depleted and evolve to the same slope as the background main belt population (e.g., O'Brien & Greenberg 2003), (ii) the $D > 10$ km population undergoes few changes, and (iii) the $0.7 \lesssim D \lesssim 5$ km objects evolve to a shallow slope that connects these two extremes, producing a shallow slope reminiscent of the slope of the main belt population in the same size range. In some test cases, we even find the slope in (iii) can be more shallow than the background population, a result consistent with some observed families (Morbidelli *et al.* 2003).

In order to compute a rough measure of how many stony meteorites should be in our collection, we combined the meteoroid decay rates taken from the evolution of our toy FSDs with CODDEM estimates of the production rates of asteroid families over the last ~ 4 Gy. We attempted to keep things as simple as possible for this calculation by assuming the following: (i) meteoroids from all parts of the main belt have an equal chance of reaching Earth, (ii) all $D > 30$ km asteroids disrupted over the last several Gy have the capability of producing a distinct class of meteorites, and (iii) once a family's meteoroid production rate drops by a factor of 100, it is unlikely to produce enough terrestrial meteorites to be noticed in our collection. The latter choice is somewhat arbitrary, but it is meant to partially balance against the possibility that our FSDs are too steep for smaller parent bodies.

Our CoDDEM results suggest that asteroid families produced by the breakup of $D \gtrsim 100$ km bodies have such slow meteoroid decay rates that all ~ 20 of them may very well be providing meteoroids today, regardless of their disruption time over the last 3-4 Gy. While this is a substantial number, it is not enough to explain the diversity of stony meteorites observed in our collection. Moreover, Nesvorný *et al.* (2004) estimate that only $\sim 40\%$ of the main belt down to a few km in diameter is comprised of asteroids from the largest families.

Among the smaller parent bodies ($30 < D < 100$ km), we find that, on average, the interval between disruption events across the main belt is short enough that many have disrupted over the last Gy or so (e.g., CoDDEM results predict a $D \sim 30$ km body disrupts once every 25 My; a $D \sim 80$ km body disrupts once every 160 My). We combine these numbers with our meteoroid decay rate values to estimate the number of parent bodies producing distinct classes of meteorites. For example, Fig. 4 shows that the meteoroid population produced by $D \sim 30$ km bodies decays by a factor of 100 over 130 My. This means that if our toy FSD is reasonable, ~ 5 stony meteorite classes (130 My / 25 My) should come from families produced by the breakup of these size bodies. Making comparable calculations for disruption events among $30 < D < 100$ km bodies, we estimate that we should see meteorites from another ~ 25 parent bodies.

Summing our values, we estimate we should find stony meteorites from ~ 45 different parent bodies. The actual value is ~ 30 parent bodies. We consider this to be surprisingly good match when all of our model uncertainties are considered. It also implies that some breakup events occurred recently enough that their FSDs at small sizes may still have power law slopes steeper than the background population.

Other than model inaccuracies, we believe there are several plausible reasons why our estimates come out 50% higher than the observed value. Two important ones are listed here:

1. Some disruption events must occur within existing families. A prominent example would be the Karin cluster, which was produced by the breakup of a $D \sim 25\text{-}30$ km asteroid inside the Koronis asteroid family (Nesvorný *et al.* 2002; 2003; Nesvorný & Bottke 2004). Because intra-family breakups are unlikely to produce distinct classes of stony meteoroids, this would decrease the total number of parent bodies represented in our model.

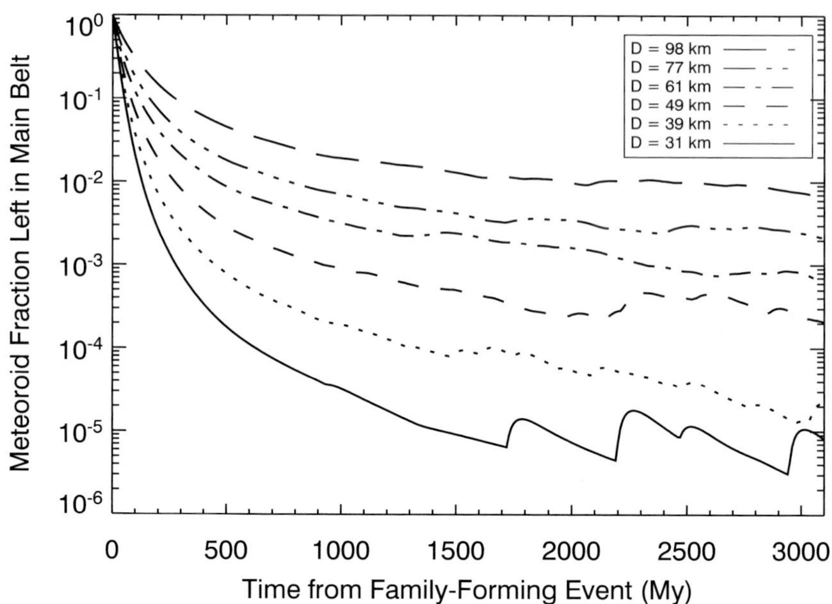

Figure 4. Decay rates of meteoroid populations from our "toy" asteroid families produced from parent bodies with $30 < D < 100$ km. All families were inserted in CoDDEM at 1.5 Gy after solar system formation. The smallest families decrease by a factor of 100 over a few 0.1 Gy while the largest take several Gy to decay by the same value.

2. Outer main belt meteoroids may have great difficulty reaching Earth. For example, Morbidelli & Gladman (1998) have shown that bodies from the ν_6 secular resonance (along the inner edge of the main belt from 2.1-2.5 AU) have a $\sim 1\%$ chance of striking the Earth, while those from the 5:2 resonance at 2.8 AU only have a $\sim 10^{-5}$ chance. The latter probability is low because most bodies placed in the 5:2 resonance readily reach Jupiter-crossing orbits, a characteristic shared by most outer main belt resonances. Moreover, many outer main belt meteoroids may also be analogous to weak carbonaceous chondritic material, which would make their survival through our atmosphere problematic. Taken together, these factors suggest that only the most prolific meteoroid producers in the outer main belt produce meteorites.

We believe these factors could reduce the number of sampled parent bodies in our model enough to match observations. To prove it, we will need to develop a next-generation meteoroid evolution model that accounts for the coupled collisional/dynamical evolution of families as they stretch toward main belt resonances (e.g., Morbidelli & Vokrouhlický 2003). This model will also need to include more realistic FSD estimates than those used here.

Before concluding, we have one additional observation to make about our results. Constraints that can be used to test our model include various meteorite CRE age distributions (e.g., H, L, and LL chondrites; Marti & Graf 1992; Morbidelli & Gladman 1998; Vokrouhlický & Farinella 2000; Eugster 2003). Our model implies that the wide range of CRE ages observed among the H, L, and LL meteorite classes (~ 10-100 My) are unlikely to come from multiple cratering events on a single parent body (e.g., Migliorini *et al.* 1997). Instead, we postulate they come from small-scale (or not so small-scale) breakup events occurring among family members, where the family itself might stretch from its point of origin all the way to multiple resonant routes leading to Earth. In this scenario, concentrations of CRE ages among meteorite classes, such as the $\sim$ 7-8 Ma cluster measured on almost half of the H chondrites (Eugster 2003), would come from distinct breakups events occurring near the resonances. The quantity of meteorites within a given CRE age cluster would then depend on the size of the disrupted immediate precursor, its proximity to the resonance, and the likelihood that material in that resonance would reach Earth.

7. Conclusions

Our main belt evolution model provides results that are consistent with the limited number of parent bodies represented in our stony meteorite collection. We believe that most stony meteorites are byproducts of a collisional cascade, with some coming from asteroid families produced by the breakup of $D > 100$ km bodies over the last several Gy and the remainder coming from smaller, more recent breakup events among $D < 100$ km asteroids that occurred over recent times (i.e., $\ll 1$ Gy).

Other notable results described in the paper include the following: (i) the wavy main belt size distribution is likely a "fossil" produced by collisions taking place in the first few tens of My of solar system history; (ii) the kink in the NEO size distribution near $D \sim 0.7$ km is likely a by-product of the Yarkovsky effect/resonances working on the wavy main belt population, and (iii) collisional evolution likely explains why so many observed asteroid families have shallow power law slopes in the size interval $0.7 < D < 5$ km (Morbidelli *et al.* 2003).

Acknowledgements

We thank Dave O'Brien for valuable discussions and input to this study. We also thank Kleomenis Tsiganis for his careful and useful review of this paper. Research funds for William Bottke were provided by NASA's Origins of Solar Systems (grant NAG5-10658) and Planetary Geology and Geophysics Programs (grant NAG5-13195).

References

Agnor, C.B., Canup, R.M. & Levison, H.F. 1999, *Icarus* 142, 219

Asphaug, E., Ryan, E.V. & Zuber, M.T. 2002, in: Bottke, W.F., Cellino, A., Paolicchi, P. & Binzel, R.P. (eds.), *Asteroids III* (Univ. Arizona Press, Tucson), p. 463.

Benz, W. & Asphaug, E. 1999, *Icarus* 142, 5

Bland, P.A., Conway, A., Smith, T.B., Berry, F.J. & Pillinger, C.T. 2000, in: Grady M. M., Hutchsion R., McCall G. J. H. & Rothery D. A. (eds) *Meteorites: Flux with Time and Impact Effects*. Geological Society, London, Special Publications 140, p.43

Bottke, W.F. & Greenberg, R. 1993, *Geophys. Res. Lett.* 20, 879

Bottke, W.F., Nolan, M.C., Greenberg, R. & Kolvoord, R.A. 1994, *Icarus* 107, 255.

Bottke, W.F., Rubincam, D.P. & Burns, J.A. 2000, *Icarus* 145, 301

Bottke, W.F., Jedicke, R., Morbidelli, A., Petit, J.-M. & Gladman, B.J. 2000, *Science* 288, 2190

Bottke, W.F., Vokrouhlický, D., Brož, M., Nesvorný, D. & Morbidelli, A. 2001, *Science* 294, 1693

Bottke, W.F., Morbidelli, A., Jedicke, R., Petit, J.-M., Levison, H.F., Michel, P. & Metcalfe, T.S. 2002a, *Icarus* 156, 399

Bottke, W.F., Vokrouhlický, D., Rubincam, D.P. & Brož, M. 2002b, in: Bottke, W.F., Cellino, A., Paolicchi, P. & Binzel, R.P. (eds.), *Asteroids III* (The University of Arizona Press, Tucson), p. 395

Bottke, W. F., Durda, D.D., Nesvorný, D., Jedicke, R. Morbidelli, A., Vokrouhlický, D. & Levison., H.L 2004a. *Icarus*, in press

Bottke, W. F., Durda, D.D., Nesvorný, D., Jedicke, R. Morbidelli, A., Vokrouhlický, D. & Levison., H.L. 2004b. *Icarus*, in preparation

Britt, D.T., Yeomans, D., Housen, K., Consolmagno, G. 2002, in: Bottke, W.F., Cellino, A., Paolicchi, P. & Binzel, R.P. (eds.), *Asteroids III* (Univ. Arizona Press, Tucson), p. 485

Burbine, T.H., McCoy, T.J., Meibom, A., Gladman, B. & Keil, K. 2002, in: Bottke, W.F., Cellino, A., Paolicchi, P. & Binzel, R.P. (eds.), *Asteroids III* (Univ. Arizona Press, Tucson), p. 653

Campo Bagatin, A., Cellino, A., Davis, D.R., Farinella, P. & Paolicchi, P. 1994, *Planet. Space Sci.* 42, 1079

Caffee, M.W., Reedy, R.C., Goswami, J.N., Hohenberg, C.M. & Marti, K. 1988, in: Kerridge, J.F. & Matthews, M.S. (eds.), *Meteorites and the Early Solar System* (Univ. Arizona Press, Tucson), p. 205.

Chambers, J.E. & Migliorini, F. 1997, *Bull. Amer. Astron. Soc.* 29, 1024

Chambers, J.E. & Wetherill, G.W. 1998, *Icarus* 136, 304

Chambers, J.E. & Wetherill, G.W. 2001, *Meteoritics Planet. Sci.* 36, 381.

Davis, D.R., Durda, D.D., Marzari, F., Campo Bagatin, A. & Gil-Hutton, R. 2002, in: Bottke, W.F., Cellino, A., Paolicchi, P. & Binzel, R.P. (eds.), *Asteroids III* (Univ. Arizona Press, Tucson), p. 545

Dohnanyi, J.W. 1969, *J. Geophys. Res.* 74, 2531

Durda, D.D., Greenberg, R. & Jedicke, R. 1998, *Icarus* 135, 431

Durda, D.D., Bottke, W.F., Enke, B.L., Merline, W.J., Asphaug, E., Richardson, D.C., & Leinhardt, Z.M. 2004, *Icarus* 170, 243

Eugster, O. 2003, *Chemie der Erde* 63, 3

Farinella, P. & Davis, D.R. 1992, *Icarus* 97, 111

Farinella, P., Froeschlé, Ch, Froeschlé, Cl., Gonczi, R., Hahn, G., Morbidelli, A. & Valsecchi, G.B. 1994. *Nature* 371, 315

Farinella, P. & Vokrouhlický, D. 1999, *Science* 283, 1507

Farinella, P., Vokrouhlický, D. & Hartmann, W.K. 1998, *Icarus* 132, 378

Fowler, J.W. & Chillemi, J.R. 1992, in: E.F. Tedesco, (ed) *The IRAS Minor Planet Survey* (Tech. Report PL-TR-92-2049., Phillips Laboratory, Hanscom Air Force Base, Massachusetts) p. 17

Gomes, R., Morbidelli, A., Tsiganis, K., & Levison, H. 2004, *AAS/Division for Planetary Sciences Meeting* 36, 1167

Grieve, R. A. F. & Shoemaker, E. M. 1994, in: Gehrels, T. & Matthews, M.S. (eds.) *Hazards Due to Comets and Asteroids* (Univ. Arizona Press, Tucson) p. 417

Guillot, T., Hueso, R. 2003, *Bull. Amer. Astron. Soc.* 35

Gladman, B.J., Migliorini, F., Morbidelli, A., Zappalà, V., Michel, P., Cellino, A., Froeschlé, C., Levison, H.F., Bailey, M. & Duncan, M. 1997, *Science* 277, 197

Grady, M. *Catalogue of meteorites* (U. Cambridge Press).

Greenberg, R. 1982, *Astron. J.* 87, 184

Hartmann, W. K., Ryder, G., Dones, L. & Grinspoon, D. 2000, in: Canup, R. & Righter, K., (eds.) *Origin of the Earth and the Moon* (Univ. Arizona Press, Tucson) p. 805

Holsapple, K., Giblin, I., Housen, K., Nakamura, A. & Ryan, E. 2002, in: Bottke, W.F., Cellino, A., Paolicchi, P. & Binzel, R.P. (eds.), *Asteroids III* (Univ. Arizona Press, Tucson), p. 443.

Inaba, S., Wetherill, G.W. & Ikoma, M. 2003, *Icarus* 166, 46

Ivezić, Z. & 31 coauthors 2001, *Astron. J.* 122, 2749

Jedicke, R., Larsen, J. & Spahr, T. 2002, in: Bottke, W.F., Cellino, A., Paolicchi, P. & Binzel, R.P. (eds.), *Asteroids III* (Univ. Arizona Press, Tucson), p. 71.

Keil, K. 2002, *Planet. Space Sci.* 48, 887.

Keil, K. 2002, in: Bottke, W.F., Cellino, A., Paolicchi, P. & Binzel, R.P. (eds.), *Asteroids III* (Univ. Arizona Press, Tucson), p. 573.

Levison, H.F. & Duncan, M.J. 1994, *Icarus* 108, 18

Love, S.G. & Ahrens, T.J. 1997, *Nature* 386, 154

Marti, K. & Graf, T. 1992, *Ann. Rev. Earth Planet. Sci.* 20, 221

Meibom A. & Clark, B.E. 1999, *Meteoritics Planet. Sci.* 34, 7.

Michel., P., Benz, W., Tanga, P. & Richardson, D. 2001, *Science* 294, 1696.

Michel., P., Tanga, P., Benz, W. & Richardson, D. 2002, *Icarus* 160, 10

Michel, P., Benz, W. & Richardson, D. 2003, *Nature* 421, 608

Morbidelli, A. & Gladman, B. 1998, *Meteoritics Planet. Sci.* 33, 999

Morbidelli, A., Petit, J.-M., Gladman, B. & Chambers, J. 2001, *Meteoritics Planet. Sci.* 36, 371

Morbidelli, A., Nesvorný, D., Bottke, W.F., Michel, P., Vokrouhlický, D. & Tanga, P. 2003, *Icarus* 162, 328.

Morbidelli, A. & Vokrouhlický, D. 2003, *Icarus* 163, 120

Nesvorný, D. & Bottke, W.F. 2004, *Icarus* 170, 324

Nesvorný, D., Bottke, W.F., Dones, L. & Levison, H.F. 2002, *Nature* 417, 720

Nesvorný, D., Bottke, W.F., Levison, H.F. & Dones, L. 2003, *Astrophys. J.* 591, 486

Nesvorný, D. & Bottke, W.F. 2004, *Icarus* 170, 324

Nesvorný, D., Jedicke, R., Whiteley, R.J. & Ivezić, Ž 2004, *Icarus*, in press.

O'Brien, D.P. & Greenberg, R. 2003, *Icarus* 164, 334

Öpik, E.J. 1951, *Proc. R. Irish Acad.* 54, 165.

Petit, J., Morbidelli, A. & Valsecchi, G.B. 1999, *Icarus* 141, 367.

Petit, J., Morbidelli, A. & Chambers, J. 2001, *Icarus* 153, 338

Petit, J.-M., Chambers, J., Franklin, F. & Nagazawa, M. 2002, in: Bottke, W.F., Cellino, A., Paolicchi, P. & Binzel, R.P. (eds.), *Asteroids III* (Univ. Arizona Press, Tucson), p. 711.

Pollack, J.B., Hubickyj, O., Bodenheimer, P., Lissauer, J., Podolak, M. & Greenzweig, Y. 1996, *Icarus* 124, 62.

Press, W.H., Flannery, B.P., Teukolsky, S.A. & Vetterling, W.T. 1989, *Numerical recipes in Fortran. The art of scientific computing* (Cambridge: University Press)

Rubincam, D. P. 2000, *Icarus* 148, 2

Stuart, J.S. & Binzel, R.P. 2004, *Icarus* 170, 295

Stokes, G.H., Yeomans, D.K., Bottke, W.F., Chesley, S.R., Evans, J.B., Gold, R.E., Harris, A.W., Jewitt, D., Kelso, T.S., McMillan, R.S., Spahr, T.B. & Worden, S.P. 2003, *Report of the Near-Earth Object Science Definition Team: A Study to Determine the Feasibility of Extending the Search for Near-Earth Objects to Smaller Limiting Diameters.* NASA-OSS-Solar System Exploration Division.

Tanga, P., Cellino, A., Michel, P., Zappalà, V., Paolicchi. P. & Dell'Oro, A. 1999, *Icarus* 141, 65

Tedesco, E.F. & Desert, F. 2002, *Astron. J.* 123, 2070

Tedesco, E.F., Noah, P.V., Noah, M. & Price, S.D. 2002, *Astron. J.* 123, 1056

Tsiganis, K., Morbidelli, A., Levison, H. F. & Gomes, R. S. 2004, *AAS/Division of Dynamical Astronomy Meeting* 35, 1177.

Vokrouhlický, D. & Farinella, P. 2000, *Nature* 407, 606

Weidenschilling, S.J. 1977, *Astrophys. Space Sci.* 51, 153

Wetherill, G.W. 1989, 2002, in: Binzel, R.P., Gehrels, T. & Matthews, M.S. (eds.), *Asteroids II* (Univ. Arizona Press, Tucson), p. 661.

Wetherill, G.W. 1967, *J. Geophys. Res.* 72, 2429.

Wetherill, G.W. & Stewart, G.R. 1989, *Icarus* 77, 330

Williams, D.R. & Wetherill, G.W. 1994., *Icarus* 107, 117

Wuchterl, G., Guillot, T. & Lissauer, J.J. 2000, in: Manning, V., Boss, A.P. & Russell, S.S. (eds.), *Protostars and Planets IV* (Univ. Arizona Press, Tucson), p. 1081

Zappalà, V., Cellino, A., dell'Oro, A. & Paolicchi, P. 2002, in: Bottke, W.F., Cellino, A., Paolicchi, P. & Binzel, R.P. (eds.), *Asteroids III* (The Univ. Arizona Press, Tucson), p. 619.

Dynamics of Populations of Planetary Systems
Proceedings IAU Colloquium No. 197, 2005
Z. Knežević and A. Milani, eds.

© 2005 International Astronomical Union
DOI: 10.1017/S1743921304008877

The dynamical structure of meteor streams and meteor shower predictions

David J. Asher

Armagh Observatory, College Hill, Armagh, BT61 9DG, United Kingdom
email: dja@arm.ac.uk

Abstract. Meteor streams that form as a result of cometary activity around perihelion consist of both structured and background components. The former are often referred to as trails. A trail is created at each perihelion passage as a result of the meteoroids' range of orbital periods. Trail locations can be precisely calculated by numerical integrations, allowing predictions of meteor outbursts and storms. The initial distribution of meteoroids, which relates to the meteor shower profile, depends on the meteoroid production rate and ejection velocity distribution as functions of heliocentric distance and on solar radiation pressure. The profile can gradually evolve owing to other radiative forces. This paper reviews such work on these aspects of shower predictions.

Keywords. Celestial mechanics, methods: numerical, meteors, meteoroids

1. Introduction

The science of meteor forecasting has more immediate practical applications than many topics in solar system dynamics, interested parties ranging from meteor observers to satellite operators. In common with other dynamical topics, stream modelling has benefitted from high speed computers, although distinctions can be drawn. Thus timescales for modelling relevant to meteor predictions are often short when compared against integration studies of various solar system populations, so that the required computation is less: apart from practical investigations of specific virtual impactors, asteroid and comet integrations using 'cloned' orbits usually cover quite long timescales. On the other hand, the practical need to map out extremely fine structure in streams can require more computation. Interest can nevertheless be focused on the fine structure in restricted regions of streams, such as those near the Earth or another planet, if computing power is limited.

Stages in stream evolution (from the early structured phases to dispersal within the same stream and thereafter beyond the original stream into the zodiacal background), and the dominant physical forces during different stages, are discussed by Williams (2002). Secular perturbation methods have been successful for constructing overall models of meteoroids gradually filling a stream (e.g., Babadzhanov & Obrubov 1987), especially when knowledge of the fine structure at precise locations is not needed.

As computing power has advanced, increasingly large scale integration studies have become possible (e.g., Brown & Jones 1998 modelled the Perseid stream, Vaubaillon 2003 the Leonids and π-Puppids). To obtain large statistical samples of particles, which greatly helps the examination of fine structure in a stream, there are also methods other than the direct integration of the equations of motion (e.g., see Ryabova 2001 on the Geminids).

Many of the most spectacular meteor storms have been found to be associated with meteoroids ejected at a single perihelion return of the parent comet. Many could be identified without building elaborate models of the meteor stream because, typically for some centuries after ejection, gravitational perturbations depend almost entirely on a single parameter (the orbital period immediately after ejection). The range of periods causes meteoroids to stretch into a long, narrow trail; a new trail is formed on each revolution of an active comet. Although recently most attention was given to the Leonids (e.g., Upton 1977, Kondrat'eva & Reznikov 1985, Kondrat'eva *et al.* 1997, Lyytinen 1999, McNaught & Asher 1999), a similar technique has been applied to many other streams (e.g., Davies & Turski 1962, Reznikov 1983, Reznikov 1993).

In what follows here, rather than the orbital period being referred to directly, the parameter a_0 is used. This is the semi-major axis when the meteoroid is released from the comet; Δa_0 is the difference from the cometary value. In cases where β (the ratio of solar radiation pressure to solar gravity) is non-zero, I define a_0 in terms of a particle with the same instantaneous position and velocity vectors moving purely gravitationally. Therefore in such cases the geometric semi-major axis (i.e., the semi-major axis of the ellipse corresponding to the path followed by the particle) differs from a_0.

To first order in computing dynamical evolution, the reason why any meteoroid has its given value of Δa_0 does not matter; while the physical cause of a_0 differing from the cometary value can be either velocities of ejection from the nucleus, or solar radiation pressure, the subsequent effect, through gravitational perturbations, is the same if Δa_0 is the same. Meteoroids with the same a_0, even if other elements are slightly different, co-move around their orbits, are therefore at a similar distance from each planet at every time, and so undergo identical perturbations. Planetary perturbations only differ significantly during close approaches to planets; it is then that substantial dispersion occurs (Asher 2002).

Usually the distance of a comet's node from the Earth's orbit is much greater than the width of a trail. Identifying parts of trails that come close to planets can be done by calculating the perturbations to the nodal position as a function of Δa_0. Without loss of generality, tangential ejection at perihelion can be assumed. The process of identifying the relevant parts of trails tends to be easy for young trails (a few revolutions), but gradually the various elements become less smooth functions of Δa_0 and detailed inspection of the values of orbital elements as functions of Δa_0 can become necessary.

2. Integrations

Modelling can be developed further by considering density distributions within trails, in particular cross sections at certain values of Δa_0. This is best illustrated using example integrations. Vaubaillon (2004, personal communication) found that meteoroids ejected from 55P/Tempel-Tuttle around its 1333 and 1733 returns would be perturbed so as to be moderately close to the Earth during the 2004 Leonid shower. I have therefore integrated particles based on those initial conditions (1333 and 1733) for illustration in this paper, although of course similar features are present in the Leonid and other streams whether or not the Earth is nearby. Integrations used the MERCURY package (Chambers 1999) implementation of the RADAU integrator (Everhart 1985) with 8 planets from Mercury to Neptune included (present day elements from JPL's DE403) and the cometary orbit from Nakano (1997). Plots shown below used isotropic ejection over the sunward hemisphere of 55P/Tempel-Tuttle but integrations with ejection over the whole surface of the nucleus gave very similar results. Meteoroid production was assumed to occur at $r < 3.4$ AU at a constant rate in true anomaly (i.e., faster near perihelion). The meteoroid ejection theory of Whipple (1951) developed by Jones (1995) predicts a dependence of ejection speed on heliocentric distance quite close to $v \propto 1/r$ and this is the relation used here. Generating sets of particles with different constants of proportionality demonstrates how cross sections relate to v (e.g., McNaught & Asher 2002). The extent of trail cross sections

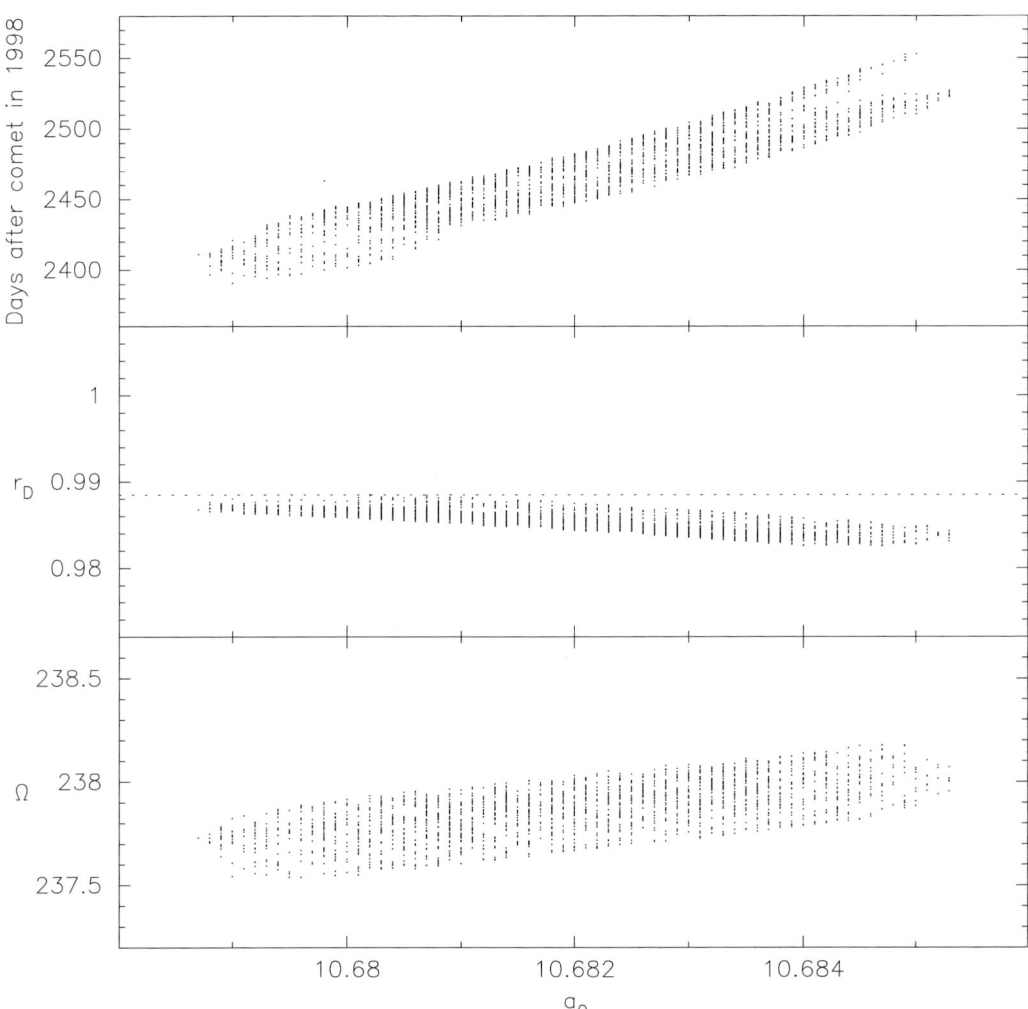

Figure 1. A short section of the 8-revolution Leonid trail in 2004. Particles that reach the ecliptic when the Earth is nearby are those whose nodal crossing is $\sim$2450 days after that of the comet in 1998; this condition implies a narrow range of initial semi-major axis a_0, and other particles were not integrated. These particles had $\beta = 0$ (gravitational perturbations only) and $v = 50/r$ m/s. Elements in 2004 are fairly linear functions of a_0.

determines which trails are and which are not encountered by the Earth, and it is notable that observations and non-observations of meteor outbursts are in accord with Whipple theory.

Figure 1 shows particles ejected during a single perihelion return, with a single value of v_1 in the relation $v = v_1/r$ and a single value of β. The parameters r_D = heliocentric distance of the descending node (Earth's distance shown as horizontal dotted line), and Ω = longitude of ascending node, correspond to the two dimensions of the trail cross section in the ecliptic. Many other sets of particles, having different combinations of v_1

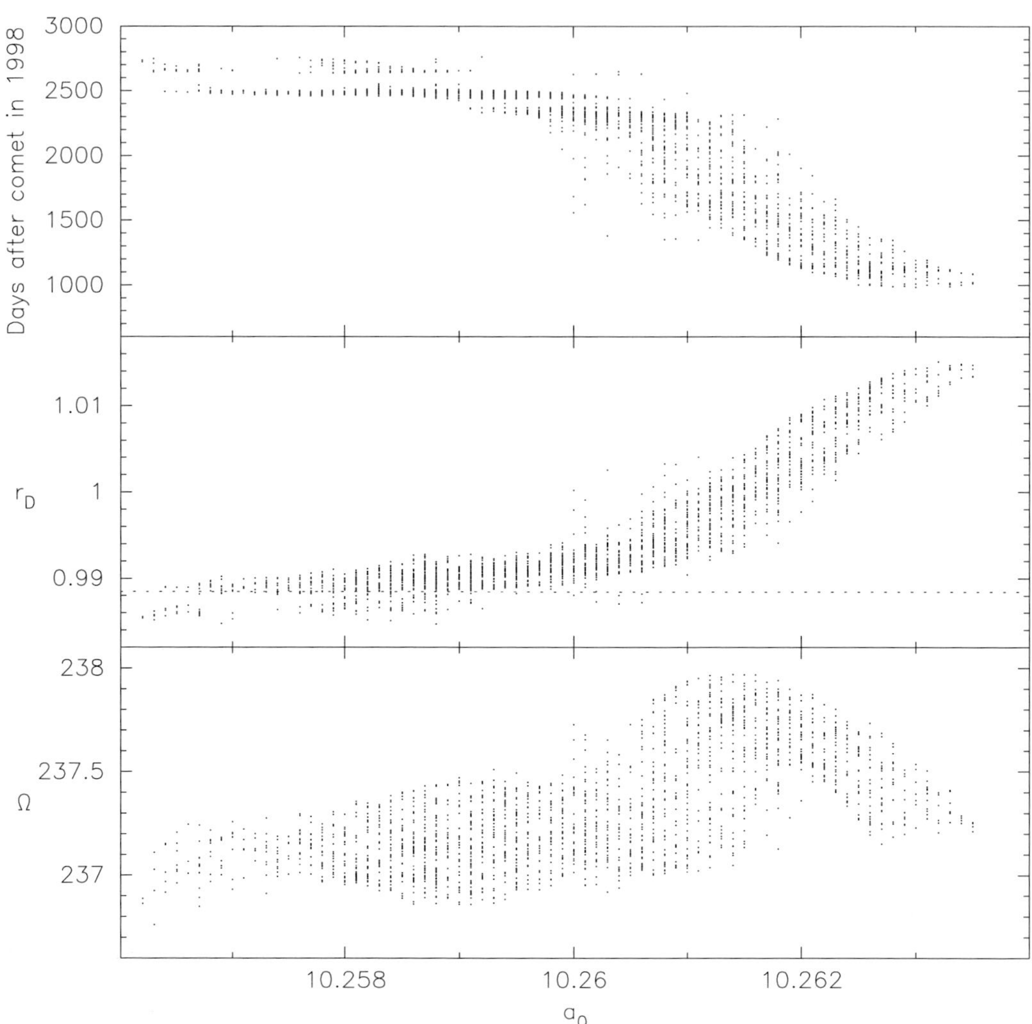

Figure 2. A short section of the 20-revolution Leonid trail in 2004. Ejection speed $v = 50/r$ m/s and radiation pressure parameter $\beta = 0.001$. The non-linearity here contrasts with a younger trail (figure 1).

and β, have been integrated, with the results similar except for the dispersion (i.e., the density distribution changes but the overall location in the ecliptic is very similar). The location of a given section of a given trail in space at a given time depends on the history of planetary perturbations starting at ejection time; that perturbation history is the same whatever the cross sectional density distribution, until that part of the trail is scattered. Figure 2 shows an older trail. The greater dispersion, and also the fact that even over a very small range the elements are not linear functions of Δa_0, can be seen. These features are present whether $\beta = 0$ or not.

Although the perturbation history determines trail nodal positions, in general planetary perturbations do not alter a trail's cross section for some time. Comparison of the

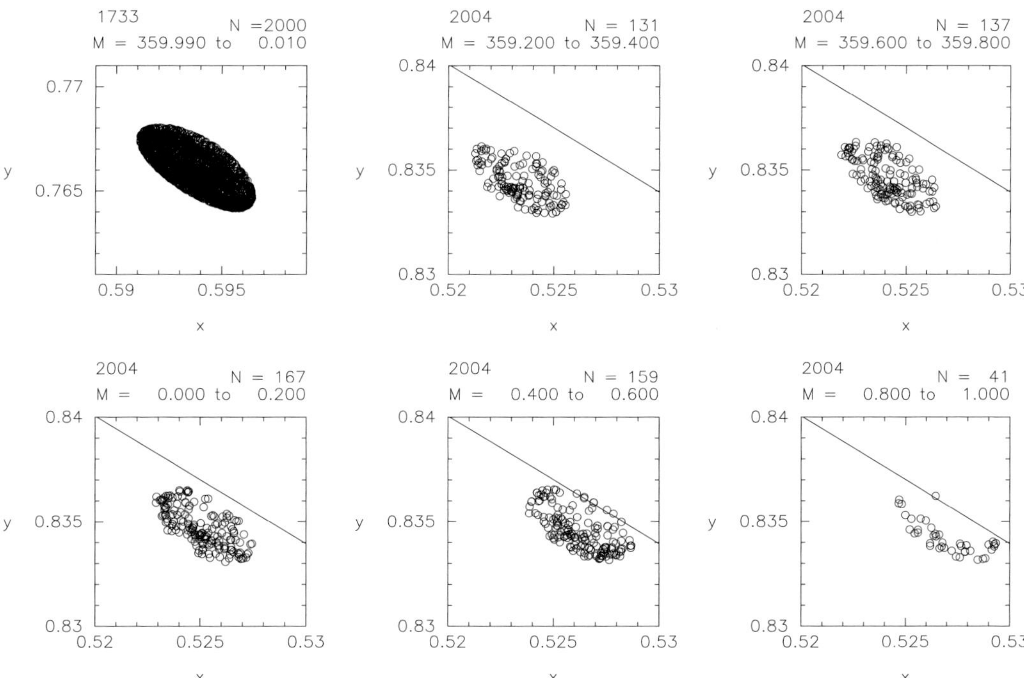

Figure 3. Cross sections in the ecliptic of the 8-revolution Leonid trail, near the part of the trail that the Earth approaches in 2004. Initial distribution in top left plot results from choice of $\beta = 0$ and $v = 50/r$. Remaining 5 plots are at roughly two week intervals in late 2004; the 3rd (i.e., left of bottom row) brackets the time when the Earth passes near. The line is the Earth's orbit.

top left panel of figure 3 with any of the remaining five panels shows that although the node has been shifted quite far (by $\sim$0.1 AU) during 8 revolutions, all particles at any single point along the trail are perturbed by almost exactly the same amount, so that the shape of the cross section is similar to the initial cross section. In a model involving gravitational perturbations and radiation pressure only, this is true for all v_1 and β. Additionally, comparison between any of those same five panels in figure 3 shows that the exact location in the ecliptic of a trail cross section is very sensitive to the position along the trail. Even a displacement a small way along a trail causes the perturbation history over several centuries to differ slightly. The lower total number of particles in the final plot of figure 3 is statistically significant: the trail has been more stretched along (but not across) the orbit at that point, diluting the number density.

Although, under gravitational perturbations alone, the dispersion eventually broadens after several centuries, cross sections within limited parts of trails can be maintained for some time. Thus in figure 4 the ecliptic crossing points of individual particles, at the same position along a trail, move relative to each other by small amounts over several centuries, up to about half the width of the cross section shown in that figure. These relative perturbations are a little more than for young trails, but insufficient to disperse that part of the trail into the Leonid background. Often resonances help to maintain compact structures in streams over substantial timescales (see Emel'yanenko 2001).

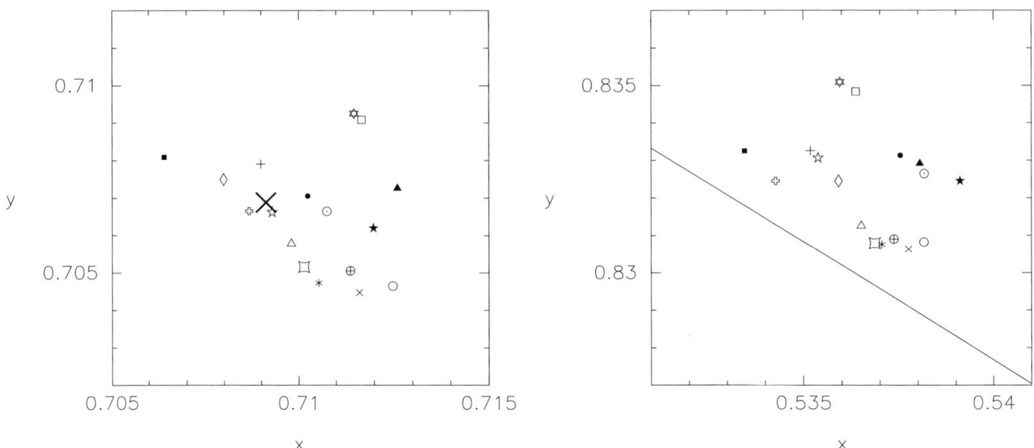

Figure 4. Example particles ($\beta = 0.001$, $v = 50/r$) in the 20-revolution Leonid trail, selected to be at a similar point along the trail, namely to cross the ecliptic within a few days of when the Earth (whose orbit is shown) is nearby in 2004. Each particle's nodal crossing point is shown with the same symbol in the left (ejection time) and right (20 revolutions later) plots. The large X in the left plot is the comet's node.

3. Discussion

Most of the sharpest meteor outbursts are due to meteoroidal material released just a few to several revolutions previously. Over these comparatively short timespans following ejection, orbital evolution under gravitational perturbations can be calculated precisely, as shown in §2 and in references given above. Conditions for Earth impact at any given time can be constrained in terms of ejection with specific velocities and at specific true anomalies (e.g., Brown & Arlt 2000, Müller $et\ al.$ 2001, Ryabova 2001, Asher & Emel'yanenko 2002).

However, a full description of a meteor activity profile requires knowledge of the initial conditions for the integrations (i.e., the distribution of particles ejected). The question also arises as to the effect of radiative forces (other than solar radiation pressure) on the orbital precession.

In principle, an activity profile can be constructed by the superposition of particles suitably distributed in v_1 and β (§2). Additionally the ejection model could be refined; for example, a distributed production model (Crifo 1995), in which fragments sublimate after being ejected from the cometary surface, may be more physically realistic. Ultimately, however, the question must be addressed of calibrating the model's prediction of activity profiles against observations. Difficulties can arise because of the effectively large number of parameters in the ejection model (even assuming any one comet shows the same activity pattern at each return); the distributions of particle sizes and ejection velocities can vary as functions of true anomaly.

Recently, the attempt by Vaubaillon (2002) to use observed meteor fluxes to calibrate predictions from dynamical simulations has met with significant success. This model uses ejection velocities given by Crifo & Rodionov (1997), and observations of the parent comet to constrain the dust production rate.

Modelling of other radiative forces has been done by Lyytinen & Van Flandern (2000) and Lyytinen $et\ al.$ (2001), who have found good evidence that such effects have measurable observational consequences on the timing and strength of meteor outbursts. The most important force is a seasonal Yarkovsky effect but a neat feature of the model is that all the relevant forces can be incorporated into a single 'A2' parameter. Each individual particle's orbit is systematically changed by radiative forces but different particles are affected to different extents (e.g., obviously not all particles have the same spin). The single parameter of the model is the width of the distribution of rates at which particles' orbits are affected. The force, acting on all particles within trails, leads to a gradual spread in a trail cross section. Additionally, the component of the force that changes the orbital period has an indirect effect on the nodal position, through gravitational perturbations (the fact that the orbital change due to radiative forces is systematic meaning that the entire perturbation history of each individual particle is slightly changed). The name 'A2' for this effect is by analogy to the well known cometary A_2 acceleration (Marsden *et al.* 1973) in terms of the dynamical result, systematically changing the period (not, of course, in terms of the physical cause, which is different).

Recently most work involving the seasonal Yarkovsky effect has focused on bodies the size of asteroids or larger meteoroids (e.g., Vokrouhlický & Farinella 1999). Lyytinen & Van Flandern (2000) and Lyytinen *et al.* (2001) have considered the smaller grains that are relevant in the present paper: the sensitivity of trail perturbations over centuries to any small changes in the orbital period means that outburst timings can in some cases (especially for slightly older trails) be expected to shift by a few hours. The size of the A2 parameter can be calibrated by comparing model fits to observations. Just as ejection velocities determined indirectly from meteor observations are of comparable size to those predicted by Whipple theory (§ 2), good fits to observed meteor outbursts are found when the A2 parameter is of a size expected theoretically from consideration of seasonal Yarkovsky.

Acknowledgements

Research at the Armagh Observatory is supported by the N. Ireland DCAL. I thank the IAU for support towards my attendance at the meeting, Esko Lyytinen and Jérémie Vaubaillon for valuable discussions, and the referee for a helpful review.

References

Asher, D.J. 2002, in: S.F. Green, I.P. Williams, J.A.M. McDonnell & N. McBride (eds.), *Dust in the Solar System and Other Planetary Systems*, IAU Colloq. No. 181 (Oxford: Pergamon), p. 61

Asher, D.J. & Emel'yanenko, V.V. 2002, *Mon. Not. R. Astron. Soc.* 331, 126

Babadzhanov, P.B. & Obrubov, Yu.V. 1987, in: Z. Ceplecha & P. Pecina (eds.), *Interplanetary Matter*, Publ. Astron. Inst. Czechoslov. Acad. Sci., vol. 67 (2), (Ondřejov: Czechoslov. Acad. Sci.) p. 141

Brown, P. & Arlt, R. 2000, *Mon. Not. R. Astron. Soc.* 319, 419

Brown, P. & Jones, J. 1998, *Icarus* 133, 36

Chambers, J.E. 1999, Mon. Not. R. Astron. Soc. 304, 793

Crifo, J.F. 1995, *Astrophys. J.* 445, 470

Crifo, J.F. & Rodionov, A.V. 1997, *Icarus* 127, 319

Davies, J.G. & Turski, W. 1962, *Mon. Not. R. Astron. Soc.* 123, 459

Emel'yanenko, V.V. 2001, in: B. Warmbein (ed.), *Proc. Meteoroids 2001 Conf.*, ESA SP–495 (Noordwijk: European Space Agency), p. 43

Everhart, E. 1985, in: A. Carusi & G.B. Valsecchi (eds.), *Dynamics of Comets: Their Origin and Evolution*, (Dordrecht: Reidel), p. 185

Jones, J. 1995, *Mon. Not. R. Astron. Soc.* 275, 773

Kondrat'eva, E.D. & Reznikov, E.A. 1985, *Sol. Syst. Res.* 19, 96

Kondrat'eva, E.D., Murav'eva, I.N. & Reznikov, E.A. 1997, *Sol. Syst. Res.* 31, 489

Lyytinen, E. 1999, *Meta Res. Bull.* 8, 33

Lyytinen, E.J. & Van Flandern, T. 2000, *Earth, Moon, Planets* 82, 149

Lyytinen, E., Nissinen, M. & Van Flandern, T. 2001, *WGN* 29, 110

Marsden, B.G., Sekanina, Z. & Yeomans, D.K. 1973, *Astron. J.* 78, 211

McNaught, R.H. & Asher, D.J. 1999, *WGN* 27, 85

McNaught, R.H. & Asher, D.J. 2002, *WGN* 30, 132

Müller, M., Green, S.F. & McBride, N. 2001, in: B. Warmbein (ed.), *Proc. Meteoroids 2001 Conf.*, ESA SP–495 (Noordwijk: European Space Agency), p. 47

Nakano, S. 1997, *Minor Planet Circ.* 29285

Reznikov, E.A. 1983, *Trudy Kazan. Gor. Astron. Obs.* 47, 131

Reznikov, E.A. 1993, *Trudy Kazan. Gor. Astron. Obs.* 53, 80

Ryabova, G.O. 2001, in: B. Warmbein (ed.), *Proc. Meteoroids 2001 Conf.*, ESA SP–495 (Noordwijk: European Space Agency), p. 77

Upton, E.K.L. 1977, *Griffith Observer* 41(5), 3

Vaubaillon, J. 2002, *WGN* 30, 144

Vaubaillon, J. 2003, Ph.D. thesis, Observatoire de Paris

Vokrouhlický, D. & Farinella, P. 1999, *Astron. J.* 118, 3049

Whipple, F.L. 1951, *Astrophys. J.* 113, 464

Williams, I.P. 2002, in: S.F. Green, I.P. Williams, J.A.M. McDonnell & N. McBride (eds.), *Dust in the Solar System and Other Planetary Systems*, IAU Colloq. No. 181 (Oxford: Pergamon), p. 3

Dynamics of Populations of Planetary Systems
Proceedings IAU Colloquium No. 197, 2005
Zoran Knežević and Andrea Milani, eds.

© 2005 International Astronomical Union
DOI: 10.1017/S1743921304008889

The origin and evolution of dust belts

Mark C. Wyatt

UK Astronomy Technology Centre, Royal Observatory, Edinburgh EH9 3HJ, UK
email: wyatt@roe.ac.uk

Abstract. Planetary systems are made up of objects with sizes ranging from gas giant planets down to asteroids and on to micron sized dust. Dust in the zodiacal cloud in the solar system originates in the break-up of asteroids and comets and then migrates in toward the Sun due to P-R drag. The dynamical evolution of the dust is also affected by the gravitational perturbations of the solar system's planets, and the consequence of those perturbations is evident in the asymmetric and clumpy structure of the zodiacal cloud. In the last couple of years features in the cloud have been identified with asteroid collisions which occurred just a few Myr ago implying that steady state models for the zodiacal cloud will have to be reconsidered. Many extrasolar systems also harbor massive dust belts and the structures of those dust belts have been linked to perturbations from unseen planets. This paper reviews the dominant physical processes affecting the evolution of dust grains and describes the techniques which have been developed to model their dynamics and identify the sources of dust structures in both the solar system and extrasolar systems.

Keywords. Celestial mechanics; methods: numerical; interplanetary medium; circumstellar matter

---

1. Dust in the Solar System

The inner solar system is populated with a large number of dust grains which lie in a disk around the Sun in which the Earth is embedded. This disk is known as the Zodiacal Cloud (ZC) and it can be seen with the naked eye: diffuse emission, known as the zodiacal light, is seen near the Sun just before sunrise or after sunset and comes from sunlight scattered by the dust grains. The ZC is even more readily detected in the infrared where it is the brightest object in the sky once outside the Earth's atmosphere (Low *et al.* 1984; Kelsall *et al.* 1998). Such wavelengths probe thermal emission from the dust grains which are heated by the Sun.

The dust has a very short lifetime and so must be constantly replenished. There are several possible sources for this replenishment. The dust could come from the break-up of large bodies in the solar system, such as from collisions between asteroids in the Asteroid Belt (AB), the sublimation or collisional disintegration of comets, or from the collisional destruction of more distant bodies such as those in the Kuiper Belt (KB). The dust could also come from outside the solar system, i.e., from the interstellar medium (ISM). All of these sources contribute to the ZC to some extent, but the contribution of the last two is thought to be low, because dust originating in the KB is scattered by Jupiter before it reaches the inner solar system (Moro-Martín & Malhotra 2002) and the Sun is currently in a tenuous region of the ISM called the local bubble (Grogan *et al.* 1996). The relative contribution of asteroids and comets to the ZC is still debated (e.g., Sykes *et al.* 1986; Kortenkamp & Dermott 1998), but at least 30% of the ZC dust has been linked to asteroids (Grogan *et al.* 2001), so the following sections explain the theory behind the structure of the ZC and how it is modeled using the AB as the main example. For a more detailed discussion the reader is referred to Dermott *et al.* (2001).

1.1. *Large Scale Structure*

The AB is a collisional cascade in which large asteroids are continually colliding and getting broken up into smaller fragments. This is the reason the size distribution of the detected members (Durda & Dermott 1997) is close to the theoretical value expected in such a cascade, $n(D) \propto D^{2-3q}$ where $q = 11/6$ (Dohnanyi 1969). While asteroids smaller than a few km have not been detected in the AB, the cascade is believed to extend down to μm-sized dust particles. It is those smallest particles that are detected in the infrared, because such observations are sensitive to the surface area of the cascade, not its mass which is in the largest asteroids. It is difficult to determine how much dust is created in this cascade, because the extrapolation of this size distribution down to small sizes depends on details of collisional processes we cannot be sure of (Durda *et al.* 1998).

For most of the cascade, gravity is the dominant force acting on its members. Small dust, however, is affected by its interaction with solar radiation (Burns *et al.* 1979; Gustafson 1994). The resulting force falls off $\propto r^{-2}$, so it is commonly defined by the ratio β, of its magnitude to that of gravity. Different size dust grains are affected to a different extent by radiation forces, and for large grains the approximation $\beta \propto 1/D$ is valid. There are two components to the force: a radial component, known as radiation pressure, which means that dust grains effectively see a smaller Sun so that their orbits are more elliptical than that of the parent asteroids and grains with $\beta > 0.5$ (corresponding to those < 1 μm in the solar system) are blown out on hyperbolic orbits as soon as they are created; and a tangential component, known as Poynting-Robertson (P-R) drag, which makes dust particle orbits spiral into the Sun, where the particles evaporate, at a rate $\dot{r}_{\mathrm{pr}} = -2\alpha/r$ for initially eccentric orbits, where $\alpha = 6.24 \times 10^{-4}\beta M_\star$ AU^2 yr^{-1}.

The large scale structure of the ZC is determined by the balance of collisions and P-R drag. The continuity equation expressing this balance can be solved analytically if you assume the dust created in the AB is of the same size, since the collisional lifetime of this dust is then just $t_{\mathrm{per}}/4\pi\tau_{\mathrm{eff}}$, where τ_{eff} is the effective optical depth of these grains and t_{per} is their orbital period (Wyatt 1999). The resulting dust distribution depends only on the factor $\eta_0 = 5000\tau_{\mathrm{eff}}(r_0)\sqrt{r_0/M_\star}/\beta$, which is determined by the rate at which the dust produced. In this model, if the AB was very dense, and the production rate was high ($\eta_0 \gg 1$), the dust would be destroyed in mutual collisions before making it into the inner regions of the solar system. However, the AB is relatively tenuous ($\eta_0 \ll 1$) implying that few collisions should occur and the dust distribution should have a surface density which is constant with distance from the Sun.

In reality dust is produced with a range of sizes, and it is clear that while the largest grains remain in the AB their whole lifetime and small grains make it to the Sun without suffering a collision, some grains only complete part of the journey and are then destroyed. While the continuity equation can be solved numerically to derive the resulting size distribution, simple arguments show that the size distribution expected in such a situation should peak at the size for which the collisional lifetime is equal to the P-R drag lifetime (Wyatt *et al.* 1999; Dermott *et al.* 2001). Since this is a few 100 μm for dust originating in the AB (particles for which collisional and P-R drag lifetimes are ~ 4 Myr, Wyatt *et al.* 1999; also Leinert & Grün 1990), this is in good agreement with the peak in the size distribution of dust accreted by the Earth found by LDEF (Love & Brownlee 1993).

1.2. *Dust Bands*

The small scale structure of the ZC is particularly telling about the sources of its dust grains. One such feature is the dust bands which were discovered by IRAS in 1984 (Low *et al.* 1984). IRAS produced scans through the ZC perpendicular to the ecliptic. As expected the brightness increases to a peak in the ecliptic, falling off on either side. However, there are also shoulders of emission at $\pm 10°$ and $\pm 2°$ ecliptic latitude. One method of isolating the dust band features is to use Fourier filtering to remove the low frequency background component. When applied to the database of IRAS scans, such analysis shows that the structure of the dust bands varies significantly throughout the year as the Earth moves around its orbit; it is also different when viewed in front of and behind the Earth as well as in different wavebands (Fig. 1a; Grogan *et al.* 1997).

The latitudes of the bands meant they were quickly identified with the Hirayama asteroid families (Dermott *et al.* 1984), because the distribution of proper inclinations (I_p) in the AB has two peaks, one at $\sim 10°$ from the Eos family and one at $\sim 2°$ from the Themis and Koronis families. The reason these families would be expected to produce dust bands is that asteroids on inclined orbits spend a disproportionate amount of time at large heights above the ecliptic plane. P-R drag has no effect on the orbital planes of dust produced by these families, so their distribution would be expected also to peak at large heights above the ecliptic at all distances from the Sun. The observing geometry means that when looking through this dust we would expect to see peaks at latitudes of $\pm I_p$.

However, that idealized model for the dust bands cannot properly account for their seasonal variation. This is because while the orbital planes of the dust grains are not affected by P-R drag, these are significantly affected by the secular perturbations of the planets. The secular perturbation equations can be solved in the presence of P-R drag to show that their effect is to make both the inclinations $y = I \exp i\Omega$ and eccentricities $z = e \exp i\tilde{\omega}$ of dust grains precess around circles centered on forced elements. However, the centers of these circles are found to be different for different particle sizes and must be determined by integrating over the particle's evolutionary history. This led Dermott *et al.* (1992) to invent the particle in a circle method of determining the orbital element distributions of different sized particles originating from different families. This method followed the evolution of waves of different sized particles started at some point in the past on circles in the z and y planes defined by the orbital elements of the family being studied. That evolution was followed up until the present epoch and circles fitted to their current distributions to find the forced and proper elements. Fig. 1b shows the outcome of several hundred runs showing the forced inclination of different sized dust from the Themis family (Kehoe *et al.* 2002). One outcome from these results was to show that while the particles are still in the AB their forced elements are essentially locked onto Jupiter, but passage through secular resonance at ~ 2 AU leads to a large dispersal in forced elements for different sized dust grains. This means that material contributing to the dust band peaks must be confined to > 2 AU.

The derived orbital element distributions were then converted into a 3D model of the spatial distribution of dust band material by choosing several hundred million orbits from these distributions and spreading the surface area around each orbit according to Kepler's laws. Observations through the dust bands were simulated by integrating the expected flux along a given line-of-sight making assumptions about the particles' emission properties and size distribution. The dust band components were compared with those observed using the same Fourier filtering methods (Fig. 1a) showing that by accounting for the full dynamical evolution of the dust grains the seasonal and wavelength variation of the bands can be accounted for (Grogan *et al.* 2001). Grogan *et al.* (2001) also found that the dust band shoulders are just the tip of the iceberg, and material in these bands contributes some 30% of the total ZC flux.

However, to fit the latitudes of the $\sim 10°$ dust bands the proper inclination of the Eos family had to be artificially reduced to $9.35°$. This has led to a revival of an interpretation

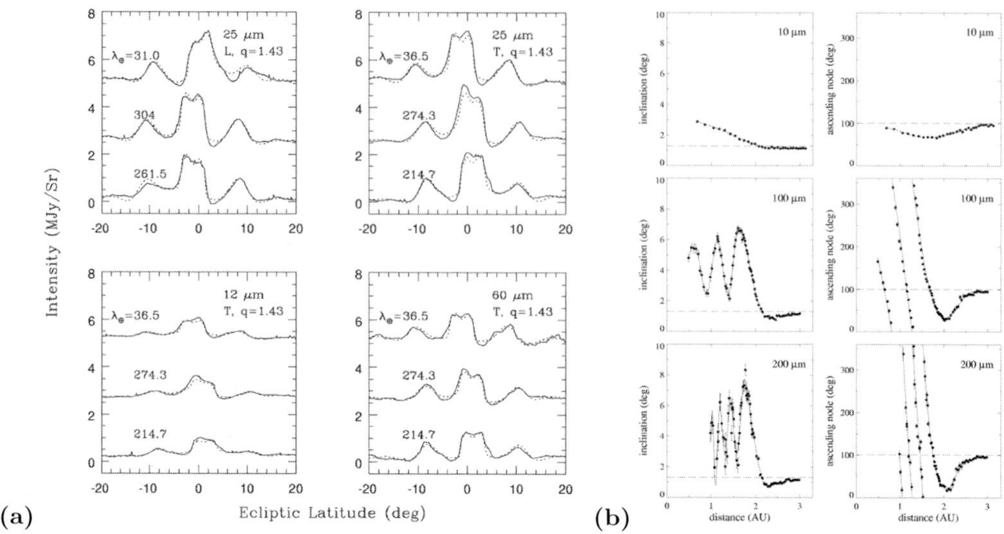

Figure 1. Dust band models. **(a)** Filtered IRAS dust band profiles (solid lines) at 12, 25 and 60 μm compared with model profiles (Grogan *et al.* 2001). All profiles were made at 90° solar elongation in the Leading (L) or Trailing (T) direction, and each figure shows the profiles at three different longitudes of the Earth. The models are made up of dust from the Eos, Themis and Koronis families with a size distribution defined by $q = 1.43$. The distribution of the orbital elements of dust from each of the families was determined using the particle in a circle method to derive the forced and proper inclinations and eccentricities on different sized dust particles at different distances from the Sun at the epoch of the IRAS observations. **(b)** shows **(left)** the forced inclination as a function of semimajor axis and **(right)** the ascending node as a function of semimajor axis for dust of different sizes from the Themis family (Kehoe *et al.* 2002); Jupiter's osculating inclination and ascending node are shown with the dashed line.

originally proposed by Sykes & Greenberg (1986), which is that the dust bands come from the recent break-up of individual asteroids (Dermott *et al.* 2002). Using new catalogues of asteroid proper elements Nesvorný *et al.* (2003) have been able to identify families (or indeed sub-families) which are at a suitable proper inclinations to form the bands: the 10° band is now believed to originate in the break-up resulting in the Veritas family and the 2° band is thought to originate in the Karin sub-family of the Koronis family (Nesvorný *et al.* 2002). Remarkably these are families believed to have formed just a few Myr ago. This new result will have a profound effect on our understanding of the ZC, since it implies that it is not in static equilibrium, rather that both its structure and brightness may vary significantly with time.

1.3. *Earth's Resonant Ring*

The other significant small scale structure in the ZC is the trailing/leading asymmetry. Both IRAS and COBE observations showed that the ZC is always brighter in the direction behind the Earth's motion than in front of it (Dermott *et al.* 1988; Reach *et al.* 1995). This asymmetry has been identified with dust trapped in resonance with the Earth (Dermott *et al.* 1994).

Resonance trapping is a well studied dynamical phenomenon (e.g., Jackson & Zook 1989; Marzari & Vanzani 1994; Beaugé & Ferraz-Mello 1994). As dust grains migrate inward due to P-R drag they encounter the Earth's exterior mean motion resonances. Resonant forces can halt that migration by providing angular momentum to the dust grains, something which also increases the eccentricities of their orbits. Trapping is a probabilistic phenomenon and usually occurs into the first order p+1:p resonances which

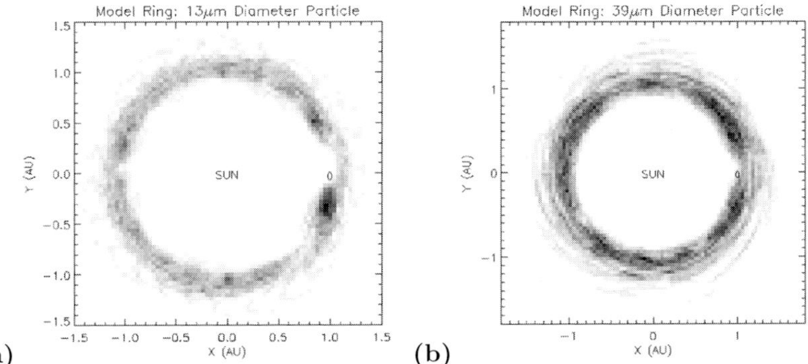

Figure 2. Face-on view of models for the Earth's resonant ring shown in the frame rotating with the Earth's mean motion (Dermott *et al.* 1994; Jayaraman & Dermott 2004). The ring models are made up of asteroidal material of diameter **(a)** 13 μm and **(b)** 39 μm which are trapped into exterior mean motion resonance with the Earth. For both sizes the Earth, whose motion is shown with an ellipse at $x = 1$ and $y = 0$, resides in a cavity in the ring. Ring model structure includes a dense clump of material just below the Earth in these plots which is most prominent for smaller particle sizes.

are the strongest. It also occurs more readily for larger particles which move slower past the resonances. Trapping lasts typically for some tens of thousands of years, long enough to cause a significant density enhancement along the Earth's orbit, and particles eventually leave resonance after a close encounter with the Earth (Marzari & Vanzani 1994). Trapping times are longer for larger particles as well as for resonances which are further from the planet (i.e., low p resonances).

Dermott *et al.* (1994) constructed a 3D model for the structure of the Earth's resonant ring based on the results of simulations which determined the trapping probabilities and trapping durations for different sized dust grains migrating in from the AB past the Earth. A quantitative analysis of how the orbital elements are affected by the resonance was used to determine the distribution of orbital elements of material in the ring and the resulting model is shown in Fig. 2. This model has two defining features: a clump which follows the Earth around its orbit, and a cavity at the location of the Earth. These are features which are more clearly defined for small rather than large particles.

To explain why there is a clump and cavity in this model, consider the geometry of a particle on a 4:3 resonant orbit. If this particle is on an elliptical orbit then even without considering its dynamical interaction with the Earth, the particle will follow a loopy pattern in the frame rotating with the Earth and spend most of its time at 3 longitudes relative to the Earth (e.g., Murray & Dermott 1999). Resonant forces from the Earth make the resonant argument of the particle, $\phi = (p+q)\lambda_{\rm r} - p\lambda_{\rm pl} - q\tilde{\omega}_{\rm r}$ (where $p+q = 4$ and $p = 3$ for the 4:3 resonance), librate about some value; this angle ϕ determines the orientation of the loopy pattern. For large bodies ϕ librates about 180°, which results in a pattern in which the loops (which correspond to locations where the particle is at pericenter) are as far from the Earth as possible and is symmetrical about the Earth-Sun line. For small dust, however, ϕ librates about an angle slightly higher than 180° so that angular momentum can be transferred to the particle to stop it migrating out of resonance by P-R drag. This means that the loopy pattern is asymmetric in the sense that the loop which is behind the Earth is closer to the Earth than the loop in front of it. The same is true for all resonances and explains why there is more material behind than in front of the Earth as well as why this asymmetry is stronger for smaller particles for which the drag force (and so asymmetry) is greater.

2. Dust in Extrasolar Systems

The Sun is not the only star to have dust around it. In fact 15% of nearby stars are known to be surrounded by dust disks (e.g., Backman & Paresce 1993). These were discovered by IRAS which showed that these stars exhibit more infrared emission than expected from the star alone (Aumann *et al.* 1984). The excess component emits at far-IR to sub-mm wavelengths and has a characteristic temperature of < 100 K. It comes from dust grains heated by the star and its temperature implies that the dust is more than 30 AU from the stars. The dust is shortlived and so must be continually replenished, a fact which implies there are also planetesimals orbiting the star (Wyatt & Dent 2002). The disks are thought to be the extrasolar equivalents to our own KB (Wyatt *et al.* 2003).

One of the main goals of subsequent studies has been to obtain images of these disks both to confirm the disk interpretation and to search for structure within them. To date some 10 or so of these disks have been imaged at wavelengths ranging from optical (Clampin *et al.* 2003), near-IR (Schneider *et al.* 1999), mid-IR (Telesco *et al.* 2000), sub-mm (Holland *et al.* 1998) and mm (Wilner *et al.* 2002). These images confirm that these are rings of dust > 30 AU from the star with holes at the center of similar size to the solar system. The images also show that the disks contain a variety of structures.

The interpretation of such structures is of fundamental importance to our understanding of the outcome of planetary formation in extrasolar systems. Not least because the identification of structure in the ZC with the solar system's planets shows that if there are planets in these disks then they would impose structure on the disks implying that the structures could be used to infer the presence of planets which would otherwise be undetectable. The challenge with modeling these structures is that there are essentially no constraints on the perturbing planetary systems meaning that detailed dynamical simulations are out of the question. Current modeling efforts focus on the ways in which different types of planetary perturbation affect the orbits of populations of planetesimals and dust. A vital prediction of any such model is how the resulting dynamical structures would be manifested in observations of the disks when parameters such as the mass and orbit of the perturbing planet are varied. The ultimate goal of a model is to be able to use observed structures to set constraints on the properties of unseen planets and to make predictions which can be tested with future observations.

2.1. *Offsets and Warps: Secular Perturbations*

Some of the ways in which planetary perturbations affect disk structure have been studied confirming not only that these perturbations are potentially observable, but that we may already have evidence of such perturbations in the current observations. The secular perturbations of a planet are a well studied example. If there is a planet on an eccentric orbit within a disk then its secular perturbations will impose a forced eccentricity on the orbits of all planetesimals near it. The effect on the planetesimal disk is to cause its center of symmetry to be offset from the star, a fact which makes one side of the disk hotter and therefore brighter than the other side (Wyatt *et al.* 1999). An asymmetry in the brightness of the lobes in the HR4796 disk has been detected (Telesco *et al.* 2000) which could be caused by a forced eccentricity as small as 0.02 (Wyatt *et al.* 1999). The planet's secular perturbations would also impose a forced inclination onto planetesimal orbits. Since this forced inclination defines the plane of symmetry of the disk, if this varies with distance, such as in a system with two or more planets on different orbital planes, then the disk will appear warped. The β Pic disk, which is seen edge-on, has been observed to be significantly warped (e.g., Heap *et al.* 2000), an observation which has been modeled as due to planets on inclined orbits (e.g., Augereau *et al.* 2001).

2.2. *Clumps: Collisions and Resonant Perturbations*

The most intriguing property of debris disks is that most of them are clumpy (Greaves *et al.* 1998; Holland *et al.* 2003). For example, a sub-mm image of the Vega disk, which has two clumps of asymmetric brightness embedded in it is shown in Fig. 3a (Holland *et al.* 1998). Taking the structure of the ZC as the basis for interpreting these clumpy structures, there are two possible models.

(a) Based on the origin of the dust bands, the first explanation is that the clumps are the products of recent individual collisions. While this remains a possibility, it is statistically unlikely. The problem is that most of these clumps have been detected in the sub-mm which means that the very fact that we are able to detect these clumps means that they contain so much mass that each clump would have to be the product of a collision between two planetesimals at least 1400 km in diameter (Wyatt & Dent 2002). The interaction between such large planetesimals (e.g., Kokubo & Ida 1995) means that there can only be ~ 100 such planetesimals in each disk, and the rate at which clumps would disperse into a ring after a collision means that they would last at most a few tens of thousands of years. Given the age of these disks of a few 100 Myr, we would have to be at a very special point in time to be witnessing the destruction of one of these planetesimals (Wyatt & Dent 2002). The same may not be true, however, for the clump recently detected in the mid-IR observations of the β Pic disk (Telesco *et al.* 2004), since such observations sample much lower dust masses and this system is only ~ 10 Myr.

(b) The other possibility is based on the Earth's resonant ring, i.e., that the clumps are resonant patterns caused by dust which has migrated into the resonances of an unseen planet. This interpretation has been pursued by several authors who have shown that structures similar to those observed in the disks around ϵ Eridani and Vega can be produced in this way if the perturbing planet is about the size of Jupiter or Saturn (Ozernoy *et al.* 2000; Wilner *et al.* 2002; Quillen & Thorndike 2002; Kuchner & Holman 2003). However, this interpretation has a serious problem which is that it ignores the effect of collisions. The collisional lifetime of particles in these disks is much shorter than their P-R drag lifetime. Using the terminology introduced in section 1.1, these are disks with $\eta_0 \gg 1$. This means that mass flows through the collisional cascade so fast that asteroids are ground into dust which is removed by radiation pressure before P-R drag has had a chance to act (Wyatt *et al.* 1999). This problem was acknowledged by Wilner *et al.* (2002) who proposed that the parent planetesimals of the dust are in resonance.

(c) This led Wyatt (2003) to propose a third possibility which is based on the structure of the KB. Many objects in the KB are trapped in resonance with Neptune (e.g., Jewitt 1999). This resonant configuration is thought to be the result of resonance sweeping of the primordial KB which occurred when Neptune migrated out to its current location from an orbit closer to the Sun due to angular momentum exchange with the residual planetesimal disk left over after planet formation (Hahn & Malhotra 1999). Such resonance sweeping may be a common feature of extrasolar planetary systems and the resulting resonant configuration in the KBs of extrasolar systems may be different to our own because of a different planet mass or migration rate.

Wyatt (2003) put together a model for the structure of planetesimal disks expected as a result of planet migration using techniques similar to those employed by Dermott *et al.* (1994). The model is based on the outcome of several hundred numerical simulations, each of which involving just 200 planetesimals, which were used to derive the probability of planetesimals becoming trapped into the 2:1, 5:3, 3:2 and 4:3 resonances as a result of the migration of different mass planets at different rates. The simulations were also used to derive the orbital element distributions of the trapped planetesimals. These results

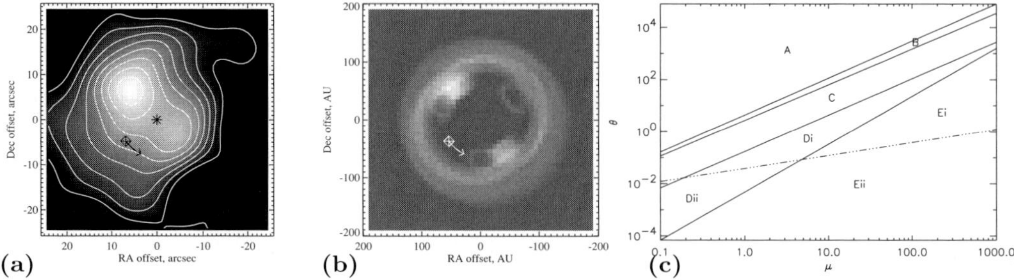

Figure 3. Planet migration model for the clumps in Vega's dust dust disk: **(a)** Sub-mm image of the Vega disk (Holland *et al.* 1998). The star contributes little to the emission at this wavelength and is at the center of the image, halfway between two clumps of unequal brightness. **(b)** Structure of the underlying planetesimal disk in the model of Wyatt (2003) in which the clumps are explained by the outward migration of a planet which traps planetesimals into its resonances. **(c)** Phase space determining which resonances can be filled for migrations defined by μ (planet mass) and θ (migration rate) (Wyatt 2003). The solid lines show 50% trapping probability for the 4:3, 3:2, 5:3 and 2:1 resonances (top to bottom); the dash-dot shows the line for which twice as many planetesimals are trapped into the 2:1(u) over the 2:1(l) resonance.

were combined to make a 3D model of the spatial distribution of material resulting from any given planet migration. Assumptions were then made about the dust emission properties and an observing geometry simulated to get an image of the disk which was then convolved with the appropriate beam size to allow the model observation to be compared directly with that observed.

When this process was applied to Vega's clumpy disk it was found that the observed structure can be explained very well by a model in which a Neptune mass planet migrated from 40-65 AU over the course of 56 Myr. Such a planet traps planetesimals primarily into its 3:2 and 2:1(u) resonances which means that the spatial distribution of those planetesimals has two clumps on either side of the star of unequal brightness; the unconvolved image of the model is shown in Fig. 3b.

The most important feature of this modeling method was that it allowed constraints to be set on the planet causing the structure. This is because the model explored the structures caused by all possible planet masses and migration rates. The four different resonances which were studied have their own clumpy signature and trapping probabilities for the different resonances were found to be determined by the two parameters $\mu = M_{\rm pl}/M_\star$ and $\theta = \dot{a}_{\rm pl}\sqrt{a/M_\star}$ in the manner described in Fig. 3c (Wyatt 2003). Thus it was possible to say that the migration causing the Vega disk structure, which must come from the superposition of the 2:1(u) and 3:2 resonances, must come from zone Di in Fig. 3c, although tighter constraints could be set on these parameters since the edges of the zones are soft and not hard as shown on this figure. So while the model proposed a Neptune mass planet taking 56 Myr to complete its migration, which is a migration strikingly similar to that proposed for our own Neptune, it is also possible that the planet was bigger and migrated faster and vice versa. The model also made testable predictions. such as that the structure will orbit the star with the planet at about 1°/year, a level which we should be able to detect in a few years.

There is also a prediction of the model which may recently have been confirmed. The Wyatt (2003) model assumed that the dust seen in the images has the same orbital distribution as the large planetesimals in the disk. This is not necessarily true because of radiation pressure. As mentioned in section 1.1, radiation pressure means that dust grains see a lower mass star than their parent planetesimals which means the dust has a slightly different orbital period. To assess the impact of this effect more simulations were performed (Wyatt, in prep.) to follow the orbital evolution of dust grains with different levels of radiation pressure resulting from the destruction of resonant planetesimals. This showed that the dust grains do stay in resonance, but at the expense of an increased width of libration of ϕ for smaller grains. Since ϕ defines the orientation of the clumpy pattern, this causes the clumps to be smeared out azimuthally. The very smallest grains are no longer in resonance and have an axisymmetric distribution.

To find out whether this has any impact on the Wyatt (2003) model the spectral energy distribution of the Vega disk emission was studied to determine that the sizes of the dust grains contributing to the 850 μm observations is between 300 μm and 20 cm in diameter (e.g., Wyatt & Dent 2002). Since the size at which dust falls out of resonance is 300 μm-2 mm depending on planet mass, the Wyatt (2003) model was correct in assuming that the dust distribution is similar to that of the planetesimals at 850 μm. However, this analysis also led to the prediction that mid- and far-IR observations of Vega will sample dust smaller than a few mm, and since that dust is no longer in resonance its distribution should be smooth. In the last couple of months the Spitzer satellite has made images of the Vega disk at 25 and 70 μm which show that the disk is indeed circularly symmetric at these wavelengths (Rieke *et al.* in prep.).

3. Conclusions

At least 30% of dust in the ZC comes from the break-up of asteroids. That dust migrates in toward the Sun by P-R drag, some of it getting broken up in collisions into smaller dust on the way. It also interacts with the planets causing structure and there is now evidence that the ZC may be sporadically replenished by the break-up of a few asteroids. Extrasolar systems also have dust belts, but these are more akin to dust coming from the KB than the AB in the solar system. The ones which have been detected so far are much more massive than the ZC, so their dynamics is different: both P-R drag and individual planetesimal collisions are much less important in shaping the disk. However, the disks do contain structures which may be evidence of unseen planets, both from the secular perturbations of those planets and clumpy resonant structure caused by the migration of planets.

Acknowledgements

I wish to acknowledge the Florida Solar System Dynamics Group (Stan Dermott *et al.*) whose work has contributed greatly to this paper.

References

Augereau, J.C., Nelson, R., Lagrange, A.M., Papaloizou, J. & Mouillet, D. 2001, *Astron. Astrophys.*, 370, 447

Aumann, H.H., Gillett, F.C., Beichman, C.A., De Jong, T., Houck, J.R., Low, F.J., Neugebauer, G., Walker, R.G. & Wesselius, P.R. 1984, *Astrophys. J.* 278, L23

Backman, D.E. & Paresce, F. 1993, in E.H. Levy & J. Lunine (eds.), *Protostars and Planets III*, (Tucson: Univ. Ariz. Press), p. 1253

Beaugé, C. & Ferraz-Mello, S. 1994, *Icarus*, 110, 239

Burns, J.A., Lamy, P.L. & Soter, S. 1979, *Icarus*, 40, 1

Clampin, M., *et al.* 2003, *Astron. J.*, 126, 385

Dermott, S.F., Nicholson, P D., Burns, J.A. & Houck, J.R. 1984, *Nature*, 312, 505

Dermott, S.F., Nicholson, P.D., Kim, Y., Wolven, B. & Tedesco, E.F. 1988, in A. Lawrance (ed.), *Comets to Cosmology*, (Berlin: Springer-Verlag), p. 3

Dermott, S.F., Gomes, R.S., Durda, D.D., Gustafson, B.Å.S., Jayaraman, S., Xu, Y.L. & Nicholson, P.D. 1992, in S. Ferraz-Mello (ed.), *Chaos, Resonance and Collective Dynamical Phenomena in the Solar System,* (Dordrecht: Kluwer Acad. Publ.), p. 333

Dermott, S.F., Jayaraman, S., Xu, Y.L., Gustafson, B.A.S. & Liou, J.-C. 1994, *Nature,* 369, 719

Dermott, S.F., Grogan, K., Durda, D.D., Jayaraman, S., Kehoe, T.J.J., Kortenkamp, S.J. & Wyatt, M.C. 2001, in E. Grun, B.Å.S. Gustafson, S.F.Dermott & H. Fechtig (eds.), *Interplanetary Dust,* (Heidelberg: Springer-Verlag), p. 569

Dermott, S.F., Kehoe, T.J.J., Durda, D.D., Grogan, K. & Nesvorný, D. 2002, in B. Warmbein (ed.), *Proceedings of Asteroids, Comets, Meteors (ACM 2002),* (Noordwijk: ESA Publications Division), p. 319

Dohnanyi, J.S. 1969, *J. Geophys. Res.* 74, 2531

Durda, D.D. & Dermott, S.F. 1997, *Icarus,* 130, 140

Durda, D.D., Greenberg, R. & Jedicke, R. 1998, *Icarus,* 135, 431

Greaves, J.S., *et al.* 1998, *Astrophys. J.,* 506, L133

Grogan, K., Dermott, S.F. & Gustafson, B.Å.S. 1996, *Astrophys. J.,* 472, 812

Grogan, K., Dermott, S.F., Jayaraman, S. & Xu, Y.L. 1997, *Planet. Space Sci.,* 45, 1657

Grogan, K., Dermott, S.F. & Durda, D.D. 2001, *Icarus,* 152, 251

Gustafson, B.Å.S. 1994, *Annual Rev. Earth Planet. Sci.,* 22, 553

Hahn, J.M. & Malhotra, R. 1999, *Astron. J.,* 117, 3041

Heap, S.R., *et al.* 2000, *Astrophys. J.,* 539, 435

Holland, W.S., *et al.* 1998, *Nature,* 392, 788

Holland, W.S., *et al.* 2003, *Astrophys. J.,* 582, 1141

Jackson, A.A. & Zook, H.A. 1989, *Icarus,* 97, 70

Jayaraman, S. & Dermott, S.F. 2004, *Icarus,* submitted

Jewitt, D.C. 1999, *Annual Rev. Earth Planet. Sci.,* 27, 287

Kehoe, T.J.J., Dermott, S.F. & Grogan, K. 2002, in S.F. Green *et al.* (eds.), *Dust in the Solar System and Other Planetary Systems,* (Amsterdam: Elsevier), p. 140

Kelsall, T., *et al.* 1998, *Astrophys. J.,* 508, 44

Kokubo, E. & Ida, S. 1995, *Icarus,* 114, 247

Kortenkamp, S.J. & Dermott, S.F. 1998, *Icarus,* 135, 469

Kuchner, M.J. & Holman, M.J. 2003, *Astrophys. J.,* 588, 1110

Leinert, C. & Grün, E. 1990, in R. Schween & E. Marsch (eds.), *Physics and Chemistry in Space — Space and Solar Physics, Vol. 20,* (Berlin: Springer), p. 207

Love, S.G. & Brownlee, D.E. 1993, *Science,* 262, 550

Low, F.J., *et al.* 1984, *Astrophys. J.,* 278, L19

Marzari, F. & Vanzani, V. 1994, *Planet. Space Sci.,* 42, 101

Moro-Martín, A. & Malhotra, R. 2002, *Astron. J.,* 124, 2305

Murray, C.D. & Dermott, S.F. 1999, *Solar System Dynamics* (Cambridge: Cambridge University Press)

Nesvorný, D., Bottke, W.F., Dones, L. & Levison, H.F. 2002, *Nature,* 417, 720

Nesvorný, D., Bottke, W.F., Levison, H.F. & Dones, L. 2003, *Astrophys. J.,* 591, 486

Ozernoy, L.M., Gorkavyi, N.N., Mather, J.C. & Taidakova, T.A. 2000, *Astrophys. J.,* 537, L147

Quillen, A.C. & Thorndike, S. 2002, *Astrophys. J.,* 578, L149

Reach, W.T., *et al.* 1995, *Nature,* 374, 521

Schneider, G., *et al.* 1999, *Astrophys. J.,* 513, L127

Sykes, M.V. & Greeberg, R. 1986, *Icarus,* 65, 51

Sykes, M.V., Lebofsky, L.A., Hunten, D.M. & Low, F.J. 1986, *Science,* 232, 1115

Telesco, C.M., *et al.* 2000, *Astrophys. J.,* 530, 329

Telesco, C.M., *et al.* 2004, *Nature,* submitted

Wilner, D.J., Holman, M.J., Kuchner, M.J. & Ho, P.T.P. 2002, *Astrophys. J.,* 569, L115

Wyatt, M.C. 1999, PhD thesis, University of Florida

Wyatt, M.C., *et al.* 1999, *Astrophys. J.,* 527, 918

Wyatt, M.C. & Dent, W.R.F. 2002, *Mon. Not. R. Astron. Soc.* 334, 589

Wyatt, M.C., Holland, W. S., Greaves, J. S. & Dent, W.R.F. 2003, *Earth, Moon, Planets,* 92, 423

Wyatt, M.C. 2003, *Astrophys. J.,* 598, 1321

Dynamics of Populations of Planetary Systems
Proceedings IAU Colloquium No. 197, 2005
Z. Knežević and A. Milani, eds.

© 2005 International Astronomical Union
DOI: 10.1017/S1743921304008890

Bolide meteor streams

Štefan Gajdoš and Vladimir Porubčan

Dept. of Astronomy, Physics of the Earth, and Meteorology
Faculty of Mathematics, Physics and Informatics
Comenius University Bratislava, Slovakia
email: gajdos@fmph.uniba.sk

Abstract. The current version of the IAU Meteor Data Centre catalogue of photographic orbits comprising the orbital and geophysical data of 4581 meteors is applied for a more representative search of the mean orbits of bolide meteor streams among photographic meteors. We have made the search based on a computerized stream search procedure utilizing the Southworth-Hawkins D-criterion. The results are compared with the previous similar analysis made by Porubčan & Gavajdová (1994).

Keywords. Meteors, meteoroids, bolide meteor streams.

1. Introduction

Porubčan & Gavajdová (1994) published a list of fireball meteoroid streams compiled from the IAU Meteor Data Centre catalogue of photographic orbits (Lindblad 1991) based on a set of 1028 fireballs ($M_{ph} \leqslant -3^m$). They found 30 fireball streams among known meteoroid streams of which 23 streams belong to regular streams listed in Cook (1973) or Kronk (1988) including their northern and southern branches and further they found additional 19 new fireball streams. As an subsidiary result of their analysis was that less than a half (46%) of the fireball population belongs to meteoroid streams population.

In our present analysis we have searched for potential fireball streams utilizing an updated version of the IAU MDC catalogue (Lindblad, Neslušan, Porubčan, *et al.* (2004)) of precise photographic meteor orbits summarizing a total of 4581 meteors. The catalogue contains a total of 1541 bolides having the ($M_{ph} \leqslant -3^m$), i.e. additional 513 bolides with respect to the previous search made by Porubčan & Gavajdová (1994). This means a substantial increase of data which can enable to define more precise mean orbits of the streams and detect additional members of some minor streams or associations.

Statistics of data available for analysis is summarized in Table 1, which lists the numbers of bolides observed in individual months of the year. As evident, the most numerous data are from August (Perseids) followed by data from autumn months, October, November and December. Table 1 shows an inhomogeneous distribution of observations throughout the year resulting from obervational campaigns carried out predominantly during the activity of major meteor showers. The last column of Table 1 lists the number of meteors without an estimate of brightness being available. The largest number of such observations is from January (Quadrantids), August (Perseids) and December (Geminids).

Of the 4581 photographic meteors listed in the current version of the IAU MDC catalogue, 1563 meteors are without an estimation of the brightness (M_{ph}). Therefore, only 1541 bolides with $M_{ph} \leqslant -3^m$ could be selected from the catalogue. Table 2 shows the statistics of various brightness groups of fireballs listed in the catalogue.

Table 1. The numbers of bolides recorded in individual months. The first column gives number of meteors brighter (or equal) -3^m, while the second one gives number of bolides, according commonly used brightness limit (-4^m). The third column contains meteors fainter than -3^m. The last column lists the number of meteors having no magnitude estimation - there are "no data" on their brightness available.

Months	$M_{ph} \leqslant -3^m$	$M_{ph} \leqslant -4^m$	$M_{ph} > -3^m$	no data
Jan	82	80	63	79
Feb	63	61	37	54
Mar	65	60	73	26
Apr	69	62	91	27
May	62	56	64	60
Jun	47	40	81	51
Jul	106	76	102	73
Aug	524	354	408	460
Sep	93	73	99	140
Oct	171	147	143	147
Nov	130	115	76	110
Dec	129	107	240	336
SUM	1541	1231	1477	1563

Table 2. The numbers of bolides recorded up to other limits of the absolute photographic magnitude M_{ph}.

M_{ph} limit	Number
$M_{ph} \leqslant -5^m$	997
$M_{ph} \leqslant -7^m$	693
$M_{ph} \leqslant -9^m$	369
$M_{ph} \leqslant -11^m$	128

2. Analysis

We have made a search for new members of the streams from the updated list of bolides in order to derive more precise mean orbits of the streams.

The analysis was based on a computerized stream search procedure utilizing the Southworth-Hawkins D-criterion (Southworth & Hawkins 1963) and an iteration procedure (Porubčan & Gavajdová 1994). For the analysis a limiting value of $D \leqslant 0.20$ was applied. This value is stricter than the value applied by Porubčan & Gavajdová (1994) in their search for fireball streams, where they have set $D \leqslant 0.25$ for the whole set of data.

The results obtained in our search are listed in Table 3 and 4, where the arithmetic mean values of the orbital elements together with the geocentric radiants and geocentric velocities are presented. The last two columns list the number of individual members of the streams obtained by Porubčan & Gavajdová (1994) and in this analysis, respectively. The largest increase of stream members is for the Perseids and Geminids as the best monitored meteor streams by almost all photographic observational campaigns.

The investigation has shown that some ecliptic low-inclination and dispersed streams (as Taurids) cannot be separated distinctly from the background and that some of their members appear also in smaller streams active either parallel or close to these streams. To separate them unequivocally the limiting D value has to be lower and carefully selected for each stream individually.

Table 3. Mean orbital parameters, geocentric velocity (in km/s) and radiants of the known meteoroid streams derived from bolides of $M_{ph} \leqslant -3^m$. In some cases, * denotes the showers, where a lower limit $D \leqslant 0.15$ had to be applied for the separation of the stream.

Shower	q	a	e	i	ω	Ω	π	V_g	R.A.	Dec	N_1	N_2
Quadrantids	0.977	3.122	0.686	72.3	170.7	284.8	95.5	41.5	231.2	48.6	6	5
Leonids-Ursids	0.843	2.309	0.634	4.0	232.0	345.7	217.8	14.4	150.4	22.7	4	5
Virginids	0.387	4.246	0.914	3.8	286.5	348.5	275.0	31.0	178.1	4.4	3	4
π Virginids	0.618	2.257	0.728	4.8	264.6	16.9	281.5	21.5	195.8	0.8	3	7
σ Leonids (S)	0.754	1.968	0.620	4.5	68.8	197.1	266.0	16.5	181.5	-10.7	5	8
Lyrids	0.925	66.509	0.986	79.8	212.8	32.7	245.5	47.0	272.8	33.7	4	9
α Scorpiids	0.324	2.546	0.872	9.7	117.3	237.5	354.8	31.0	247.1	-28.7	3	3
θ Ophiuch. (N)	0.566	2.367	0.761	4.5	271.4	92.2	3.6	22.9	272.4	-17.3	6	6
θ Ophiuch. (S)	0.657	1.949	0.660	2.7	83.0	285.4	8.4	18.9	282.4	-27.3	5	6
ι Aquarids (N)	0.238	3.599	0.929	5.8	306.6	130.0	76.6	34.4	327.5	-9.3	3	3
α Capricor. (N)	0.589	2.427	0.758	8.2	268.3	128.1	36.5	22.5	307.3	-8.1	15	27
α Capricor. (S)	0.519	2.149	0.762	4.0	97.2	327.0	64.3	23.2	334.0	-15.6	6	5
δ Aquarids (N)	0.085	3.002	0.971	20.0	329.3	136.9	106.2	40.7	342.7	-1.1	5	6
δ Aquarids (S)	0.095	3.206	0.969	26.5	147.3	313.1	100.5	40.6	344.4	-15.2	10	11
Perseids	0.951	27.396	0.965	112.9	151.1	138.7	289.9	59.2	46.0	57.9	193	*230
Perseids-2	0.911	1.995	0.542	108.3	135.6	137.2	272.8	53.1	45.0	57.3	6	4
κ Cygnids	0.983	3.535	0.721	32.7	200.0	141.6	341.7	21.6	279.5	50.8	13	21
Sept. Perseids	0.734	-44.510	1.016	140.6	242.4	166.9	49.4	65.6	47.4	39.0	3	3
Piscids (S)	0.544	3.314	0.844	2.2	90.0	12.2	102.3	24.4	12.0	2.4	7	6
Andromedids	0.611	2.550	0.760	5.5	264.2	189.9	94.2	21.3	4.0	10.2	5	6
Andromedids-2	0.760	2.409	0.679	14.2	245.1	207.9	93.0	18.1	3.9	31.8	3	3
Oct. Draconids	0.993	2.338	0.575	22.9	180.6	203.0	23.6	15.4	278.5	48.2	7	12
Taurids (N)	0.332	2.071	0.839	2.8	297.3	222.8	160.1	28.8	53.1	21.4	12	*11
Taurids (S)	0.364	2.232	0.836	5.0	112.8	43.6	156.4	28.2	53.7	14.7	19	*27
Orionids	0.578	27.478	0.978	163.7	81.2	27.8	109.0	66.7	94.5	15.6	12	17
Leonids	0.985	7.641	0.894	162.6	174.5	236.0	50.6	70.6	153.8	21.7	5	9
σ Hydrids	0.261	76.920	0.994	128.1	118.3	77.4	195.7	59.2	126.2	2.0	4	2
χ Orionids (N)	0.430	2.143	0.798	2.5	285.5	255.8	181.3	26.1	83.1	25.8	7	*4
χ Orionids (S)	0.529	2.215	0.761	5.0	94.0	75.3	169.4	23.4	77.6	16.6	6	*5
Geminids	0.138	1.375	0.899	24.1	324.6	261.7	226.3	34.7	113.2	32.3	19	38

Fig. 1 shows the distribution of the radiants (right ascension and declination) of all bolides and the mean stream radiants listed in Tables 3 and 4, respectively.

3. Discussion

Based on the updated version of the IAU MDC catalogue of photographic orbit, we have derived more precise orbits of fireball meteor streams (meteors with the absolute photographic magnitude $M_{ph} \leqslant -3^m$) by applying a limiting value of $D \leqslant 0.20$ to all the streams. The search has shown that each stream has to be analyzed individually and equal limiting value of D can be used only as a first approximation.

The statistics of meteors in the catalogue shows that the distribution of meteors all over the year is not homogeneous, while the Perseids are by far the richest stream in the majority of the photographic catalogues. This fact is due to a selection effect caused by observations carried out predominantly in the summer months. Eliminating the Perseids, the distribution of fireballs in the IAU MDC catalogue is more homogeneous. For the majority of the known meteor streams an increase in the shower numbers is recorded (see Table 3). On the contrary, in some cases a decrease in the number of members is

Table 4. Mean orbital parameters, geocentric velocity and radiants of minor bolide streams and their number N_1, as stated by Porubčan & Gavajdová (1994), and number N_2 resulting from our analysis.

Shower	q	a	e	i	ω	Ω	π	V_g	R.A.	Dec	N_1	N_2
α Cancrids	0.482	1.160	0.582	7.2	113.0	124.1	237.1	19.3	135.7	6.2	3	3
β Cancrids	0.800	2.115	0.614	4.1	59.9	136.1	196.0	14.8	120.0	10.9	3	5
Lynxids	0.966	1.925	0.493	6.9	202.4	343.4	185.9	8.9	124.7	49.3	3	4
March Cassiopeids	0.939	2.406	0.610	15.8	149.9	349.4	139.3	13.6	358.7	54.0	4	5
α Coma Berenicids	0.858	2.532	0.660	9.9	229.5	23.2	252.7	15.4	189.9	20.8	5	7
γ Corvids	0.832	2.116	0.610	4.5	55.3	208.3	263.7	14.4	182.2	-13.2	5	9
β Librids	0.508	2.623	0.808	10.1	276.4	49.8	326.2	25.9	233.8	-7.3	3	7
η Ursa Maiorids	0.939	2.308	0.592	12.0	212.5	32.2	244.7	13.1	192.7	33.7	3	9
μ Ursa Maiorids	0.977	2.154	0.546	8.8	198.7	48.2	247.0	10.5	186.5	32.0	3	8
λ Aquilids	0.943	1.840	0.485	4.1	215.8	150.7	6.5	9.2	290.4	-5.7	4	6
δ Piscids	0.257	1.881	0.864	4.3	307.3	173.6	121.0	30.6	9.1	7.1	3	5
Sept. ι Aquarids	0.908	2.361	0.613	1.6	39.4	1.3	40.8	11.2	328.4	-19.0	3	5
λ Cygnids	0.947	2.532	0.626	11.2	206.7	202.7	49.5	12.1	321.7	23.8	3	7
α Taurids	0.792	2.205	0.641	3.0	60.0	85.3	145.3	14.9	67.9	14.8	4	12
Dec. Aurigids	0.665	2.153	0.692	7.2	257.7	270.7	168.5	19.5	85.5	35.5	5	5
τ Geminids (N)	0.385	1.446	0.734	4.0	296.4	272.8	209.2	24.5	108.3	26.5	4	4
τ Geminids (S)	0.536	2.036	0.733	4.6	94.2	100.8	195.0	22.8	104.1	16.7	6	6
Dec. β Perseids	0.929	2.238	0.584	4.9	208.5	275.5	124.1	10.3	39.9	35.2	3	8

revealed, mainly due to more rigorous limit for D (e.g. narrow Perseids-2, σ Hydrids, both the χ Orionids).

A representative stream search desires to take into account all the factors influencing geometry of encounter of the potential meteoroid streams with the Earth, allowing for the size and form of the radiant, its daily motion and dispersion of geocentric velocities of the stream members.

It is apparent that these factors strongly influence the possibility of detection, especially for meteoroid streams moving in low-inclination, typical asteroidal orbits and this affects the posibility of detection of meteoroid streams of asteroidal origin.

The Earth approaching asteroids are moving in prograde orbits with low geocentric velocities, then due to geometry of the encounter with the Earth their potential streams will be dispersed and large radiant areas; they frequently split into several "sub-streams" and this greatly hampers their detection and recognition as single meteoroid streams (Kresák 1968). Therefore the above factors have to be taken into account, at least in a study or in the selection of meteoroid streams moving in low inclination short-period orbits.

In a search for meteoroid stream and their parent bodies besides the orbital similarity also the orbital evolution of both the parent body and of the members of the stream has to be taken into account.

According to the principles of meteoroid stream formation (Babadzhanov 1996), meteor showers can be produced also by parents which are, at present, on orbits at a distance from the Earth's orbit > 0.3 AU, but which crossed it in the past. Therefore a study also of the orbital evolution of the stream members is desirable.

For comets as parent bodies of meteoroid streams, resonant binding of cometary dust particles provides a new possibility in the field of meteoroid stream research. As recently revealed, mean-motion resonance with Jupiter is responsible, at least in the Leonids and the June Bootids, for dust trails evolution (e.g. Asher, Bailey & Emel'yanenko (1999), Asher & Emel'yanenko (2002)). Also, a detailed study of dust filaments formation resulted

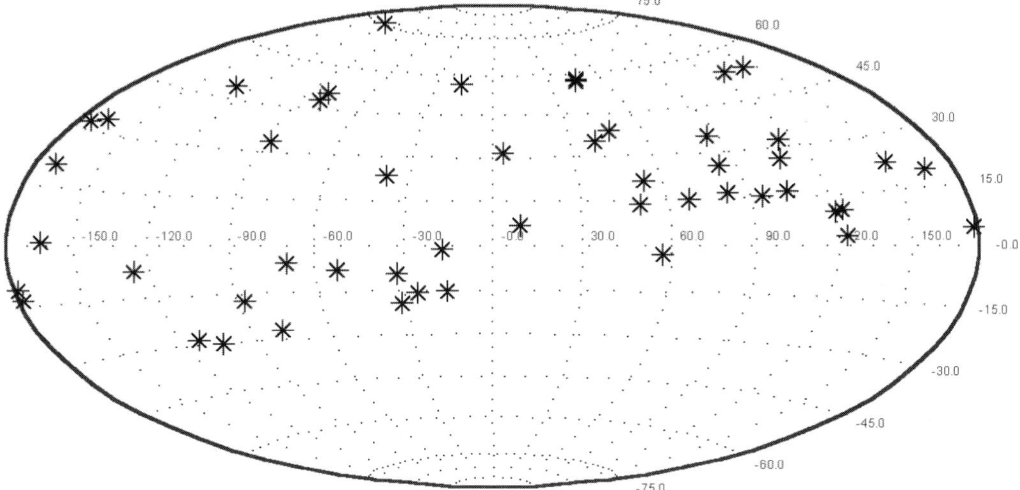

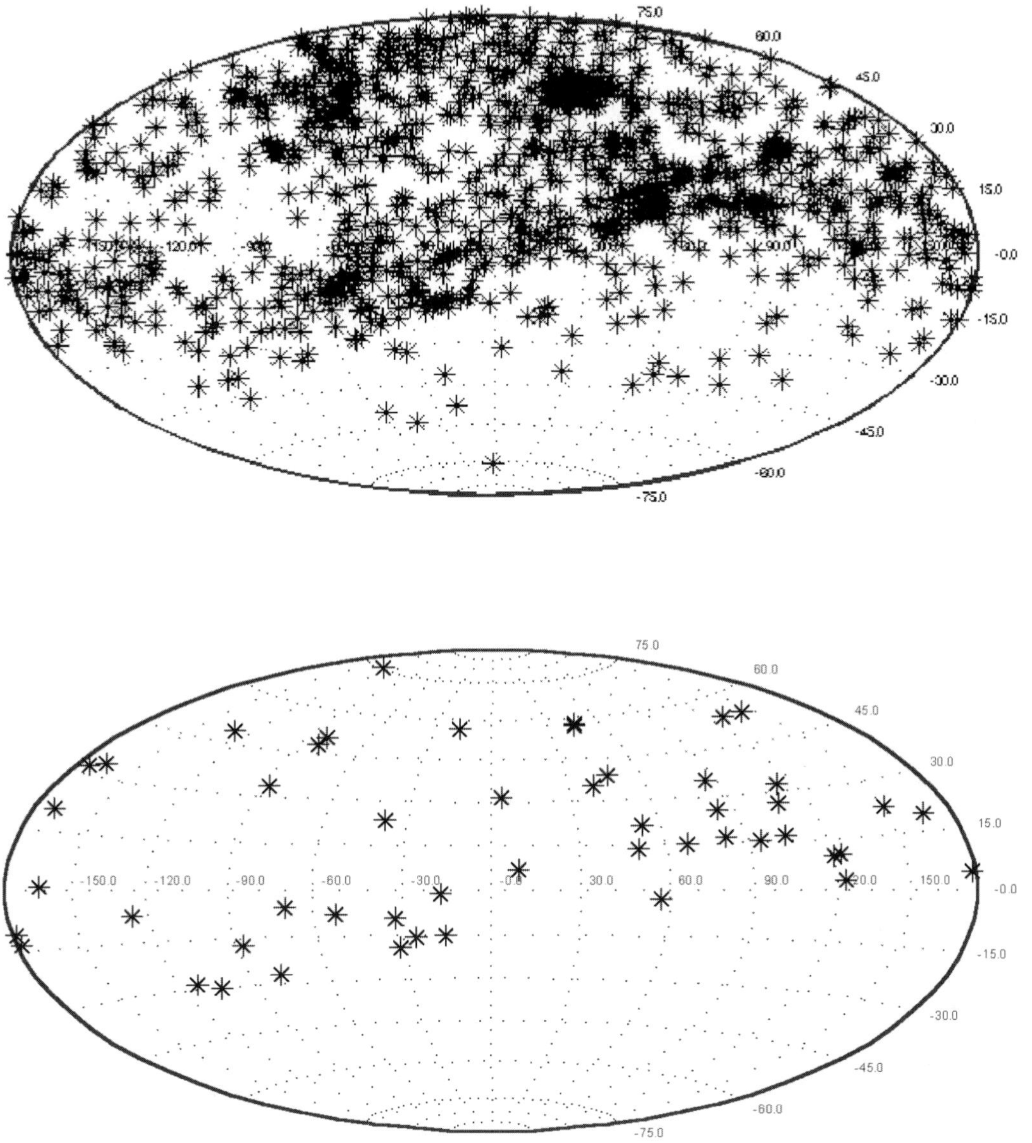

Figure 1.
Up - Distribution of radiants (α,δ) of all 1541 bolides separated from the IAU MDC catalogue. Down - Distribution of the mean radiants (α,δ) of the streams listed in Table 3 and 4.

in successful predictions of enhanced activity of the Leonids starting from the last years of the previous millennium (Asher 1999; Lyytinen 1999). At the same time, as during recent returns of Leonids rich in bolides, especially in 1998, many new orbits were obtained, a great increase in the number of Leonid orbits also in photographic catalogues is to be expected rather soon.

Acknowledgements

The authors acknowledge support of the research by the VEGA grants Nos. 01/0204/03 and 01/0206/03.

                                  Š. Gajdoš and V. Porubčan

References

Asher, D.J. 1999, *Mon. Not. R. Astron. Soc.* 307, 919

Asher, D.J., Bailey, M.E. & Emel'yanenko, V.V. 1999, *Mon. Not. R. Astron. Soc.* 304, L53

Asher, D.J. & Emel'yanenko, V.V. 2002, *Mon. Not. R. Astron. Soc.* 331, 126

Babadzhanov, P.B. 1996, *Sol. Sys. Res.* 30, 499

Cook, A.F. 1973, *Working List of Meteor Streams*, NASA SP 319, 183

Kresák, L. 1968, in: P.M. Millman and L. Kresák (eds.), *Physics and Dynamics of Meteors*, (Dordrecht: D. Reidel Publ. Co.), p. 391

Kronk, G.W. 1988, *Meteor Showers: A Descriptive Catalogue*, (Enslow Publishers, Hillside, USA), p. 291

Lindblad, B.A. 1991, in: A.C. Levasseur-Regourd and H.Hasegawa (eds.), *Origin and Evolution of Interplanetary Dust*, (Dordrecht: Kluwer Academic Publishers), p. 311

Lindblad, B.A., Neslušan, L., Porubčan, V., Svoreň, J. 2004, *Earth, Moon, Planets*, in press

Lyytinen, E. 1999, *Meta Res. Bull.* 8, 33

Porubčan, V. & Gavajdová, M. 1994, *Planet. Space Sci.* 42, 151

Southworth, R.B. & Hawkins, G.S. 1963, *Smithson. Contrib. Astrophys.* 7, 261

Dynamics of Populations of Planetary Systems
Proceedings IAU Colloquium No. 197, 2005
Z. Knežević and A. Milani, eds.

© 2005 International Astronomical Union
DOI: 10.1017/S1743921304008907

Migration of small bodies and dust to the terrestrial planets

Sergei I. Ipatov[1,2] and John C. Mather[3]

[1]Catholic University of America, USA
email: siipatov@hotmail.com
[2]Institute of Applied Mathematics, Moscow, Russia
[3]LASP, NASA/Goddard Space Flight Center, Greenbelt, USA
email: John.C.Mather@nasa.gov

Abstract. We integrated the orbital evolution of 30,000 Jupiter-family comets, 1300 resonant asteroids, and 7000 asteroidal, trans-Neptunian, and cometary dust particles. For initial orbital elements of bodies close to those of Comets 2P, 10P, 44P, and 113P, a few objects got Earth-crossing orbits with semi-major axes $a<2$ AU and moved in such orbits for more than 1 Myr (up to tens or even hundreds of Myrs). Three objects (from 2P and 10P runs) even got inner-Earth orbits (with aphelion distance $Q<0.983$ AU) and Aten orbits for Myrs. Our results show that the trans-Neptunian belt can provide a significant portion of near-Earth objects, or the number of trans-Neptunian objects migrating inside the solar system can be smaller than it was earlier considered, or most of 1-km former trans-Neptunian objects that had got near-Earth object orbits for millions of years disintegrated into mini-comets and dust during a smaller part of their dynamical lifetimes. The probability of a collision of an asteroidal or cometary particle during its lifetime with the Earth was maximum at diameter $d\sim100$ μm. At $d<10$ μm such probability for trans-Neptunian particles was less than that for asteroidal particles by less than an order of magnitude, so the fraction of trans-Neptunian particles with such diameter near Earth can be considerable.

Keywords. Comets, asteroids, Kuiper Belt

---

1. Introduction

Celestial bodies and dust particles migrate to near-Earth space from different regions of the solar system (from the main asteroid and Edgeworth-Kuiper belts, the Oort and Hills clouds, etc.). Some scientists (Farinella *et al.* 1993; Bottke *et al.* 2002) considered that most near-Earth objects (NEOs) are asteroidal fragments, others (Wetherill 1988) supposed that half of NEOs are former short-period comets. We studied the migration of small bodies and dust particles based on extensive integrations, considering more Jupiter-crossing objects than before. The interval of considered diameters of dust particles was wider than in previous publications. Our previous work on these problems can be found in (Ipatov 2001, Ipatov & Mather 2004a-b, Ipatov *et al.* 2004a).

2. Migration of Jupiter-family comets to the Earth

As migration of trans-Neptunian objects (TNOs) to Jupiter's orbit was investigated by several authors (e.g., Levison & Duncan, 1997), Ipatov & Mather (2003, 2004a-b) integrated the orbital evolution of 30,000 Jupiter-family comets (JFCs) under the gravitational influence of planets, and based on the results of these integrations they investigated the migration of TNOs to the near-Earth space. Also the orbital evolution of 1300 resonant main-belt asteroids was studied. We omitted the influence of Mercury (except for

Comet 2P/Encke) and Pluto; we used the Bulirsch-Stoer and symplectic methods (BUL-STO and RMVS3 codes) from the integration package of Levison and Duncan (1994). In the first series of runs (denoted as $n1$), we calculated the evolution of 3100 Jupiter-crossing objects (JCOs) moving in initial orbits close to those of 20 real comets with period $5<P_a<9$ yr, and in the second series of runs (denoted as $n2$) we considered 13,500 JCOs moving in initial orbits close to those of 10 real comets with period $5<P_a<15$ yr. In other series of runs, initial orbits were close to those of a single comet (2P, 9P, 10P, 22P, 28P, 39P, or 44P). We investigated the orbital evolution during the dynamical lifetimes of objects (at least until all the objects reached perihelion distance $q>6$ AU).

In our runs, planets were considered as material points, so literal collisions did not occur. However, based on the orbital elements sampled with a 500 yr step, we calculated the mean probability P of collisions during lifetime of a JCO. The results can differ considerably depending on the initial orbits of comets. The values of P for the Earth were about $(1\text{-}4)\cdot10^{-6}$ for Comets 9P, 22P, 28P, and 39P. They were $(6\text{-}10)\cdot10^{-6}$ for Comet 10P and exceeded 10^{-4} for Comet 2P. The ratio of the mean probability of a JCO with semi-major axis $a>1$ AU with a planet to the mass of the planet was greater for Mars by a factor of several than that for Earth and Venus. Using $P=4\cdot10^{-6}$ (this value is smaller than the mean value of P obtained in our runs) and assuming that the total mass of planetesimals that ever crossed Jupiter's orbit is about $100m_E$ (Ipatov 1987, 1993), where m_E is the mass of the Earth, we obtained that the total mass of water delivered from the feeding zone of the giant planets to the Earth could be about the mass of water in Earth's oceans. The ancient Venus and Mars could have had large oceans.

For series $n1$ with RMVS3, the probability of a collision with the Earth for one object with initial orbit close to that of Comet 44P was 88.3% of the total probability for 1200 objects from this series, and the total probability for 1198 objects was only 4%. The probabilities of collisions of one object with Earth and Venus could be greater than for thousands other objects combined.

A few JFCs got Earth-crossing orbits with aphelion distance $Q<4.2$ AU (and sometimes with $a<2$ AU) and moved in such orbits for more than 1 Myr (up to tens or even hundreds of Myrs). The probability of a collision of one of such objects, which move for millions of years inside Jupiter's orbit, with a terrestrial planet can be greater than analogous total probability for thousands other objects. With BULSTO, six and nine objects, respectively from 10P and 2P series, moved into Apollo orbits with $a<2$ AU for at least 0.5 Myr each, and five of them remained in such orbits for more than 5 Myr each. Only two objects in series $n2$ got Apollo orbits with $a<2$ AU during more than 1 Myr.

Three considered former JFCs even got inner-Earth object (IEO) orbits (with $Q<0.983$ AU) or Aten orbits for Myrs. Note that Ipatov (1995) obtained migration of JCOs into IEOs using the method of spheres to consider the gravitational influence of planets. Now lets return to our recent results of integration. One former JCO got Aten orbits ($a<1$ AU, $Q>0.983$ AU) during >3 Myr, but a probability of its collision with the Earth was greater than that for 10^4 other considered former JCOs. This object also moved during about 10 Myr (before its collision with Venus) in IEO orbits, and during this time interval a probability of its collision with Venus was even greater than the total probability of collisions of 10^4 considered JCOs with Earth.

After 40 Myr one considered object with initial orbit close to that of Comet 88P got $Q<3.5$ AU, and it moved in orbits with $a=2.60\text{-}2.61$ AU, $1.7<q<2.2$ AU, $3.1<Q<3.5$ AU, eccentricity $e=0.2\text{-}0.3$, and inclination $i=5\text{-}10°$ for 650 Myr. Another object (with initial orbit close to that of Comet 94P) moved in orbits with $a=1.95\text{-}2.1$ AU, $q>1.4$ AU, $Q<2.6$ AU, $e=0.2\text{-}0.3$, and $i=9\text{-}33°$ for 8 Myr (and it had $Q<3$ AU for 100 Myr). So JFCs can rarely get typical asteroid orbits and move in them for Myrs. In our opinion, it can be possible that Comet 133P (Elst-Pizarro) moving in a typical asteroidal orbit was earlier a JFC and it circularized its orbit also due to non-gravitational forces. JFCs got typical asteroidal orbits less often than NEO orbits.

Results of the orbital evolution of JCOs show that many Earth-crossing objects can move in highly eccentrical ($e>0.6$) orbits and, probably, most of 1-km objects in such orbits have not yet been discovered. The obtained results show that during the accumulation of the giant planets the total mass of icy bodies delivered to the Earth could be about the mass of water in Earth's oceans.

The results obtained by the Bulirsh-Stoer method (BULSTO code) with an integration step error $\epsilon_r \sim 10^{-9}$-10^{-8}, with 10^{-12}, and by a symplectic method (RMVS3 code) with an integration step $d_s \leqslant 10$ days were usually similar, and the difference at these three series was about the difference at small variation of ϵ_r or d_s. In the case of close encounters with the Sun, the values of the collision probability P_S with the Sun obtained by BULSTO and RMVS3 and at different ϵ_r or d_s could be different, but all other results were usually similar. For Comet 2P, Comet 96P, and the 3:1 resonance with Jupiter, close encounters with the Sun took place, and with a symplectic method we got greater mean probabilities of collisions with the Sun than with BULSTO. The fraction of asteroids migrated from the 3:1 resonance with Jupiter that collided with the Earth was greater by a factor of several than that for the 5:2 resonance.

3. Trans-Neptunian objects in near-Earth object orbits

Using our results of the orbital evolution of JCOs and the results of migration of TNOs obtained by Duncan *et al.* (1995) and considering the total of $5 \cdot 10^9$ 1-km TNOs with $30<a<50$ AU, Ipatov & Mather (2003, 2004a-b) estimated the number of 1-km former TNOs in NEO orbits. Results of our runs testify in favor of at least one of these conclusions: 1) the portion of 1-km former TNOs among NEOs can exceed several tens of percents, 2) the number of TNOs migrating inside the solar system could be smaller by a factor of several than it was earlier considered, 3) most of 1-km former TNOs that had got NEO orbits disintegrated into mini-comets and dust during a smaller part of their dynamical lifetimes if these lifetimes are not small.

Our runs showed that if one observes the former JFCs in NEO orbits, then most of them could have already moved in such orbits for millions (or at least hundreds of thousands) of years (if they didn't disintegrate). Some former comets that have moved in typical NEO orbits for millions or even hundreds of millions of years, and might have had multiple close encounters with the Sun, could have lost their typically dark surface material, thus brightening their low albedo and assuming the aspect typical of an asteroid (for most observed NEOs, the albedo is greater than that for comets). The number of extinct comets can exceed the number of active comets by several orders of magnitude if most of comets do not disintegrate into mini-comets and dust during their dynamical lifetimes.

4. Migration of dust particles to the terrestrial planets

Using the Bulirsh-Stoer method of integration, Ipatov *et al.* (2004a) integrated the orbital evolution of 7000 asteroidal, trans-Neptunian, and cometary dust particles, under the gravitational influence of planets, the Poynting-Robertson drag, radiation pressure, and solar wind drag. The values of the ratio β between the radiation pressure force and the gravitational force were varied from 0.0004 to 0.4. For silicates, such values correspond to particle diameters d between 1000 and 1 microns; for water ice, the diameters are greater by a factor of 3 than those for silicates. After the above publication, we have made more calculations for small values of β and more analysis of the runs. Considered asteroidal and trans-Neptunian (kuiperoidal) dust particles started with zero relative velocity from the numbered main-belt asteroids and known trans-Neptunian objects, respectively. We also studied particles started from Comet Encke. Several hundred of particles were usually considered at each β. The simulations continued until all of the particles either collided with the Sun or reached 2000 AU from the Sun. In our runs, planets were considered as material points, but using orbital elements obtained with a step dt of $\leqslant 20$ yr, we calculated the mean probability P of a collision of a particle with a planet during the lifetime of the particle. We also made similar runs without planets to investigate the role of planets in interplanetary dust migration.

The mean probabilities P of collisions of the asteroidal and cometary dust particles with Earth and Venus during the lifetimes of the particles were maximum at $\beta \sim 0.002$-0.004 (i.e., at $d \sim 100$-200 microns for silicate particles). At $\beta \geqslant 0.01$ the values of P and the mean times T spent by particles in orbits with $q < 1$ AU quickly decrease with an increase of β (usually P and T are proportional to $1/\beta$). For asteroidal dust particles at $\beta \sim 0.0004$-0.001 the values of P and T were smaller than those at $\beta \sim 0.002$-0.004, though the maximum times elapsed until collisions of particles with the Sun were greater for smaller β (greater times were needed for larger particles to migrate to the orbits of the terrestrial planets). The probability of a collision of a migrating dust particle with the Earth for $\beta \leqslant 0.01$ is greater by a factor > 200 than that for $\beta = 0.4$. Cratering records in lunar material and on the panels of the Long Duration Exposure Facility have shown that the mass distribution of dust particles encountering Earth peaks at $d = 200$ μm. The probability of collisions of cometary particles with the Earth is smaller than that for asteroidal particles at the same β, and this difference is greater for larger particles.

Almost all asteroidal particles with $d \leqslant 4$ μm collided with the Sun. At $\beta \geqslant 0.02$ and $\beta \leqslant 0.001$ some asteroidal particles migrated beyond Jupiters orbit. A few asteroidal particles can collide with the Sun after moving outside Jupiter's orbit for a long time.

At $\beta \geqslant 0.05$ (i.e., at $d < 10$ μm) the probability of a collision of a trans-Neptunian particle with the Earth and the mean time T spent in orbits with $q < 1$ AU were usually less than those for a typical asteroidal particle by less than an order of magnitude (only at $\beta = 0.05$ values of T differed by a factor of 20). The total mass of the Edgeworth-Kuiper belt is at least two orders of magnitude greater than that of the main asteroid belt. So the fraction of trans-Neptunian particles with $d < 10$ μm (runs with greater d have not yet been finished) among all particles near the orbit of the Earth can be considerable. At $\beta \sim 0.05$-0.4, the fraction P_S of trans-Neptunian particles collided with the Sun was 0.1-0.2, i.e., it was smaller by a factor of 5 than that of asteroidal particles for the same β.

Several plots of the distribution of migrating asteroidal particles in their orbital elements and the distribution of particles with their distance R from the Sun and their height h above the initial plane of the Earth's orbit were presented by Ipatov et $al.$ (2004a). For $\beta = 0.01$ the local maxima of the mean time t_a (the total time divided by the number N of particles) during which an asteroidal dust particle had a semi-major axis a in an interval of fixed width corresponding to the 6:7, 5:6, 3:4, and 2:3 resonances with the Earth are greater than the maximum at 2.4 AU. There are several other local maxima corresponding to the n:(n+1) resonances with Earth and Venus (e.g., the 7:8 and 4:5 resonances with Venus). The trapping of dust particles in the n:(n+1) resonances causes Earth's dust ring. The greater the β, the smaller the local maxima corresponding to these resonances. At $\beta \leqslant 0.1$ there are gaps with a a little smaller than the semi-major axes of Venus and Earth that correspond to the 1:1 resonance for each; the greater the β, the smaller the corresponding values of a. A small gap for Mars is seen only at $\beta \leqslant 0.01$. There are also gaps corresponding to the 3:1, 5:2, and 2:1 resonances with Jupiter.

Spatial density n_s of considered trans-Neptunian particles near ecliptic at distance from the Sun $R=1$ AU was greater than at $R>1$ AU. At $0.1 \leqslant \beta \leqslant 0.4$ and $2<R<45$ AU (at $\beta=0.05$ for $11<R<50$ AU), n_s varied with R by less than a factor of 4. For asteroidal and cometary particles, n_s quickly decrease with an increase of R, e.g., for $\beta=0.2$ at $R=5$ AU n_s was smaller than at $R=1$ AU by a factor of 70 and 50 for asteroidal and cometary particles, respectively.

In contrast to the asteroidal dust particles, for cometary dust particles the values of a mean time in a planet-crossing orbit did not differ much between Venus, Earth, and Mars. Collision probabilities with Earth were greater by a factor of 10-20 than those with Mars and greater for particles starting at perihelion than aphelion. For the same values of β, the probability of cometary dust particles colliding with a terrestrial planet was several times smaller than for asteroidal dust particles, mainly due to the greater eccentricities and inclinations of the cometary particles. This difference is greater for larger particles. In almost all cases, inclinations $i<50°$, and at $a>10$ AU the maximum i was less for larger asteroidal particles.

Based on the obtained distributions of migrating dust particles, Ipatov *et al.* (2004b) investigated how the solar spectrum is changed by scattering by the zodiacal cloud grains. As positions of particles were stored with a step of 10 yr, for each such stored position we calculated many ($>10^3$) different positions of a particle and the Earth considering that orbital elements do not vary during a period of revolution of the particle around the Sun. One of the obtained results is that the results of modeling are relatively insensitive to the scattering function considered. In some models, the scattering function depended on a scattering angle θ in such a way: $1/\theta$ for $\theta<c$, $1+(\theta-c)^2$ for $\theta>c$, where $c=2\pi/3$ radian. In other runs, we added the same dependence on elongation ϵ (considered westward from the Sun). In the third runs, the scattering function didn't depend on these angles at all.

The plots of the obtained spectrum are in general agreement with the observations made by Reynolds *et al.* (2004). They measured the profile of the scattered solar Mg Iλ5184 absorption line in the zodiacal light. Unlike results by Clarke *et al.* (1996), our modeled spectra do not exhibit strong asymmetry. As these authors, we obtained that minima in the plots of dependencies of the intensity of light on its wavelength near 5184 Angstrom are not so deep as those for the initial solar spectrum. The details of plots depend on diameters, inclinations, and a source of particles. For example, at inclination $i=0$ and $\epsilon=90°$, for kuiperoidal and cometary particles the shift of the plot to the blue and its minimum were greater than those for asteroidal particles. Our preliminary models and comparison with observational data indicate that for more precise observations it will be possible to distinguish well the sources of the dust and impose constrains on the particle size.

For asteroidal and cometary particles at $\beta=0.2$, about 65-80% of brightness is due to particles at distance from the Earth $r<1$ AU (90-95% for $r<1.5$ AU), and at most values of ϵ (not close to 180°) 75-90% of brightness is from particles with $R<1.5$ AU. For trans-Neptunian particles, these fractions were smaller. At $\beta=0.2$ for $\epsilon=270°$ about half of brightness was from particles with $R>4$ AU and $r>4$ AU; for $\epsilon=90°$ 30-40 % (depending on the considered scattering function) of brightness was from particles with such values of R and r. Velocities of dust particles relative to the Earth that mainly contributed to brightness were different for different ϵ. At $i=0$ they were between -25 and 25 km/s.

5. Conclusions

Our results show that some former JFCs got Earth-crossing orbits with aphelion distance $Q<4.2$ AU and even inner-Earth object orbits (with $Q<0.983$ AU), Aten orbits, or typical asteroidal orbits for Myrs. The probability of a collision of one former JFC moving inside Jupiter's orbit for millions of years with a terrestrial planet can be greater than the analogous total probability for thousands other JFCs. Results obtained by the Bulirsch-Stoer method were mainly similar (except for probabilities of close encounters with the Sun when they were high) to those obtained by a symplectic method. The total mass of icy bodies delivered to the Earth during the accumulation of the giant planets could be about the mass of water in Earth's oceans.

We concluded that the trans-Neptunian belt can provide a significant portion of Near-Earth objects, or the number of trans-Neptunian objects migrating inside solar system could be smaller than it was earlier considered, or most of 1-km former trans-Neptunian objects that had got near-Earth object orbits disintegrated into mini-comets and dust during a smaller part of their dynamical lifetimes if these lifetimes are not small.

The probability of a collision of an asteroidal or cometary particle with the Earth was maximum at diameter $d\sim100$ μm. At $d<10$ μm such probability for a trans-Neptunian particle was less than that for an asteroidal particle by less than an order of magnitude, and the fraction of trans-Neptunian particles with such diameter near Earth can be considerable. The peaks in the distribution of migrating asteroidal dust particles with semi-major axis corresponding to the n:(n+1) resonances with Earth and Venus and the gaps associated with the 1:1 resonances with these planets are more pronounced for larger particles.

Acknowledgements

This work was supported by INTAS (00-240) and NASA (NAG5-12265).

References

Bottke, W.F., Morbidelli, A., Jedicke, R., Petit, J.M., Levison, H.F., Michel, P. & Metcalfe, T.S. 2002, *Icarus* 156, 399

Clarke D., Matthews, S.A., Mundell, C.G. & Weir, A.S., 1996, *Astron. Astrophys.* 308, 273

Duncan, M.J., Levison, H.F. & & Budd, S.M. 1995, *Astron. J.* 110, 3073

Farinella, P., Gonczi, R., Froeschle, Ch. & Froeschle, C. 1993, *Icarus* 101, 174

Ipatov, S.I. 1987, *Earth, Moon, Planets* 39, 101

Ipatov, S.I. 1993, *Solar System Res.* 27, 65

Ipatov, S.I. 1995, *Solar System Res.* 29, 261

Ipatov, S.I. 2001, *Adv. Sp. Res.* 28, 1107

Ipatov S.I. & Mather J.C. 2003, *Earth, Moon, Planets* 92, 89

Ipatov, S.I. & Mather, J.C. in: E. Belbruno, D. Folta, & P. Gurfil (ed.) 2004a, *Astrodynamics, Space Missions, and Chaos*, Annals of the New York Academy of Sciences, vol. 1017, 46

Ipatov, S.I. & Mather, J.C. 2004b, *Adv. Sp. Res.* 33, 1524

Ipatov, S.I., Mather, J.C., & Taylor, P.A. in: E. Belbruno, D. Folta, & P. Gurfil (ed.) 2004a, *Astrodynamics, Space Missions, and Chaos*, Annals of the New York Academy of Sciences, vol. 1017, 66

Ipatov, S.I., Kutyrev, A.S., Madsen, G.J., Mather, J.C., Moseley, S.H. & Reynolds, R.J. 2004b, *Bull. Amer. Astron. Soc.*, vol. 36, n. 4, in press.

Levison, H.F. & Duncan, M.J. 1994, *Icarus* 108, 18

Levison, H.F.& Duncan, M.J. 1997, *Icarus* 127, 13

Reynolds, R.J., Madsen, G.J., & Moseley, S.H. 2004, *Astrophys. J.* in press

Wetherill, G.W. 1988, *Icarus* 76, 1

Dynamics of Populations of Planetary Systems
Proceedings IAU Colloquium No. 197, 2005
Z. Knežević and A. Milani, eds.

© 2005 International Astronomical Union
DOI: 10.1017/S1743921304008919

Forward-scattering of meteors: a digital way

Saša Nedeljković[1] and Vjera Miović[2]

University of Toronto, 60 St. George Street, Toronto, M5S 1A7, Canada
[1]email: sasa@physics.utoronto.ca
[2]email: miovic@physics.utoronto.ca

Abstract. The new cutting edge technologies have opened the possibility of using digital spectrometers to probe the sky. With the sampling rate of $\sim$400 Msamples/s and the capability to digitally process data in real-time it is possible to get wideband spectra covering the frequencies in the range 0-200 MHz with the resolution of a few kHz. This presentation will show how such a spectrometer can be applied to the forward-scattering method of meteor detection.

Keywords. Meteors, atmospheric effects, instrumentation: spectrographs, radio lines: general

1. Introduction

Advances in high speed digital electronics and data acquisition have enabled the development of the new radio frequency spectrometers with the capability of exploring exciting astrophysical questions. Modern analog/digital (A/D) converters are capable of transforming the analog signals of up to 1GHz into a digital form. At the same time the processing power of inexpensive computers is becoming sufficient for the real time data analysis.

A digital spectrometer can be described as a system composed of an antenna, an amplifier, AD converter and a computer. At the University of Toronto, we are currently building such an instrument, known as TREX (21 cm Reionization EXperiment) to both probe the Epoch of Reionization and to characterise the micrometeorite arrival statistics.

TREX is designed to work with frequencies up to 200MHz. To achieve 200MHz Nyquist frequency we need to sample data using 400MHz A/D converter. This exceeds the transfer rate of all commercially available A/D converters, which typically are only able to transfer a few MB/sec, leading to very inefficient data acquisition. An ongoing project to develop a faster A/D converter and acquisition system is under way at the Physics Electronics Resource Centre at the University of Toronto. 64-bit PCI-X bus will be used to transfer all data to the computer's memory. Such a design will permit streaming of all data to the computer's memory in real time. After the data is received in the main memory, fast Fourier transform (FFT) needs to be performed to obtain the power spectrum. FFT calculation is a computationally very intense task. N-point FFT uses $\frac{1}{2}N\log_2 N$ operations. Even for the fastest personal computers today, this is a challenging computational task. In the next section we will discuss a possible way of speeding up the FFT calculation by using Intel's SIMD (Single Instruction Multiple Data) technology.

The meteor detection via forward-scattering needs a properly chosen location for the receiver. The forward-scattering technique uses transmitters which are at a different location than the receiver. The transmitter-receiver system must not be in the line-of-sight. In section 3 we discuss how to find a proper place for the receiver.

Figure 1. TREX schematics

2. Digital Spectrometer

2.1. *Power Spectrum*

Obtaining the data for a single antenna system is quite simple: from the input signal the power spectrum must be calculated. This is the most time costly operation, so a carefully chosen algorithm is crucial. Two well known FFT packages are FFTW (Fastest Fourier Transform in the West) and Intel's IPP (Integrated Performace Primitives). These algorithms are optimised for working with the floating-point numbers. For the purpose of our digital spectrometer we will use integer instead of the floating-point arithmetics to additionally speed up the calculations. Intel's processor streaming SIMD extensions can be used to speed up the FFT algorithm by adapting it to the specific hardware such as Pentium 4.

2.2. *FFT code*

New Intel's microprocessors such as Pentium 4 and Itanium have implemented a set of instructions for calculations with multiple data in a single instruction - SIMD processing. An excellent reference for the interested reader is Intel C++ Compiler for Linux Systems User's Guide (2003). Primarily designed for speeding up multimedia applications, these instructions can be used for speeding up almost any programming task. Implementation is done trough extensions to previously defined/implemented instructions. From the programmer's point of view, the usage of SIMD can be implemented using the inlined assembly code, or even simpler, by using Intel's C++ API extension sets, better known as intrinsics. Intrinsics are the special coding extensions that allow using the syntax of C function calls and C variables instead of the hardware registers. Using these intrinsics frees programmers from having to program in the assembly language and to manage registers manually. The compiler optimization is automated so the instructions are scheduled to maximise the running speed.

SIMD extensions use eight 128-bit registers capable of processing the data elements in parallel. Since each of the registers can store more than one data element, the processor can operate with more data simultaneously.

An algorithm is written to use SIMD2 extensively with 16-bit data (short). Each 128-bit SIMD register can hold and simultaneously operate with eight 16-bit integer values. We are computing eight FFTs simultaneously by packing eight 16-bit data in one 128-bit SIMD type. Which means that our algorithm performs eight FFTs at the cost of one.

We have tested the FFT performance of this new algorithm and compared the results with the FFTW code working with 64-bit data and 32-bit data. The results are presented on Figure 2.

Our algorithm shows a faster execution time by a factor of two. The drop in speed at the array size of 2^{14} is due to the processor cache memory. FFTW is also affected by cache limitation but with the larger array sizes. We are initially using eight 16-bit data arrays compared to only one 32 or 64-bit data array. One might expect a greater difference in the execution speed knowing that 8 FFTs are being executed at the cost of one. However, FFTW also uses the SIMD technology to speed up the execution time. FFTW parallelises groups of instructions if possible. For example if there are two consecutive additions they can be done in parallel. Such an optimisation is limited by compiler/programmer ability to transfer the linear code into parallel. A good example of instructions which can not

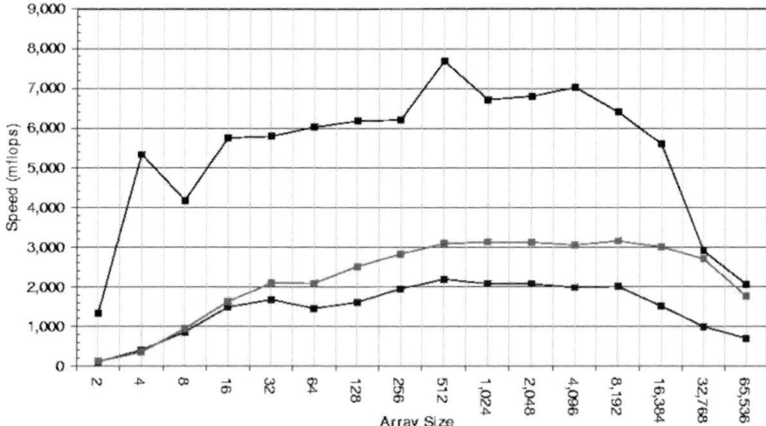

Figure 2. Speed comparison of SIMD2 optimised FFT code (top line) with the FFTW algorithm working with 32-bit floating point data (middle line) and 64-bit double floating point data (bottom line). The tests were performed on a 2.4GHz Intel Xeon machine.

be done in parallel is

$$x = 3 + 5 \tag{2.1}$$
$$y = x + 3, \tag{2.2}$$

because the input of the second equation depends on the result of the first one.

There are two limitations to our code. The first one is data overflow caused by 2 byte data size limitation. If the result of any operation is longer than 2 bytes, the overflow will be generated. There is no cure for this, since the overflow testing would slow down the algorithm by a factor of ~ 10. The second one is due to the arithmetic data shifting. Since the division in the set of short numbers is not hardware defined, one has to use data shifting to perform this operation. To divide a binary number by 2 it is enough to arithmetically shift the binary number one place to the right. The shifting also carries the signum, so that for example, $1/2 = 0$, but $-1/2 = -1$. It causes the error in the FFT result. However, despite the cost in the result precision, this algorithm can be used for the detection of individual spectral lines.

The comparison in precision is shown on Figure 3. At the two 16-bit sets in the Figure 3 it is easy to notice the "ghost" line at 155.9MHz. The origin of this "ghost line" is the data overflow. At frequencies at about 70-80MHz in the bottom spectra there are ghost lines originating from the division error discussed above. Despite these two types of errors the spectra calculated using our algorithm can be used for scientific data analysis. These errors are "constant" in time if the spectra do not change. However, a change in spectra might affect the errors. In the case of spectral lines caused by meteor forward-scattering, the error is negligible.

3. Where to observe from?

Man-made interference can be the source of contamination of the astronomical signals. Radio signals of astronomical origin are by far weaker than the signals used by terrestrial communication systems. The interference can result in "bad data" and even in completely wrong scientific interpretation.

FM stands for "Frequency Modulated". The FM radio band goes from 88MHz up to 108MHz. To resolve different stations the FM signals must be at least 200kHz apart.

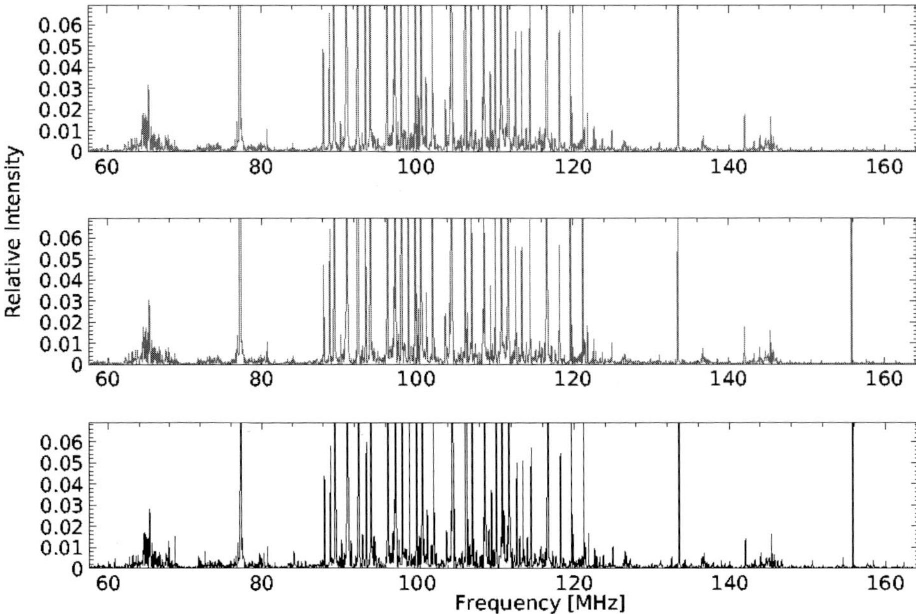

Figure 3. Comparison of spectra calculated from the same dataset using different data types: 32-bit floating point (top), 16-bit integer (middle) and 16-bit integer using SIMD2 instructions (bottom). For the two bottom spectra, FFT is done with 16-bit precision and the final power spectra are calculated as 32-bit integers. Data were obtained using a rectangular approximation of two arm conical spiral antenna at the Department of Physics, University of Toronto on July 28th, 2004.

Knowing that the total FM bandwidth is 20MHz, there can not be more than 100 stations visible in the whole FM band.

There is no reflection from the Earth's ionosphere in the FM band, which means that the receiver must be in the line-of-sight of the transmitter in order to receive data. However, one should keep in mind that the signals can be reflected from mountains and buildings. The maximum distance one can receive the FM signal from depends on the curvature of the Earth, terrain configuration and the position of the receiver and the transmitter.

The maximum distance l a receiver (at the sea level) can receive a line-of-sight signal from a transmitter at the altitude h (above the sea level) is given by the expression:

$$l = R \arctan \frac{\sqrt{h^2 + 2Rh}}{R}, \tag{3.1}$$

where $R = 6380$ km is the Earth's radius and h is the height of the transmitter. To avoid interference coming from the line-of-sight terrestrial receivers at height h one must be further than the distance l away.

For example, the famous CN Tower is often stated as the tallest free-standing structure in the world. Located in Toronto, Canada, completed in 1976, the tower stands at 553 meters (including antennas). Using the equation above, the maximum distance at which we can hear the signal emitted from the top of the CN Tower is about 85 km.

4. Meteor Detection

When a meteoroid enters the region of the ionosphere in the altitude range of 80–120 kilometers, it produces a trail of free electrons due to collisions with air molecules. The high kinetic energy involved in the process determines the transformation of a solid body into a plasma, which can scatter radio waves and emit light (meteor). Signals produced by FM radio and TV stations, being reflected from a meteor trail, might be used to extract all vital information describing the meteor.

When the transmitter (radio or TV station) is located far enough from the receiver, so that direct contact is impossible due to the curvature of the Earth, the signal reflected from a meteor trail can be detected. Such technique is known as forward meteor scattering. Even by knowing all characteristics of the transmitter and the receiver beams, when the receiver operates only on one fixed frequency, the orbital elements of a meteor can not be resolved. One solution commonly used is to have one transmitter and more than one receiver in order to collect enough information so that the meteor parameters can be determined by comparing the data from different receivers looking at the same trail.

A technically simpler system is composed of only one receiver: 2ACSA (2-Arm Conical Spiral Antenna) optimised for 80-250MHz coupled with a digital spectrometer. Such a system with a resolution of a few kHz can resolve different terrestrial VHF signals. All non-field-of-view VHF sources can be used as transmitters for forward scattering method. By knowing the frequency, directivity and the power of transmitters, and keeping in mind that there is a large number of suited sources, it is possible to extract all meteor parameters by having only one receiver. 2ACSA is an excellent choice for the receiver antenna because of its frequency independent characteristics.

The detected meteors could be coming from comet debris within the solar system, or they could be the micron size extrasolar dust particles. The proper method of detection could distinguish between solar and extra solar origin of meteoritic grains, once the orbital elements of the particles are known. The duration of the trail can give information about mass. Knowing the velocity of entry and height from the observations one can get the orbital elements of the meteor. This detection can yield valuable information for astronomy.

If electrons in plasma are offset from the equilibrium position, they execute simple harmonic motion. The equation of motion for the electrons gives a characteristic angular frequency of oscillation:

$$\omega_p = \sqrt{\frac{n_e e^2}{\epsilon_0 m_e}}, \tag{4.1}$$

where n_e and m_e are electron number density and mass, respectively. Since electrons move faster than ions, they mainly contribute to the plasma frequency, which is the collective characteristic of the plasma and is referred to as the Langmuir frequency.

There are two recognisable classes of meteors based on their Langmuir frequency: underdense and overdense. In underdense meteors, the electron density is small enough to allow the radio wave to propagate through the meteor and each individual electron scatters the incident electromagnetic wave. In the overdense meteor, the incident wave is totally reflected, and a simple model can be used in which the meteor trail is treated as a cylinder. In the overdense scatter, the Langmuir frequency is lower than that of the radio wave. In our observations, we can distinguish between overdense and underdense meteors by observing the time evolution of the signal.

Using the plasma physics approach, a simple model of an overdense meteor can lead to obtaining the approximate height of a meteor, as shown by Foschini (1999):

$$h \approx \frac{L}{2}\left(\frac{1}{\tan\phi} - \frac{L}{4R_\oplus}\right),$$

(4.2)

where $L = 2R_\oplus sin\alpha$, and the angles ϕ and α are the half angular distance between the receiver and the transmitter, from the meteor trail and the centre of the Earth, respectively.

5. Conclusion

The advances in digital electronics allow us to develop a digital spectrometer which can be used to probe the epoch of reionisation in the universe and to do meteor physics. We have shown how such a system can be used to characterise meteor trail orbits and heights. A new algorithm has been developed to perform FFT in order to obtain power spectra of forward-scatter detected meteors. It is implemented on the Intel Pentium 4 SIMD processors. Because it performs 8 FFT calculations at once and works only with integers, it is significantly faster than the conventional algorithms, without the loss of performance for our purposes. The spectrometer developed at the University of Toronto digitally processes data in real-time producing spectra in the range of frequencies covering up to 200MHz.

Acknowledgements

The authors would like to thank Ue-Li Pen, Barth Netterfield and Anna Lubchenko for their involvement in this project.

References

Intel C++ Compiler for Linux Systems User's Guide, 2003, Intel Corporation
Foschini, L. 1999, *Astron. Astrophys.* 341, 634

Dynamics of Populations of Planetary Systems
Proceedings IAU Colloquium No. 197, 2005
Z. Knežević and A. Milani, eds.

© 2005 International Astronomical Union
DOI: 10.1017/S1743921304008920

On the dynamical consequences of the Poynting-Robertson drag caused by solar wind

Galina O. Ryabova

Research Institute of Applied Mathematics and Mechanics of Tomsk State University, 634050
Tomsk, Russia
email: ryabova@niipmm.tsu.ru

Abstract. To correctly take into account the solar wind drag equations for secular variations in the meteoroid orbital semimajor axes and eccentricities were obtained. The equations are similar to those derived for the Poynting- Robertson radiation drag (Wyatt and Whipple 1950). An estimation for the Geminid meteoroid stream shows that the ratios of corpuscular to radiation drags are $0.4 - 0.7$ for particles of size $> 10\mu m$, i.e. the effect is much stronger than was assumed before.

Keywords. Meteoroids, solar wind, methods: analytical, celestial mechanics

1. Introduction

Mukai & Yamamoto (1982) derived an improved formula for the drag generated by motion of dust grains through solar wind. As a rule, the influence of the corpuscular analogue is taken into account by introducing the proper coefficient equal to the ratio between the corpuscular and radiation drags. The estimates of the different authors (Whipple 1967a; Whipple 1967b; Burns, Lamy & Soter 1979; Polyakhova 1979) give close values of the order of 0.2. The estimate of Misconi (1976) is slightly higher and equal to 0.26. Mukai & Yamamoto (1982) found all these estimates to be underestimated. Their results were applied to study the orbital evolution of dust particles in the region of exterior mean motion resonances by Marzari & Vanzani (1994). But they did not find application in numerous works devoted to study of mass separation in meteoroid streams (for details see Ryabova 1999) leading to overestimation of meteoroid stream's age.

Based on the results of Mukai & Yamamoto (1982), the author derived in this paper equations for secular variations in semimajor axes a and eccentricities e of the particles' orbits. The influence of the corpuscular analogue of the Poynting-Robertson effect was confirmed to be stronger than it was assumed before.

Below, unless otherwise specified, we shall use CGS units.

2. Solar wind force

The force acting on meteoroid and caused by its interaction with the solar wind is

$$\mathbf{F}_w = f_w \frac{\mathbf{U}}{U} = \left(\sum_{i=p,\alpha} \frac{m_i n_i}{2} U^2 A\, C_{D,i} \right) \times \tag{2.1}$$

$$\times\, [(W\cos\phi - \dot{r})\hat{\mathbf{r}} + (\mp W\sin\phi - r\dot\theta)\hat{\boldsymbol{\theta}}] \frac{1}{U}.$$

Here m_i and n_i are the mass and the number density of the ith component of the solar wind, respectively; $C_{D,i}$ is the drag coefficient caused by this component; indices "p" and

"α" denote proton and alpha particles; $\mathbf{U} = \mathbf{W} - \mathbf{V}$, where $\mathbf{W}$ is the average velocity of the solar wind flow and $\mathbf{V}$ is the meteoroid velocity; $\hat{\mathbf{r}}$ and $\hat{\boldsymbol{\theta}}$ are the radial and transverse unit vectors, respectively; $\dot{r}$ and $r\dot{\theta}$ are the radial and transverse components of the meteoroid velocity respectively; θ is the orbital longitude; A is the area of the cross-section of the meteoroid; and ϕ is the angle between the vectors $\overline{\mathbf{W}}$ and $\hat{\mathbf{r}}$.

Solar wind plasma consists of protons, electrons, alpha particles, and heavy ions. In equation (2.1), only protons and alpha particles are taken into account, because the contribution of electrons and heavy ions is negligible. Following Mukai & Yamamoto (1982), we take, for the heliocentric distance r (Eyni & Steinitz 1980),

$$n_p = 8.1 \left(\frac{r_0}{r}\right)^2 \left(\frac{400\,km/s}{W}\right)^2 \; cm^{-3},$$

where $r_0 = 1$ AU. The mean velocity of the solar wind is constant in the range $0.3 < r < 10$ AU and equal to 400 km/s, and $n_\alpha/n_p = 0.05$ (Svalgaard 1977). As a consequence $n_{p,0}m_p + n_{\alpha,0}m_\alpha = 1.2 n_{p,0}m_p = 1.626 \times 10^{-23}$, where $n_{p,0}$ and $n_{\alpha,0}$ are the values at r_0. The average direction from the Sun of the interplanetary plasma flow is deflected from the radial. But, since the terms with $\sin\phi$ and $\cos\phi$ in (2.1) do not contribute to $< da/dt >$, we will not discuss ϕ. In the limit, when the dispersion in the velocity of the solar wind particles is small in comparison with the flow velocity and the sputtering process is also omitted, we have $C_{D,i} \to 2$ and $f_w \to f_{w0}$. Mukai & Yamamoto (1982) calculated $\psi = f_w/f_{w0}$ in the range $0.1 < r < 5$ AU for the average parameters of the solar wind flow and for three grain materials : water ice, magnetite, and obsidian). The value ψ proved to be nearly constant for these materials, and the following values can be used for estimations: 1.6 (water ice), 1.4 (magnetite), and 1.1 (obsidian). Taking all this into account, we replace f_w in (2.1) by its approximate expression:

$$f_w \approx \psi f_{w0} = 1.626 \times 10^{-23}\psi \left(\frac{r_0}{r}\right)^2 U^2 A. \tag{2.2}$$

3. Secular variations in semimajor axis and eccentricity

Now we can find the time-averaged secular changes $< da/dt >_w$ and $< da/dt >_w$ by means of the procedure used by Burns, Lamy & Soter (1979, Sec. VII). We use the Gaussian equations for the derivatives of orbital elements in the form adopted by Burns, Lamy & Soter (1979):

$$\frac{da}{dt} = \frac{2a^2}{mH}\left[F_r e \sin v + F_t(1 + e\cos v)\right], \tag{3.1}$$

$$\frac{de}{dt} = \frac{H}{m\mu}\left[F_r \sin v + F_t \frac{e + 2\cos v + e\cos^2 v}{1 + e\cos v}\right],$$

here m is the meteoroid mass; $H = [\mu p]^{1/2}$ is the meteoroid angular momentum per unit mass; $p = a(1 - e^2)$; μ is the solar gravitational constant; v is the true anomaly; F_r and F_t are the radial and transverse components of the perturbing force $\mathbf{F}_w$, respectively. Let us average the variation in the semimajor axis over the revolution period, assuming that a and e are nearly constant over this time interval and that the angular momentum is conserved. Then

$$\left\langle \frac{da}{dt} \right\rangle = \frac{1}{PH} \oint \frac{da}{dt} r^2 \, dv, \tag{3.2}$$

where P is the orbital period.

Next, we replace f_w in (2.1) by its approximate expression (2.2). Obtaining the new expression for $\mathbf{F}_w$, we substitute the resulting components F_r and F_t into (3.1), and then (3.1) into (3.2). In F_r and F_t only terms depending on velocity are taken into account, because the circular integral of terms independent on velocity is equal to zero. Since $\dot{r} = \mu e \sin v / H$, and $r\dot{\theta} = \mu(1 + e \cos v)/H$, we have

$$\left\langle \frac{da}{dt} \right\rangle = \frac{1.626 \times 10^{-23}\psi r_0^2}{\pi a(1 - e^2)^{3/2}} \frac{A}{m} \oint U[e^2 \sin^2 v + (1 + e \cos v)^2]dv. \qquad (3.3)$$

In the right-hand side of (3.3) we see value U under the integral. This value changes along the meteoroid's orbit. It is obvious from (2.1) that

$$\mathbf{U} = \left[(W \cos \phi - \dot{r})\hat{\mathbf{r}} + (\mp W \sin \phi - r\dot{\theta})\hat{\boldsymbol{\theta}} \right].$$

For our estimation it is quite enough to use U averaged over the orbit. Let us use the same technique as above. Then

$$\langle U^2 \rangle = \frac{1}{PH} \int_0^\pi U^2 r^2 \, dv = \pi\mu a^{3/2}(1 - 2p^2)p^{-5/2} - 4Wae(\mu\pi)^{1/2}. \qquad (3.4)$$

The equation for the eccentricity is derived in a similar way. Then after taking integrals and transformations the final equations are

$$\left\langle \frac{da}{dt} \right\rangle_w = -3.65 \times 10^3 \psi\hat{U}\frac{A}{m}\frac{2 + 2e^2}{a(1 - e^2)^{3/2}} \, ,$$

$$\left\langle \frac{de}{dt} \right\rangle_w = -3.65 \times 10^3 \psi\hat{U}\frac{A}{m}\frac{2e}{a^2(1 - e^2)^{1/2}} \, . \qquad (3.5)$$

The values are obviously comparable with those, derived for the Poynting-Robertson effect, and should be taken into account in estimating the stream age.

4. Estimation for the Geminid meteoroid stream

For the orbit of asteroid Phaethon, the parent body of the Geminids (with $a = 1.323$ AU and $e = 0.891$), and a solar wind velocity W (directed radially for simplicity) of 400 km/s, U varies from 355 to 450 km/s, the averaged value over the orbit $\hat{U}$, calculated from (3.4), being 386 km/s. Because the uncertainties in n_p and W are much larger, we take $U/W \approx 1$ for the Geminids.

The ratios between corpuscular and radiation drags for water ice, magnetite, and obsidian are equal to about 0.55, 0.48, and 0.38 respectively, for the coefficient of Mie scattering $Q_{pr} = 1$. If we take into account, for example, that $Q_{pr} \approx 0.53$ for obsidian grains of 10 to 100 μm in size (Mukai & Yamamoto 1982), then the above ratio will increase to 0.71.

What consequences these results have for the Geminid stream evolution study? Estimations of the Geminid's age on secular variations in a and e due to the Poynting-Robertson effect only are in the range from 4700 to 6700 years (a review can be found in Ryabova 1999) for spherical particles. Correction for the solar wind decreases the estimated age to $3.1 - 4.8$ thousand of years, and correction for nonspherical shape (Ryabova 1999) gives the final value $2.1 - 3.2$ thousand of years.

Acknowledgements

The author thanks IAU and the organizers of the Colloquium for their support.

References

Burns, J.A., Lamy, Ph.L. & Soter, S. 1979, *Icarus* 40, 1

Eyni, M. & Steinitz, R. 1980, in: M.Dryer & E.Tandberg-Hanssen (eds.), *Solar and Interplanetary Dynamics*, Proc. IAU Symp. No. 91, p. 147

Marzari, F. & Vanzani, V. 1994, *Astron. Astrophys.* 283, 275

Misconi, N.Y. 1976, *Astron. Astrophys.* 51, 357

Mukai, T. & Yamamoto, T. 1982, *Astron. Astrophys.* 107, 97

Polyakhova, E.H. 1979, *Uchenye zapiski LGU* 400, 161 (in Russian)

Ryabova, G.O. 1999, *Solar System Research* 33, 258

Svalgaard, L. 1977, in: A.Bruzek & C.J. Durrant (eds.), *Illustrated glossary for solar and solar-terrestrial physics* (Astrophys. and Space Sci. Lib. Vol.69), p. 149

Whipple, F.L. 1967a, *Smithsonian Astrophys. Obs. Spec. Rept.* 239, 1

Whipple, F.L. 1967b, in: J. Weinberg (ed.), *The Zodiacal Light and the Interplanetary Medium* (NASA SP-150), p. 409

Wyatt, S.P. & Whipple, F.L. 1950, *Astrophys. J.* 111, 134

Dynamics of Populations of Planetary Systems
Proceedings IAU Colloquium No. 197, 2005
Z. Knežević and A. Milani, eds.

© 2005 International Astronomical Union
DOI: 10.1017/S1743921304008932

Invariant of motion for interstellar dust captured in the solar system

J. Klačka[1] and M. Kocifaj[2]

[1]Department of Astronomy, Physics of the Earth, and Meteorology, Faculty of Mathematics, Physics and Informatics, Comenius University, 842 48 Bratislava, Slovak Republic, email: klacka@fmph.uniba.sk

[2]Astronomical Institute, Department of Interplanetary Matter, Slovak Academy of Sciences, Dúbravská cesta 9, 845 04 Bratislava, Slovak Republic, e-mail: kocifaj@astro.savba.sk

Abstract. Interstellar dust grains approaching the Sun are influenced mainly by solar gravity, solar electromagnetic radiation and Lorentz force due to the existence of interplanetary magnetic field. These interactions together with the effect of the solar wind on dust grain and the effect of solar cycle are taken into account when modelling behaviour of interstellar dust in the vicinity of the Sun. As a consequence, nonspherical dust grains can be captured and survive in the solar system – they can orbit the Sun in sufficiently large distances from the Sun not to be thermally destroyed. On the other hand, captured spherical dust grains are practically all destroyed. Detailed numerical simulations showed an interesting behaviour of a quantity of the dimension of length cubed divided by time squared. The quantity behaves practically as an invariant of motion: it is a constant during the process when surviving captured interstellar grain is orbiting the Sun. The constancy is fulfilled with an accuracy better than 1%.

Keywords. Dust grain, electromagnetic radiation, Celestial Mechanics

---

1. Introduction

Particles that enter the Solar System from interstellar space have been identified in the data from their high impact speed and from their impact directions (Grün *et al.* 1994). The β Pictoris system is assumed to be one of the potential sources of interstellar meteoroids. Baggaley (2000) found that the most significant stream of interstellar meteor particles arrives from a direction that is compatible with being released about 10^6 years ago from β Pictoris at an escape speed of 20 $km.s^{-1}$. Recently, the dust particles within the Solar System were systematically monitored by the Ulysses and Galileo spacecrafts. The typical speed of the detected interstellar dust was found about to be 26 $km\,s^{-1}$. Frisch *et al.* (1999) reported on the flow of interstellar grains arriving from the upstream direction with the heliocentric ecliptic longitude $259°$ and latitude of $+8°$.

The dynamics of dust particles in the Solar System is dominated by the solar gravity, solar electromagnetic radiation and by Lorentz force due to interaction of charged dust grains with the interplanetary magnetic field (Grün and Landgraf 2001). One of the most recent papers (Jackson 2001) deals with the capture of interstellar dust due to the Poynting-Robertson (P-R) effect. Interstellar particles of the spherical shape vanish in the vicinity of the Sun (they either hit the Sun, or evaporate in the sublimation region close the Sun) (Gulyaev and Shcheglov 1999, Jackson 2001, Kocifaj and Klačka 2003). Morever, from observations of the extinction and polarization by the interstellar medium, it is clear that interstellar particles themselves are not spherical but rather have an elongated structure (Wurm and Schnaiter 2002). Since the P-R effect (Robertson 1937, Klačka 2004) represents only a very special form of equation of motion, which is

415

mainly applicable to particles of spherical shape (see Eq. (120) or Eq. (122) in Klačka 1992; compare Eqs. (40) and (48) in Klačka 2004), we use more general equation of motion. Once the non-sphericity of interstellar particles is established, one can see that their motion considerably differs from the P-R effect (Kocifaj and Klačka 2003, 2004). Trajectories have to be calculated in detail and the final form of the orbit depends on the shape of the particle, on its size and optical properties.

In this paper the motion of interstellar dust particles (ISDPs) in the Solar System is investigated. Gravitational force of the Sun, solar electromagnetic and corpuscular radiation and interplanetary magnetic field are considered. The effect of solar electromagnetic radiation plays an important role in the sense that non-spherical ISDPs can be captured (and survive) much more effectively than spherical particles. It turns out that particles of effective radii $\approx 0.4~\mu m$, moving initially near the solar equatorial plane and with impact parameter $400~R_S \lesssim b \lesssim 500~R_S$ (solar radii) exhibit a high probability of capture and survival in the Solar System (Kocifaj and Klačka 2004). Only a very small number of spherical particles can be captured. An invariant of motion is presented for the motion of non-spherical interstellar grains.

2. Model and equation of motion

All our numerical simulations are based on equation of motion written in the form (see Klačka and Kocifaj 2001, for more details)

$$\frac{d\vec{v}}{dt} = -\frac{4\,\pi^2}{r^2}\,\vec{e}_R + \frac{q}{m}\,(\vec{v} - \vec{v}_{sw}) \times \vec{B} +$$

$$\beta\,\frac{4\,\pi^2}{r^2} \sum_{j=1}^{3} \frac{Q'_{prj}}{Q'_{pr1}} \left\{ \left(1 - 2\,\frac{\vec{v}\cdot\vec{n}_1}{c} + \frac{\vec{v}\cdot\vec{n}_j}{c} \right)\,\vec{n}_j - \frac{\vec{v}}{c} \right\},$$

$$\vec{n}_1 \equiv \vec{e}_R \equiv \vec{r}/|\vec{r}|; \quad \beta = 7.6 \times 10^{-4}\,Q'_{pr1}\,\frac{A'\,[m^2]}{m\,[kg]}, \qquad (2.1)$$

where $\vec{v}$ (measured in units $[AU\,year^{-1}]$) is the dust grain's heliocentric velocity, $\vec{r}$ (measured in AU) is the position vector of the particle with respect to the Sun, A' is the geometrical cross section of a sphere of volume identical to the volume of the particle ($A' = \pi a'^2_{eff}$, where a'_{eff} is the "effective radius"), m is the mass of the particle, Q'_{pr1}, Q'_{pr2}, Q'_{pr3} are components of the radiation pressure efficiency factor vector; see Eq. (27) in Klačka and Kocifaj (2001). As for the Lorentz acceleration, the charge of the grain is $q = 4\pi\varepsilon_0 U a'_{eff}$; ε_0 is the permitivity of vacuum, U is the surface potential given by Kimura and Mann (1998), radially expanding solar wind is given by

$$\vec{v}_{sw} = v_{sw}\hat{\vec{e}}_R, \qquad (2.2)$$

magnetic field is considered in the form (Parker 1958, Grün *et al.* 1994)

$$\vec{B} = B_R\hat{\vec{e}}_R + B_T\hat{\vec{e}}_T,$$
$$B_R = B_{R0}\,(r_0/r)^2 \cos(\pi * t[years]/11 + \varphi_0),$$
$$B_T = B_{T0}\,(r_0/r) \cos\vartheta \cos(\pi * t[years]/11 + \varphi_0),$$
$$B_{R0} = B_{T0} = 3nT,$$
$$r_0 = 1AU, \quad \hat{\vec{e}}_T = \hat{\vec{\omega}} \times \hat{\vec{e}}_R, \qquad (2.3)$$

where $\hat{\omega}$ defines magnetic axis of the Sun, ϑ is the altitude from the solar equatorial plane; different orientations of magnetic field for northern and southern hemispheres are also considered.

We simulated the continuous flow of interstellar dust by the particle stream with time-independent volume concentration and random distribution of tilts of particle rotation axes (uniform in solid angles) at large distances from the Sun. Particles enter the Solar System from interstellar space from their upstream direction (Grün *et al.* 1994). We initially consider the particle to approach the Sun from infinity with velocity $v_\infty = 26$ km/s (Landgraf *et al.* 1999, Jackson 2001) with an impact parameter b. We assume spatial dependency of dust surface potential according to Kimura and Mann (1998) with value +3V in the central part of the Solar System. The slow solar wind component of $400 \ km.s^{-1}$ is taken into account. The phase angle $\varphi_0 = 0$ of magnetic field is assumed.

As for the model of the cosmic dust particle we have used particle U2015B10 (Clanton *et al.* 1984; Kocifaj *et al.* 1999). The particle shape appears to be representative of cosmic dust, since its aspect aspect ratio (equal to 1.4) coincides with the results obtained by means of the mid-infrared spectropolarimetry (Hildebrand and Dragovan 1995): the interstellar grains were identified to be oblate and the best fit of the aspect ratio lies between 4/3 and 3/2. The characteristics of radiation scattered by the cosmic dust particle were calculated using the discrete dipole approximation (DDA; Draine 1988). The advantage of the DDA method is its easy application to various geometries and material configurations. Our calculations are limited to magnesium-rich silicate particles, as these could be representative for interstellar dust grains (Dorschner *et al.* 1995). We are interested in particles of the radius 0.4 μm moving from the infinity in plane $\alpha = 90°$, where α is the particle position angle within dust stream (Kocifaj and Klačka 2004). The slope of particle rotation axis is fixed during simulation. Impact parameters vary between 150 and 500 solar radii. Such input parameters corespond to particles which survive in the Solar System with high probability: it was shown in Kocifaj and Klačka (2004) that charged grains can be captured only in a very narrow belt near ecliptic plane and surviving charged grains (grains are not sublimated, i.e. their perihelion distances from the Sun are large enough) exist mainly for impact parameters $b > 150$ solar radii $\equiv$ 0.7 AU.

3. Invariant of motion

In spite of the fact that real motion of ISDPs in the Solar System is very complicated, we have succeeded in finding a quantity which can be considered to be a constant of motion for the case when the ISDPs revolve around the Sun. The motivation comes from Kepler's third law which holds for two-body problem: $(a[AU])^3/(T[years])^2 = 1 - \beta$, where a is semi-major axis, T is period of revolution around the Sun and β is a nondimensional quantity and is often called the ratio of radiation pressure force (it's radial component independent of velocity) to the gravitational force. However, our physical situation is much more complicated and it is evident that the semi-major axis is not conserved during the revolution. Fig. 1 depicts two possibilities which are of the same dimension as the third Kepler's law: $< (r[AU])^3 > /(T[years])^2$ and $< r[AU] >^3 /(T[years])^2$, where r is the actual distance of the ISDP from the center of the Sun and the symbol $< x >$ denotes time averaging of the quantity x over the period T. One can see from Fig. 1 that $< r[AU] >^3 /(T[years])^2$ exhibits a large dispersion (from 0.15 up to 0.36) during the motion of the ISDP. However, surprisingly, $< (r[AU])^3 > /(T[years])^2$ is approximately constant (the values vary from 0.64 up to 0.69), i. e. it behaves as an

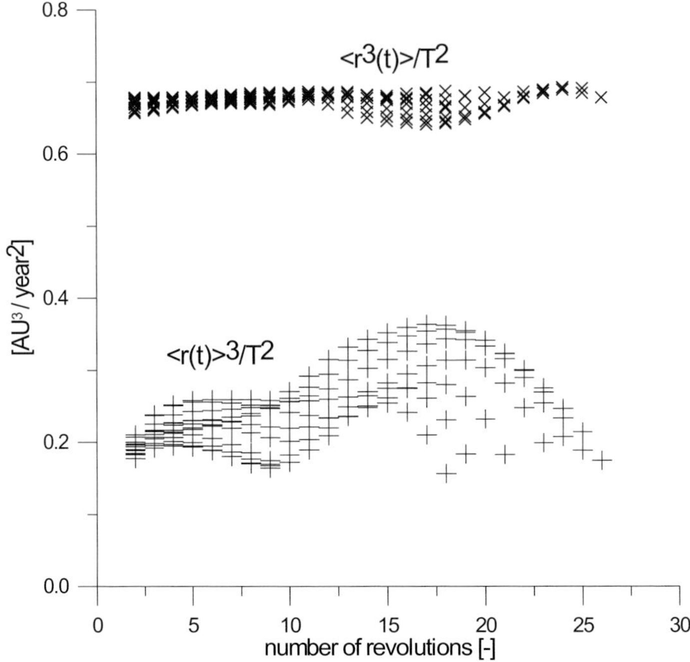

Figure 1. Dependency of quantities $\langle r^3(t)\rangle/T^2$ and $\langle r(t)\rangle^3/T^2$ on number of revolutions around the Sun. The distribution contains averaged data over each revolution for particles moving in region $b \in (300, 500)$ solar radii. Model characteristics: particle radius: $a'_{eff} = 0.4\ \mu m$, incident slope of particle rotation axis with respect to the orbital plane: $\theta = 40°$, $\phi = 0°$.

approximate invariant. Indeed, the upper part of Fig. 2 confirms this statement, and, moreover, $< (r[AU])^3 > /(T[years])^2 = 0.673 \pm 0.002$; the values lie in the interval 0.671 to 0.675. However, this numerical value is not consistent with the value $1 - \beta$, as it is in the case of exact Kepler's third law: in our case $\beta = 0.4847$. Moreover, while the obtained value of $< r^3 > /T^2$ corresponds to Keplerian motion for eccentricity $e = 0.318$, the small values of $< r >^3 /T^2$ cannot be understood in terms of Keplerian motion:

$$\frac{\langle r^3 \rangle}{T^2} = (1 - \beta)\ (1 - e^2)^{9/2}\ \frac{1}{2\ \pi} \int_0^{2\pi} \frac{dx}{(1 + e \cos x)^5},$$

$$\frac{\langle r \rangle^3}{T^2} = (1 - \beta)\ (1 - e^2)^{15/2} \left\{ \frac{1}{2\ \pi} \int_0^{2\pi} \frac{dx}{(1 + e \cos x)^3} \right\}^3. \tag{3.1}$$

(Central acceleration $- 4\pi^2(1 - \beta)\ \vec{e}_R/r^2$ is used for calculations of orbital elements.)

The question how the approximate invariant depends on impact parameter is answered in Fig. 2: the relation $< (r[AU])^3 > /(T[years])^2 \approx 0.673$ may be considered as an expression independent of impact parameter. Relative errors of the considered quantities are presented in Fig. 3. The reduction of error with an increase of initial impact parameter b for non-spherical particles is evident. Such behaviour can be observed for both functions: $< r(t)^3 > /T^2$, and $< r(t) >^3 /T^2$. Moreover, there is a significant gap between the curves (one order in magnitude). In general, the relative error has a decreasing trend with b and seems to be independent of number of revolutions of ISDP around the Sun.

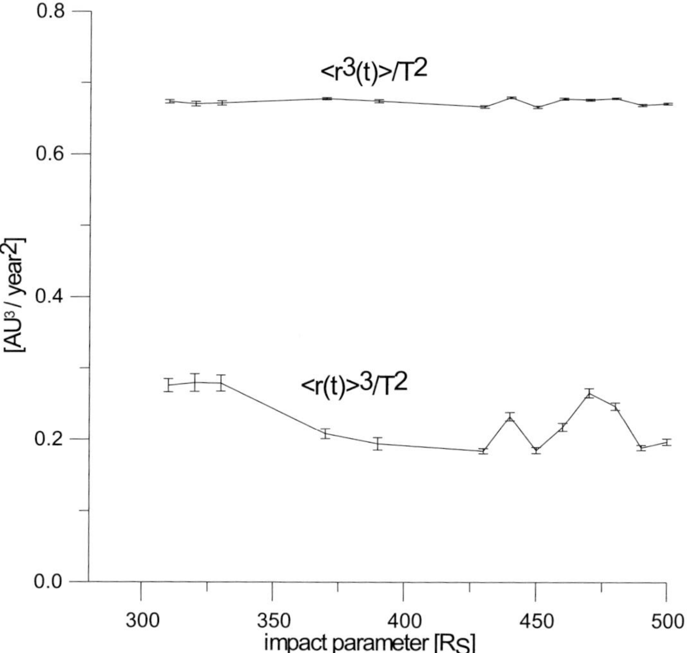

Figure 2. Dependency of quantities $\langle r^3(t)\rangle/T^2$ and $\langle r(t)\rangle^3/T^2$ (averaged over total number of revolutions) on initial impact parameter b. The error bars define mean square deviations. Model characteristics are the same as in Fig. 1.

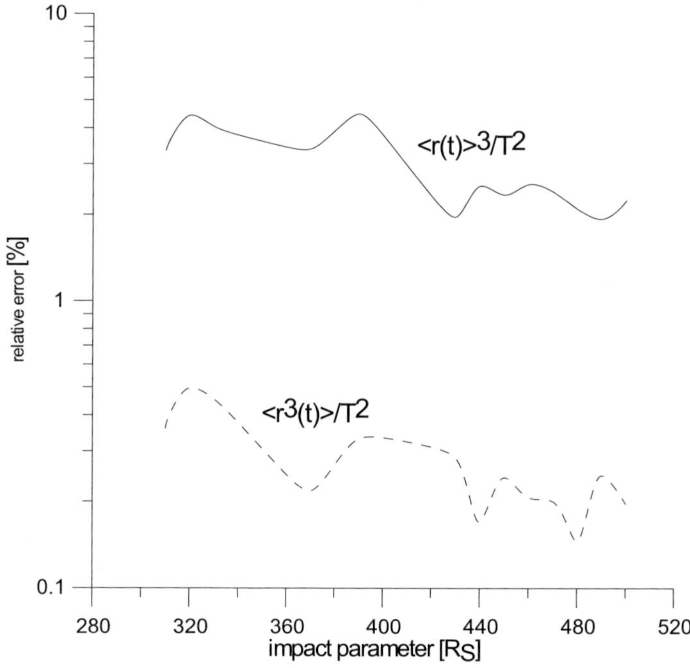

Figure 3. Relative error of $\langle r^3(t)\rangle/T^2$ and $\langle r(t)\rangle^3/T^2$ as a function of initial impact parameter b. Model characteristics are the same as in Fig. 1.

4. Conclusion

The characteristics of motion of the interstellar dust particles (ISDPs) in the Solar System were investigated. Gravitational force of the Sun, solar electromagnetic and corpuscular radiation and interplanetary magnetic field were considered. The effect of solar electromagnetic radiation plays an important role in the sense that non-spherical ISDPs can be captured (and survive) much more effectively than spherical particles. Surviving non-spherical ISDPs orbiting around the Sun are characterized by a nearly constant value of $< r^3 > /T^2$, where T is orbital period and $< r^3 >$ is time average of cubed distance from the Sun – time average over the period T: $< r^3 > /T^2 = 0.673 \pm 0.002$. The validity of quasi-constant profile of $< (r[AU])^3 > /(T[years])^2$ needs to be analysed for different types of cosmic dust particles in the future works. In any case, such a fact could be efficiently utilized in modelling of ISDPs dynamics in the circumsolar region.

Acknowledgements

The paper was partially supported by the Scientific Grant Agency VEGA (grants Nos. 1/0206/03 and 2/3024/23). J. K. would like to thank to organizers of the colloquium and IAU for financial support.

References

Baggaley, W. J. 2000, *J. Geophys. Res.* 105, 10353

Clanton, V. S., Gooding, J. L., Mckay, D. S., Robinson, G. A., Warren, J. L. & Watts, L. A. 1984, *Cosmic dust catalog (particles from collection flag U2015)*, NASA, Johnson Space Center, (Houston, Texas), vol. 70, No. 1, p. 10

Dorschner, J., Begemann, B., Henning, Th., Jäger, C. & Mutschke, H. 1995, *Astron. Astrophys.* 300, 503

Draine, B. T. 1988, *Astrophys. J.* 333, 848

Frisch, P., Dorschner, J., Geiss, J., Greenberg, M., Grűn, E., Landgraf, M., Hoppe, P., Jones, A., Krätschmer, W., Linde, T., Morfill, G., Reach, W., Slavin, J., Švestka, J., Witt, A. & Zank, G. 1999, *Astrophys. J.* 525, 492

Fuller, K. A. 1994, *Opt. Lett.* 19, 1272

Grűn, E., Gustafson, B. Å, Mann, I., Baguhl, M., Morfill, G. E., Staubach, P., Taylor, A. & Zook, H. A. 1994, *Astron. Astrophys.* 286, 915

Grűn, E. & Landgraf, M. 2001, *Space Sci. Rev.* 99, 151

Gulyaev, R. A. & Shcheglovm, P. V. 1999, *Contrib. Astron. Obser. Skalnaté Pleso* 28, 237

Hildebrand, R. H. & Dragovan, M. 1995, *Astrophys. J.* 450, 663

Jackson, A. A. 2001, *Planet. Space Sci.* 49, 417

Kimura, H. & Mann, I. 1998, *Astrophys. J.* 499, 454

Klačka, J. 1992, *Earth, Moon, Planets* 59, 41

Klačka, J. 2004, *Cel. Mech. Dyn. Astron.* 89, 1

Klačka, J. & Kocifaj, M. 2001, *J. Quant. Spectr. Radiat. Transfer* 70, 595

Kocifaj, M. & Klačka, J. 2003, *Planet. Space Sci.* 51, 617

Kocifaj, M. & Klačka, J. 2004, *Planet. Space Sci.* 52, 839

Kocifaj, M., Kapišinský, I. & Kundracík, F. 1999, *J. Quant. Spectr. Radiat. Transfer* 63, 1

Landgraf, M., Műller, M. & Grűn, E. 1999, *Planet Space Sci.* 47, 1029

Parker, E. N. 1958, *Astrophys. J.* 128, 664

Perrin, J. M. & Sivan, J. P. 1991, *Astron. Astrophys.* 247, 497

Robertson, H. P. 1937, *Mon. Not. Roy. Astron. Soc.* 97, 423

Wurm, G. & Schnaiter, M. 2002, in: G. Videen & M. Kocifaj (eds.), *Optics of Cosmic Dust*, Kluwer Academic Publishers, Dordrecht, 89

Dynamics of Populations of Planetary Systems
Proceedings IAU Colloquium No. 197, 2005
Z. Knežević and A. Milani, eds.

© 2005 International Astronomical Union
DOI: 10.1017/S1743921304008944

Nonspherical dust in exterior resonances with Neptune

M. Kocifaj[1], J. Klačka[2] and H. Horvath[1]

[1]Institute for Experimental Physics, University of Vienna, Boltzmangasse 5, 1090 Wien, Austria, email: kocifaj@ap.univie.ac.at

[2]Department of Astronomy, Physics of the Earth, and Meteorology, Faculty of Mathematics, Physics and Informatics, Comenius University, 842 48 Bratislava, Slovak Republic, email: klacka@fmph.uniba.sk

Abstract. Nonspherical dust grains orbiting the Sun are influenced by solar electromagnetic radiation. Interaction of electromagnetic radiation with nonspherical grains is complex and analogy with spherical grains may not be physically justified. As a consequence, equation of motion for nonspherical grain is more general than the equation of motion for spherical particle. Application of this more general interaction to possible trapping of the grain in resonances with planet Neptune is investigated. Approximation of the planar circular restricted three-body problem with action of solar electromagnetic radiation on nonspherical grain is used.

The orbital evolution of nonspherical dust grains of radius ≈ 2 micrometers is numerically calculated. Attention is payed to exterior resonances with the Neptune, and the numerical experiments are concentrated on their possible high capture efficiency. Physical difference between possibilities for resonant captures for spherical and nonspherical dust particles is pointed out.

Keywords. Dust grain, electromagnetic radiation, Celestial Mechanics

1. Introduction

Scientists have intensively studied the effects of resonances with solar system planets during the last two decades. The orbital evolution of dust grains near resonances with planets was also investigated. In addition to gravitational forces, the influence of solar electromagnetic radiation in the form of the Poynting-Robertson effect is usually taken into account (Robertson 1937, Klačka 2004), see also Jackson & Zook (1992), Marzari & Vanzani (1994), Šidlichovský & Nesvorný (1994), Liou & Zook (1995), Liou *et al.* (1995).

In reality Poynting-Robertson effect (P-R effect) holds only if a special condition is fulfilled (see Eq. (120) or Eq. (122) in Klačka 1992; compare Eqs. (40) and (48) in Klačka 2004). The nonspherical particles do not fulfill this condition. Thus, we have to take into account effect of solar electromagnetic radiation in an appropriate form.

2. Model and equation of motion

All our numerical simulations are based on equation of motion written in the form (see Klačka & Kocifaj 2001 for more details)

$$\frac{d\vec{v}}{dt} = -\frac{4\,\pi^2}{r^2}\,\vec{e}_R - 4\,\pi^2\,\frac{m_P}{M_\odot}\left\{\frac{\vec{r}-\vec{r}_P}{|\vec{r}-\vec{r}_P|^3} + \frac{\vec{r}_P}{|\vec{r}_P|^3}\right\} +$$

$$\beta\,\frac{4\,\pi^2}{r^2}\sum_{j=1}^{3}\frac{Q'_{prj}}{Q'_{pr1}}\left\{\left(1 - 2\,\frac{\vec{v}\cdot\vec{n}_1}{c} + \frac{\vec{v}\cdot\vec{n}_j}{c}\right)\vec{n}_j - \frac{\vec{v}}{c}\right\},$$

$$\vec{n}_1 \equiv \vec{e}_R \equiv \vec{r}/|\vec{r}|\;;\quad \beta = 7.6\times10^{-4}\,Q'_{pr1}\,\frac{A'[m^2]}{m\,[kg]}, \tag{2.1}$$

where $\vec{r}$ is the position vector of the particle and $\vec{r}_P$ is the position vector of the planet (Neptune) which moves in circular orbit (m_P is mass of the planet, $M_\odot$ is mass of the Sun). As for the model of the cosmic dust particle we have used particle morphologically identical to U2015B10 (Clanton *et al.* 1984, Kocifaj *et al.* 1999). The particle shape appears to be representative of cosmic dust, since its aspect ratio (equal to 1.4) coincides with results obtained by means of the mid-infrared spectropolarimetry (Hildebrand & Dragovan 1995). The characteristics of radiation scattered by the cosmic dust particle were calculated using the discrete dipole approximation (DDA; Draine (1988)). We have made simulations for effective radius of the particle corresponding to 2 microns. Several samples of materials with different optical properties were considered: ice (mass densities 1 g/cm^3 and 2 g/cm^3), magnesium-rich silicate (mass densities 2 g/cm^3 and 4 g/cm^3), and, iron-rich silicate (mass densities 6 g/cm^3 and 8 g/cm^3). Central acceleration $-\,4\pi^2(1-\beta)\,\vec{e}_R/r^2$ is used in calculations of orbital elements.

3. The most important results

Simulations for various optical properties were done by sampling the initial positions of nonspherical particle with respect to the Neptune in all angles. It was assumed that initial position of the grain was in the plane of the planet's orbit. In our computations we sampled the initial distances from the Sun in the range ($\approx$ 1.0, 1.5)-times the distance between the Sun and the planet. Various initial orientations of the particle's rotation axis were considered (rotation axis was fixed during the particle's motion).

Application of Eq. (2.1) for the case $Q'_{pr2} = Q'_{pr3} = 0$ yielded standard positions for resonances (see Eq. (4.1) below: the fraction of orbital periods T/T_P defines type of the resonance). However, real particles exhibit nonzero values for Q'_{pr2} and Q'_{pr3} and their values may be much more important than the ratio v/c present in the P-R effect (Klačka & Kocifaj 2001). Our simulations for orbital evolution of micron-sized dust particles show several temporary captures of grains in exterior resonances with the planet Neptune (capture with time larger than 4×10^4 years was not found for ice particles). Results for 2 micron sized dust grains are presented in Table 1. One has to take into account that particle's orientation (with respect to the incident radiation) change during its motion and the presented values of β may change within a few percent.

Table 1. Capture times for nonspherical particle of effective radius of 2 microns.

material	mass density [g/cm^3]	β [10^{-2}]	resonance	capture times [10^3 years]
magnesium-rich silicate	4.0	3.60	7/6	> 40
magnesium-rich silicate	4.0	3.60	9/8	> 50
iron-rich silicate	6.0	2.43	3/2	> 40
iron-rich silicate	6.0	2.43	8/7	> 50
iron-rich silicate	6.0	2.43	5/3	> 160

Fig. 1 illustrates time evolution of semimajor axis and eccentricity for nonspherical iron-rich silicate dust grain. While spherical particle is characterized by a decrease of semimajor axis outside resonance (P-R effect), nonspherical dust grain may exhibit also an increase of semimajor axis outside the resonance; analogous situation holds for eccentricity. The resonance 5/3 corresponds to semimajor axes 41.81 AU $\leqslant a \leqslant 42.02$ AU, while its theoretical value for $\beta = 0.0243$ equals to $a = 41.908$ AU, according to Eq. (4.1).

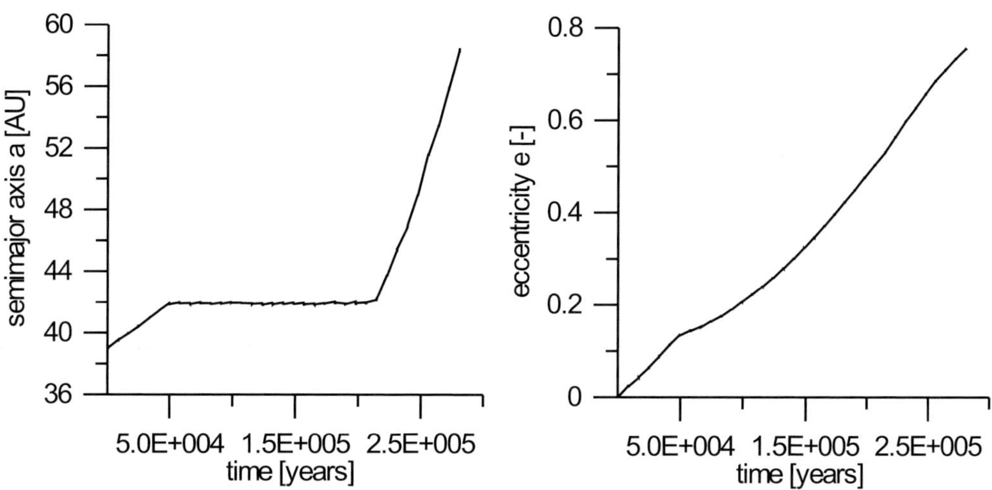

Figure 1. Time evolution of semimajor axis and eccentricity for nonspherical iron-rich silicate dust grain of effective radius 2 microns. The plots show exterior resonance 5/3 with Neptune.

4. Spherical/nonspherical particles and resonances – comparison

There are important differences in orbital evolution for spherical and nonspherical particles.

While spherical particles can be described by the P-R effect, this may not hold for micron-sized nonspherical particles. While P-R effect exhibits decrease of semi-major axis before trapping in a resonance, interaction of electromagnetic radiation with nonspherical grain may initiate also an increase of semi-major axis close to the resonant regions.

Spherical particles always move in the plane of the planetary orbit, if the initial orbital angular momentum vector of the particle was normal to the plane of the planetary orbit. This does not hold for nonspherical particles – inclination of the particle's orbit is not conserved.

Another important physical difference between the orbital evolution of spherical and nonspherical particles concerns the values of parameter β. This parameter is a constant of motion for spherical particles. Thus, the third Kepler's law yields

$$\frac{a}{a_P} = (1 - \beta)^{1/3} \left(\frac{T}{T_P}\right)^{2/3} \left(1 + \frac{m_P}{M_\odot}\right)^{-1/3} , \qquad (4.1)$$

for the semimajor axes and periods of revolution around the Sun of the particle and the planet. Defining any resonance by the ratio T/T_P, Eq. (4.1) yields immediately the ratio a/a_P. However, parameter β is not conserved during the motion of nonspherical particle. If its relative change is ϱ_β, then (approximately)

$$\varrho_a = \frac{\beta}{3\,(1 - \beta)}\,\varrho_\beta , \qquad (4.2)$$

where $\varrho_a = \sigma_a/a$, $\varrho_\beta = \sigma_\beta/\beta$, σ_a and σ_β are standard deviations for mean values a and β. The consequence of this equation is, that several first and higher order resonances may overlap for larger values of β. Moreover, Eq. (4.2) shows that the change of semimajor axis a in orbital resonance caused by the change of parameter β may be relatively large. Thus, one should expect that resonant capture times for nonspherical particles are smaller than the capture times for spherical grains.

5. Conclusion

Our numerical simulations have shown that resonant captures of dust grains occur for exterior resonances with the planet Neptune for micron-sized nonspherical particles in circular restricted three-body problem with solar electromagnetic radiation (see Table 1 and Fig. 1). The existence of resonant capture can be influenced by three facts:

i) interaction of electromagnetic radiation with nonspherical particle changes inclination of particle's orbit,

ii) resonances are not defined in a unique way – several resonances may overlap due to the changing value of parameter β for its greater values (and, even for spherical grains, capture times are very small for similar values of T and T_P), and,

iii) the change of β during particle's motion can cause significant change of semimajor axis (see Eq. 4.2), and, thus, also decrease of capture time.

Interaction of electromagnetic radiation with nonspherical grain may initiate an increase of semi-major axis close to the resonant regions. Resonant trapping is not normally expected for diverging orbits such as the trapping presented in Fig. 1.

Acknowledgements

The paper was supported by the Fonds FWF (Project M772-N02) and by the Scientific Grant Agency VEGA (grant No. 1/0206/03). J. K. would like to thank to organizers of the colloquium and IAU for financial support.

References

Clanton, V. S., Gooding, J. L., Mckay, D. S., Robinson, G. A., Warren, J. L. & Watts, L. A. 1984, *Cosmic dust catalog (particles from collection flag U2015)*, NASA, Johnson Space Center, (Houston, Texas), vol. 70, No. 1, p. 10

Draine, B. T. 1988, *Astrophys. J.* 333, 848

Hildebrand, R. H. & Dragovan, M. 1995, *Astrophys. J.* 450, 663.

Jackson, A. A. & Zook, H. A. 1992, *Icarus* 97, 70

Klačka, J. 1992, *Earth, Moon, Planets* 59, 41

Klačka, J. 2004, *Cel. Mech. Dyn. Astron.* 89, 1

Klačka, J. & Kocifaj, M. 2001, *J. Quant. Spectrosc. Radiat. Transfer* 70, 595

Kocifaj, M., Kapišinský, I. & Kundracík, F. 1999, *J. Quant. Spectrosc. Radiat. Transfer* 63, 1

Liou, J-Ch. & Zook, H. A. (1995) 1995, *Icarus* 113, 403

Liou, J-Ch., Zook, H. A. & Jackson, A. A. 1995, *Icarus* 116, 186

Marzari, F. & Vanzani, V. 1994, *Astron. Astrophys.* 283, 275

Robertson, H. P. 1937, *Mon. Not. R. Astron. Soc.* 97, 423

Šidlichovský, M. & Nesvorný, D. 1994, *Astron. Astrophys.* 289, 972

Part 6

SATELLITE AND STELLAR SYSTEMS

Dynamics of Populations of Planetary Systems
Proceedings IAU Colloquium No. 197, 2005
Z. Knežević and A. Milani, eds.

© 2005 International Astronomical Union
DOI: 10.1017/S1743921304008956

Population models of space debris

Alessandro Rossi

ISTI/CNR, Via Moruzzi 1, 56127 Pisa, Italy
email: Alessandro.Rossi@isti.cnr.it

Abstract. More than 300 000 artificial debris particles with diameter larger than 1 cm are orbiting the Earth. The space debris population is similar to the asteroid belt, since it is subject to a process of high-velocity mutual collisions that affects the long-term evolution of its size distribution. The near–Earth space can be divided in three major regions where orbital debris is of concern: Low Earth Orbits (LEOs), below about 2000 km, Geosynchronous Orbits (GEOs), at an altitude of about 36000 km and the Medium Earth Orbits (MEOs) in between. The issues are in principle the same in the three regions, nevertheless they require different approaches and solutions. The space debris are composed by several different populations according to their source and their orbital region. A description of the nature and dynamics of the different populations in the low, medium and high orbital regimes is given. The impact risk posed by these debris is then briefly outlined.

The long term evolution of the whole debris population can be studied with computer models allowing the simulation of all the known source and sinks mechanisms. One of these codes is described and the evolution of the debris environment, over the next 100 years, under different traffic scenarios is shown, pointing out the possible measures to mitigate the growth of the orbital debris population.

Keywords. Space vehicles, celestial mechanics, n-body simulations

1. Introduction

When in October 1957 the first artificial satellite, Sputnik 1, was launched by the USSR, nobody could even imagine that after less than 50 years we would be facing an environmental problem in the near-Earth space. A few years later, on June 29, 1961 the first known break–up in orbit, the explosion of the *Transit 4A* rocket body happened. From then on, the repetition of these two events (launches of new satellites and break-up of in orbit spacecraft) contributed to build up a huge population of objects that are now polluting, perhaps in an irreversible way, the space around us. Many years later, on July 24, 1996, the danger posed by the space debris to all the human activities in space has been clearly showed by the first recorded accidental collision between an operational satellite and a piece of debris: the French micro-satellite *Cerise* has been hit, at the relative velocity of 14.77 km/s, by a fragment, of about 10 cm^2, coming from the explosion of an Ariane rocket upper stage, that happened ten years before (Alby *et al.* 1997). Though still very unlikely from a statistical point of view, collisions with space debris represent now a threat for all the space missions, especially the manned ones, and are deemed to become the most important source of debris in a not too distant future.

2. The space debris population

The near–Earth space can be roughly divided in three main regions. The Low Earth Orbit (LEO) region, below about 2 000 km, is where most of the orbiting objects of any size can be found. Due to the high objects density and the large impact velocities (larger than 9 km/s on average (Rossi & Farinella 1992)) it is in this region where the space debris issue is presently most urgent and where it has been historically more studied.

The Medium Earth Orbit (MEO) between 2 000 km and about 36 000 km is emerging as a new issue in the space debris studies, since it is the home of the so-called navigation constellations, i.e., the constellations of satellites, like the Global Positioning System (GPS), used to locate with high accuracy the position of a receiver on the ground. The vital role that this navigation services is acquiring in the air and terrestrial transportation make these constellations and the region of space they occupy correspondingly important.

The geostationary ring can be defined roughly as a toroidal region of space close to the equatorial plane at about 36 000 km of altitude. A satellite placed in the Geosynchronous Orbit (GEO) region will orbit the Earth in about one sidereal day, remaining almost fixed with respect to a given ground station. For this reason most of the telecommunication satellites are launched in the geostationary ring; hence its uniqueness and importance.

In Fig. 1, the spatial density of objects with diameters larger than 1 mm, 1 cm and 10 cm in Earth Orbit is plotted. The peaks of density corresponding to LEO, MEO and GEO are immediately apparent from the figure. In particular it should be noted that there are mainly two regions of the near Earth space where the debris "pollution" is particularly high: one around 900 km and one around 1400 km.

The space debris issues in the three regions are in principle the same, nevertheless the orbital dynamics and the existing population are different and therefore they require different approaches and solutions.

From the mm–sized particles up to the largest satellites, tens of meters across, the population of objects of artificial origin orbiting in near–Earth space spans many orders of magnitude in size and mass. Below about 1 μm in diameter the natural meteoroid population still dominates the particle flux, whereas above this threshold artificial space debris accounts for the majority of the solid bodies present in the near–Earth environment.

This huge ensemble of particles includes many different populations produced by various mechanisms. In the following a brief review of these populations is given.

Since the Sputnik I about 5 000 payloads have been launched. Among these objects, some 3 000 have re–entered in the atmosphere. The others (about 2 000) are still orbiting and represent most of the large (> 1 m) objects in orbit. It should be remembered that every launch often implies that more than one object is injected into Earth orbit since, beyond the payload(s), we have the upper stages of the rocket and the so-called mission related debris (e.g., sensors caps, yo-yo masses used to slow down the spacecraft spin, etc).

All the un-classified spacecraft currently in orbit are cataloged by the United States Space Command in the so called Two–Line Element (TLE) catalog. In this catalog about 10 000 objects are listed along with their current orbital parameters. The limiting size of the objects included in the catalog (due to limitations in sensors power and in observation and data processing procedures) is about 5 to 10 cm in LEO and about 0.5 - 1 m in GEO. The orbits of the TLE catalog objects are maintained thanks to the observations performed by the Space Surveillance Network. The network is composed by 25 sensors, both radars (providing the vast majority of information) and optical sensors. Only about 6% of the objects in the catalog are operative satellites. Approximately 24% is composed by non-operative spacecraft; around 17% by the upper stages of the rockets used to place the satellite in orbit. Then about 13% is composed by mission related debris. Finally some 40% are debris generated mostly by about 170 explosions and 2 collisions which have involved rocket upper stages or spacecraft in orbit (Bendish *et al.* 2004). About 99% of the mass in orbit is due to the large objects included in the catalog. In Fig. 2 (left panel) the cataloged objects are plotted in the semi-major axis versus inclination

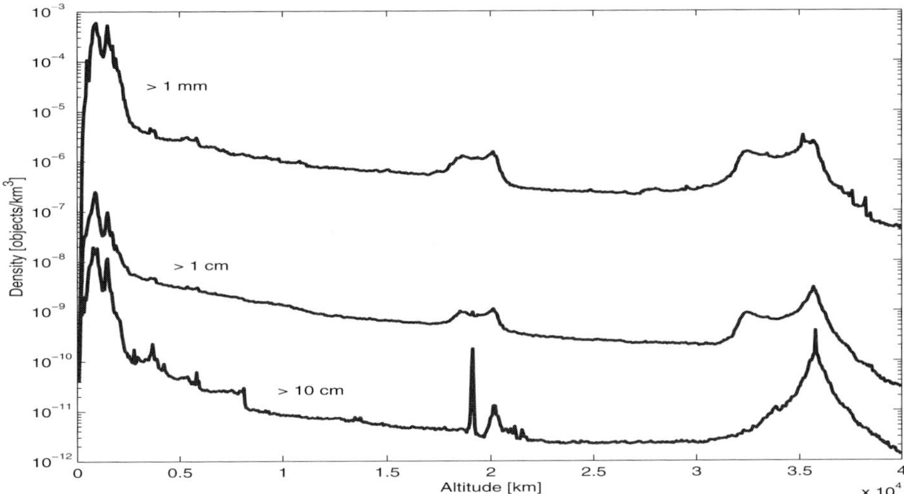

Figure 1. Density of objects as a function of altitude for three different size thresholds: objects with diameter larger than 1 mm, 1 cm and 10 cm.

space. This representation clearly highlights some features in the distribution of objects in space with the spacecraft (and the resulting debris) being clearly grouped in "families" or constellations, according to their different purposes and to the different launching bases: e.g., we can distinguish the US GPS (Global Positioning System) satellites and their Russian analogues GLONASS ($a \simeq 26,000$ km, $i \simeq 55°$ and $i \simeq 63°$, respectively), the Russian communication satellites in Molniya–type orbits ($a \simeq 26,000$ km, $e \simeq 0.7$, $i \simeq 63°$), the geosynchronous satellites ($a \simeq 42,000$ km, $e \simeq 0$, $15° \geqslant i \geqslant 0°$), the satellites in Sun-synchronous orbits ($i \simeq 100°$) , the satellites in polar orbits ($i \simeq 90°$), some families of Russian COSMOS satellites between $i \simeq 60°$ and $i \simeq 80°$, the LEO satellites launched from the Kennedy Space Center (at $i \simeq 27°$) and the families of objects in geosynchronous transfer orbits (GTO) (mostly upper stages) launched from Kourou (ESA Ariane rockets, $i \simeq 7°$), from the Kennedy Space Center ($i \simeq 27°$) and from Baikonour ($i \simeq 48°$).

To get data on the smaller objects not included in the catalog, different sensors, or the same sensors but operated in a different way, are needed. Radar campaigns have been carried out to detect objects of 1 cm and below in LEO by putting the radar in a "beam park" mode, where the radar stares in a fixed direction and the debris randomly passing through the field of view are detected. This allows a counting of the number of objects, i.e., the determination of the objects flux and density, but only a rough determination of their orbits.

These radar campaigns gave an explanation of the prominent peak of density of objects around 900 km of altitude (see Fig. 1). It is mainly due to the presence in this altitude band of a large number of sodium-potassium liquid metal droplets leaked outside a number of Russian ocean surveillance satellites (RORSAT) (Foster *et al.* 2003). This liquid was used as a coolant for the nuclear reactor which generated the power on board and was dispersed in space after the core of the reactor was ejected from the spacecraft in order to prevent possible risks due to its reentry into the Earth atmosphere. About 70 000 drops with diameter between 0.5 mm and about 5.5 cm have been estimated to orbit the observed region.

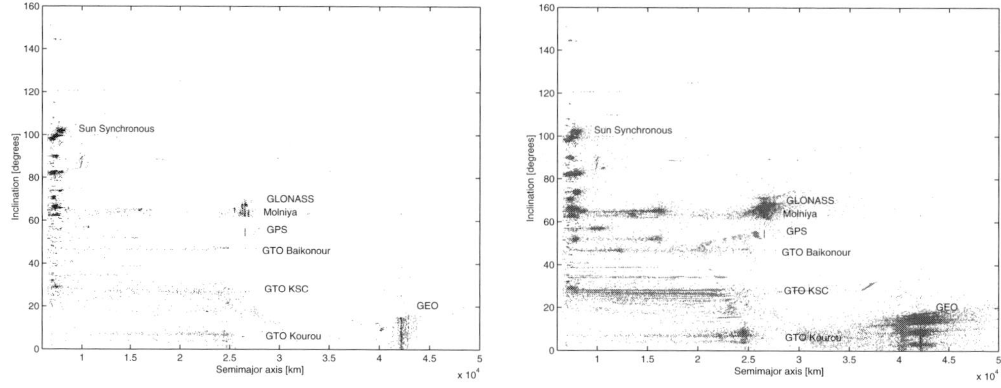

Figure 2. Distribution in the semimajor axis - inclination space of the cataloged objects (left) and of the whole population of objects larger than 1 cm inside MASTER 2001 (right)

The Haystack observations were instrumental also to point out the importance of another unexpected source of space debris, the aluminum oxide particles coming from the burns of the rocket motors with solid propellant. During these burns a large number of sub-millimeter sized particles are ejected. As a matter of fact the solid rocket motor (SRM) exhaust are probably the main contributors to the debris population between $10\,\mu$ and $100\,\mu$. Between $100\,\mu$ and 1 cm the SRM exhaust are again one of the main components of the population, together with fragments and paint flakes that detached from spacecraft surfaces exposed to the space environment effects. In some cases (particularly toward the end of the burn) also slugs of this propellant are released from the SRM, which are of centimetric dimensions (Jackson *et al.* 1997); particularly at low inclinations, where the SRM firings have been more frequent, these slag particles could be responsible for a significant portion of the centimetric debris and even dominate the 1 cm population below about 400 km and above 2 500 km of altitude.

Another previously unknown debris population, around 2900 km of altitude, consisting of the so-called *West Ford Needles* has been detected by radar surveys. Using the powerful Goldstone radar, Goldstein *et al.* (1998) found the remnants of the copper dipoles, 1.77 cm long, which were released in space in 1961 and 1963 by the American satellites Midas 3 and Midas 6, for telecommunication experiments. They were conceived to reenter the atmosphere in about 5 years, but apparently some of them stuck together after the release so lowering their area over mass ratio and therefore augmenting their orbital lifetime. According to the Goldstone observations, a population of about 40 000 such clusters is orbiting between 2400 and 3100 km of altitude.

With circular orbits and an altitude of 20 181 km and 19 100 km respectively, the American GPS and the Russian GLONASS are the two currently deployed navigation constellations. Along with the Russian telecommunication spacecraft in Molnyia orbits, they are the most sensible objects orbiting the MEO region and can be clearly distinguished in Fig. 1 and Fig. 2. According to the MASTER 2001 population model (see later), in the MEO region there are about 60 000 objects larger than 1 cm that are possibly crossing the orbits of the navigation constellations. Actually most of the objects in MEO are clustered about the Molnyia orbits and have therefore a minimal interaction with the navigation constellations. But, even if we exclude the objects close to Molnyia orbits, about 16 000 objects with diameter larger than 1 cm have orbits potentially crossing the navigation constellations. In particular, the GPS orbit appears within reach of several thousand objects, due to the non-zero eccentricity of most of the debris in the

MEO zone. The risk of an impact in this region and its possible consequences on the navigation constellations is investigated in Rossi *et al.* (2003).

Notwithstanding its paramount importance, the picture of the debris environment in the GEO region is still very uncertain, mainly due to the physical distance which prevent its mapping by radars. The peculiarities of the GEO region are mainly the absence of any natural decay mechanism, such as air drag (see Sec. 4) and the fact that each satellite in geostationary orbit is assigned an "orbital slot" of about $0.1°$ of width in longitude. For these reasons, though huge in physical terms, the useful space in the GEO region is actually operationally limited since an orbital slot not freed by a "dead" satellite (or a debris) is not any more usable by other spacecraft. Moreover any debris created in the region will stay there almost for ever. Note that, during their operative lifetime, the satellites are periodically maneuvered to keep them inside the slot, counteracting the perturbations that would tend to change their orbital parameters. In particular perturbations due to the Sun, the Moon and the Earth oblateness would induce a precessional motion of the orbital plane inclination, inducing an oscillation of i of $15°$, with a period of about 53 years. Then the solar radiation pressure would induce small periodic variations in e. Therefore, once the satellite is no more operational its inclination and eccentricity will tend to deviate from the nominal zero values. This means that they will start crossing the operational orbits with relative velocities of several hundreds of m/s, much higher than those common for operative coorbiting GEO satellites. Dedicated optical observation campaigns are performed to characterize the environment in this orbital region. The European Space Agency has installed for this purpose a 1 m Schmidt telescope in the Canary Islands. The limiting detection size in GEO for this telescope is about 20 – 30 cm. Since 1999 routine observations have been performed leading to unexpected and worrying results (Flury *et al.* 2000; Schildknecht *et al.* 2004). About 1040 objects have been detected near GEO. Only 340 are active satellites, while the rest are debris, mostly uncataloged objects. The source of this objects remains still uncertain since only two explosions have been recorded in GEO and these two events cannot account for all the observed debris. Probably ten more unrecorded fragmentation events must have happened in GEO (Bendish *et al.* 2004). Of course the number of non-trackable objects, smaller than the telescope detection threshold, should be much larger than 1000. A large uncertainty remains obviously in the geostationary region and international efforts are under way to fill this important gap.

The ground based observations are then supplemented by the data, mainly about mm and sub-mm particles, obtained from the analysis of the surfaces of spacecraft returned to Earth after some time spent in orbit (e.g. the Long Duration Exposure Facility (LDEF), a satellite released and then retrieved by the Space Shuttle, the Hubble Space Telescope solar panels, the Space Shuttle external surface itself, etc) and from impact sensors on board a few satellites.

The observations and in-situ data allow to calibrate the models of the space environment. There are models produced mainly by fitting the observational data to derive values of the flux of debris as a function of the altitude and on a given orbit (Liou *et al.* 2002). Another kind of models reconstruct the environment by reproducing all the known source and sink mechanisms (such as the atmospheric drag) with ad-hoc computer models. One of these last models has been developed by the European Space Agency and is called MASTER 2001 (Bendish *et al.* 2004). Figs. 1 – 2 have been produced by using the MASTER 2001 population of objects.

The right panel of Fig. 2 shows the distribution of the objects larger than 1 cm from the MASTER 2001 population. The families of orbits, described in the left panel of the same figure, are still recognizable. Nonetheless they are now covered by large clouds of smaller objects. Note the spread of the orbital elements (mostly in semi-major axis, since it is very expensive to impart changes in inclination) due to the energy imparted to the fragments in case of breakups. The long stripes of objects at the inclination of the different GTOs are mainly due to the release of the slag by the SRM of the upper stages. Also noticeable, with respect to the situation in the left panel, is the large number of objects in the GEO vicinity due both to slag from upper stage burns and to the large number of uncataloged fragments.

The current estimate, derived from the observations and the simulated populations, is that the total number of non–trackable particles of 1 cm and greater is around 350 000, while those larger than 1 mm could be more than 3×10^8.

3. Risk of collision for orbiting spacecraft

The overcrowding of the space around the Earth makes collisions a serious threat and, as pointed out by the *Cerise* event, a reality. Also the Space Shuttle has already performed several maneuvers to avoid pieces of junk which might have crossed its path.

The average impact velocity for orbiting bodies in LEO, of about 9.7 km/s (Rossi & Farinella 1992), means that even centimeter sized particles are delivering a large kinetic energy on the target. If the mass of the projectile exceeds about 1/1000 of the target's mass, then the target can be completely shuttered producing a cloud of fragmentation debris (see Sec. 4); this corresponds to a threshold for the impact specific energy (i.e the ratio between the projectile kinetic energy and the target mass) of the order of some 10^4 J/kg. Below this value a localized crater–like damage occurs.

To highlight the collision risk, another interesting way of representing the space debris population is shown in Fig. 3, following Valsecchi *et al.* (1999). In this representation, the space debris population is plotted as a function of the square of the impact velocity, U^2 (in units of the target orbital velocity, i.e., about 7 km/s in LEO), of every object with respect to a selected target in a given circular orbit. In the left panel of Fig. 3 the population of the cataloged objects of Fig. 2, is plotted with respect to a target in an equatorial ($i = 0°$) circular orbit with semi-major axis $a_{ref} = 6\,828$ km. The different families of objects are still recognizable, but now the additional information about the impact velocity against the selected target is available. Moreover, all the objects found on the left of the line tagged "Tangency condition" are not crossing the target orbit, and therefore are not potential projectiles. Note also how, in the same 2-dimensional plot, the information about the eccentricity and the inclination of the projectile orbit, with respect to the target one, is included. In the right panel the same population is plotted with respect to a target in a circular orbit with the same a but with $i = 52°$ (i.e., an orbit similar to the one of the International Space Station (ISS)). Note how the the projectile population spreads and mixes and especially how the impact velocities can become significantly larger since the relative inclination between the target and the projectile orbit has to be taken into account. Therefore also almost head-on collision at very high velocity can take place. This target centered risk analysis, based on Öpik's studies of impacts in the Solar System (Öpik 1976), has been applied to the case of multi-plane satellite constellations in LEO (Rossi *et al.* 1999) and MEO (Rossi *et al.* 2003) and to the ISS (Valsecchi & Rossi 2002), showing the dependence of the collision risk to the complex dynamics of these systems. The results of these studies cannot be reported here and the interested reader can refer to the cited papers.

To protect the space assets against impacts with small debris ($\leqslant$ 1 cm) multi-wall bumper shields have been devised and installed on some modules of the ISS. Yet for larger debris the shields are not enough to prevent the penetration of the target or

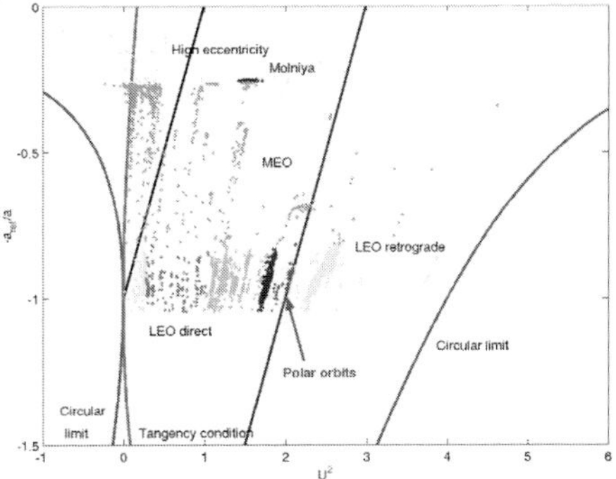
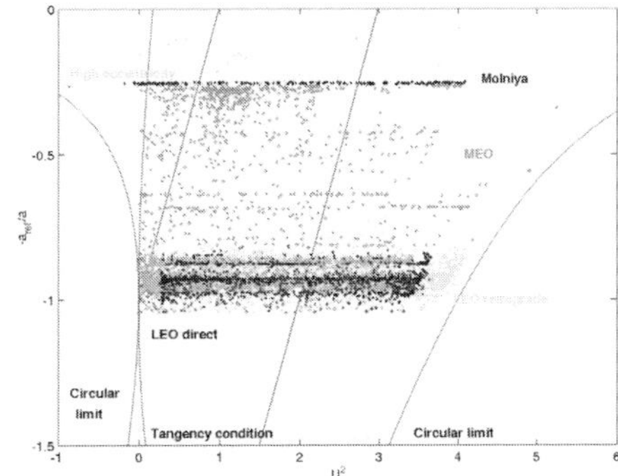

Figure 3. Representation of the TLE catalog population in the space U^2 vs $E = -a_{target}/a_{projectile}$, with respect to a target in a circular orbit at 450 km of altitude. In the left panel the target orbit has equatorial inclination $i = 0°$, while in the right panel the target has $i = 52°$. For a thorough explanation of the lines drawn in the plot, refer to Valsecchi *et al.* (1999). Note that the admissible region is included between the two circular limits and that the objects possibly crossing the orbit of the target lie to the right of the tangency condition. The objects to the right of the inclined line starting from the target position have higher probability of hitting the target on the front; on the left of this line impacts from the back predominate (the more so as we approach the tangency condition). Finally, the straight line from $(1.5, -1.5)$ to $(3, 0)$ is the locus of orbits with inclination equal to $90°$ with respect to the selected reference plane.

even its complete fragmentation; in this case an avoidance maneuver, if the projectile is trackable from the ground, is the only solution to save the Station module. Note moreover that, as specified in the previous section, most of the debris between 1 and 10 cm are not cataloged. In addition, most of the operational satellites cannot carry the heavy bumper shields so they can rely only on avoidance maneuvers or on their "good luck" to survive the harsh debris environment.

4. Long–term evolution of the debris population

As a matter of fact, the low–orbiting Earth debris population is similar to the asteroid belt, since it is subject to a process of high–velocity mutual collisions that affects the long-term evolution of its size distribution. However, the situation is more complex than for the asteroids, because here the source and sink mechanisms are (partially) subject to human control (e.g. launches, explosions and retrievals), the number density of objects is a sensitive function of the altitude (and so is the sink mechanism due to atmospheric drag) and the relative speeds are dominated by mutual inclinations, which are much larger than typical orbital eccentricities and unevenly distributed (whereas among the asteroids eccentricities and sine of inclinations have similar, fairly broad distributions, with average values ≈ 0.15).

In the end of the '70s Donald Kessler (Kessler & Cour–Palais 1978) first pointed out the possibility that a process of mutual collisions between the objects presently in orbit could lead to the creation of a debris belt surrounding our planet and jeopardizing, if not preventing, all the space activities. Mathematical models and large numerical codes have been developed to simulate the interplay of all the physical processes involved in the evolution of the debris population. In these codes the main source and sink mechanisms of debris that have to be modelled are:

- the launches: being the only source that adds mass to the population in orbit it is of great importance to be able to predict, in a reliable way, the future space traffic. On the other hand, it is extremely difficult to make a reliable forecast since the traffic will depend on several technical, economical and political factors. The best way to deal with this problem is to produce models able to simulate, in an efficient way, different traffic scenarios and to compare the results of the various cases to identify significant trends.

- the explosions: the fragmentations due to explosions of on-orbit spacecraft represent the major source of cataloged objects. After an explosion, a single object produces a swarm of fragments with a mass distribution that approximately matches exponential laws.

- the collisions: although only one accidental collision has been recorded up to now, the collisions are going to represent the most important source of debris for the long term evolution of the space environment. The energy involved in a collision at about 10 km/s is huge, of the order of 10^3 Joules even for a centimetric projectile. The mass distribution of the fragments follows a power law:

$$N(> m) \propto m^{-b}$$

where b is a suitable positive exponent (< 1 to be consistent with a finite total mass). This mass distribution means that more small fragments are produced with respect to an explosion. In our model the exponent b has typically an energy-dependent value (Petit & Farinella 1993). The few laboratory experiments, with non-classified results, make it very difficult to estimate, in a fully reliable way, the outcome of a hypervelocity collision (i.e. a collision where the impact velocity is larger than the velocity of sound inside the materials, typically around 5 km/s) between space objects. This remains one of the most important uncertainties in modelling the space debris evolution.

- as described in Sec. 2, other non-fragmentation debris sources played an important role in determining the present population of orbiting debris. The SRM slag production has to be modelled as well to have a good picture of the future debris environment, whereas the RORSAT drops should not represent a significant source of debris in the future.

On the other side there are the sink mechanisms, that is the processes that tend to remove objects from the orbit. Several natural perturbations act on an orbiting spacecraft, altering its motion from a pure two body orbit:

- gravitational perturbations, due to the non-spherical shape of the Earth and to a third body, i.e. the Moon or the Sun. These perturbations do not affect the semi-major axis of an orbiting object (i.e. do not change the orbital energy), therefore they are not efficient in removing debris form space. Actually, lunisolar perturbation, coupled with non-gravitational perturbations, may play a role in speeding the orbital decay of certain classes of highly eccentric orbits.

- non-gravitational perturbations such as the solar radiation pressure and the atmospheric drag (Milani *et al.* 1987). The latter is the most important one since it subtracts energy from an orbiting object causing its decay into the atmosphere; it represents, therefore, the main sink process. Unfortunately the atmosphere density is decreasing exponentially with the altitude, so that this perturbation is efficient only up to about 800 km above the surface of the Earth. Above this level the air drag takes several hundreds of years to remove a typical satellite from orbit.

Another "non-natural" way to remove objects from space is represented by the de-orbiting of the spacecraft after they completed their mission; we will discuss this issue in the following.

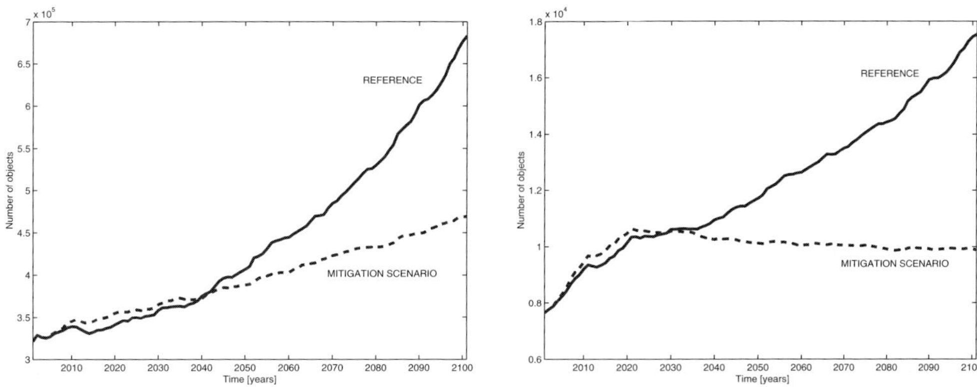

Figure 4. Number of objects larger than 1 cm between 0 and 40 000 km (left) and number of objects larger than 10 cm in LEO (right), with two different simulation scenarios.

The Semi Deterministic Model (SDM) is a software package for the simulation of the long term evolution of the debris population developed in Pisa since the early '90s. In SDM, along with a modelling of all the above mentioned source and sink mechanisms, the actual orbital evolution of all the larger objects is followed by means of an *ad-hoc* fast orbit propagator (Rossi *et al.* 1998). SDM can simulate many complex scenarios allowing to test the influence on the evolution of different fragmentation models, launch traffic, explosion patterns and mitigation measures.

In particular SDM has been used in the last years to study the effect on the environment of the measures proposed at international level to control and, if possible, reduce the number of debris in Earth orbit. The main international committee established to study and face the space debris issue is the Inter-Agency Debris Coordination Committee (IADC) that periodically gathers the representatives of all the main space agencies to share the results of the researches on the different aspects of the field. The most favored *mitigation measures* include:

• change the design of the spacecraft in order to prevent the release of mission related debris;

• prevent on-orbit explosions: this includes venting the upper stages of the residual fuel to avoid over-pressurization and discharging any power system on board after the end of the operational life;

• de-orbit all the upper stages and the satellites at the end-of-life (EOL), sending them to a fiery reentry into the atmosphere. The preferable solution would be, of course, to send them to an immediate reentry at EOL. Nonetheless, depending on the operational orbit, it has been shown how the required maneuver might be too expensive in terms of fuel consumption (Rossi 2002). Therefore an alternative solution may be a delayed reentry, where the spacecraft is de-orbited to an orbit that will naturally decay, under the effect of air drag, within a fixed amount of years (a common accepted value is 25 years). If the operational orbit is still such that this cannot be realistically accomplished (e.g. in the case of the GEO satellites), the spacecraft should be sent to a "graveyard" orbit, i.e. into a zone of space where no other operative spacecraft are orbiting (Rossi 2002).

Fig. 4 summarizes the results of a typical long term evolution study performed by SDM. For the sake of simplicity only two scenarios are shown, out of the several recently studied ones (Rossi *et al.* 2004). The scenario tagged REFERENCE is the so-called business-as-usual evolution. Namely, it is characterized by a launch activity deduced from the traffic

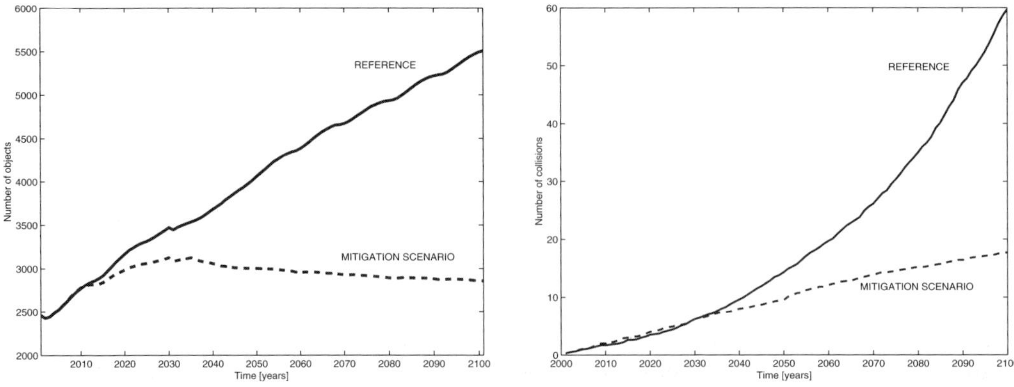

Figure 5. Number of objects larger than 1 m in LEO (left) and cumulative number of catastrophic collisions (right) with two different simulation scenarios.

observed over the last five years (1999-2003), adjusted by taking into account the phasing out of obsolete launchers and the introduction of new rocket families, with different hardware and mission characteristics. Mission related objects are released according to the current practices and no de-orbiting or re-orbiting of spacecraft and upper stages is performed at the EOL. The explosion statistics is based over the last 5 years (1999-2003) events, with an average of 2.4 explosions/year. Nonetheless, the introduction of mitigation measures on several classes of old and new upper stages systems (e.g., passivation of upper stages after burn), the existence, in orbit, of a large number of old upper stages prone to explode for at least a few decades, the progressive introduction, over the coming 5-25 years, of explosion prevention measures on the systems currently in use are taken into account. All these assumptions lead to a complete stop of explosions after 2030. As far as the production of slag is concerned, a minimum use of SRM is envisaged, based on current and planned practices. Two solid rocket motor firings are simulated for each GPS mission (perigee and apogee burns) and one of them is considered as well for each Chinese rocket Long March 3 GEO injection.

The left panel of Fig. 4 shows the number of objects larger than 1 cm as a function of time, between 0 and 40 000 km of altitude. In the REFERENCE scenario an almost linear growth is observed in the first 20-30 years. This growth is sustained by the remaining on-orbit explosions and by a progressively larger number of collisions. After 2030, collisions remain the only substantial source of centimetre sized objects and a more than linear pace takes place, leading to a number of objects, after 100 years, more than twice the initial population.

The right panel of Fig. 4 shows the number of objects larger than 10 cm in LEO. The initial growth rate levels off after 2020 due the cessation of most of the on-orbit explosions and the growth rate never returns to the values experienced in the first 10-20 years, even though the absolute increase is still about a factor 2 in one century for the REFERENCE scenario.

The population of the objects larger than 1 metre is dominated by the intact spacecraft and upper stages and the linear trend observed in the REFERENCE case is just the result of the net accumulation due to new launches in the considered time span (Fig. 5, left panel).

The picture changes considerably if some mitigation measures are introduced.

In the simulated mitigated scenario first the MRO are no more released starting from the year 2020, then the re-orbiting of spacecraft at EOL is performed. In particular, the GEO satellites are re-orbited at EOL to a circular orbit about 300 km above GEO (according to the IADC recommendation). Starting from the year 2010, all the spacecraft with perigee altitude $h_p < 1400$ km, or in high eccentricity orbits crossing the LEO region, are maneuvered to orbits with a residual lifetime $T_{res} = 25$ years. The spacecraft with $h_p \geqslant 1400$ km are re-orbited in a super-LEO storage zone above 2000 km (with a width of 100 km).

In Fig. 4 and Fig. 5 the effect of the mitigation measures are shown in comparison with the REFERENCE case. The measures seem to be able to maintain approximately stable the growth of centimetre sized debris, with a moderate linear trend. Moreover, the adoption of a combination of re-orbiting and de-orbiting at the EOL is able to stabilize the population of objects larger than 10 cm and 1 m in LEO, and even lead to slightly decreasing trends. These conclusions are supported by the cumulative number of catastrophic collisions in LEO (Figure 5, right panel). It is worth remembering that, due to the typical impact velocities in LEO, the fragmentation of a satellites requires a projectile larger than about 10 cm. In the REFERENCE scenario there is a progressive rise of the yearly number of catastrophic collisions, but in the mitigation scenario the collision rate remains approximately constant, with a difference of a factor 3 between the two cases.

These results suggest that the use of disposal orbits might be the solution to adopt in the near future to stabilize the debris environment and to guarantee the continued exploitation of the circumterrestrial space. Nonetheless it is again worth stressing that the compliance to the recommendations of international committees, such as the IADC, and the actual implementation of these measures sometimes may collide with technical, political or economical reasons that can slow down and reduce their efficiency. As an example, in the last few years only about 10 % of the GEO satellites have followed the de-orbiting procedures recommended by the IADC. In fact the operators often tend to privilege the immediate return of a few more months of operations, performed with the fuel that should be used for the de-orbiting maneuvers, as opposed to the long term benefit of all the space environment.

5. Acknowledgements

The studies on the long term evolution of the debris population and the development of the SDM software have been performed under ESA funding.

References

Alby, F., Lansard, E. & Michal, T. 1997, *Second European Conference on Space Debris*, Proceedings ESA SP-393 (Darmstadt), 589

Bendisch, J., Bunte, K., Klinkrad, H., Krag, H., Martin, C., Sdunnus, H., Walker, R. & Wegener, P. 2004, *Advances in Space Research* 34, 959

Flury, W., Massart, A., Schildknecht, T., Hugentobler, U., Kuusela, J. & Sodnik, Z. 2000, ESA Bulletin 104, 92

Schildknecht, T., Musci, R., Ploner, M., Beutler, G., Flury, W., Kuusela, J., de Leon Cruz, J. & de Fatima Dominguez Palmerod, L. 2004, *Advances Space Res.* 34, 901

Foster, J., Krisko, P., Matney, M. & Stansbery, E. 2003, 54th *International Astronautical Congress*, Paper IAC-03-IAA.5.2.02 (Brehmen, Germany)

Goldstein R.M., Goldstein S.J. & Kessler D.J. 1998, *Planet. Space Sci.* 46, 1007

Jackson, A., P. Eichler, Potter, R., Reynolds A. & Johnson, N. 1997, *Second European Conference on Space Debris*, Proceedings ESA SP-393 (Darmstadt), p. 279

Kessler D.J. & Cour–Palais B.G. 1978, *J. Geophys. Res.* 83, 2637

Liou, J.-C., Matney M.J., Anz-Meador P.D., Kessler D., Jansen M. & Theall R. 2002, *NASA/TP-2002-210780*

Milani A., Nobili A. & Farinella P. 1987, *Non gravitational perturbations and satellite geodesy*, Adam Hilger Ltd., Bristol and Boston.

Öpik, E.J. 1976 *Interplanetary Encounters*, Elsevier, New York, USA

Petit, J.–M. & Farinella, P. 1993, *Cel. Mech. Dyn. Astron.* 57, 1

Rossi A. & Farinella P. 1992, *ESA Journal* 16, 339

Rossi, A., Cordelli A. Farinella P. & Anselmo L. 1994, *J. Geophys. Res.* 99, No. E11, 23 195

Rossi, A., Anselmo L., Cordelli A., Farinella P. & Pardini C. 1998, *Planet. Space Sci.* 46, 1583

Rossi, A., Valsecchi, G.B. & Farinella, P. 1999, *Nature*, 399, 743

Rossi, A. 2002 *Journal of Spacecraft and Rockets* 39, No. 4, 540

Rossi A., Valsecchi G.B. & and Perozzi E. 2003, *13th AAS/AIAA Spaceflight Mechanics Meeting*, Paper AAS 03-185 (Ponce, Puerto Rico).

Rossi, A., Anselmo L., Pardini C., Valsecchi G.B. & Jehn R. 2004, *55th International Astronautical Congress*, Paper IAC-04-IAA.5.12.1.06 (Vancouver, Canada)

Valsecchi G.B., Rossi A. & Farinella P. 1999, *Space Debris* 1, 143

Valsecchi G.B. & A. Rossi 2002, *Cel. Mech. Dyn. Astron.* 83, 63

Dynamics of Populations of Planetary Systems
Proceedings IAU Colloquium No. 197, 2005
Z. Knežević and A. Milani, eds.

© 2005 International Astronomical Union
DOI: 10.1017/S1743921304008968

The use of the two-body energy to study problems of escape/capture

Ernesto Vieira Neto[1], O.C. Winter[1,2] and C.F. Melo[2]

[1]Grupo de Dinâmica Orbital e Planetologia – UNESP†.
email: ernesto@feg.unesp.br

[2]Instituto Nacional de Pesquisas Espaciais.

Abstract. The problem of escape/capture is encountered in many problems of the celestial mechanics – the capture of the giants planets irregular satellites, comets capture by Jupiter, and also orbital transfer between two celestial bodies as Earth and Moon. To study these problems we introduce an approach which is based on the numerical integration of a grid of initial conditions. The two-body energy of the particle relative to a celestial body defines the escape/capture. The trajectories are integrated into the past from initial conditions with negative two-body energy. The energy change from negative to positive is considered as an escape. By reversing the time, this escape turns into a capture. Using this technique we can understand many characteristics of the problem, as the maximum capture time, stable regions where the particles cannot escape from, and others. The advantage of this kind of approach is that it can be used out of plane (that is, for any inclination), and with perturbations in the dynamics of the n-body problem.

Keywords. Celestial mechanics, planets and satellites: general.

---

1. Introduction

A precise definition of gravitational capture is not yet established. There are several papers which define some kind of capture, as Heppenheimer (1975) with the passage of the particle by the inner Lagrangian point (L_1). In this work the definition given by Yamakawa (1992) is adopted, and it reads: if the local two-body energy of the particle relative to a celestial body changes from positive (hyperbolic motion) to negative (elliptic motion) it is considered a capture. The local two-body energy can be formulated as:

$$C3 = v^2 - \frac{2\mu}{r} \tag{1.1}$$

where v is the velocity of the particle relative to the celestial body, μ is the mass parameter of the body, and r is the distance from the body to the particle.

In the two-body problem $C3$ is constant, but any perturbation to the two-body problem will change this energy. For negatives values of $C3$ the orbit of a particle is closed (elliptic motion), for positives values the orbit is open (hyperbolic motion). Thus, in the course of the integration of the particle's orbit, we monitor the value of $C3$ and when a change of sign occurs, we record an escape or a capture.

In the restricted three-body problem, if a particle has a Jacobi constant C_J greater than the $C_J(L_1)$, and it is in the vicinity of the secondary, $C3$ will vary, but never assume a positive value. Thus this particle will never escape from the secondary, as expected. If the particle has C_J smaller than $C_J(L_1)$ (the bottle neck near L_1 is open), and it is in the vicinity of the secondary, it can have a negative value of $C3$, which may vary to a positive value at some time, that is, it can escape (see Figure 1).

† Address: Av. Ariberto Pereira da Cunha, 333, CEP 12516-410, Guaratinguetá, S.P. Brazil.

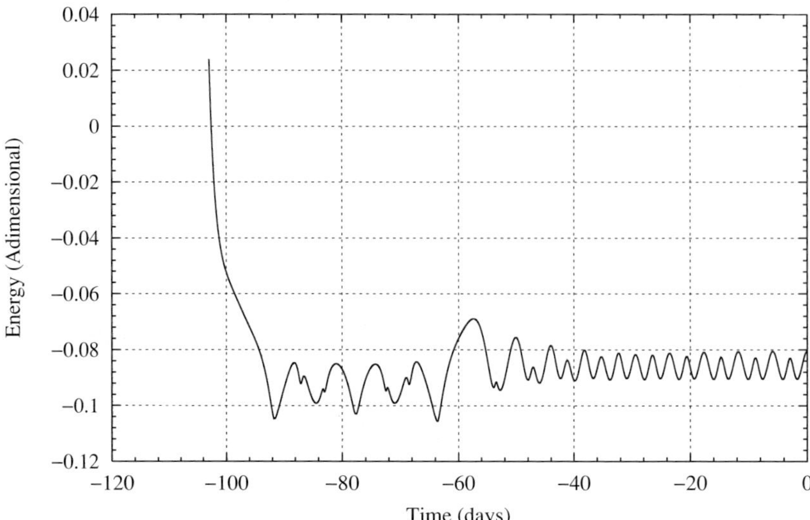

Figure 1. Variation of the energy ($C3$) for the restricted three body problem. The energy of the particle is measured relative to the secondary body (or planet).

We use this simple rule to study regions near a celestial body. There are other tools to make this kind of study, such as Poincaré Surface of section (see for example Hénon 1970), but they have some limitations. For example, it is very difficult to use the Poincaré Surface of Section to study the four-body problem, or inclined orbits in the three-body problem.

2. The method

To use this method it is necessary to choose a region in the vicinity of the celestial body, the variables to represent the region, and the dynamical model to be used. For example, the orbital elements of a particle relative to a planet can be used to study the stability of the trajectories. Orbits with eccentricities lower than one (elliptic orbits) have negative two-body energy; if we introduce another planet (or the Sun) the perturbations on the trajectory can make the particle crash into the planet, or escape from the planet. We define these trajectories as unstable, to distinguish them from the others which stay around the planet.

As an example, lets study a region around the Moon. The dynamical model was the circular restricted three-body problem. The variables were the selenocentric orbital elements. The initial values of the angular variables were fixed at zero, that is $i = 0°, \Omega = 0°, \omega = 0°, M = 0°$. The value of the semi-major axis and the eccentricity varied from $20\,000$ km to $35\,000$ km, and from 0.00 to 0.99, respectively. The trajectories were integrated from time $t = 0$ to $t = -5\,000$ days. The integration was made to the past because in reversing the time the escape becomes a capture. The result for this grid is shown in Figure 2 where the gray scale is associated with the escape time, that is, the time when $C3$ changes its sign from negative to positive.

On the left hand side of Figure 2 there is a region in white whose trajectories have the C_J value greater than $C_J(L_1)$. As expected, these trajectories do not escape in the time span covered by the integration. However, there are some trajectories which do not escape, but have C_J value smaller than $C_J(L_1)$; these are the trajectories on the right hand side of the curve $C_J = C_J(L_1)$. These trajectories are associated with the family of

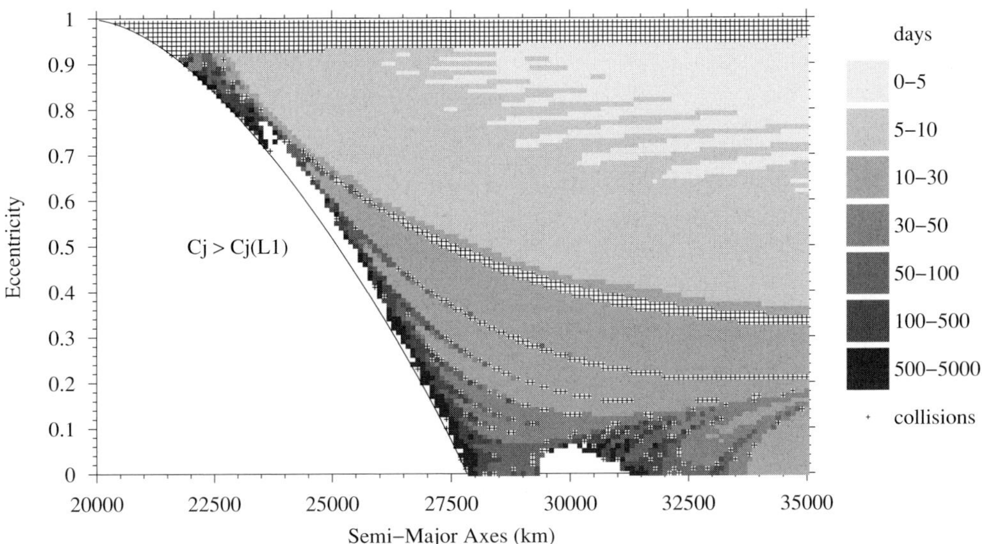

Figure 2. Study of the stable region around the Moon. The escape time is shown in gray scale. Each point is a trajectory with initial conditions $i = 0°, \Omega = 0°, \omega = 0°, M = 0°$, and a and e as on the axes. The white regions are trajectories that did not escape in $5\,000$ days.

direct periodic orbits denoted h_1 by Broucke (1968) (or family g_1 by Hénon 1970). The other white region, with semi-major axis about $30\,000$ km is associated with family h_2 by Broucke (1968) (or g_2 of Hénon 1970).

There are other remarkable characteristics in Figure 2. For example, on the top we see a region with crosses, this region is filled with trajectories which collide with the Moon due to the high initial eccentricity of the orbits. It is also possible to see these collision lines in the figure with trajectories with low initial eccentricity.

The h_2 family is better seen when we change the pericenter to $\omega = 180°$, as it is shown in Figure 3. In Figure 2 the particles have initial conditions at opposition with respect to the Moon and in Figure 3 the trajectories initiate at conjunction with respect to the Moon. Thus, with simple change of the initial values of the relevant variable we can have better information about the characteristics of the region.

In Winter & Vieira Neto (2002) these regions were discussed in detail. In that paper the Poincaré Surface of Section method was used, and it was shown that the white regions are associated with periodic orbits.

As stated before, the advantage of the two-body energy method is the possibility to study regions out of plane and to use more complex dynamics. In Figure 4 we explore this possibility by using the four-body dynamical model to study the stability of the same region as of Figure 3. In this case Earth, Moon and Sun interact with each other, but all moving in the same plane. Here it is not possible to make use of the usual Poincaré Surface of Section method because there is not a Jacobi constant for this problem, thus only the use of the two-body energy in this problem can reveal the location of the trajectories which will not escape in the time span of integration. Although the area of stability is reduced in Figure 4, we see that the solar perturbation on the region does not destroy it.

With the two-body energy method it is also possible to study other effects on the trajectories, like the effect of the Moon's orbital inclination, or that of the Earth's and the Moon's orbital eccentricities, one can use dynamical models with variation of mass in the three-body problem, as in Vieira Neto *et al.* (2004). In this case, due to the change of the gravitational influence of the planet, the particles integrated with a negative step

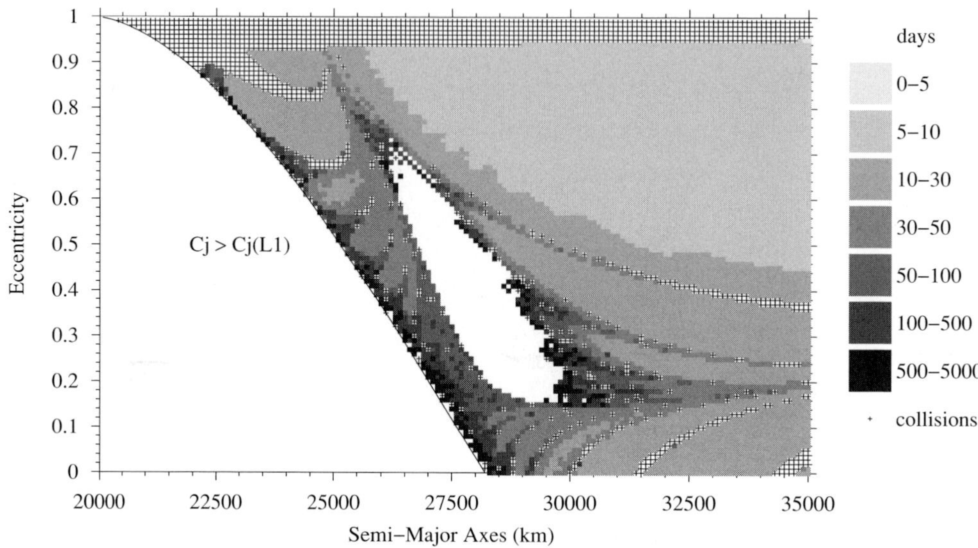

Figure 3. Study of the stable region around the Moon. The escape time is shown in gray scale. Each point is a trajectory with initial conditions $i = 0°, \Omega = 0°, \omega = 180°, M = 0°$, and a and e as on the axes. The white regions are trajectories that did not escape in 5 000 days.

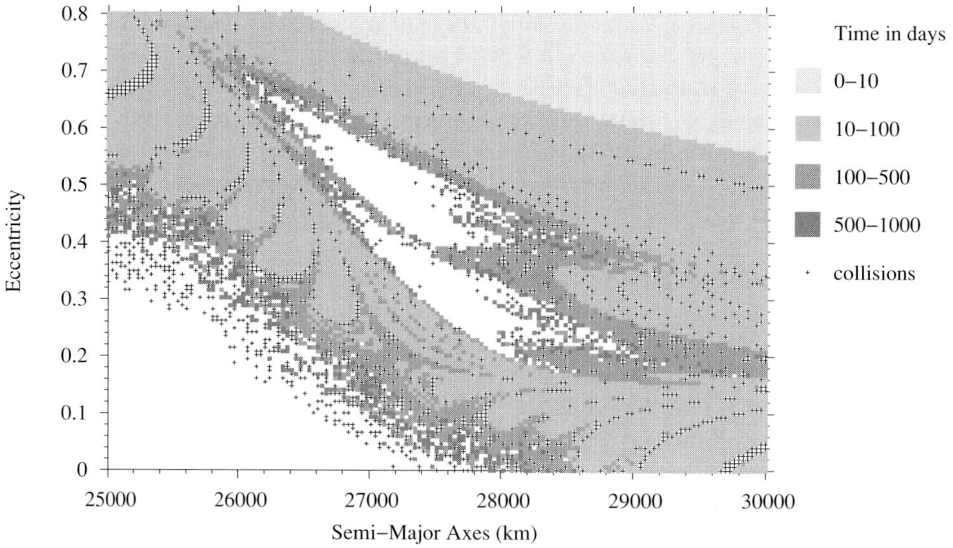

Figure 4. Study of the stable region around the Moon. Each point is a trajectory with initial conditions $i = 0°, \Omega = 0°, \omega = 180°, M = 0°$, and a and e as on the axes. The white regions are trajectories that did not escape in 5 000 days. The dynamical model used in this case is the four body problem with the particle, Earth, Moon and the Sun.

in time escaped. In this kind of analysis, the mass of the planet when the escape occurs is used as indicative for the analysis instead of the escape time. Reversing the time we found that particle in heliocentric orbit approached the planet and, due to the mass variation of the planet, became a satellite.

Another approach is to study the lowest energy of the capture around the Moon. In Figure 5 we vary the energy and the pericenter distance of the particle relative to the Moon and we measure the time for a trajectory to escape (Winter *et al.* 2003). As in

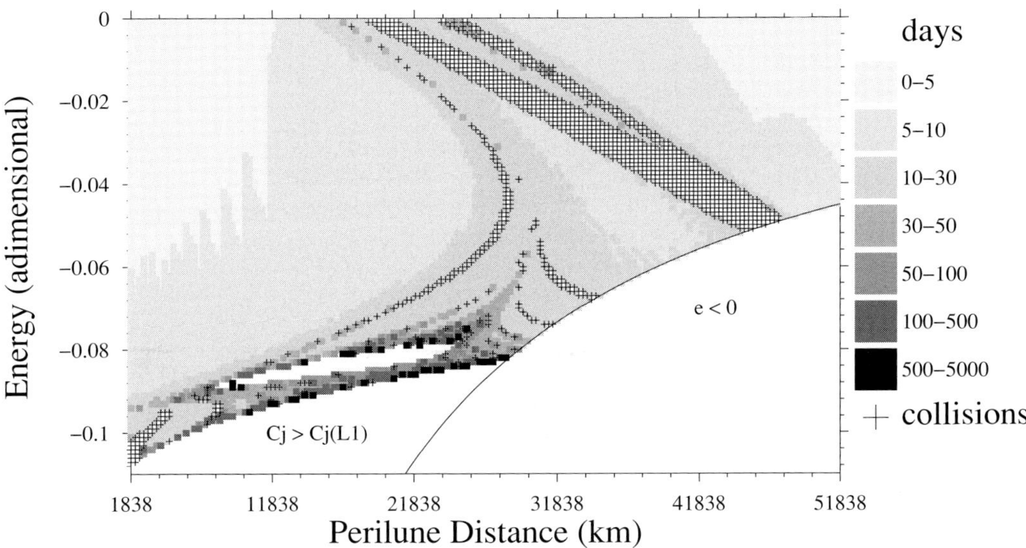

Figure 5. The minimum orbital energy of a particle as a function of the perilune position. The gray scale represents the time, in days, that the trajectory takes to escape from the Moon. The pericenter is pointed towards the Earth.

the previous figures, there is a region where the particle can not escape from the Moon due to the value of the Jacobi constant. In the lower right corner of the figure there is a region where the eccentricity of the particle is negative for the given values of energy and pericenter distance. The time for the capture has an exponential growth, and the longest capture times are detected at the border defined by the value of the Jacobi constant $C_J(L_1)$. The white island inside the grey region, is a region where the trajectories do not escape during the integration time of 5 000 days. Basically we have the same structures as on the other figures but this one gives us information of the lowest energy possible to achieve certain perilune distance and the time that the particle takes to achieve that position and energy.

3. Conclusion

The two-body energy of a particle, relative to some celestial body, can be used to study escape/capture problems. This approach can give us ample information as discussed in this work. This method have some difficulties, as the extensive consumption of computational time when integrating trajectories in the stable regions. This computational time depends on the total time used for the trajectory integration, and for some systems the integration time must therefore be long. It is, however, easier than with other methods to study the stability, and it can also be used in more general conditions such as out of the orbital plane and with dynamics which includes mass variation, or gas drag.

Acknowledgements

We would like to acknowledge the support of FAPESP under grant number 02/00344-7 and CNPq. Also we are grateful to an anonymous referee for his useful comments.

References

Broucke, R. A. 1968, *JPL-NASA, Technical Report* 32-1168
Hénon, M. 1970, *Astron. Astrophys.* 9, 24
Heppenheimer, T. A. 1975. *Icarus* 24, 172
Vieira Neto, E., Winter, O. C. and Yokoyama, T. 2004, *Astron. Astrophys.* 414, 727
Winter, O. C. and Vieira Neto, E. 2002, *Astron. Astrophys.* 393, 661
Winter, O.C., Vieira Neto, E. and Prado, A.F.B.A. 2003, *Advances Space Res.* 31, 2005
Yamakawa, H. 1992, PhD Thesis, University of Tokyo.

Dynamics of Populations of Planetary Systems
Proceedings IAU Colloquium No. 197, 2005
Z. Knežević and A. Milani, eds.

© 2005 International Astronomical Union
DOI: 10.1017/S174392130400897X

Numerical modelling of the paleotidal evolution of the Earth-Moon System

Eugene Poliakow

Central(Pulkovo) Astronomical Observatory
Russian Academy of Sciences
St.Petersburg, Russia
email: poliakow@rol.ru

Abstract. The results of a numerical simulation of the tidal evolution of the Earth-Moon system during the Phanerozoic epoch (the last 600 million years) are given. In most of the researches devoted to the solution of the problem the authors simplified and parametrized very complicated tidal phenomena to a primitive integral hump on the Earth's surface. As distinct from these the numerical model of the ocean tides in its most complete form is the core of the present study: the problem is solved for a viscous liquid in a paleoocean with variable outlines and depth allocations stimulated by the drift of the lithospheric platforms; the global interaction between the ocean and earth tides and the fluctuations of the gravitational field of the planet caused by them are taken into account. The astronomical component of the model is simplified. It is assumed that the Earth-Moon system is isolated, the Moon's orbit circular and the moment of inertia of the Earth constant during the Phanerozoic epoch. It is shown that the evolution of the Earth-Moon system during the Phanerozoic was nonuniform and that the primary role in this process belongs to the geodynamic factor.

Keywords. Paleotides, Earth-Moon system, evolution, energy dissipation, geodynamics

---

1. Introduction

Over a hundred years ago G. Darwin made a hypothesis about formation of an Earth-Moon system and constructed, speaking in modern terms, a mathematical model of its tidal evolution (Darwin G.H. (1978), Ball P.S. (1900)). According to Darwin, the Moon was formed from substance of the Earth which has been thrown out into an orbit by the resonant forces. The satellite appearing in close orbit induced high tides in the planet's body. Because the Moon revolved about the Earth in forward direction and the Earth's angular rate of rotation about its axis was greater than the Moon's angular orbital velocity, the tidal hump due to action of the friction forces "ran out" from under the Moon forward. Mutual attraction of hump and Moon imparted to the latter, acceleration in the orbit and slowed-down the Earth's diurnal rotation. The Moon gradually moved away from the Earth until a synchronization of the planet's rotation and satellite's revolution took place. Further on, however, the tidal influence of the Sun on the Earth slowed down its rate of rotation to a still greater extent, the tidal hump appeared to be behind the Moon and the process of evolution of the Moon's orbit continued with opposite sign. Finally, the Moon will approach the Earth so closely that the forces of attraction will disintegrate it. This, in general terms, is the picture of evolution of the Earth-Moon system calculated by Darwin.

The subsequent century did not bring fundamental changes the in theory of tidal evolution for this system, but it proved the failure of Darwin's hypothesis of the formation of our satellite. Two theories appeared: capture of the Moon by the Earth (Alfven X. & Arrhenius G. (1976), MacDonald G.J.F. (1975)) and accumulation of the Moon in the near-Earth swarm. The first theory finds it difficult to explain what happened to the energy liberated due to capture in the near-Earth orbit of a body that massive and why on the Earth no traces of this extraordinary event have been found anywhere. The theory of accumulation experiences difficulties in an attempt to correlate the length of the evolution with the Earth's age exceeding 4.5 bln years (Russkol E.A. (1975)).

The length of the evolution is defined by the magnitude of tidal friction, in other words, how high the rate of tidal energy dissipation is: the higher the dissipation, the more intensive the evolutionary process. The tidal friction can be characterized by the so-called effective lag angle, δ, indicating how far the tidal hump is carried away from under the Moon and by what angle the Moon lags compared with the tide as the meridian is passed. At the present time, the magnitude of is estimated to be $2.5 - 3^o$. The data have been obtained both in a numerical simulation of the tides in the World Ocean (Gordeev P.G., Kagan B.A. & Polyakov E.V. (1976)) and on the basis of observation data for the solar eclipses in the ancient world (Morrison L.V. (1978)). Using the present-day magnitude of δ in calculations of the tidal evolution yields its length equal to 1.5-1.75 bln years (Russkol E.A. (1975)), which is a serious argument in favor of the capture hypothesis and constitutes the main difficulty of the accumulation hypothesis. Attempts to overcome the latter are usually associated with the assumption that in the past the lag angle was smaller than the present value although justification of this assumption appears to be insufficiently convincing.

2. Construction of equations

In the present paper, in contrast with previous investigations of the tidal evolution for the Earth-Moon system, the core of the problem is a mathematical model of the tides in a most complete presentation, to which we connected a simple evolution model. Thus, in order to calculate the evolution rather than use an a prioriassigned lag angle δ, a complex tidal problem is solved and. In the tidal problem itself some quantities usually taken as constants, namely the angular rate of rotation of the Earth and the frequency and amplitude of the tide-generating force, become variable and are computed in the evolution model. A mathematical description of the problem formulation can be found in Gordeev P.G., Kagan B.A. & Polyakov E.V. (1976), Polyakov E.V. (1986).

The model in question describes an Earth-Moon system isolated in space. The Moon's orbit is assumed to be circular coplanar with the plane of the Earth's equator. The latter plane is considered as an elastic-viscous body covered by a gravitational fluid shell, the ocean. The continents jutting in the ocean are in permanent movement, so that the ocean configuration changes continuously, the volume of water in the ocean remaining constant. The Moon, a point mass, induces tides in the terrestrial crust and ocean where allowance is made for bottom friction and horizontal turbulent exchange. Moving above the bottom the tidal wave causes its deformations, i.e. deflections of the crust under the crest and a rise under the wave trough. As a result of changes in the planet's form under the action of the tide-generating forces and the effects of loading and self-attraction and also redistribution of the water masses the planet's gravitational field experiences disturbances which, in turn, take part in formation of the tides in the ocean and Earth's body. In this way the effect of self-attraction of the tides is manifested.

The calculation of the Earth-Moon system evolution has been performed for the Phanerozoic epoch (to 600 mln years into the past), a short time interval compared with the system age. This limitation is connected with the fact that as of today the reconstruction of the position of the continents had been carried out only for the period specified (Zonnenshain L.P. & Gorodnitsky A.M. (1979)).

An integro-differential problem has been solved. With allowance for the assumption on the harmonic character of the oscillations, the linearized equations of tidal dynamics are written in the form (Gordeev P.G., Kagan B.A. & Polyakov E.V. (1976))

$$(r - i\sigma)\overline{\mathbf{w}} + A\overline{\mathbf{w}} - k_l \Delta\overline{\mathbf{w}} + gH\nabla\xi = gH\nabla(\gamma_L\bar{\xi}^+ + \bar{\xi}^\oplus), \tag{2.1}$$

$$-i\sigma\bar{\xi} + div\overline{\mathbf{w}} = 0. \tag{2.2}$$

Here, $\overline{\mathbf{w}}$ is the integral transport vector, g is the gravity acceleration, H is the depth, $\bar{\xi}$ is the height of the ocean tide with respect to the bottom, $\bar{\xi}^+$ is the static tide in the ocean for a rigid Earth, γ_L is the Love factor for an elastic Earth, r and k_l are the coefficients of bottom friction and turbulent exchange, σ is the frequency of the driving force, A is the Coriolis matrix,

$$A = \begin{pmatrix} 0 & -\lambda \\ \lambda & 0 \end{pmatrix},$$

where $\lambda = 2\Omega\cos\theta$; Ω is the angular rate of rotation of the Earth. The origin of the coordinates is situated in the North Pole, θ is the colatitude, φ is the longitude. The integral term $\bar{\xi}^\oplus$ describes the deformations of the Earth's crust and disturbances of the gravitational potential as a response to the ocean level oscillations

$$\bar{\xi}^\oplus = \frac{1}{2\pi} \int\limits_0^{2\pi}\int\limits_0^{\pi} \bar{\xi}(\theta', \varphi') \sum_n \gamma_n'\alpha_n \sum_n N_{nq}^{-1} P_n^q(\cos\theta) P_n^q(\cos\theta') \begin{pmatrix} \cos q\varphi \cos q\varphi' \\ \sin q\varphi \sin q\varphi' \end{pmatrix} \sin\theta' d\theta' d\varphi' \tag{2.3}$$

where γ_n' and α_n are coefficients for calculating induced disturbances of the static tide, $P_n^q(\cos\theta)$ is the Legendre connected function of the n order and to the q power; and N_{nq} is a normalizing factor. Set (1-3) is supplemented by the nonslip condition on the contour β of the region in question

$$\overline{\mathbf{w}}|_\beta = 0. \tag{2.4}$$

For epochs of the geological past, in addition, it is necessary to define the parameters ξ^+ and σ of the driving force. For the M_2 wave the following relations are valid (Marchuk G.I. & Kagan B.A. (1983)):

$$\xi^+ = (3Cm/4M)a^4 R^{-3} \sin^2\theta \cos(\sigma t - 2\varphi), \tag{2.5}$$

$$\sigma = 2(\Omega - \omega). \tag{2.6}$$

Here, C is the amplitude coefficient † of the M_2 harmonic; m and M are the masses of the Moon and Earth, respectively, a is the mean radius of the Earth, R is the radius of the Moon's orbit, and ω is the Moon's angular orbital velocity. The values of R, Ω and

---

† According to Marchuk G.I. & Kagan B.A. (1983) $C = (1 - 5e^2/2)\cos^4(i/2)$, where e is the eccentricity and i is the inclination of the Moon's orbit. Although the set of evolution equations has been written for the case of the satellite's circular orbit ($e = 0$) lying in the plane of the planet's equator ($i = 0$), the magnitude of C is assumed to be equal to the present magnitude $C_0 = 0,9081$ in order to obtain results corresponding precisely to the M_2 wave and subsequently compare them with those of further investigations.

ω are defined using the equation for impulse moment conservation in the Earth-Moon system

$$I\frac{d\Omega}{dT} + \frac{Mm}{M+m}\frac{d}{dT}(R^2\omega) = 0, \tag{2.7}$$

Kepler's third law is

$$R^3\omega^2 = G(M+m) \tag{2.8}$$

and the expression for moment L of the tidal force zonal component [8]

$$L = I\frac{d\Omega}{dT} = 6.4\pi\rho_0\gamma_L\frac{3}{4}CGma^4R^{-3}D_{22}^-\sin\varepsilon_{22}^-, \tag{2.9}$$

where T is the time in the geological scale directed into the past, I is the Earth's moment of inertia equal to $0,3315Ma^2$ (Marchuk G.I. & Kagan B.A. (1983)), G is the gravitational constant, ρ_0 is the mean density of water, D_{22}^- and ε_{22}^- are parameters obtained from the coefficients for expansion of the amplitude of level ξ into a series using the spherical function in a system of coordinates with westerly direction of the longitudinal axis φ, therefore here, in contrast to Marchuk G.I. & Kagan B.A. (1983) in expression (11), use is made of parameters D_{22}^- and ε_{22}^-, in place of D_{22}^+ and ε_{22}^+.

Using (8) and (9), Eq. (7) takes the form

$$\frac{d}{dT}R^{1/2} = -\frac{L}{Mm}\left(\frac{M+m}{G}\right)^{1/2}. \tag{2.10}$$

Set (1)–(6), (8)–(10) is complemented with initial conditions corresponding to the present epoch: $\Omega_0 = 7.2921\cdot10^{-5}s^{-1}$, $\omega_0 = 2.6617\cdot10^{-6}s^{-1}$. The proposed formulation of the tide problem makes it possible to find the solution based only on the masses of the Moon and the Earth, the mean radius of the latter, and the depth of the ocean bottom relief: $m = 7.348\cdot10^{22}$ kg, $M = 5.976\cdot10^{24}$ kg, $a = 6.371\cdot10^6$ m, $H = H(\theta,\varphi)$.

3. Results

It should be noted that the tidal model in question has been worked out for solving an independent problem on interaction between the ocean and Earth tides, precalculation of the tides in an ocean with an elastic bottom and investigation of the tidal variations in the force of gravity. Having been used for its direct purpose it has recommended itself quite well. The results of calculations of the tidal harmonics M_2, S_2, K_1 and O_1 for the present–day epoch (Fig. 1) are in satisfactory agreement with data by other authors (Accad Y. & Pekeris C.L. (1978), Zahel W. (1978)) and observations (Schwiderski E.W. (1978)). This makes it possible to acheive some credibility for the solutions of the tidal problem also for past epochs (Fig. 2–4).

A note on the angle δ: clearly, the tides in the ocean have nothing in common with the hump and are distinguished by a complex spatial structure. Adding together the moments of the attraction forces which act between the Moon and each of the particles in the ocean we obtain integral moment L of the tide-generating forces. One can imagine a tidal hump producing a moment equal to L in magnitude. The crest of the hump producing moment L must be spaced from the point under the Moon by the angle δ.

Application of the model for calculating the evolution of the Earth–Moon system has led to unexpected results which may be used as a reason for revision of long-standing concepts about the character of secular changes in the system parameters and eliminate

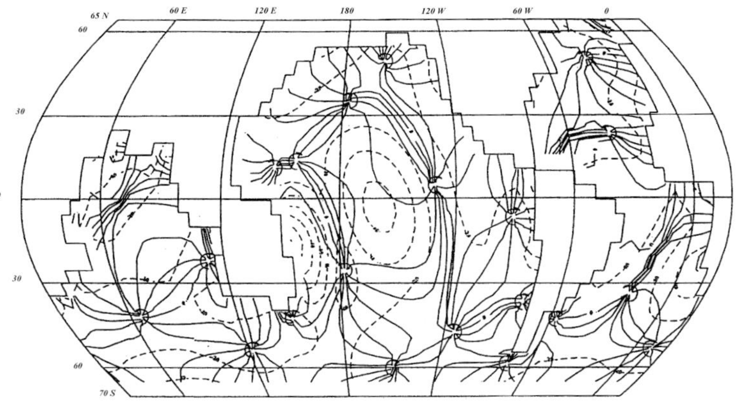

Figure 1. Tidal map. M_2 wave. Present epoch.

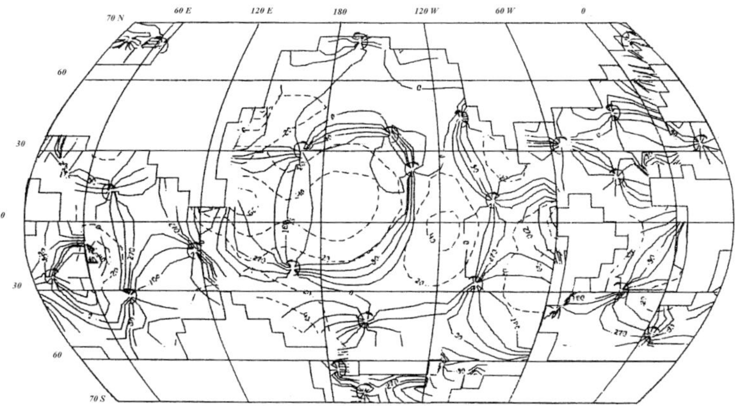

Figure 2. Tidal map. M_2 wave. 70 Myr ago.

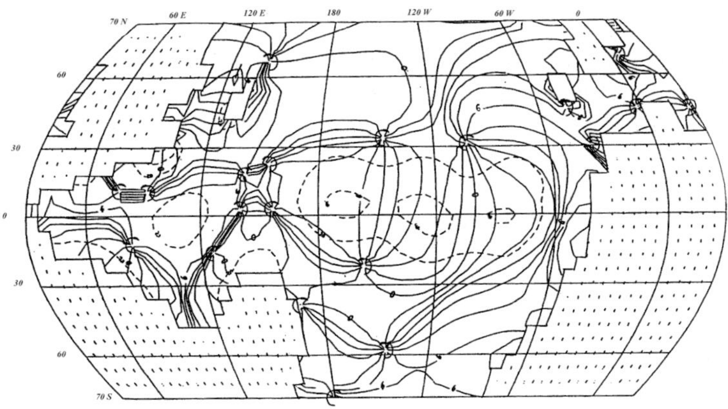

Figure 3. Tidal map. M_2 wave. 240 Myr ago.

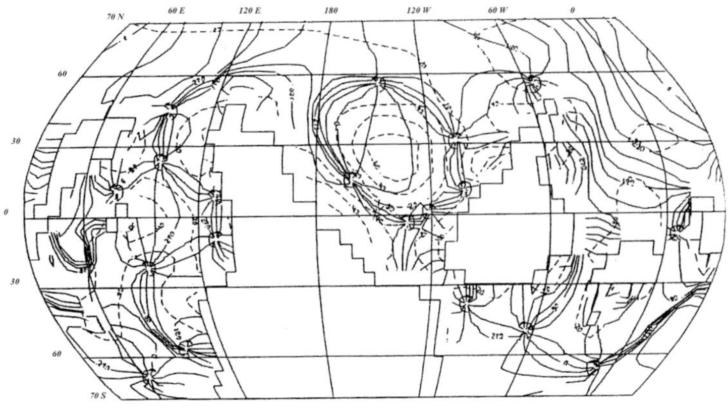

Figure 4. Tidal map. M_2 wave. 570 Myr ago.

Main characteristics of tidal evolution of the Earth–Moon system. M_2 wave

$T,$ 10^6 yr	$\Omega,$ $10^{-5}\ s^{-1}$	$\omega,$ $10^{-6}\ s{-}1$	$\sigma,$ $10^{-4}\ s^{-1}$	$A+,$ cm	$L,$ $10^{16}\ n\cdot m$	$\delta,$ deg	Q	$\dot{R},$ cm/yr	$\dot{\omega},$ "/cent^2	$\dot{\tau},$ s/cent
0	7.292	2.662	1.405	24.23	3.42	5.18	11.0	2.91	19.7	1.59
10	7.306	2.665	1.408	24.28	3.14	4.75	12.0	2.68	18.4	1.46
50	7.349	2.675	1.416	24.46	2.17	3.27	17.5	1.83	12.4	0.98
100	7.382	2.682	1.423	24.60	0.94	1.43	40.0	0.81	5.6	0.43
200	7.407	2.688	1.428	24.71	0.40	0.61	93.9	0.34	2.3	0.18
300	7.423	2.691	1.431	24.77	0.53	0.81	70.7	0.46	3.2	0.24
350	7.439	2.695	1.434	24.85	0.93	1.41	40.6	0.80	5.5	0.42
400	7.452	2.698	1.436	24.89	0.66	1.01	56.7	0.57	3.9	0.30
450	7.476	2.703	1.441	24.99	1.80	2.70	21.2	1.54	10.7	0.80
500	7.527	2.715	1.451	25.21	2.95	4.48	12.8	2.53	17.7	1.31
570	7.617	2.736	1.469	25.60	2.54	3.82	15.0	2.19	15.5	1.11

the contradiction, cited above, between the age of the system and the length of its tidal evolution.

It appears that moment L of the tide-generating forces and the lag angle δ (see table) changed in the Phanerozoic in a complicated manner. Thus, 500 mln years ago they had maximum magnitudes ($2.95\cdot 10^{16}$ nm and 2.24^o, respectively) close to present magnitude ($3.42\cdot 10^{16}$ nm and 2.59^o) and during periods from the Devonian to the Cretaceous (400 – 100 mln year) their magnitudes were a factor of $3-8$ below the present values. The table, apart from L and δ, shows the mechanical quality Q of the Earth's ocean-crust oscillatory, the system rate of separation $\dot{R}$ and secular acceleration of the Moon's $\dot{\omega}$, and the lengthening of the terrestrial day $\dot{\tau}$. The magnitudes of the above parameters obtained from an analysis of movement of the satellites and calculated on a model for the present epoch are in satisfactory agreement.

Figure 5 shows graphs for the number of days in one year N and for the length of the terrestrial day τ in the past. It is not difficult to see that these parameters changed unevenly in the Phanerozoic. Their smallest changes correspond to the period with low rate of dissipation $-\dot{E}$. The curve of relative dissipation $-\dot{E}/\dot{E}_o$, where $-\dot{E}_o$ is the

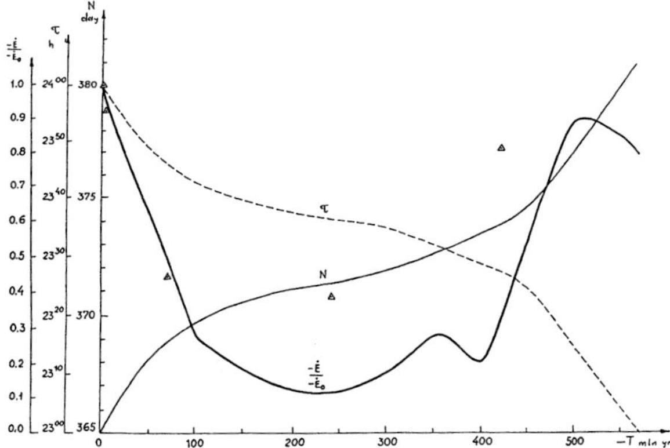

Figure 5. Changes during the Phanerozoic in the dissipation $-\dot{E}$ of the tidal energy normalized by its present-day (calculated) magnitude $-\dot{E}_o = 2.41 \cdot 10^{12}W$, length τ of the mean solar days and number N of days in the year

present-day rate of dissipation, is nearly coincident in shape with curves for L and δ (not given in Fig. 5).

4. Analysis

This complex behavior of the dissipation (or L or δ) in the past cannot be explained only by a change in the Moon's orbit. It remains to assume that the character of the evolution of the Earth–Moon system is also dependent on the ocean configuration varying as a result of the movement of the lithospheric plates.

Figure 6 shows the results of checking this assumption. In order to construct the figure, 81 numerical experiments have been performed, each representing a solution of the problem about tides in the paleo-ocean (and subsequent calculation of tidal energy dissipation) for all possible combinations of nine ocean configurations (one, present-day, plus eight paleoreconstructions borrowed from Zonnenshain L.P. & Gorodnitsky A.M. (1979)) and nine different radii of the Moon's orbit. It can be clearly seen that the movement of the continents has a stronger influence on the dissipation than the increase of the Moon's orbital radius.

In Fig. 6 are compared two curves characterizing the separation of the Moon from Earth. Curve 1 has been calculated in a traditional manner on the assumption of the constant lag angle; curve 2 has been obtained as the problem in question has been solved. Calculation of the evolution based on the tidal model in which the continental drift has been allowed for leads to a more than two-fold reduction of the Moon's separation. This strong influence of the ocean configuration on the tidal energy dissipation should be sought for in resonant nature of the excitation of the ocean tides. If we assume that the dissipation is proportional to the tidal energy, then based on Fig. 6 one may come to some conclusion about evolution of the ocean resonant properties. Namely: at the present time and 500 mln years ago the ocean configuration contributed to "adjustment" of the tides to frequency of the disturbing force and the amplitude of the tidal waves increased, resulting in growth of their gravitational interaction with the Moon and intensification of the evolutionary process. Vice versa, mismatching of the frequencies in a range of 400-150 mln years ago led to slowing down of the evolution.

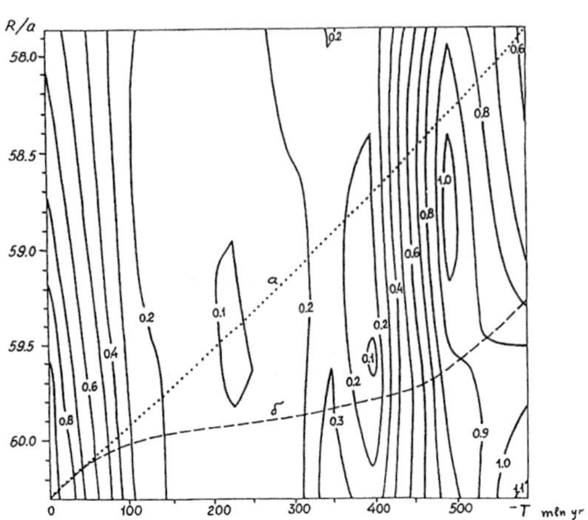

Figure 6. Dependence of tidal energy dissipation $(-\dot{E}/\dot{E}_o)$ on the evolution of the Earth-Moon system (along the ordinate axis) and the drift of the lithospheric plates (along the abscissa axis). Superimposed two plots of the separation of Moon from Earth as a function of the time ine past: 1 - with constant lag angle δ; 2 - based on a calculation of the paleotides

5. Conclusions

The results presented indicate that reaching a consistent and noncontradictory solution to the problem of Earth-Moon system evolution is only possible in the case of joint consideration of the astronomical, geophysical and geodynamic influencing factors, the latter two, as it has become clear, having an important role.

References

Accad, Y. & Pekeris, C.L. 1978, *Phil. Trans. Roy. Soc.* A290, 235

Alfven, X. & Arrhenius, G. 1976, *NASA Tech.*, p. 512

Ball, P.S. 1900, *Mathesis*, Odessa, p. 104, (in Russian)

Darwin, G.H. 1978, *Phil. Roy. Soc.* A170, 447

Goldreich, P. 1975, *Mir Publ.*, Moscow, 97, (in Russian)

Gordeev, P.G., Kagan, B.A. & Polyakov 1976, *Dokl. Akad. Nauk SSSR.* 228(4), 817, (in Russian)

Hansen & Kirk S. 1982, *Rev. Geophys. Space Phys.* 20(3), 457

Kagan, B.A. & Maslova, N.B. 1994, *Earth, Moon, Planets* 66,173

MacDonald, G.J.F. 1975, *Mir Publ.* Moscow, 9, (in Russian)

Marchuk, G.I. & Kagan, B.A. 1983, *Gidrometeoizdat Publ.* Leningrad, p. 360,(in Russian)

Morrison, L.V. 1978, in: P. Brosche & J. Sundermann (eds.) *Tidal Friction and the Earth Rotation*, Springer-Verlag, p. 35

Polyakov, E.V. 1986, *Izv. Akad. Nauk SSSR. Oceanic and Atmospheric Physics.* 22(5), 505, (in Russian)

Ray, R.D., Bills, B.G. & Chao, B.F. 1999, *J. Geophys. Res. - Solid Earth* 104(B8), 17653

Russkol, E.A. 1975, *Nauka Publ*, Moscow, p. 189, (in Russian)

Schwiderski, E.W. 1978, Naval Surface Weapons Center. - Dahlgren Lab. - Dahlgren. p. 88

Zahel, W. 1978, in: P.Brosche & J.Sunderman (eds.) *Tidal Friction and the Earth's Rotation*, Springer-Verlag, p. 132.

Zonnenshain, L.P. & Gorodnitsky, A.M. 1979, *Nauka Publ.*, Moscow, 327, (in Russian)

Dynamics of Populations of Planetary Systems
Proceedings IAU Colloquium No. 197, 2005
Z. Knežević and A. Milani, eds.

© 2005 International Astronomical Union
DOI: 10.1017/S1743921304008981

The k:k+4 resonances in planetary systems

Benoît Noyelles and Alain Vienne

IMCCE - Laboratoire d'Astronomie de Lille, UMR 8028 du CNRS, 1 impasse de
l'Observatoire, 59000 Lille, France
e-mail : Benoit.Noyelles@imcce.fr

Abstract. In this study we give a first description of De Haerdtl's 3:7 inequality between the
Jovian satellites Ganymede and Callisto and 1:5 inequality between the Saturnian Titan and
Iapetus and the resonant arguments associated. For each inequality, 19 arguments are associated.
The overlapping of resonant zones induces stochasic layers that the system might have crossed
in the past thanks to tidal dissipation.

Keywords. Celestial mechanics, planets and satellites: individual: Jupiter - Saturn

1. Introduction

The problem of high order resonances has often been studied in the case of resonances
between asteroides and Jupiter (see for instance Moons & Morbidelli (1995)) but there
might have been such resonances in satellite systems, since there is a 3:7 great inequality
(De Haerdtl's inequality, see De Haerdtl (1892)) between the Jovian satellites Ganymede
and Callisto and 1:5 inequality between the Saturnian ones Titan and Iapetus. The case
of satellites is very different from asteroides because there are not 3-bodies restricted
problems, so there are much more arguments involved, but there are fewer bodies. This
study is based on Ganymede - Callisto and Titan - Iapetus problems.

2. Physical description of the two systems

Physical and dynamical data related to Ganymede-Callisto and Titan-Iapetus are listed
in Tables 1 and 2. It is important to notice that the ratio of the mean motions of
Ganymede and Callisto is very near from $\frac{3}{7}$ and the one of Titan and Iapetus very near
from $\frac{1}{5}$. The mass ratios are important too. The one of Ganymede-Callisto is near 1, so
this problem is very different from an asteroide-Jupiter problem, which can be modeled by
a 3-bodies restricted problem. The one of Titan-Iapetus is here 76 but could be evaluated
between 45 and 125 because of a large uncertainty on Iapetus' mass.

Table 1. Physical data related to Ganymede and Callisto. The mean motions, semimajor axis
and inclinations come from E5 ephemerides (Lieske (1998)), the eccentricities from Burns (1986),
the mean radii from the IAU recommandations (Seidelmann *et al.* (2002)) and the masses from
Pioneer and Voyager spacecrafts (Campbell & Synnott (1985)).

	Ganymede	Callisto
Mean motion n $(rad.\,d^{-1})$	0.878132432119	0.376443069266
Semi-major axis a (km)	1070427.5	1882758.6
Eccentricity e	0.0021	0.007
Inclination i	$11'8"$	$15'10"$
Mean radius r (km)	2632.345	2409.3
Mass m (M_j)	7.80431×10^{-5}	5.66832×10^{-5}

Table 2. Physical data related to Titan and Iapetus. The mean motions, semi-major axis, eccentricities, inclinations as well as Iapetus' mass come from TASS1.6 (Vienne & Duriez (1995)), whereas Titan's mass comes from Pioneer and Voyager data (Campbell & Anderson (1989)) and the mean radii from the IAU recommmamdations (Seidelmann *et al.* (2002)).

	Titan	Iapetus
Mean motion n $(rad.d^{-1})$	0.3940956849	0.07924104005
Semi-major axis a (km)	1221728.7	3559387.7
Eccentricity e	0.0289	0.029
Inclination i	$27' - 50'$	$11° - 18°$
Mean radius r (km)	2575	718
Mass m (M_s)	2.36638×10^{-4}	3.1×10^{-6}

3. Location of the resonances

The common point between two groups of resonances of the same order (here 4) is that they have the same number of arguments, every argument being likely to generate a resonance if considered isolated. For k:k+4 resonances, there are 19 arguments of the form:

$$\phi = k\lambda - (k+4)\lambda' + q_1\varpi + q_2\varpi' + q_3\Omega + q_4\Omega' \qquad (3.1)$$

where $q_1 + q_2 + q_3 + q_4 = 4$ and $q_3 + q_4$ is even, so as to respect D'Alembert law. At the middle of the resonance zone, $< \phi >= 0$, what gives:

$$kn - (k+4)n' + q_1\dot\varpi + q_2\dot\varpi' + q_3\dot\Omega + q_4\dot\Omega' = 0 \qquad (3.2)$$

A knowledge of the velocities of nodes and pericentres could allow us to determine a ratio α between the semi-major axis, deduced from the mean motions. Velocities of the nodes and pericentres (Table 3) come from Lainey's ephemerides (Lainey, Arlot & Vienne (2004)) for Ganymede and Callisto and TASS1.6 (Vienne & Duriez (1995)) for Titan and Iapetus.

Table 3. Estimations of velocities of pericentres and nodes of the involved satellites in the laplacian plane

	Pericentre	Node
Ganymede	$1.27274 \times 10^{-4}d^{-1}$	$-1.24913 \times 10^{-4}d^{-1}$
Callisto	$3.20651 \times 10^{-5}d^{-1}$	$-3.05609 \times 10^{-5}d^{-1}$
Titan	$2.44596 \times 10^{-5}d^{-1}$	$-2.44524 \times 10^{-5}d^{-1}$
Iapetus	$5.40641 \times 10^{-6}d^{-1}$	$-5.27184 \times 10^{-6}d^{-1}$

So as to really model the location of resonances, we have to evaluate their width thanks to Eq. (3.3) from Champenois & Vienne (1999) giving the α's amplitude for an isolated resonance:

$$\Delta\alpha = 4k^{\frac{2}{3}}(k+4)^{-\frac{5}{3}}y\sqrt{\frac{1}{3}[k^2\alpha_r^{-2}m' + (k+4)^2m]}|f(\alpha_r)| \qquad (3.3)$$

where

$$y = \sqrt{e^{|q_1|}e'^{|q_2|}\gamma^{|q_3|}\gamma'^{|q_4|}} \text{ with } \gamma = \sin\frac{i}{2} \qquad (3.4)$$

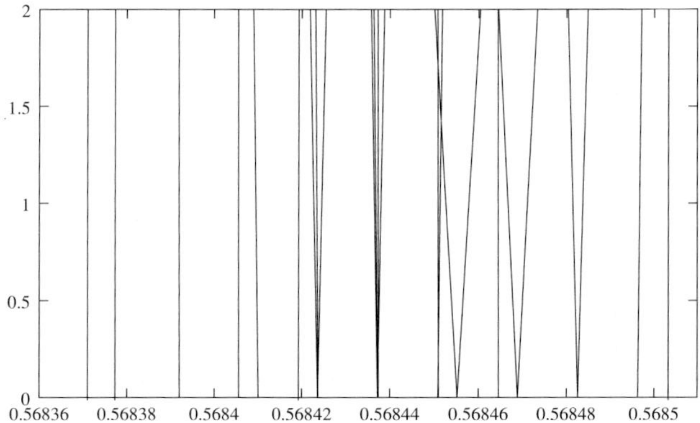

Figure 1. Locations of the resonance for the case Ganymede - Callisto. The x-axis is α, the y-axis is $\frac{y}{y_1}$ where y_1 is y (see Eq.3.4) evaluated with $e = 0.0021$, $e' = 0.007$, $\gamma = 0.0016$ and $\gamma' = 0.0022$ (actual values of these parameters).

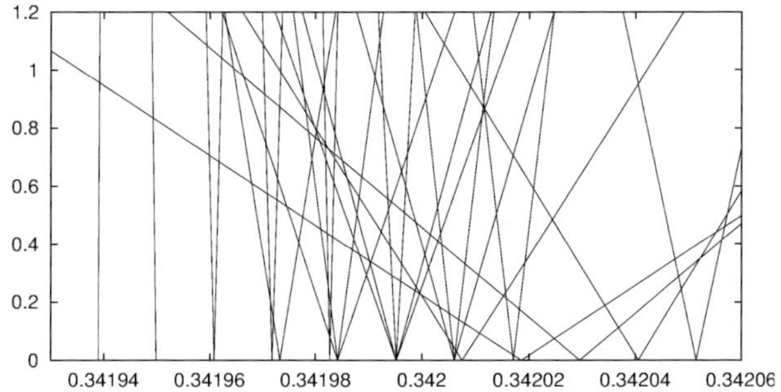

Figure 2. Locations of the resonance for the case Titan-Iapetus. Here y_1 is evaluated with $e = 0.289$, $e' = 0.029$, $\gamma = 0.055$ and $\gamma' = 0.012$. These values correspond to actual values except of Iapetus' inclination which is nowadays 10 times bigger; its actual value maybe due to the crossing of this chaotic zone.

Moreover, α_r is α's value at the exact resonance (say, $< \dot{\phi} >= 0$, see also Eq. (3.2), and f is a function of α appearing in the developpement of the disturbing function, see for instance Champenois & Vienne (1999).

When two resonances overlap, as in Figures 1 and 2, a stochastic layer appears (see Chirikov (1979)). In our cases, there are always at least three overlaps due to the quasi equality $< \dot{\varpi} + \dot{\Omega} >= 0$, but more may appear due to significant values of eccentricities and inclinations that enlarge the resonance zones.

4. A first exploration of the resonance zones

4.1. *Migration due to tidal effects*

The two systems are not actually in the resonance zones described above, the actual values of α being respectively 0.56854 for Ganymede-Callisto and 0.34324 for Titan-Iapetus. Migration of the systems is mostly due to tidal dissipation in the satellites and planets, which provokes change in semi-major axes and so in the mean motions as in Eq. (4.1) taken from Kaula (1964) and Peale & Cassen (1978):

$$\frac{\dot{n}}{n} = -\frac{9k_2 nm}{2QM}\left(\frac{r}{a}\right)^5\left[1 + \frac{51e^2}{4}\right] + \frac{63k_2^* nM}{2Q^* m}\left(\frac{r^*}{a}\right)^5 e^2 \tag{4.1}$$

where $*$ refers to the satellite, k_2 and k_2^* are Love numbers, Q and Q^* are dissipation functions, and r and r^* are the mean radii. This formula works for Callisto, Titan and Iapetus but not for Ganymede since the laplacian resonance with Io and Europa provokes energy transfer to Ganymede (see Yoder (1979)), in fact Ganymede's secular acceleration depends strongly on Io's, as shown by Lainey's numerical simulations (Lainey (2002)). So, we can estimate Ganymede's secular acceleration as $-10^{-11} \cdot y^{-1}$, Callisto's as $\pm 2 \times 10^{-14} \cdot y^{-1}$ and Iapetus' as $10^{-19} - 10^{-18} \cdot y^{-1}$. The value of Titan's is very important because it determines whether the system has crossed the 1:5 stochastic zone or not. With Eq. (4.1) this quantity can be estimated as $\pm 10^{-12} \cdot y^{-1}$ but only a negative value is compatible with the hypothesis that the system has crossed the stochastic layer, which gives a condition on geophysical parameters of Saturn and Titan, more precisely:

$$0.0178 \frac{k_2^*}{Q^*} < \frac{k_2}{Q} \tag{4.2}$$

4.2. Analytical study of some resonances

To describe analytically the systems, we used the second fundamental model of the resonance (see Henrard & Lemaître (1983)). First of all, we expressed the hamiltonian of the system in Jacobi variables while keeping only the terms associated to the studied inequality and the secular terms, which gave :

$$H(S_1, S_2, S_3, S_4, \Phi_1, \Phi_2, s_1, s_2, s_3, s_4, \phi_1, \phi_2) = \Gamma_0 + \sum_{i=1}^{4} \Gamma_i S_i + \sum_{i=1}^{4} \Gamma_{ii} S_i^2$$

$$+ \sum_{j=i+1}^{4} \sum_{i=1}^{4} \Gamma_{ij} S_i S_j + K_{4000} S_1^2 \cos(4s_1) + K_{3100} S_1^{\frac{3}{2}} \sqrt{S_2} \cos(3s_1 + s_2)$$

$$+ K_{2200} S_1 S_2 \cos(2s_1 + 2s_2) + K_{1300} \sqrt{S1} S_2^{\frac{3}{2}} \cos(s_1 + 3s_2) + K_{0400} S_2^2 \cos(4s_2)$$

$$+ K_{0040} S_3^2 \cos(4s_3) + K_{0022} S_3 S_4 \cos(2s_3 + 2s_4) + K_{0004} S_4^2 \cos(4s_4)$$

$$+ K_{1111} \sqrt{S_1 S_2 S_3 S_4} \cos(s_1 + s_2 + s_3 + s_4) + K_{2020} S_1 S_3 \cos(2s_1 + 2s_3)$$

$$+ K_{2011} S_1 \sqrt{S_3 S_4} \cos(2s_1 + s_3 + s_4) + K_{2002} S_1 S_4 \cos(2s_1 + 2s_4)$$

$$+ K_{0220} S_2 S_3 \cos(2s_2 + 2s_3) + K_{0211} S_2 \sqrt{S_3 S_4} \cos(2s_2 + s_3 + s_4)$$

$$+ K_{0202} S_2 S_4 \cos(2s_2 + 2s_4) + K_{1120} \sqrt{S_1 S_2} S_3 \cos(s_1 + s_2 + 2s_3)$$

$$+ K_{1102} \sqrt{S_1 S_2} S_4 \cos(s_1 + s_2 + 2s_4) + K_{0031} S_3^{\frac{3}{2}} \sqrt{S_4} \cos(3s_3 + s_4)$$

$$+ K_{0013} \sqrt{S_3} S_4^{\frac{3}{2}} \cos(s_3 + 3s_4) + \sqrt{S_1 S_2} \left(C_0 + \sum_{i=1}^{4} C_i S_i \right) \cos(s_1 - s_2)$$

$$+ \sqrt{S_3 S_4} \left(D_0 + \sum_{i=1}^{4} D_i S_i \right) \cos(s_3 - s_4) + E S_1 S_2 \cos(2s_1 - 2s_2) + F S_3 S_4 \cos(2s_3 - 2s_4)$$

$$+ G \sqrt{S_1 S_2 S_3 S_4} \cos(s_1 + s_2 - s_3 - s_4) + I \sqrt{S_1 S_2 S_3 S_4} \cos(s_1 - s_2 - s_3 + s_4)$$

$$+ J \sqrt{S_1 S_2 S_3 S_4} \cos(s_1 - s_2 + s_3 - s_4) + K S_1 S_3 \cos(2s_1 - 2s_3)$$

$$+ L S_1 S_4 \cos(2s_1 - 2s_4) + M S_2 S_3 \cos(2s_2 - 2s_3) + N S_2 S_4 \cos(2s_2 - 2s_4) \tag{4.3}$$

where S_1 and S_2 are proportional to the squares of eccentricities, S_3 and S_4 to the squares of inclinations, the 19 first angular terms are related to the great inequality, and the others are secular terms.

Our first study consists in studying a resonance as isolated, so all angular terms disappear except one. The case of arguments like $4s_i$ is the easiest to study and has already been studied by Lemaître (1984) because it is a reliable modelization of the 3-bodies restricted problem where only one argument is significant.

In such a case, the hamiltonian becomes (second fundamental model of the resonance):

$$H = \alpha S + \beta S^2 + 4\epsilon S^2 \cos 4s \tag{4.4}$$

which becomes

$$K = R^2 - \delta R + bR^2 \cos 4r \tag{4.5}$$

after the canonical transformation

$$R = S$$

$$r = (sign\beta)s \text{ if } \beta\epsilon > 0 \quad r = (sign\beta)s - \frac{\pi}{4} \text{ if } \beta\epsilon < 0$$

with

$$\delta = -\frac{\alpha}{\beta}$$

$$b = 4\left|\frac{\epsilon}{\beta}\right|$$

The exact resonance is reached when $\delta = 0$. When $\delta < 0$, the argument circulates and when $\delta > 0$ the phase space is divided into 3 zones: 2 circulation zones (one prograde and one retrograde) and a resonance zone. Since the resonance zone and the retrograde circulation zone appear at the same time, the capture into resonance is always probabilistic, the probability depending on the evolution of the areas of the critical zones (see Figure 3).

When the hypothesis of the adiabatic invariant is reliable, the probability of capture can be expressed as

$$P = \frac{2\pi - 2\arccos\frac{3b-1}{1+b}}{\pi(2 - b^2) - \arccos\frac{3b-1}{1+b}} \tag{4.6}$$

when δ is increasing. If this expression gives a negative result, $P = 0$, and if it gives something bigger than 1, $P = 1$. When δ is decreasing, capture is impossible.

In our cases, P can be estimated at a few percents.

5. Conclusion

This study shows that 4-order resonances in satellites systems might have played a significant role in the dynamical history of the systems and it is worth to continue the study, while using proper elements instead of nodes and pericentres which are much more representative of the resonant arguments. We also have to check the hypothesis of the adiabatic invariant. If this hypothesis is wrong, probabilities of capture may change (see Gomes (1997)). An exhaustive study of every arguments and their overlaps, as we plan, will precise the role of these inequalities, and maybe will explain Iapetus' actual inclination.

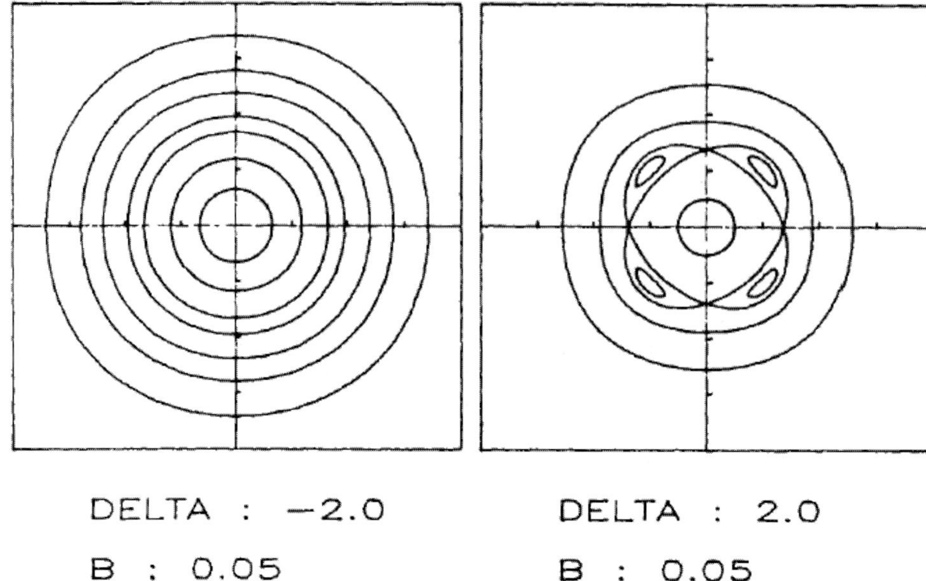

Figure 3. The phase space in the case of an isolated resonance of the form 4s, taken from Lemaître (1984). On the left there is only one circulation zone, whereas on the right the resonance zone has appeared with its 4 stable equilibria. These figures have been obtained using the Poincaré's variables $x = \sqrt{2R}\cos r$ and $y = \sqrt{2R}\sin r$.

References

Burns, J.A. 1986, in: J.A. Burns & Matthews, eds. *Satellites*, Univ. Arizona Press
Campbell, J.K. & Anderson, J.D. 1989, *Astron. J.* 97, 1485
Campbell, J.K. & Synnott, S.P. 1985, *Astron. J.* 1549, 364
Champenois, S. & Vienne A. 1999, *Cel. Mech. Dyn. Astr.* 74, 111
Chirikov, B.V. 1979, *Physics Reports* 52, 263
De Haerdtl, E. 1892, *Bulletin Astronomique* 9, 2121
Gomes, R.S. 1997, *Astron. Astrophys.* 121, 960
Henrard, J. & Lemaître, A. 1983, *Celest. Mech.* 30, 197
Kaula, W.M. 1964, *Review of Geophysics and Space Physics* 2, 467
Lainey, V. 2002, PhD thesis, Observatoire de Paris
Lainey, V., Arlot, J.-E. & Vienne, A. 2004, *Astron. Astrophys.*, 427, 371
Lemaître, A. 1984, *Cel. Mech. Dyn. Astr.* 32, 109
Lieske, J.H. 1998 *Astron. Astrophys. Suppl.*, 129, 205
Moons, M. & Morbidelli, A. 1995, *Icarus* 114, 33
Peale S.J. & Cassen P. 1978, *Icarus* 36, 245
Seidelmann, P.K., Abalakin V.K., Bursa M. *et al.* 2002, *Cel. Mech. Dyn. Astr.* 82, 83
Vienne, A. & Duriez, L. 1995, *Astron. Astrophys.* 297, 588
Yoder, C.F. 1979, *Nature* 279, 767

Dynamics of Populations of Planetary Systems
Proceedings IAU Colloquium No. 197, 2005
Z. Knežević and A. Milani, eds.

© 2005 International Astronomical Union
DOI: 10.1017/S1743921304008993

Fine-scale density wave structure of Saturn's A, B, and C rings

Evgeny Griv

Department of Physics, Ben-Gurion University of the Negev, Beer-Sheva 84105, Israel
email: griv@bgumail.bgu.ac.il

Abstract. In view of the possibility of employing *Cassini*'s experiments for the diagnosis of the Saturnian ring system, local N-body simulations of low and moderately high optical depth regions of Saturn's main rings are presented. A special emphasis is made on fine-scale spiral structures (irregular cylindric wave-type structures of the order of 100 m or so) of Saturn's A, B, and C rings. It is predicted that *Cassini* spacecraft high-resolution images of Saturn's rings will reveal this kind of small-scale irregular density wave structure.

Keywords. Planets, Saturn's rings, waves and instabilities, N-body simulations

1. Introduction

Voyager images of Saturn's main A, B, and C rings have shown evidence of radial structures ranging from a few kilometres down to the several hundrends metres resolution of the spacecraft's camera (Smith *et al.* 1982). The best resolution demonstrated structures at all scales in the rings, down to the limit of resolution ~ 5 km. Most of the structures are irregularly spaced and do not correspond to resonances with known satellities. It is important that the *Voyager*'s stellar occultation data revealed some indirect evidence for structuring in the densest central parts of Saturn's B ring down to 100 m length scale. One cannot exclude the existence of a fine irregular structure of this kind in other regions of the Saturnian ring system of mutually gravitating and colliding particles.

Among all mechanisms, the self-excited nonresonant Jeans instability of gravity disturbances (e.g., those produced by a spontaneous perturbation or a satellite system) has long been suspected to be the key one that determines the ubiquitous irregular structure of Saturn's rings, with the appearance of record-grooves (e.g., Esposito 1993, Fig. 5 therein; Cuzzi *et al.* 2002, Fig. 2b therein). Accordingly, a system of mutually gravitating particles of Saturn's rings exhibits collective, gravitationally unstable modes of motions. A quasi-linear kinetic theory of the almost aperiodic Jeans instability in Saturn's rings has been developed by Griv *et al.* (2000, 2003a, 2003b) and Griv & Gedalin (2003). The theory predicts that as a direct result of the instability of small-amplitude gravity disturbances the main parts of A, B, and C rings are divided into numerous irregular spiral ringlets of the order of 2π times the local thickness $h = 5 - 30$ m. Below I describe N-body simulations in order to verify the validities of the theory.

2. Local simulations

The dynamical behavior of planetary rings has already studied via simplified N-body simulations of an orbiting patch of the ring by Salo (1992, 1995), Richardson (1994), Osterbart & Willerding (1995), Sterzik *et al.* (1995), Griv (1998), Daisaka & Ida (1999), and others. See Griv & Gedalin (2003) as a review of the problem. In these N-body experiments in a local, or Hill's approximation dynamics of particles in small regions of the disk are assumed to be statistically independent of dynamics of the particles in other regions. The local numerical model thus simulates only a small part of the system and more distant parts are represented as copies of the simulated region. The system of Newton's equations of three-dimensional motion in local approximation for $N \gg 1$ identical particles,

$$\frac{d^2 x}{dt^2} - 2\Omega r_0 \frac{d\Omega}{dr} x - 2\Omega \frac{dy}{dt} = F_x \,, \tag{2.1}$$

$$\frac{d^2 y}{dt^2} + 2\Omega \frac{dx}{dt} = F_y \,, \tag{2.2}$$

$$\frac{d^2 z}{dt^2} + \Omega^2 z = F_z \,, \tag{2.3}$$

was integrated by the Runge–Kutta method of the fourth order. In Eqs. (2.1)–(2.3),

$$x = r - r_0 \,, \quad y = r_0(\varphi - \Omega t) \,,$$

r_0 is the reference radius, $\Omega = \Omega(r_0)$, and $A_0 = -(r_0/2)(d\Omega/dr)_0 \approx (3/4)\Omega$ is the first Oort constant of the differential rotation which is a measure of the shear strength. In general, F_x, F_y, and F_z are the forces due to interactions with other particles. The gravitational forces are

$$\vec{F}_i = -G \sum_{j \neq i}^{N} \frac{(\vec{r}_i - \vec{r}_j)}{[(\vec{r}_i - \vec{r}_j)^2]^{3/2}} \,, \tag{2.4}$$

where $\vec{r}_i$ is the position of the i-th particle and $\vec{r}_j$ is the position of the j-th particle. Following Wisdom & Tremaine (1988), Salo (1995), Richardson (1995), Daisaka & Ida (1999), and Ohtsuki & Emori (2000), I adopted the standard hard-sphere collision model. A collision changes only impact velocity in normal direction, $v'_n = -\epsilon v_n$ and $v'_t = v_t$, where v_n, v_t and v'_n, v'_t, respectively, are relative velocities of colliding particles in the normal direction and the tangential direction.

A rotating Cartesian coordinate system with origin at the reference position r_0 was chosen, the x axis pointing radially outward and the y axis pointing in the direction of the rotation (Wisdom & Tremaine 1988; Salo 1995; Sterzik $et\ al.$ 1995). The particles were initially placed on nearly circular orbits with an anisotropic Maxwellian (Schwarzschild) distribution of small random velocities. The initial distribution of particles was generated by means of pseudo-random number generator placing particles uniformly in the box in real space. The box should be thought of as being embedded in Saturn's ring disk which has a constant angular velocity gradient in the x direction.

I present simulations which are distinguished primarily by their values of the radial c_r and azimuthal c_φ velocity dispersions. The "cool" model has a value of $c_r \equiv 2c_\varphi = 0.5c_T$, and so I expect the model to be initially violently unstable to both radial and spiral gravity disturbances. Here, $c_T = 3.4G\Sigma_0/\Omega$ is the well-known Safronov–Toomre (Safronov 1960; Toomre 1964) random velocity dispersion to suppress the instability of axisymmetric (radial) disturbances and Σ_0 is the mean (local) surface density of the disk. The "warm" model has $c_r \equiv 2c_\varphi = c_T$, and so it is expected to be unstable only to spiral disturbances. In turn, the "hot" model with $c_r \equiv 2c_\varphi = 2c_T$ is expected to be at best only marginally unstable to the growth of spiral waves. See Griv $et\ al.$ (2000, 2003a, 2003b) and Griv & Gedalin (2003) for an explanation. In all experiments, initially $c_z = 0.2c_r$.

To maintain the system under the shearing stress in a steady state, the cyclic boundary conditions were used in the form suggested by Wisdom & Tremaine (1988), Toomre (1990), Salo (1995), and others. The direction of the disk rotation was taken to be

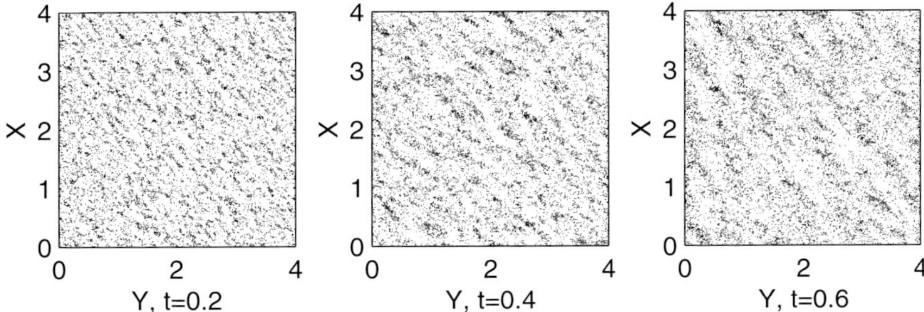

Figure 1. Time-development of the differentially rotating three-dimensional model of particles (face-on view) distributed over the $L_x \times L_y \times L_z = 4\lambda_{\rm J} \times 4\lambda_{\rm J} \times \lambda_{\rm J}$ unit cell; $c_r = 0.5c_{\rm T}$. The model is violently unstable against spontaneous trailing gravity perturbations.

clockwise. Time $t = 1$ corresponds to a single revolution of the disk, and the orbital period was $T_{\rm orb} = 2\pi/\Omega$. All the particles moved with the same fixed time step $\Delta T = 0.001T_{\rm orb}$. In Saturn's rings, a measure of the fundamental vibration period, or the dynamical time is of the order of $T_{\rm dyn} \sim (G\rho_{\rm ring})^{-1/2} \sim 1$ h if we assume the volume density $\rho_{\rm ring} \sim 1$ g/cm^3. This means that $T_{\rm orb} \sim 10T_{\rm dyn}$, so a choice of the stepsize $\Delta T = 0.001T_{\rm orb}$ gives about 100 steps per particle dynamical time, which should be sufficient to accurately resolve particle–particle interactions.

In experiments reported below the following physical parameters for a simulated patch of the ring were chosen: the orbital period $T_{\rm orb} \equiv 2\pi/\Omega = 7.027$ h which corresponds to a typical orbital period of the C ring's particle at the distance $r = 85\,000$ km from the planet, the surface density $\Sigma_0 = 15$ g/cm^2, and the total number of particles $N = 12\,000$ in all models. The particle radius $r_{\rm p} = 4.2$ cm in $L_x \times L_y \times L_z = 4\lambda_{\rm J} \times 4\lambda_{\rm J} \times \lambda_{\rm J}$ three-dimensional models. The corresponding Jeans–Toomre wavelength $\lambda_{\rm J} \approx 6.4$ m and the optical depth $\tau \approx \pi r_{\rm p}^2 n \approx 0.1$ in all models (n is the number density per unit area). The constant coefficient of restitution was $\epsilon = 0.8$. A very popular model of the particles in Saturn's rings is a smooth ice sphere, whose restitution coefficient is quite high, exceeding 0.63, and decreases as the collision velocity increases (Goldreich & Tremaine 1982; Bridges *et al.* 1984; Kerr 1985). Saturnian ring system is populated primarily by centimeter- to a few meter-sized mutually gravitating and physically colliding particles (Zebker *et al.* 1985). A particle size distribution function exhibits approximately inverse-cubic power-law behavior. In addition, Saturn has extensive but much more tenuous rings containing mainly micrometer-sized particles (Lissauer & Cuzzi 1985). Because low-velocity collisions of ice particles will always involve some dissipation of acoustic energy, some source of energy must be available to keep them from total collapse to a featureless monolayer.

3. Fine-scale structure

In Fig. 1 I show a series of snapshots from a run with the cool model, i.e., the nonuniformly rotating model, in which initially particles all move along almost circular orbits and the radial dispersion of the random velocities is smaller than the critical Safronov–Toomre one, namely $c_r = 0.5c_{\rm T}$, or Toomre's Q-value, $Q = c_r/c_{\rm T}$ is equal to 0.5, respectively. As has been predicted in the theory, the Jeans instability develops quickly in the system. Figure 1 clearly shows that in a disk with the Keplerian shear profile a spiral pattern (more accurately, a chainlike structure or irregular "wakes") develops spontaneously in the initially featureless disk on a dynamical time scale $< \Omega^{-1}$. The wakes are created if self-gravity is included; only collisions do not create the structure.

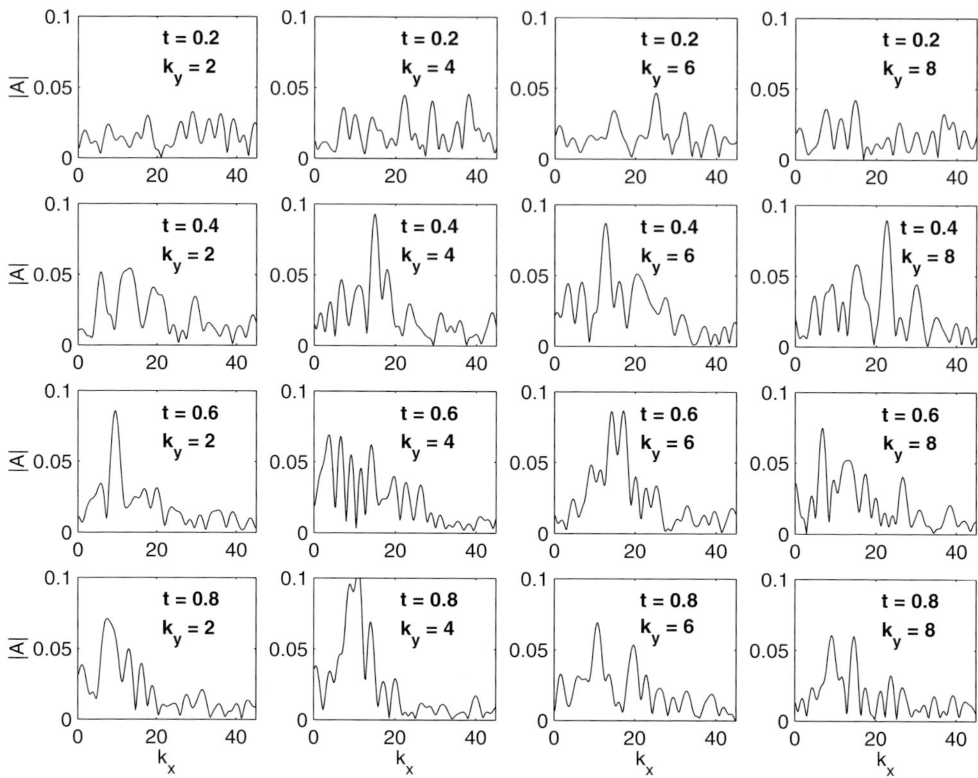

Figure 2. Density spectrum $|A(k_x, k_y)|$ of the particle distribution shown in Fig. 1 for different azimuthal k_y and radial k_x mode numbes at the calculation times $t = 0.2 - 0.8$.

Their pattern speed $\Omega_{\rm p}$ is zero, because the system is spatially homogeneous (Griv *et al.* 2003b). The Lin–Shu type density wave structure (Lin & Shu 1966; Lin *et al.* 1969; Shu 1970; Lin & Lau 1979; Griv & Gedalin 2003) is time dependent and transient.

Early in the evolution, multiple spirals interfere with each other and produce a complicated set of density concentrations. At most, eight or nine individual high-density wakes can be seen (Fig. 1, $t = 0.4 - 0.6$). The structure consists of elongated trailing filaments with a definite pitch angle $\psi_{\rm crit}$ with respect to the local shear flow. The pitch angle of spiral wakes $\psi_{\rm crit} \approx 20°$ can be obtained by examining Fig. 1 directly.

Figure 1 gives a feeling for the evolutionary process, but in order to trace and quantify the growth of instabilities in the disk, it is necessary to compute Fourier decompositions of the surface density distribution for various mode numbers. Shown in Fig. 2 are the sequences of density-spectrum evolution in the wavenumber space for the case shown in Fig. 1 at the calculation times $t = 0.2 - 0.8$. The power spectrum is constructed by employing a technique where one regards the particle, labeled by j, as a discrete δ-function to calculate the density Fourier component, i.e.,

$$A(k_x, k_y) = \frac{1}{N} \sum_{j=1}^{N} \exp(\vec{k} \cdot \vec{r}_j). \tag{3.1}$$

In addition, the azimuthal wavenumber k_y assumes discrete values compatible with the finite boundaries in the streamwise direction, that is, in the y-direction, whereas the radial wavenumber k_x assumes continuous values because of the background flow shear. The

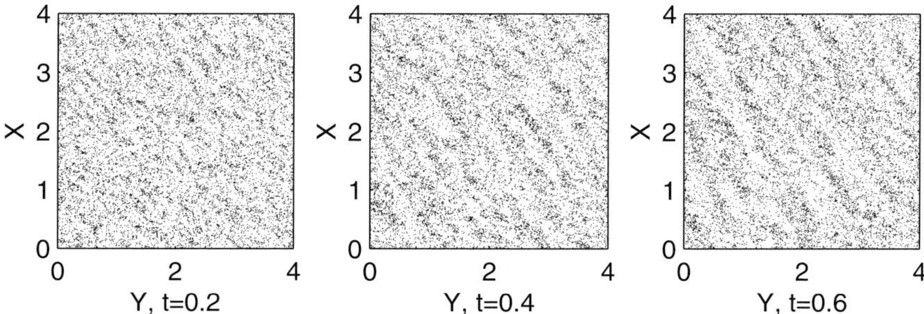

Figure 3. Time-development of the differentially rotating model; $c_r = c_{\rm T}$. In agreement with the theory, even though the initial radial velocity dispersion c_r is equal to the Safronov–Toomre one $c_{\rm T}$, the model is still unstable against nonaxisymmetric perturbations.

pitch angle of a spiral ψ is given by $\psi = \arctan(k_y/k_x)$, and positive k_x corresponds to trailing spirals and negative k_x to leading. The wavenumber is normalized to π/L_y. Thus, the quantity k_y gives the number of halfwaves of a spiral mode of collective oscillations in the y-direction.

No particular set of k_x, k_y dominates at the beginning of calculations ($t = 0.2$), which corresponds to the initial noise. It is evident from Fig. 2 that the $c_r < c_{\rm T}$ case produces rigorous instabilities. One can clearly see that at times $t \gtrsim 0.6$ a dominant k_y is equal about to $(2L_y/\lambda_{\rm J}) \tan \psi \approx 4$, which is consistent with the linear stability analysis for the marginally unstable mode (Griv *et al.* 2003b; Griv & Gedalin 2003). That is, a wavelength of the dominant mode in the streamwise direction $\lambda_y = \lambda_{\rm crit}/\tan \psi \approx 2\lambda_{\rm J}$, where $\psi = 20° - 25°$. The latter fact convincingly indicates that we have dealt with a collective-type instability rather than with a random process (as advocated by Toomre 1990 and Toomre & Kalnajs 1991). It is natural to attribute the observed instability to the Jeans instability so far discussed in the paper. Also, we see that there are a number of $k_y = 2$, $k_y = 6$, and $k_y = 8$ discrete harmonics present.

In the second set of experiments with the Keplerian disk, I simulated a system which is stable according to Safronov (1960) and Toomre (1964): the initial radial velocity dispersion is equal to $c_r = c_{\rm T}$, or Toomre's parameter $Q = 1$, respectively. The evolution of the model is shown in Fig. 3. As is seen, in a Safronov–Toomre stable disk with the differential rotation a spiral instability develops rapidly on a dynamical time scale $\lesssim \Omega^{-1}$. This fierce instability of the system with $c_r = c_{\rm T}$ indicates that in a *differentially* rotating system the Jeans instability of *nonaxisymmetric* ($\psi \neq 0$) gravity perturbations cannot be suppressed by the ordinary Safronov–Toomre critical velocity spread $c_{\rm T} \approx 3.4G\Sigma_0/\Omega$, in line with theoretical expectations (Griv 1996, 1998; Griv & Gedalin 2003).

The initial unstable growth is clear from Figs. 1–3. The simulated wake structure rapidly at times $t \gtrsim 0.6$ reaches nonlinearity, and is thus beyond the scope of our theory (or any corresponding linear analysis of shearing modes). I then simulated a differentially rotating disk which is stable in accordance with the modified stability criterion (Griv *et al.* 2000, 2003a, 2003b; Griv & Gedalin 2003). Figure 4 shows the observed evolution of the dynamically hot model with $c_r = 2c_{\rm T}$ (or $Q = 2$, respectively). As one can see, in such a system this relatively high "temperature" c_r almost eliminates completely the growth rate of the gravitational instability. In sharp contrast to the previous simulations, the model is now almost gravitationally stable. The contrast between Figs. 1–3 and Fig. 4 establishes experimental evidence to support the stability theory developed by Griv and co-workers. Note that Salo (1995), Richardson (1994), Osterbart & Willerding (1995), Sterzik *et al.* (1995), and Daisaka & Ida (1999) have already found that the stability

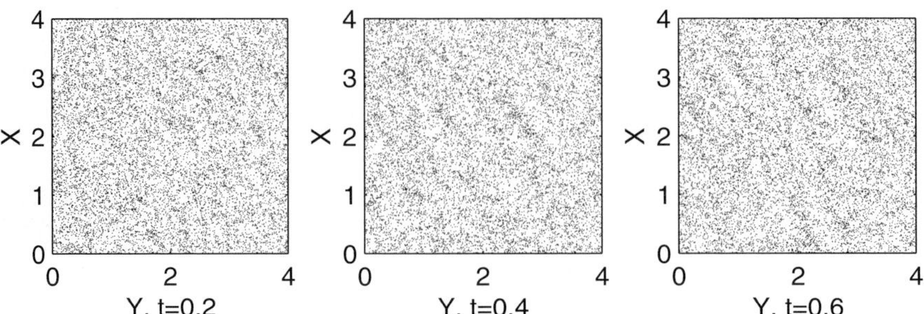

Figure 4. Time-development of the differentially rotating model; $c_r = 2c_{\mathrm{T}}$. All perturbations are almost suppressed, including the most unstable nonaxisymmetric ($\psi \neq 0$) ones.

number Q of Toomre in relaxed equilibrium disks does not fall below a critical value, which lies about $Q_{\mathrm{crit}} = 2 - 2.5$. However, no adequate explanation of the latter has been presented. See also Griv *et al.* (1999) for a discussion of the problem.

4. Vertical structures

Figures 1–3 show the density distributions of unstable models of Saturn's rings in the plane (face-on view). As has been mentioned by Salo (1992, 1995) and others, another feature of the simulations is the radical changes in vertical structures of unstable systems that result from the instabilities. Figure 5 shows distributions of particles and isodensity contours in the (z, η)-plane for the three-dimensional model at the simulation times $t = 0$ and $t = 0.4$. (I introduced a new coordinate system (z, η); η is perpendicular to the spiral wakes, while z gives the normal to the plane position.) The disk surface generally evolves from a smooth height profile with η to complex structures with peaks and valleys that result from instabilities. The results presented in Fig. 5 are consistent with the hypothesis that we have dealt with the even Jeans perturbations, because the perturbed density is an even function of z. This type of vertical motions does not deform the horizontal disk plane $z = 0$, because the vertical velocity v_z in a density wave is odd in z: $v_z(-z) = -v_z(z)$. Such "sausage-like" perturbations (Bertin & Casertano 1982) can release gravitational energy and are subject to classical gravitational Jeans-type instability.

5. Summary

The most puzzling features of Saturn's rings, revealed by *Voyager* fly-bys, are the radial density variations seen on all scales down to the resolution limit of few kilometers. At the present time, their origin is far from being understood (Tremaine 2003). This paper reports on an investigation of the significance of self-excited (that is, intrinsic), off-resonant, almost aperiodic Jeans instability in Saturn's main rings to the formation of the fine-scale structures. I explored the linear regime of Jeans instabilities in Saturn's rings by means of sliding N-body patch model (e.g., Salo 1992, 1995) and compared those numerical results with the quasi-linear kinetic stability theory by Griv et al. (2000, 2003a, 2003b) and Griv & Gedalin (2003). The aim of the work is to discuss specific astronomical implications of the study to Saturn's rings in view of the possibility of employing *Cassini*'s experiments. Summarizing, I claim that

1. In order to suppress the instability of arbitrary but not only axisymmetric Jeans-type perturbations in a differentially rotating disk, including the most unstable nonaxisymmetric ones, the value of the radial dispersion of random velocities of particles must

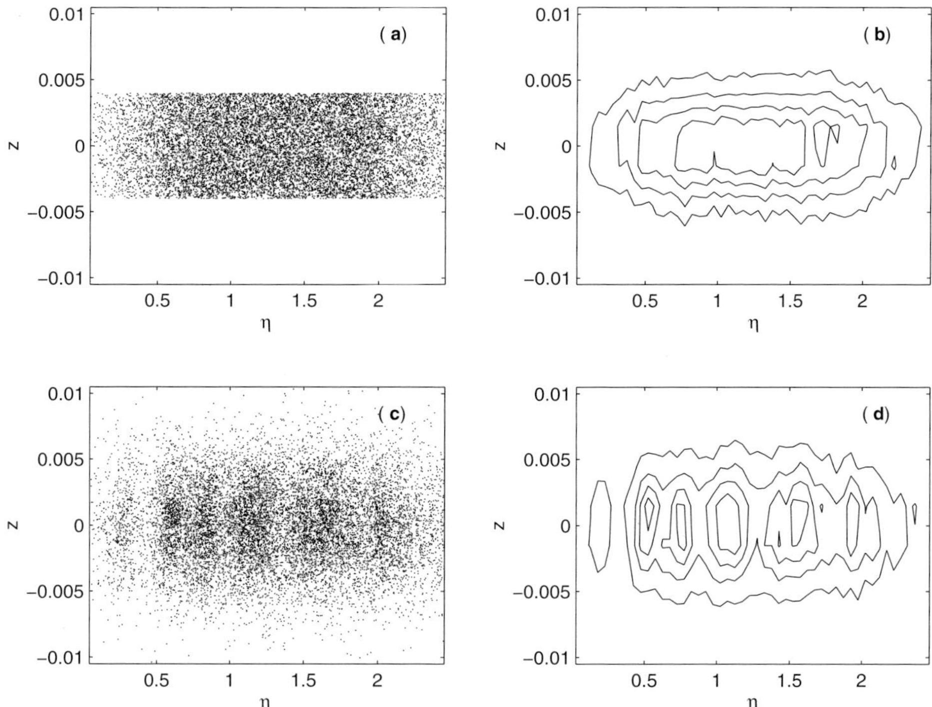

Figure 5. Comparison of vertical structure. Shown are both distributions of particles and isodensity contours in the (z, η)-plane for the three-dimensional model shown in Fig. 3 at the calculation times $t = 0$ and $t = 0.4$: (a) distribution of particles at $t = 0$, (b) isodensity contours at $t = 0$, (c) distribution of particles at $t = 0.4$, and (d) isodensity contours at $t = 0.4$. Figures 5c and 5d give the end-on edge-on views of the positions and isocontours of the density of the disk particles at $t = 0.4$, respectively. The unstable spiral disturbances significantly alter the disk's vertical surface.

exceed $c_{\mathrm{crit}} \approx 2c_{\mathrm{T}}$, or Toomre's stability parameter $Q \gtrsim 2$. It is expected that in the main rings of Saturn $Q \approx 2$ (cf. Lane *et al.* 1982, p. 543). I again argue that sufficient velocity dispersion ($Q \gtrsim 2$) prevents the Jeans instability from occurring but inelastic particle collisions reduce the relative particle velocities so that the Jeans instability may be an effective generating mechanism for the recurrent fine structure of the ring system.

2. In Saturn's rings, this almost aperiodic gravitational instability manifests itself as trailing cylindric density-wave enhancements, forming $\approx 20°$ angle with respect to tangential direction.

3. By local simulations of a particle model, we see that at the limit of stability with respect to all gravity perturbations of a differentially rotating disk the critical radial wavelength becomes approximately equal to $\lambda_{\mathrm{crit}} = 2\lambda_{\mathrm{J}}$. Modern observations indicate that the ring thickness ranges from $1 - 2$ m in the C ring to $1 - 5$ m in the B ring and $5 - 30$ m in the A ring (Esposito 2002). Then, in the C ring estimations give the value of $\lambda_{\mathrm{crit}} = 6 - 20$ m, in the B ring $\lambda_{\mathrm{crit}} = 6 - 50$ m, and in the A ring $\lambda_{\mathrm{crit}} = 30 - 300$ m. Thus, both the theory and the simulations forecast the existence of the fine-scale ~ 100 m or even less radial structures in Saturn's main rings.

4. Seen edge-on with the fine spirals seen end-on (with the line of sight along the spiral local major axis), Saturn's rings will show even with respect to the equatorial plane "sausage" structures.

Acknowledgements

The author would like to thank Tzi-Hong Chiueh, David Eichler, Michael Gedalin, Edward Liverts, Yury Lyubarsky, Irena Shuster, Raphael Steinitz, and Chi Yuan for several valuable discussions and constructive criticism of the manuscript. Part of this work has been conducted under the auspices of the Israeli Ministry of Immigrant Absorption program "KAMEA." Support was also received from the Israel Science Foundation and the Binational U.S.–Israel Science Foundation.

References

Bertin, & Casertano, S. 1982, *Astron. Astrophys.* 106, 274

Bridges, F.G., Hatzes, A.P. & Lin, D.N.C. 1984, *Nature* 309, 333

Cuzzi, J.N., Colwell, J.E., Esposito, L.W., *et al.* 2002, *Space Sci. Rev.* 118, 209

Daisaka, H. & Ida, S. 1999, *Earth Planets Space* 51, 1195

Esposito, L.W. 1993, *ARE&PS* 21, 487

Esposito, L.W. 2002, *Rep. Prog. Phys.* 65, 1741

Goldreich, P. & Tremaine, S. 1982, *ARA&A* 20, 249

Griv, E. 1996, *Planet. Space Sci.* 44, 579

Griv, E. 1998, *Planet. Space Sci.* 46, 615

Griv, E. & Gedalin, M. 2003, *Planet. Space Sci.* 51, 899

Griv, E., Gedalin, M., Eichler, D. & Yuan, C. 2000, *Planet. Space Sci.* 48, 679

Griv, E., Rosenstein, B., Gedalin, M. & Eichler, D. 1999, *Astron. Astrophys.* 347, 821

Griv, E., Gedalin, M. & Yuan, C. 2003a, *Astron. Astrophys.*, 400, 375

Griv, E., Gedalin, M. & Yuan, C. 2003b, *Mon. Not. R. Astron. Soc.* 342, 1102

Kerr, R.A. 1985, *Science* 220, 1376

Lane, A.L., Hord, C.W., West, R.A., *et al.* 1982, *Science* 215, 537

Lin, C.C. & Shu F.H. 1966, *Proc. Natl. Acad. Sci.* 55, 229

Lin, C.C. & Lau, Y.Y. 1979, *SIAM Stud. Appl. Math.* 60, 97

Lin, C.C., Yuan, C. & Shu, F.H. 1969, *Astrophys. J.* 155, 721

Lissauer, J.J. & Cuzzi, J.N. 1985, in: D.C. Black & M.S. Matthews (Eds.), *Protostars and Planets, II*, Univ. of Arizona Press, p. 920

Ohtsuki, K. & Emori, H. 2000, *Astrophys. J.* 119, 403

Osterbart, R. & Willerding, E. 1995, *Planet. Space Sci.* 43, 289

Richardson, D.C. 1994, *Mon. Not. R. Astron. Soc.* 269, 493

Safronov, V.S. 1960, *Ann. d'Astrophys.* 23, 979

Salo, H. 1992, *Nature* 359, 619

Salo, H. 1995, *Icarus* 117, 287

Shu, F.H. 1970, *Astrophys. J.* 160, 99

Smith, B.A., Soderblom, L., Batson, R., *et al.* 1982, *Science* 215, 504

Sterzik, M.F., Herold, H., Ruder, H. & Willerding, E. 1995, *Planet. Space Sci.* 43, 259

Toomre, A. 1964, *Astrophys. J.* 139, 1217

Toomre, A. 1990, in: R. Wielen (Ed.), *Dynamics and Interactions of Galaxies*, Springer-Verlag, p. 292

Toomre, A. & Kalnajs, A.J. 1991, in: B. Sundelius (Ed.), *Dynamics of Disc Galaxies*, Göteborg Univ. Press, p. 341

Tremaine, S. 2003, *Astron. J.* 125, 894

Wisdom, J. & Tremaine, S. 1988, *Astron. J.* 95, 925

Zebker, H.A., Marouf, E.A. & Tyler, G.L. 1985, *Icarus* 64, 531

Dynamics of Populations of Planetary Systems
Proceedings IAU Colloquium No. 197, 2005
Z. Knežević and A. Milani, eds.

© 2005 International Astronomical Union
DOI: 10.1017/S1743921304009007

Symmetric periodic orbits in proto-stellar systems

Vasile Mioc[1], Mira-Cristiana Anisiu[2] and Magda Stavinschi[1]

[1]Astronomical Institute of the Romanian Academy, Str. Cuțitul de Argint 5, RO-040557 Bucharest, RO
email: vmioc@aira.astro.ro, magda@aira.astro.ro

[2]T. Popoviciu Institute of Numerical Analysis of the Romanian Academy, P.O. Box 68, RO-400110 Cluj-Napoca, RO
email: mira@math.ubbcluj.ro

Abstract. We approach the dynamics in proto-stellar systems via the two-body problem associated to an anisotropic Schwarzschild-type potential. On the basis of the natural symmetries of the characteristic vector field, and using variational methods (particularly the classical lower-semicontinuity method), we prove the existence of infinitely many families of symmetric periodic orbits.

Keywords: Celestial Mechanics, methods: analytical

1. Introduction

Dynamics in proto-stellar systems is a topic that can be tackled via anisotropic two-body problems (Saslaw 1978). Actually, astronomy provides a much larger class of problems tractable by this model (e.g., Mioc *et al.* 2003).

The physical framework of our model is a central (proto-)star in fast rotation, hence oblate, but still axisymmetric, surrounded by a nonuniformly dense accretion disk. We are interested in the dynamics of a body that moves in the plane of the disk, without gravitationally interacting with the disk particles, but being influenced by the nonuniform (due to the presence of the disk) radiation of the central star that falls on the body. The oblateness of the star, created by the fast rotation, leads to the existence of a Schwarzschild-type potential. The influence of the disk makes the Newtonian-type term of the potential to be anisotropic.

Our previous attmepts (Mioc *et al.* 2003) to prove the existence or nonexistence of periodic orbits within such a Schwarzschild-type problem were unsuccessful. Here we solve this problem, even if within a little bit restricted framework. To this end, we use the natural symmetries of the problem, and a variational principle: the extrema of the action integral are genuine periodic solutions.

In this paper we present results without proofs. A complete presentation will be published elsewhere.

2. Basic equations and properties

The problem is described by a two-degrees-of-freedom system of ODE with the Hamiltonian $H(\mathbf{q}, \mathbf{p}) = |\mathbf{p}|^2/2 - W(\mathbf{q})$, in which the potential $W : \mathbb{R}^2 \backslash \{(0,0)\} \to \mathbb{R}$ is (cf.

Mioc *et al.* 2003):

$$W\left(q_1, q_2\right) = \left(\mu q_1^2 + q_2^2\right)^{-1/2} + b\left(q_1^2 + q_2^2\right)^{-3/2}, \tag{2.1}$$

where $\mu > 0$ and $b > 0$ are parameters. The presence of the disk and the oblateness of the star lead to this expression of the potential inside the disk. The corresponding Lagrangian, $L\left(\mathbf{q}, \mathbf{p}\right) = |\mathbf{p}|^2/2 + W\left(\mathbf{q}\right)$, is always positive.

The equations of motion define the two-body problem dynamics in an anisotropic plane, in which, for $\mu > 1$, the attraction is the strongest in the q_1-direction and the weakest in the q_2-direction; for $\mu < 1$ the situation is inverse. We consider, without loss of generality, that $\mu > 1$.

The model admits the integral of energy, but not the angular momentum integral (because of the anisotropy).

An important property of the potential W is that it generates a *strong force* according to Gordon's (1975) definition (using an alternative definition, one immediately proves that $W\left(\mathbf{q}\right) \geqslant b/|\mathbf{q}|^2$ for $0 < |\mathbf{q}| < 1$). This makes the variational methods easier to apply.

Another important property is that the vector field that characterizes the problem benefits of eight natural symmetries $S_i = S_i\left(q_1, q_2, p_1, p_2, t\right)$, $i = \overline{0,7}$, as follows:

$$\begin{aligned}
&S_0 = \left(q_1, q_2, p_1, p_2, t\right) = I \text{ (identity)}, \ S_1 = \left(q_1, q_2, -p_1, -p_2, -t\right), \\
&S_2 = \left(q_1, -q_2, -p_1, p_2, -t\right), \ S_3 = \left(-q_1, q_2, p_1, -p_2, -t\right), \\
&S_4 = \left(q_1, -q_2, p_1, -p_2, t\right), \ S_5 = \left(-q_1, q_2, -p_1, p_2, t\right), \\
&S_6 = \left(-q_1, -q_2, -p_1, -p_2, t\right), \ S_7 = \left(-q_1, -q_2, p_1, p_2, -t\right)
\end{aligned} \tag{2.2}$$

It is easy to translate these symmetries in physical terms. For instance, S_1 implies that, for every solution, there is another solution with the same coordinates and with inverse velocities, all in reversed time, and so forth.

3. Main steps

The anisotropy of the potential is a strongly destabilizing factor for the motion we study. Within a slightly more general model (Mioc et al. 2003), we found collision/ejection-type or escape/capture-type orbits, but did not succeed in proving the existence or nonexistence of periodic orbits. Here, taking into account those previous results, we use some symmetries and topological constraints in connection with a variational principle to get periodic orbits as extrema of the action. To this end, we resort to results concerning periodic solutions of fixed period for symmetric, singular, Lagrangian systems (Ambrosetti and Coti Zelati 1993), as it is the case for the system associated to (2.1).

We first choose a value $T > 0$, and dwell upon the space of T-periodic C^∞ cycles $f : [0, T] \rightarrow \mathbb{R}^2$. Let L^2 be the space of square integrable functions, and let H^1 be the Sobolev space of all absolutely continuous T-periodic functions with L^2 derivatives defined almost everywhere.

The potential (2.1) is singular at $(0,0)$ (collision). Let $\Lambda = \{f \in H^1 \mid f(t) \neq (0,0), \forall t \in [0, T]\}$ be the open subset of noncollisional cycles in H^1. We define the *winding number* $w(f)$, which shows how many times the continuous cycle f winds around the origin. It easily follows that $\Lambda = \cup_{k \in \mathbb{Z}} \Lambda_k$, with $\Lambda_k = \{f \in \Lambda \mid w(f) = k\}$. In other words, the set of noncollisional cycles is partitioned by the winding number.

Coming back to the natural symmetries of the system, denote the subsets of H^1 formed by S_i-symmetric cycles by Σ_i.

As a first step, we proved that Σ_i, $i \in \{1, 2, 3, 7\}$, are Sobolev spaces, which are the natural framework for finding periodic solutions by variational methods. Moreover, we proved that $H^1 = \Sigma_2 \oplus \Sigma_3 = \Sigma_1 \oplus \Sigma_7$ (orthogonal decomposition), hence such a couple is sufficient to cover the whole loop space H^1.

In the second step, in order to avoid cycles that pass through origin (collision) or have zero winding number (escape or quasiperiodic orbits), we examined all subsets Σ_i. Only Σ_2 and Σ_3 fulfil these requirements. With the above results, Σ_2 and Σ_3 can provide periodic solutions for the whole H^1.

We define now the action integral (whose extremal values will provide periodic orbits) $A_T : \Lambda \to \mathbb{R}$ between the instants 0 and T, along a cycle f whose Euclidean coordinate representation is $\mathbf{q} = (q_1, q_2)$ as

$$A_T(f) = \int_0^T L(\mathbf{q}(t), \mathbf{p}(t)) \, dt, \tag{3.1}$$

with the positive Lagrangian specified in Section 2. To obtain periodic solutions, we are forced to minimize A_T on subsets Λ_k of Λ, chosen via symmetries). After selecting a suitable subset, we use the *lower-semicontinuity method* (e.g., Struwe 1996) to get a minimizer in that subset, which we prove to be an extremal value of A_T..

A specification from the astronomer's standpoint is necessary here: why the minimization of the action leads to periodic orbits? Recall that the Lagrangian of our problem is the sum of two positive terms: the kinetic energy K and the force function W (the negative of the potential energy). Also note that K and W are not independent each other, being related by the energy integral. Since $K, W > 0$, any minimization of their sum involves the minimization of both K and W; both push the trajectory away from the field-generating centre. But the limit imposed by the fixed energy-level and by the fixed value of T stops the orbit expansion to a finite value, which can lead to a periodic orbit.

A new step concerns the connection between solutions of our problem and the extremals (critical points) of A_T. We proved that, if a cycle f is a critical point of A_T on Λ, then f is a classical periodic solution of the problem.

Next we showed that the elements for which the action is bounded are bounded away from zero. This prevents critical points from being collisional solutions. Another property of the action, the *coercivity*, avoids critical points at infinity. The fact that our procedure also avoids quasiperiodic solutions makes our results lead to genuine periodic solutions.

4. Main results

Provided all previous results (Σ_i, $i \in \{1, 2, 3, 7\}$, are Sobolev spaces; $H^1 = \Sigma_2 \oplus \Sigma_3 = \Sigma_1 \oplus \Sigma_7$; only the cycles belonging to Σ_2 and Σ_3 are noncollisional, nonescape, and do not represent quasiperiodic orbits; Σ_2 and Σ_3 can provide periodic solutions for the whole H^1; the minimizer in a Λ_k subset is an extremal of the action; a cycle that is aa critical point of A_T is a classical periodic solution of the problem), we proved the following:

THEOREM 1. *For any $T > 0$ and any $k \in \mathbb{Z} \backslash \{0\}$, there exists at least one S_i-symmetric ($i = 2, 3$) periodic orbit with period T and winding number k.*

THEOREM 2. *By the existence theorems of a minimal period τ of an autonomous system, and of a period $\tau/2$ (e.g., Amann 1990), there exist infinitely many families of distinct T-periodic orbits.*

Remark 3. Consider the present model to be a perturbation of the isotropic case (Stoica and Mioc 1997) via the parameter $\mu > 1$. This anisotropy, no matter how large its size, deforms the S_i-symmetric ($i = 2, 3$) periodic orbits of the isotropic problem, but does not destroy them. This makes the symmetries S_i ($i = 2, 3$) constitute an indicator of the robustness of the system to perturbations.

5. Conclusions

From the astronomical point of view, a central proto-star in fast rotation (hence oblate), surrounded by a nonhomogeneous accretion disk, and a body moving under this combined influence constitute a good concrete situation to be studied via a two-body problem associated to the anisotropic potential (2.1).

Our results prove that initially far orbits can collide with the central star or escape from the system (cf. Mioc *et al.* 2003), but genuine periodic orbits exist, too.

Even if outside our framework, problems as dynamics and structure of planetary rings, or satellite dynamics under the influence of the re-emitted solar/stellar radiation, can be tackled via the same mathematical tools.

The results we presented add new features as regards structure of proto-stellar systems, but can also serve to the understanding of the dynamics in some concrete situations in the Solar system.

References

Amann, H. 1990, *Ordinary Differential Equations: an Introduction to Nonlinear Analysis* (Walter de Gruyer, Berlin. New York).

Ambrosetti, A. and Coti Zelati, V. (1993) *Periodic Solutions of Singular Lagrangian Systems*, Progresses in Nonlinear Differential Equations and their Applications, No. 10, Birkhäuser, Boston.

Gordon, W. B. 1975, *Trans. Amer. Math. Soc.* 204, 113

Mioc, V., Pérez-Chavela, E. and Stavinschi, M. (2003) *Cel. Mech. Dyn. Astron.* 86, 81

Saslaw, W. C. 1978, *Astrophys. J.* 226, 240

Stoica, C. & Mioc, V. 1997, *Astrophys. Space Sci.* 249, 161

Struwe, M. 1996, *Variational Methods. Applications to Nonlinear Partial Differential Equations and Hamiltonian Systems* (Springer-Verlag, Berlin).

Dynamics of Populations of Planetary Systems
Proceedings IAU Colloquium No. 197, 2005
Z. Knežević and A. Milani, eds.

© 2005 International Astronomical Union
DOI: 10.1017/S1743921304009019

Improved proper motions in declination of some Hipparcos stars via long–term classical observations

Goran Damljanović[1] and Nada Pejović[2]

[1]Astronomical Observatory, Volgina 7, 11160 Belgrade – 74, Serbia and Montenegro
email: gdamljanovic@aob.bg.ac.yu

[2]Dept. of Astronomy, Faculty of Mathematics, University of Belgrade,
Studentski trg 16, 11000 Belgrade, Serbia and Montenegro
email: nada@matf.bg.ac.yu

Abstract. In this paper we present the new possibilities of classical ground – based observations for the purpose of improving the proper motions of some HIPPARCOS stars. The coordinates of asteroids, comets and other objects are calculated mostly by using the relative method and presented in a relevant reference frame via nearby stars (at the moment of observation) which materialize that reference frame. To improve the proper motions of stars we can use the latitude/universal time variations of ground – based data. Here, the results of improved proper motions in declination of some HIPPARCOS stars are presented.

Keywords. Stellar catalogues, proper motion

---

1. Introduction

Two catalogues appear as products of ESA mission, the Hipparcos one and the Tycho one. The former is the primary representation of the ICRS (International Celestial Reference System) in the optical wavelengths (ESA 1997). It contains 118218 stars with precise parallaxes, positions (better than 1 mas), proper motions and photometrical data. The median precision of positions for 1058332 Tycho Catalogue stars is 25 mas. The coordinates of the other objects, (asteroids, comets, etc.) calculated by applying the relative method and using the positions of nearby stars (at the moment of observation) are presented in a catalogue which materializes the reference frame. Thus the quality of the coordinates of the observed solar system objects depends upon the accuracy of the star positions.

During the last years, problems were noted for the proper motion of some Hipparcos stars, because the mission lasted < 4 years and was too short to provide satsisfactory accuracy in proper motion. The proper motions of some Hipparcos stars have comparatively low accuracy, mostly double and multiple stars (Vondrák & Ron 2004).

On the other hand, we can see that ground–based observations with a long history have a good potential to materialize a better reference frame, more stable in time. During last few years some catalogues including ground–based data were made : TYCHO-2, ACT, FK6(I), FK6(III), GC+HIP, TYC2+HIP , ARIHIP and the most recent Earth Orientation Catalogue – EOC (Vondrák & Ron 2003).

There is a large data set of ground–based measurements of latitude/universal time variations. The Hipparcos Catalogue comprises several thousand stars which were observed from the ground for few decades and for which a large number of observations is available (Vondrák *et al.* 1998). We used some of these data to improve the proper

471

motions in declination for some Hipparcos stars. Our results give a contribution to the investigations of possible improvement of the reference frame.

2. Data

In this paper we present our computational procedure by using the observations of several Photographic Zenith Tubes (PZT):
- PIP data, Punta Indio PZT (Argentina), observation time span 1971.6 – 1984.5,
- MS ones, Mount Stromlo PZT (Australia), 1957.8 – 1985.7,
- OJP ones, Ondřejov PZT (Czech Republic), 1973.1 – 1992.0.

We received the data from J. Vondrák, and decided to apply our procedure star by star, because the data contain star by star latitude observations (the data of each star are independent from each other). The data are referred to the Hipparcos Catalogue, corrected for instrumental constants (micrometer screw, plate scale, etc.), refraction, and errors in proper motions of some Hipparcos stars (but this correction is available for only about 20 % of the stars).

The precession–nutation model IAU 1980 and the MERIT Standards were used to compute the apparent places of the stars. The mean latitudes, corrections of conventional longitudes and tectonic plate motions (from the NUVEL–1 NNR geophysical model) were subtracted from the results in order to bring them all to a homogeneous system.

For each star we prepared a separate file with all observations made with PIP, MS or OJP. There are 165 Hipparcos stars observed with PIP instrument, 184 with MS, and 285 with OJP (157 stars are common to both PIP and MS). We used the polar motion coordinates from the file EOPOA00.dat by Vondrák.

3. Results

The latitude data we received mostly contain time variations due to polar motion changes with time, but we want to investigate the proper motion. Thus, at first it was necessary to calculate and remove the polar motion component.

We took the coordinates of the polar motion (x and y) from the file EOPOA00.dat and adapted x and y by linear interpolation to the moment of interest. Also, we repeated that calculation for each of the mentioned instruments, and removed the calculated polar motion part from the received latitude data for each star. For this purpose, we used the Kostinski's formula (Kulikov 1962) where λ (longitude of the instrument) is calculated from the zero–meridian to the west. The values of λ and φ (latitude) for mentioned stations are from (Vondrák et al. 1998).

After removing the polar motion part, for each star we are left with residuals (polar motion value minus mean value of n latitudes of some Hipparcos star observed during the subperiod, approximately one year long). Thus for each star we have got nearly one data point (residual) per year. Each data point contains: MJD, residual, standard error, and number of latitudes n (used in the mean value calculation of the mean value). A few examples are presented in the figures: in Fig. 1 we show the residuals as a function of time (MJD) of the common PIP and MS Hipparcos star H68419 (MS is the longer curve); Fig. 2 shows the same for OJP H80585. The curves are almost straight lines; thus the residuals are well prepared. The calculation of the free term a, of the linear one b and of its standard errors was made for each star by using the Lest-Squares Method (LSM) and linear model

$$Y(i) = a + b * (X(i) - 1991.25),$$

where: $X(i)$ is time in years, $Y(i)$ – mentioned residuals, i is the number of residuals (points). The linear term b is our correction of μ_δ (proper motion in declination).

For the cases presented in the figures, the results are (LSM, linear model, no weights):
- PIP, H68419 , $a = -0.''071 \pm 0.''043$, $b = -0.''0021/year \pm 0.''0032/year$,
- MS, H68419 , $a = -0.''103 \pm 0.''028$, $b = -0.''0042/year \pm 0.''0013/year$,
- OJP, H80585 , $a = -0.''411 \pm 0.''013$, $b = -0.''0103/year \pm 0.''0013/year$.

The results for the star H68419, as obtained from the PIP and from the MS data set, are in good agreement (thus our procedure is correct), but the values for the star H80585 (OJP data) are too high and questionable (the same occurs for the other OJP stars).

To explain the results for the star H80585 it was necessary to identify the trends of all PIP, MS and OJP data. We determined the values of a and b (LSM, the same model, no weights), but for the points of all PIP stars (similar for MS and OJP):
- PIP, $a = -0.''069 \pm 0.''005$, $b = -0.''0018/year \pm 0.''0004/year$,
- MS, $a = -0.''019 \pm 0.''003$, $b = -0.''0008/year \pm 0.''0002/year$,
- OJP, $a = -0.''426 \pm 0.''002$, $b = -0.''0118/year \pm 0.''0002/year$.

Evidently, the results concerning PIP and MS are still in a good agreement (the values of b are small for both instruments), but the results for OJP show the reason why the values for a and b are so large in case of Hipparcos star H80585. Similar results are obtained for most of the other PIP, MS and OJP stars. This means that these data still contain some systematic errors which we need to determine and remove before our final calculations, but the procedure of calculation is correct, which was our purpose to show here. Also, after finishing the calculations for the other instruments (here, we present only three of them) we are going to send the results to Dr. Vondrák, and expect new Vondrák's solutions of the Earth Orientation Parameters (EOP) and full comparison of our results, in line with the long–term Earth rotation studies (Vondrák & Ron 2004). At present, we can conclude that our calculated correction is justified if our procedure for the same star observed at different sites (see Fig. 1) gives us similar results. It means that the procedure can extract the effect of proper motion affecting latitude changes with time.

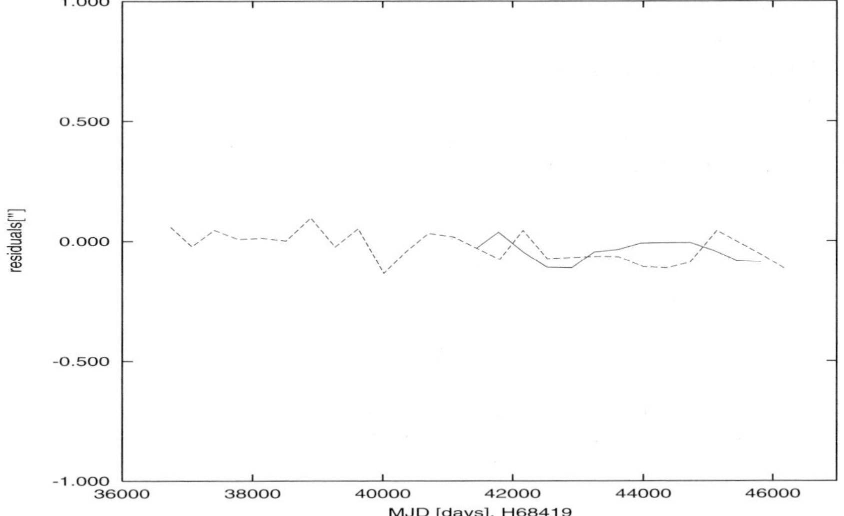

Figure 1. The residuals as a function of time (MJD) of the common PIP (shorter curve) and MS (longer one) Hipparcos star H68419.

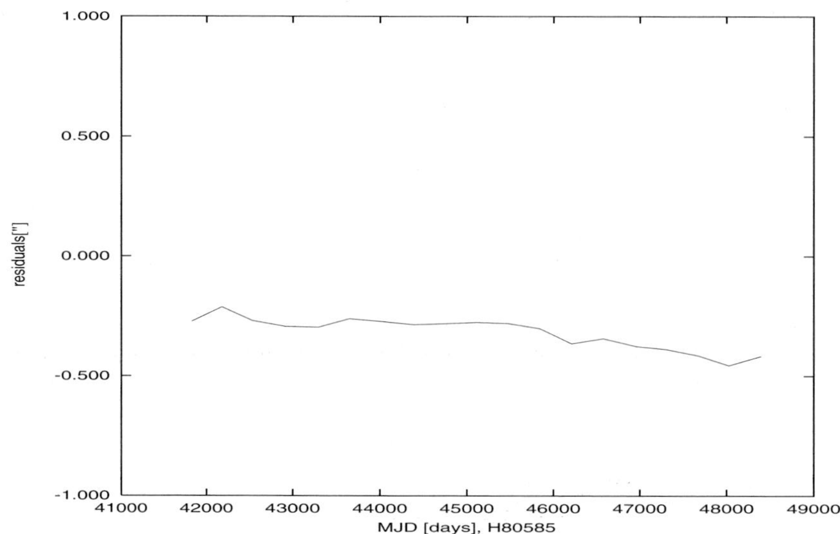

Figure 2. The residuals with MJD of OJP Hipparcos star H80585.

It is necessary to be careful with PZT latitude data because of their level of formal errors (much larger than the Hipparcos ones) and of different kinds of systematic ones, but these data with a long history still can give us useful informations. One possibility is to improve proper motion for some Hipparcos stars as shown here.

4. Conclusions

We used the latitude data of PIP, MS, and OJP PZTs to improve the proper motions in declination for some Hipparcos stars. The main features of the data and our calculation procedure are explained. A few examples (see the Fig. 1 and Fig. 2) are given. This kind of investigation is the way, step by step, to obtain more accurate data than the Hipparcos ones, in a long–term sense, and the combination of ground – based optical observations with space ones can give us a better reference frame.

Acknowledgements

One of the authors (GD) has done this work as a part of the project No. 1468 "Structure, Kinematics and Dynamics of the Milky Way" supported by the Ministry of Science and Environmental Protection of Serbia.

References

ESA 1997, *The Hipparcos and Tycho Catalogues* ESA SP-1200

Kulikov A.K. 1962, *Izmenyaemost' sirot i dolgot*, Moskva

Vondrák, J., Pešek, I., Ron, C. & Čepek, A. 1998, *Astron. Inst. Acad. Sci. Czech R., Publ. No.,* 87

Vondrák, J. & Ron, C. 2003 in: N. Capitaine (ed.), Observatoire de Paris, *Journées 2002 Systèmes de Référence Spatio – Temporels*, p. 49

Vondrák, J. & Ron, C. 2004 in: R.Gaume, D.McCarthy and J.Souchay (eds.), *in Proc. IAU XXV Joint Discussion 16: The ICRS Maintenance and Future Realizations*, in press

Dynamics of Populations of Planetary Systems
Proceedings IAU Colloquium No. 197, 2005
Z. Knežević and A. Milani, eds.

© 2005 International Astronomical Union
DOI: 10.1017/S1743921304009020

A mass estimate for a sample of nearby stars

Slobodan Ninković and Veselka Trajkovska

Astronomical Observatory, Volgina 7, 11160 Belgrade 74, Serbia and Montenegro
email: sninkovic@aob.bg.ac.yu
email: vtrajkovska@aob.bg.ac.yu

Abstract. We form a sample of nearby single stars. Their masses are estimated on the basis of the mass-luminosity relation. Our aim is to specify a reasonable space limit, i.e. a heliocentric volume, as small as possible, but to contain a sufficient number of stars so that an application of statistical laws becomes possible. Among the quantities calculated by us a special attention is paid to the mean mass of a single star and, also, to the local value of Agekyan's factor.

Keywords. Stellar mass, nearby stars

1. Introduction

It is sufficiently well known that for a star its mass is, practically, its most important physical property, but also that to determine reliably the mass of an individual star is a difficult task. The data sources for stellar masses are binary stars. However, this case is not without difficulties. Even if the question, whether a stellar pair considered as a binary is really bound, is put aside, nevertheless, problems are still present; for some kinds of binaries, say visual, the total mass can be found, whereas in the case of, say spectroscopic binaries, the first quantity to be determined is the mass ratio. Despite of all of this masses of many stars have been determined (better to say, estimated) and, as a consequence, an empirical rule was obtained, according to which for, at least, the stars of the main sequence of the HR diagram the higher the luminosity is, the larger is the mass (for details see Angelov 1993, and the references therein). Such a relationship is to be expected bearing in mind the energy sources of stars.

Once the masses for a sufficiently large number of stars are known, it is possible to analyse the mass distribution of stars. There are mathematical models developed for this purpose (e.g. Marochnik & Suchkov 1984, p. 200, and the references therein). Among the questions inestigated by one of the present authors (Ninković 1995), of importance is the value of Agekyan's factor in stellar systems. In the present paper an attempt is done to answer these questions.

2. Theory

Agekyan's factor is a dimensionless quantity appearing in the formula for a criterion aimed at estimating the role of regular, i.e. irregular forces in stellar systems. The criterion was proposed by Agekyan (1962). Such criteria are normally based on a characteristic ratio indicating the importance of irregular, i.e. regular forces. The concept of dividing gravitational forces into these two groups is almost fundamental in stellar dynamics. The regular forces are described by means of the gravitational potential which has its origin in the mass distribution within a given stellar system, whereas the irregular forces have a stochastic nature and they are usually manifested through stellar encounters. Agekyan estimates the volume of the part of a stellar system within which the irregular forces prevail over the regular ones. As the unit he uses the total volume of that stellar system. In this way we have the following expression:

$$A = \frac{2}{\sqrt{N}} \frac{\langle m^{3/2} \rangle}{\langle m \rangle^{3/2}}$$

As easily seen, the first factor on the right hand side of this equation is a simple decreasing function of N (total number of stars in the stellar system). This is to be expected because the higher the total number of stars, the less is the importance of irregular forces; however the factor depending on star masses is something new, in which Agekyan's criterion is different from other similar criteria, since it indicates that in addition to the total number of stars the mass distribution (mass scatter) is also important in estimating the importance of irregular forces. For this reason this factor is called Agekyan's one, and it is treated in the present paper.

It should be mentioned that the action of irregular forces is also characterised by the amount of the so-called relaxation time (time interval within which the moduli of the average change of mechanical energy per star and of the average mechanical energy per star become equal to each other). The best solution is to express it in terms of the so-called crossing time (time interval needed on the average for a star to cross the system diameter). This ratio also depends on N (e.g. Binney & Tremaine 1987 - eq. (8.1), p. 489), also being a simple function of it and leading to the same conclusions as the first factor in Agekyan's formula.

3. Mass Estimates

The data source is the Hipparcos Catalogue (ESA 1997). It contains the parallaxes, apparent magnitudes and other data for a large number of stars (more than 100 000). We expect these parallaxes to be accurate enough so that they enable deriving of absolute magnitudes for all stars with known apparent magnitudes. Since the Hipparcos stars are sufficiently close to the Sun, the interstellar absorption is practically negligible. As the relation yielding the mass of a star on the basis of its luminosity, we use the one proposed by Angelov 1993. The input data for this relation are visual absolute magnitudes and, consequently, there is no need to calculate the corresponding bolometric magnitudes. We form the input parameters (star magnitudes, in fact the apparent magnitudes corrected for the distance, i.e. parallax), and as the output we have the masses. Our previous experience (Trajkovska & Ninković 1997) shows that the star masses determined in this way can be reliable enough. For this reason we insist on single stars only because the astrophysical data necessary for our estimation are more reliable in the case of single stars.

4. Results

Our first step is to separate those Hipparcos stars with parallaxes exceeding 0".01. Of course, of interest are only the cases where all the data necessary for our work are available. In this set of stars we look for the ones with designation V in the Catalogue which indicates the main sequence (MS), because the relation for mass estimation used by us is valid for MS only. The corresponding HR diagram is presented in Fig. 1.

Within this sample (Sample 1) we form a subsample containing very close stars (Sample 2), with parallaxes greater than 0".1. The sample containing stars with parallaxes between 0".01 and 0".1 is referred to as Sample 3. The parallax distribution is presented

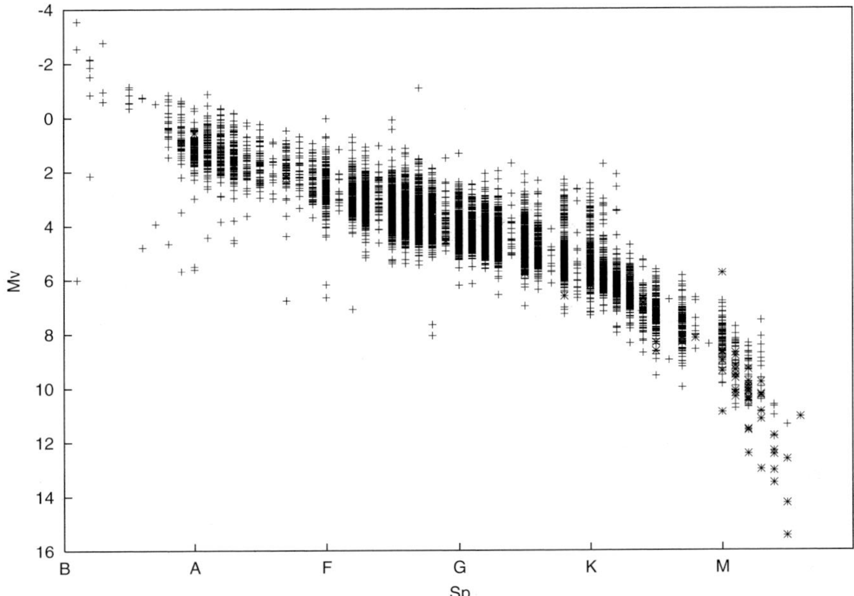

Figure 1. HR Diagram for Stars of Sample 1.

Table 1. The mean mass of a single star, standard deviation and the value of Agekyan's factor for all samples.

$Sample$	n	$\langle m \rangle$	$\sigma_{\langle m \rangle}$	A
Sample 1	9496	1.288	0.582	1.071
Sample 2	96	0.602	0.495	1.221
Sample 3	9400	1.295	0.579	1.069

in Fig. 2 for Sample 1 (above), and for Sample 2 (below). As for the parallax distribution for Sample 3, it is very similar to that for Sample 1. The reason is very simple, Sample 2 contains a very small number of stars as will be shown below. Thus, separate plots for Sample 3 are not given.

For each star we calculate the mass by applying Angelov's (1993) relation. The mass distribution is presented in Fig. 3 for Sample 1 (above), and for Sample 2 (below).

For all these samples we calculate the mean mass of a single star, its standard deviation and the value of Agekyan's factor. These values are given in Table 1. The number of stars in each sample is given in the first column. Note that the number of stars within a sample is in no way connected with the total number of stars in a stellar system (N in the formula of Agekyan's criterion). Here the stellar system under study is the Galaxy, thus N would be of order of 10^{11}. The unit for the mean mass and its standard deviation is the mass of the Sun.

5. Discussion and Conclusion

First, it should be emphasized that in the case of Sample 2 the mean mass of a single star is significantly different from the values obtained for the other two samples which are very close to each other. The same can be said for the case of Agekyan's factor, a dimensionless quantity characterising the scatter in the star masses. As for the standard

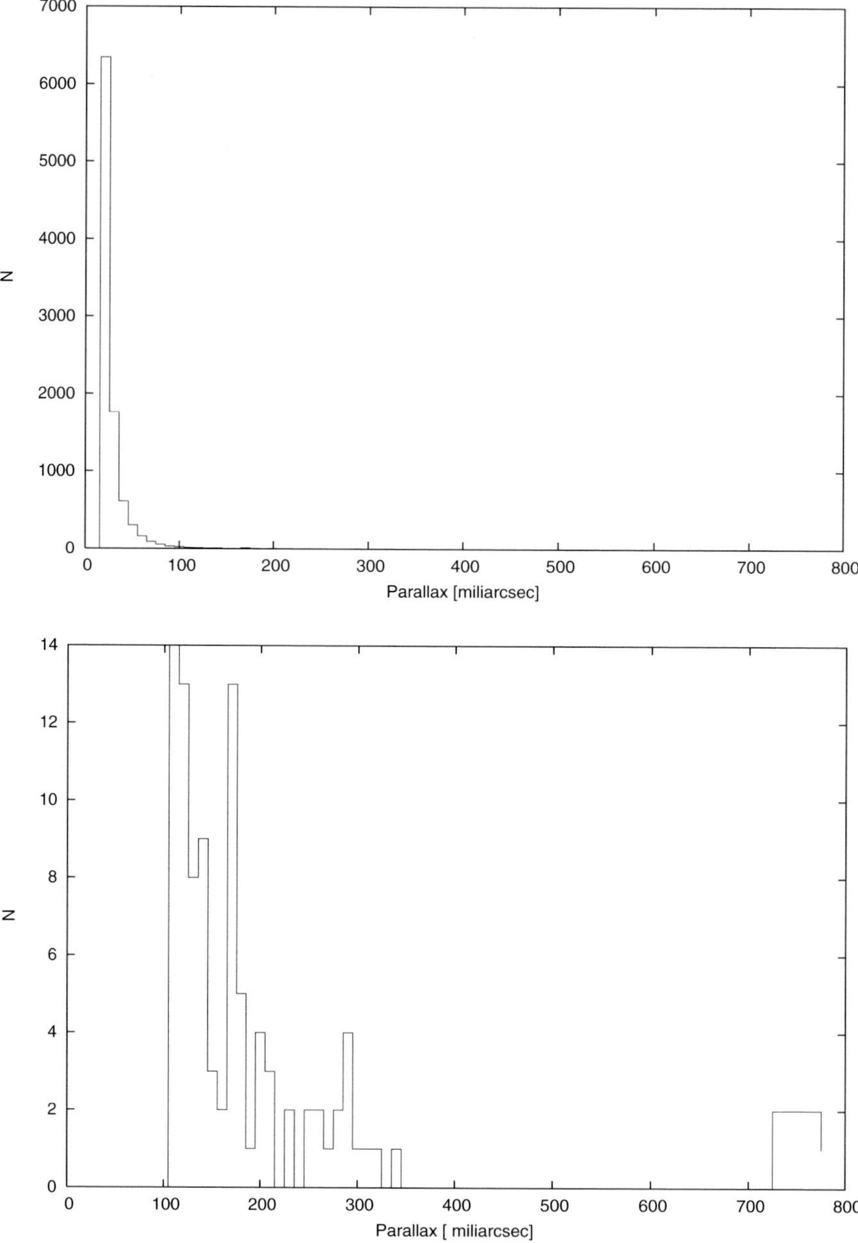

Figure 2. The parallax distribution
Above - Sample 1, Below - Sample 2.

deviation, the difference is not so prominent. This fact is interpreted by us as due to the observational selection. The best way to demonstrate its influence is to present the dependence mass-parallax (Fig. 4).

As clearly seen, the low-mass stars prevail at high parallaxes and vice versa. It is well known that high-mass stars are less frequent than low-mass ones. Therefore, within a small volume like that corresponding to a lower parallax limit of 0".1 the probability of finding massive stars is low. On the other hand low-mass stars as less luminous (we are

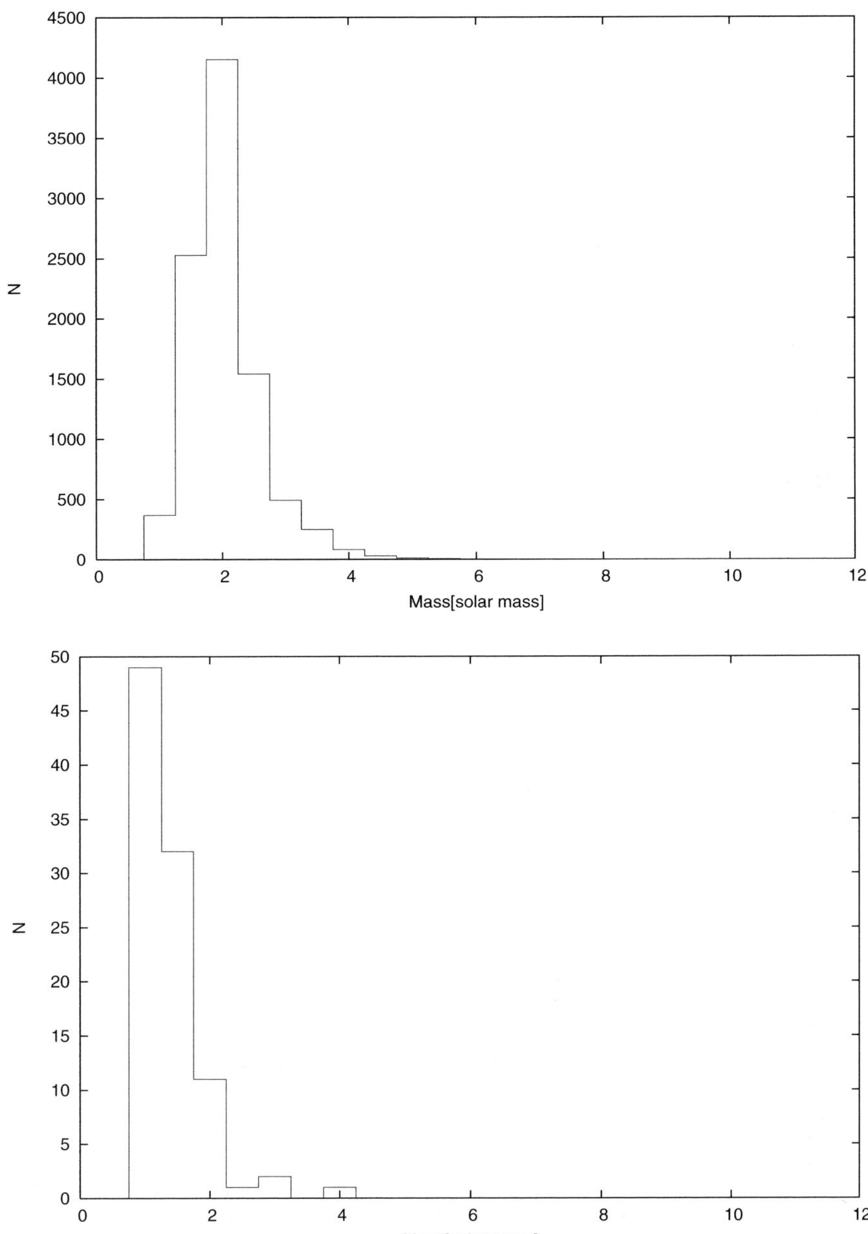

Figure 3. The mass distribution
Above - Sample 1, Below - Sample 2.

on MS!) are difficult to be discovered. The same is true for Agekyan's factor where the lack of low-mass stars produces an artificial equalling of star masses, i.e. diminishes the mass scatter. Nevertheless, it is even for very close stars still significantly smaller than the value expected on the basis of high numbers (Ninković 1995). As a possible explanation we offer the influence of the accuracy with which the parallaxes used here are obtained. The estimates of the number density of stars and their mean mass performed at the same

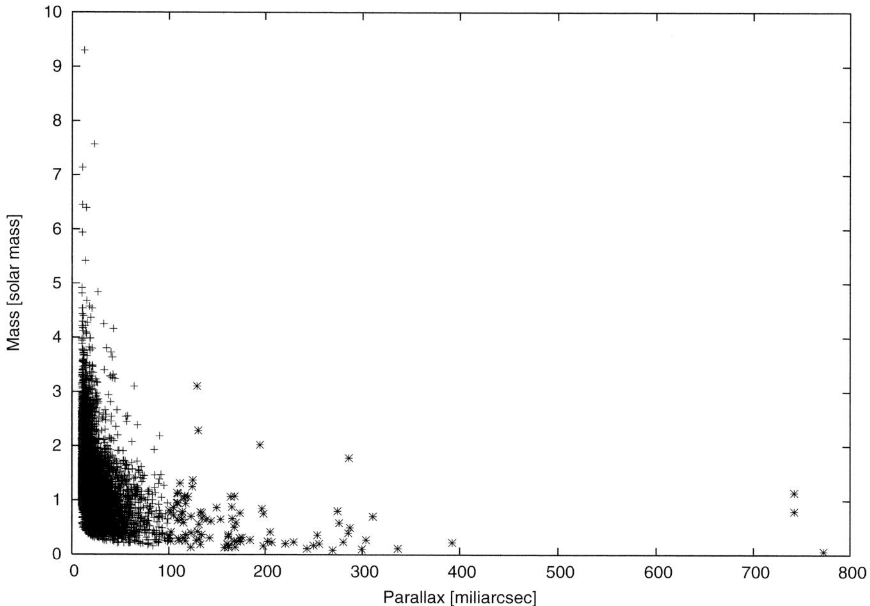

Figure 4. Dependence Mass-Parallax for Stars of Sample 1.

location are correlated; this indicates problems with the fainter end of MS, irrespectively of the precision of our results.

Acknowledgements

This work is a part of Project 1468 supported by the Ministry for Science and Environmental Protection of the Republic of Serbia.

References

Agekyan, T.A. 1962, *in Kurs astrofiziki i zvezdnoj astronomii, tom vtoroj, ed. A. A. Mikhajlov*, Gos. izd. fiz. - mat. lit., Moskva, p. 528

Angelov, T. 1993, *Bull. Astron. Belgrade* 148, 1

Binney, J. & Tremaine, S. 1987, *Galactic Dynamics*, "Princeton University Press", Princeton, New Jersey

ESA (The Hipparcos Science Team: U. Bastian, P.L. Bernacca, M. Creze, F. Donati, M. Grenon, M. Grewing, E. Hg (TDAC Consortium Leader), J. Kovalevsky, (FAST Consortium Leader), F. van Leeuwen, L. Lindegren, (NDAC Consortium Leader), H. van der Marel, F. Mignard, C.A. Murray, M.A.C. Perryman, ESA (Chairman) Hipparcos Project Scientist, R.S. Le Poole, H. Schrijver, C. Turon (INCA Consortium Leader)) 1997, *The Hipparcos and Tycho Catalogues ESA SP - 1200*

Marochnik, L.S. & Suchkov, A.A. 1984, *Galaktika*, "Nauka", glav. red. fiz. - mat. lit., Moskva

Ninković, S. 1995, *Bull. Astron. Belgrade* 151, 1

Trajkovska, V. & Ninković, S. 1997, *Bull. Astron. Belgrade* 155, 35

Part 7
ENDING THE COLLOQUIUM

Concluding remarks

At the meeting we had 18 invited and 38 contributed oral presentations in 13 scientific sessions; 35 posters were presented in 2 poster sessions. To summarize the meeting results is necessarily unfair to some of the many very interesting contributions. This volume is fairly representative of the scientific content of the meeting.

At the meeting and the following few weeks 70 papers were submitted. These papers covered 16 of the 18 invited talks (extended abstracts were requested for the other two), 30 contributed oral presentations and 24 posters. All the papers have been submitted to a referee. The referee reports received have been forwarded to the authors with the request to apply modifications, in the majority of cases minor ones. However, in some cases major modifications were required, and 12 papers have been rejected on the basis of the referee report and the editors' own judgment: thus this book contains 58 papers and 2 extended abstracts. The revised versions of the papers have been checked for agreement with the referee's comments, for the correct format of the references and for the use of the IAU LaTeX style. In many cases the English had to be significantly improved: the editors felt that this was a duty, to give some real meaning to the commitment requested from the organizers by the IAU of accepting participants from every country, including those in which very different languages are spoken. The editors whish to acknowledge the essential contribution of Ivana Damjanov, of the Astronomical Observatory of Belgrade, in all the stages of this editorial process.

All the above may sound like a normal procedure, but in this case the entire process, from the end of the meeting to submission of the manuscript, took place in three months. The editors wish to express the feeling that to complete the manuscript with such a tight deadline, without skipping any of the stages of the process, has been an achievement. The editors hope that the quality of this book will testify both of the very high scientific level of the meeting and of the rigorous procedure used to publish the proceedings.

Author Index

Object Index